Big Data in Omics and Imaging: Association Analysis

CHAPMAN & HALL/CRC
Mathematical and Computational Biology Series

Aims and Scope:
This series aims to capture new developments and summarize what is known over the entire spectrum of mathematical and computational biology and medicine. It seeks to encourage the integration of mathematical, statistical, and computational methods into biology by publishing a broad range of textbooks, reference works, and handbooks. The titles included in the series are meant to appeal to students, researchers, and professionals in the mathematical, statistical and computational sciences, fundamental biology and bioengineering, as well as interdisciplinary researchers involved in the field. The inclusion of concrete examples and applications, and programming techniques and examples, is highly encouraged.

Series Editors

Proposals for the series should be submitted to one of the series editors above or directly to:
CRC Press, Taylor & Francis Group
3 Park Square, Milton Park
Abingdon, Oxfordshire OX14 4RN
UK

Published Titles

An Introduction to Systems Biology: Design Principles of Biological Circuits
Uri Alon

Glycome Informatics: Methods and Applications
Kiyoko F. Aoki-Kinoshita

Computational Systems Biology of Cancer
Emmanuel Barillot, Laurence Calzone, Philippe Hupé, Jean-Philippe Vert, and Andrei Zinovyev

Python for Bioinformatics, Second Edition
Sebastian Bassi

Quantitative Biology: From Molecular to Cellular Systems
Sebastian Bassi

Methods in Medical Informatics: Fundamentals of Healthcare Programming in Perl, Python, and Ruby
Jules J. Berman

Chromatin: Structure, Dynamics, Regulation
Ralf Blossey

Computational Biology: A Statistical Mechanics Perspective
Ralf Blossey

Game-Theoretical Models in Biology
Mark Broom and Jan Rychtář

Computational and Visualization Techniques for Structural Bioinformatics Using Chimera
Forbes J. Burkowski

Structural Bioinformatics: An Algorithmic Approach
Forbes J. Burkowski

Spatial Ecology
Stephen Cantrell, Chris Cosner, and Shigui Ruan

Cell Mechanics: From Single Scale-Based Models to Multiscale Modeling
Arnaud Chauvière, Luigi Preziosi, and Claude Verdier

Bayesian Phylogenetics: Methods, Algorithms, and Applications
Ming-Hui Chen, Lynn Kuo, and Paul O. Lewis

Statistical Methods for QTL Mapping
Zehua Chen

An Introduction to Physical Oncology: How Mechanistic Mathematical Modeling Can Improve Cancer Therapy Outcomes
Vittorio Cristini, Eugene J. Koay, and Zhihui Wang

Normal Mode Analysis: Theory and Applications to Biological and Chemical Systems
Qiang Cui and Ivet Bahar

Kinetic Modelling in Systems Biology
Oleg Demin and Igor Goryanin

Data Analysis Tools for DNA Microarrays
Sorin Draghici

Statistics and Data Analysis for Microarrays Using R and Bioconductor, Second Edition
Sorin Drăghici

Computational Neuroscience: A Comprehensive Approach
Jianfeng Feng

Mathematical Models of Plant-Herbivore Interactions
Zhilan Feng and Donald L. DeAngelis

Biological Sequence Analysis Using the SeqAn C++ Library
Andreas Gogol-Döring and Knut Reinert

Gene Expression Studies Using Affymetrix Microarrays
Hinrich Göhlmann and Willem Talloen

Handbook of Hidden Markov Models in Bioinformatics
Martin Gollery

Meta-Analysis and Combining Information in Genetics and Genomics
Rudy Guerra and Darlene R. Goldstein

Differential Equations and Mathematical Biology, Second Edition
D.S. Jones, M.J. Plank, and B.D. Sleeman

Knowledge Discovery in Proteomics
Igor Jurisica and Dennis Wigle

Introduction to Proteins: Structure, Function, and Motion
Amit Kessel and Nir Ben-Tal

Published Titles (continued)

RNA-seq Data Analysis: A Practical Approach
Eija Korpelainen, Jarno Tuimala, Panu Somervuo, Mikael Huss, and Garry Wong

Introduction to Mathematical Oncology
Yang Kuang, John D. Nagy, and Steffen E. Eikenberry

Biological Computation
Ehud Lamm and Ron Unger

Optimal Control Applied to Biological Models
Suzanne Lenhart and John T. Workman

Clustering in Bioinformatics and Drug Discovery
John D. MacCuish and Norah E. MacCuish

Spatiotemporal Patterns in Ecology and Epidemiology: Theory, Models, and Simulation
Horst Malchow, Sergei V. Petrovskii, and Ezio Venturino

Stochastic Dynamics for Systems Biology
Christian Mazza and Michel Benaïm

Statistical Modeling and Machine Learning for Molecular Biology
Alan M. Moses

Engineering Genetic Circuits
Chris J. Myers

Pattern Discovery in Bioinformatics: Theory & Algorithms
Laxmi Parida

Exactly Solvable Models of Biological Invasion
Sergei V. Petrovskii and Bai-Lian Li

Computational Hydrodynamics of Capsules and Biological Cells
C. Pozrikidis

Modeling and Simulation of Capsules and Biological Cells
C. Pozrikidis

Cancer Modelling and Simulation
Luigi Preziosi

Computational Exome and Genome Analysis
Peter N. Robinson, Rosario M. Piro, and Marten Jäger

Introduction to Bio-Ontologies
Peter N. Robinson and Sebastian Bauer

Dynamics of Biological Systems
Michael Small

Genome Annotation
Jung Soh, Paul M.K. Gordon, and Christoph W. Sensen

Niche Modeling: Predictions from Statistical Distributions
David Stockwell

Algorithms for Next-Generation Sequencing
Wing-Kin Sung

Algorithms in Bioinformatics: A Practical Introduction
Wing-Kin Sung

Introduction to Bioinformatics
Anna Tramontano

The Ten Most Wanted Solutions in Protein Bioinformatics
Anna Tramontano

Combinatorial Pattern Matching Algorithms in Computational Biology Using Perl and R
Gabriel Valiente

Managing Your Biological Data with Python
Allegra Via, Kristian Rother, and Anna Tramontano

Cancer Systems Biology
Edwin Wang

Stochastic Modelling for Systems Biology, Second Edition
Darren J. Wilkinson

Big Data in Omics and Imaging: Association Analysis
Momiao Xiong

Big Data Analysis for Bioinformatics and Biomedical Discoveries
Shui Qing Ye

Bioinformatics: A Practical Approach
Shui Qing Ye

Introduction to Computational Proteomics
Golan Yona

Big Data in Omics and Imaging: Association Analysis

Momiao Xiong

CRC Press is an imprint of the
Taylor & Francis Group, an **informa** business
A CHAPMAN & HALL BOOK

CRC Press
Taylor & Francis Group
6000 Broken Sound Parkway NW, Suite 300
Boca Raton, FL 33487-2742

First issued in paperback 2021

ISBN 13: 978-1-03-209598-1 (pbk)
ISBN 13: 978-1-4987-2578-1 (hbk)

Publisher's Note
The publisher has gone to great lengths to ensure the quality of this reprint but points out that some imperfections in the original copies may be apparent.

Visit the Taylor & Francis Web site at
http://www.taylorandfrancis.com

and the CRC Press Web site at
http://www.crcpress.com

To Ping

Contents

Preface

THE NEXT GENERATION OF genomic, sensing, and imaging technologies has generated a deluge of DNA sequencing, transcriptomes, epigenomic, metabolic, physiological (ECG, EEG, EMG, and MEG), image (CT, MRI, fMRI, DTI, PET), behavioral, and clinical data with multiple phenotypes and millions of features. Analysis of increasingly larger, deeper, more complex, and more diverse genomic, epigenomic, molecular, and spatiotemporal physiological and anatomical imaging data provides invaluable information for the holistic discovery of the genetic and epigenetic structure of disease and has the potential to be translated into better understanding of basic biomedical mechanisms and to enhance diagnosis of disease, prediction of clinical outcomes, characterization of disease progression, management of health care, and development of treatments. Big data sparks machine learning and causal revolutions and rethinking the entire health and biomedical data analysis process. The analysis of big data in genomics, epigenomics, and imaging that covers fundamental changes in these areas is organized into two books: (1) *Big Data in Omics and Imaging: Association Analysis* and (2) *Big Data in Omics and Imaging: Integrated Analysis and Causal Inference.*

The focus of this book is association analysis and machine learning. The standard approach to genomic association analysis is to perform analysis individually, one trait and one variant at a time. The traditional analytic tools were originally designed for analyzing homogeneous, single phenotype, and common variant data. They are not suitable to cope with big heterogeneous genomic data due to both methodological and performance issues. Deep analysis of high-dimensional and heterogeneous types of genomic data in the sequencing era demands a paradigm shift in association analysis from standard multivariate data analysis to functional data analysis, from low-dimensional data analysis to high-dimensional data analysis, and from individual PC to multicore cluster and cloud computing.

There has been rapid development of novel and advanced statistical methods and computational algorithms for association analysis with next-generation sequencing (NGS) in the past several years. However, very few books have covered their current development. This book intends to bridge the gap between the traditional statistical methods and computational tools for small genetic data analysis and the modern, advanced statistical methods, computational algorithms, and cloud computing for sequencing-based association analysis. This book will bring technologies for statistical modeling, functional data analysis, convex optimization, high-dimensional data reduction, machine learning, and multiple phenotype data analysis together. This book also aims to discuss interesting real data analysis and applications.

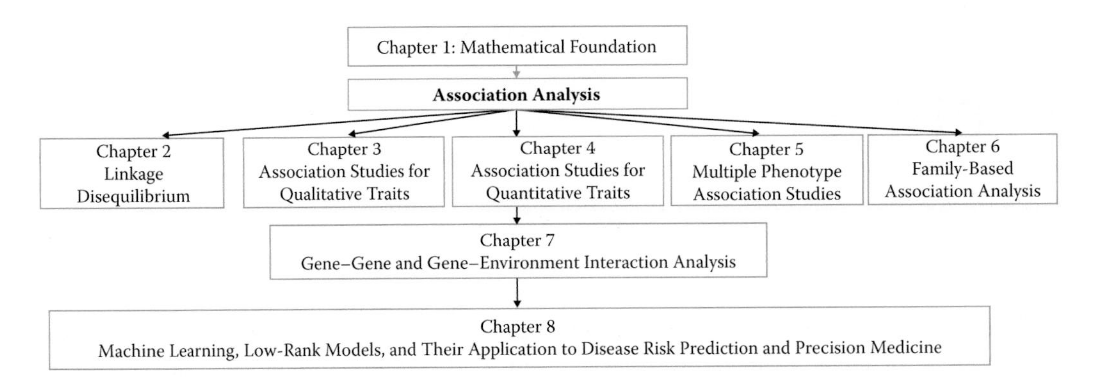

FIGURE P.1 Outline of this book.

As shown in Figure P.1, this book is organized into eight chapters. The following is a description of each chapter.

Chapter 1, "Mathematical Foundation," covers (1) sparsity-inducing norms, dual norms, and Fenchel conjugate, (2) subdifferential, (3) proximal methods, (4) matrix calculus, (5) principal component analysis, (6) functional principal component analysis, (6) canonical correlation analysis, and (7) functional canonical correlation analysis. Various norms, subdifferentials and subgradients, and matrix calculus are powerful tools for developing efficient statistical methods and computational algorithms in genomic, epigenomic, and imaging data analysis. Detailed descriptions of these topics are not included in the standard statistical and genetics books. The coverage of these materials in this book will facilitate the training of a new generation of statistical geneticists and computational biologists with modern mathematics. Proximal methods have recently been developed for convex optimization, which provide foundations for regularized statistical methods, low-rank models, machine learning, and causal inferences. Both multivariate and functional principal component analysis and canonical correlation analysis are major high-dimension data reduction methods and offer powerful tools for big genomic data analysis.

Chapter 2, "Linkage Disequilibrium (LD)," includes the standard two-locus LD measures, composite LD measures, haplotype reconstruction, and multilocus LD measures. Chapter 2 also introduces mutual information for two-locus measures, multi-information for multilocus measures of LD, and interaction information measures of genetic variation. Particularly, Chapter 2 presents canonical correlation as a measure of LD between two genomic regions and establishes the relationship between canonical correlation and joint information measure of LD. The LD patterns across the genomes in three populations, Africa, Europe, and Asia, using the 1000 Genome Project data are presented in Chapter 2.

Chapter 3, "Association Studies for Qualitative Traits," addresses the major issues for association analysis of qualitative traits. The Hardy–Weinberg equilibrium, odds ratio, and disease models are first discussed in this chapter. Then, after introducing the widely used statistics for testing the association of common variants with a disease, we focus on the paradigm shift of association analysis from single marker analysis to the joint analysis of multiple variants in a genomic region for coping with next-generation sequencing data. Chapter 3 covers (1) multivariate group tests, including collapsing method, combined multivariate and collapsing (CMC) method, weighted sum method, score test and logistic regression, and sequencing kernel association test (SKAT), and (2) functional association analysis, including function principal component analysis for association tests and smoothed functional principal component analysis for association tests.

Chapter 4, "Association Studies for Quantitative Traits," moves association analysis for qualitative traits to association analysis to quantitative traits. Basic concepts, models, and theories for quantitative genetics, which are genetic additive and dominance effects, genetic variance, and linear regression as a major model for genetic studies of a quantitative trait are first introduced. Then the discussion progresses from a simple linear regression model with a single marker to a multiple regression model with multiple markers. Both simple and multiple regression models are used to test the association of common variants with a quantitative trait. We show that these classical quantitative genetic models are not suitable for rare variants. To cope with next-generation sequencing data, we shift our attention to gene-based quantitative trait association analysis. Three approaches to gene-based quantitative trait analysis, functional linear models, canonical correlation analysis, and kernel approach, are introduced. To develop a general framework for quantitative genetic studies, we formulate an association analysis problem as an independent test in a Hilbert space and use a dependence measure to quantify the level of association of the genetic variant with the quantitative trait.

Chapter 5, "Multiple Phenotype Association Studies," focuses on an association analysis of multiple traits. Three major approaches are commonly used to explore the association of genetic variants with multiple correlated phenotypes: multiple regression methods, integration of P-values of univariate analysis, and dimension reduction methods. Chapter 5 will cover pleiotropic additive and dominance genetic models of multiple traits, multivariate marginal regressions, multivariate analysis of variance, multivariate multiple linear regression models, multivariate functional linear models for gene-based genetic analysis of multiple traits, both multivariate and functional canonical correlation analysis for gene-based genetic pleiotropic analysis, principal component analysis for phenotype dimension reduction, and dependence measures for association tests of multiple traits. Chapter 5 also presents two novel statistics: quadratically regularized canonical correlation analysis and quadratically regularized principal component analysis for genetic analysis of multiple traits. In Chapter 5, the connection among regression, canonical correlations, and kernel methods for association analysis of multiple traits are explained.

Chapter 6, "Family-Based Association Analysis," considers family-based designs for association analysis. Population-based sample design is the current major study design for many association studies. However, many rare variants are from recent mutations in pedigrees. The inability of common variants to account for most of the supposed heritability and the low power of population-based tests for the association of rare variants have led to a renewed interest in family-based design with enrichment for risk alleles to detect the association of rare variants. It is increasingly recognized that analyzing samples from populations and pedigrees separately is highly inefficient. It is natural to unify population and family study designs for association studies. This chapter focuses on the statistical methods for a unified approach to the genetic analysis of both qualitative and quantitative traits.

Chapter 6 will cover (1) kinship coefficient, genome similarity matrix, and heritability; (2) mixed linear models for both single quantitative trait and multiple quantitative traits with common variants; (3) mixed functional linear models for both single and multiple quantitative traits with rare variants; (4) a unified general framework for sequence-based association studies with pedigree structure and unrelated individuals; and (5) family-based genome information content and functional principal component analysis statistics for pathway analysis.

Chapter 7, "Gene–Gene and Gene–Environment Interaction Analysis," investigates statistical methods for the detection of gene–gene and gene–environment interaction. Disease development is a dynamic process of gene–gene and gene–environment interactions within a complex biological system. Gene–gene and gene–environment interactions are ubiquitous. The interactions hold a key for dissecting the genetic structure of complex diseases and elucidating the biological and biochemical pathway underlying the diseases. The current paradigm for gene–gene and gene–environment interaction is to study pairwise interaction between two single markers and is designed for identifying interaction between common variants. Chapter 7 will first review the odds ratio, disequilibrium, and information measure of gene–gene and gene–environment interaction and introduce the relative risk, odds ratio, and disequilibrium and information measure–based statistics for testing the gene–gene and gene–environment interaction. The classical statistical methods for gene–gene and gene–environment interaction detection are designed for common variants; they are not suitable for rare variants and next-generation sequencing data. To deal with next-generation sequencing data, Chapter 7 will also cover the current development of gene-based gene–gene and gene–environment interaction analysis: function regression models for single quantitative trait, multivariate function regression models for multiple quantitative traits, canonical correlation analysis for both single and multiple quantitative traits, and functional logistic regression for qualitative traits.

Chapter 8, "Machine Learning, Low-Rank Models, and Their Application to Disease Risk Prediction and Precision Medicine," covers the core part of data science and machine learning. It includes both discriminant analysis and penalized discriminant analysis, support vector machine, kernel support vector machine, sparse support vector machine, multitask and multiclass support vector machine, classical penalized and network-penalized logistic regression, low rank models with both generalized cost functions and generalized constraints, generalized canonical correlation analysis, canonical correlation analysis for classification, unsupervised and supervised dimension reduction, and sufficient dimension reduction. Chapter 8 will also investigate the application of machine learning and feature selection to disease risk prediction and precision medicine.

Overall, this book introduces state-of-the-art studies and practice achievements in genomic and multiple phenotype data analysis. It sets the basis and analytical platforms for further research in this challenging and rapidly changing field. The expectation is that the presented concepts, statistical methods, computational algorithms, and analytic platforms will facilitate the training of next-generation statistical geneticists, bioinformaticians, and computational biologists.

I thank Sara A. Barton for editing this book. I am deeply grateful to my colleagues and collaborators Li Jin, Eric Boerwinkle, Ruzong Fan, and others whom I have worked with for many years. I especially thank my former and current students and postdoctoral fellows for their strong dedication to the research and scientific contributions to this book: Jinying Zhao, Li Luo, Shenying Fang, Yun Zhu, Long Ma, Nan Lin, Hua Dong, Futao Zhang, Pengfei Hu, and Hoicheong Siu. Finally, I must thank my editor, David Grubber, for his encouragement and patience during the process of creating this book.

MATLAB® is a registered trademark of The MathWorks, Inc. For product information, please contact:

The MathWorks, Inc.
3 Apple Hill Drive
Natick, MA 01760-2098 USA
Tel: 508-647-7000
Fax: 508-647-7001
E-mail: info@mathworks.com
Web: www.mathworks.com

Author

Momiao Xiong, PhD is a professor in the Department of Biostatistics, University of Texas School of Public Health, and a regular member of the Genetics & Epigenetics (G&E) Graduate Program at the University of Texas MD Anderson Cancer Center, UTHealth Graduate School of Biomedical Sciences.

CHAPTER 1

Mathematical Foundation

1.1 SPARSITY-INDUCING NORMS, DUAL NORMS, AND FENCHEL CONJUGATE

In this section, we introduce several important concepts, including norms, dual norms, and Fenchel conjugate and its dual and subdifferential, which play an important role in optimization with sparsity-inducing penalties and genomic and epigenomic data analysis.

Definition 1.1 (Norm)

A norm $||x||$ of an element x in a space is a mapping from the space to R with the following three properties:

(1) $||x|| > 0$, if $x \neq 0$; $||x|| = 0$, if $x = 0$.
(2) $||\lambda x|| = |\lambda| \, ||x||$, for any $\lambda \in R$.
(3) $||x + y|| \leq ||x|| + ||y||$ for any elements x, y in the space.

As examples, we introduce several widely used norms.

Example 1.1 L_2 norm

Consider a vector, $x = [x_1, x_2, \dots, x_n]^T$. The Euclidean norm ($L_2$ norm) is defined as

$$\|x_2\| = \sqrt{\sum_{i=1}^{n} x_i^2} = \sqrt{x^T x}.$$

We can show

(1) Positivity: $||x||_2 \geq 0$, and when $x \neq 0$, $||x||_2 > 0$.
(2) Positive scalability: $||\lambda x||_2 = \sqrt{(\lambda x)^T \lambda x} = \sqrt{\lambda^2 x^T x} = |\lambda| \sqrt{x^T x} = |\lambda| ||x||_2$.
(3) Triangular inequality: Using the Cauchy–Schwarz inequality $x^T y \leq ||x||_2 ||y||_2$, we have

$$||x+y||_2 = \sqrt{(x+y)^T (x+y)} = \sqrt{x^T x + 2x^T y + y^T y} \leq \sqrt{(||x||_2 + ||y||_2)^2} = ||x||_2 + ||y||_2.$$

Example 1.2 L_1 norm

An L_1 norm of x is defined as

$$\|x\|_1 = \sum_{i=1}^{n} |x_i|.$$

We can show that L_1 is a norm. In fact, we have

(1) $||x||_1 = \sum_{i=1}^{n} |x_i| > 0$ if $x \neq 0$.
(2) $||\lambda x||_1 = \sum_{i=1}^{n} |\lambda x_i| = |\lambda| \sum_{i=1}^{n} |x_i| = |\lambda| ||x||_1$.
(3) $||x+y||_1 = \sum_{i=1}^{n} ||x_i + y_i|| \leq \sum_{i=1}^{n} (|x_i| + |y_i|) = \sum_{i=1}^{n} |x_i| + \sum_{i=1}^{n} |y_i| = ||x||_1 + ||y||_1$.

Example 1.3 L_∞ and L_P norms

Other popular norms include the L_∞ norm and L_P norm. The L_∞ norm and L_P norm are respectively defined as

$$||x||_\infty = \max\{x_1, \ldots, x_n\} \quad \text{and} \quad ||x||_P = \left(\sum_{i=1}^{n} x_i^p \right)^{\frac{1}{p}}.$$

We can similarly show that L_∞ and L_P satisfy the requirements of norm definition (Exercise 1.1).

Now we use a simple example to compare the magnitudes of the above defined norms.

Example 1.4

Let $x = [3, 3, \ldots, 3]^T$. Then, we have

$$||x||_1 = 3n,$$

$$||x||_2 = 3\sqrt{n},$$

$$||x||_p = 3n^{\frac{1}{p}}, \quad \text{and}$$

$$||x||_\infty = 3.$$

Next we introduce the vector norm–induced matrix norm.

Definition 1.2 (Induced matrix norm)

The vector norm $||\cdot||$ induced matrix norm is defined as

$$||A|| = \sup\left\{\frac{||Ax||}{||x||}\right\}. \tag{1.1}$$

We first check that the induced matrix norm defined in Equation 1.1 is a norm:

(1) $||A|| > 0$ if $A \neq 0$.
(2) $||\lambda A|| = \sup\left\{\frac{||\lambda Ax||}{||x||}\right\} = \sup\left\{\frac{|\lambda|\,||Ax||}{||x||}\right\} = |\lambda| \sup\left\{\frac{||Ax||}{||x||}\right\} = |\lambda|\,||A||$.
(3) $||A+B|| = \sup\left\{\frac{||(A+B)x||}{||x||}\right\} \leq \sup\left\{\frac{||Ax||+||Bx||}{||x||}\right\} = ||A|| + ||B||$.

The induced matrix norm has very useful properties:

(1) $||I|| = 1$.
(2) $||Ax|| \leq ||A||\,||x||$.
(3) $||AB|| \leq ||A||\,||B||$.

In fact, we can show that

(1) $||I|| = \sup\left\{\frac{||Ix||}{||x||}\right\} = \sup\left\{\frac{||x||}{||x||}\right\} = 1$.

(2) By definition, $||A|| = \sup\left\{\frac{||Ax||}{||x||}\right\} \geq \frac{||Ax||}{||x||}$, which implies that $||A||\,||x|| \geq ||Ax||$.

(3) $||AB|| = \sup\left\{\frac{||ABx||}{||x||}\right\} \leq \sup\left\{\frac{||A||\,||Bx||}{||x||}\right\} = ||A|| \sup\left\{\frac{||Bx||}{||x||}\right\} = ||A||\,||B||$.

Now we illustrate how to calculate the L_1, L_2, L_p, and L_∞ induced matrix norms.

Example 1.5

(1) $||A||_1 = \max_j \sum_i |a_{ij}|$.

(2) $||A||_\infty = \max_i \sum_j |a_{ij}|$.

(3) $||A||_2 = \max_j \sqrt{\lambda_j}$,

where λ_j are the eigenvalues of the matrix $A^T A$.

We first prove (1). By definition, we have

$$||A||_1 = \sup\left\{\frac{||Ax||_1}{||x||_1}\right\} = \sup\left\{\frac{\sum_i \left|\sum_j a_{ij}x_j\right|}{||x||_1}\right\} \le \sup\left\{\frac{\sum_i \sum_j |a_{ij}||x_j|}{||x||_1}\right\}$$

$$\le \max_j \sum_i |a_{ij}| \sup\left\{\frac{\sum_j |x_j|}{||x||_1}\right\} = \max_j \sum_i |a_{ij}|.$$

Suppose that $\sum_i |a_{ij_0}| = \max_j \sum_i |a_{ij}|$. Take $x_0 = [0, \ldots, 1, \ldots, 0]^T$, where 1 is located at the j_0 position. Then, we have $||A||_1 \ge \frac{||Ax_0||_1}{||x_0||_1} = \frac{\sum_i |a_{ij_0}|}{1} = \sum_i |a_{ij_0}| = \max_j \sum_i |a_{ij}|$.

Therefore, we have $||A||_1 = \max_j \sum_i |a_{ij}|$.

Formula (2) can be similarly proved. Now we prove (3). By definition, $||A||_2$ is defined as

$$||A||_2 = \sup\left\{\frac{||Ax||_2}{||x||_2}\right\} = \sup\left[\frac{\sqrt{x^T A^T A x}}{||x||_2}\right] = \sup\left\{\frac{\sqrt{\lambda x^T x}}{||x||_2}\right\} \le \sqrt{\lambda_{\max}}.$$

Suppose that $A^T A x_{\max} = \lambda_{\max} x_{\max}$. Then, we have $||A||_2 \ge \frac{||Ax_{\max}||_2}{||x_{\max}||_2} = \frac{\sqrt{x_{\max}^T A^T A x_{\max}}}{||x_{\max}||_2} = \frac{\sqrt{\lambda_{\max}}||x_{\max}||_2}{||x_{\max}||_2} = \sqrt{\lambda_{\max}}$, which implies that

$$||A||_2 = \sqrt{\lambda_{\max}}.$$

Next we introduce other widely used matrix norms.

1.1.1 "Entrywise" Norms

Entrywise norm of the matrix is defined as

$$||A||_p = \left(\sum_i \sum_j |a_{ij}|^p\right)^{\frac{1}{p}}.$$

1.1.1.1 $L_{2,1}$ Norm

The $L_{2,1}$ norm is defined as the sum of the Euclidean norm of the columns of the matrix

$$||A||_{2,1} = \sum_j \left(\sum_i |a_{ij}|^2 \right)^{\frac{1}{2}}.$$

The $L_{2,1}$ norm can be generalized to $L_{p,q}$ norm.

1.1.1.2 $L_{p,q}$ Norm

The $L_{p,q}$ norm of the matrix is defined as

$$||A||_{p,q} = \left[\sum_j \left(\sum_i |a_{ij}|^p \right)^{q/p} \right]^{\frac{1}{q}}.$$

1.1.2 Frobenius Norm

The Frobenius norm of the matrix, a very useful norm in numerical analysis, is defined as

$$||A||_F = \sqrt{\sum_i \sum_j |a_{ij}|^2} = \sqrt{\text{Trace}\left(A^T A\right)}. \tag{1.2}$$

1.1.2.1 l_1/l_2 Norm

A gene consists of multiple SNPs and a pathway consists of multiple genes. The multiple SNPs or multiple genes can form groups. A gene or a pathway can be taken as a unit in the analysis. We can select or remove jointly all the variables in the group. To explicitly explore the group structure, we define a l_1/l_2 norm as a group norm. Let a set of indices, $\{1, \ldots, p\}$, be divided into a set if nonoverlapping groups G and $(w_g)_{g \in G}$ be some strictly positive weights. An l_1/l_2 norm is defined as (Bach et al. 2011)

$$\Omega(x) = \sum_{g \in G} w_g ||x_g||_2 = \sum_{g \in G} w_g \sqrt{\sum_{gj} x_{gj}^2}. \tag{1.3}$$

We can show that the l_1/l_2 norm is a norm. In fact, it satisfies the following conditions:

(i) If $x \neq 0$, then there exists at least one $g_0 \in G$ such that $||x_{g0}||_2 > 0$. Thus, $\Omega(x) \geq w_{g0} ||x_{g0}||_2 > 0$.

(ii) $\Omega(\lambda x) = \sum_g w_g ||\lambda x_g||_2 = |\lambda| \sum_g w_g ||x_g||_2 = |\lambda| \Omega(x)$.

(iii) $\Omega(x+y) = \sum_g w_g ||(x+y)_g||_2 \leq \sum_g w_g (||x_g||_2 + ||y_g||_2) = \Omega(x) + \Omega(y)$.

The l_1/l_2 norm can be generalized to an l_1/l_q **norm**:

$$\Omega(x)=\sum_{g\in G} w_g \|x_g\|_q. \quad (1.4)$$

1.1.3 Overlapping Groups

Rich structures exist in the data. In the previous section, we introduce nonoverlapping group structure. Now we consider overlapping group structure. Overlapping groups are defined as groups that share at least one variable. Figure 1.1 shows one-dimensional overlapping group, two-dimensional overlapping group, and hierarchical and directed acyclic graph (DAG) structures.

In Figure 1.1a, group 1 overlaps with group 2, group 2 overlaps with group 3, and group 3 overlaps with groups 2 and 4. The data can be organized on a two-dimensional space. Figure 1.1b shows that group 3 covers groups 1 and 2, and group 2 covers group 1. In the tree and graph structure, the variables correspond to the nodes of the tree or graph. The node and its ancestors are overlapped.

The l_1/l_2 and l_1/l_q norms defined for nonoverlapping groups can also be used for overlapping groups. When some of groups overlap, Equations 1.3 and 1.4 still define a norm because all covariates should be included in at least one group (Obozinski et al. 2011). The l_1/l_2 and l_1/l_q norms can be used for hierarchical variable selection and structure sparsity (Bach 2010; Jenatton et al. 2011). The precise effect of the penalty for group norm is to set the groups of variables to zero and not to select them. When groups are not overlapped, setting the norms of the subgroups to zero implies to leave the other full groups of variables to nonzero. However, in the presence of group overlapping, setting the norm of one group to zero shrinks all its variates in the group to zero, even if some variables belong to other groups. We plot Figure 1.2a with four overlapping groups of variables to illustrate this. We assume that the norms $\|X_{G_1}\|_2=0$ and $\|X_{G_3}\|_2=0$. These penalties remove all variables in groups 1 and 3 and maintain the remaining variables in groups 2 and 4, which are neither in group 1 nor in group 3. In other words, we cannot select entire groups of variables. We only can select the variables in the set

$$(G_1 \cup G_3)^c = (G_1)^c \cap (G_3)^c.$$

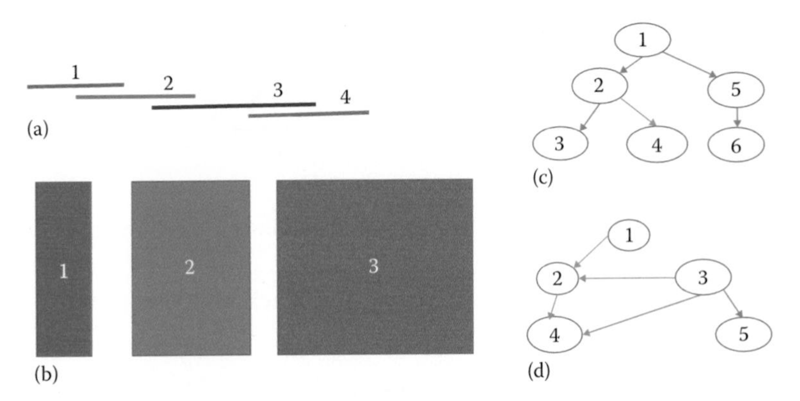

FIGURE 1.1 (a) One-dimensional overlapping sequences. (b) Two-dimensional overlapping structure. (c) Tree structure. (d) Directed acyclic graph structure.

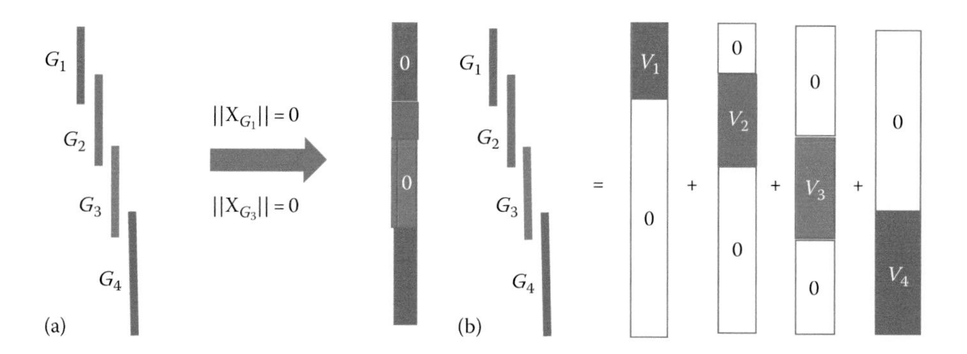

FIGURE 1.2 (a) Effect of penalty for group norms. (b) Latent decomposition of the data.

In practice, we often want to select entire groups of variables. For example, as Figure 1.2b shows, the variable x belongs to both groups G_1 and G_2. We want to remove the group G_2 and select the group G_1; the variable x in the group G_1 can still be selected, even if the group G_2 is to be removed. To achieve this, we need to introduce latent variables to independently represent variables in the different groups. Let $\bar{V} = \left(v^g\right)_{g \in G}$ be a set of latent variables such that $v^g \in R^p$ and support of $v^g \subset g$. In other words, we assume

$$\forall g \in G, \quad v_j^G = 0, \quad \text{if } j \notin g.$$

We decompose the vector space X as a sum of latent variables whose support is included in each group:

$$X = \sum_{g \in G} v^g. \tag{1.5}$$

The formula (1.4) is then reduced to

$$\Omega_{\cup}\left(x\right) = \sum_{g \in G} w_g ||v^g||_q. \tag{1.6}$$

Unlike the penalty $\Omega(x)$ enforcing the group G_2 of variables to be zero, the variable x belonging to the group G_2 should be zero; for the penalty $\Omega_{\cup}(x)$ enforcing the group G_2 of variables to be zero, the variable x in the group G_1 can be nonzero (Figure 1.2). Since decomposition (1.5) is not unique, we minimize the $\Omega_{\cup}(x)$ over $\bar{V}$:

$$\Omega_{\cup}\left(x\right) = \min_{v \in R^{p \times |G|}} \sum_{g \in G} w_g ||v^g||_q, \quad \text{s.t.} \quad X = \sum_{g \in G} v^g. \tag{1.7}$$

$\Omega_{\cup}(x)$ defined in Equation 1.7 is referred to as the **latent group lasso norm.**

Next we prove that $\Omega_U(x)$ is a norm. Indeed, we can prove this through the following:

(i) If $x \neq 0$, then there exists a g_0 such that $v^{g_0} \neq 0$, which implies that $\Omega_U(x) \geq w_{g_0} \|v^{g_0}\|_q > 0$.

(ii) $\Omega_U(\lambda x) = \min_{v \in R^{p \times |G|} \sum_{g \in G} v^g = X} \sum_{g \in G} \left\| \lambda v^g \right\|_q = \min_{v \in R^{p \times |G|} \sum_{g \in G} v^g = X} \sum_{g \in G} |\lambda| \left\| v^g \right\|_q = |\lambda| \Omega_U(x).$

(iii) $\Omega_U(x+y) = \min_{v \in R^{p \times |G|} \sum_{g \in G} v^g = X} \sum_{g \in G} \left\| v_x^g + v_y^g \right\|_q \leq \min_{v \in R^{p \times |G|} \sum_{g \in G} v^g = X} \sum_{g \in G} \left\| v_x^g \right\|_q + \left\| v_y^g \right\|_q$

$\leq \Omega_U(x) + \Omega_U(y).$

1.1.4 Dual Norm

For a given norm $||.||$, a dual norm, denoted by $\|\cdot\|_*$, is defined as

$$||y||_* = \max \frac{y^T x}{||x||}.$$

We can show that the dual norm satisfies three conditions of the norm:

(i) $||y||_* \geq \frac{y^T y}{||y||} > 0$, if $y \neq 0$.

(ii) If $\lambda > 0$, then $||\lambda y||_* = \max \frac{\lambda y^T x}{||x||} = \lambda \max \frac{y^T x}{||x||} = \lambda ||y||_*$.

(iii) $||y+z||_* = \max \frac{(y+z)^T x}{||x||} = \max \left(\frac{y^T x}{||x||} + \frac{z^T x}{||x||} \right) \leq \max \frac{y^T x}{||x||} + \max \frac{z^T x}{||x||} = ||y||_* + ||z||_*$.

By definition of dual norm, we have

$$\frac{y^T x}{||x||} \leq \max \frac{y^T x}{||x||} = ||y||_*.$$

Multiplying both sides of the above inequality by $\|x\|$, we obtain $y^T x \leq \|x\| \, \|y\|_*$, which is a generalization of the Cauchy–Schwartz inequality.

Example 1.6

(1) The norm dual to the L_2 norm is itself. In fact, by the traditional Cauchy–Schwartz inequality, we have $||y||_* = \max \frac{y^T x}{||x||_2} \leq \frac{||x||_2 ||y||_2}{||x||_2} = ||y||_2$. Taking $x_0 = y$, we have $||y||_* \geq \frac{y^T y}{||y||_2} = ||y||_2$. Therefore, we obtain $\|y\|_* = \|y\|_2$.

(2) The norm dual to the L_∞ norm is the L_1 norm. Similarly, we have $y^T x = \sum_j y_j x_j \le \sum_j |x_j||y_j| \le \max(|x_j|)\sum_j |y_j| \le ||x||_\infty ||y||_1$, i.e., $||y||_1 \ge \max \frac{y^T x}{||x||_\infty} = ||y||_*$. Taking $x = sign(y)$, we obtain $||y||_* \ge \frac{y^T sign(y)}{||sign(y)||_\infty} = \frac{\sum_j |y|_j}{1} = ||y||_1$. Combining the above two inequalities, we have $||y||_* = ||y||_1$.

(3) The norm dual to the L_1 norm is the L_∞ norm. It is clear that $y^T x = \sum_j y_j x_j \le \sum_j |x_j||y_j|| \le \max(|y_j|)\sum_j |x_j| \le ||x||_1 ||y||_\infty$, which implies that $||y||_* \le y_\infty$. On the other hand, suppose that at j_0, we have $||y||_\infty = |y_{j0}|$. Taking $x_0 = [0, \ldots, y_{j0}, \ldots, 0]^T$, we can obtain $||y||_* \ge \frac{y^T x_0}{||x_0||_1} = \frac{y_{j0}^2}{|y_{j0}|} = |y_{j0}| = |y|_\infty$. Combining the above two inequalities, we obtain $||y||_* = ||y||_\infty$.

(4) The norm dual to L_p is the L_q norm. Now we prove this.

It is well known that $y^T x \le ||x||_q ||y||_p$, which implies that $||y||_* \le ||y||_p$. On the other hand, taking $x = \sum_j y_j^{\frac{1}{q-1}}$, we obtain $||y||_* \ge \frac{y^T x}{||x||_q} = \frac{\sum_j y_j^p}{\left(\sum_j y_j^p\right)^{\frac{1}{q}}} = \left(\sum_j y_j^p\right)^{\frac{1}{p}} = ||y||_p$.

Thus, combining the above two inequalities, we prove that the norm dual to L_p is the L_q norm.

1.1.4.1 The Norm Dual to the Group Norm

Proposition 1.1

The dual norm $\Omega^*(y)$ of the group norm $\Omega(x)$ satisfies

$$\Omega^*(y) = \max_{g \in G} \frac{1}{w_g} \Omega^*(y_g), \tag{1.8}$$

where $\Omega^*(y_g) = ||y_g||_p$ if the group norm is defined by the $l_1//l_q$ norm in Equation 1.4.

It is clear that

$$y^T x \le \sum_{g \in G} y_g^T x_g \le \sum_{g \in G} ||y_g||_* ||x_g|| = \sum_{g \in G} \frac{||y_g||_*}{w_g} w_g ||x_g||$$
$$\le \max_{g \in G} \frac{||y_g||_*}{w_g} \sum_{g \in G} w_g ||x_g|| = \max_{g \in G} \frac{\Omega^*(y_g)}{w_g} \Omega(x),$$

which implies that $\Omega^*(y) \le \max_{g \in G} \frac{\Omega^*(y_g)}{w_g}$.

Let g_0 denote $\frac{\Omega^*\left(y_{g0}\right)}{w_{g0}}=\max_{g\in G}\frac{1}{w_g}\Omega^*\left(y_g\right)$ and $x=[0,\ldots,x_{g_0},\ldots,0]^T$. Then, we have $\Omega^*\left(y\right)\geq\frac{y_{g0}^T x_{g0}}{\Omega\left(x_{g0}\right)}=\frac{y_{g0}^T w_{g0}x_{g0}}{w_{g0}\Omega\left(x_{g0}\right)}=\max_{g\in G}\frac{1}{w_g}\Omega^*\left(y_g\right)\frac{\Omega\left(x_{g0}\right)}{\Omega\left(x_{g0}\right)}=\max_{g\in G}\frac{1}{w_g}\Omega^*\left(y_g\right)$. Therefore, combining the above two inequalities, we obtain $\Omega^*\left(y\right)=\max_{g\in G}\frac{1}{w_g}\Omega^*\left(y_g\right)$.

The results can be extended to the latent group lasso norm. Similarly, we can prove Proposition 1.2.

Proposition 1.2

The dual norm $\Omega^*(y)$ of the latent group lasso norm $\Omega(x)$ satisfies (Exercise 1.3)

$$\Omega^*\left(y\right)=\max_{g\in G}\frac{1}{w_g}\Omega^*\left(y_g\right). \tag{1.9}$$

Proposition 1.3

Dual to the dual norm is the original norm. Let $\Omega(x)$ be the original norm, $\Omega^*(y)$ be the dual norm of the norm $\Omega(x)$, and $\Omega^{**}(x)$ be the dual to the dual norm $\Omega^*(y)$. Then, $\Omega^{**}(x)=\Omega(x)$.

Proof.

We sketch the proof as follows. By definition of dual norm of the norm $\Omega(x)$, we have

$$x^T y\leq\Omega\left(x\right)\Omega^*\left(y\right),\quad\forall y,x,$$

which implies that

$$\Omega^{**}\left(x\right)=\sup_y\left\{\frac{x^T y}{\Omega^*\left(y\right)}\right\}\leq\Omega\left(x\right). \tag{1.10}$$

On the other hand, suppose that $\Omega^*\left(y\right)=\frac{x_0^T y}{\Omega\left(x_0\right)}$. Then, we have

$$\Omega^{**}\left(x_0\right)\geq\frac{x_0^T y}{\Omega^*\left(y\right)}=\frac{x_0^T y}{x_0^T y/\Omega\left(x_0\right)}=\Omega\left(x_0\right).$$

1.1.5 Fenchel Conjugate

A function is often represented as a locus of points $(x,f(x))$. However, in some cases, it is more convenient to represent the function as an envelope of tangents. A tangent is parameterized by the slope z and the intercept $f^*(z)$, which is referred to as Fenchel conjugate. Formally, the Fenchel conjugate of a function is defined as

$$f^*(z) = \sup_{x \in \mathrm{dom}\, f} \{x^T z - f(x)\}. \tag{1.11}$$

Example 1.7 Norm function

Let $f(x) = ||x||$. Then, $f^*(z) = I_{||z||_* \le 1}(z)$, where I is an indicator function. In fact, by definition, we have

$$f^*(z) = \sup_{x \in \mathrm{dom}\, f} \{x^T z - ||x||\}.$$

Consider two cases: (1) $||z||_* \le 1$ and (2) $||z||_* > 1$.

First, we consider $||z||_* \le 1$.

By Cauchy–Schwarz inequality, we have

$$x^T z \le ||z||_* ||x|| \quad \text{and} \quad \sup\{x^T z\} = ||z||_* ||x||.$$

Therefore, (1.11) can be reduced to

$$f^*(z) = \sup_{x \in \mathrm{dom}\, f} \{||x||(||z||_* - 1)\} \le 0$$

when $||z||_* \le 1$, which implies

$$f^*(z) = 0, \quad \text{when } ||z||_* \le 1. \tag{1.12}$$

Now we consider $||z||_* > 1$. By definition of dual norm, we have

$$||z||_* = \sup_x \left\{\frac{z^T x}{||x||}\right\} = \frac{z^T x_0}{||x_0||}.$$

Let $x = \lambda x_0$. Then, we have

$$\begin{aligned} f^*(z) &= \sup_{x \in \mathrm{dom}\, f} \{x^T z - ||x||\} \\ &= \lambda \sup_{x \in \mathrm{dom}\, f} \left\{||x_0|| \left(\frac{x_0^T z}{||x_0||} - 1\right)\right\} \\ &= \sup_{x \in \mathrm{dom}\, f} \{\lambda ||x_0||(||z||_* - 1)\} \to \infty. \end{aligned} \tag{1.13}$$

Combining Equations 1.12 and 1.13, we obtain

$$f^*(z) = \begin{cases} 0 & ||z||_* \le 1 \\ \infty & \text{otherwise.} \end{cases} \tag{1.14}$$

Example 1.8

A general quadratic curve $f(x) = \frac{1}{2} x^T A x$ where A is a symmetric matrix and its inverse exists.

Let $F = z^T x - \frac{1}{2} x^T A x$. Then, differentiating F with respect to x, we obtain

$$\frac{\partial F}{\partial x} = z - Ax = 0,$$

which implies that $x = A^{-1}z$. Therefore, we have

$$f^*(z) = z^T A^{-1} z - \frac{1}{2} z^T A^{-1} A A^{-1} z = \frac{1}{2} z^T A^{-1} z.$$

Example 1.9 l_p Norms $f(x) = \frac{1}{p} ||x||^p$

Let $F = x^T z - \frac{1}{p} ||x||^p$. Differentiating F with respect to x yields

$$\frac{\partial F}{\partial x} = z - ||x||^{p-1} \frac{x}{||x||} = 0,$$

which implies that

$$z = x||x||^{p-2}. \tag{1.15}$$

Taking $||.||$ on both sides of the above equation, we obtain

$$||z|| = ||x||^{p-1}$$

or

$$||x|| = ||z||^{\frac{1}{p-1}}. \tag{1.16}$$

Substituting Equation 1.16 into Equation 1.15 yields

$$x = z||z||^{-\frac{p-2}{p-1}}.$$

Thus, we obtain

$$f^*(z) = \sup_x \left\{ x^T z - \frac{1}{p} ||x||^p \right\}$$
$$= ||z||^{-\frac{p-2}{p-1}} ||z||^2 - \frac{1}{p} ||z||^{\frac{p}{p-1}}$$
$$= \left(1 - \frac{1}{p}\right) ||z||^{\frac{1}{1-\frac{1}{p}}}$$
$$= \frac{1}{q} ||z||^q, \tag{1.17}$$

where $q = \frac{1}{1 - \frac{1}{p}}$.

1.1.6 Fenchel Duality

In many cases, the Fenchel conjugate of the objective functions has closed forms and can be easily computed. The primal problems are often transformed to the dual problems that can be much easily solved.

Consider this primal problem:

$$\begin{aligned} &\text{minimize} \quad f_1(x) - f_2(x) \\ &\text{subject to} \quad x \in X_1 \cap X_2, \end{aligned}$$

where $f_1 : R^n \to (-\infty, \infty]$ and $f_2 : R^n \to [-\infty, \infty)$.

Suppose that $y = u$; the primal problem can be further transformed to

$$\begin{aligned} &\text{minimize} \quad f_1(y) - f_2(u) \\ &\text{subject to} \quad y = u. \end{aligned}$$

Multiplying the Lagrange multiplier to the constraint $y = u$ and adding it to the objective function, we obtain

$$q(z) = \min_{y,u} \left\{ f_1(y) - f_2(u) + (u - y)^T z \right\}$$
$$= \min_u \left\{ z^T u - f_2(u) \right\} - \max_y \left\{ z^T y - f_1(y) \right\}$$
$$= f_{2*}(z) - f_1^*(z).$$

The dual problem is then defined as

$$f^* = \max_z q(z) = \max_z \left\{ f_{2*}(z) - f_1^*(z) \right\}.$$

In general, by duality we have

$$\max_z \left\{ f_{2*}(z) - f_1^*(z) \right\} \le \min_x \left\{ f_1(x) - f_2(x) \right\}.$$

Under an optimal condition that there is no duality gap, we have

$$\max_z \left\{ f_{2*}(z) - f_1^*(z) \right\} = \min_x \left\{ f_1(x) - f_2(x) \right\}.$$

Let $f_1(x) = f(x)$ and $f_2(x) = -\lambda\|x\|$. By definition of Fenchel conjugate, we obtain

$$\begin{aligned} f_{2*}(z) &= \inf_x \left\{ z^T x - (-\lambda\|x\|) \right\} \\ &= \inf_x \left\{ -\lambda \left[\left(\frac{z}{-\lambda} \right)^T x - \|x\| \right] \right\} \\ &= -\lambda \sup_x \left\{ \left[\left(\frac{z}{-\lambda} \right)^T x - \|x\| \right] \right\} \\ &= -\lambda \left\| \frac{z}{\lambda} \right\|_* \\ &= \begin{cases} 0 & \|z\|_* \le \lambda \\ -\infty & \|z\|_* > \lambda \end{cases} \end{aligned}$$

which implies that

$$\max_z \left\{ f_{2*}(z) - f_1^*(z) \right\} = \max_{\|z\|_* \le \lambda} - f^*(z) \le \min_x \{ f_1(x) - f_2(x) \}.$$

We prove the following proposition.

Proposition 1.4 Fenchel Duality theorem

If $f^*(z)$ and $\|z\|_*$ are the Fenchel conjugate of a convex and the differentiable function $f(x)$ and the dual norm of $\|x\|$, respectively, then we have

$$\max_{\|z\|_* \le \lambda} - f^*(z) \le \min_x \left\{ f(x) + \lambda\|x\| \right\}.$$

Quality holds when the domain of function $f(x)$ has a nonempty interior.

The difference between the left and right terms in the above inequality is referred to as a duality gap. The duality gap denotes the difference between the value of the primal objective function $f(x)+\lambda||x||$ and a dual objective function $-f^*(z)$. Duality gaps are important in convex analysis. It offers an upper bound between the current value, a primal objective function, and the optimal value, which can be used as a criterion for stopping iteration.

Given a current iterative x, calculating a duality gap requires information on the value of dual variable z. Recall that $f^*(z)=\sup_x\{x^Tz-f(x)\}$. $z^*=\nabla f(x^*)$ is the unique solution to the dual problem. Making $||z||_*$ to satisfy $||z||_*\leq\lambda$, we set $z=\min\left(1,\frac{\lambda}{||\nabla f(x)||_*}\right)\nabla f(x)$. Then, we calculate a duality gap:

$$\text{Gap}=f(x)+\lambda||x||+f^*(z).$$

When a gap is less than a prespecified error ε, i.e., $\text{Gap}\leq\varepsilon$, iteration stops.

The function $f(f)$ often takes the form $f(x)=\varphi(Wx)$, where $\varphi:R^n\rightarrow R$ and W is a design matrix. For example, the cost function for linear regression is of the form

$$\varphi(Wx)=||Y-Wx||_2^2,$$

where Y and W are observations and Wx are predictions. The Fenchel conjugate of φ is easy to compute. In this case, the primal optimization problem can be rewritten as

$$\begin{aligned}&\min_{x\in R^P,u\in R^n}\ \varphi(u)+\lambda||x||\\&\text{s.t.}\quad u=Wx.\end{aligned}$$

Similar to Proposition 1.4, Proposition 1.5 specifies the Fenchel dual problem for the above specific case.

Proposition 1.5

Let $\varphi^*(\alpha)=\sup_u\{\alpha^Tu-\varphi(u)\}$. The Fenchel dual of the above primal problem is

$$\begin{aligned}&\max_{\alpha}-\varphi^*(\lambda\alpha)\\&\text{s.t.}\ \ ||W^T\alpha||_*\leq 1.\end{aligned}$$

Proof.

Adding the product of the Lagrangian multiplier and the constraint to the objective function can obtain the Lagrangian dual of the primal problem:

$$\min_{x\in R^P,u\in R^n}\ \max_{\alpha\in R^n}\varphi(u)+\lambda||x||+\lambda\alpha^T(Wx-u)$$

or

$$\min_{x\in R^p, u\in R^n} \max_{\alpha\in R^n} \varphi(u)-\lambda\alpha^T u+\lambda\left(||x||+\alpha^T Wx\right).$$

It is clear that

$$\begin{aligned}
&\min_{x\in R^p, u\in R^n} \max_{\alpha\in R^n} \varphi(u)-\lambda\alpha^T u \\
&= \max_{\alpha\in R^n} \min_{u\in R^n} -\left\{(\lambda\alpha)^T u-\varphi(u)\right\} \\
&= \max_{\alpha\in R^n} - \max_{u\in R^n}\left\{(\lambda\alpha)^T u-\varphi(u)\right\} \\
&= \max_{\alpha\in R^n} -\varphi^*(\lambda\alpha).
\end{aligned}$$

Next we consider $\min_{x\in R^p, u\in R^n} \max_{\alpha\in R^n} \lambda\left(||x||+\alpha^T Wx\right)$, which can be reduced to

$$\begin{aligned}
&\max_{\alpha\in R^n} \min_{x\in R^p, u\in R^n} \lambda\left(||x||+\alpha^T Wx\right) \\
&= \max_{\alpha\in R^n} \min_{x\in R^p} \lambda\left(\left(\alpha^T Wx\right)-\left(-||x||\right)\right) \\
&= \max_{\alpha\in R^n} \min_{x\in R^p} -\lambda\left\{(-\alpha)^T Wx-||x||\right\} \\
&= \max_{\alpha\in R^n} -\lambda \max_{x\in R^p}\left\{\left(-W^T\alpha\right)^T x-||x||\right\} \\
&= \max_{\alpha\in R^n} \lambda I_{\|W^T\alpha\|_*\le 1}(z),
\end{aligned}$$

where $I_{||W^T\alpha||_*\le 1}(z)=\begin{cases} 0 & ||W^T\alpha||_*\le 1 \\ -\infty & \text{otherwise} \end{cases}$.

Therefore, we obtain

$$\min_{x\in R^p, u\in R^n} \max_{\alpha\in R^n} \varphi(u)-\lambda\alpha^T u+\lambda\left(||x||+\alpha^T Wx\right)=\max_{\alpha\in R^n} -\varphi^*(\lambda\alpha)$$

when $||W^T\alpha||_*\le 1$.

1.2 SUBDIFFERENTIAL

The concept of subdifferential for nonsmooth convex functions was introduced and developed by Moreau and Rockafellar (Mordukhovich and Nam 2017). A subdifferential is an extension of the traditional differential of a smooth convex function to a nonsmooth convex function. The major difference between the standard derivative of a differentiable function

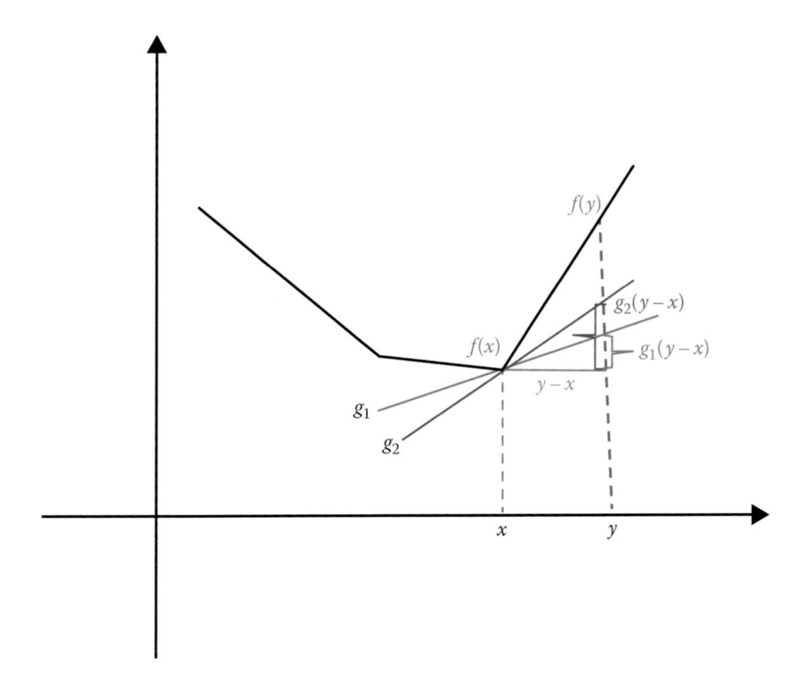

FIGURE 1.3 Concept of a subdifferential of a nondifferential function.

and the subdifferential of a convex function at a given point is that the subdifferential is a set of "derivatives," which reduces to a standard derivative if the function is differentiable.

1.2.1 Definition of Subgradient

A vector $g \in R^n$ is a subgradient of function $f: R^n \rightarrow R$ at x if for all y (Figure 1.3)

$$f(y) \geq f(x) + g^T(y - x). \tag{1.18}$$

Figure 1.3 shows that the function at point x is not differentiable; there is more than one subgradient, $g_1, g_2, \ldots$, at point x.

Subdifferential: The set of subgradients of $f(x)$ at point x is referred to as the subdifferential of $f(x)$ at x and is denoted by $\partial f(x)$.

Example 1.10 Absolute value: $f(x) = |x|$

For $x > 0$, $\partial f(x) = 1$, and for $x < 0$, $\partial f(x) = -1$. At $x = 0$, for $y > 0$ and $g \leq 1$, we have $gy \leq y = |y|$, and for $y \leq 0$ and $g \geq -1$, we have $gy \leq -y = |y|$. Combining the above two inequalities yields $|y| \geq gy$, where $|g| \leq 1$. In other words, $\partial f(0) = [-1, 1]$. Summarizing the above results, we have

$$\partial |x| = \begin{cases} 1 & x > 0 \\ -1 & x < 0 \\ [-1, 1] & x = 0. \end{cases} \tag{1.19}$$

Example 1.11 L_2 norm $\|x\|_2$

For $x \neq 0$, by calculus, we have $\frac{\partial \|x\|_2}{\partial x} = \frac{x}{\|x\|_2}$. For $x = 0$, by Cauchy inequality, we have

$$g^t y \le \|g\|_2 \|y\|_2 \le \|y\|_2, \quad \forall \|g\|_2 \le 1.$$

Therefore, the subdifferential of the L_2 norm $\|x\|_2$ is given by

$$\partial \|x\|_2 = \begin{cases} \frac{x}{\|x\|_2} & x \neq 0 \\ \left\{ g \,\middle|\, \|g\|_2 \le 1 \right\} & x = 0. \end{cases} \tag{1.20}$$

Example 1.12

Maximum of functions $f(x) = \max(f_1(x), f_2(x))$, where both $f_1(x)$ and $f_2(x)$ are convex and differentiable (Figure 1.4).

The subdifferential of function $f(x)$ is given by

$$\partial f(x) = \begin{cases} f_1'(x) & f_1(x) > f_2(x) \\ f_2'(x) & f_1(x) < f_2(x) \\ \left[f_2'(x), f_1'(x) \right] & f_1(x) = f_2(x). \end{cases} \tag{1.21}$$

1.2.2 Subgradients of Differentiable Functions

If $f(x)$ is convex and differentiable at x, then $\partial f(x) = \nabla f(x)$. Conversely, if $f(x)$ is convex and its subgradient is unique, i.e., $\partial f(x) = \{g\}$, then $f(x)$ is differentiable at x and $g(x) = \nabla f(x)$.

1.2.3 Calculus of Subgradients

Similar to the traditional calculus, we can develop calculus of subgradients that describe the rules for calculate subgradients.

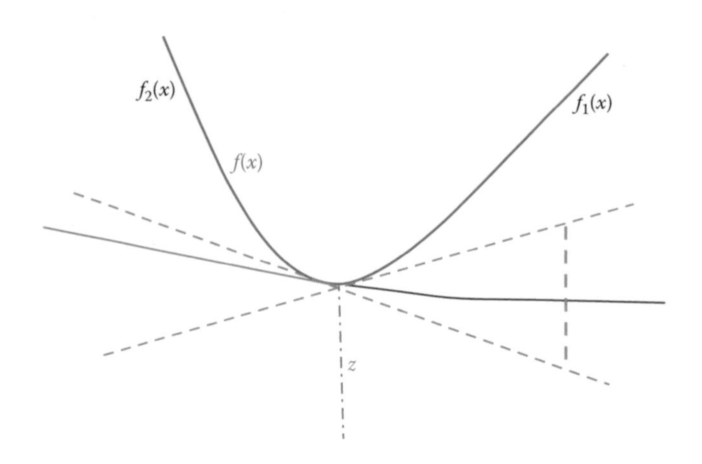

FIGURE 1.4 Subdifferential of maximum of functions.

1.2.3.1 Nonnegative Scaling

For $\alpha>0$, we have $\alpha f(y)\geq \alpha f(x)+(\alpha g(x))^T(y-x)$, where $g(x)\in\partial f(x)$. Therefore, $\partial(\alpha f(x))=\alpha\partial f(x)$.

1.2.3.2 Addition

Suppose that $f(x)=f_1(x)+\ldots+f_k(x)$, where $f_1(x),\ldots,f_k(x)$ are convex functions. Then, we have

$$\partial f(x)=\partial f_1(x)+\ldots+\partial f_k(x).$$

1.2.3.3 Affine Transformation of Variables

Suppose that $f(x)$ is convex and $h(x)=f(Ax+b)$. Then, $\partial h(x)=A^T\partial f(Ax+b)$.

In fact, by definition of subgradient, we have

$$\begin{aligned} f(Ay+b)&\geq f(Ax+b)+g^T(Ay+b-Ax-b)\\ &\geq f(Ax+b)+\left(A^Tg\right)^T\left(A(y-x)\right), \end{aligned}$$

where $g\in\partial f$. Thus, we have $\partial h(x)=A^T\partial f(Ax+b)$.

1.2.3.4 Pointwise Maximum

Suppose that $f_1(x),\ldots,f_k(x)$ are convex functions. Define function $f(x)=\max\{f_1(x),\ldots,f_k(x)\}$. Then, a subdifferential of function $f(x)$ at x is given by

$$\partial f(x)=\mathrm{Co}\bigcup\left\{\partial f_i(x)\,|\,f_i(x)=f(x)\right\}, \tag{1.22}$$

i.e., convex hull of union of subdifferentials of "active" functions where $f_i(x)=f(x)$ at x.

Assume that $f_m(x)=f(x)$ and $g\in\partial f_m(x)$. By definition of subgradient, we have

$$f(y)\geq f_m(y)\geq f_m(x)+g^T(y-x)=f(x)+g^T(y-x), \tag{1.23}$$

i.e., $g\in\partial f(x)$.

Suppose $I=\{i|f_i(x)=f(x)\}$. Then, it follows from Equation 1.23 that

$$\begin{aligned} f(y)&=\sum_{i\in I}\lambda_i f(y)\geq\sum_{i\in I}\lambda_i f_i(y)\geq\\ &=\sum_{i\in I}\lambda_i\left[f_i(x)+g_i^T(y-x)\right]=f(x)+\sum_{i\in I}\left(\lambda_i g_i\right)^T(y-x). \end{aligned}$$

Let $g=\sum_{i\in I}\lambda_i g_i$. Then, g is a subdifferential of the function $f(x)$ at x. In other words, a convex hull of union of subdifferentials of "active" functions is subdifferential of the function $f(x)$ at x.

Example 1.13 Maximum of differentiable functions

Suppose that $f_1(x), \ldots, f_k(x)$ are convex and differentiable functions and $f(x) = \max\{f_1(x), \ldots, f_k(x)\}$. Then,

$$\partial f(x) = \text{Co}\{\nabla f_i(x) | f_i(x) = f(x)\}.$$

Example 1.14 l_1-norm

The l_1-norm

$$\|x\|_1 = |x_1| + \ldots + |x_k|$$

is a nondifferentiable convex function. The function $\|x\|_1$ can be expressed as

$$\|x\|_1 = \max\{S^T x\},$$

where $S = [s_1, \ldots, s_k]^T$, $s_i = \{-1, 1\}$, $i = 1, \ldots, k$.

To find an active function $S^T x = \|x\|_1$, we can chose

$$s_i = \begin{cases} 1 & x_i > 0 \\ -1 & x_i < 0 \\ -1,1 & x_i = 0. \end{cases}$$

The function $S^T x$ is differentiable. Its unique subgradient is S. Let $g = [g_1, \ldots, g_k]^T = S$. The subdifferential that is the convex hull of all subgradients can be generated by

$$\partial\|x\|_1 = \{g \,|\, \|g\|_\infty \le 1,\ g^T x = \|x\|_1\}.$$

Figure 1.5 plots the subdifferential of the function $f(x) = \|x\|_1 = |x_1| + |x_2|$ at three points $(0, 0)$, $(0, 1)$, and $(2, 2)$.

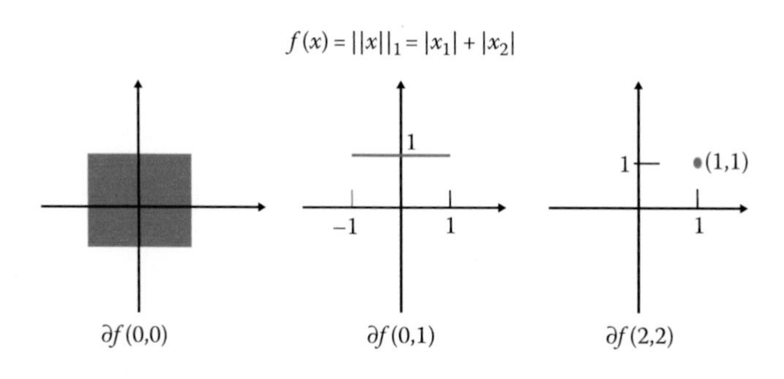

FIGURE 1.5 Subdifferential of function $\|x\|_1$.

1.2.3.5 Pointwise Supremum

The results in Section 1.2.3.4 can be extended to the supremum over an infinite number of convex functions. We consider

$$f(x)=\sup_{\alpha\in A}\{f_\alpha(x)\},$$

where the function $f_\alpha(x)$ is subdifferentiable. Suppose that $\beta\in A$ and $f_\beta(x)=f(x)$. Let $g\in\partial f_\beta(x)$. Then, $g\in\partial f(x)$. The subdifferential of function $f(x)$ can be represented by

$$\partial f(x)=\mathrm{Co}\cup\{\partial f_\alpha(x)|f_\alpha(x)=f(x)\}. \quad (1.24)$$

Example 1.15 Maximum eigenvalue of a symmetric matrix function

Consider a symmetric matrix function: $A(x)=A_0+x_1A_1+\ldots+x_kA_k$. Let $f(x)=\lambda_{\max}(A(x))$. The function $f(x)$ can be expressed as the pointwise supremum of a convex function:

$$f(x)=\lambda_{\max}(A(x))=\sup_{\|y\|_2=1}y^TA(x)y.$$

Now calculation of a subdifferential of the maximum eigenvalue of the symmetric matrix function is transformed as the calculation of the subdifferential for the pointwise supremum of the convex function. Define an index set $A=\{y\|y\in R^k,\|y\|_2\le1\}$. Function $f_y(x)=y^TA(x)y$ can be expressed as

$$f_y(x)=y^TA(x)y=y^TA_0y+x_1y^TA_1y+\ldots+x_ky^TA_ky. \quad (1.25)$$

The function $f_y(x)$ is differentiable with gradient

$$\nabla_xf_y(x)=\left[y^TA_1y,\ldots,y^TA_ky\right]^T.$$

To identify active function $f_y(x)=y^TA(x)y$, we compute an eigenvector y with the largest eigenvalue $\lambda_{\max}(A(x))$, normalized to have a unit norm. The subdifferential of the function $f(x)=\lambda_{\max}(A(x))$ is given by

$$\partial f(x)=\mathrm{Co}\{\nabla_xf_y(x)|,A(x)y=\lambda_{\max}y|,\|y\|_2=1\}. \quad (1.26)$$

1.2.3.6 Expectation

Consider convex function $f(x,z)$ and its expectation, $f(x)=E_z[f(x,z)]$. Let $g(x,z)\in\partial f(x,z)$. Then, $g(x)=E_z[g(x,z)]\in\partial f(x)$. To see this, by definition of subgradient of the function $f(x,z)$ with respect to x, we have

$$f(y,z)\ge f(x,z)+g(x,z)^T(y-x). \quad (1.27)$$

Taking expectation on both sides of the inequality (1.27), we obtain

$$E_z\left[f(y,z)\right] \geq E_z\left[f(x,z)\right] + E_Z\left[g(x,z)\right]^T (y-x). \tag{1.28}$$

Let $f(x) = E_z[f(y,z)]$, $f(x) = E_z[f(x,z)]$, and $g(x) = E_z[g(x,z)]$. Then, inequality (1.28) is reduced to

$$f(y) \geq f(x) + g(x)^T (y-x),$$

which shows that $g(x) \in \partial f(x)$.

1.2.3.7 Chain Rule

Similar to calculus, chain rule is also an important tool in subdifferential calculus. We consider the following composite function:

$$f(t) = \begin{bmatrix} f_1(t_1,\ldots,t_n) \\ \vdots \\ f_m(t_1,\ldots,t_n) \end{bmatrix}, \quad k(x) = \begin{bmatrix} k_1(x_1,\ldots,x_p) \\ \vdots \\ k_n(x_1,\ldots,x_p) \end{bmatrix}, \quad \text{and} \quad h(x) = f(k(x)).$$

Then, the chain rule for subdifferential is given by

$$\partial h(x) = (\partial f)(\partial k),$$

where

$$\partial f = \begin{bmatrix} \partial f_1(t_1,\ldots,t_n) \\ \vdots \\ \partial f_m(t_1,\ldots,t_n) \end{bmatrix} \quad \text{and} \quad \partial k = \begin{bmatrix} \partial k_1(x_1,\ldots,x_p) \\ \vdots \\ \partial k_n(x_1,\ldots,x_p) \end{bmatrix}.$$

We briefly show the chain rule as follows. By definition of subgradient, we have

$$\begin{aligned} h(y) = f(k(y)) &\geq f(k(x)) + (\partial f)(k(y) - k(x)) \\ &\geq h(x) + (\partial f)(\partial k)(y-x). \end{aligned}$$

1.2.3.8 Subdifferential of the Norm

Let $\Omega(w)$ be a norm. Then, its differential is given by

$$\partial\Omega(w) = \begin{cases} \left\{z \in R^p;\ \Omega^*(z) \leq 1\right\} & \text{if } w = 0, \\ z \in R^p;\ \Omega^*(z) = 1 \text{ and } z^T w = \Omega(w) & \text{otherwise.} \end{cases} \tag{1.29}$$

Proof.

There are two cases: (i) $w=0$ and (ii) $w \neq 0$. We first consider (i) $w=0$. Using $1 \geq \Omega^*(z)$ and Cauchy inequality, we obtain

$$\begin{aligned}\Omega(u) &\geq \Omega^*(z)\Omega(u)\\ &\geq z^T u\\ &\geq \Omega(w)+z^T(u-w),\end{aligned}$$

which implies $\{z|\,\Omega^*(z) \leq 1\} \in \partial\Omega(w)$.

Next consider case (ii) $w \neq 0$. Let $z^T w = \Omega(w)$ and $\Omega^*(z) = 1$. By Cauchy inequality, we obtain

$$\begin{aligned}z^T u &\leq \Omega^*(z)\Omega(u)\\ &\leq \Omega(u),\end{aligned}$$

which implies that

$$\Omega(u) \geq \Omega(w) - z^T w + z^T u = \Omega(w) + z^T(u-w).$$

Therefore, by definition of subdifferential, $\{z|\,\Omega^*(z)=1 \text{ and } z^T w = \Omega(w)\}$ is the subdifferential $\partial\Omega(w)$ of $\Omega(w)$.

1.2.3.9 Optimality Conditions: Unconstrained

In standard calculus, it is well known that the necessary and sufficient condition for a differentiable function $f(x)$ to reach its optimal value at point x_* is $\nabla f(x_*)=0$. This optimality condition can be extended to a nondifferentiable convex function. Let

$$f(x_*) = \inf_x f(x).$$

A necessary and sufficient optimality condition is

$$0 \in \partial f(x_*). \tag{1.30}$$

Proof.

By definition of subdifferential, we have

$$\begin{aligned}f(x) &\geq f(x_*) + g^T(x-x_*)\\ &= f(x_*) + 0^T(x-x_*),\end{aligned}$$

which implies $0 \in \partial f(x_*)$.

1.2.3.10 Application to Sparse Regularized Convex Optimization Problems

We consider widely used convex optimization problems:

$$\min_{w \in R^p} f(w) + \lambda \Omega(w), \tag{1.31}$$

where $f: R^p \to R$ is a convex differentiable function and $\Omega(w): R^p \to R$ is a sparsity-inducing norm such as $\|w\|_1$, $\|w\|_2$, $\|w\|_p$, and $\|w\|_\infty$.

The optimality conditions discussed in Section 1.2.3.9 can be applied to solving the convex optimization problem.

Proposition 1.6

Necessary and sufficient conditions for convex optimization problems (1.30) (Bach et al. 2012).

A vector w for problems (1.30) is optimal if and only if $-\frac{1}{\lambda}\nabla f(w) \in \partial\Omega(w)$, where

$$\partial\Omega(w) = \begin{cases} \{z \in R^p;\ \Omega^*(z) \le 1\} & \text{if } w = 0, \\ z \in R^p;\ \Omega^*(z) = 1 \text{ and } z^T w = \Omega(w) & \text{otherwise.} \end{cases} \tag{1.32}$$

To apply optimality conditions (1.30), we first calculate the subdifferential of the function $F(w) = f(w) + \lambda\Omega(w)$. Since we assume that the function $f(w)$ is differentiable, we have $\partial f(w) = \nabla f(w)$, which implies $\partial F(w) = \nabla f(w) + \lambda \partial\Omega(w)$, where $\partial\Omega(w)$ is given by Equation 1.29. Using optimality condition (1.30), we obtain

$$0 \in \partial F(w) = \nabla f(w) + \lambda \partial \Omega(w),$$

which leads to

$$-\frac{1}{\lambda}\nabla f(w) \in \partial\Omega(w). \tag{1.33}$$

Proposition 1.6 can be easily applied to the lasso for regression. Let $X \in R^{n \times p}$ be the observed predictor matrix such as an SNP matrix, or a gene expression matrix, and $y \in R^n$ be a vector of phenotypes. The regression function for fitting the data X and Y is $\|y - Xw\|_2^2$, where $w \in R^p$ is a vector of regression coefficients. The lasso is to seek the vector of regression coefficients w that minimizes

$$\min_w \frac{1}{2n}\|y - Xw\|_2^2 + \lambda\|w\|_1. \tag{1.34}$$

Applying Proposition 1.6, we can obtain the optimality conditions for the lasso.

Proposition 1.7 Optimality conditions for the lasso

A vector w is an optimal solution to the lasso (1.34) if and only if for every component w_j of the vector w, $j=1, \dots, p$,

$$\begin{cases} \left|X_j^T(y-Xw)\right| \le n\lambda & \text{if } w_j = 0 \\ X_j^T(y-Xw) = n\lambda \, \text{sgn}(w_j) & \text{if } w_j \ne 0, \end{cases} \tag{1.35}$$

where X_j is the jth column vector of the matrix X.

Proof.

Let $f(w)=\frac{1}{2n}\|y-Xw\|_2^2$. To apply Proposition 1.6, we first calculate $\nabla f(w) = -\frac{1}{n}X^T(y-Xw)$. Recall that the dual norm of $\Omega(w)=\|w\|_1$ is $\Omega^*(z)=\|z\|_\infty$ and

$$\partial\Omega(w) = \begin{cases} z,\ \|z\|_\infty \le 1 & \text{if } w = 0 \\ z,\ \|z\|_\infty = 1,\ z^T w = \|w\|_1 & \text{if } w \ne 0. \end{cases}$$

Applying Proposition 1.6, we obtain

$$\frac{1}{n\lambda}X^T(y-Xw) \in \partial\Omega(w),$$

or

(i) if $w_j=0$, $\left|X_j^T(y-Xw)\right| \le \|X^T(y-Xw)\|_\infty \le n\lambda$ and

(ii) if $w_j \ne 0$, then we have $\|X^T(y-Xw)\|_\infty = n\lambda$ and $w^T X^T(y-Xw) = n\lambda\|w\|_1$.

Note that $\frac{w_j}{|w_j|} = \text{sgn}(w_j)$. Thus, for (ii), we have $X_j^T(y-Xw) = n\lambda \, \text{sgn}(w_j)$.

It is well known that if a and b are nonnegative real numbers and p and q are positive real numbers such that $\frac{1}{p}+\frac{1}{q}=1$, then $ab \le \frac{a^p}{p}+\frac{b^q}{q}$. This inequality can be extended to a high-dimensional and functional space, which will be very useful for duality study (Borwein and Lewis, 2006).

Proposition 1.8 Fenchel–Young inequality

Let $w \in R^p$ be a vector, $f: R^p \to R$ be a function, and z be a vector in the domain of f^*. The following inequality holds:

$$f(w) + f^*(z) \ge w^T z \tag{1.36}$$

To see this, by the definition of Fenchel conjugate, we have

$$f^*(z) \geq w^T z - f(w).$$

Moving $f(w)$ from the right side of the above inequality to its left side yields $f(w) + f^*(z) \geq w^T z$.

1.3 PROXIMAL METHODS

1.3.1 Introduction

In modern genomic and epigenomic analysis, we often need to consider convex optimization problems of the forms

$$\min_{w} f(w) + \lambda \Omega(w), \tag{1.37}$$

where $f(w)$ is a convex differentiable function of p-dimensional vector of variables w and $\Omega(w)$ is a nonsmooth function, typically a nonsmooth norm. The objective function is a sum of a generic convex differentiable function and a nonsmooth function. The traditional Newton's method is an efficient tool for solving an unconstrained smooth optimization problem but is not suited for solving a large nonsmooth convex problem. Proximal methods can be viewed as an extension of Newton's method from solving smooth optimization problems to nonsmooth optimization problems (Bach et al. 2012; Parikh and Boyd 2014). In this section, we briefly introduce the basic principle and algorithms of the proximal methods. For details, we refer the readers to the monographs (Bach et al. 2012; Parikh and Boyd 2014).

To solve the optimization problem (1.37), at each iteration, we often expand the function $f(w)$ in a neighborhood of the current iterate w^t by a Taylor expansion:

$$f(w) = f(w^t) + \nabla f(w^t)^T (w - w^t) + \frac{L}{2} \|w - w^t\|_2^2, \tag{1.38}$$

where L is an upper bound on the Lipschitz constant of gradient ∇f. Substituting Equation 1.38 into the optimization problem (1.37), we obtain the reduced optimization problem:

$$\min_{w \in R^p} f(w^t) + \nabla f(w^t)^T (w - w^t) + \frac{L}{2} \|w - w^t\|_2^2 + \lambda \Omega(w). \tag{1.39}$$

The Taylor expansion $f(w^t) + \nabla f(w^t)^T (w - w^t) + \frac{L}{2} \|w - w^t\|_2^2$ can be reformulated as

$$f(w^t) + \nabla f(w^t)^T (w - w^t) + \frac{L}{2} \|w - w^t\|_2^2 = L \left\{ \frac{1}{2} \left\| w - \left(w^t - \frac{1}{L} \nabla f(w^t) \right) \right\|_2^2 \right\}$$
$$+ f(w^t) - \frac{1}{2L} \|\nabla f(w^t)\|_2^2. \tag{1.40}$$

Using Equation 1.40, the optimization problem (1.39) can be reduced to

$$\min_{w \in R^p} \frac{1}{2} \left\| w - \left(w^t - \frac{1}{L} \nabla f\left(w^t\right) \right) \right\|_2^2 + \frac{\lambda}{L} \Omega(w). \tag{1.41}$$

In the absence of nonsmooth term $\Omega(w)$, the approximate solution to the optimization problem (1.37) is $w = w^t - \frac{1}{L} \nabla f\left(w^t\right)$. In the presence of nonsmooth term $\Omega(w)$, we use the optimization problem (1.41) to approximate the optimization problem (1.37). For the sake of convenience, we name the solution to the problem (1.41) as a proximal operator.

1.3.2 Basics of Proximate Methods

1.3.2.1 Definition of Proximal Operator

The proximal operator $\mathrm{Prox}_\Omega(u)$ of the function $\Omega(w)$ is defined by

$$\mathrm{Prox}_\Omega(u) = \arg\min_{w \in R^p} \left(\Omega(w) + \frac{1}{2} \|w - u\|_2^2 \right). \tag{1.42}$$

We often need to consider the penalty parameter for the penalty function $\Omega(w)$. Therefore, we extend the definition of the proximal operator to the scale function $\lambda\Omega(w)$:

$$\mathrm{Prox}_{\lambda\Omega}(u) = \arg\min_{w \in R^p} \left(\Omega(w) + \frac{1}{2\lambda} \|w - u\|_2^2 \right). \tag{1.43}$$

The corresponding proximal operator for the optimization problem (1.41) is therefore denoted as $\mathrm{Prox}_{\frac{\lambda}{L}\Omega}\left(w^t - \frac{1}{L} \nabla f\left(w^t\right) \right)$.

Consider an indicator function

$$\Omega_C(w) = \begin{cases} 0 & w \in C \\ +\infty & w \notin C, \end{cases}$$

where C is a closed nonempty convex set. The proximal operator of $\Omega_C(w)$ is

$$\mathrm{Prox}_C(u) = \arg\min_{w \in C} \frac{1}{2} \|w - u\|_2^2 = \Pi_C(u),$$

which is a Euclidean projection of point u onto C (Figure 1.6).

If we assume that $\Omega(w)$ is differentiable, then the proximal operator can be obtained by solving the following equation:

$$\nabla\Omega(u) + \frac{1}{\lambda}(w - u) = 0,$$

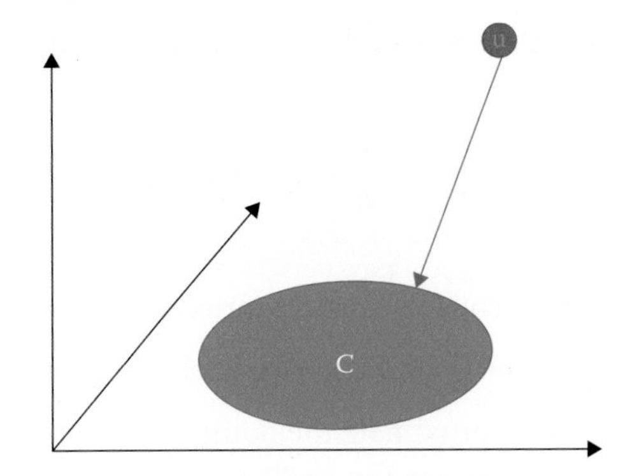

FIGURE 1.6 Euclidean projection of point u onto the region C.

which implies that

$$w = u - \lambda \nabla \Omega(u).$$

This suggests that the proximal operator can be searched by gradient methods with λ as a role similar to step size in a gradient method.

1.3.3 Properties of the Proximal Operator

We briefly introduce the major properties of proximal operators, which are useful for evaluating the proximal operator of a given function (Parikh and Boyd 2014).

1.3.3.1 Separable Sum

If Ω is separable across two sets of variables, i.e., $\Omega(x, y) = \phi(x) + \psi(y)$, then

$$\text{Prox}_\Omega(u, v) = \left[\text{Prox}_\phi(u), \text{Prox}_\psi(v)\right]. \tag{1.44a}$$

Proof.

To see this, by definition of the proximal operator of a function, we have

$$\begin{aligned}
\text{Prox}_\Omega(u,v) &= \underset{(x,y)}{\arg\min}\left\{\phi(x) + \psi(y) + \frac{1}{2}\left[(x-u)^T, (y-v)^T\right]\begin{bmatrix} x-u \\ y-v \end{bmatrix}\right\} \\
&= \underset{(x,y)}{\arg\min}\left\{\phi(x) + \frac{1}{2}\|x-u\|_2^2 + \psi(y) + \frac{1}{2}\|y-v\|_2^2\right\} \\
&= \left[\underset{x}{\arg\min}\left\{\phi(x) + \frac{1}{2}\|x-u\|_2^2\right\}, \underset{y}{\arg\min}\left\{\psi(y) + \frac{1}{2}\|y-v\|_2^2\right\}\right] \\
&= \left[\text{Prox}_\phi(u), \text{Prox}_\psi(v)\right].
\end{aligned}$$

In general, the formula (1.44) can be extended to a sum of n functions. Suppose that $\Omega(x)=\sum_{i=1}^{n}\Omega_i(x_i)$. Then, we have

$$\left(\text{Prox}_{\Omega}(u)\right)_i=\text{Prox}_{\Omega_i}(v_i). \tag{1.44b}$$

If a complex function can be decomposed into a number of separable simple functions, evaluating the proximal operator of a separable can be carried out by independently and easily evaluating the proximal operators of each simple function.

1.3.3.1.1 Postcomposition   If $\Omega(w)=\alpha\varphi(w)+b$ and $\alpha>0$, then $\text{Prox}_{\lambda\Omega}(u)=\text{Prox}_{\alpha\lambda\varphi}(u)$. Indeed, by definition of proximal operator of function $\Omega(w)$, we have

$$\begin{aligned}\text{Prox}_{\lambda\Omega}(u)&=\arg\min_{w}\left\{\Omega(w)+\frac{1}{2\lambda}\|w-u\|_2^2\right\}\\&=\arg\min_{w}\left\{\alpha\varphi(w)+b+\frac{1}{2\lambda}\|w-u\|_2^2\right\}\\&=\arg\min_{w}\left\{\alpha\left[\varphi(w)+\frac{1}{2\alpha\lambda}\|w-u\|_2^2\right]\right\}\\&=\arg\min_{w}\left\{\varphi(w)+\frac{1}{2\alpha\lambda}\|w-u\|_2^2\right\}\\&=\text{Prox}_{\alpha\lambda\varphi}(u).\end{aligned} \tag{1.45}$$

1.3.3.1.2 Precomposition (1)   If $\Omega(w)=\varphi(\alpha w+b)$ with $\alpha\neq 0$, then

$$\text{Prox}_{\lambda\Omega}(u)=\frac{1}{\alpha}\left[\text{Prox}_{\alpha^2\lambda\varphi}(\alpha u+b)-b\right]. \tag{1.46}$$

Proof.

$$\begin{aligned}\text{Prox}_{\lambda\Omega}(u)&=\arg\min_{w}\left\{\Omega(w)+\frac{1}{2\lambda}\|w-u\|_2^2\right\}\\&=\arg\min_{w}\left\{\varphi(\alpha w+b)+\frac{1}{2\lambda}\|w-u\|_2^2\right\}\\&=\arg\min_{w}\left\{\varphi(x)+\frac{1}{2\lambda}\left\|\frac{x-b}{\alpha}-u\right\|_2^2\right\}\quad(\text{let } x=\alpha w+b)\\&=\frac{1}{\alpha}\left[\arg\min_{x}\left\{\varphi(x)+\frac{1}{2\alpha^2\lambda}\|x-\alpha u-b\|_2^2\right\}-b\right]\\&=\frac{1}{\alpha}\left[\text{Prox}_{\alpha^2\lambda\varphi}(\alpha u+b)-b\right].\end{aligned}$$

1.3.3.1.3 Precomposition (2) If $\Omega(w)=\varphi(Qw)$, where Q is orthogonal, then we have

$$\text{Prox}_{\lambda\Omega}(u)=Q^T\text{Prox}_{\lambda\varphi}(Qu). \tag{1.47}$$

Proof.

Let $x=Qw$, then $w=Q^Tx$ because Q is an orthogonal matrix. By definition of proximal operator, we have

$$\begin{aligned}
\text{Prox}_{\lambda\Omega}(u) &= \arg\min_w\left\{\Omega(w)+\frac{1}{2\lambda}||w-u||_2^2\right\} \\
&= \arg\min_w\left\{\varphi(Qw)+\frac{1}{2\lambda}||w-u||_2^2\right\} \\
&= \arg\min_w\left\{\varphi(x)+\frac{1}{2\lambda}||Q^Tx-u||_2^2\right\} \\
&= Q^T\arg\min_x\left\{\varphi(x)+\frac{1}{2\lambda}||Q^T(x-Qu)||_2^2\right\} \\
&= Q^T\arg\min_x\left\{\varphi(x)+\frac{1}{2\lambda}||(x-Qu)||_2^2\right\} \\
&= Q^T\text{Prox}_{\lambda\varphi}(Qu).
\end{aligned}$$

1.3.3.1.4 Affine Addition If $\Omega(w)=\varphi(w)+a^Tw+b$, then

$$\text{Prox}_{\lambda\Omega}(u)=\text{Prox}_{\lambda\varphi}(u-\lambda a). \tag{1.48}$$

Proof.

It is clear that

$$\begin{aligned}
\text{Prox}_{\lambda\Omega}(u) &= \arg\min_w\left\{\Omega(w)+\frac{1}{2\lambda}||w-u||_2^2\right\} \\
&= \arg\min_w\left\{\varphi(w)+a^Tw+b+\frac{1}{2\lambda}||w-u||_2^2\right\} \\
&= \arg\min_w\left\{\varphi(w)+\frac{1}{2\lambda}\left[||w-u||_2^2+2\lambda a^Tw\right]\right\} \\
&= \arg\min_w\left\{\varphi(w)+\frac{1}{2\lambda}\left[||w-u+\lambda a||_2^2\right]\right\} \\
&= \text{Prox}_{\lambda\varphi}(u-\lambda a).
\end{aligned}$$

1.3.3.1.5 Regularization Let $\tilde{\lambda} = \frac{\lambda}{1+\rho\lambda}$. If $\Omega(w) = \varphi(w) + \frac{\rho}{2}\|w-a\|_2^2$, then

$$\text{Prox}_{\lambda\Omega}(u) = \text{Prox}_{\tilde{\lambda}\varphi}\left(\frac{\tilde{\lambda}}{\lambda}u + \rho\tilde{\lambda}a\right). \tag{1.49}$$

Proof.

By definition of the proximal operator of $\Omega(w) = \varphi(w) + \frac{\rho}{2}\|w-a\|_2^2$, we have

$$\begin{aligned}
\text{Prox}_{\lambda\Omega}(u) &= \arg\min_w \left\{\Omega(w) + \frac{1}{2\lambda}\|w-u\|_2^2\right\} \\
&= \arg\min_w \left\{\varphi(w) + \frac{\rho}{2}\|w-a\|_2^2 + \frac{1}{2\lambda}\|w-u\|_2^2\right\} \\
&= \arg\min_w \left\{\varphi(w) + + \frac{1}{2\lambda}\left[\|w-u\|_2^2 + \rho\lambda\|w-a\|_2^2\right]\right\} \\
&= \arg\min_w \left\{\varphi(w) + + \frac{1}{2\lambda}(1+\rho\lambda)\left\|w - \frac{u+\rho\lambda a}{1+\rho\lambda}\right\|_2^2\right\} \\
&= \arg\min_w \left\{\varphi(w) + + \frac{1}{2\tilde{\lambda}}\left\|w - \frac{\tilde{\lambda}}{\lambda}u - \rho\tilde{\lambda}a\right\|_2^2\right\} \\
&= \text{Prox}_{\tilde{\lambda}\varphi}\left(\frac{\tilde{\lambda}}{\lambda}u + \rho\tilde{\lambda}a\right).
\end{aligned}$$

Fixed point theory plays a fundamental role in devising algorithms for optimization. To employ fixed point theory for developing proximal operator-based algorithms for convex optimization, we need to explore the following property that connects the proximal operator with the fixed point.

1.3.3.1.6 Fixed Points The necessary and sufficient conditions that the point w^* minimizes the function $\Omega(w^*)$ are if and only if

$$w^* = \text{Prox}_f(w^*). \tag{1.50}$$

Proof (Parikh and Boyd 2014).

We first show that if w^* minimizes $\Omega(w)$, then $w^* = \text{Prox}_f(w^*)$. Suppose that w^* is the minimizer of the function $\Omega(w)$. Then, we have

$$\Omega(w) + \frac{1}{2}\|w-w^*\|_2^2 \geq \Omega(w^*) + \frac{1}{2}\|w-w^*\|_2^2 \geq \Omega(w^*) + \frac{1}{2}\|w^*-w^*\|_2^2 = \Omega(w^*),$$

which implies w^* minimizes $\text{Prox}_f(w^*)$, i.e., $w^* = \text{Prox}_f(w^*)$.

On the contrary, $w^* = \text{Prox}_f(w^*)$, i.e., w^* minimizes

$$\Omega(w) + \frac{1}{2}\|w - w^*\|_2^2. \tag{1.51}$$

Assume that $\Omega(w)$ is subdifferentiable. Then, w^* minimizes Equation 1.51, which implies

$$0 \in \partial\Omega(w^*) + (w^* - w^*).$$

Therefore, we show that

$$0 \in \partial\Omega(w^*). \tag{1.52}$$

Equation 1.52 indicates that $\Omega(w) \geq \Omega(w^*)$ for all w. Therefore, w^* minimizes $\Omega(w)$.

1.3.3.1.7 Moreau Decomposition We introduce Moreau decomposition as a generalization of orthogonal decomposition. Let $\Omega^*(z)$ be the Fenchel conjugate of the function $\Omega(x)$. We have

$$v = \text{Prox}_{\Omega}(v) + \text{Prox}_{\Omega^*}(v). \tag{1.53}$$

Proof.

Let

$$u = \text{Prox}_{\Omega}(v). \tag{1.54}$$

In other words, u is the minimizer of the function

$$\Omega(w) + \frac{1}{2}\|w - v\|_2^2.$$

Therefore, we obtain

$$0 \in \partial\Omega(u) + u - v$$

or

$$v - u \in \partial\Omega(u). \tag{1.55}$$

Using Fenchel–Young inequality, we obtain the following equality:

$$\Omega(u)+\Omega^*(v-u)=(v-u)^T u. \tag{1.56}$$

Taking a subdifferential on both sides of the equality (1.56), we obtain

$$u \in \partial\Omega^*(v-u). \tag{1.57}$$

Recall that

$$\text{Prox}_{\Omega^*}(v)=\arg\min_{w}\left\{\Omega^*(w)+\frac{1}{2}\|w-v\|_2^2\right\}.$$

From Equation 1.57, we obtain the following equation:

$$0 \in \partial\Omega^*(v-u)-u,$$

which implies that

$$v-u=\text{Prox}_{\Omega^*}(v). \tag{1.58}$$

Adding Equations 1.54 and 1.58 yields

$$v=\text{Prox}_{\Omega}(v)+\text{Prox}_{\Omega*}(v).$$

1.3.3.2 Moreau–Yosida Regularization

A very useful tool for proximal operator analysis is the **infimal convolution** of closed convex functions $\Omega(x)$ and $\Psi(x)$ (Parikh and Boyd 2014). It is defined as

$$(\Omega\oplus\Psi)(v)=\inf_{x}\{\Omega(x)+\Psi(v-x)\}. \tag{1.59}$$

Infimal convolution can also be extended to a multiple-convex function:

$$\Omega(x)=(\Omega_1\otimes\Omega_2\otimes\ldots\otimes\Omega_m)(x)=\inf_{x_k}\left\{\sum_{k=1}^{m}\Omega_k(x_k)|\sum_{k=1}^{m}x_k=x\right\}. \tag{1.60}$$

The Fenchel conjugate of infimal convolution satisfies the addition

$$(\Omega_1\otimes\Omega_2\otimes\ldots\otimes\Omega_m)^*=\Omega_1^*+\ldots+\Omega_m^*. \tag{1.61}$$

To see this, by definition of Fenchel conjugate, we have

$$
\begin{aligned}
(\Omega_1 \otimes \Omega_2 \otimes \ldots \otimes \Omega_m)^*(z) &= \sup_x \left\{ z^T x - \inf\left(\sum_{k=1}^m \Omega_k(x_k) \mid \sum \sum_{k=1}^m x_k = x \right) \right\} \\
&= \sup_x \sup_{x_k \mid \sum_{k=1}^m x_k = x} \left\{ z^T x - \sum_{k=1}^m \Omega_k(x_k) \right\} \\
&= \sup_x \sup_{x_k \mid \sum_{k=1}^m x_k = x} \left\{ z^T \left(\sum_{k=1}^m x_k \right) - \sum_{k=1}^m \Omega_k(x_k) \right\} \\
&= \sup_{x_k \mid \sum_{k=1}^m x_k = x} \left\{ \sum_{k=1}^m \left(z^T x_k - \Omega_k(x_k) \right) \right\} \\
&= \sum_{k=1}^m \Omega_k^*(z).
\end{aligned}
$$

A quadratic function is an important function in convex optimization. Its convolution with other nonsmooth functions will smooth a nonsmooth function, and it plays an important role in convex optimization. **Moreau–Yosida regularization** or the **Moreau envelope** $M_{\lambda\Omega}$ of the function $\lambda\Omega$ is defined as its informal convolution with the quadratic function $\frac{1}{2}\|\cdot\|_2^2$:

$$
M_{\lambda\Omega}(v) = \lambda\Omega \otimes \left(\frac{\|\cdot\|_2^2}{2} \right) = \inf_x \left(\Omega(x) + \frac{1}{2\lambda} \|x - v\|_2^2 \right). \tag{1.62}
$$

When a function is nonsmooth, we can construct its Moreau envelope to transform the nonsmooth function $\Omega(x)$ to a smoothed form $M_{\lambda\Omega}(v)$.

It follows from Equation 1.62 that

$$
M_\Omega(v) = \Omega(\text{Prox}_f(v)) + \frac{1}{2} \|\text{Prox}_f(v) - v\|_2^2,
$$

which states that the prox(v) gives the unique point reaching the infimum that defines the Moreau envelope.

Applying Equation 1.61 to the Moreau envelope $M_{\lambda\Omega}(v)$, we obtain

$$
M_{\lambda\Omega}^*(v) = \Omega^*(v) + \frac{1}{2\lambda} \|x - v\|_2^2. \tag{1.63}
$$

Since $M_{\lambda\Omega} = M_{\lambda\Omega}^{**}$, taking a Fenchel conjugate on both sides of Equation 1.63 yields

$$M_{\lambda\Omega}(v) = \left(\Omega^*(v) + \frac{1}{2\lambda}||x - v||_2^2\right)^*. \tag{1.64}$$

In general, the conjugate function of a closed convex function is smooth; Equation 1.64 shows that the Moreau envelope is smooth. Now we calculate the gradient of the Moreau envelope $M_{\lambda\Omega}(v)$.

Suppose that $x = \arg\inf_y \left\{\Omega(y) + \frac{1}{2\lambda}||y - v||_2^2\right\}$. By definition of the Moreau envelope $M_{\lambda\Omega}(v)$, we have

$$M_{\lambda\Omega}(v) = \Omega(x) + \frac{1}{2\lambda}||x - v||_2^2.$$

Therefore, taking a derivative of $M_{\lambda\Omega}(v)$ with respect to v yields

$$\begin{aligned}\nabla_v M_{\lambda\Omega}(v) &= \frac{1}{\lambda}(v - x) \\ &= \frac{1}{\lambda}(v - \text{prox}_{\lambda\Omega}(v)),\end{aligned} \tag{1.65}$$

or

$$\text{prox}_{\lambda\Omega}(v) = v - \lambda\nabla_v M_{\lambda\Omega}(v).$$

A proximal operator is to seek points that minimize $\Omega(x) + \frac{1}{2\lambda}||x - v||_2^2$, which can be obtained by taking the subdifferential

$$0 \in \partial_x\left(\Omega(x) + \frac{1}{2\lambda}||x - v||_2^2\right)$$

or

$$0 \in \partial\Omega(x) + \frac{x - v}{\lambda},$$

which implies that

$$v \in (\lambda\partial\Omega + I)(x),$$

where variable x minimizes $\Omega(x) + \frac{1}{2\lambda}||x - v||_2^2$, i.e., $x = \text{prox}_{\lambda\Omega}(v)$.

Therefore, we obtain

$$\text{prox}_{\lambda\Omega}(v) = x = (\lambda\partial\Omega + I)^{-1}(v). \tag{1.66}$$

1.3.3.3 Gradient Algorithms for the Calculation of the Proximal Operator

Using Equation 1.66, we can develop a gradient algorithm to calculate the proximal operator. When λ is small, by expansion, Equation 1.66 can be approximated by

$$\begin{aligned}\text{prox}_{\lambda\Omega}(v) &= (\lambda\partial\Omega + I)^{-1}(v)\\ &= (\lambda\nabla\Omega + I)^{-1}(v)\\ &\approx v - \lambda\nabla\Omega(v).\end{aligned} \tag{1.67}$$

Taylor expansion is a useful tool for developing algorithms for the calculation of the proximal operator. The second-order Taylor expansion of the function $\Omega(v)$ is given by

$$\Omega(x) \approx \Omega(v) + \nabla\Omega(v)^T (x - v) + \frac{1}{2}(x - v)^T \nabla^2\Omega(v)(x - v). \tag{1.68}$$

Then, the function $\Omega(x) + \frac{1}{2\lambda}||x - v||_2^2$ can be approximated by

$$\Omega(x) + \frac{1}{2\lambda}||x - v||_2^2 \approx \Omega(v) + \nabla\Omega(v)^T (x - v) + \frac{1}{2}(x - v)^T \nabla^2\Omega(v)(x - v). \tag{1.69}$$

To minimize $\Omega(x) + \frac{1}{2\lambda}||x - v||_2^2$, we take its gradient and set it equal to zero:

$$\nabla\Omega(v) + \nabla^2\Omega(v)(x - v) + \frac{1}{\lambda}(x - v) = 0. \tag{1.70}$$

Solving Equation 1.70 for x yields the proximal operator of the second-order approximation:

$$\text{prox}_{\lambda\Omega}(v) = v - \left(\nabla^2\Omega(v) + \frac{1}{\lambda}I\right)^{-1} \nabla\Omega(v). \tag{1.71}$$

The formula (1.71) is usually referred to as a Tikhonov-regularized Newton update. Equations 1.67 and 1.71 provide proximal operators of first- and second-order approximations of function Ω.

1.3.4 Proximal Algorithms

Recall that we often need to solve the following optimization problem:

$$\min_w f(w) + \lambda\Omega(w), \tag{1.72}$$

whose objective function is the sum of a differentiable function and a nondifferentiable function. In Section 1.3.1, we showed that the optimization problem (1.72) can be reduced to

$$\min_{w \in R^p} \frac{1}{2}\left\| w - \left(w^t - \frac{1}{L}\nabla f\left(w^t\right)\right)\right\|_2^2 + \frac{\lambda}{L}\Omega(w),$$

which leads to introducing the proximal operator

$$\mathrm{Prox}_{\lambda\Omega}(u) = \underset{w \in R^p}{\mathrm{argmin}}\left(\Omega(w) + \frac{1}{2\lambda}\|w - u\|_2^2 \right). \tag{1.73}$$

The corresponding proximal operator for the optimization problem (1.72) is therefore denoted as $\mathrm{Prox}_{\frac{\lambda}{L}\Omega}\left(w^t - \frac{1}{L}\nabla f\left(w^t\right)\right)$. Now we introduce several algorithms, which solve the optimization problems via proximal operators (Parikh and Boyd 2014).

1.3.4.1 Proximal Point Algorithm

The proximal point algorithm is defined as

$$u^{k+1} = \mathrm{prox}_{\lambda\Omega}\left(u^k\right), \tag{1.74}$$

where

k is the number of iterations

u^k represents the kth iterate of the variable u

The proximal point algorithm can be simply viewed as a fixed point iteration algorithm or the standard gradient method that is applied to the Moreau envelope M_Ω.

1.3.4.2 Proximal Gradient Method

Consider a general optimization problem (1.72) in which the objective function can be split into one differentiable function $f(w)$ and one nonsmooth function $\Omega(w)$. As we discussed before, the differentiable function $f(w)$ can be expanded in a Taylor series around the current solution. Then, the optimization problem (1.72) can be reduced to compute the following proximal operator:

$$\mathrm{Prox}_{\frac{\lambda}{L}\Omega}\left(w^t - \frac{1}{L}\nabla f\left(w^t\right)\right).$$

This motivates the proximal gradient method:

$$u^{k+1} = \mathrm{prox}_{\lambda^k\Omega}\left(u^k - \lambda^k \nabla f\left(u^k\right)\right), \tag{1.75}$$

where $\lambda^k > 0$ is a step size.

In theory, the step size can be chosen by Lipschitz constant. When ∇f is Lipschitz continuous with constant L, the fixed step size $\lambda^k = \lambda \in \left(0, \frac{1}{L}\right)$ is selected. But, in general, the Lipschitz constant L is unknown. In practice, the step size is selected by a "majorization-minimization method" and a line search.

Majorization-minimization algorithms mean that we iteratively majorize the upper bound of the objective function and then minimize the majorization. Consider

$$\hat{f}_\lambda(u, y) = f(y) + \nabla f(y)^T (u - y) + \frac{1}{2\lambda} \|u - y\|_2^2 \quad \text{with } \lambda > 0. \tag{1.76}$$

After some calculations, we know that

$$\hat{f}_\lambda(u, u) = f(u), \quad \nabla \hat{f}(u, u) = 0 \quad \text{and} \quad \nabla \hat{f}_\lambda(u, u)\left(\nabla \hat{f}_\lambda(u, u)\right)^T = -\frac{1}{\lambda} I$$

are a negative definite matrix, which implies $\hat{f}_\lambda(u, y) \geq f(u)$ and the function $\hat{f}_\lambda(u, y)$ is the upper bound of the function $f(u)$.

The line search algorithm for selecting λ is given as (Parikh and Boyd 2014).

Algorithm

Step 1: Given u^k, λ^{k-1}, and parameter $\alpha \in (0, 1)$, set $\lambda = \lambda^{k-1}$.
Step 2: Repeat

$$\text{(1) Set } z = \text{prox}_{\lambda\Omega}\left(u^k - \lambda \nabla f\left(u^k\right)\right). \tag{1.77}$$

(2) Break if $f(z) \leq \hat{f}_\lambda\left(z, u^k\right)$.

(3) Update $\lambda = \alpha\lambda$.

Step 3: Return $\lambda^k = \lambda$, $u^{k+1} = z$.

1.3.4.3 Accelerated Proximal Gradient Method

To accelerate the proximal gradient algorithm, we can add an additional step to the algorithm, which extrapolates the solution in the previous step. Specifically, we modify the proximal gradient algorithm as follows:

$$\begin{aligned} y^{k+1} &= u^k + \omega^k\left(u^k - u^{k-1}\right) \\ u^{k+1} &= \text{prox}_{\lambda^k\Omega}\left(y^{k+1} - \lambda^{k+1} \nabla f\left(y^{k+1}\right)\right), \end{aligned} \tag{1.78}$$

where $\omega^k(\omega^0=0)\in[0,1]$ is an extrapolation parameter and λ^k is the step size, which can be similarly determined by algorithm (1.77). A simple choice for the extrapolation parameter is

$$\omega^k = \frac{k}{k+3}.$$

The software package TFOCS includes several implementations of the accelerated proximal gradient methods.

1.3.4.4 Alternating Direction Method of Multipliers

The alternating direction method of multipliers (ADMM) attempts to split the objective function into two major parts: the differential objective function and the nondifferentiable objective function (Parikh and Boyd 2014). The ADMM is to combine dual decomposition and augmented Lagrangian methods for constrained optimization. Consider the optimization problem (1.72):

$$\min_w f(w)+\lambda\Omega(w).$$

To separate the differentiable objective function $f(w)$ from the nonsmooth objective function $\Omega(w)$, we introduce new variables, z, and rewrite the unconstrained optimization problem (1.72) as a constrained optimization problem:

$$\begin{array}{ll} \min & f(w)+g(z) \\ \text{subject to} & w-z=0, \end{array} \tag{1.79}$$

where $g(z)=\lambda\Omega(z)$. In the problem (1.72), the variables are split into two sets of variables, w and z, and the consensus constraint that they must agree is introduced.

The augmented Lagrangian methods can be used to solve the problem (1.79):

$$L_\rho(w,z,y)=f(w)+g(z)+y^T(w-z)+\frac{\rho}{2}\|w-z\|_2^2, \tag{1.80}$$

where $\rho>0$ is a penalty parameter and $y\in R^n$ is a vector of dual variables associated with the constraint. The optimization problem (1.80) can be iteratively and separately solved as follows:

$$w^{k+1}=\arg\min_w L_\rho\left(w,z^k,y^k\right), \tag{1.81a}$$

$$z^{k+1}=\arg\min_z L_\rho\left(w^{k+1},z,y^k\right), \tag{1.81b}$$

$$y^{k+1}=y^k+\rho\left(w^{k+1}-z^{k+1}\right). \tag{1.81c}$$

In (1.81a), L_ρ is minimized over the primal variable w and is reduced to

$$w^{k+1}=\arg\min_{w}\left(f(w)+\left(y^k\right)^T(w-z)+\frac{\rho}{2}\|w-z^k\|_2^2\right). \tag{1.82a}$$

Similarly, the optimization problem (1.81b) can be reformulated as

$$z^{k+1}=\arg\min_{z}\left(g(z)+\left(y^k\right)^T\left(w^{k+1}-z\right)+\frac{\rho}{2}\|w^{(k+1)}-z\|_2^2\right). \tag{1.82b}$$

Note that

$$\begin{aligned}\left(y^k\right)^T(w-z)+\frac{\rho}{2}\|w-z^k\|_2^2&=\frac{\rho}{2}\left(\|w-z^k\|_2^2+\frac{2}{\rho}\left(y^k\right)^T(w-z)\right)\\&=\frac{\rho}{2}\left\|w-z^k-\left(\frac{y^{(k)}}{\rho}\right)\right\|_2^2-\frac{2}{\rho}\|y^{(k)}\|_2^2.\end{aligned}$$

Therefore, the optimization problems (1.82a) and (1.82b) can be reduced to

$$w^{k+1}=\arg\min_{w}\left(f(w)+\frac{\rho}{2}\left\|w-z^k+\frac{y^k}{\rho}\right\|_2^2\right), \tag{1.83a}$$

$$z^{k+1}=\arg\min_{z}\left(g(z)+\frac{\rho}{2}\left\|w^{(k+1)}-z-\frac{y^k}{\rho}\right\|_2^2\right). \tag{1.83b}$$

Let $u=\frac{y}{\rho}$ and $\tau=\frac{1}{\rho}$. The optimization problems (1.81a), (1.81b), and (1.81c) will be reduced to

$$w^{k+1}=\arg\min_{w}\left(f(w)+\frac{1}{2\tau}\|w-z^k+u^k\|_2^2\right), \tag{1.84a}$$

$$z^{k+1}=\arg\min_{z}\left(g(z)+\frac{1}{2\tau}\|w^{(k+1)}-z+u^k\|_2^2\right), \tag{1.84b}$$

$$u^{k+1}=u^k+w^{k+1}-z^{k+1}. \tag{1.84c}$$

We use a proximal operator to rewrite algorithms (1.84a), (1.84b), and (1.84c) as

$$w^{k+1}=\text{prox}_{\tau f}\left(z^k-u^k\right), \tag{1.85a}$$

$$z^{k+1}=\text{prox}_{\tau g}\left(w^{k+1}+u^k\right), \tag{1.85b}$$

$$u^{k+1}=u^k+w^{k+1}-z^{k+1}. \tag{1.85c}$$

1.3.4.5 Linearized ADMM

Consider the problem

$$\begin{aligned} &\text{minimize} \quad f(w)+g(z) \\ &\text{subject to} \quad Aw-z=0. \end{aligned} \tag{1.86}$$

Again, we use the augmented Lagrangian to rewrite the problem (1.86) as

$$L_\rho(w,z,y)=f(w)+g(z)+y^T(Aw-z)+\frac{\rho}{2}||Aw-z||_2^2. \tag{1.87}$$

Linearizing the quadratic term, we obtain

$$\begin{aligned} \frac{\rho}{2}||Aw-z||_2^2 &\approx \rho\left(Aw^k-z^k\right)^T Aw \\ &=\rho\left[A^T\left(Aw^k-z^k\right)\right]^T w. \end{aligned} \tag{1.88}$$

Adding a new quadratic regularization term $\frac{1}{2\mu}||w-w^k||_2^2$ into Equation 1.88 yields

$$\rho\left[A^T\left(Aw^k-z^k\right)\right]^T w+\frac{1}{2\mu}||w-w^k||_2^2. \tag{1.89}$$

Replacing $\frac{\rho}{2}||Aw-z||_2^2$ in Equation 1.87 by Equation 1.89, we obtain

$$\begin{aligned} &\min_w\left\{f(w)+g(z)+y^T(Aw-z)+\rho\left[A^T\left(Aw^k-z^k\right)\right]^T w+\frac{1}{2\mu}||w-w^k||_2^2\right\} \\ &=\min_w\left\{f(w)+\frac{1}{2\mu}\left[||w-w^k||^2+2\rho\mu\left(A^T\left(Aw^k-z^k\right)\right)^T w+2\mu\left(A^Ty\right)^T w\right]\right\} \\ &=\min_w\left\{f(w)+\frac{1}{2\mu}\left[||w||_2^2-2w^kw+2\rho\mu\left(A^T\left(Aw^k-z^k\right)\right)^T w+2\mu\left(A^Ty\right)^T w\right]\right\} \\ &=\min_w\left\{f(w)+\frac{1}{2\mu}\left[||w||_2^2-2(w^k-\rho\mu\left(A^T\left(Aw^k-z^k+\frac{1}{\rho}y\right)^T\right)w\right]\right\} \\ &\min_w\left\{f(w)+\frac{1}{2\mu}\left[\left\|w-\left(w^k-\frac{\mu}{\lambda}A^T\left(Aw^k-z^k+u^k\right)\right)\right\|_2^2\right]\right\}, \end{aligned}$$

where $u=\frac{y}{\rho}$ and $\lambda=\frac{1}{\rho}$.

Therefore, we obtain

$$w^{k+1}=\text{prox}_{\mu f}\left(w^k-\frac{\mu}{\lambda}A^T\left(Aw^k-z^k+u^k\right)\right).$$

Now we consider

$$\begin{aligned}
&\min_z L_\rho\left(w^{k+1}, z, y^k\right) \\
&= \min_z \left\{ g(z) + \left(y^k\right)^T \left(Aw^{k+1} - z\right) + \frac{\rho}{2}\|z - Aw^{k+1}\|_2^2 \right\} \\
&= \min_z \left\{ g(z) + \frac{\rho}{2}\left\| z - Aw^{k+1} - \frac{y^k}{\rho} \right\|_2^2 \right\} \\
&= \min_z \left\{ g(z) + \frac{1}{2\lambda}\left\| z - Aw^{k+1} - u^k \right\|_2^2 \right\},
\end{aligned}$$

which implies

$$z^{k+1} = \operatorname{prox}_{\lambda g}\left(Aw^{k+1} + u^k\right).$$

In summary, the linearized ADMM is given by

$$w^{k+1} = \operatorname{prox}_{\mu f}\left(w^k - \frac{\mu}{\lambda} A^T \left(Aw^k - z^k + u^k \right) \right), \tag{1.90a}$$

$$z^{k+1} = \operatorname{prox}_{\lambda g}\left(Aw^{k+1} + u^k\right), \tag{1.90b}$$

$$u^{k+1} = u^k + Aw^{k+1} - z^{k+1}. \tag{1.90c}$$

1.3.5 Computing the Proximal Operator

An essential step for proximal algorithms is to compute the proximal operators. Computing the proximal operators involves solving a convex optimization problem. We often can obtain the closed form for the solution to the convex optimization problem. However, even if the closed form for computing the proximal operator is not available, a generic optimization algorithm for the proximal operator computation is still useful (Parikh and Boyd 2014).

1.3.5.1 Generic Function

In general, the problem we consider is

$$\begin{array}{ll} \text{minimize} & \Omega(w) + \dfrac{1}{2\lambda}\|w - v\|_2^2 \\ \text{subject to} & w \in C, \end{array} \tag{1.91}$$

where $w \in R^n$ and C is the set that defines the function $\Omega(w)$.

If $C = R^n$, then the problem (1.91) is unconstrained. The algorithms used for solving the problem (1.91) depend on the properties of the function $\Omega(w)$. If $\Omega(w)$ is differentiable, then a wide range of methods such as gradient and Newton's methods can be used. If $\Omega(w)$ is nonsmooth, then subgradient methods will be used to solve the problem.

If C is a set in a n-dimensional space and the problem (1.91) is constrained, the methods for unconstrained optimization can be adapted to projection methods. For example, if the function $\Omega(w)$ is differentiable, then the projected gradient methods can be used. For the nondifferentiable function $\Omega(w)$, the projected subgradient methods can be used. Next we briefly introduce methods for solving problem (1.91) with several specific functions.

1.3.5.1.1 Quadratic Functions We assume a quadratic function $\Omega(w)=\frac{1}{2}w^T Aw+w^T b+c$, with the nonnegative definite matrix A. To compute the proximal operator, we need to solve the problem

$$\min_w \ F(w)=\frac{1}{2}w^T Aw+w^T b+c+\frac{1}{2\lambda}\|w-v\|_2^2.$$

Taking gradient $\nabla_w F(w)$ and setting it to zero yields

$$Aw+b+\frac{w-v}{\lambda}=0. \tag{1.92}$$

Solving Equation 1.92 for w, we obtain

$$w=(I+\lambda A)^{-1}(v-\lambda b).$$

Thus, the proximal operator is

$$\text{prox}_{\lambda\Omega}(v)=(I+\lambda A)^{-1}(v-\lambda b). \tag{1.93}$$

We discuss three special cases:

(i) Assume $A=0$. In this case, Equation 1.93 is reduced to

$$\text{prox}_{\lambda\Omega}(v)=v-\lambda b. \tag{1.94}$$

(ii) If $\Omega(w)=c$, then we have

$$\text{prox}_{\lambda\Omega}(v)=v. \tag{1.95}$$

(iii) Consider $\Omega(w)=\frac{1}{2}\|w\|_2^2$. Its proximal operator is given by

$$\text{prox}_{\lambda\Omega}(v)=\frac{1}{1+\lambda}v. \tag{1.96}$$

1.3.5.1.2 Smooth Functions The optimization problems with smooth functions are easy to solve. We can use gradient or Newton's methods to compute the proximal operators. The major step for Newton's method to compute the proximal operators is to solve the following linear equation:

$$Hw = -g, \tag{1.97}$$

where $H = \nabla^2\Omega(w)$ and $g = \nabla\Omega(w)$. The classical method for solving a system of linear equation (1.97) is to first factorize the Hessian matrix H into $H = LL^T$ and then transform Equation 1.97 to

$$LL^T w = -g,$$

which leads to the solution $w = -L^{-T}L^{-1}g$.

Now we consider the specific structure of the matrix H:

$$H = D + zz^T, \tag{1.98}$$

which is the sum of a diagonal matrix $D \in R^{n\times n}$ and a rank-one matrix zz^T. We can easily compute the inverse of the matrix H:

$$H^{-1} = D^{-1} - \frac{D^{-1}zz^T D^{-1}}{1 + z^T D^{-1} z}. \tag{1.99}$$

Next we introduce a class of smooth functions whose Hessian matrix structure is the sum of a diagonal and a rank-one matrix:

$$\Omega(w) = f\left(\sum_{i=1}^{k} \alpha_i(w_i) + b) + \sum_{i=1}^{k} \beta_i(w_i)\right). \tag{1.100}$$

Let

$$\alpha(w) = \begin{bmatrix} \alpha_1(w_1) \\ \vdots \\ \alpha_k(w_k) \end{bmatrix}, \quad \beta(w) = \begin{bmatrix} \beta_1(w_1) \\ \vdots \\ \beta_k(w_k) \end{bmatrix}, \quad \nabla_w\alpha(w) = \begin{bmatrix} \alpha_1'(w_1) \\ \vdots \\ \alpha_k'(w_k) \end{bmatrix}, \quad \nabla_w\beta(w) = \begin{bmatrix} \beta_1'(w_1) \\ \vdots \\ \beta_k'(w_k) \end{bmatrix}.$$

Then, the gradient and the Hessian matrix of $\nabla_w\Omega(w)$ are respectively given by

$$\nabla_w\Omega(w) = f'\left(\sum_{i=1}^{k} \alpha_i(w_i) + b\right)\nabla_w\alpha(w) + \nabla_w\beta(w) \tag{1.101}$$

and

$$H = \frac{\partial \Omega(w)}{\partial w \partial w^T}$$
$$= f''\left(\sum_{i=1}^{k} \alpha_i(w_i) + b\right) \nabla_w \alpha(w) \nabla_w^T \alpha(w) + f'\left(\sum_{i=1}^{k} \alpha_i(w_i) + b\right) \text{diag}\left(\alpha_i''(w_i)\right) + \text{diag}\left(\beta_i''(w_i)\right),$$

where

$$\nabla_w \alpha(w) \nabla_w^T \alpha(w) = \begin{bmatrix} \left(\alpha_1'(w_1)\right)^2 & \cdots & \alpha_1'(w_1)\alpha_k'(w_k) \\ \vdots & \vdots & \vdots \\ \alpha_k'(w_k)\alpha_1'(w_1) & \cdots & \left(\alpha_k'(w_k)\right)^2 \end{bmatrix}. \quad (1.102)$$

Let

$$z = \sqrt{f''\left(\sum_{i=1}^{k} \alpha_i(w_i) + b\right)} \nabla_w \alpha(w) \quad \text{and} \quad D = \text{diag}\left(f'\left(\sum_{i=1}^{k} \alpha_f(w_i) + b\right) \alpha_i''(w_i) + \beta_i''(w_i)\right).$$

Then, the Hessian matrix H can be expressed as

$$H = D + zz^T.$$

Consider a fully separable function $\Omega(w) = \sum_{i=1}^{k} \Omega_i(w_i)$. The proximal operator of the separable function $\Omega(w)$ is

$$\text{prox}_{\lambda\Omega}(v) = \left[\text{prox}_{\lambda\Omega_1}(v_1), \ldots, \text{prox}_{\lambda\Omega_k}(v_k)\right]^T.$$

If $\Omega_i(w_i) = -\log(w_i)$, then it is clear that

$$\text{prox}_{\lambda\Omega_i}(v_i) = \frac{v_i + \sqrt{v_i^2 + 4\lambda}}{2}. \quad (1.103)$$

For a nonsmooth function $\Omega_i(w_i) = |w_i|$, we have

$$\text{prox}_{\lambda\Omega_i}(v_i) = \begin{cases} v - \lambda & \text{if } v \geq \lambda \\ 0 & |v|\lambda \\ v + \lambda & \text{if } v \leq -\lambda. \end{cases} \quad (1.104)$$

1.3.5.1.3 Linear Constraints Consider a quadratic problem:

$$\begin{array}{ll} \text{minimize} & \frac{1}{2}\|x-v\|_2^2 \\ \text{subject to} & Ax=b,\ Cx\le d, \end{array} \tag{1.105}$$

where $A\in R^{m\times n}$, $C\in R^{p\times n}$, and $\Xi=\{x\in R^n|\,Ax=b, Cx\le d\}$.

Dual method can be used to solve the quadratic problem (1.105). If we have a few constraints, the number of dual problems will be small. The dual method is to transform the high-dimensional primal problem to a low-dimensional problem with a few dual variables corresponding to the constraints in Ξ.

Dual method is to add the constraints in Ξ to the objective function, which leads to the following unconstrained optimization problem:

$$F=\min_x \frac{1}{2}\|x-v\|^2+(Ax-b)^T\mu+(Cx-d)^T\eta. \tag{1.106}$$

Taking derivative $\frac{\partial F}{\partial d}$ and setting it to zero, we obtain

$$\frac{\partial F}{\partial x}=x-v+A^T\mu+C^T\eta=0.$$

Solving this equation yields

$$x=v-\begin{bmatrix} A \\ C \end{bmatrix}^T\begin{bmatrix} \mu \\ \eta \end{bmatrix}. \tag{1.107}$$

Substituting x into Equation 1.106, we have

$$L(\mu,\eta)=-\frac{1}{2}\|A^T\mu+C^T\eta\|_2^2+(Av-b)^T\mu+(Cv-d)^T\eta,$$

where $\mu\in R^m$ and $\eta\in R^p$ are dual variables. Then, the primal problem (1.105) is transformed to the dual problem:

$$\begin{array}{ll} \text{maximize} & L(\mu,\eta) \\ \text{subject to} & \eta\ge 0. \end{array} \tag{1.108}$$

Let μ^* and η^* be the solution to the quadratic problem (1.108). The original solution x^* in the problem (1.105) is given by

$$x^*=v-A^T\mu^*-C^T\eta^*. \tag{1.109}$$

1.3.5.1.4 Equality Constraints As a special case, we consider equality constraints. Assume $\Xi = \{x \in R^n | Ax = b\}$. In this case, the dual problem (1.108) is reduced to

$$\text{maximize} \quad L(\mu), \tag{1.110}$$

where $L(\mu) = -\frac{1}{2}||A^T\mu||_2^2 + (Av - b)^T \mu$.

Taking derivative $\frac{\partial L(\mu)}{\partial \mu}$ and setting it to zero, we obtain

$$-AA^T\mu = Av - b. \tag{1.111}$$

If $m < n$ and A has full rank, we can solve Equation 1.111:

$$\mu = -(AA^T)^{-1}(Av - b).$$

Equation 1.107 is then reduced to

$$\Pi_\Xi(v) = x = v - A^T(AA^T)^{-1}(Av - b). \tag{1.112}$$

Next we introduce their applications to several special cases.

For the hyperplane, we only have one equality constraint:

$$\Xi = \{w | a^T w = b\}.$$

Applying Equation 1.112 to the hyperplane, we obtain the projection onto the hyperplane:

$$\begin{aligned} \Pi_\Xi &= v - a(a^Ta)^{-1}(a^Tv - b) \\ &= v + \frac{b - a^T v}{||a||_2^2} a. \end{aligned} \tag{1.113}$$

For the half space, we have one inequality constraint $\Xi = \{w | a^T w \leq b\}$. In this case, the Lagrangian function is

$$L(\eta) = \min_w \frac{1}{2}||w - v||_2^2 + \eta(a^T w - b). \tag{1.114}$$

Taking a derivative of its objective function and setting it to zero, we obtain

$$w - v + \eta a = 0. \tag{1.115}$$

Solving Equation 1.115 for w yields

$$w = v - \eta a, \tag{1.116}$$

which leads to the Lagrangian function

$$L(\eta) = -\frac{a^T a}{2}\eta^2 + (a^T v\eta) - b\eta.$$

To solve the problem (1.108), we first take derive $\frac{dL(\eta)}{d\eta}$ and set it to zero:

$$\frac{dL(\eta)}{d\eta} = a^T a\eta + (a^T v - b) = 0.$$

The solution to the problem (1.108) is

$$\eta = -\frac{a^T v - b}{||a||_2^2} \quad \text{and} \quad \eta \geq 0.$$

In other words, we have solution

$$\eta = -\frac{(a^T v - b)_+}{||a||_2^2} a. \tag{1.117}$$

Thus, the solution follows from Equations 1.107 and 1.117:

$$\Pi_\Xi(v) = v - \frac{(a^T v - b)_+}{||a||_2^2} a. \tag{1.118}$$

Finally, we consider the intervals $\Xi = \{w | l \leq w \leq u\}$. Its Lagrangian function is

$$L(\eta, \mu) = \min_w \frac{1}{2}||w - v||_2^2 + \eta^T (w - u) + \mu^T (l - w). \tag{1.119}$$

Solving this problem, we have

$$w = v - \eta + \mu, \tag{1.120a}$$

$$L(\eta, \mu) = \frac{1}{2}||\mu - \eta||_2^2 + \eta^T (v - u + \mu - \eta) + \mu^T (l - v + \eta - \mu). \tag{1.120b}$$

The dual problem is

$$\begin{array}{ll} \text{maximize} & L(\eta,\mu) \\ \text{subject to} & \eta \geq 0,\ \mu \geq 0. \end{array} \tag{1.121}$$

We calculate derivatives of the Lagrangian function with respect to η and μ:

$$\frac{\partial L(\eta,\mu)}{\partial \eta} = \eta - \mu + v - u + \mu - 2\eta + \mu = -\eta + \mu + v - u = w - u$$

and

$$\frac{\partial L(\eta,\mu)}{\partial \mu} = \mu - \eta + \eta + l - v + \eta - 2\mu = l - v + \eta - \mu = l - w.$$

The conditions for the optimality include

$$\eta_i(w_i - u_i) = 0,\quad \mu_i(l_i - w_i) = 0,\quad \eta_i \geq 0,\quad \mu_i \geq 0. \tag{1.122}$$

The solutions to the problem (1.121) can be grouped into three cases:

(1) $v_i \leq l_i$. If $w_i > l_i$, which implies $\mu_i = 0$ and $w_i = v_i - \eta_i \leq v_i \leq l_i$, this leads to contradiction. Therefore, $w_i \leq l_i$. If $w_i < l_i$, then $\eta_i = 0$, $\mu_i = 0$, which implies $w_i = v_i$ and $\frac{\partial L(\eta,\mu)}{\partial \eta_i} < 0$, $\frac{\partial L(\eta,\mu)}{\partial \mu_i} < 0$. In this case, $L(\eta,\mu)$ will not reach maximum. If $w_i = l_i$, then $\frac{\partial L(\eta,\mu)}{\partial \mu_i} = 0$ and $\mu_i > 0$. $L(\eta,\mu)$ will reach maximum. Therefore, the original solution is $w_i = l_i$.

(2) $v_i \geq u_i$. If $w_i \neq u_i$, then $\eta_i = 0$, $\frac{\partial L(\eta,\mu)}{\partial \eta_i} \neq 0$, and $w_i = v_i + \mu_i \geq v_i \geq u_i > l_i$. Thus, we have $\frac{\partial L(\eta,\mu)}{\partial \mu_i} \neq 0$. Therefore, $L(\eta,\mu)$ will not reach maximum. $w_i = u_i$ should hold.

(3) $l_i < v_i < u_i$. If $w_i \neq v_i$, from Equation 1.120a, we obtain that $\eta_i \neq \mu_i$, which implies that either $\eta_i \neq 0$ or $\mu_i \neq 0$ or both $\eta_i \neq 0$, $\mu_i \neq 0$. Suppose that $\eta_i \neq 0$, then it follows from Equation 1.122 that $w_i = u_i > v_i > l_i$. Thus, again from Equation 1.122, we obtain that $\mu_i = 0$. Recall Equation 1.120a that $w_i = v_i - \eta_i + \mu_i = v_i - \eta_i < v_i < u_i$. This contradicts the assumption $w_i = u_i$. Therefore, we have $w_i = v_i$.

In summary, we prove that

$$\left(\Pi_{\Xi}(v)\right)_i = \begin{cases} l_i & v_i \leq l_i \\ v_i & l_i < v_i < u_i \\ u_i & v_i \geq u_i. \end{cases} \tag{1.123}$$

1.3.5.2 Norms

The norm is a widely used class of function. Computing the proximal operator of the norm function is a key step for optimization with sparsity-inducing penalties. Suppose that $\Omega(w) = ||w||$ is a general norm on R^n. In the Fenchel conjugate section, we showed that the Fenchel conjugate of $||w||$ is an indicator function:

$$I_{||z||_* \le 1} = \begin{cases} 0 & ||z||_* \le 1 \\ \infty & \text{otherwise.} \end{cases} \tag{1.124}$$

Before we compute the proximal operator of the norm function, we first extend Moreau decomposition (1.53) to a more general case:

$$\text{prox}_{\lambda\Omega}(v) = v - \text{prox}_{\Omega^*/\lambda}(v/\lambda). \tag{1.125}$$

Proof.

Now we give a brief proof.

Let

$$u = \text{Prox}_{\lambda\Omega}(v). \tag{1.126}$$

In other words, u is the minimizer of the function

$$\Omega(w) + \frac{1}{2\lambda}||w - v||_2^2.$$

Therefore, we obtain

$$0 \in \partial\Omega(u) + (u - v)/\lambda$$

or

$$\frac{v-u}{\lambda} \in \partial\Omega(u). \tag{1.127}$$

Using Fenchel–Young inequality, we obtain the following equality:

$$\Omega(u) + \Omega^*\left(\frac{v-u}{\lambda}\right) = \left(\frac{v-u}{\lambda}\right)^T u. \tag{1.128}$$

Taking a subdifferential on both sides of the equality (1.128), we obtain

$$\partial\Omega(u) - \frac{1}{\lambda}\partial\Omega^*\left(\frac{v-u}{\lambda}\right) \ni \frac{v-u}{\lambda} - \frac{u}{\lambda}. \tag{1.129}$$

It follows from Equations 1.127 and 1.128 that

$$u \in \partial\Omega^*\left(\frac{v-u}{\lambda}\right). \tag{1.130}$$

Recall that

$$\text{Prox}_{\Omega^*/\lambda}\left(\frac{v}{\lambda}\right) = \arg\min_{w}\left\{\Omega^*(w) + \frac{\lambda}{2}\left\|w - \frac{v}{\lambda}\right\|_2^2\right\}. \tag{1.131}$$

Let $w_0 = \text{Prox}_{\Omega^*/\lambda}\left(\frac{v}{\lambda}\right)$. Then, w_0 must satisfy the equation

$$0 \in \partial\Omega^*(w_0) + \lambda\left(w_0 - \frac{v}{\lambda}\right). \tag{1.132}$$

From Equation 1.31, we obtain the following equation:

$$0 \in \partial\Omega^*\left(\frac{v-u}{\lambda}\right) - u. \tag{1.133}$$

Taking $w_0 = \frac{v-u}{\lambda}$, Equation 1.132 is reduced to

$$0 \in \partial\Omega^*\left(\frac{v-u}{\lambda}\right) + \lambda\left(\frac{v-u}{\lambda} - \frac{v}{\lambda}\right) = \partial\Omega^*\left(\frac{v-u}{\lambda}\right) - u,$$

which implies Equation 1.133 ensures Equation 1.132 holds when we take $w_0 = \frac{v-u}{\lambda}$.

$$\frac{v-u}{\lambda} = \text{Prox}_{\Omega^*/\lambda}\left(\frac{v}{\lambda}\right). \tag{1.134}$$

Multiplying by λ on both sides of Equation 1.134 and adding Equation 1.126 results in

$$v = \text{prox}_{\lambda\Omega}(v) + \lambda\text{Prox}_{\Omega^*/\lambda}\left(\frac{v}{\lambda}\right). \tag{1.135}$$

Since Ω^* is an indicator function, the proximal operator of Ω^* is the projection of the point to the set

$$B = \left\{w \,\middle|\, \|w\|_* \leq 1\right\}.$$

Therefore, the proximal operator $\text{prox}_{\lambda\Omega}(v)$ is

$$\text{prox}_{\lambda\Omega}(v) = v - \lambda \Pi_B\left(\frac{v}{\lambda}\right). \tag{1.136}$$

Equation 1.136 provides a tool to calculate the proximal operator of various norms.

1.3.5.2.1 l_1-Norm Recall that dual norm of the l_1 norm is the l_∞ norm. The l_∞ norm is defined as $\|w\|_\infty = \max(|w_1|, \ldots, |w_p|)$. The unit ball is $B = \{w_i \| w_i\| \leq 1, \forall i \leq p\}$, i.e., the unit ball of the l_∞ norm is an n-dimensional hypercube of sidelength 2 $[-1, 1]^p$. Using Equation 1.123, we obtain the projection onto a hypercube:

$$\left(\Pi_B(v)\right)_i = \begin{cases} 1 & v_i \geq 1 \\ v_i & 1 > v_i > -1 \\ -1 & v_i \leq -1. \end{cases}$$

Using Equation 1.124, we obtain the proximal operator of the l_1 norm:

$$\begin{aligned}
\left(\text{prox}_{\lambda\|w\|_1}(v)\right)_i &= v - \lambda \Pi_B\left(\frac{v}{\lambda}\right) \\
&= v_i - \begin{cases} \lambda & v_i \geq \lambda \\ v_i & \lambda > v_i > -\lambda \\ -\lambda & v_i \leq -\lambda \end{cases} \\
&= \begin{cases} v_i - \lambda & v_i \geq \lambda \\ 0 & \lambda > v_i > -\lambda \\ v_i + \lambda & v_i \leq -\lambda \end{cases} \\
&= sign(v_i)\left(|v_i| - \lambda\right)_+ .
\end{aligned} \tag{1.137}$$

The proximal operator $\text{prox}_{\lambda\|w\|_1}(v)$ is an elementwise soft-thresholding operator.

1.3.5.2.2 l_2-Norm Consider the Euclidean norm in R^p. The Euclidean unit ball B is defined as $B = \left\{w \| \|w\|_2^2 \leq 1\right\}$. The projection of point v onto the Euclidean unit ball B is given by

$$\begin{aligned} \min_w \quad & \|w - v\|_2^2 \\ \text{subject to} \quad & \|w\|_2^2 \leq 1. \end{aligned} \tag{1.138}$$

To derive the Lagrangian function of the problem (1.138), we first minimize

$$\min_w F = \|w - v\|_2^2 + \eta\left(\|w\|_2^2 - 1\right). \tag{1.139}$$

Taking the derivative $\frac{\partial F}{\partial w}$ and setting it to zero, we obtain

$$\frac{\partial F}{\partial w} = 2(w - v + \eta w) = 0. \tag{1.140}$$

Solving Equation 1.140 for w, we obtain

$$w = \frac{v}{1+\eta}. \tag{1.141}$$

Consider two cases:

(i) $||v||_2 \le 1$. In this case

$$||w||_2 = \frac{||v||_2}{1+\eta} \le \frac{1}{1+\eta}. \tag{1.142}$$

We claim that $\eta = 0$. Otherwise, from the optimal condition for the problem (1.138) that $\eta\left(||w||_2^2 - 1\right) = 0$, it must be $||w||_2 = 1$. However, from inequality (1.142), we conclude that when $\eta > 0$, we have $||w||_2 < 1$, which leads to the contradiction. Therefore, when $||v||_2 \le 1$, the solution is $w = v$.

(ii) $||v||_2 > 1$. In this case, the Lagrangian function of the problem (1.138) is

$$L(\eta) = \eta\left(\frac{||v||_2^2}{1+\eta} - 1\right). \tag{1.143}$$

The dual problem is

$$\max_{\eta \ge 0} L(\eta).$$

Taking derivative $\frac{\partial L(\eta)}{\partial \eta}$ and setting it to zero yields

$$\frac{\partial L(\eta)}{\partial \eta} = \frac{||v||_2^2}{(1+\eta)^2} - 1 = 0. \tag{1.144}$$

Since $1 + \eta > 0$, solving Equation 1.144, we obtain

$$1 + \eta = ||v||_2^2. \tag{1.145}$$

Substituting Equation 1.145 into Equation 1.141, we obtain the projection

$$w = \frac{v}{||v||_2}.$$

In summary, the projection of point v onto the Euclidean unit ball B is

$$\Pi_B(v) = \begin{cases} \dfrac{v}{\|v\|_2} & \|v\|_2 > 1 \\ v & \|v\|_2 \le 1. \end{cases} \tag{1.146}$$

Using Equation 1.136, we obtain the proximal operator of the Euclidean norm:

$$\begin{aligned} \operatorname{prox}_{\lambda\|w\|_2}(v) &= v - \lambda \Pi_B\left(\frac{v}{\lambda}\right) \\ &= \begin{cases} \left(1 - \dfrac{\lambda}{\|v\|_2}\right) v & \|v\|_2 > \lambda \\ 0 & \|v\|_2 \le \lambda \end{cases} \\ &= \left(1 - \frac{\lambda}{\|v\|_2}\right)_+ v. \end{aligned} \tag{1.147}$$

The proximal operator of the Euclidean norm is often referred to as block soft thresholding.

1.3.5.2.3 $l_1 + l_2$-Norm Consider a combination of the l_1-norm and the Euclidean norm. It can be written as

$$\Omega(w) = \|w\|_1 + \frac{\gamma}{2}\|w\|_2^2, \tag{1.148}$$

where $\gamma > 0$. Equation 1.148 is the elastic net regularization (Zou and Hastie 2005). Let $\bar{\lambda} = \dfrac{\lambda}{1+\gamma\lambda}$. Using Equation 1.49, we obtain

$$\begin{aligned} \operatorname{prox}_{\lambda\Omega} &= \operatorname{prox}_{\bar{\lambda}\|w\|_1}\left(\frac{\bar{\lambda}}{\lambda} v\right) \\ &= sign\left(\frac{\bar{\lambda}}{\lambda} v\right)\left(\left|\frac{\bar{\lambda}}{\lambda} v\right| - \frac{\lambda}{1+\gamma\lambda}\right)_+ \\ &= sign(v)\left(\left|\frac{1}{1+\gamma\lambda} v\right| - \frac{1}{1+\gamma\lambda}\lambda\right)_+ \\ &= \frac{1}{1+\gamma\lambda} sign(v)\left(|v| - \lambda\right)_+ \\ &= \frac{1}{1+\gamma\lambda} \operatorname{prox}_{\lambda\|w\|_1}(v). \end{aligned} \tag{1.149}$$

1.3.5.2.4 Group Lasso Group lasso is a sum of l_2-norm regularization. Let Φ be a partition of $\{1, \ldots, p\}$. Consider the function:

$$\Omega(w) = \sum_{g \in \Phi} \|w_g\|_2. \tag{1.150}$$

Since the function $\Omega(w)$ is fully separable, using Equation 1.44b, we obtain

$$\begin{aligned}\left(\text{Prox}_{\lambda\Omega}(u)\right)_{gi} &= \text{Prox}_{\lambda\Omega_g}(v_g) \\ &= \text{Prox}_{\lambda\|w_g\|_2}(v_g) \\ &= \left(1 - \frac{\lambda}{\|v_g\|_2}\right)_+ v_g \quad (\text{by Equation 1.147}).\end{aligned} \tag{1.151}$$

1.4 MATRIX CALCULUS

In many situations, it is necessary to obtain the partial derivatives of a function with respect to a vector or matrix of variables.

1.4.1 Derivative of a Function with Respect to a Vector

Definition 1.3

Let $f(x_1, \ldots, x_k)$ be a function of k real variables $x_1, \ldots, x_k$. Define a vector:

$$x = \begin{bmatrix} x_1 \\ \vdots \\ x_k \end{bmatrix}.$$

The derivative of the function $f(x_1, \ldots, x_k)$ with respect to the vector x is defined as

$$\nabla_x f = \frac{\partial f}{\partial x} = \begin{bmatrix} \frac{\partial f}{\partial x_1} \\ \vdots \\ \frac{\partial f}{\partial x_k} \end{bmatrix}.$$

Example 1.16

Let f be a linear function of k real variables defined by $f(x) = \sum_{i=1}^{k} a_i x_i = a^T x$, where

$$a = \begin{bmatrix} a_1 \\ \vdots \\ a_k \end{bmatrix}.$$

Then, we have

$$\frac{\partial f}{\partial x} = \begin{bmatrix} a_1 \\ \vdots \\ a_k \end{bmatrix} = a. \tag{1.152}$$

Example 1.17

Let f be a quadratic form in the k real variables $x_1, \ldots, x_k$ defined by

$$f(x) = x^T Ax,$$

where

$$A = \begin{bmatrix} a_{11} & \cdots & a_{1k} \\ \vdots & \ddots & \vdots \\ a_{k1} & \cdots & a_{kk} \end{bmatrix}.$$

Then, we have (1.153)

$$\frac{\partial f}{\partial x} = 2Ax. \tag{1.153}$$

1.4.2 Derivative of a Function with Respect to a Matrix

Definition 1.4

Let f be a function of the $m \times n$ matrix defined by

$$X = \begin{bmatrix} x_{11} & \cdots & x_{1n} \\ \vdots & \ddots & \vdots \\ x_{m1} & \cdots & x_{mn} \end{bmatrix}.$$

We assume that each partial derivative $\partial f / \partial x_{ij}$ exists. The derivative of function f with respect to the matrix X, denoted by $\partial f / \partial X$, is defined by

$$\frac{\partial f}{\partial X} = \begin{bmatrix} \frac{\partial f}{\partial x_{11}} & \cdots & \frac{\partial f}{\partial x_{1n}} \\ \vdots & \ddots & \vdots \\ \frac{\partial f}{\partial x_{m1}} & \cdots & \frac{\partial f}{\partial x_{mn}} \end{bmatrix}.$$

Example 1.18

Consider the function $f(X) = a^T Xb$, where a is an m-dimensional vector of constants, b is an n-dimensional vector of constants, and X is an $m \times n$ dimensional matrix of the variables. Then, we have

$$\frac{\partial f}{\partial X} = ab^T. \tag{1.154}$$

Proof.

The function $f(X)$ can be rewritten as

$$f(X) = \sum_{i=1}^{n} \sum_{j=1}^{n} a_i x_{ij} b_j.$$

By definition, we have $\frac{\partial f}{\partial x_{ij}} = a_i b_j$, which implies that

$$\frac{\partial f}{\partial X} = \begin{bmatrix} a_1 b_1 & \cdots & a_1 b_n \\ \vdots & \ddots & \vdots \\ a_m b_1 & \cdots & a_m b_n \end{bmatrix} = \begin{bmatrix} a_1 \\ \vdots \\ a_m \end{bmatrix} \begin{bmatrix} b_1 & \cdots & b_n \end{bmatrix} = ab^T.$$

Example 1.19 (Exercise 1.5)

Let function $f(X)$ be defined as $f(X) = a^T Xa$, where a is a k-dimensional vector of constants and X is a $k \times k$ dimensional matrix of variables. Then

$$\frac{\partial f}{\partial X} = 2aa^T - \text{diag}\left(a_1^2, \ldots, a_k^2\right). \tag{1.155}$$

1.4.3 Derivative of a Matrix with Respect to a Scalar

Definition 1.5

Let Y be a $k \times k$ matrix with elements being a function of a scalar. The derivative of the matrix Y with respect to the scalar x is defined as

$$\frac{\partial Y}{\partial x} = \begin{bmatrix} \frac{\partial y_{11}}{\partial x} & \cdots & \frac{\partial y_{1k}}{\partial x} \\ \vdots & \ddots & \vdots \\ \frac{\partial y_{k1}}{\partial x} & \cdots & \frac{\partial y_{kk}}{\partial x} \end{bmatrix}.$$

Example 1.20

Let Y be a nonsingular $k \times k$ matrix of function of scalar β. Then, we have

$$\frac{\partial Y^{-1}}{\partial \beta} = -Y^{-1}\frac{\partial Y}{\partial \beta}Y^{-1}. \tag{1.156}$$

Proof.

Recall that

$$YY^{-1} = I.$$

Taking a derivative of matrix Y with respect to the scalar β on both sides of the above equation, we obtain

$$\frac{\partial Y}{\partial \beta}Y^{-1} + Y\frac{\partial Y^{-1}}{\partial \beta} = 0,$$

which implies that

$$\frac{\partial Y^{-1}}{\partial \beta} = -Y^{-1}\frac{\partial Y}{\partial \beta}Y^{-1}.$$

Definition 1.6

Similarly, we can define the derivative of a vector function of a scalar as

$$\frac{\partial Y}{\partial x} = \begin{bmatrix} \frac{\partial y_1}{\partial x} \\ \vdots \\ \frac{\partial y_k}{\partial x} \end{bmatrix},$$

where $Y = \begin{bmatrix} y_1 \\ \vdots \\ y_k \end{bmatrix}$.

1.4.4 Derivative of a Matrix with Respect to a Matrix or a Vector

Definition 1.7

Let

$$Y = \begin{bmatrix} y_{11} & \cdots & y_{1k} \\ \vdots & \ddots & \vdots \\ y_{k1} & \cdots & y_{kk} \end{bmatrix}, \quad X = \begin{bmatrix} x_{11} & \cdots & x_{1n} \\ \vdots & \ddots & \vdots \\ x_{m1} & \cdots & x_{mn} \end{bmatrix}, \quad \text{and} \quad \frac{\partial Y}{\partial x_{ij}} = \begin{bmatrix} \frac{\partial y_{11}}{\partial x_{ij}} & \cdots & \frac{\partial y_{1k}}{\partial x_{ij}} \\ \vdots & \ddots & \vdots \\ \frac{\partial y_{k1}}{\partial x_{ij}} & \cdots & \frac{\partial y_{kk}}{\partial x_{ij}} \end{bmatrix}.$$

The derivative of matrix Y with respect to X, denoted by $\frac{\partial Y}{\partial X}$, is defined as

$$\frac{\partial Y}{\partial X} = \begin{bmatrix} \frac{\partial Y}{\partial x_{11}} & \cdots & \frac{\partial Y}{\partial x_{1n}} \\ \vdots & \ddots & \vdots \\ \frac{\partial Y}{\partial x_{m1}} & \cdots & \frac{\partial Y}{\partial x_{mn}} \end{bmatrix}.$$

Example 1.21

Let $Y = \begin{bmatrix} x_1^2 & 2x_1 + x_2 & x_2 \\ x_2^2 & x_1 & x_1 + 2x_2^2 \end{bmatrix}$ and $X = \begin{bmatrix} x_1 \\ x_2 \end{bmatrix}$.

Then, we have

$$\frac{\partial Y}{\partial x} = \begin{bmatrix} \frac{\partial Y}{\partial x_1} \\ \frac{\partial Y}{\partial x_2} \end{bmatrix} = \begin{bmatrix} 2x_1 & 2 & 0 \\ 0 & 1 & 1 \\ 0 & 1 & 1 \\ 2x_2 & 0 & 4x_2 \end{bmatrix}.$$

1.4.5 Derivative of a Vector Function of a Vector

Let $Y = [y_1(x_1, \ldots, x_m), \ldots, y_n(x_1, \ldots, x_m)]^T$. The derivative of a vector function with respect to a vector is defined as

$$\frac{\partial Y^T}{\partial x} = \begin{bmatrix} \frac{\partial y_1}{\partial x_1} & \cdots & \frac{\partial y_n}{\partial x_1} \\ \vdots & \ddots & \vdots \\ \frac{\partial y_1}{\partial x_m} & \cdots & \frac{\partial y_n}{\partial x_m} \end{bmatrix}.$$

1.4.6 Chain Rules

Next we introduce chain rules that are very useful in application of matrix calculus.

1.4.6.1 Vector Function of Vectors

Define two vector functions of a vector:

$$z = \begin{bmatrix} z_1 \\ \vdots \\ z_k \end{bmatrix} = h(y) = \begin{bmatrix} h_1(y) \\ \vdots \\ h_k(y) \end{bmatrix} \quad \text{and} \quad y = \begin{bmatrix} y_1 \\ \vdots \\ y_m \end{bmatrix} = f(x) = \begin{bmatrix} f_1(x) \\ \vdots \\ f_m(x) \end{bmatrix},$$

where $x = \begin{bmatrix} x_1 \\ \vdots \\ x_n \end{bmatrix}$.

Then, we have

$$\frac{\partial z^T}{\partial x}=\begin{bmatrix}\frac{\partial h_1(y)}{\partial x_1} & \cdots & \frac{\partial h_k(y)}{\partial x_1}\\ \vdots & \ddots & \vdots \\ \frac{\partial h_1(y)}{\partial x_n} & \cdots & \frac{\partial h_k(y)}{\partial x_n}\end{bmatrix}=\begin{bmatrix}\sum_{l=1}^{m}\frac{\partial h_1}{\partial y_l}\frac{\partial y_l}{\partial x_1} & \cdots & \sum_{l=1}^{m}\frac{\partial h_k}{\partial y_l}\frac{\partial y_l}{\partial x_1}\\ \vdots & \ddots & \vdots \\ \sum_{l=1}^{m}\frac{\partial h_1}{\partial y_l}\frac{\partial y_l}{\partial x_n} & \cdots & \sum_{l=1}^{m}\frac{\partial h_k}{\partial y_l}\frac{\partial y_l}{\partial x_n}\end{bmatrix}$$

$$=\begin{bmatrix}\frac{\partial y_1}{\partial x_1} & \cdots & \frac{\partial y_m}{\partial x_1}\\ \vdots & \ddots & \vdots \\ \frac{\partial y_1}{\partial x_n} & \cdots & \frac{\partial y_m}{\partial x_n}\end{bmatrix}\begin{bmatrix}\frac{\partial h_1}{\partial y_1} & \cdots & \frac{\partial h_k}{\partial y_1}\\ \vdots & \ddots & \vdots \\ \frac{\partial h_1}{\partial y_m} & \cdots & \frac{\partial h_k}{\partial y_m}\end{bmatrix}$$

$$=\frac{\partial y^T}{\partial x}\frac{\partial z^T}{\partial y}. \tag{1.157}$$

1.4.6.2 *Scalar Function of Matrices*

Let $z=h(y)$ be a scalar function of matrix and $y=f(x)$ be a matrix function of a scalar, where $y=\begin{bmatrix}y_{11} & \cdots & y_{1n}\\ \vdots & \ddots & \vdots \\ y_{m1} & \cdots & y_{mn}\end{bmatrix}$.

The derivative $\frac{dz}{dx}$ is given by

$$\frac{dz}{dx}=\sum_{i=1}^{m}\sum_{j=1}^{n}\frac{\partial z}{\partial y_{ij}}\frac{\partial y_{ij}}{\partial x}$$

$$=\text{Trace}\left[\frac{\partial z}{\partial y}\left(\frac{\partial y}{\partial x}\right)^T\right]. \tag{1.158}$$

1.4.7 Widely Used Formulae

We extend some formulae from standard calculus to matrix calculus that are often used in practice. Let $A, B, C,$ and D be matrices and $a, b, c, d, x, y,$ and z be vectors.

1.4.7.1 *Determinants*

Let $|A|$ be the determinant of matrix A. Then,

$$\frac{\partial |A|}{\partial A}=|A|A^{-T}. \tag{1.159}$$

In fact, we can write

$$|A| = \sum_{p=1}^{n} a_{pj} A_{pj}, \quad j = 1, \ldots, n,$$

where A_{ij} is the cofactor of a_{ij}. By definition of the derivative of scale function with respect to matrix, we obtain

$$\frac{\partial |A|}{\partial A} = \begin{bmatrix} A_{11} & \cdots & A_{1n} \\ \vdots & \ddots & \vdots \\ A_{m1} & \cdots & A_{mn} \end{bmatrix} = |A| \frac{1}{|A|} \begin{bmatrix} A_{11} & \cdots & A_{1n} \\ \vdots & \ddots & \vdots \\ A_{m1} & \cdots & A_{mn} \end{bmatrix} = |A| A^{-T}.$$

Using Equation 1.159, we can easily obtain

$$\frac{\partial \log |A|}{\partial A} = A^{-T}. \tag{1.160}$$

1.4.7.2 Polynomial Functions

(1) Linear functions

$$\frac{\partial x^T ay}{\partial x} = \frac{\partial x^T}{\partial x} ay = ay. \tag{1.161}$$

(2) Quadratic functions

Let $F = (Ax + b)^T C(Dx + e)$. Then

$$\begin{aligned} \frac{\partial F}{\partial x} &= \frac{\partial x^T}{\partial x} A^T C(Dx+e) + \frac{\partial x^T}{\partial x} D^T C^T (Ax+b) \\ &= A^T C(Dx+e) + D^T C^T (Ax+b) \end{aligned}$$

$$\frac{\partial}{\partial A} x^T Ay = xy^T. \tag{1.162}$$

1.4.7.3 Trace

$$(1)\ \frac{\partial}{\partial A} \text{Trace}(A) = \begin{bmatrix} \frac{\partial}{\partial a_{11}} \sum_{i=1}^{n} a_{ii} & \cdots & 0 \\ \vdots & \ddots & \vdots \\ 0 & \cdots & \frac{\partial}{\partial a_{nn}} \sum_{i=1}^{n} a_{ii} \end{bmatrix} = I. \tag{1.163}$$

(2) $\frac{\partial}{\partial B}\text{Trace}(ABC) = A^T C^T.$ (1.164)

To see this, we first calculate $F = \text{Trace}(ABC) = \sum_i \sum_k \sum_j a_{ik} b_{kj} c_{ji}$. By definition of the derivative of a scale function with respect to a matrix, we obtain

$$\frac{\partial F}{\partial B} = \begin{bmatrix} \frac{\partial F}{\partial b_{11}} & \cdots & \frac{\partial F}{\partial b_{1n}} \\ \vdots & \ddots & \vdots \\ \frac{\partial F}{\partial b_{n1}} & \cdots & \frac{\partial F}{\partial b_{nn}} \end{bmatrix}$$

$$= \begin{bmatrix} \sum_i a_{i1} c_{1i} & \cdots & \sum_i a_{i1} c_{ni} \\ \vdots & \ddots & \vdots \\ \sum_i a_{in} c_{1i} & \cdots & \sum_i a_{in} c_{ni} \end{bmatrix} = A^T C^T.$$

(3) $\frac{\partial}{\partial B}\text{Trace}(AB^T) = A.$ (1.165)

Proof.
Let $F = \text{Trace}(AB^T) = \sum_i \sum_j a_{ij} b_{ij}$. Then, we have

$$\frac{\partial F}{\partial B} = \begin{bmatrix} a_{11} & \cdots & a_{1n} \\ \vdots & \ddots & \vdots \\ a_{n1} & \cdots & a_{nn} \end{bmatrix} = A.$$

Similarly, we have

(4) $\frac{\partial}{\partial B}\text{Trace}(B^T A) = A.$ (1.166)

Combining Equations 1.165 and 1.166, we obtain

(5) $\frac{\partial}{\partial B}\text{Trace}(B^T AB) = (A + A^T)B.$ (1.167)

Now we introduce a formula involving inverse matrix.

(6) $\frac{\partial}{\partial A}\text{Trace}(A^{-1}B) = -A^{-T} B^T A^{-T}.$ (1.168)

Let Δ_{ij} be the matrix with 1 in the i,j position and zero elsewhere and $W = A^{-1}$. By definition of the derivative of trace with respect to matrix, we have

$$\frac{\partial}{\partial A}\text{Trace}(A^{-1}B) = \begin{bmatrix} \frac{\partial}{\partial a_{11}}\text{Trace}(A^{-1}B) & \cdots & \frac{\partial}{\partial a_{1n}}\text{Trace}(A^{-1}B) \\ \vdots & \ddots & \vdots \\ \frac{\partial}{\partial a_{n1}}\text{Trace}(A^{-1}B) & \cdots & \frac{\partial}{\partial a_{nn}}\text{Trace}(A^{-1}B) \end{bmatrix}$$

$$= \begin{bmatrix} \text{Trace}\left(\frac{\partial}{\partial a_{11}}(A^{-1})B\right) & \cdots & \text{Trace}\left(\frac{\partial}{\partial a_{1n}}(A^{-1})B\right) \\ \vdots & \ddots & \vdots \\ \text{Trace}\left(\frac{\partial}{\partial a_{n1}}(A^{-1})B\right) & \cdots & \text{Trace}\left(\frac{\partial}{\partial a_{nn}}(A^{-1})B\right) \end{bmatrix}$$

$$= \begin{bmatrix} \text{Trace}((-A^{-1}\Delta_{11}A^{-1})B) & \cdots & \text{Trace}((-A^{-1}\Delta_{1n}A^{-1})B) \\ \vdots & \ddots & \vdots \\ \text{Trace}((-A^{-1}\Delta_{n1}A^{-1})B) & \cdots & \text{Trace}((-A^{-1}\Delta_{nn}A^{-1})B) \end{bmatrix}$$

$$= \begin{bmatrix} -\text{Trace}(W_{.1}W_{1.}^{T}B) & \cdots & -\text{Trace}(W_{.1}W_{n.}^{T}B) \\ \vdots & \ddots & \vdots \\ -\text{Trace}(W_{..n}W_{1.}^{T}B) & \cdots & -\text{Trace}(W_{.n}W_{n.}^{T}B \end{bmatrix}$$

$$= -\begin{bmatrix} \sum_i w_{i1}W_{1.}^{T}B_{.i} & \cdots & \sum_i w_{i1}W_{n.}^{T}B_{.i} \\ \vdots & \ddots & \vdots \\ \sum_i w_{in}W_{1.}^{T}B_{.i} & \cdots & \sum_i w_{in}W_{n.}^{T}B_{.i} \end{bmatrix}$$

$$= -A^{-T} = B^{T}A^{-T}.$$

Specifically, we have

$$\frac{\partial}{\partial A}\text{Trace}\left(A^{-1}\right) = -\left(A^{-2}\right)^{T}. \tag{1.169}$$

1.5 FUNCTIONAL PRINCIPAL COMPONENT ANALYSIS (FPCA)

Due to advances in sequencing technologies, sensing and communications, three major types of biological information—the digital information of the genomes, epigenomes, and environmental signals—are generated. Fast and cheaper next-generation sequencing (NGS) technologies will generate unprecedentedly massive (thousands or even tens of thousands of individuals) and high-dimensional (up to hundreds of millions) genomic and epigenomic variation data. The application of mobile and new sensing technologies in health and wellness produces a deluge of physiological, imaging, and environmental data. Analyses of these extremely big and diverse types of datasets provide invaluable information for holistic discovery of the genetic and epigenetic structure of disease and for the prediction, prevention, diagnosis, and treatment of disease but also pose great conceptual, analytical, and computational challenges.

Functional data analysis techniques can be used to meet these challenges. In this section, we introduce FPCA for high-dimensional functional data reduction. We first review traditional multivariate principal component analysis (PCA) and functional principal component analysis (FPCA). Then, we extend FPCA to smooth FPCA (SFPCA).

1.5.1 Principal Component Analysis (PCA)

1.5.1.1 Least Square Formulation of PCA

PCA is one of the oldest and most widely used tools for dimension reduction and visualization. PCA intends to project high-dimensional data to a low-dimensional space with a few directions (axes) along which data variation is maximized. We consider p random variables $X_1, \ldots, X_p$. Assume that n samples are taken. The generated data matrix is denoted by $X=[x_1, \ldots, x_n]$, where $x_i=[x_{i1}, \ldots x_{ip}]^T$ and n is the number of samples. The goal of PCA is to seek the projection $\tilde{V} \in R^{d\times n}$ of the matrix X in a d-dimensional linear space with an orthonormal basis, $\tilde{U} \in R^{p\times d}$. In other words, we approximate the matrix X by $L=\tilde{U}\tilde{V}$, where $\tilde{U}$ is called a matrix of principal component scores and $\tilde{V}$ is called a matrix of principal components.

Let the rank of the matrix L be d and the rank of the matrix X be r. The singular value decomposition (SVD) of X is given by $X=U\Lambda V^T$, where $U\in R^{m\times r}$, $V\in R^{n\times r}$, and $\Lambda=\text{diag}\,(\sigma_1, \ldots, \sigma_r)$ with $\sigma_1 \geq \ldots \sigma_r > 0$. Let $\Lambda_{(d)}=\text{diag}\,(\sigma_1, \ldots, \sigma_d)$.

Theorem 1.1

$L=\tilde{U}\tilde{V}$ minimizes.

$$F=\|X-\tilde{U}\tilde{V}\|_F^2 \tag{1.170}$$

with rank of $L=d$, where $\tilde{U}=(U)_d\Lambda_{(d)}^{1/2}$ and $\tilde{V}=\Lambda_{(d)}^{1/2}\left(V^T\right)_d$ (Appendix 1A).

Theorem 1.1 shows that the matrix X can be best approximated by $(U)_d\Lambda_{(d)}(V^T)_d$ under the Frobenius norm.

We define

$$\xi=U^TX.$$

Then, its variance is given by

$$\text{Var}(\zeta)=\Lambda^2.$$

This shows that the variance of the first principal component is the square of the largest single value of the matrix X, the variance of the second principal component is the square of the second largest single value of the matrix X, and so on, while all pairs of the principal components are uncorrelated.

1.5.1.2 Variance-Maximization Formulation of PCA

PCA can also be interpreted as finding linear combinations of the variables called principal components, maximizing variance in the data. Consider the linear combinations

$$\begin{aligned} Y_1 &= e_{11}X_1 + \ldots + e_{1p}X_p = e_1^T X \\ Y_2 &= e_{21}X_1 + \ldots + e_{2p}X_p = e_2^T X \\ &\vdots \\ Y_p &= e_{p1}X_1 + \ldots + e_{pp}X_p = e_p^T X, \end{aligned} \tag{1.171}$$

where $e_i^T = \left[e_{i1},\ldots,e_{ip}\right]^T$, $X = [X_1, \ldots, X_p]^T$, and $Y_i = [y_{i1}, \ldots, y_{in}]$. Since $X_1, \ldots, X_p$ are random variables, $Y_1, \ldots, Y_p$ are also random variables.

Define

$$Y = \begin{bmatrix} Y_1 \\ \vdots \\ Y_p \end{bmatrix} \quad \text{and} \quad E = \begin{bmatrix} e_{11} & \cdots & e_{1p} \\ \vdots & \ddots & \vdots \\ e_{p1} & \cdots & e_{pp} \end{bmatrix} = \begin{bmatrix} e_1^T \\ \vdots \\ e_p^T \end{bmatrix}.$$

Equation 1.171 can be written in a matrix form:

$$Y = EX. \tag{1.172}$$

Denote the variance–covariance matrix of the random vector X by

$$\Sigma = \text{var}\left(X\right) = \Sigma = \begin{bmatrix} \sigma_{11} & \cdots & \sigma_{1p} \\ \vdots & \ddots & \vdots \\ \sigma_{p1} & \cdots & \sigma_{pp} \end{bmatrix}.$$

The variance $\text{var}(Y_i)$ of the random variable Y_i is then given by

$$\text{var}\left(Y_i\right) = e_i^T \Sigma e_i, \quad i = 1,\ldots,p.$$

We want that the transformed variables $y_1, \ldots, y_p$ are uncorrelated, i.e.,

$$e_j^T \Sigma e_i = 0. \tag{1.173}$$

The goal of the PCA is to seek $e_1, \dots, e_p$ that maximize $\text{var}(Y_1), \dots, \text{var}(Y_p)$ subject to the constraints:

$$e_i^T e_i = 1, \quad e_j^T e_i = 0, \quad j < i, \; i = 1, \dots, p. \tag{1.174}$$

The summation of the variances $\text{var}(Y_1), \dots, \text{var}(Y_p)$ can be expressed as

$$\sum_{i=1}^{p} \text{var}(Y_i) = \sum_{i=1}^{p} e_i^T \Sigma e_i = \text{Trace}\left(E \Sigma E^T\right). \tag{1.175}$$

The constraints (1.174) can be written in a matrix form:

$$E^T E = I. \tag{1.176}$$

Thus, the PCA can be mathematically formulated as the following optimization problem:

$$\begin{aligned} \max_{E} \quad & \text{Trace}\left(E \Sigma E^T\right) \\ \text{subject to} \quad & E E^T = I. \end{aligned} \tag{1.177}$$

By Lagrange multiplier method, the constrained optimization problem (1.177) can be transformed into the following unconstrained optimization problem:

$$\max_{E} \quad \text{Trace}\left(E \Sigma E^T\right) + \text{Trace}\left(\Lambda\left(I - E E^T\right)\right), \tag{1.178}$$

Using formula (1.164), the optimal conditions for solving the optimization problem (1.178) are

$$\Sigma E^T = E^T \Lambda \tag{1.179}$$

or

$$\Sigma e_i = \lambda_{i1} e_1 + \dots + \lambda_{ii-1} e_{i-1} + \lambda_{ii} e_i + \lambda_{ii+1} e_{i+1} + \dots + \lambda_{ip} e_p, \quad i = 1, \dots, p. \tag{1.180}$$

Recall that

$$e_j^T \Sigma e_i = 0,$$

which implies

$$\lambda_{ij} = 0, \quad \forall j \neq i.$$

Therefore, the matrix Λ is diagonal and denoted by $\Lambda = \text{diag}(\lambda_1, \dots, \lambda_p)$, and Equation 1.180 is reduced to the following eigenequation:

$$\Sigma e_i = \lambda_i e_i, \quad i = 1, \dots, p. \tag{1.181}$$

Equation 1.181 implies that

$$\text{var}\left(Y_i\right) = e_i^T \Sigma e_i = \lambda_i, \quad i = 1, \ldots, p. \tag{1.182}$$

The above discussions show that the variance of the first principal component is the largest eigenvalue value of the covariance matrix Σ of X, the variance of the second principal component is the second largest eigenvalue value of the covariance matrix Σ of, and so on, while all pairs of the principal components are uncorrelated.

It follows from Equation 1.175 that the total population variance is equal to the summation of eigenvalues of the covariance matrix Σ. A sample measure of how well the first k principal components represent the k original variable is given by

$$\frac{\lambda_1 + \lambda_2 + \ldots + \lambda_k}{\lambda_1 + \lambda_2 + \ldots + \lambda_p}.$$

It is easy to see that the sampling variance–covariance matrices var(Y) and var(X) are given by

$$\hat{\text{var}}\left(Y\right) = YY^T \quad \text{and} \quad \hat{\Sigma} = \hat{\text{var}}\left(X\right) = XX^T.$$

Therefore,

$$\hat{\Sigma} = U\Lambda^2 U^T.$$

For the sampling data, the matrix E is equal to U. The principal components Y defined in Equation 1.172 is the same as ξ defined as $\xi = U^T X$ in the least square formulation of the PCA.

Example 1.22 Real Example

To illustrate PCA, we apply it to the low coverage pilot dataset in the 1000 Genomes Project, which was released in July 2010 (ftp://ftp.1000genomes.ebi.ac.uk). The dataset included 179 unrelated individuals from four populations: Yoruba in Ibadan, Nigeria (YRI, 59 individuals); Utah residents with ancestry from Northern and Western Europe (CEU, 60 individuals); Han Chinese in Beijing, China (CHB, 30 individuals); and Japanese in Tokyo, Japan (JPT, 30 individuals). In this study, CHB and JPT populations were combined as one population (ASI). These samples were sequenced on an average cover rate of 4X. A total of 14,397,437 SNPs on the 22 autosomes were identified. We plot the first two PC scores for 179 individuals from four populations YRI, CEU, CHB, and JPT on 14,397,437 SNPs in Figure 1.7 to study the power of the popular PCA for detecting the population structure. Figure 1.7 shows that the individuals from CEU, YRI, and ASI were well separated, but individuals from the CHB and JPT populations were not separated very well by the PCA.

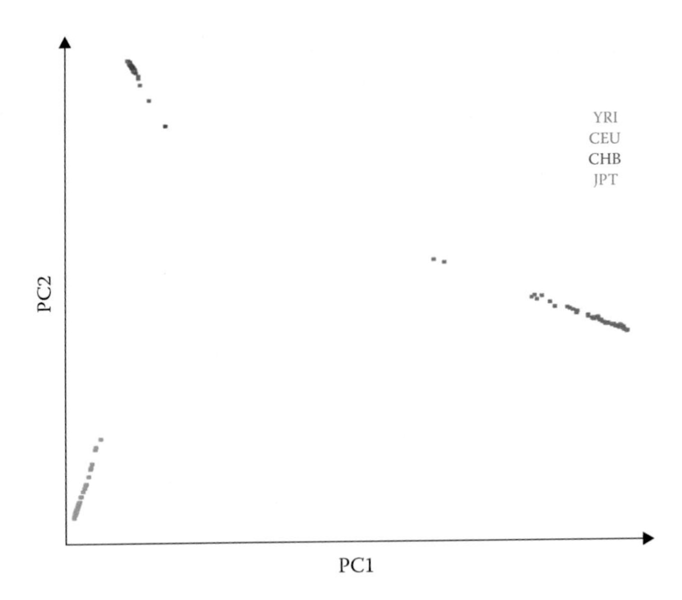

FIGURE 1.7 Two-PC score plot for 179 individuals from four populations YRI, CEU, CHB, and JPT on 14,397,437 SNPs.

1.5.2 Basic Mathematical Tools for Functional Principal Component Analysis

1.5.2.1 Calculus of Variation

Calculus of variation is a very useful mathematical tool that deals with maximizing or minimizing functionals, which map a set of functions to the real numbers. Let $y(x)$ be a function and $J[y(x)]$ be its corresponding functional. The variation of the function $\delta y(x)$ is defined as the difference between two functions (Figure 1.8)

$$\delta y(x) = y(x) - y_0(x).$$

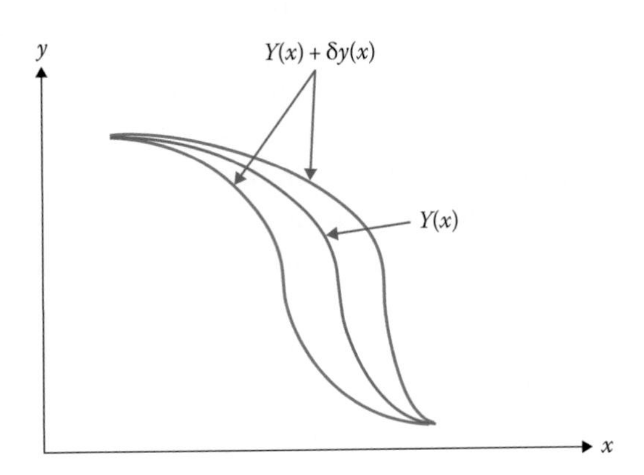

FIGURE 1.8 Illustration of function variation.

If the functional $J(y(x))$ for any function $y(x)$ that is close to $y^*(x)$ is larger than or equal to $J(y^*(x))$, then $J(y(x))$ reaches its minimum $J(y^*(x))$ at the function $y^*(x)$. Let $\Delta J[h] = J[y+h] - J[y]$ be its change due to an increment $h(x)$ of the "variable" $y = y(x)$.

When $y(x)$ is fixed, $\Delta J[h]$ is a function of $h(x)$.

Definition 1.8

$\delta J[h]$ is called the first variation of $J[y]$ at $y(x) = y_0(x)$ if for $t \in R$ we have that

$$\delta J[h] = \frac{d}{dt} J[y_0(x) + th(x)]_{t=0} \tag{1.183}$$

exists for all $h(x) \in$ a normal linear space S.

Definition 1.9

Let $\Psi \subset S$ represent a space of competing functions. $J[h]$ reaches its relative maximum at $y_0(x) \in \Psi$ if

$$J[y] - J[y_0] \le 0 \tag{1.184}$$

for all $y \in \Psi$ for which $\|y - y_0\| < \varepsilon$ for some $\varepsilon > 0$.

A key issue for functional data analysis is to search the optimum of a functional. Next we introduce a tool for finding a maximum of a functional. Similar to the necessary condition for a relative optimal of a function that $\frac{dy(x)}{dx} = 0$, the necessary condition for reaching the optimum of a functional is $\delta J[h] = 0$.

Theorem 1.2

First necessary condition for a relative maximum of a functional. If the functional $J[y]$ is assumed to have a relative maximum at $y(x) = y_0(x)$, then it is necessary that

$$\delta J[h] = 0. \tag{1.185}$$

Intuitively, if $\delta J[h] \neq 0$, suppose $\delta J[h] > 0$. Then, $\delta J[h] = \frac{d}{dt} J[y_0 + th]_{t=0} > 0$, which implies that $J[y_0 + th] - J[y_0] > 0$ for small t. Thus, this violates the assumption that $J[y]$ reaches its relative maximum at $y_0(x)$. If $\delta J[h] < 0$, by the similar argument, we can lead to the contradiction that $J[y]$ reaches its relative maximum at $y_0(x)$.

1.5.2.2 Stochastic Calculus

Functional data analysis is to study variation in stochastic process. Similar to the standard calculus that is a powerful tool for multivariate analysis, stochastic calculus provides a useful tool for stochastic process analysis.

1.5.2.2.1 Definition of Stochastic Integrals Let $X(t)$ be a stochastic process and $f(t)$ a function. We partition an integral, $[a, b]$, into a number of subintervals: $a = t_0 < t_1 < \ldots < t_n = b$. Let $\Delta t_k = t_k - t_{k-1}$ and $\Delta_n = \max_{1 \le k \le n} \Delta t_k$. Then, the integral $\int_a^b f(t)X(t)dt$ is defined as

$$\int_a^b f(t)X(t)dt = \lim_{\Delta_n \to 0} \sum_{k=1}^n f(u_k)X(u_k)\Delta t_k. \tag{1.186}$$

Theorem 1.3

Let $\operatorname{cov}(X(s), X(t)) = R(s, t)$. We have

$$(1)\ E\left[\int_a^b f(t)X(t)dt\right] = \int_a^b f(t)E[X(t)]dt$$

and

$$(2)\ Cov\left(\int_a^b f(s)X(s)ds, \int_a^b f(t)X(t)dt\right) = \int_a^b \int_a^b f(s)R(s,t)f(t)dsdt. \tag{1.187}$$

Intuitively, by definition, we can observe

$$E\left[\int_a^b f(t)X(t)dt\right] = E\left[\lim_{\Delta_n \to 0} \sum_{k=1}^n f(u_k)X(u_k)\Delta t_k\right]$$

$$= \lim_{\Delta_n \to 0} \sum_{k=1}^n f(u_k)E[X(u_k)]\Delta t_k = \int_a^b f(t)E[X(t)]dt$$

and

$$\operatorname{Cov}\left(\int_a^b f(s)X(s)ds, \int_a^b f(t)X(t)dt\right) = \operatorname{Cov}\left\{\lim_{\Delta_n \to 0} \sum_{k=1}^n f(u_k)X(u_k)\Delta s_k \lim_{\Delta_n \to 0} \sum_{j=1}^n f(u_j)X(u_j)\Delta t_j\right\}$$

$$= \lim_{\Delta_n \to 0} \sum_{k=1}^n \sum_{j=1}^n f(u_k)Cov(X(u_k), X(u_j))f(u_j)\Delta s_k \Delta t_j$$

$$= \lim_{\Delta_n \to 0} \sum_{k=1}^n \sum_{j=1}^n f(u_k)R(u_k, u_j)f(u_j)\Delta s_k \Delta t_j$$

$$= \int_a^b \int_a^b f(s)R(s,t)f(t)dsdt.$$

1.5.3 Unsmoothed Functional Principal Component Analysis

Similar to multivariate PCA where we consider a linear combination of variables to capture the variations contained in the entire dataset (see Equation 1.171), we can consider a linear combination of functional values (Ramsay and Silverman 2005):

$$f = \int_0^1 \beta(t) X(t) dt, \tag{1.188}$$

where

$\beta(t)$ is a weight function
$X(t)$ is a centered function

Mathematical representation of a linear combination of functional value in Equation 1.188 is a limit of linear combination of variables in Equation 1.171. To capture the variation of process $X(t)$, we chose weight function $\beta(t)$ to maximize the variance of f. By the formula for the variance of stochastic integral (1.187), we have

$$Var(f) = \int_0^1 \int_0^1 \beta(s) R(s,t) \beta(t) dsdt, \tag{1.189}$$

where $R(s,t)$ is the covariance function of the process $X(t)$. Since multiplying $\beta(t)$ by a constant will not change the maximizer of the variance $Var(f)$, we impose a constraint to make the solution unique:

$$\int_0^1 \beta^2(t) dt = 1 \tag{1.190}$$

Therefore, to find the weight function, we seek to solve the following optimization problem:

$$\begin{aligned} \max \quad & \int_0^1 \int_0^1 \beta(s) R(s,t) \beta(t) dsdt \\ \text{s.t.} \quad & \int_0^1 \beta^2(t) dt = 1. \end{aligned} \tag{1.191}$$

By the Lagrange multiplier, we reformulate the constrained optimization problem (1.191) into the following nonconstrained optimization problem:

$$\max_{\beta} \quad \frac{1}{2} \int_0^1 \int_0^1 \beta(s) R(s,t) \beta(t) dsdt + \frac{1}{2} \lambda \left(1 - \int_0^1 \beta^2(t) dt \right), \tag{1.192}$$

where λ is a parameter.

We define the functional

$$J[\beta]=\frac{1}{2}\int_0^1\int_0^1\beta(s)R(s,t)\beta(t)dsdt+\frac{1}{2}\lambda\left(1-\int_0^1\beta^2(t)dt\right).$$

Similar to multivariate calculus where the necessary condition for a maximum of a function $f(x)$ is $\frac{\partial f(x)}{\partial x}=0$, by variation calculus (Sagan 1969), the necessary condition for a relative maximum of a functional is $\delta J(h)=0$. It follows from Equation 1.183 that its first variation is given by

$$\begin{aligned}\delta J[h]&=\frac{d}{d\varepsilon}J[\beta(t)+\varepsilon h(t)]\\&=\frac{d}{d\varepsilon}\left\{\frac{1}{2}\int_0^1\int_0^1[\beta(s)+\varepsilon h(s)]R(s,t)[\beta(t)+\varepsilon h(t)]dsdt+\frac{1}{2}\lambda\left(1-\int_0^1[\beta(t)+\varepsilon h(t)]^2dt\right)\right\}\Bigg|_{\varepsilon=0}\\&=\int_0^1\int_0^1\beta(s)R(s,t)h(t)dsdt-\lambda\int_0^1\beta(t)h(t)dt\\&=\int_0^1\left[\int_0^1R(s,t)\beta(s)ds-\lambda\beta(t)\right]h(t)dt\quad\left(\text{taking } h(t)=\int_0^1R(s,t)\beta(s)ds-\lambda\beta(t)\right)\\&=\int_0^1\left[\int_0^1R(s,t)\beta(s)ds-\lambda\beta(t)\right]^2dt=0,\end{aligned}$$

which implies the following integral equation:

$$\int_0^1R(s,t)\beta(s)ds=\lambda\beta(t)\qquad(1.193a)$$

for an appropriate eigenvalue λ. The left side of the integral equation (1.192) defines an integral transform R of the weight function β. Therefore, the integral transform of the covariance function $R(s,t)$ is referred to as the covariance operator R. The integral equation (1.193a) can be rewritten as

$$R\beta=\lambda\beta,\qquad(1.193b)$$

where $\beta(t)$ is an eigenfunction and referred to as a principal component function. Equation 1.193b is also referred to as an eigenequation. Clearly, the functional eigenequation (1.193b) looks the same as the eigenequation for the multivariate PCA if the covariance operator and eigenfunction are replaced by a covariance matrix and eigenvector.

Since the number of function values is theoretically infinity, we may have an infinite number of eigenvalues. Provided the functions X_i and Y_i are not linearly dependent, there will be only $N-1$ nonzero eigenvalues, where N is the sample size. Eigenfunctions satisfying the eigenequation are orthonormal (Ramsay and Silverman 2005). In other words, Equation 1.193b generates a set of principal component functions

$$R\beta_k = \lambda_k \beta_k, \quad \text{with } \lambda_1 \geq \lambda_2 \geq \cdots.$$

These principal component functions satisfy

(1) $\int_0^1 \beta_k^2(t)\,dt = 1$ and

(2) $\int_0^1 \beta_k(t)\beta_m(t)\,dt = 0 \quad \text{for all } m < k.$

The principal component function β_1 with the largest eigenvalue is referred to as the first principal component function, the principal component function β_2 with the second largest eigenvalue is referred to as the second principal component function, etc.

1.5.4 Smoothed Principal Component Analysis

The observed genotype profiles and other functional data are often not smooth, which will lead to substantial variability in the estimated functional principal component curves. To improve the smoothness of the estimated functional principal component curves, we impose the roughness penalty on the functional principal component weight functions. We balance the goodness-of-fit and the roughness of the estimated functional principal component curves.

We often penalize the roughness of the functional principal component curve by its integrated squared second derivative. The balance between the goodness-of-fit and the roughness of the function is controlled by a smoothing parameter, μ. We implement roughness penalty by defining an extended inner product of two functions as follows:

$$(f, g)_\mu = \int f(t)g(t)\,dt + \mu \int D^2 f(t) D^2 g(t)\,dt, \tag{1.194}$$

where $D^2 f(t) = \frac{d^2 f(t)}{dt^2}$. Similar to Equation 1.189, the penalized sample variance is defined as

$$F = \frac{Var\left(\int_0^1 x(t)\beta(t)\,dt\right)}{\|\beta(t)\|_\mu^2}, \tag{1.195}$$

where $\|\beta(t)\|_\mu^2 = \int_0^1 \beta^2(t)\,dt + \mu \int_0^1 \left[D^2 \beta(t)\right]^2 dt.$

Therefore, to find the functional principal components, we seek to solve the following optimization problem:

$$\begin{aligned} \max \quad & \int_0^1\int_0^1 \beta(s)R(s,t)\beta(t)dsdt \\ \text{s.t.} \quad & ||\beta(t)||_\mu^2 = \int_0^1 \beta^2(t)dt + \mu\int_0^1 \left[D^2\beta(t)\right]^2 dt = 1. \end{aligned} \tag{1.196}$$

Replacing the norm of the function $\beta(t)$, defined as its inner product $||\beta(t)||^2 = \int_0^1 \beta^2(t)dt$ by its new norm $||\beta(t)||_\mu^2$, Equation 1.191 is reduced to Equation 1.196. By the Lagrange multiplier, we reformulate the constrained optimization problem (1.196) into the following nonconstrained optimization problem:

$$\max_\beta J(\beta) = \int_0^1\int_0^1 \beta(s)R(s,t)\beta(t)dsdt + \lambda\left(1 - \int_0^1 \beta^2(t)dt - \mu\int_0^1 \left[D^2\beta(t)\right]^2 dt\right), \tag{1.197}$$

where λ and μ are parameters. Similar to the standard FPCA, its first variation is given by

$$\begin{aligned} \delta J[h] &= \frac{d}{d\varepsilon} J\left[\beta(t) + \varepsilon h(t)\right] \\ &= \frac{d}{d\varepsilon}\int_0^1\int_0^1 \left[\beta(s) + \varepsilon h(s)\right] R(s,t)\left[\beta(t) + \varepsilon h(t)\right] dsdt \\ &\quad + \lambda\left\{1 - \int_0^1 \left[\beta(t) + \varepsilon h(t)\right]^2 dt - \mu\int_0^1 \left[D^2(\beta + \varepsilon h)\right]^2\right\}\Bigg|_{\varepsilon=0} \\ &= 2\left\{\int_0^1\int_0^1 \beta(s)R(s,t)h(t)dsdt - \lambda\left[\int_0^1 \beta(t)h(t)dt + \mu\int_0^1 D^2\beta(t)D^2h(t)dt\right]\right\} \\ &= 2\int_0^1 \left\{\int_0^1 R(s,t)\beta(s)ds - \lambda\left[\beta(t) + \mu D^4\beta(t)\right]\right\} h(t)dt \\ &\quad \left(\text{taking } h(t) = \int_0^1 R(s,t)\beta(s) - \lambda\left[\beta(t) + \mu D^4\beta(t)\right]\right) \\ &= 2\int_0^1 \left\{\int_0^1 R(s,t)\beta(s)ds - \lambda\left[\beta(t) + \mu D^4\beta(t)\right]^2\right\} dt = 0, \end{aligned}$$

which implies the following integral function:

$$\int_0^1 R(s,t)\beta(s)ds = \lambda\left[\beta(t) + \mu D^4\beta(t)\right]. \tag{1.198}$$

Note that when $\mu = 0$, integral functional eigenequation (1.198) is reduced to Equation 1.193a. In other words, the smoothed functional principal components analysis is reduced to unsmoothed functional principal component analysis when $\mu = 0$. Unsmoothed FPCA is a special case of the smoothed FPCA.

1.5.5 Computations for the Principal Component Function and the Principal Component Score

The eigenfunction is an integral function and difficult to solve in closed form. A general strategy for solving the eigenfunction problem in (1.198) is to convert the continuous eigen-analysis problem to an appropriate discrete eigen-analysis task (Ramsay and Silverman 2005). A popular method is to use basis function expansion methods to achieve this conversion.

Let $\{\phi_j(t)\}$ be the series of Fourier functions. For each j, define $\omega_{2j-1} = \omega_{2j} = 2\pi j$. We expand each function or genetic variant profile $X_i(t)$ as a linear combination of the basis function ϕ_j:

$$X_i(t) = \sum_{j=1}^{T} C_{ij}\phi_j(t). \tag{1.199}$$

The expansion coefficients C_{ij} are calculated by Fourier series analysis (Ramsay and Silverman 2005).

Define the vector-valued function $X(t) = [X_1(t), \cdots, X_N(t)]^T$ and the vector-valued function $\phi(t) = [\phi_1(t), \cdots, \phi_T(t)]^T$. The joint expansion of all N functions or genetic variant profiles can be expressed as

$$X(t) = C\phi(t), \tag{1.200}$$

where the matrix C is given by

$$C = \begin{bmatrix} C_{11}, & \cdots & C_{1T} \\ \cdots & \cdots & \cdots \\ C_{N1} & \cdots & C_{NT} \end{bmatrix}.$$

In matrix form, we can express the sampling variance–covariance function of the genetic variant profiles, gene expression, methylation variation profiles, or other observed function curves as

$$\begin{aligned} R(s,t) &= \frac{1}{N} X^T(s)X(t) \\ &= \frac{1}{N}\phi^T(s)C^T C\phi(t). \end{aligned} \tag{1.201}$$

Similarly, the eigenfunction $\beta(t)$ can be expanded as

$$\beta(t)=\sum_{j=1}^{T} b_j\phi_j(t) \quad \text{and} \quad D^4\beta(t)=\sum_{j=1}^{T}\omega_j^4 b_j\phi_j(t)$$

or

$$\beta(t)=\phi(t)^T b \quad \text{and} \quad D^4\beta(t)=\phi(t)^T S_0 b, \tag{1.202}$$

where $b=[b_1,\ldots,b_T]^T$ and $S_0=\text{diag}\left(\omega_1^4,\ldots,\omega_T^4\right)$. Let $S=\text{diag}\left(\left(1+\mu\omega_1^4\right)^{-\frac{1}{2}},\ldots,\left(1+\mu\omega_T^4\right)^{-\frac{1}{2}}\right)$. Then, we have

$$\beta(t)+\mu D^4\beta(t)=\phi(t)^T S^{-2}b. \tag{1.203}$$

Substituting expansions (1.201) and (1.203) of variance–covariance $R(s, t)$ and eigenfunction $\beta(t)$ into the functional eigenequation (1.188), we obtain

$$\phi(t)^T \frac{1}{N}C^TCb=\lambda\phi^T(t)S^{-2}b. \tag{1.204}$$

Since Equation 1.204 must hold for all t, we obtain the following eigenequation:

$$\frac{1}{N}C^TCb=\lambda S^{-2}b, \tag{1.205}$$

which can be rewritten as

$$\left[S\left(\frac{1}{N}C^TC\right)S\right]\left[S^{-1}b\right]=\lambda\left[S^{-1}b\right],$$

or

$$S\left(\frac{1}{N}C^TC\right)Su=\lambda u, \tag{1.206}$$

where $u=S^{-1}b$. Thus, $b=Su$ and $\beta(t)=\varphi(t)^Tb$ is a solution to functional eigenequation (1.188). We can easily check (Exercise 1.17)

$$\left\langle\beta_j,\beta_k\right\rangle_\mu=0 \quad \text{for all } k\neq j. \tag{1.207}$$

The vector of functions $\beta(t)$ forms a set of orthonormal functional principal components under the extended inner product. The set of orthonormal functional principal components can be used as a set of new basis functions. We can expand any functions in terms of principal component functions. Let $x_i(t)$ be a function. We can expand $x_i(t)$ as

$$x_i(t) = \sum_{j=1}^{T} \xi_{ij}\beta_j(t), \tag{1.208}$$

where

$$\xi_{ij} = \langle x_i(t), \beta_j(t) \rangle_{\mu}. \tag{1.209}$$

1.6 CANONICAL CORRELATION ANALYSIS

The goal of canonical correlation analysis is to seek linear combinations of two sets of variables, which maximize the correlation between two sets of variables. Specifically, it first identifies the pair of linear combinations that have the largest correlation. Next we identify the pair of linear combinations having the largest correlation among all pairs uncorrelated with the initially selected pair and so on. The pairs of linear combinations are called canonical variates, and their correlations are called canonical correlations (Anderson 1984).

1.6.1 Mathematical Formulation of Canonical Correlation Analysis

First we quantify the measure of association between two groups of variables. Consider two groups of variables. The first group of p variables is denoted by X and the second group of q variables is denoted by Y. We assume that $p+q$ variables X and Y jointly have the mean

$$\begin{bmatrix} \mu_x \\ \mu_y \end{bmatrix}$$

and covariance matrix

$$\Sigma = \begin{bmatrix} \Sigma_{xx} & \Sigma_{xy} \\ \Sigma_{yx} & \Sigma_{yy} \end{bmatrix}.$$

Linear combinations are simple summary measures of a set of variables. Let

$$U = a^T X$$

and

$$V = b^T Y,$$

where a and b are a pair of vectors of coefficients. The variances of the variables U and V and their covariance are given by

$$\mathrm{Var}(U) = a^T \Sigma_{xx} a,$$

$$\mathrm{Var}(V) = b^T \Sigma_{yy} b,$$

and

$$\mathrm{Cov}(U, V) = a^T \Sigma_{xy} b.$$

We shall seek coefficient vectors a and b to make

$$\mathrm{Corr}(U, V) = \frac{a^T \Sigma_{xy} b}{\sqrt{a^T \Sigma_{xx} a}\sqrt{b^T \Sigma_{yy} b}} \tag{1.210}$$

as large as possible.

We define the first pair of canonical variables as the pair of linear combinations U_1, V_1 having unit variance, which maximizes correlation (1.210). We define the second pair of canonical variables as the pair of linear combinations U_2, V_2 having unit variance, which maximizes the correlation (1.210) among all linear combinations that are uncorrelated with the first pair of canonical variables and so on.

1.6.2 Correlation Maximization Techniques for Canonical Correlation Analysis

For simplicity, we assume that the random variables U and V have unit variances:

$$\begin{aligned} a^T \Sigma_{xx} a &= 1 \\ b^T \Sigma_{yy} b &= 1. \end{aligned} \tag{1.211}$$

The first step is to find the vectors of correlation coefficients a and b such that the random variables U and V have maximum correlation:

$$\mathrm{Corr}(U, V) = a^T \Sigma_{xy} b. \tag{1.212}$$

By a Lagrangian multiplier, to find a and b to maximize (1.212) under constraints (1.211), we set

$$f(a, b) = a^T \Sigma_{xy} b + \frac{\lambda}{2}\left(1 - a^T \Sigma_{xx} a\right) + \frac{\mu}{2}\left(1 - b^T \Sigma_{yy} b\right), \tag{1.213}$$

where λ and μ are Lagrangian multipliers. Taking derivatives of the objective function $f(a, b)$ and setting it equal to zero

$$\begin{aligned}\frac{\partial f(a,b)}{\partial a} &= \Sigma_{xy}b - \lambda\Sigma_{xx}a = 0\\ \frac{\partial f(a,b)}{\partial b} &= \Sigma_{yx}a - \mu\Sigma_{yy}b = 0.\end{aligned} \tag{1.214}$$

Multiplying a^T on the left side of the first equation in (1.214), we obtain

$$\lambda = a^T\Sigma_{xy}b. \tag{1.215}$$

Multiplying b^T on the left side of the second equation in (1.214), we obtain

$$\mu = b^T\Sigma_{yx}a = \lambda. \tag{1.216}$$

Thus, Equation 1.214 can be rewritten as

$$\Sigma_{xy}b - \lambda\Sigma_{xx}a = 0, \tag{1.217a}$$

$$\Sigma_{yx}a - \lambda\Sigma_{yy}b = 0. \tag{1.217b}$$

Premultiplying Equation 1.217a by $\Sigma_{yx}\Sigma_{xx}^{-1}$, then substituting Equation 1.217b into the result yields

$$\Sigma_{yx}\Sigma_{xx}^{-1}\Sigma_{xy}b = \lambda^2\Sigma_{yy}b, \tag{1.218}$$

which implies that

$$\Sigma_{yy}^{-1/2}\Sigma_{yx}\Sigma_{xx}^{-1}\Sigma_{xy}\Sigma_{yy}^{-1/2}g = \lambda^2 g, \tag{1.219}$$

where $g = \Sigma_{yy}^{1/2}b$.

Equation 1.219 implies that the maximum correlation between U and V can be achieved by taking the largest eigenvalue λ_1 of matrix R:

$$R = \Sigma_{yy}^{-1/2}\Sigma_{yx}\Sigma_{xx}^{-1}\Sigma_{xy}\Sigma_{yy}^{-1/2}. \tag{1.220}$$

Let e_1 be the eigenvector of the matrix R associated with the largest eigenvalue λ_1^2. Then, the vectors of coefficients a and b are given by

$$a_1 = \Sigma_{xx}^{-1}\Sigma_{xy}\Sigma_{yy}^{-1/2}e_1 \quad \text{and} \quad b_1 = \Sigma_{yy}^{-1/2}e_1, \tag{1.221}$$

and the first canonical correlation is

$$\max_{a,b} Corr(U, V) = \lambda_1.$$

Given $U_1 = a_1^T X$ and $V_1 = b_1^T Y$. Let $U = a^T X$ and $V = b^T Y$ be a second pair of linear projection with unit variances. We seek U and V to have a maximum correlation among all linear combinations with unit variances, which are also uncorrelated with U_1 and V_1. In other words, it requires that

$$Cov(U, U_1) = a^T \Sigma_{xx} a_1 = 0, \tag{1.222a}$$

$$Cov(V, V_1) = b^T \Sigma_{yy} b_1 = 0, \tag{1.222b}$$

$$Cov(U, V_1) = a^T \Sigma_{xy} b_1 = 0, \tag{1.222c}$$

$$Cov(V, U_1) = b^T \Sigma_{yx} a_1 = 0. \tag{1.222d}$$

Again, by Lagrangian multipliers, we set

$$f(a, b) = a\Sigma_{xy} b + \frac{\lambda}{2}\left(1 - a^T \Sigma_{xx} a\right) + \frac{\mu}{2}\left(1 - b^T \Sigma_{yy} b\right) + \eta a^T \Sigma_{xx} a_1 + \xi b^T \Sigma_{yy} b_1, \tag{1.223}$$

where λ, μ, η, and ξ are Lagrangian multipliers. Differentiate $f(a, b)$ with respect to a and b, and set its derivatives equal to zero:

$$\frac{\partial f(a, b)}{\partial a} = \Sigma_{xy} b - \lambda \Sigma_{xx} a + \eta \Sigma_{xx} a_1 = 0, \tag{1.224a}$$

$$\frac{\partial f(a, b)}{\partial b} = \Sigma_{yx} a - \mu \Sigma_{yy} b + \xi \Sigma_{yy} b_1 = 0. \tag{1.224b}$$

Multiplying a^T and b^T on the left sides of Equations 1.224a and 1.224b, respectively, we have

$$\lambda = \mu = a^T \Sigma_{xy} b. \tag{1.225}$$

Multiplying a_1^T and b_1^T on the left sides of Equations 1.224a and 1.224b, respectively, we obtain

$$\eta = \xi = 0. \tag{1.226}$$

Combining Equations 1.224a, 1.224b, 1.225, and 1.226 gives

$$\begin{aligned} -\lambda\Sigma_{xx}a+\Sigma_{xy}b &= 0 \\ \Sigma_{yx}a-\lambda\Sigma_{yy}b &= 0 \end{aligned} \tag{1.227}$$

which is similar to Equations 1.217a and 1.217b. Using the second eigenvalue λ_2^2 of the matrix R ($R=\Sigma_{yy}^{-1/2}\Sigma_{yx}\Sigma_{xx}^{-1}\Sigma_{xy}\Sigma_{yy}^{-1/2}$) and associated eigenvector e_2, we obtain the second pair of canonical variants:

$$a_2=\Sigma_{xx}^{-1}\Sigma_{xy}\Sigma_{yy}^{-1/2}e_2 \quad \text{and} \quad b_2=\Sigma_{yy}^{-1/2}e_2. \tag{1.228}$$

The second canonical correlation between U_2 and V_2 is given by λ_2.

This procedure continues until all the canonical variants are found.

Example 1.23

Assume that the covariance matrix of $\begin{bmatrix} X \\ Y \end{bmatrix}$ is given by

$$\Sigma=\begin{bmatrix} 1 & 0.4 & 0.5 & 0.6 \\ 0.4 & 12 & 0.3 & 0.4 \\ 0.5 & 0.3 & 1 & 0.2 \\ 0.6 & 0.4 & 0.2 & 1 \end{bmatrix}.$$

Then,

$$\Sigma_{yy}^{-1/2}=\begin{bmatrix} 1.0155 & -0.1026 \\ -0.1026 & 1.0155 \end{bmatrix}, \quad \Sigma_{xx}^{-1}=\begin{bmatrix} 1.1905 & -0.4762 \\ -0.4762 & 1.1905 \end{bmatrix}.$$

Thus,

$$R=\begin{bmatrix} 0.2077 & 0.2644 \\ 0.2644 & 0.3389 \end{bmatrix}.$$

The eigenvalues and eigenvectors are given by

$$\lambda_1=0.5457, \quad \lambda_2=0.0009;$$

$$e_1=\begin{bmatrix} 0.6161 \\ 0.7877 \end{bmatrix}, \quad e_2=\begin{bmatrix} -0.7877 \\ 0.6161 \end{bmatrix};$$

$$a_1=\Sigma_{xx}^{-1}\Sigma_{xy}\Sigma_{yy}^{-1/2}e_1=\begin{bmatrix} 0.6323 \\ 0.2052 \end{bmatrix} \quad \text{and} \quad b_1=\Sigma_{yy}^{-1/2}e_1=\begin{bmatrix} 0.5448 \\ 0.7367 \end{bmatrix};$$

$$a_2 = \Sigma_{xx}^{-1}\Sigma_{xy}\Sigma_{yy}^{-1/2}e_2 = \begin{bmatrix} 0.0204 \\ 0.0318 \end{bmatrix} \quad \text{and} \quad b_2 = \Sigma_{yy}^{-1/2}e_2 = \begin{bmatrix} 0.8631 \\ 0.7064 \end{bmatrix}.$$

Therefore, we have

$$U_1 = 0.6323x_1 + 0.2052x_2, \quad V_1 = 0.5448y_1 + 0.7367y_2$$
$$U_2 = 0.0204x_1 + 0.0318x_2, \quad V_2 = 0.8631y_1 + 0.7064y_2.$$

1.6.3 Single Value Decomposition for Canonical Correlation Analysis

Canonical correlation analysis can be performed by single value decomposition (SVD). Consider two data matrices: $X \in R^{n\times p}$ and $Y \in R^{n\times q}$. Let $A \in R^{p\times d}$ and $B \in R^{q\times d}$ be the matrices of coefficients of the linear combinations of the data matrices X and Y, respectively. We denote linear combinations of the matrices X and Y by

$$u = XA, \quad v = YB. \tag{1.229}$$

Suppose that the SVD of the data matrices X and Y are respectively given by

$$X = u_1 s_1 v_1^T, \quad Y = u_2 s_2 v_2^T. \tag{1.230}$$

Next we calculate the SVD of the matrix $u_1^T u_2$ as

$$u_1^T u_2 = U\Lambda V^T. \tag{1.231}$$

To identify the matrices A and B, we impose $u^T v = \Lambda$. Therefore, we require

$$u^T v = A^T X^T Y B = \Lambda. \tag{1.232}$$

Substituting Equations 1.230 into Equation 1.232, we obtain

$$A^T X^T Y B = A^T v_1 s_1 u_1^T u_2 s_2 v_2^T B. \tag{1.233}$$

Again, substituting Equation 1.231 for $u^T v$ into Equation 1.232, we have

$$A^T X^T Y B = A^T v_1 s_1 U \Lambda V^T s_2 v_2^T B. \tag{1.234}$$

To make Equation 1.232 to hold, we must have

$$A^T v_1 s_1 U = I, \quad V^T s_2 v_2^T B = I. \tag{1.235}$$

Solving Equation 1.235 for the matrices A and B, we obtain

$$A = v_1 s_1^{-1} U, \quad B = v_2 s_2^{-1} V. \tag{1.236}$$

Using Equations 1.231 and 1.236, we can confirm that

$$u^T v = \Lambda.$$

From Equation 1.230, we can calculate the covariance matrices $X^T X$ and $Y^T Y$:

$$\begin{aligned} X^T X &= v_1 s_1 u_1^T u_1 s_1 v_1^T = v_1 s_1^2 v_1^T, \\ Y^T Y &= v_2 s_2 u_2^T u_2 s_2 v_2^T = v_2 s_2^T v_2^T. \end{aligned} \tag{1.237}$$

Using Equations 1.236 and 1.237, we can easily check

$$u^T u = A^T X^T X A = U^T s_1^{-1} v_1^T v_1 s_1^2 v_1^T v_1 s_1^{-1} U = I.$$

Similarly, we can show $V^T V = I$.

To establish the relationships between correlation-maximum techniques and SVD for canonical correlation analysis, we show that

$$\text{Trace}\left(\Lambda^2\right) = \text{Trace}\left(R\right). \tag{1.238}$$

From Equation 1.220, it follows that

$$\begin{aligned} \text{Trace}\left(R\right) &= \text{Trace}\left(\Sigma_{yy}^{-1/2} \Sigma_{yx} \Sigma_{xx}^{-1} \Sigma_{xy} \Sigma_{yy}^{-1/2}\right) \\ &= \text{Trace}\left(\left[\Sigma_{xx}^{-1/2} \Sigma_{xy} \Sigma_{yy}^{-1/2}\right]^2\right). \end{aligned}$$

1.6.4 Test Statistics

To develop statistics for testing the null hypothesis that X and Y are independent, which is equivalent to testing the hypothesis that each variable in the set X is uncorrelated with each variable in the set Y, we need to first calculate the likelihood ratio (Anderson 1984). Let

$$Z = \begin{bmatrix} X \\ Y \end{bmatrix}, \quad \mu = \mu_z = \begin{bmatrix} \mu_x \\ \mu_y \end{bmatrix}, \quad \text{and} \quad \Sigma = \Sigma_z = \begin{bmatrix} \Sigma_{xx} & \Sigma_{xy} \\ \Sigma_{yx} & \Sigma_{yy} \end{bmatrix}.$$

We assume that both X and Y are normally distributed with density functions:

$$L\left(x \middle| \mu_x, \Sigma_{xx}\right) = \frac{1}{\left(2\pi\right)^{\frac{p}{2}} \left|\Sigma_{xx}\right|^{\frac{1}{2}}} e^{-\frac{1}{2}\left(x-\mu_x\right)^T \Sigma_{xx}^{-1} \left(x-\mu_x\right)},$$

$$L\left(y \middle| \mu_y, \Sigma_{yy}\right) = \frac{1}{\left(2\pi\right)^{\frac{q}{2}} \left|\Sigma_{yy}\right|^{\frac{1}{2}}} e^{-\frac{1}{2}\left(y-\mu_y\right)^T \Sigma_{yy}^{-1} \left(x-\mu_y\right)},$$

and

$$L(z|\mu,\Sigma)=\frac{1}{(2\pi)^{\frac{p+q}{2}}|\Sigma|^{\frac{1}{2}}}e^{-\frac{1}{2}(z-\mu)^T\Sigma^{-1}(z-\mu)}.$$

Under the normal assumption, the null hypothesis that X and Y are independent is equivalent to the hypothesis $H_0:\Sigma_{xy}=0$, which implies that Σ is of the form

$$\Sigma_0=\begin{bmatrix}\Sigma_{xx} & 0\\ 0 & \Sigma_{yy}\end{bmatrix}.$$

Given a sample $\{x_1, y_1, \ldots, x_N, y_N\}$ of N observations, the likelihood functions are

$$L(x_1,\ldots,x_N\mid\mu_x,\Sigma_{xx})=\prod_{i=1}^{N}L(x_i\mid\mu_x,\Sigma_{xx}),\quad L(y_1,\ldots,y_N\mid\mu_y,\Sigma_{yy})=\prod_{i=1}^{N}L(y_i\mid\mu_y,\Sigma_{yy}),$$

and

$$L(z_1,\ldots,z_N\mid\mu,\Sigma)=\prod_{i=1}^{N}L(z_i\mid\mu,\Sigma).$$

The likelihood ratio is defined as

$$\lambda=\frac{\max_{\mu,\Sigma_0}L(z_1,\ldots,z_N\mid\mu,\Sigma_0)}{\max_{\mu,\Sigma}L(z_1,\ldots,z_N\mid\mu,\Sigma)},\tag{1.239}$$

where $L(z_1, \ldots, z_N| \mu, \Sigma_0) = L(x_1, \ldots, x_N| \mu_x, \Sigma_{xx})L(y_1, \ldots, y_N| \mu, \Sigma_{xx})$.

Let

$$A=\sum_{i=1}^{N}(z_i-\bar{z})(z_i-\bar{z})^T,\quad A_{xx}=\sum_{i=1}^{N}(x_i-\bar{x})(x_i-\bar{x})^T,\quad A_{yy}=\sum_{i=1}^{N}(y_i-\bar{y})(y_i-\bar{y})^T,$$

$$\bar{z}=\frac{1}{N}\sum_{i=1}^{N}z_i,\quad \bar{x}=\frac{1}{N}\sum_{i=1}^{N}x_i,\quad \bar{y}=\frac{1}{N}\sum_{i=1}^{N}y_i.$$

Then, the maximum likelihood estimates of the parameters Σ_{xx}, Σ_{yy}, and Σ are

$$\hat{\Sigma}_{xx}=\frac{1}{N}A_{xx},\quad \hat{\Sigma}_{yy}=\frac{1}{N}A_{yy},\quad\text{and}\quad \hat{\Sigma}=\frac{1}{N}A.\tag{1.240}$$

Substituting Equation 1.240 into a likelihood function yields

$$L\left(z_1,\ldots,z_N|\hat{\mu},\hat{\Sigma}\right)=\prod_{i=1}^{N}\frac{1}{(2\pi)^{\frac{(p+q)}{2}}\left|\hat{\Sigma}\right|^{\frac{1}{2}}}e^{-\frac{1}{2}\mathrm{Trace}\left(\hat{\Sigma}^{-1}A\right)}$$

$$=\frac{1}{(2\pi)^{\frac{N(p+q)}{2}}\left|\hat{\Sigma}\right|^{\frac{N}{2}}}e^{-\frac{N(p+q)}{2}},$$

$$L\left(x_1,\ldots,x_N|\hat{\mu}_x,\hat{\Sigma}_{xx}\right)=\frac{1}{(2\pi)^{\frac{Np}{2}}\left|\hat{\Sigma}_{xx}\right|^{\frac{N}{2}}}e^{-\frac{Np}{2}},$$

$$L\left(y_1,\ldots,y_N|\hat{\mu}_y,\hat{\Sigma}_{yy}\right)=\frac{1}{(2\pi)^{\frac{Nq}{2}}\left|\hat{\Sigma}_{yy}\right|^{\frac{N}{2}}}e^{-\frac{Nq}{2}}. \tag{1.241}$$

Thus, using Equation 1.241, we obtain

$$\lambda=\frac{\max_{\mu,\Sigma_0}L\left(z_1,\ldots z_N|\mu,\Sigma_0\right)}{\max_{\mu,\Sigma}L\left(z_1,\ldots,z_N|\mu,\Sigma\right)}=\frac{\frac{1}{(2\pi)^{\frac{Np}{2}}\left|\hat{\Sigma}_{xx}\right|^{\frac{N}{2}}}e^{-\frac{NP}{2}}\frac{1}{(2\pi)^{\frac{Nq}{2}}\left|\hat{\Sigma}_{yy}\right|^{\frac{N}{2}}}e^{-\frac{Nq}{2}}}{\frac{1}{(2\pi)^{\frac{N(p+q)}{2}}\left|\hat{\Sigma}\right|^{\frac{N}{2}}}e^{-\frac{N(P+q)}{2}}}$$

$$=\frac{\left|\hat{\Sigma}\right|^{\frac{N}{2}}}{\left|\hat{\Sigma}_{xx}\right|^{\frac{N}{2}}\left|\hat{\Sigma}_{yy}\right|^{\frac{N}{2}}}$$

$$=\frac{\frac{1}{N^{(p+q)}}|A|^{\frac{N}{2}}}{\frac{1}{N^p}|A_{xx}|^{\frac{N}{2}}\frac{1}{N^q}|A_{yy}|^{\frac{N}{2}}}=\frac{|A|^{\frac{N}{2}}}{|A_{xx}|^{\frac{N}{2}}|A_{yy}|^{\frac{N}{2}}}.$$

The test statistics based on likelihood ratio is defined as

$$T_{CCA}=-2\log\lambda=-N\log\frac{|A|}{|A_{xx}||A_{yy}|}. \tag{1.242}$$

Under the null hypothesis $H_0: \sum_{xy} = 0$, T_{CCA} is asymptotically distributed as a central χ^2_{pq} distribution (Serfling 1980).

Now we show that we can use canonical correlation coefficients to calculate T_{CCA}. Recall that the canonical variables are defined as

$$U_i = a_i^T X \quad \text{and} \quad V_j = b_j^T Y.$$

We define the matrices

$$W = \begin{bmatrix} a_i^T \\ \vdots \\ a_p^T \end{bmatrix}, \quad \Gamma = \begin{bmatrix} b_1^T \\ \vdots \\ b_q^T \end{bmatrix}.$$

Suppose that the data are centered. Then, the vectors of canonical variables can be expressed as

$$U = \begin{bmatrix} U_1 \\ \vdots \\ U_p \end{bmatrix} = WX, \quad V = \begin{bmatrix} V_1 \\ \vdots \\ V_q \end{bmatrix} = \Gamma Y.$$

Canonical variates should satisfy the following conditions:

$$E\left[UU^T\right] = W\Sigma_{xx}W^T \approx WA_{xx}W^T = I_p, \tag{1.243}$$

$$E\left[VV^T\right] = \Gamma\Sigma_{yy}\Gamma^T \approx \Gamma A_{yy}\Gamma^T = I_q, \tag{1.244}$$

and

$$E\left[UV^T\right] = W\Sigma_{xy}\Gamma^T \approx WA_{xy}\Gamma^T = \left[\Lambda \quad 0\right], \tag{1.245}$$

where $\Lambda = \text{diag}(\lambda_1, \ldots, \lambda_p)$.

Combining Equations 1.243, 1.244, and 1.245, we obtain

$$\begin{bmatrix} W & 0 \\ 0 & \Gamma \end{bmatrix} \begin{bmatrix} A_{xx} & A_{xy} \\ A_{yx} & A_{yy} \end{bmatrix} \begin{bmatrix} W^T & 0 \\ 0 & \Gamma^T \end{bmatrix} = \begin{bmatrix} I_p & \Lambda & 0 \\ \Lambda & I_p & 0 \\ 0 & 0 & I_{q-p} \end{bmatrix}, \tag{1.246}$$

$$WW^T = I_p, \quad \Gamma\Gamma^T = I_q.$$

Thus,

$$\frac{|A|}{|A_{xx}||A_{yy}|}=\frac{\begin{vmatrix} W & 0 \\ 0 & \Gamma \end{vmatrix}}{|W||\Gamma|}\frac{\begin{vmatrix} A_{xx} & A_{xy} \\ A_{yx} & A_{yy} \end{vmatrix}}{|A_{xx}||A_{yy}|}\frac{\begin{vmatrix} W^T & 0 \\ 0 & \Gamma^T \end{vmatrix}}{|W^T||\Gamma^T|}$$

$$=\frac{\begin{vmatrix} I_p & \Lambda & 0 \\ \Lambda & I_p & 0 \\ 0 & 0 & I_{q-p} \end{vmatrix}}{|I_p||I_q|}$$

$$=\begin{vmatrix} I_p & \Lambda \\ \Lambda & I_p \end{vmatrix}=|I_p-\Delta^2|$$

$$=\begin{vmatrix} 1-\lambda_1^2 & \cdots & 0 \\ \vdots & \ddots & \\ 0 & \cdots & 1-\lambda_p^2 \end{vmatrix}=\prod_{i=1}^{p}\left(1-\lambda_i^2\right). \tag{1.247}$$

Substituting Equation 1.247 into Equation 1.242 yields

$$T_{CCA}=-N\sum_{i=1}^{p}\log\left(1-\lambda_i^2\right). \tag{1.248}$$

When the sample size is large, Bartlett (1939) suggests using the following statistic T_{CCA} for hypothesis testing:

$$T_{CCA}=-\left[N-\frac{(p+3)}{2}\right]\sum_{i=1}^{p}\log\left(1-\lambda_i^2\right). \tag{1.249}$$

1.6.5 Functional Canonical Correlation Analysis

Multivariate CCA can be easily extended to functional canonical correlation analysis (FCCA). We consider two sets of functions, $(x_i(t), y_i(t))$, $i=1, \ldots, N$. Define the set of functions:

$$Z(t)=\begin{bmatrix} X(t) \\ Y(t) \end{bmatrix},\quad X(t)=\begin{bmatrix} x_1(t) \\ \vdots \\ x_N(t) \end{bmatrix},\quad Y(t)=\begin{bmatrix} y_1(t) \\ \vdots \\ y_N(t) \end{bmatrix}.$$

The concept of canonical variables as a linear combination of the variables a^TX (inner product between two vectors) can be extended as an inner product between two functions,

$\int a^T(t)X(t)dt$, where $a(t)=[a_1(t), \ldots, a_N(t)]^T$. Let $U=\int a^T(t)X(t)dt$ and $V=\int b^T(t)Y(t)dt$. The correlation coefficient between U and V is given by

$$\rho=\frac{\operatorname{cov}(U,V)}{\sqrt{\operatorname{var}(U)\operatorname{var}(V)}}. \tag{1.250}$$

Making the results of FCCA meaningful, we should ensure that the functions $a(t)$ and $b(t)$ are smooth. Our goal is to find functions $a(t)$ and $b(t)$ such that the correlation coefficient ρ is maximized, i.e.,

$$\max_{a(t),b(t)} \frac{\operatorname{cov}\left(\int a^T(t)X(t)dt, \int b^T(t)Y(t)dt\right)}{\sqrt{\operatorname{var}\left(\int a^T(t)X(t)dt\right)\operatorname{var}\left(\int b^T(t)Y(t)dt\right)}}. \tag{1.251}$$

We use FPCA to solve the problem (1.251). Using Equations 1.248 and 1.250, we expand the functions $x_i(t)$ and $y_i(t)$ as

$$x_i(t)=\sum_{j=1}^{T_x}\xi_{ij}\beta_j(t), \quad y_i(t)=\sum_{j=1}^{T_y}\eta_{ij}\beta_j(t), \tag{1.252}$$

where

$$\xi_{ij}=\left\langle x_i(t),\beta_j(t)\right\rangle_\mu \quad \text{and} \quad \eta_{ij}=\left\langle y_i(t),\beta_j(t)\right\rangle_\mu. \tag{1.253}$$

Define the matrices of functional principal component scores:

$$\xi=\begin{bmatrix}\xi_{11} & \cdots & \xi_{1T_x}\\ \vdots & \ddots & \vdots\\ \xi_{N1} & \cdots & \xi_{NT_x}\end{bmatrix}, \quad \eta=\begin{bmatrix}\eta_{11} & \cdots & \eta_{1T_y}\\ \vdots & \ddots & \vdots\\ \eta_{N1} & \cdots & \eta_{NT_y}\end{bmatrix}. \tag{1.254}$$

The multivariate CCA is then applied to the matrices ξ and η.

Ramsay and Silverman (2005) take a slightly different approach. Equation 1.251 is replaced by

$$\max_{a(t),b(t)} \frac{\operatorname{cov}\left(\int a^T(t)X(t)dt, \int b^T(t)Y(t)dt\right)}{\sqrt{\left[\operatorname{var}\left(\int a^T(t)X(t)dt\right)+\lambda_1\int\|\ddot{a}(t)\|^2dt\right]\left[\operatorname{var}\left(\int b^T(t)Y(t)dt\right)+\lambda_2\int\|\ddot{b}(t)\|^2dt\right]}}. \tag{1.255}$$

We expand functions $a_i(t), b_i(t), x_i(t)$, and $y_i(t)$ in terms of basis functions as

$$x_i(t)=\sum_{j=1}^{T_x}\xi_{ij}\beta_j(t)=\beta^T(t)\xi,\quad y_i(t)=\sum_{j=1}^{T_y}\eta_{ij}\beta_j(t)=\beta^T(t)\eta,$$

$$a_i(t)=\sum_{j=1}^{T_x}a_{ij}\beta_j(t)=\beta^T(t)a,\quad b_i(t)=\sum_{j=1}^{T_y}b_{ij}\beta_j(t)=\beta^T(t)b. \tag{1.256}$$

Then, we have

$$\begin{aligned}\int \|\ddot{a}(t)\|^2dt &= \int a^T\ddot{\beta}(t)\ddot{\beta}^T(t)adt\\ &= a^T\int \ddot{\beta}(t)\ddot{\beta}^T(t)dta\\ &= a^TKa,\end{aligned}$$

$$\begin{aligned}\int \|\ddot{b}(t)\|^2dt &= b^T\int \ddot{\beta}(t)\ddot{\beta}^T(t)dtb\\ &= b^TKb,\end{aligned} \tag{1.257}$$

$$\int a^T(t)X(t)dt=\int a^T\beta(t)\beta^T(t)\xi dt=a^TJ\xi,$$

$$\int b^T(t)Y(t)dt=\int b^T\beta(t)\beta^T(t)\eta dt=b^TJ\eta,$$

where

$$K=\int \ddot{\beta}(t)\ddot{\beta}^T(t)dt,\quad J=\int \beta(t)\beta^T(t)dt.$$

Substituting Equation 1.257 into Equation 1.255, we obtain

$$\max_{a,b}\quad \frac{a^TJ\xi^T\eta Jb}{\sqrt{\left(a^TJ\xi^T\xi Ja+\lambda_1a^TKa\right)\left(b^TJ\eta^T\eta b+\lambda_2b^TKb\right)}}. \tag{1.258}$$

Let

$$V_{11}=\xi^T\xi,\quad V_{12}=\xi^T\eta,\quad V_{22}=\eta^T\eta.$$

We can transform the optimal problem (1.258) into the following optimization problem:

$$\max_{a,b} \quad a^T J V_{12} J b \tag{1.259}$$

subject to

$$a^T \left(JV_{11}J + \lambda_1 K\right)a = 1 \quad \text{and} \quad b^T \left(JV_{22}J + \lambda_2 K\right)b = 1. \tag{1.260}$$

By Lagrangian multiplier, to find a and b to maximize (1.259) under constraints (1.260), we set

$$f(a,b) = a^T JV_{12}Jb + \frac{\rho}{2}\left(1 - a^T\left(JV_{11}J + \lambda_1 K\right)a\right) + \frac{\mu}{2}\left(1 - b^T\left(JV_{22}J + \lambda_2 K\right)b\right),$$

where λ and μ are Lagrangian multipliers. Taking derivatives of the objective function $f(a, b)$ and setting it equal to zero

$$\begin{aligned} &JV_{12}Jb - \rho\left(JV_{11}J + \lambda_1 K\right)a = 0, \\ &JV_{21}J - \mu\left(JV_{22}J + \lambda_2 K\right)b = 0. \end{aligned} \tag{1.261}$$

Let

$$\Sigma_{xy} = JV_{12}J, \quad \Sigma_{xx} = JV_{11}J + \lambda_1 K, \quad \Sigma_{yx} = JV_{21}J, \quad \Sigma_{yy} = JV_{22}J + \lambda_2 K.$$

Equation 1.261 is reduced to Equation 1.214. The FCCA problem is then transformed to the standard CCA problem. The methods for CCA discussed in Section 1.6.2 can be used to solve Equation 1.261.

However, we should point out that the denominator in Equation 1.255 is no longer a pure variance. It involves the penalty terms. The quantity, which we want to maximize in Equation 1.255, is not exactly a canonical correlation coefficient. We suggest that we first use FPCA to obtain the functional principal scores and apply the multivariate CCA to the matrices of functional principal scores ξ and η defined in Equation 1.254.

APPENDIX 1A

Now we will prove Theorem 1.1.

Recall $F = \text{Trace}\left(\left(X - \tilde{U}\tilde{V}\right)^T \left(X - \tilde{U}\tilde{V}\right)\right)$. The optimality conditions for minimizing $\|X - \tilde{U}\tilde{V}\|_F^2$ are $\frac{\partial F}{\partial \tilde{U}} = 0$ and $\frac{\partial F}{\partial \tilde{V}} = 0$.

From the matrix calculus, we know that

$$\frac{\partial \,\text{Trace}\left(A^T BC\right)}{\partial B} = AC^T \quad \text{and} \quad \frac{\partial \,\text{Trace}\left(A^T BC\right)}{\partial C} = B^T A.$$

Let $A=\left(X-\tilde{U}\tilde{V}\right)$, $B=\tilde{U}$, and $C=\tilde{V}$. Then, we have

$$\frac{\partial F}{\partial \tilde{U}}=\frac{-\partial \operatorname{Trace}\left(A^T BC\right)}{\partial B} \quad \text{and} \quad \frac{\partial F}{\partial \tilde{V}}=\frac{-\partial \operatorname{Trace}\left(A^T BC\right)}{\partial C}.$$

Therefore, we have

$$\frac{\partial F}{\partial \tilde{U}}=-2\left(X-\tilde{U}\tilde{V}\right)\tilde{V}^T=0 \tag{1A.1}$$

and

$$\frac{\partial F}{\partial \tilde{V}}=-2\tilde{U}^T\left(X-\tilde{U}\tilde{V}\right)=0,$$

which implies

$$\frac{\partial F}{\partial \tilde{V}}=-2\left(X-\tilde{U}\tilde{V}\right)^T\tilde{U}=0. \tag{1A.2}$$

Equations 1A.1 and 1A.2 can be rewritten as

$$\begin{bmatrix} 0 & X \\ X^T & 0 \end{bmatrix}\begin{bmatrix} \tilde{U} \\ \tilde{V}^T \end{bmatrix}=\begin{bmatrix} \tilde{U}\tilde{V}\tilde{V}^T \\ \tilde{V}^T\tilde{U}^T\tilde{U} \end{bmatrix}. \tag{1A.3}$$

Let $U_*=\left(U\right)_d\Lambda_{(d)}^{1/2}$ and $V_*=\Lambda_{(d)}^{1/2}\left(V^T\right)_d$.

Then, we have

$$X\left(V_*\right)^T=U\Sigma V^T\left(V^T\right)_d\Lambda_{(d)}=U\Sigma\begin{bmatrix} I_d \\ 0 \end{bmatrix}\Lambda_{(d)}^{1/2}=U\begin{bmatrix} \Lambda_{(d)} \\ 0 \end{bmatrix}\Lambda_{(d)}^{1/2}=\left(U\right)_d\Lambda_{(d)}^{3/2} \tag{1A.4}$$

and

$$X^TU_*=V\Sigma U^T\left(U\right)_d\Lambda_{(d)}^{1/2}=V\Sigma\begin{bmatrix} I_d \\ 0 \end{bmatrix}\Lambda_{(d)}^{1/2}=V\begin{bmatrix} \Lambda_{(d)} \\ 0 \end{bmatrix}\Lambda_{(d)}^{1/2}=\left(V\right)_d\Lambda_{(d)}^{3/2}. \tag{1A.5}$$

Note that

$$U_*V_*V_*^T=\left(U\right)_d\Lambda_{(d)}\left(V\right)_d^T\left(V\right)_d\Lambda_{(d)}^{1/2}=\left(U\right)_d\Lambda_{(d)}^{3/2} \tag{1A.6}$$

and

$$V_*^TU_*^TU_*=\left(V\right)_d\Lambda_{(d)}^{1/2}\Lambda_{(d)}^{1/2}\left(U^T\right)_d\left(U\right)_d\Lambda_{(d)}^{1/2}=\left(V\right)_d\Lambda_{(d)}^{3/2}. \tag{1A.7}$$

Combining Equations 1A.5 through 1A.7, we show that U_* and V_* satisfy Equation 1A.3.

EXERCISES

Exercise 1.1 Show that L_∞ and L_P define a norm.

Exercise 1.2 Find the dual norm of the matrix norm $||A||_F = \sqrt{\text{Trace}\left(A^T A\right)}$.

Exercise 1.3 Show that the dual norm $\Omega^*(y)$ of the latent group lasso norm $\Omega(x)$ is $\Omega^*\left(y\right) = \max_{g \in G} \frac{1}{w_g} \Omega^*\left(y_g\right)$.

Exercise 1.4 Prove Equation 1.153:

$$\frac{\partial f}{\partial x} = 2Ax.$$

Exercise 1.5 Let function $f(X)$ be defined as $f(X) = a^T X a$, where a is a k-dimensional vector of constants and X is a $k \times k$ dimensional matrix of variables. Show

$$\frac{\partial f}{\partial X} = 2aa^T - \text{diag}\left(a_1^2, \ldots, a_k^2\right).$$

Exercise 1.6 Find the Fenchel conjugate of the nuclear norm of the matrix A: $||A||_* = \sum_{i=1}^{k} \lambda_i$, where λ_i is a singular value of the matrix A.

Exercise 1.7 Find the Fenchel conjugate of exponential function e^x.

Exercise 1.8 Calculate the subdifferential $\partial e^{|x|}$.

Exercise 1.9 Define a point maximize function $f(x) = \max\{|x|, |x|^{1/2}\}$. Find a subdifferential $\partial f(x)$.

Exercise 1.10 Define a composite function $h(x) = f(k(x))$ where $f\left(t\right) = \begin{bmatrix} t_1 + 2t_2 \\ t_1^2 + e^{t_2} \end{bmatrix}$ and $t = k\left(x\right) = \begin{bmatrix} t_1 \\ t_2 \end{bmatrix} = \begin{bmatrix} |x_1| + x_2^2 \\ \left(x_1^2 + x_2^2\right)^{1/2} \end{bmatrix}$. Find a subdifferential $\partial h(x)$.

Exercise 1.11 Find an optimal of the function $x^2 + |x|$.

Exercise 1.12 Find a solution to the problem:

$$\min_w \frac{1}{2n} ||y - Xw||_2^2 + \lambda ||w||_2.$$

Exercise 1.13 Find the proximal operator $\text{Prox}_\Omega(u)$ of the function $\Omega(w) = ||w||_2$.

Exercise 1.14 Find the proximal operator $\text{Prox}_\Omega(u)$ of the function $\Omega\left(x\right) = \sum_{i=1}^{n} |x_i|_1$.

Exercise 1.15 Let $V=\sigma^2 A^T DA$ and A and D be constant matrices. Find a derivative of function Trace (V^{-1}) with respect to σ^2: $\frac{\partial}{\partial \sigma^2}\left(\text{Trace}\left(V^{-1}\right)\right)$.

Exercise 1.16 Let $J\left[h\right]=\int_0^1 \sqrt{1+\dot{y}^2\left(x\right)}dx$. Calculate $\delta J[h]$ at $y_0=e^x$.

Exercise 1.17 Prove equality (1.207)

$$\left\langle \beta_j, \beta \right\rangle_{k\mu}=0 \quad \text{for all } k \neq j.$$

Exercise 1.18 Let $f(x)=x+e^x$. Find the Fourier series expansion of the function $f(x)$.

Exercise 1.19 Show that functional principal scores ξ_{ij} are independent variables with the variances of $\text{var}(\xi_{ij})=\lambda_j$, $j=1, 2, \ldots$.

Exercise 1.20 Let $u=XA$, $v=YB$ as defined in Equation 1.229. Establish the relationship between $u^T v$ and canonical correlation analysis of the matrices X and Y.

CHAPTER 2

Linkage Disequilibrium

GENETIC VARIANTS include single nucleotide polymorphisms (SNPs), indels, and structural variants such as copy number variants. The 1000 Genome Project has sequenced 2504 individuals from 26 populations. In total, 88 million variants, of which 84.7 million are SNPs, 3.6 million short insertions/deletions (indels), and 60,000 structural variants, are observed (The 1000 Genome Project Consortium 2015). This provides rich resources for the investigation of LD across the genome and genome-wide association studies.

2.1 CONCEPTS OF LINKAGE DISEQUILIBRIUM

Linkage disequilibrium (LD) refers to the nonrandom association of alleles at different marker loci, also called "allelic association" or "gametic disequilibrium." LD is of fundamental importance in genetic studies of complex diseases (Horikawa et al. 2000; Vilhjálmsson et al. 2015). Linkage disequilibrium is due to evolutionary forces in the history of populations such as mutations, selection, population bottleneck, recombination, and random genetic drift.

As an example, we consider two loci: A and B. At locus A, there are two alleles A and a with frequencies $P_A = 0.6$ and $P_a = 0.4$, respectively (Figure 2.1). At locus B, there is only one allele B. Consider a haplotype spanned by alleles A and B. The frequency of the haplotype AB is $P_{AB} = 0.6$. If we view the frequencies of the alleles and haplotypes as the probability of observing alleles and haplotypes in the population, then $P_A P_B = 0.6 * 1 = P_{AB}$ indicates that observing an allele at locus A is independent of observing allele at locus B. When mutation at B occurs, there are now two alleles B and b at locus B. Suppose that the frequencies of allele b and haplotype AB are $P_b = 0.1$ and $P_{AB} = 0.5$, respectively. Now, $P_A P_B = 0.6 * 0.9 = 0.54 \neq P_{AB} = 0.5$, which implies that the event of observing alleles at the loci A and B are no longer independent and that dynamics of the alleles at two loci are in linkage disequilibrium (LD).

One of major evolutionary forces for changing LD is recombination. Suppose that a new mutation arises on individual chromosomes. Over years of transmission of the mutation, through multiple meioses to successive generations, recombination separates the mutation from the original alleles at the loci that are unlinked to the mutations (Figure 2.2). At very closely linked loci, the likelihood of recombination with the disease mutation is low, and the

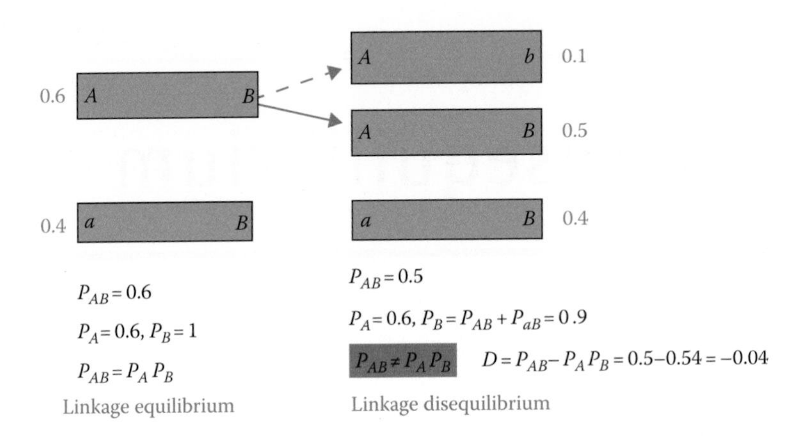

FIGURE 2.1 Linkage disequilibrium due to mutation.

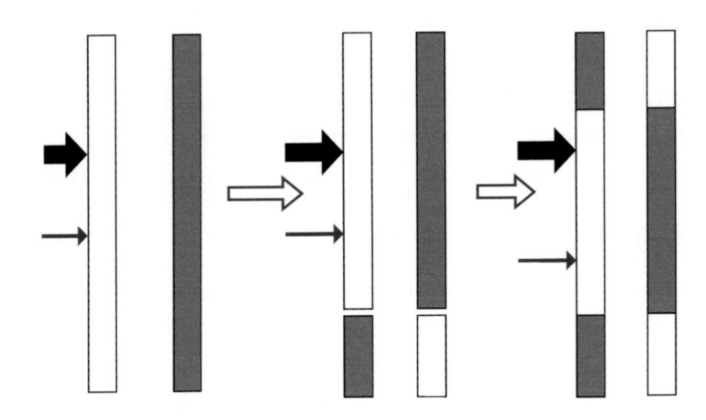

FIGURE 2.2 Evolution of linkage disequilibrium.

original alleles will remain in linkage with the mutation for many generations. By examining the haplotypes at many loci within a large region that does not exhibit recombination, it is sometimes possible to identify a smaller region that appears to be in "linkage disequilibrium" with the mutation because the same alleles are present in different families with the diseases. Linkage disequilibrium provides an indication of which part of the chromosome to study first.

2.2 MEASURES OF TWO-LOCUS LINKAGE DISEQUILIBRIUM

Because of the important implications of LD in association studies of complex diseases and population genetics, several quantities have been proposed to measure the level of LD between loci, which quantifies the dependence of alleles at two loci. Here, we review several widely used measures of linkage disequilibrium.

2.2.1 Linkage Disequilibrium Coefficient D

Consider two marker loci with two alleles D_1 and d_1 at the first locus and two alleles D_2 and d_2 at the second locus. Let $P_{D_1}, P_{d_1}, P_{D_2}$, and P_{d_2} be the frequencies of the alleles D_1,

d_1, D_2, and d_2, respectively. Assume the population is mating randomly. The disequilibrium coefficient for alleles D_1 and D_2 at two loci is defined as the difference between the haplotype frequency and the product of allele frequencies: $D_{D_1D_2} = P_{D_1D_2} - P_{D_1}P_{D_2}$, where $P_{D_1D_2}$ denotes the frequency of haplotype D_1D_2. The maximum likelihood estimate (MLE) of $D_{D_1D_2}$ is estimated by $\hat{D}_{D_1D_2} = \hat{P}_{D_1D_2} - \hat{P}_{D_1}\hat{P}_{D_2}$, where $\hat{P}_{D_1}$ and $\hat{P}_{D_2}$ are the estimated frequency of allele D_1 and D_2, respectively. $\hat{P}_{D_1D_2}$ is the estimated frequency of haplotype D_1D_2.

2.2.2 Normalized Measure of Linkage Disequilibrium D'

The linkage disequilibrium coefficient $D_{D_1D_2}$ depends on the frequencies of haplotype and alleles, making comparisons between two populations difficult. For the convenience of comparison, Lewontin (1964) normalized the above measure of LD by dividing the coefficient D by its maximum value D_{max}, which is given by (Exercise 2.2)

$$|D|_{max} = \min\left[P_{D_1}P_{D_2}, \left(1 - P_{D_1}\right)\left(1 - P_{D_2}\right)\right] \text{ if } D_{D_1D_2} < 0,$$

$$|D|_{max} = \min\left[P_{D_1}P_{d_2}, P_{d_1}P_{D_2}\right] \text{ if } D_{D_1D_2} > 0.$$

This normalized LD measure D' is therefore defined as

$$D'_{D_1D_2} = \frac{D_{D_1D_2}}{|D|_{max}}. \tag{2.1}$$

The normalized LD measure lies between −1 and +1, achieving these values, −1 and +1, when two loci are in complete linkage disequilibrium.

2.2.3 Correlation Coefficient r

Pearson's correlation coefficient r^2 between two loci is another commonly used measure of the LD. Consider two loci, D_1 and D_2. Define two indicator variables:

$$X = \begin{cases} 1 & D_1 \\ 0 & d_1 \end{cases} \quad \text{and} \quad Y = \begin{cases} 1 & D_2 \\ 0 & d_2 \end{cases}. \tag{2.2}$$

Pearson's correlation coefficient r between two loci can be defined in terms of two indicator variables as follows:

$$r = \frac{\text{cov}(X, Y)}{\sqrt{\text{var}(X)\text{var}(Y)}}, \tag{2.3}$$

where

$$\begin{aligned}\operatorname{cov}(X,Y) &= E[XY]-E[X]E[Y] \\ &= P_{D_1D_2} = D^2_{D_1D_2},\end{aligned} \tag{2.4}$$

$$\begin{aligned}\operatorname{var}(X) &= E\left[X^2\right]-\left(E[X]\right)^2 \\ &= P_{D_1} - P^2_{D_1} = P_{D_1}P_{d_1},\end{aligned} \tag{2.5}$$

and

$$\begin{aligned}\operatorname{var}(Y) &= E\left[Y^2\right]-\left(E[Y]\right)^2 \\ &= P_{D_2} - P^2_{D_2} = P_{D_2}P_{d_2}.\end{aligned} \tag{2.6}$$

Substituting Equations 2.4 through 2.6 into Equation 2.3 results in the square of correlation coefficient:

$$r^2 = \frac{D^2_{D_1D_2}}{P_{D_1}P_{d_1}P_{D_2}P_{d_2}}. \tag{2.7}$$

r^2 is often used to eliminate the arbitrary sign introduced. When two loci are in linkage equilibrium, r^2 is reduced to zero. There is a simple inverse relationship between this measure and the sample size required to detect association (Jorde 2000).

Although LD levels quantified by three measures are different, when two loci are in linkage equilibrium they are equal to zero, e.g.,

$$D = 0,$$
$$D' = 0,$$
$$r = 0.$$

Example 2.1 (Figure 2.3)

Denote "A" by allele A and "G" by allele a at the first locus and "T" by allele B and "C" by allele b. The frequencies of haplotype AB and alleles A and B, $|D|_{\max}$ are respectively given by

$$P_{AB} = 0.1425,$$
$$P_A = 0.1425,$$
$$P_B = 0.1425 + 0.7521 = 0.8937.$$

$$|D|_{\max} = \min\{0.1425 * (1-0.8937), (1-0.1425) * 0.8937\} = 0.0151.$$

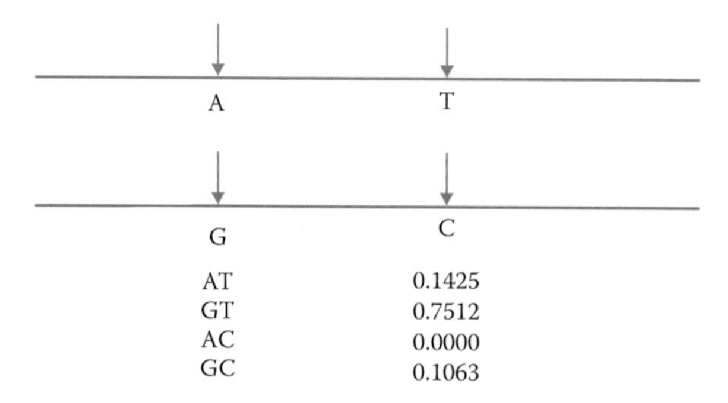

FIGURE 2.3 Figure for Example 2.1.

Three LD measures are then given by

$$D = 0.1425 - 0.1425 * 0.8937 = 0.0151,$$

$$D' = \frac{D}{|D|_{max}} = \frac{0.0151}{0.0151} = 1,$$

$$r = \frac{0.0151}{0.0177} = 0.1401.$$

Example 2.2

Consider four genes: *DEF6*, *KCNK5*, *MRPS18A*, and *PRDM13* that are located in chromosome 6 with physical distance between neighboring genes 1.87M, 6.5M, and 56M, respectively, and are sampled from the NHLBI's Exome Sequencing Project dataset. "Haploview," a software that provides a comprehensive set of tools for haplotype analysis, LD level calculation, and LD pattern visualization (Barrett et al. 2005), is used to analyze and plot the pattern of LD. Figure 2.4 shows the map of LD between SNPs with MAF > 0.0005 within the above four genes where the red color indicates the strong magnitude of LD, and white and gray colors indicate the week LD.

Example 2.3

We analyze the low-coverage pilot with whole-genome sequencing of 179 individuals from four populations and the exon pilot with exon-targeted sequencing of 697 individuals from seven populations in the 1000 Genomes Project to examine LD patterns in humans. Large-scale surveys of genome-wide LD patterns using data generated in the 1000 Genomes Project will reveal the full complexity of empirical patterns of LD. The squared correlation coefficient between the two SNPs is used to measure the levels of pair-wise LD. If r^2 between two SNPs is larger than or equal to 0.8, then the LD between two SNPs is viewed as strong. We used an intermarker distance of 50kb to calculate LD between SNPs. The proportions of pair-wise SNPs with r^2 in five intervals for the intermarker distance of 50kb based on a low-coverage pilot dataset and HapMap phase II (r22) dataset are shown in Figure 2.5. We observed that the LD

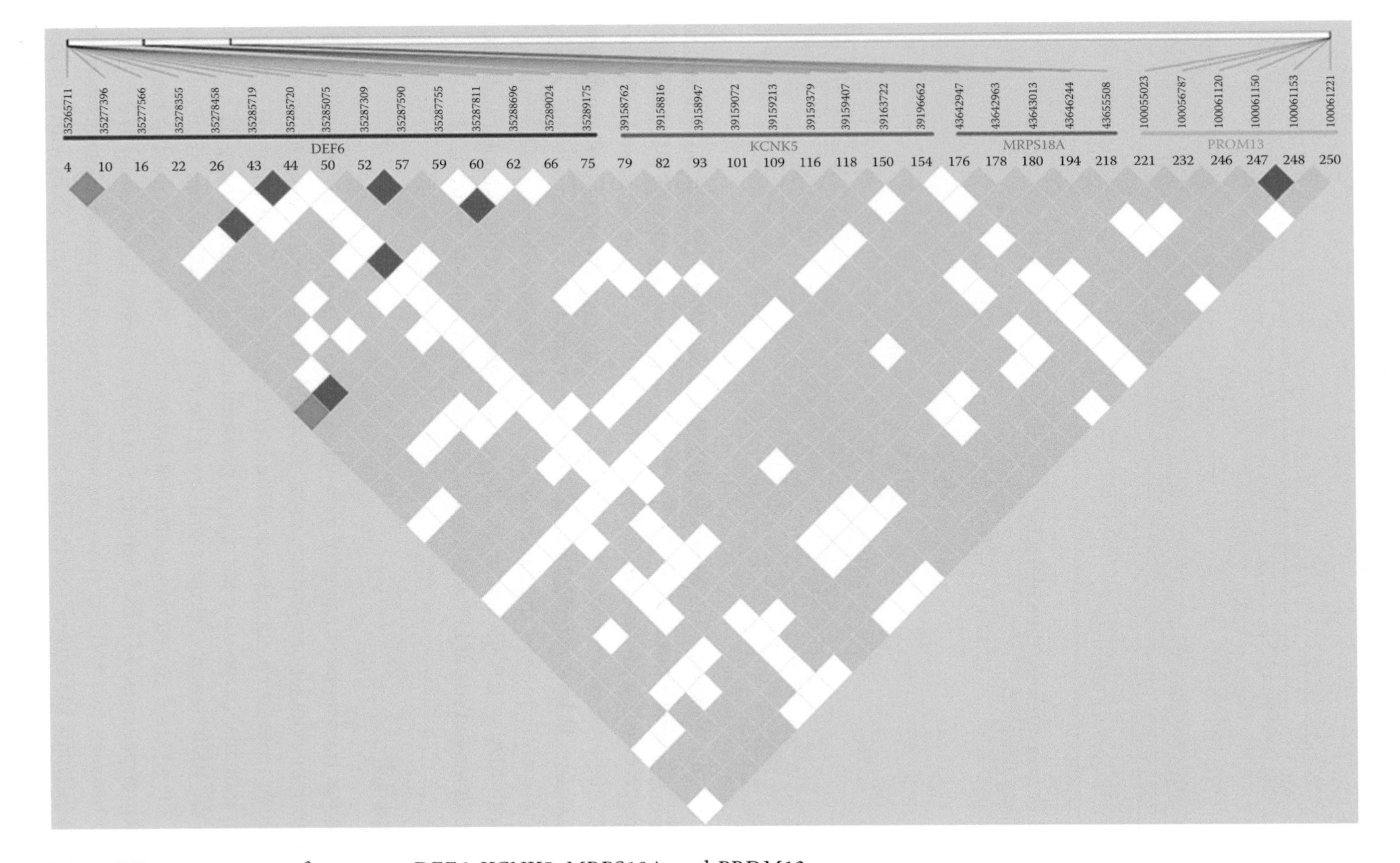

FIGURE 2.4 LD pattern among four genes: *DEF6*, *KCNK5*, *MRPS18A*, and *PRDM13*.

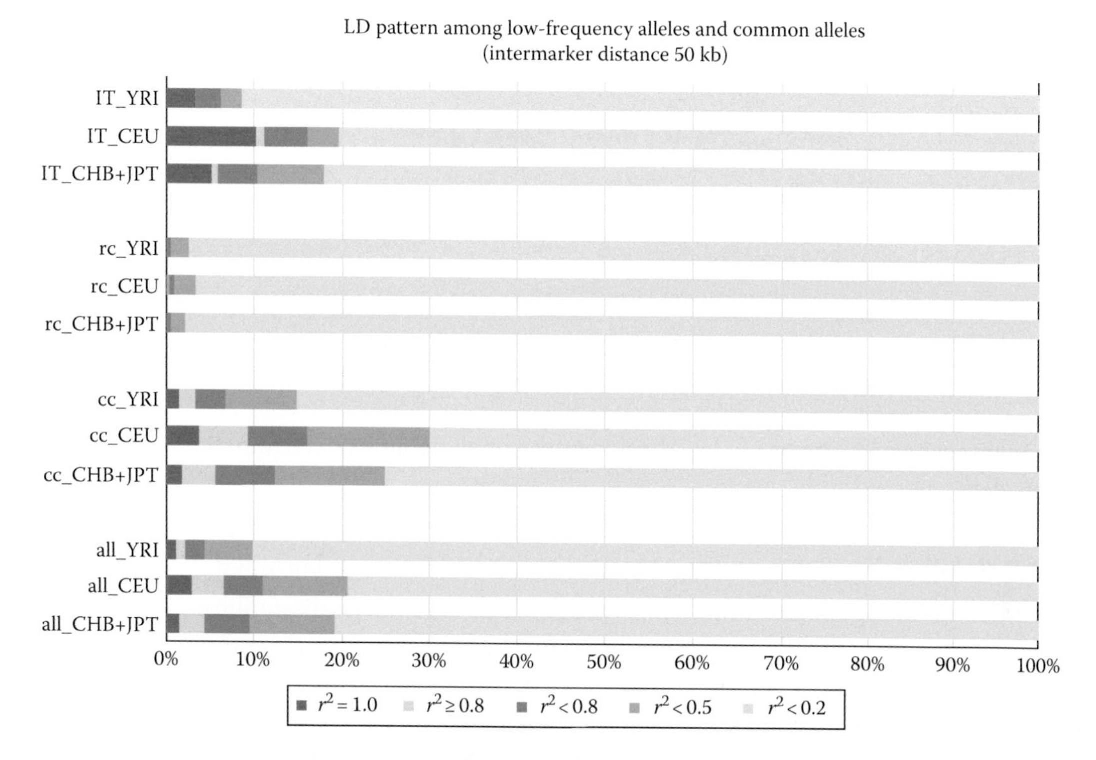

FIGURE 2.5 LD pattern among low-frequency alleles and common alleles. The proportions of pair-wise SNPs with r^2 between common and common SNPs (cc), low frequency and common SNPs (rc), and low frequency and low-frequency SNPs (rr) in five intervals of r^2 within each category of the MAF of SNPs (i.e., cc, rc, and rr categories) for low-coverage pilot dataset. We plotted graphs under three intermarker distances of 50 kb where r^2 between the target SNP and all its nearby SNPs within the distance was calculated.

between common and common SNPs is much stronger than that between low frequency and low-frequency SNPs, and low frequency and common SNPs. In general, we observed only less than 10% of the pair-wise low-frequency SNPs have strong LD.

2.2.4 Composite Measure of Linkage Disequilibrium

LD measures D, D', and r; assume that individuals mate at random. Under the assumption of random mating, the frequency of the genotype is the product of frequencies of haplotypes, and the previously discussed measures of LD can be calculated by estimations of frequencies of alleles and haplotypes, which are obtained by maximum likelihood estimation. However, this assumption is not always satisfied. When only genotypic data are available and random mating cannot be assumed, the measures of gametic disequilibrium introduced previously cannot be calculated directly. Weir (1979) and Weir and Cockerham (1989) introduced the following composite measure of LD, which combines gametic and nongametic digenic disequilibrium coefficients and uses only genotype data:

$$\Delta_{AB} = D_{AB} + D_{A/B} = P_{AB} + P_{A/B} - 2P_A P_B.$$

This can be calculated as

$$\Delta_{AB} = \frac{1}{n}\left(2n_{AABB} + n_{AABb} + n_{AaBB} + \frac{1}{2}n_{AaBb}\right) - 2\hat{P}_A\hat{P}_B,$$

where

n is the number of individuals sampled

n_{AABB}, n_{AABb}, n_{AaBB}, and n_{AaBb} are the numbers of individuals carrying corresponding genotypes at two loci

$\hat{P}_A$ and $\hat{P}_B$ are the sample frequencies for allele A and B, respectively

The composite measure of LD has the advantage of allowing its determination with genotypic data.

2.2.5 Relationship Between the Measure of LD and Physical Distance

Let t denote the age of the mutation, which creates the linkage disequilibrium. Let θ be the recombination fraction between the two marker loci and $P_{ij}(t)$ be the frequency of the haplotype A_iB_j at t generations after the mutation causing linkage disequilibrium. The haplotype in the next generation is produced either by transmission without recombination or by transmission with recombination between two loci (Figure 2.6). Thus, we have on average

$$P_{ij}(t+1) = (1-\theta)P_{ij}(t) + \theta P_{A_i}P_{B_j}. \tag{2.8}$$

Recall that the LD coefficient at t generation is given by

$$D_{ij}(t) = P_{ij}(t) - P_{A_i}P_{B_j}. \tag{2.9}$$

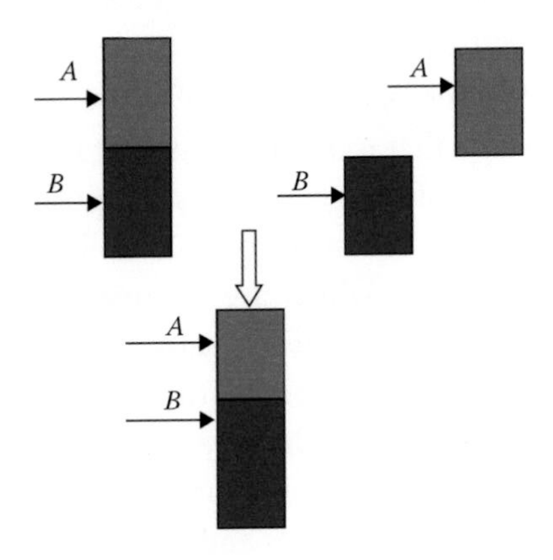

FIGURE 2.6 Scheme of haplotype evolution.

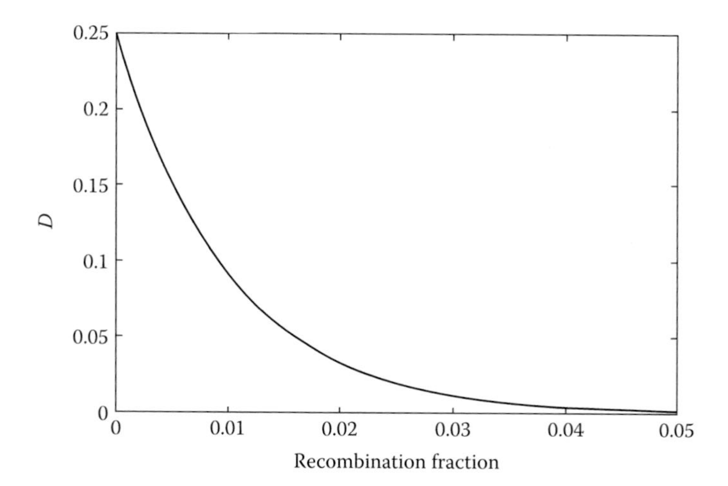

FIGURE 2.7 The LD coefficient decreases exponentially as a function of recombination.

Combining Equations 2.8 and 2.9 yields the following recursive formula for the calculation of the expectation of the measure of the LD:

$$D_{ij}(t+1)=(1-\theta)D_{ij}(t). \tag{2.10}$$

Therefore, recursively, we have

$$\begin{aligned} D_{ij}(t) &= (1-\theta)D_{ij}(t-1) \\ &= (1-\theta)^2 D_{ij}(t-2) \\ &= \cdots \\ &= (1-\theta)^t D_0, \end{aligned} \tag{2.11}$$

where $D_0 = P_{ij}(0) - P_{A_i}P_{B_j}$ is an initial measure of LD. Equation 2.11 implies that the LD coefficient is a function of mutation age, recombination fraction, and the initial measure of LD.

Example 2.4

Assume that the age of mutation is $t=20$ generations and the initial LD coefficient D_0 is equal to 0.25. The LD coefficient curve $D(t)$ as a function of recombination fraction is shown in Figure 2.7. The $D(t)$ decreases exponentially to zero at $\theta=0.05$.

2.3 HAPLOTYPE RECONSTRUCTION

Haplotypes are, in general, not directly observable. Phase-unknown multilocus genotype data are the primary data sources available. Although several experimental technologies for molecular haplotyping have been developed (Reich et al. 2003), these methods are labor intensive, low throughput, and costly. Therefore, experimental haplotyping methods are not practically useful for large-scale population studies. Analyzing family data with many relatives is another method to infer haplotypes, but (1) collecting family data is costly and (2) ambiguity still exists, especially as the number of markers increases. Therefore, computational methods for estimating haplotypes using phase-unknown genotype data offer practical and cost-effective solutions. Haplotype estimation is often one of the first stages in genetic studies.

2.3.1 Clark's Algorithm

Clark was the earliest to propose an algorithm based on maximum parsimony to reconstruct haplotypes among unrelated individuals using genotype data. This algorithm first determines the haplotypes from all individuals with no haplotype ambiguity, i.e., the individuals who are complete homozygotes and single-site heterozygotes, assuming Hardy–Weinberg equilibrium, the basic model of a stable frequency distribution among haplotypes in the presence of random mating. Then the remaining individuals with ambiguous haplotypes are sequentially screened for the possible occurrence of previously recognized haplotypes; the complementary haplotype was then added to the list of resolved haplotypes. Clark's algorithm is straightforward but does not give unique solutions and does not explicitly assume Hardy–Weinberg equilibrium.

2.3.2 EM algorithm

The expectation–maximization (EM) algorithm infers haplotypes based on maximum likelihood that optimizes the likelihood of occurrence of molecular haplotype frequencies from the observed data, assuming Hardy–Weinberg equilibrium (HWE). The advantages of the EM algorithm include its solid theory, good performance for large samples, and relative robustness to the departure from HWE. However, since the optimization method is greedy, its performance is sensitive to the initial solution. Inappropriate initial solutions may lead to a wrong local maximum, which is serious when there are many distinct haplotypes. Therefore, to ensure finding the global maximum likelihood estimate of haplotype frequencies, the EM algorithms should start to multiply with several initial solutions. Further, a standard application of the EM algorithm may not be feasible when a large number of markers are analyzed simultaneously since the number of haplotypes and hence the computation time increase exponentially with the number of markers.

2.3.3 Bayesian and Coalescence-Based Methods

Stephens et al. (2001) proposed to use a Bayesian approach, either using a simple Dirichlet-prior distribution or a prior distribution, which approximates the coalescent, to reconstruct haplotypes from genotype data. The algorithm has been implemented in the program PHASE and its modified versions. This algorithm infers haplotypes based on the following logic: a haplotype that is more similar to the commonly observed haplotype patterns has a higher probability to be present in the population than the less similar haplotypes. The principle for Bayesian haplotype reconstruction methods is to treat the unknown haplotypes as random quantities and to calculate the posterior distribution of the unobserved haplotypes given the observed genotype data using prior information and the likelihood. The haplotypes (or haplotype frequencies) can then be estimated from maximizing this posterior distribution. The prominent feature of this algorithm is the incorporation of coalescence theory into the algorithms and outperforms two previously introduced algorithms. The disadvantages include the lack of a measure of overall quality of the inferred haplotypes, slow computation, and unclear performance in admixed or rapidly expanding populations when the coalescent model does not hold.

Several other coalescence-based and Bayesian algorithms as well as the modified EM algorithms have also been developed to facilitate haplotype reconstruction for multiple markers.

2.4 MULTILOCUS MEASURES OF LINKAGE DISEQUILIBRIUM

While pair-wise LD measure is widely used in association and evolution studies over the past several decades, there are several reasons where the joint study of multilocus LD may be helpful. First, using the multilocus measure of LD for association studies may improve the power to detect association. Multilocus LD will provide not only pair-wise LD information between the marker and the functional loci but also additional high-order LD information involving functional sites. Second, complex diseases are often caused by the combined mutations at multiple sites. To capture the contributions of multiple mutations to the disease, we need to compare differences in the pattern of multilocus LD between affected and unaffected individuals. Developing commonly accepted high-order and global LD measures at multiple loci is a challenging but indispensable task when NGS is popular.

Motivated by equivalence of LD among markers and stochastic dependency of random variables, information theory can be used as a general framework for developing multilocus LD measures. We introduce three types of multilocus measure of LD: global measures of LD, high-order measures of LD, and the measure of LD between two sets of markers. The concept of entropy and mutual information that were proposed by Shannon (1948) can serve as a general measure of correlation between two systems. There are two directions to extend mutual information to a finite number of variables. One direction for extension of mutual information is to characterize high-order interaction information. Mutual information can also be extended to measure the degree of stochastic dependence among multiple variables and will be called multi-information.

2.4.1 Mutual Information Measure of LD

Mutual information is used to measure stochastic dependency of variables. Similar to LD where we use LD information to predict the status of other marker(s) from the status of the allele at one marker locus, the mutual information maximizes the amount of information about variables, which we can obtain from observing other variables. Therefore, the mutual information can be used to characterize the LD between markers (Liu and Lin 2005).

Consider two SNP markers, M_1 and M_2. The marker M_1 has two alleles, A_1 and A_2, with frequencies P_{A_1} and P_{A_2}, respectively. The marker M_2 has two alleles, B_1 and B_2, with frequencies P_{B_1} and P_{B_2}, respectively. Let $P_{A_iB_j}$ be the frequency of the haplotype A_iB_j ($i = 1, 2$, $j = 1, 2$). The mutual information between two marker loci M_1 and M_2 is defined as

$$I(M_1; M_2) = \sum_{i=1}^{2}\sum_{j=1}^{2} P_{A_iB_j} \log \frac{P_{A_iB_j}}{P_{A_i} P_{B_j}}, \tag{2.12}$$

which can be approximated under the assumption of weak LD between two markers by

$$I(M_1; M_2) \approx \frac{1}{2}\sum_{i=1}^{2}\sum_{j=1}^{2} \frac{\left(P_{A_iB_j} - P_{A_i} P_{B_j}\right)^2}{P_{A_i} P_{B_j}} \tag{2.13}$$

or

$$I(M_1;M_2) \approx \frac{1}{2} r^2, \tag{2.14}$$

where $r^2 = \dfrac{\delta^2}{P_{A_1} P_{A_2} P_{B_1} P_{B_2}}$ and $\delta = P_{A_1B_1} - P_{A_1} P_{B_1}$.

Equation 2.14 shows that the mutual information $I(M_1;M_2)$ is half of a square of the correlation coefficient between two markers and hence has close relationship with the traditional two-locus measure of LD.

The entropy of a marker (Zhao et al. 2005), which is a measure of uncertainty of the marker, is defined by

$$H(M_1) = -\sum_{i=1}^{2} P(A_i) \log\left(P_{A_i}\right).$$

The joint entropy between two markers is defined as

$$H(M_1, M_2) = -\sum_{i=1}^{2}\sum_{j=1}^{2} P_{A_iB_j} \log\left(P_{A_iB_j}\right).$$

The conditional entropy of marker M_1 given marker M_2 is defined as

$$H(M_1|M_2) = -\sum_{i=1}^{2}\sum_{j=1}^{2} P_{A_iB_i} \log P\left(A_i|B_j\right).$$

Thus, the mutual information can be rewritten as

$$I(M_1, M_2) = H(M_1) - H(M_1|M_2).$$

This indicates that the mutual information $I(M_1, M_2)$ is the reduction in the uncertainty of the marker M_1 due to the information of marker M_2.

It can be shown that (Cover and Thomas 1991)

$$0 \le I(M_1, M_2) \le \min\left(H(M_1), H(M_2)\right).$$

Mutual information between two markers is equal to zero if and only if two markers are in linkage equilibrium. The above inequality also shows that the mutual information can be normalized by the smallest entropy of two markers. The mutual information is a measure of dependency between two markers, which is equivalent to the standard measure of LD between two markers. Mutual information can clearly characterize the pattern of LD and has a close relationship with the traditional r^2 measure of LD. Therefore, we define the mutual information between two markers as a measure of LD between them.

When the haplotype phase is unknown, the mutual information between two marker loci can be defined in terms of genotypes. Let the genotypes at the marker locus M_1 be indexed by G_i and the genotypes at the marker locus M_2 be indexed by G_j. Let G_{ij} be the genotype G_i at the marker M_1 and the genotype G_j at the marker M_2. For example, G_{11} will be the genotype A_1A_2 at the marker M_1 and the genotype B_1B_2 at the marker M_2. The mutual information between two markers M_1 and M_2 is then defined as

$$I(G_1; G_2) = \sum_{i=1}^{3}\sum_{j=1}^{3} P_{G_{ij}} \log \frac{P_{G_{ij}}}{P_{G_i} P_{G_j}}, \tag{2.15}$$

where P_{G_i} and P_{G_j} are the frequencies of the genotype G_i at the marker M_1 and the genotype G_j at the marker M_2, respectively, and $P_{G_{ij}}$ the frequency of the genotype G_{ij}.

The mutual information $I(G_1; G_2)$ can also be approximated by

$$I(G_1; G_2) \approx \frac{1}{2}\sum_{i=1}^{3}\sum_{j=1}^{3} \frac{\left(P_{G_{ij}} - P_{G_i} P_{G_j}\right)^2}{P_{G_i} P_{G_j}}. \tag{2.16}$$

In Appendix 2A, we have shown that assuming Hardy–Weinberg equilibrium (HWE), we have

$$I(G_1; G_2) \approx \frac{1}{2} r^2 \left(r^2 + 1\right), \tag{2.17}$$

where r is defined as before.

2.4.2 Multi-Information and Multilocus Measure of LD

The mutual information can be generalized to multiple marker loci for measuring stochastic dependency among a finite number of markers. Consider a sequence of marker loci $M_1 \cdots M_k$. Each locus M_i has two alleles, M_{1_i} and M_{2_i}, with frequencies $P(M_1)$ and $P(M_2)$, respectively. Each locus M_i also has three genotypes: $M_{1_i}M_{1_i}$, $M_{1_i}M_{2_i}$, and $M_{2_i}M_{2_i}$, labeled as G_{1_i}, G_{2_i}, and G_{3_i}, respectively. Let $P(G_{j_i})$ be the frequency of the genotype G_{j_i} ($j = 1, 2, 3$). Let $P(M_{j_1} \cdots M_{j_k})$ be the frequency of the haplotype $M_{j_1} \cdots M_{j_k}$ ($j=1,2$), and $P(G_{j_1} \cdots G_{j_k})$ be the frequency of the sequence of the genotypes $G_{j_1} \cdots G_{j_k}$ ($j=1,2,3$). We first study haplotype multi-information among k marker loci, which is defined as

$$MI(M_1,\ldots,M_k)=\sum_{j1}\cdots\sum_{jk}P(M_{j1}\cdots M_{jk})\log\frac{P(M_{j1}\cdots M_{jk})}{P(M_{j1})\cdots P(M_{jk})}. \tag{2.18}$$

Similar to the mutual information, multi-information $MI(M_1, \ldots, M_k)$ is also bounded by

$$0\le MI(M_1,\ldots,M_k)\le H(M_1)+\cdots+H(M_k),$$

where $H(M_i)$ is the entropy of marker $M_i(i, \ldots, k)$. The multi-information $MI(M_1, \ldots, M_k)$ is equal to zero if and only if the set of markers $M_1, \ldots, M_k$ are independent (or in linkage equilibrium), i.e., $P(M_{j_1} \cdots M_{j_k}) = P(M_{j_1}) \cdots P(M_{j_k})$. The multi-information measures the departure of a set of markers from equilibrium. Therefore, $MI(M_1, \ldots, M_k)$ can be defined as a k-locus measure of LD at the marker loci $M_1, \ldots, M_k$.

By the same argument as that for mutual information, $MI(M_1, \ldots, M_k)$ can be approximated under the assumption of weak dependency among the markers by

$$MI(M_1,\ldots,M_k)\approx\frac{1}{2}\sum_{j1}\cdots\sum_{jk}\frac{\left[P(M_{j1}\cdots M_{jk})-P(M_{j1})\cdots P(M_{jk})\right]^2}{P(M_{j1})\cdots P(M_{jk})}. \tag{2.19}$$

In Appendix 2B, we show that

$$MI(M_1,M_2,M_3)\approx\frac{1}{2}\left(\frac{\delta^2_{M_1M_2M_3}}{\prod_{i=1}^{3}\prod_{j=1}^{2}P_{M_{ji}}}+\frac{\delta^2_{M_1M_2}}{\prod_{i=1}^{2}\prod_{j=1}^{2}P_{M_{ji}}}+\frac{\delta^2_{M_1M_3}}{\prod_{i=1,3}\prod_{j=1}^{2}P_{M_{ji}}}+\frac{\delta^2_{M_2M_3}}{\prod_{i=2}^{3}\prod_{j=1}^{2}P_{M_{ji}}}\right), \tag{2.20}$$

where $\delta_{M_1M_2M_3}$ is a three-locus measure of LD and defined as

$$\begin{aligned}\delta_{M_1M_2M_3}&=P(M_{1_1}M_{1_2}M_{1_3})-P(M_{1_1})\delta_{M_2M_3}-P(M_{1_2})\delta_{M_1M_3}-P(M_{1_3})\delta_{M_1M_2}\\&\quad-P(M_{1_1})P(M_{1_2})P(M_{1_3})\end{aligned}$$

(Weir 1990); $\delta_{M_1M_2}$, $\delta_{M_1M_3}$, and $\delta_{M_2M_3}$ are pair-wise LD between the markers M_1 and M_2, M_1 and M_3, and M_2 and M_3, respectively. Let r_{12}, r_{13}, and r_{23} be the correlation coefficients between the markers M_1 and M_2, M_1 and M_3, and M_2 and M_3, respectively, i.e., $r_{12}^2 = \frac{\delta_{M_1M_2}^2}{P_{M_{11}}P_{M_{21}}P_{M_{12}}P_{M_{22}}}$, $r_{13}^2 = \frac{\delta_{M_1M_3}^2}{P_{M_{11}}P_{M_{21}}P_{M_{13}}P_{M_{23}}}$, and $r_{23}^2 = \frac{\delta_{M_2M_3}^2}{P_{M_{12}}P_{M_{22}}P_{M_{13}}P_{M_{23}}}$. The concept of correlation relations between two marker loci can be extended to multiple loci. Define

$$r_{123}^2 = \frac{\delta_{M_1M_2M_3}^2}{P_{M_{11}}P_{M_{21}}P_{M_{12}}P_{M_{22}}P_{M_{13}}P_{M_{23}}},$$

which can be viewed as a multilocus generalization of the correlation coefficient r^2. For the convenience of presentation, r_{123} is called a three-locus correlation coefficient.

Then Equation 2.20 can be rewritten as

$$MI(M_1, M_2, M_3) \approx \frac{1}{2}\left(r_{123}^2 + r_{12}^2 + r_{13}^2 + r_{23}^2\right). \tag{2.21}$$

Equations 2.20 and 2.21 show that three-locus multi-information involves a three-locus high-order measure of LD and all pair-wise LD of three markers. Therefore, the multi-information captures all pair-wise and three-order LD information. By the same arguments as that for Equation 2.20, the k-locus multi-information can be approximated by

$$\begin{aligned} MI(M_1,\cdots,M_k) &\approx \frac{1}{2}\left(\frac{\delta_{M_1\cdots M_k}^2}{\prod_{i=1}^{k}\prod_{j=1}^{2}P_{M_{ji}}} + \cdots + \frac{\delta_{M_1M_2}^2}{\prod_{i=1}^{2}\prod_{j=1}^{2}P_{M_{ji}}} + \cdots + \frac{\delta_{M_{k-1}M_k}^2}{\prod_{i=k-1}^{k}\prod_{j=1}^{2}P_{M_{ji}}}\right) \\ &\approx \frac{1}{2}\left(r_{1\cdots k}^2 + \cdots + r_{12}^2 + \cdots + r_{k-1k}^2\right) \end{aligned} \tag{2.22}$$

where $\delta_{M_1\cdots M_k}^2$ is a k-order measure of LD, and other high-order LD measures can be similarly defined as $r_{1\cdots k}^2 = \frac{\delta_{M_1\cdots M_k}^2}{\prod_{i=1}^{k}\prod_{j=1}^{2}P_{M_{ji}}}$ a generalization of r^2 to k markers, and

$$r_{ij}^2 = \frac{\delta_{M_iM_j}^2}{P_{1_i}P_{2_i}P_{1_j}P_{2_j}}, \quad (i = 1,\ldots,k,\ j = i+1,\ldots,k).$$

2.4.3 Joint Mutual Information and a Measure of LD between a Marker and a Haplotype Block or between Two Haplotype Blocks

The two important concepts of mutual information are conditional mutual information and joint mutual information. For ease of understanding, we first consider three markers: M_1, M_2, and M_3. The joint mutual information between the marker M_3 and the combination of markers M_1 and M_2 is defined as

$$I(M_1, M_2; M_3) = \sum_{j1} \sum_{j2} \sum_{j3} P(M_{j1}, M_{j2}, M_{j3}) \log \frac{P(M_{j1}, M_{j2}, M_{j3})}{P(M_{j3}) P(M_{j1}, M_{j2})}, \quad (2.23)$$

which can be approximated by

$$I(M_1, M_2; M_3) \approx \frac{1}{2} \sum_{j1} \sum_{j2} \sum_{j3} \frac{\left(P(M_{j1}, M_{j2}, M_{j3}) - P(M_{j3}) P(M_{j1}, M_{j2})\right)^2}{P(M_{j3}) P(M_{j1}, M_{j2})}, \quad (2.24)$$

Define a measure of LD between the allele M_{j3} and the combination of the markers M_{j1} and M_{j2} (Xiong et al. 2003) as

$$\delta_{j1,j2,j3} = P(M_{j1}, M_{j2}, M_{j3}) - P(M_{j1}, M_{j2}) P(M_{j3}).$$

Clearly, $I(M_1, M_2; M_3)$ is a monotonic function of $\delta_{j1 j2, j3}$. We also can show that

$$I(M_1, M_2; M_3) \geq 0$$

with equality if and only if the marker M_3 is independent of (or in linkage equilibrium with) the block formed by the markers M_1 and M_2, which satisfies the requirement that the measure of LD between a marker and a block be zero when the marker is in linkage equilibrium with the block. Thus, we define the joint information $I(M_1, M_2; M_3)$ as a measure of LD between a marker and a block.

The conditional mutual information between markers M_2 and M_3, given marker M_1 is given by

$$I(M_2; M_3 | M_1) = \sum_{j1} \sum_{j2} \sum_{j3} P(M_{j1}, M_{j2}, M_{j3}) \log \frac{P(M_{j2}, M_{j3} | M_{j1})}{P(M_{j2} | M_{j1}) P(M_{j3} | M_{j1})}. \quad (2.25)$$

The multi-information can be decomposed as the summation of mutual information and joint mutual information as the following chain rule of multi-information

$$MI(M_1, M_2, M_3) = I(M_1; M_2) + I(M_1, M_2; M_3). \quad (2.26)$$

Similarly, the joint mutual information can be decomposed as summation of the mutual information and conditional mutual information.

$$I(M_1, M_2; M_3) = I(M_1; M_3) + I(M_2; M_3 | M_1). \quad (2.27)$$

The concept of joint mutual information and conditional mutual information can be extended to more general cases. The joint mutual information between the two blocks, $X_1, \ldots, X_l$ and $Y_1, \ldots, Y_k$, can be defined as

$$I(X_1,\ldots,X_l,Y_1,\ldots,Y_k) = E\left[\log\frac{P(X_1,\ldots,X_l,Y_1,\ldots,Y_k)}{P(X_1,\ldots,X_l)P(Y_1,\ldots,Y_k)}\right]. \tag{2.28}$$

Similar to other information measures, it can be shown that

$$I(X_1,\ldots,X_l;Y_1,\ldots,Y_k) \geq 0,$$

with equality if and only if two blocks of markers are independent (in linkage equilibrium), which can be interpreted that the joint frequencies of the two blocks of markers are equal to the product of the frequencies of each block of markers. It is natural to define the joint information $I(X_1, \ldots, X_l; Y_1, \ldots, Y_k)$ as a measure of LD between two blocks.

The conditional mutual information between two blocks, $X_1, \ldots, X_l$ and $Y_1, \ldots, Y_k$, given the third block, $Z_1, \ldots, Z_n$, is given by

$$\begin{aligned} &I(X_1,\ldots,X_l;Y_1,\ldots,Y_k|Z_1,\ldots,Z_n) \\ &= E\left[\log\frac{P(X_1,\ldots,X_l;Y_1,\ldots,Y_k|Z_1,\ldots,Z_n)}{P(X_1,\ldots,X_l|Z_1,\ldots,Z_n)P(Y_1,\ldots,Y_k|Z_1,\ldots,Z_n)}\right]. \end{aligned} \tag{2.29}$$

The distribution of the markers in Equations 2.28 and 2.29 can be either haplotype/allele frequencies or genotype frequencies.

Now we are in a position to study a general chain rule of multi-information. By definition of multi-information, we have

$$\begin{aligned} MI(M_1,\ldots,M_k) &= E\left[\log\frac{P(M_1,\ldots,M_k)}{P(M_1)\cdots P(M_k)}\right] \\ &= E\left[\log\frac{P(M_2|M_1)}{P(M_2)}\frac{P(M_3|M_1|M_2)}{P(M_3)}\cdots\frac{P(M_k|M_1,\ldots,M_{k-1})}{P(M_k)}\right] \\ &= E\left[\log\frac{P(M_2|M_1)}{P(M_2)}\right]+E\left[\log\frac{P(M_3|M_1,M_2)}{P(M_3)}\right]+\cdots \\ &\quad E\left[\log\frac{P(M_1,\ldots,M_k)}{P(M_1,\ldots,M_{k-1})P(M_k)}\right] \\ &= I(M_1;M_2)+I(M_1,M_2;M_3)+\cdots+I(M_1,\ldots,M_{k-1};M_k). \end{aligned} \tag{2.30}$$

Next, we decompose the joint mutual information into a series of the conditional mutual information. By definition of the joint mutual information, we have

$$\begin{aligned} I(M_1,\ldots,M_{k-1};M_k) &= E\left[\log\frac{P(M_1,\ldots,M_{k-1},M_k)}{P(M_k)P(M_1,\ldots,M_{k-1})}\right] \\ &= I(M_1;M_k)+I(M_2;M_k|M_1)+I(M_3;M_k|M_1,M_2) \\ &\quad +\cdots+I(M_{k-1};M_k|M_1,\ldots,M_{k-2}) \\ &= \sum_{i=1}^{k-1} I(M_i;M_k|M_{i-1},\ldots,M_1). \end{aligned} \tag{2.31}$$

Thus, $MI(M_1,\ldots,M_k)$ can be decomposed as

$$\begin{aligned} MI(M_1,\ldots,M_k) &= I(M_1;M_2)+I(M_1;M_3)+I(M_2;M_3|M_1) \\ &\quad +I(M_1;M_4)+I(M_2;M_4|M_1)+I(M_3;M_4|M_1,M_2) \\ &\quad +\cdots+I(M_1;M_k)+\cdots+I(M_{k-1};M_k|M_1,\ldots,M_{k-2}). \end{aligned} \tag{2.32}$$

Multi-information can also be expressed by the entropy of the markers:

$$\begin{aligned} MI(M_1,\ldots,M_k) &= \sum_{j_1}\cdots\sum_{j_k} P(M_{j_1},\ldots,M_{j_k})\log P(M_{j_1},\ldots,M_{j_k}) - \sum_{i=1}^{k} P(M_{j_i})\log P(M_{j_i}) \\ &= -S(M_1,\ldots,M_k)+S(M_1)+\cdots+S(M_k). \end{aligned} \tag{2.33}$$

If k markers are in linkage equilibrium, then

$$S(M_1,\ldots,M_k) = S(M_1)+\cdots+S(M_k).$$

Let $S_E(M_1,\ldots,M_k)$ denote the entropy of the markers $M_1,\ldots,M_k$ in the equilibrium. Then, Equation 2.33 is reduced to

$$MI(M_1,\ldots,M_k) = S_E(M_1,\ldots,M_k) - S(M_1,\ldots,M_k). \tag{2.34}$$

2.4.4 Interaction Information

Interaction among loci (or genes) is a fundamental concept we often use in genetic epidemiology but rarely specify with precision. Interaction can be interpreted as inseparable genetic effects of the multiple loci. It implies some sort of correlation and association with the phenotype. From the information point of view, the interaction can be understood as sharing common information causing disease among loci (or genes). The amount of information about causing disease shared among loci (genes) is defined as a measure of interaction.

Mutual information is to quantify common information shared by two variables. Mutual information between two markers (variables) can be generalized to multiple markers to measure shared information among multiple markers, which will provide a powerful tool for the detection of interactions among markers (Matsuda 2000). Common information shared by k markers is defined as k-locus interaction information and calculated by

$$I_k\left(M_1,\ldots,M_k\right)=\sum_{n=1}^{k}(-1)^{n+1}\sum_{i_1<\cdots<i_n}S\left(M_{i_1},\ldots,M_{i_n}\right). \tag{2.35}$$

Particularly,

$$\begin{aligned}I_3\left(M_1M_2,M_3\right)=&S\left(M_1,M_2,M_3\right)-S\left(M_1,M_2\right)-S\left(M_1,M_3\right)-S\left(M_2,M_3\right)\\&+S\left(M_1\right)+S\left(M_2\right)+S\left(M_3\right).\end{aligned}$$

k-locus interaction information can also be calculated by multi-information (Appendix 2C):

$$\begin{aligned}I_k\left(M_1,\ldots,M_k\right)&=\sum_{n=1}^{k}(-1)^{n}\sum_{i_1<\cdots<i_n}MI\left(M_{i_1},\ldots,M_{i_n}\right)\\&\approx(-1)^k\frac{1}{2}r_{1\cdots k}^2\end{aligned} \tag{2.36}$$

k-locus interaction information can be directly expressed by the frequencies of the haplotypes and alleles. Matsuda (2000) showed that

$$I_k\left(M_1,\ldots,M_k\right)=(-1)^k\sum_{j_1}\cdots\sum_{j_k}P\left(M_{j_1},\ldots,M_{j_k}\right)\log\frac{P\left(M_{j_1},\ldots,M_{j_k}\right)}{\hat{P}\left(M_{j_1},\ldots,M_{j_k}\right)}, \tag{2.37}$$

where $\hat{P}\left(M_{j_1},\ldots,M_{j_k}\right)$ is the Kirkwood superposition (Matsuda 2000) and given by

$$\hat{P}\left(M_{j_1},\ldots,M_{j_k}\right)=\prod_{j_1<\cdots<j_{k-1}}P\left(M_{j_1},\ldots,M_{j_{k-1}}\right)/\prod_{j_1<\cdots<j_{k-2}}P\left(M_{j_1},\ldots,M_{j_{k-2}}\right)/\cdots/\prod_{j}P\left(M_j\right). \tag{2.38}$$

When $k=3$, we have

$$I_3\left(M_1,M_2,M_3\right)=-\sum_{j_1}\sum_{j_2}\sum_{j_3}P\left(M_{j_1},M_{j_2},M_{j_3}\right)\log\frac{P\left(M_{j_1},M_{j_2},M_{j_3}\right)}{\hat{P}\left(M_{j_1},M_{j_2},M_{j_3}\right)},$$

where

$$\hat{P}(M_{j1}, M_{j2}, M_{j3}) = \frac{P(M_{j1}, M_{j2})P(M_{j2}, M_{j3})P(M_{j1}, M_{j3})}{P(M_{j1})P(M_{j2})P(M_{j3})}.$$

It is interesting to note that when the haplotype frequency $P(M_{j1},\ldots, M_{jk}) = \hat{P}(M_{j1},\ldots, M_{jk})$, k-locus interaction information $I_k(M_1, \ldots, M_k)$ will be equal to zero. This implies that k-locus interaction information measures the intrinsic k-way correlation among k markers, which is not influenced by $(k-1)$-way, …, 2-way correlation among the subset of k markers.

It follows from Equations 2.26 and 2.27 that

$$MI(M_1,\ldots, M_k) = \sum_{i<j} I_2(M_i, M_j) - \sum_{i<j<l} I_3(M_i, M_j, M_l) + \cdots, \tag{2.39}$$

where $I_2(M_i, M_j) = J(M_i, M_j)$.

Equation 2.39 implies that an approximation for the multi-information can be made by neglecting high-order interaction information.

2.4.5 Conditional Interaction Information

The conditional $(k-1)$-locus mutual information among markers $M_1, \ldots, M_{k-1}$, given the marker M_k, is defined as

$$I_{k-1}(M_1,\ldots, M_{k-1}|M_k) = E\left(\log\frac{P(M_1,\ldots,M_{k-1}|M_k)}{\hat{P}(M_1,\ldots,M_{k-1}|M_k)}\right),$$

where

$$\hat{P}(M_1,\ldots, M_{k-1}|M_k) = \prod_{i_1<\cdots<i_{k-2}} P(M_{i_1},\ldots, M_{i_{k-2}}|M_k) / \prod_{i_1<\cdots<i_{k-3}} P(M_{i_1},\ldots, M_{i_{k-3}}|M_k)/\cdots/\prod_i P(M_i|M_k). \tag{2.40}$$

It follows from Equation 2.40 that

$$I_{k-1}(M_1,\ldots, M_{k-1}|M_k) = \sum_{n=1}^{k-1}(-1)^{n+1}\sum_{i_1<\cdots<i_n} S(M_{i_1},\ldots, M_{i_n}|M_k) \tag{2.41}$$

and

$$I_{k-1}(M_1,\ldots, M_{k-1}|M_k) = \sum_{n=2}^{k}(-1)^{n}\sum_{i_1<\cdots<i_n} MI(M_{i_1},\ldots, M_{i_n}|M_k). \tag{2.42}$$

2.4.6 Normalized Multi-Information

To allow for comparisons in LD between different sets of marker loci, we normalize multi-information as follows from Liu and Lin (2005)

$$MI_N\left(M_1,\ldots,M_k\right)=\frac{MI\left(M_1,\ldots,M_k\right)}{\sum_{i=1}^{k}S\left(M_i\right)-\max_i S\left(M_i\right)}. \tag{2.43}$$

From Equation 2.43, we can see that

$$0\le MI_N\left(M_1,\ldots,M_k\right)\le 1.$$

Normalized $MI_N(M_1,\ldots,M_k)$ is equal to ε in Nothnagel et al. (2002), which is used to define haplotype blocks.

2.4.7 Distribution of Estimated Mutual Information, Multi-Information, and Interaction Information

Since the allele and haplotype frequencies are unknown, the four information measures, mutual information, the joint mutual information, multi-information, and interaction information, need to be estimated from a number of observed samples. These information measures are nonlinear functions of allele and haplotype frequencies. Finding the exact distribution of the estimators of the information measures is a nontrivial problem. Fortunately, large sample theory can be used to derive asymptotical distribution of these information measures quantifying the LD.

In general, the four information measures can be written as the minimum discrimination information $I(*,H_0)$ that minimizes the following Kullback Leibler distance (Kullback 1959):

$$I\left(H,H_0\right)=\sum_{x\in X}f_1\left(x\right)\log\frac{f_1\left(x\right)}{f_2\left(x\right)}$$

for a given $f_2(x)(H_0)$ and all $f_1(x)(H)$ such that

$$\theta=E_H\left[T\left(x\right)\right],$$

where

θ is a multidimensional parameter
$T(x)$ a measurable statistic

Kullback (1959) showed that

$$2N\hat{I}\left(*,H_0\right)=-2\log\lambda,$$

where

N is the number of observed samples
λ is the likelihood ratio

Thus, $2N\hat{I}(*, H_0)$ is asymptotically distributed as a χ^2 distribution with $m-r$ degrees of freedom under the null hypothesis that the parameters lie on an r-dimensional hyperplane of m-dimensional space.

Using the results of distribution of the minimum discrimination information, we first study the asymptotic distribution of estimators of the multi-information statistic that includes a mutual information statistic as its special case. The multi-information $M\hat{I}(M_1,\ldots, M_k)$ can be used to test the null hypothesis of independence (or linkage equilibrium) of the variables $M_1, \ldots, M_k$:

$$H_0: P(M_{j1},\ldots,M_{jk}) = P(M_{j1})\cdots P(M_{jk}).$$

The multi-information $M\hat{I}(M_1,\ldots,M_k)$ can be decomposed into additive components. These components correspond to a hypothesis $H_0(R_1)$ specifying the values of the $P(M_{j_1})$, a hypothesis $H_0(R_2)$ specifying the values of the $P(M_{j_2}), \ldots$, a hypothesis $H_0(R_k)$ specifying the values of the $P(M_{j_k})$, and an independence hypothesis $H_0\left(\prod_{i=1}^{k} R_i\right)$. There are the (2 – 1) independent parameter $P_{M_{j1}}$, the (2 – 1) independent parameter $P_{M_{jk}}$, and 2^k-k-1 independent parameters $P(M_{j_1}, \ldots, M_{j_k})$. Thus, $2NM\hat{I}(M_1,\ldots,M_k)$ is asymptotically distributed as a χ^2 distribution with degrees of freedom 2^k-k-1. Under the alternative hypothesis of dependence (or linkage disequilibrium), $2N\, M\hat{I}(M_1,\ldots,M_k)$ is distributed as a noncentral χ^2 distribution with the noncentrality parameter $2N\, MI(M_1, \ldots, M_k)$. Therefore, the mean and variance of $M\hat{I}(M_1,\ldots,M_k)$ are given by

$$E\left[M\hat{I}(M_1,\ldots,M_k)\right] = \frac{2^k-k-1}{2N} + MI(M_1,\ldots,M_k)$$

and

$$Var\left[M\hat{I}(M_1,\ldots,M_k)\right] = \frac{2^k-k-1}{2N^2} + \frac{2MI(M_1,\ldots,M_k)}{N}.$$

The joint information statistic $\hat{I}(X_1,\ldots,X_k,Y_1,\ldots,Y_l)$ can be decomposed into additive components: $H_0(R_1)$ specifying the values of the $P(X_{j_1}, \ldots, X_{j_k})$, $H_0(R_2)$ specifying the values of the $P(Y_{j_1}, \ldots, Y_{j_l})$, and an independence hypothesis $H_0(R_1 \cap R_2)$. There are (2^k-1) independent parameters $P(X_{j_1}, \ldots, X_{j_k})$, (2^l-1) independent parameters $P(Y_{j_1}, \ldots, Y_{j_l})$, and $(2^k-1)(2^l-1)$ independent parameters $P(X_{j_1}, \ldots, X_{j_k}, Y_{j_1}, \ldots, Y_{j_l})$. Therefore, $2N\,\hat{I}(X_1,\ldots,X_k,Y_1,\ldots,Y_l)$ is asymptotically distributed as a χ^2 distribution with degrees of freedom $(2^k-1)(2^l-1)$. Under the alternative hypothesis of dependence between two sets of variables (or LD between the blocks $X_1, \ldots, X_k$ and $Y_1, \ldots, Y_l$), $2N\,\hat{I}(X_1,\ldots,X_k,Y_1,\ldots,Y_l)$ is asymptotically distributed as a noncentral $\chi^2_{(2^k-1)(2^l-1)}$ distribution with the noncentrality parameter $2N\,I(X_1,\ldots,X_k,Y_1,\ldots,Y_l)$. The mean and variance of $\hat{I}(X_1,\ldots,X_k,Y_1,\ldots,Y_l)$ are given by

$$E\left[\hat{I}(X_1,\ldots,X_k,Y_1,\ldots,Y_l)\right]=\frac{(2^k-1)(2^l-1)}{2N}+I(X_1,\ldots,X_k,Y_1,\ldots,Y_l)$$

and

$$V\left[\hat{I}(X_1,\ldots,X_k,Y_1,\ldots,Y_l)\right]=\frac{(2^k-1)(2^l-1)}{2N^2}+\frac{2I(X_1,\ldots,X_k,Y_1,\ldots,Y_l)}{N}.$$

By similar argument, $2N\,\hat{I}_k(M_1,\ldots,M_k)$ is asymptotically distributed as a χ^2 distribution with 1 degrees of freedom. Under the alternative hypothesis of k-order interaction (or k-order LD), $2N\,\hat{I}_k(M_1,\ldots,M_k)$ is distributed as a noncentral $\chi^2_{(1)}$ distribution with the non-centrality parameter $2N\,I_k(M_1,\ldots,M_k)$. Thus, the mean and variance of $\hat{I}_k(M_1,\ldots,M_k)$ are

$$E\left[\hat{I}_k(M_1,\ldots,M_k)\right]=\frac{1}{2N}+I_k(M_1,\ldots,M_k)$$

and

$$V\left[\hat{I}_k(M_1,\ldots,M_k)\right]=\frac{1}{2N^2}+\frac{2I_k(M_1,\ldots,M_k)}{N}.$$

Example 2.5

Mutual information, interaction information, multi-information, and pair-wise r^2 and multilocus r^2 for measuring LD were calculated using chromosome 22 HapMap phase II data (Altshuler and Clark 2005). The analysis was restricted to the common SNPs. To illustrate the utility of the proposed LD measures, we present Table 2.1 summarizing the averaged pair-wise and multilocus r^2, multilocus interaction information, mutual information, and multi-information over chromosome 22. Table 2.1 indicated that variation of LD across populations quantified by pair-wise and multilocus LD measures was well matched with the "out of Africa" hypothesis of human evolution. All these LD measures almost consistently showed that when modern human populations were out of Africa and moved to Europe and to Asia, the level of LD was clearly increased.

Figure 2.8 plotted the mutual information and half of two-locus r^2 as a function of SNPs' positions along chromosome 22. From these figures, we can see that all the mutual

TABLE 2.1 Averaged LD and Information Measures in Africa (YRI), North America (CEU), China (CHB), and Japan (JPT)

Average	YRI	CEU	CHB + JPT
Pair-wise R^2	0.3070	0.3796	0.4065
Three-locus R^2	0.3201	0.4931	0.5812
Four-locus R^2	0.5657	1.1576	1.6645
Mutual information	0.1649	0.2037	0.2163
Three-locus interaction information	0.0908	0.1333	0.1476
Four-locus interaction information	0.3778	0.3779	0.4236
Three-locus multi-information	0.4073	0.4886	0.5091
Four-locus multi-information	0.6950	0.8132	0.8386

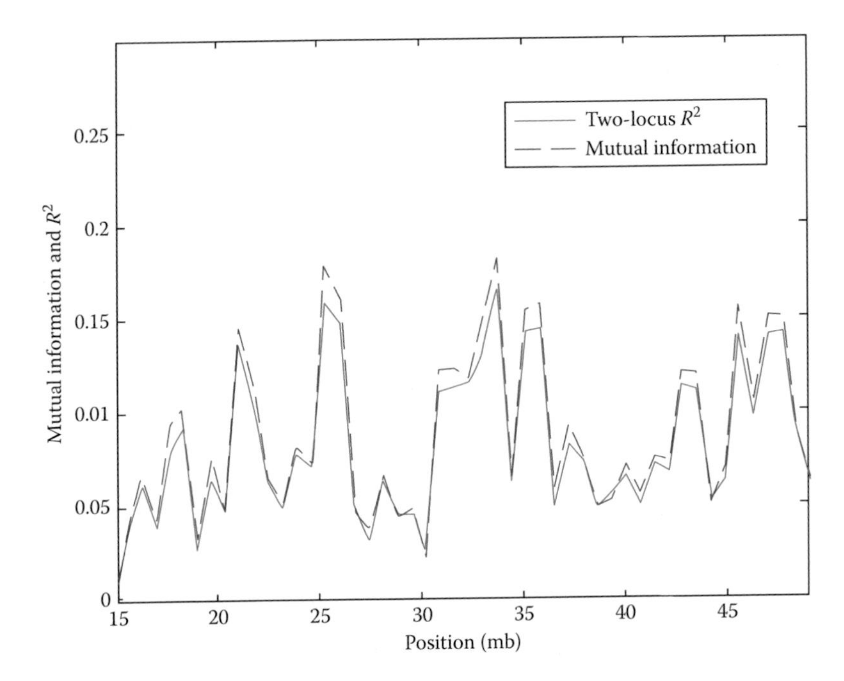

FIGURE 2.8 Average mutual information and half of an average two-locus r^2 with a window size 700 kb as a function of SNPs' positions along chromosome 22 in YRI.

information across the populations is very close to half of the two-locus r^2. This strongly implied that information measure can be used to characterize the patterns of LD and to quantify the level of LD.

The patterns of mutual information, three-locus interaction information, and multi-information characterizing pair-wise LD and joint multilocus LD across chromosome 22 for CEU were shown in Figure 2.9. Figure 2.9 demonstrated that for the fixed number of loci, the MI is much larger than the mutual information. This suggested that ignoring high-order LD will cause the loss of a large amount of information.

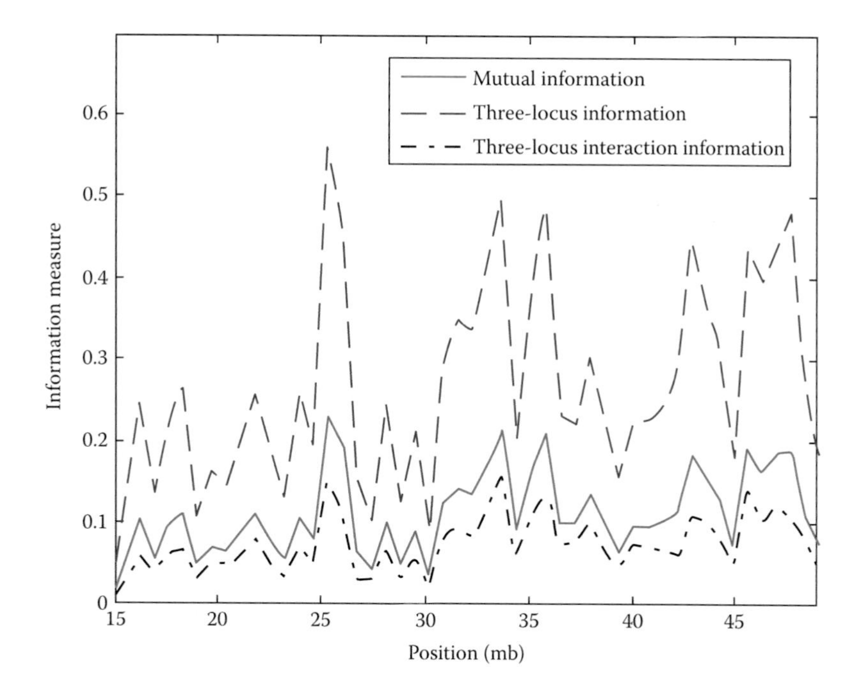

FIGURE 2.9 Average mutual information, three-locus interaction information, and three-locus multi-information with a window size of 700 kb as a function of SNPs' position along chromosome 22 in CEU.

2.5 CANONICAL CORRELATION ANALYSIS MEASURE FOR LD BETWEEN TWO GENOMIC REGIONS

Alternative to information measures of multilocus LD, canonical correlation analysis (CCA) is another tool for investigating dependent relationships or LD between two genomic regions. When a growing number of genetic variants becomes available, multi-information and interaction information are difficult to calculate for a large number of genetic variants. As we introduced in Section 1.6, CCA is a standard statistical tool for identifying linear relationships between two sets of variables and hence can be used to study LD between two sets of SNPs in two genomic regions.

2.5.1 Association Measure between Two Genomic Regions Based on CCA

Consider two genomic regions (or genes). The first genomic region contains p SNPs, each having two alleles, M_j and m_j, $j=1,\ldots,p$, and the second genomic region contains q SNPs, each SNP having alleles G_j and g_j, $j=1,\ldots,q$ (Figure 2.10). For the convenience of discussion, we assume $p \le q$. Indicator variables for SNPs can be defined for haplotype and genotype. For haplotype, the indicator variable for an SNP is defined as

$$x_j = \begin{cases} 1 & M_j \\ 0 & m_j \end{cases} \quad \text{and} \quad y_j = \begin{cases} 1 & G_j \\ 0 & g_j \end{cases}. \tag{2.44}$$

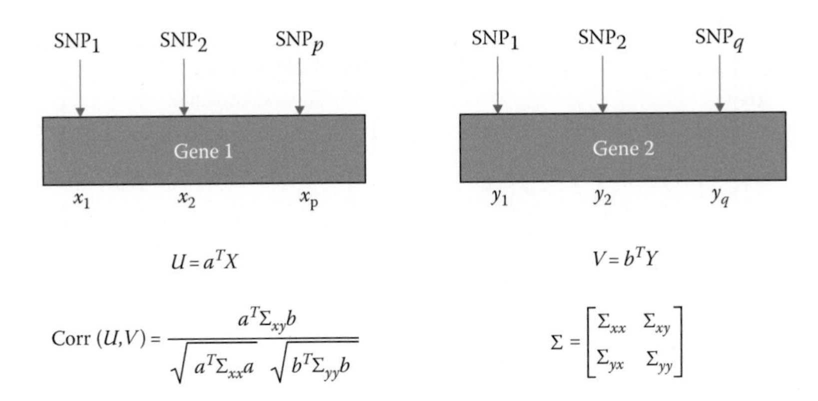

FIGURE 2.10 Schematic picture of CCA for LD between two genes.

For genotype, we define the indicator variable for an SNP as

$$x_j = \begin{cases} 2 & M_jM_j \\ 1 & M_jm_j \\ 0 & m_jm_j \end{cases} \quad \text{and} \quad y_j = \begin{cases} 2 & G_jG_j \\ 1 & G_jg_j \\ 0 & g_jg_j \end{cases}. \tag{2.45}$$

For the haplotype representation, the covariance matrix has simple biological interpretation:

$$\Sigma_{xx} = \begin{bmatrix} P_{M_1}P_{m_1} & D_{12}^x & \cdots & D_{1p}^x \\ D_{21}^x & P_{M_2}P_{m_2} & \cdots & D_{2p}^x \\ \vdots & \vdots & \ddots & \vdots \\ D_{p1}^x & D_{p2}^x & \vdots & P_{M_p}P_{m_p} \end{bmatrix},$$

$$\Sigma_{yy} = \begin{bmatrix} P_{G_1}P_{g_1} & D_{12}^y & \cdots & D_{1q}^y \\ D_{21}^y & P_{G_2}P_{g_2} & \cdots & D_{2q}^y \\ \vdots & \vdots & \ddots & \vdots \\ D_{q1}^y & D_{q2}^y & \cdots & P_{G_q}P_{g_q} \end{bmatrix},$$

$$\Sigma_{xy} = \begin{bmatrix} D_{11}^{xy} & D_{12}^{xy} & \cdots & D_{1p}^{xy} \\ D_{21}^{xy} & D_{22}^{xy} & \cdots & D_{2p}^{xy} \\ \vdots & \vdots & \ddots & \vdots \\ D_{q1}^{xy} & D_{q2}^{xy} & \cdots & D_{qp}^{xy} \end{bmatrix},$$

$$\Sigma = \begin{bmatrix} \Sigma_{xx} & \Sigma_{xy} \\ \Sigma_{yx} & \Sigma_{yy} \end{bmatrix},$$

where $P_{M_j}, P_{m_j}, P_{G_k}$ and P_{g_k} are frequencies of alleles M_j, m_j, G_k, and g_k, respectively, and D^x_{jk}, D^y_{jk}, and D^{xy}_{jk} are the LD coefficient between SNPs in the first genomic region, between SNPs in the second genomic region, and between the SNP in the first genomic region and the SNP in the second genomic region, respectively. It is clear that the elements in the covariance matrix are the pair-wise LD coefficients between SNPs and allele frequencies.

CCA is to seek linear combinations of the original variable to construct a pair of canonical variables, $U = a^T X$ and $V = b^T Y$, which maximize

$$\mathrm{Corr}(U,V) = \frac{a^T \Sigma_{xy} b}{\sqrt{a^T \Sigma_{xx} a}\sqrt{b^T \Sigma_{yy} b}}. \tag{2.46}$$

Using the method introduced in Section 1.6, we obtain p canonical correlations $\lambda_1, \lambda_2, \ldots, \lambda_p$. We expect that the measure derived from CCA can capture the LD or association relations between SNPs in the two genomic regions.

We define

$$R_{CCA} = -\sum_{i=1}^{p} \log\left(1-\lambda_i^2\right), \tag{2.47}$$

where λ_j are the canonical correlations.

Genomic data analysis was traditionally dominated by the paradigm of multivariate statistics. This paradigm is less effective in the era of next-generation sequencing (NGS) because the SNPs generated by NGS are densely distributed and the position nature of the genomic data was not included in the analysis. To overcome the limitations of the traditional CCA, we employ the functional CCA that is discussed in Section 1.6.5. The widely used functional CCA is to solve the following optimization problem:

$$\max_{a(t),b(t)} \frac{\mathrm{cov}\left(\int a^T(t)X(t)dt, \int b^T(t)Y(t)dt\right)}{\sqrt{\left[\mathrm{var}\left(\int a^T(t)X(t)dt\right)+\lambda_1\int \|\ddot{a}(t)\|^2 dt\right]\left[\mathrm{var}\left(\int b^T(t)Y(t)dt\right)+\lambda_2\int \|\ddot{b}(t)\|^2 dt\right]}}. \tag{2.48a}$$

However, the denominator in Equation 2.48a is no longer pure variance. It involves the penalty terms. The quantity, which we want to maximize in Equation 2.48a, is not exactly canonical correlation coefficient. We suggest that we first use FPCA to obtain the functional principal scores and apply the multivariate CCA to the matrices of functional principal scores ξ and η defined in Equation 1.254. Specifically, we define a new dataset that consists of the functional principal scores:

$$Z = \begin{bmatrix} \xi \\ \eta \end{bmatrix}.$$

Its covariance matrix is

$$\Sigma_{zz} = \begin{bmatrix} \Sigma_{\xi\xi} & \Sigma_{\xi\eta} \\ \Sigma_{\eta\xi} & \Sigma_{\eta\eta} \end{bmatrix}.$$

Now the functional CCA is transformed to the following multivariate CCA problem: seeking a pair of canonical variables, $U = a^T\xi$ and $V = b^T\eta$, which maximize

$$\mathrm{Corr}(U,V) = \frac{a^T\Sigma_{\xi\eta}b}{\sqrt{a^T\Sigma_{\xi\xi}a}\sqrt{b^T\Sigma_{\eta\eta}b}}.$$

2.5.2 Relationship between Canonical Correlation and Joint Information

In general, functional principal scores follow a normal distribution. Assume that the density functions for ξ, η, Z are

$$f(\xi) = \frac{1}{(2\pi)^{P/2}|\Sigma_{\xi\xi}|^{1/2}} e^{-\frac{1}{2}(\xi-\mu_\xi)^T\Sigma_{\xi\xi}^{-1}(\xi-\mu_\xi)},$$

$$f(\eta) = \frac{1}{(2\pi)^{q/2}|\Sigma_{\eta\eta}|^{1/2}} e^{-\frac{1}{2}(\eta-\mu_\eta)^T\Sigma_{\eta\eta}^{-1}(\eta-\mu_\eta)},$$

$$f(Z) = \frac{1}{(2\pi)^{(p+q)/2}|\Sigma_{zz}|^{1/2}} e^{-\frac{1}{2}(Z-\mu_z)^T\Sigma_{zz}^{-1}(Z-\mu_z)}.$$

Using Equation 2.28a, we obtain the joint information

$$\begin{aligned}
I(\xi;\eta) &= E\left[\log\frac{f(Z)}{f(\xi)f(\eta)}\right] \\
&= \log\frac{|\Sigma_{\xi\xi}|^{1/2}|\Sigma_{\eta\eta}|^{1/2}}{|\Sigma_{zz}|^{1/2}} - \frac{1}{2}E\left[\mathrm{Trace}\left(\Sigma_{zz}^{-1}(Z-\mu_z)(Z-\mu_z)^T\right)\right] \\
&\quad + \frac{1}{2}E\left[\mathrm{Trace}\left(\Sigma_{\xi\xi}^{-1}(\xi-\mu_\xi)(\xi-\mu_\xi)^T\right)\right] + \frac{1}{2}E\left[\mathrm{Trace}\left(\Sigma_{\eta\eta}^{-1}(\eta-\mu_\eta)(\eta-\mu_\eta)^T\right)\right] \\
&= \log\frac{|\Sigma_{\xi\xi}|^{1/2}|\Sigma_{\eta\eta}|^{1/2}}{|\Sigma_{zz}|^{1/2}} - \frac{1}{2}\mathrm{Trace}\left(\Sigma_{zz}^{-1}E\left[(Z-\mu_z)(Z-\mu_z)^T\right]\right) \\
&\quad + \frac{1}{2}\mathrm{Trace}\left(\Sigma_{\xi\xi}^{-1}E\left[(\xi-\mu_\xi)(\xi-\mu_\xi)^T\right]\right) + \frac{1}{2}\mathrm{Trace}\left(\Sigma_{\eta\eta}^{-1}E\left[(\eta-\mu_\eta)(\eta-\mu_\eta)^T\right]\right) \\
&= \log\frac{|\Sigma_{\xi\xi}|^{1/2}|\Sigma_{\eta\eta}|^{1/2}}{|\Sigma_{zz}|^{1/2}} - \frac{(p+q)}{2} + \frac{p}{2} + \frac{q}{2} \\
&= \log\frac{|\Sigma_{\xi\xi}|^{1/2}|\Sigma_{\eta\eta}|^{1/2}}{|\Sigma_{zz}|^{1/2}} = \frac{1}{2}\log\frac{|\Sigma_{\xi\xi}||\Sigma_{\eta\eta}|}{|\Sigma_{zz}|}.
\end{aligned} \tag{2.48b}$$

Using Equation 1.247, we obtain

$$\log \frac{|\Sigma_{zz}|}{|\Sigma_{\xi\xi}||\Sigma_{\eta\eta}|} = \sum_{i=1}^{p} \log\left(1-\lambda_i^2\right). \tag{2.49}$$

Substituting Equation 2.49 into Equation 2.48b yields

$$I(\xi,\eta) = -\frac{1}{2}\sum_{i=1}^{p} \log\left(1-\lambda_i^2\right). \tag{2.50}$$

Comparing Equations 2.47 and 2.50, we establish the following relationship between the CCA measure and joint information measure of LD between two genomic regions:

$$R_{CCA} = 2I(\xi,\eta). \tag{2.51}$$

In other words, CCA measure of LD between two genomic regions is twice the joint information measure of LD between two genomic regions. Canonical correlation between two genomic regions captures the information on DNA variation between two genomic regions. Therefore, CCA measure is a well-defined measure of LD between two genomic regions.

SOFTWARE PACKAGE

PHASE and fast PHASE packages are developed for haplotype reconstruction and recombination rate estimation. Beagle 4.1, minimac, MACH, and IMPUTE2 are software packages that perform imputation of ungenotyped markers. Haploview is designed to simplify and expedite the process of haplotype analysis and LD visualization (Barrett et al. 2005); DISEQ is developed for multilocus LD estimation (http://linkage.rockefeller.edu/soft/diseq.html). CRAN estimates the entropy, mutual information, and other information measures (http://strimmerlab.org/software/entropy/); CCA is an R package to extend canonical correlation Analysis (https://cran.r-project.org/).

BIBLIOGRAPHICAL NOTES

Composite measure of linkage disequilibrium was introduced by Weir (1979) and Weir and Cockerham (1989). Algorithms for haplotype estimation can be found in Clark (1990), Delaneau et al. (2013), Qin et al. (2002), Stephens et al. (2001), and Lin et al. (2002). Information theory and its application for measuring LD level can be found in Shannon (1948), Dawy et al. (2006), Cover and Thomas (1991), and Zhao et al. (2005), and Weir (1990) briefly introduces high-order LD. High-order information measure can be found in Matsuda (2000). Taking gene as information can be found in Sherwin (2015). Excellent theoretic treatment of CCA can be found in Anderson (1984), and its application to genomics is investigated in Lin et al. (2013).

APPENDIX 2A

There are nine genotypes at two loci. The marker M_1 has three genotypes: $G_{1.}(A_1A_1)$, $G_{2.}(A_1A_2)$, and $G_{3.}(A_2A_2)$. The marker M_2 has three genotypes: $G_{.1}(B_1B_1)$, $G_{.2}(B_1B_2)$, and $G_{.3}(B_2B_2)$. We first calculate $\frac{\left(P_{G11}-P_{G1.}P_{G.1}\right)^1}{P_{G1.}P_{G.1}}$.

Note that assuming HWE, we have

$$\begin{aligned}P_{G11}-P_{G1.}P_{G.1} &= P_{A_1B_1}^2-P_{A_1}^2P_{B_1}^2\\ &=\left(P_{A_1B_1}+P_{A_1}P_{B_1}\right)\left(P_{A_1B_1}-P_{A_1}P_{B_1}\right)\\ &=\delta\left(\delta+2P_{A_1}P_{B_1}\right),\end{aligned}$$

where δ is the LD coefficient between the markers M_1 and M_2. Thus, we obtain

$$\frac{\left(P_{G11}-P_{G1.}P_{G.1}\right)^2}{P_{G1.}P_{G.1}}=\frac{\delta^2\left(\delta+2P_{A_1}P_{B_1}\right)^2}{P_{A_1}^2P_{B_1}^2}. \tag{2A.1}$$

Similarly, we have

$$\frac{\left(P_{G12}-P_{G1.}P_{G.2}\right)^2}{P_{G1.}P_{G.2}}=\frac{2\delta^2\left[\delta+P_{A_1}\left(P_{B_1}-P_{B_2}\right)\right]^2}{P_{A_1}^2P_{B_1}P_{B_2}}, \tag{2A.2}$$

$$\frac{\left(P_{G13}-P_{G1.}P_{G.3}\right)^2}{P_{G1.}P_{G.3}}=\frac{\delta^2\left(\delta-2P_{A_1}P_{B_2}\right)^2}{P_{A_1}^2P_{B_2}^2}, \tag{2A.3}$$

$$\frac{\left(P_{G21}-P_{G2.}P_{G.1}\right)^2}{P_{G2.}P_{G.1}}=\frac{2\delta^2\left[\delta+P_{B_1}\left(P_{A_1}-P_{A_2}\right)\right]^2}{P_{A_1}P_{A_2}P_{B_1}^2}, \tag{2A.4}$$

$$\frac{\left(P_{G22}-P_{G2.}P_{G.2}\right)^2}{P_{G2.}P_{G.2}}=\frac{\delta^2\left[2\delta+\left(P_{A_1}-P_{A_2}\right)\left(P_{B_1}-P_{B_2}\right)\right]^2}{P_{A_1}P_{A_2}P_{B_1}P_{B_2}}, \tag{2A.5}$$

$$\frac{\left(P_{G23}-P_{G2.}P_{G.3}\right)^2}{P_{G2.}P_{G.3}}=\frac{2\delta^2\left[\delta-P_{B_2}\left(P_{A_1}-P_{A_2}\right)\right]^2}{P_{A_1}P_{A_2}P_{B_2}^2}, \tag{2A.6}$$

$$\frac{\left(P_{G31}-P_{G3.}P_{G.1}\right)^2}{P_{G3.}P_{G.1}}=\frac{\delta^2\left(\delta-2P_{A_2}P_{B_1}\right)^2}{P_{A_2}^2P_{B_1}^2}, \tag{2A.7}$$

$$\frac{\left(P_{G32}-P_{G3.}P_{G.2}\right)^2}{P_{G3.}P_{G.2}}=\frac{2\delta^2\left[\delta-P_{A_2}\left(P_{B_1}-P_{B_2}\right)\right]^2}{P_{A_2}^2P_{B_1}P_{B_2}}, \tag{2A.8}$$

and

$$\frac{\left(P_{G33}-P_{G3.}P_{G.3}\right)^2}{P_{G3.}P_{G.3}}=\frac{\delta^2\left(\delta+2P_{A_2}P_{B_2}\right)^2}{P_{A_2}^2P_{B_2}^2}. \tag{2A.9}$$

Thus, substituting Equations 2A.1 through 2A.9 into Equation 2A.5, we obtain

$$\begin{aligned}
I\left(G_1;G_2\right)&=\frac{1}{2}\left\{\frac{\delta^2}{P_{A_1}^2}\left(\frac{\delta^2}{P_{B_1}^2P_{B_2}^2}+\frac{2P_{A_1}^2}{P_{B_1}P_{B_2}}\right)+\frac{\delta^2}{P_{A_1}P_{A_2}}\left[\frac{2\delta^2}{P_{B_1}^2P_{B_2}^2}+\frac{\left(P_{A_1}-P_{A_2}\right)^2}{P_{B_1}P_{B_2}}\right]\right.\\
&\quad\left.+\frac{\delta^2}{P_{A_2}^2}\left(\frac{\delta^2}{P_{B_1}^2P_{B_2}^2}+\frac{2P_{A_2}^2}{P_{B_1}P_{B_2}}\right)\right\}\\
&=\frac{1}{2}\left[\frac{\delta^4}{P_{A_1}^2P_{A_2}^2P_{B_1}^2P_{B_2}^2}+\frac{\delta^2}{P_{A_1}P_{A_2}P_{B_1}P_{B_2}}\right]\\
&=\frac{1}{2}r^2\left(r^2+1\right).
\end{aligned}$$

APPENDIX 2B

Consider three marker loci: M_1, M_2, and M_3. From the definition of LD at three loci, it follows that

$$\begin{aligned}
P_{M_{11}M_{12}M_{13}}-P_{M_{11}}P_{M_{12}}P_{M_{13}}&=\delta_{M_1M_2M_3}+P_{M_{11}}\delta_{M_2M_3}+P_{M_{12}}\delta_{M_1M_3}+P_{M_{13}}\delta_{M_1M_2}\\
P_{M_{11}M_{12}M_{23}}-P_{M_{11}}P_{M_{12}}P_{M_{23}}&=-\delta_{M_1M_2M_3}-P_{M_{11}}\delta_{M_2M_3}-P_{M_{12}}\delta_{M_1M_3}+P_{M_{23}}\delta_{M_1M_2}\\
P_{M_{11}M_{22}M_{13}}-P_{M_{11}}P_{M_{22}}P_{M_{13}}&=-\delta_{M_1M_2M_3}-P_{M_{11}}\delta_{M_2M_3}+P_{M_{22}}\delta_{M_1M_3}-P_{M_{13}}\delta_{M_1M_2}\\
P_{M_{11}M_{22}M_{23}}-P_{M_{11}}P_{M_{22}}P_{M_{23}}&=\delta_{M_1M_2M_3}+P_{M_{11}}\delta_{M_2M_3}-P_{M_{22}}\delta_{M_1M_3}-P_{M_{23}}\delta_{M_1M_2}\\
P_{M_{21}M_{12}M_{13}}-P_{M_{21}}P_{M_{12}}P_{M_{13}}&=-\delta_{M_1M_2M_3}+P_{M_{21}}\delta_{M_2M_3}-P_{M_{12}}\delta_{M_1M_3}-P_{M_{13}}\delta_{M_1M_2}\\
P_{M_{21}M_{12}M_{23}}-P_{M_{21}}P_{M_{12}}P_{M_{23}}&=\delta_{M_1M_2M_3}-P_{M_{21}}\delta_{M_2M_3}+P_{M_{12}}\delta_{M_1M_3}-P_{M_{23}}\delta_{M_1M_2}\\
P_{M_{21}M_{22}M_{13}}-P_{M_{21}}P_{M_{22}}P_{M_{13}}&=\delta_{M_1M_2M_3}-P_{M_{21}}\delta_{M_2M_3}-P_{M_{22}}\delta_{M_1M_3}+P_{M_{13}}\delta_{M_1M_2}\\
P_{M_{21}M_{22}M_{23}}-P_{M_{21}}P_{M_{22}}P_{M_{23}}&=-\delta_{M_1M_2M_3}+P_{M_{21}}\delta_{M_2M_3}+P_{M_{22}}\delta_{M_1M_3}+P_{M_{23}}\delta_{M_1M_2}
\end{aligned} \tag{2B.1}$$

We first calculate the coefficient of $\delta_{M_1M_2M_3}^2$ in $MI(M_1,M_2,M_3)$:

$$\begin{aligned}
&\frac{1}{P_{M_{11}}P_{M_{12}}P_{M_{13}}}+\frac{1}{P_{M_{11}}P_{M_{12}}P_{M_{23}}}+\frac{1}{P_{M_{11}}P_{M_{22}}P_{M_{13}}}+\frac{1}{P_{M_{11}}P_{M_{22}}P_{M_{23}}}\\
&+\frac{1}{P_{M_{21}}P_{M_{12}}P_{M_{13}}}+\frac{1}{P_{M_{21}}P_{M_{12}}P_{M_{23}}}+\frac{1}{P_{M_{21}}P_{M_{22}}P_{M_{13}}}+\frac{1}{P_{M_{21}}P_{M_{22}}P_{M_{23}}}\\
&=\frac{1}{P_{M_{11}}P_{M_{21}}P_{M_{12}}P_{M_{22}}P_{M_{13}}P_{M_{23}}}
\end{aligned} \tag{2B.2}$$

We then calculate the coefficients of $D^2_{M_2M_3}$:

$$
\begin{aligned}
& P_{M_{11}}\left(\frac{1}{P_{M_{12}}P_{M_{13}}}+\frac{1}{P_{M_{12}}P_{M_{23}}}\right)+P_{M_{11}}\left(\frac{1}{P_{M_{22}}P_{M_{13}}}+\frac{1}{P_{M_{22}}P_{M_{23}}}\right) \\
& +P_{M_{21}}\left(\frac{1}{P_{M_{12}}P_{M_{13}}}+\frac{1}{P_{M_{12}}P_{M_{23}}}\right)+P_{M_{21}}\left(\frac{1}{P_{M_{22}}P_{M_{13}}}+\frac{1}{P_{M_{22}}P_{M_{23}}}\right) \\
& =\frac{P_{M_{11}}}{P_{M_{12}}P_{M_{22}}P_{M_{13}}P_{M_{23}}}+\frac{P_{M_{21}}}{P_{M_{12}}P_{M_{22}}P_{M_{13}}P_{M_{23}}}=\frac{1}{P_{M_{12}}P_{M_{22}}P_{M_{13}}P_{M_{23}}}
\end{aligned}
\tag{2B.3}
$$

Similarly, we can obtain the coefficient of $D^2_{M_1M_3}$ and $D^2_{M_1M_2}$

$$
\frac{1}{P_{M_{11}}P_{M_{21}}P_{M_{13}}P_{M_{23}}} \quad \text{and} \quad \frac{1}{P_{M_{11}}P_{M_{21}}P_{M_{12}}P_{M_{22}}}. \tag{2B.4}
$$

After lengthy calculations, we obtain that the other terms in $MI(M_1, M_2, M_3)$ are zero. Thus, putting above terms together, we have

$$
\begin{aligned}
MI(M_1,M_2,M_3) = & \frac{\delta^2_{M_1M_2M_3}}{P_{M_{11}}P_{M_{21}}P_{M_{12}}P_{M_{22}}P_{M_{13}}P_{M_{23}}}+\frac{\delta^2_{M_2M_3}}{P_{M_{12}}P_{M_{22}}P_{M_{13}}P_{M_{23}}} \\
& +\frac{\delta^2_{M_1M_3}}{P_{M_{11}}P_{M_{21}}P_{M_{13}}P_{M_{23}}}+\frac{\delta^2_{M_1M_2}}{P_{M_{11}}P_{M_{21}}P_{M_{12}}P_{M_{22}}}.
\end{aligned}
$$

APPENDIX 2C

It follows from Equations 2.34 and 2.35 that

$$
\begin{aligned}
I_k(M_1,\ldots,M_k) &= \sum_{n=1}^{k}(-1)^{n+1}\sum_{i_1<\cdots<i_n} S(M_{i_1},\ldots,M_{i_n}) \\
&= \sum_{n=1}^{k}(-1)^{n+1}\sum_{i_1<\cdots<i_n}\left[-MI(M_{i_1},\ldots,M_{i_n})+S(M_{i_1})+\cdots+S(M_{i_n})\right] \\
&= \sum_{n=2}^{k}(-1)^{n}\sum_{i_1<\cdots<i_n} MI(M_{i_1},\ldots,M_{i_n})+ \\
&= \sum_{n=1}^{k}(-1)^{n+1}\sum_{i_1<\cdots<i_n}\left[S(M_{i_1})+\cdots+S(M_{i_n})\right].
\end{aligned}
\tag{2C.1}
$$

However,

for $n = 1$

$$
\sum_{i_1<\cdots<i_n}\left[S(M_{i_1})+\cdots+S(M_{i_n})\right]=S(M_1)+\cdots+S(M_k),
$$

for $n = 2$

$$\sum_{i_1<\cdots<i_n}\left[S\left(M_{i_1}\right)+\cdots+S\left(M_{i_n}\right)\right]=(k-1)\left[S\left(M_1\right)+\cdots+S\left(M_k\right)\right],$$

for $n = 3$

$$\sum_{i_1<\cdots<i_n}\left[S\left(M_{i_1}\right)+\cdots+S\left(M_{i_n}\right)\right]=\frac{(k-1)(k-2)}{2}\left[S\left(M_1\right)+\cdots+S\left(M_k\right)\right],$$

$$\vdots$$

for $n = k - 2$

$$\sum_{i_1<\cdots<i_n}\left[S\left(M_{i_1}\right)+\cdots+S\left(M_{i_n}\right)\right]=\frac{(k-1)(k-2)}{2}\left[S\left(M_1\right)+\cdots+S\left(M_k\right)\right],$$

for $n = k - 1$

$$\sum_{i_1<\cdots<i_n}\left[S\left(M_{i_1}\right)+\cdots+S\left(M_{i_n}\right)\right]=(k-1)\left[S\left(M_1\right)+\cdots+S\left(M_k\right)\right],$$

for $n = k$

$$\sum_{i_1<\cdots<i_n}\left[S\left(M_{i_1}\right)+\cdots+S\left(M_{i_n}\right)\right]=S\left(M_1\right)+\cdots+S\left(M_k\right).$$

Thus,

$$\begin{aligned}
&\sum_{n=1}^{k}(-1)^{n+1}\sum_{i_i<\cdots<i_n}\left[S\left(M_{i_1}\right)+\cdots+S\left(M_{i_n}\right)\right] \\
&=S\left(M_1\right)+\cdots+S\left(M_k\right)-(k-1)\left[S\left(M_1\right)+\cdots+S\left(M_k\right)\right]+\binom{k-1}{2} \\
&\quad\left[S\left(M_1\right)\cdots+S\left(M_k\right)\right]+\cdots+(-1)^{k-1}\binom{k-1}{k-3}\left[S\left(M_1\right)+\cdots+S\left(M_k\right)\right] \\
&\quad+(-1)^{k}\binom{k-1}{k-2}\left[S\left(M_1\right)+\cdots+S\left(M_k\right)\right]+(-1)^{k+1}\left[S\left(M_1\right)+\cdots+S\left(M_k\right)\right] \\
&=(1-1)^{k-1}\left[S\left(M_1\right)+\cdots+S\left(M_k\right)\right] \\
&=0.
\end{aligned}$$

Therefore, we have

$$
\begin{aligned}
I_k\left(M_1,\cdots,M_k\right) &= \sum_{n=2}^{k}(-1)^n \sum_{i_1<\cdots<i_n} MI\left(M_{i_1},\cdots,M_{i_n}\right) \\
&\approx \frac{1}{2}\sum_{i<j} r_{ij}^2 - \frac{k-2}{2}\sum_{i<j} r_{ij}^2 - \frac{1}{2}\sum_{i_1<i_2<i_3} r_{i_1i_2i_3}^2 + \cdots + (-1)^{k-1}\frac{k-2}{2}\sum_{i<j} r_{ij}^2 \\
&\quad + (-1)^{k-1}\frac{k-3}{2}\sum_{i_1<i_2<i_3} r_{i_1i_2i_3}^2 + \cdots + (-1)^{k-1}\frac{1}{2}\sum_{i_1<\cdots+i_{k-1}} r_{i_1i_2\cdots i_{k-1}}^2 + (-1)^k\frac{1}{2}\sum_{i<j} r_{ij}^2 \\
&\quad + (-1)^k\frac{1}{2}\sum_{i_1<i_2<i_3} r_{i_1i_2i_3}^2 + \cdots + (-1)^k\frac{1}{2}\sum_{i_1<\cdots i_{k-1}} r_{i_1i_2\cdots i_{k-1}}^2 + (-1)^k\frac{1}{2} r_{1\cdots k}^2 \\
&= \frac{1}{2}(1-1)^{k-2}\sum_{i<j} r_{ij}^2 + \frac{1}{2}(1-1)^{k-3}\sum_{i_1<i_2<i_3} r_{i_1i_2i_3}^2 + \cdots + (1-1)\sum_{i_1<\cdots i_{k-1}} r_{i_1i_2\cdots i_{k-1}}^2 \\
&\quad + (-1)^k\frac{1}{2} r_{1\cdots k}^2 \\
&= (-1)^k\frac{1}{2} r_{1\cdots k}^2
\end{aligned}
$$

EXERCISES

Exercise 2.1 Consider two loci, A and B. At locus A, there are two alleles, A and a, with frequencies P_A and P_a, respectively. At locus B, there are two alleles, B and b, with frequencies P_B and P_b, respectively. Let P_{AB}, P_{Ab}, P_{aB}, and P_{ab} be the frequencies of the haplotype AB, Ab, aB, and ab, respectively. Define linkage disequilibrium coefficients

$$D_{AB} = P_{AB} - P_A P_B, \quad D_{Ab} = P_{Ab} - P_A P_b, \quad D_{aB} = P_{aB} - P_a P_B, \quad \text{and} \quad D_{ab} = P_{ab} - P_a P_b.$$

Show that $D_{AB} = D_{ab}$, $D_{Ab} = D_{aB} = -D_{AB}$.

Exercise 2.2 Show

$$\min\left[P_{A_i}P_{B_j}, \left(1-P_{A_i}\right)\left(1-P_{B_j}\right)\right] \quad \text{if } D_{D_1D_2} < 0,$$

$$\min\left[P_{A_i}\left(1-P_{B_j}\right), \left(1-P_{A_i}\right)P_{B_j}\right] \quad \text{if } D_{D_1D_2} > 0.$$

Exercise 2.3 Consider two loci, A and B, with haplotype frequencies:

$$P_{AB} = 0.1240, \quad P_{Ab} = 0.7834, \quad P_{aB} = 0.0011, \quad \text{and} \quad P_{ab} = 0.0915.$$

Calculate three LD measures: D, D', and r^2.

Exercise 2.4 Show

$$I(M_1; M_2) \approx \frac{1}{2} r^2$$

and

$$MI(M_1, M_2, M_3) \approx \frac{1}{2}\left(r_{123}^2 + r_{12}^2 + r_{13}^2 + r_{23}^2\right).$$

Exercise 2.5 Show

$$0 \le MI(M_1, \cdots, M_k) \le H(M_1) + \cdots + H(M_k).$$

Exercise 2.6 Show

$$(M_1, M_2; M_3) \ge 0.$$

Exercise 2.7 Show

$$I(M_1, M_2; M_3) = I(M_1; M_3) + I(M_2; M_3|M_1).$$

Exercise 2.8 Consider two loci, A and B, with haplotype frequencies

$$P_{AB} = 0.1240, \quad P_{Ab} = 0.7834, \quad P_{aB} = 0.0011, \quad \text{and} \quad P_{ab} = 0.0915.$$

Calculate the mutual information between two loci.

Exercise 2.9 Assume that X, Y, and $Z = [X, Y]^T$ follow normal distributions $N(0, \sigma_x^2), N(0, \sigma_y^2)$ and $N(0, \Sigma)$, where $\Sigma = \begin{bmatrix} \sigma_x^2 & \sigma_x\sigma_y\rho \\ \sigma_x\sigma_y\rho & \sigma_y^2 \end{bmatrix}$. Calculate the mutual information $I(X, Y)$.

CHAPTER 3

Association Studies for Qualitative Traits

3.1 POPULATION-BASED ASSOCIATION ANALYSIS FOR COMMON VARIANTS

3.1.1 Introduction

Most complex diseases including obesity, diabetes, cardiovascular disease, hypertension, asthma, bipolar, schizophrenia, Alzheimer's disease, cancer, and inflammatory disease are common diseases and hence pose great public health concerns (Collins 2004). Health states of individuals are a complex, multidimensional phenomenon. Clinical manifestations arise from integrated actions of multiple genetic and environmental factors, through an epigenetic and regulatory mechanism (Sing et al. 2003). Therefore, clinical phenotypes can be thought of as a synthesis of genes, gene–gene interactions, and gene–environment interactions (Carlson et al. 2004). A general disease model is represented in Figure 3.1. The general disease model assumes that multiple modules of phenotypes, a set of genes, and a set of environments contribute to the outcome of the disease. A module of phenotypes consists of a number of phenotypes, which are influenced by the genes and environments. The genes and environments will be classified into four categories: (1) the genes and environments directly influencing a phenotype, (2) the genes and environments influencing several phenotypes in a module of phenotypes, (3) the genes and environments simultaneously influencing several modules of phenotypes, and (4) the genes and environments directly influencing the outcome of the disease. Therefore, the genes and environments will have direct and indirect effects on the disease. The genes and environments that affect the disease through influencing the phenotypes in the modules will have only indirect effects on the disease. Therefore, the proposed disease model is a hierarchically organized network of phenotypes, genes, and environments. In this chapter, we will introduce statistical methods for both testing the association of genetic variants with each trait separately and with multiple traits simultaneously.

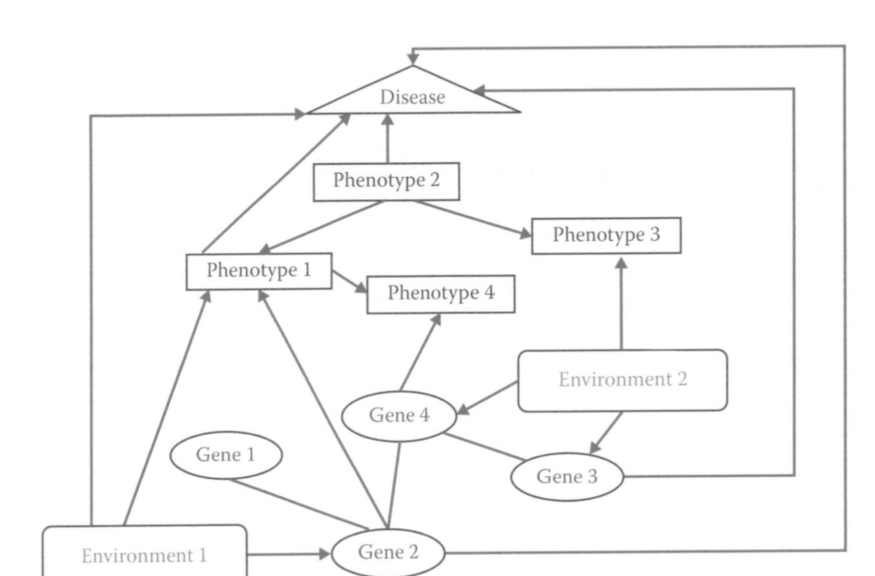

FIGURE 3.1 Scheme of a general disease model.

Resequencing of genomes will generate unprecedentedly high-dimensional genetic variation data that allow nearly complete evaluation of the genetic variation including dozens of millions of common (>5% population frequency), low-frequency (1%< and <5% population frequency), and rare variants (<1% population frequency) in the typical human genomes (http://www.1000genomes.org/). In the past decades, linkage analyses have been the primary method for genetic studies of diseases. Linkage analysis tests for the cosegregation of a genetic marker and a disease phenotype using family data. A significant linkage result implies that a marker and a susceptibility gene are genetically linked. Linkage analysis has been highly successful for many rare single-gene disorders (Jimenez-Sanchez et al. 2001).

However, the fact that many diseases are caused by multiple mutations and genes that individually contribute only modestly to disease risk limits the power of linkage studies. Furthermore, linkage analysis requires multiplex families with multiple affected relatives, which are not feasible for many occasions. An alternative method to linkage analysis for genetic studies of diseases is association studies, which examines the co-occurrence of a marker and disease at the population level and establish the functional and pathogenic significance of genetic variants (Brookes and Robinson 2015; Pritchard and Cox 2002; Risch and Merikangas 1996). Association analysis has higher power than linkage studies to detect small effects.

This chapter begins with an introduction of the Hardy–Weinberg equilibrium and genetic models that are the basis of underlying test statistics. Then, we study statistical methods for single marker and multimarker association analyses that are widely used to test the association of common variants with the diseases. Next-generation sequencing emerges as a major genotyping technique. The currently used statistical methods for testing the association of rare variants with the qualitative disease will be included in this chapter. Finally, statistical methods unifying both population-based and family-based genetic analysis are presented in this chapter.

3.1.2 The Hardy–Weinberg Equilibrium

The Hardy–Weinberg equilibrium predicts how gene frequencies will be transmitted from generation to generation given the assumptions of an infinitely large population, random mating, and absent outside evolutionary forces (immigration, mutation, and natural selection). Consider a locus with two alleles, A and a, that generate three genotypes AA, Aa, and aa. The frequencies of the genotypes AA, Aa, and aa are denoted by P, $2Q$, and R, respectively, with $P+2Q+R=1$. The transition of three genotypes to the offspring is shown in Table 3.1.

The frequencies of three genotypes AA, Aa, and aa in the offspring are, respectively, given by

$$\begin{aligned} P_1 &= P\times P\times 1+2\times P\times 2Q\times\frac{1}{2}+2Q\times 2Q\times\frac{1}{4}=P^2+2PQ+Q^2=(P+Q)^2, \\ 2Q_1 &= 2\times P\times 2Q\times\frac{1}{2}+2\times P\times R\times 1+2Q\times 2Q\times\frac{1}{2}+2\times 2Q\times R\times\frac{1}{2}=2(P+Q)(Q+R), \\ R_1 &= 2Q\times 2Q\times\frac{1}{4}+2\times 2Q\times R+R\times R\times 1=(Q+R)^2. \end{aligned} \tag{3.1}$$

If we require that the genotype frequencies be the same from one generation to the next, the following equalities must hold:

$$\begin{aligned} P &= P_1=(P+Q)^2, \\ 2Q &= 2Q_1=2(P+Q)(Q+R), \\ R &= R_1=(Q+R)^2, \end{aligned} \tag{3.2}$$

subject to $P+2Q+R=1=P_1+2Q_1+R_1$.

To satisfy the equality (3.1), we must have

$$(P+Q)(Q+R)=Q, \tag{3.3}$$

which implies that

$$PQ+PR+Q^2+QR=Q. \tag{3.4}$$

TABLE 3.1 Illustration of the Hardy–Weinberg Equilibrium

		Progeny		
Mating Type	**Frequency**	***AA***	***Aa***	***aa***
$AA\times AA$	$P\times P$	1		
$AA\times Aa$	$2\times P\times 2Q$	1/2	1/2	
$AA\times aa$	$2\times P\times R$		1	
$Aa\times Aa$	$2Q\times 2Q$	1/4	1/2	1/4
$Aa\times aa$	$2\times 2Q\times R$		1/2	1/2
$aa\times aa$	$R\times R$			1

Dividing both sides of the equality (3.4) by Q yields

$$P+\frac{PR}{Q}+Q+R=1,$$

or

$$\frac{PR}{Q}=1-P-Q-R=Q. \tag{3.5}$$

Multiplying both sides of the equality (3.5) by Q, we obtain

$$Q^2=PR. \tag{3.6}$$

Substituting Equation 3.6 into Equation 3.2 results in

$$\begin{aligned} P_1 &= P^2+2PQ+Q^2=P^2+2PQ+PR=P(P+2Q+R)=P, \\ 2Q_1 &= 2(PQ+PR+Q^2+QR)=2(PQ+Q^2+Q^2+QR)=2Q(P+2Q+R)=2Q, \\ R_1 &= Q^2+2QR+R^2=PR+2QR+R^2=R(P+2Q+R)=R, \end{aligned} \tag{3.7}$$

which show that the frequencies of genotypes under condition (3.6) will not change from generation to generation. This equilibrium is referred to as the Hardy–Weinberg equilibrium.

The **Hardy–Weinberg equilibrium** states that allele and genotype frequencies in a population will remain constant from generation to generation in the absence of other evolutionary influences.

Let $p=P+Q$ and $q=Q+R$, then have

$$P=p^2,\quad R=q^2,\quad 2Q=2pq \tag{3.8}$$

and

$$p+q=1. \tag{3.9}$$

We define p and q as frequencies of alleles A and a, respectively. For the convenience of discussion, we let $P=P(AA)$, $2Q=P(Aa)$, and $R=P(aa)$. Equation 3.8 can be written as

$$P(AA)=p^2,\quad P(Aa)=2pq,\quad P(aa)=q^2. \tag{3.10}$$

Therefore, under the Hardy–Weinberg equilibrium, the genotype frequencies satisfy the equality (3.10). When the genotype frequencies are available, we can estimate the allele frequencies as follows:

$$p=P(AA)+\frac{1}{2}P(Aa) \tag{3.11a}$$

and

$$q = P(aa) + \frac{1}{2} P(Aa). \tag{3.11b}$$

Deviation of genotype frequencies from their equilibrium is defined as the Hardy–Weinberg disequilibrium coefficient:

$$D_A = P(AA) - p^2. \tag{3.12}$$

We can show that

$$\begin{aligned} P(Aa) - 2pq &= -2D_A, \\ P(aa) - q^2 &= D_A. \end{aligned} \tag{3.13}$$

Evolutionary forces such as mutations, population substructure, and natural selection and genotype errors will change the Hardy–Weinberg equilibrium to disequilibrium. Deviations from the Hardy–Weinberg equilibrium (HWE) can be used to detect the presence of inbreeding, population stratification, and selection and genotype errors. In this section, we introduce the widely used goodness-of-fit χ^2 test and likelihood ratio test for HWE. Formally, we test for the null hypothesis:

$$H_0 : D = 0. \tag{3.14}$$

We first study the goodness-of-fit test. The observed and expected genotype frequencies under HWE are listed in Table 3.2. The goodness-of-fit χ^2 statistic is

$$\begin{aligned} \chi^2_{HW} &= \sum_{genotypes} \frac{(Observed - Expected)^2}{Expected} \\ &= \frac{\left(n_{AA} - n\hat{p}^2\right)}{n\hat{p}^2} + \frac{\left(n_{Aa} - 2n\hat{p}\hat{q}\right)^2}{2n\hat{p}\hat{q}} + \frac{\left(n_{aa} - n\hat{q}^2\right)}{n\hat{q}^2}, \end{aligned} \tag{3.15}$$

where

$$\hat{p} = \frac{2n_{AA} + n_{Aa}}{2n} = \frac{n_A}{2n}, n_A = 2n_{AA} + n_{Aa}$$

$$\hat{q} = \frac{2n_{aa} + n_{Aa}}{2n} = \frac{n_a}{2n}, n_a = 2n_{aa} + n_{Aa}$$

Under the null hypothesis of HWE $H_0 : D = 0$, χ^2_{HW} is distributed as the central $\chi^2_{(1)}$ distribution.

TABLE 3.2 Test for HWE

	AA	**Aa**	**aa**	**Total**
Observed	n_{AA}	n_{Aa}	n_{aa}	$n = n_{AA} + n_{Aa} + n_{aa}$
Expected	np^2	$2npq$	nq^2	

We observe that the expected values appear in the denominator of the test statistic; the small minor allele frequency will cause numerical instability. χ^2_{HW} will not work well for the rare variants. In general, likelihood ratio test provides a general framework for testing the null hypothesis. The likelihood ratio test can also be used to test for HWE. It is clear that the observed counts of genotypes follow a multinomial distribution. The likelihood function for the observed counts of genotypes is

$$L = \frac{n!}{n_{AA}!n_{Aa}!n_{aa}!} \frac{(n_{AA})^{n_{AA}} (n_{Aa})^{n_{Aa}} (n_{aa})^{n_{aa}}}{n^n}. \tag{3.16}$$

Under the null hypothesis of HWE, the likelihood function is

$$L_0 = \frac{n!}{n_{AA}!n_{Aa}!n_{aa}!} (\hat{p}^2)^{n_{AA}} (2\hat{p}\hat{q})^{n_{Aa}} (\hat{q}^2)^{n_{aa}}. \tag{3.17}$$

The likelihood ratio statistic for testing HWE is defined as

$$\begin{aligned} T_{HWE} &= -\log \frac{L_0}{L} \\ &= -\log \frac{(n_A)^{n_A} (n_a)^{n_a}}{2^{2n - n_{Aa}} n^n (n_{AA})^{n_{AA}} (n_{Aa})^{n_{Aa}} (n_{aa})^{n_{aa}}}. \end{aligned} \tag{3.18}$$

Under the null hypothesis of HWE, the test statistic T_{HWE} is asymptotically distributed as a central $\chi^2_{(1)}$ distribution.

3.1.3 Genetic Models

Genetic disease models specify the contributions of genotypes to phenotypes. We consider a marker, typically an SNP, with two alleles D and d having frequencies p and q, respectively. We denote the genotypes $G_0 = DD$, $G_1 = Dd$, and $G_2 = dd$. For the simplicity of discussion, we assume that disease is caused by a disease risk allele at one locus. The disease penetrance is associated with a given genotype. The penetrance is defined as the conditional probability of an individual being affected, given a certain genotype. Specifically, we define

$$f_0 = P(\text{being affected} | G_0),$$

$$f_1 = P(\text{being affected} | G_1),$$

and

$$f_2 = P(\text{being affected} | G_2).$$

Let K_p be the population prevalence of the disease. Under the assumption of HWE, we have

$$K_p = p^2 f_0 + 2pqf_1 + q^2 f_2. \quad (3.19)$$

Genotype relative risk is defined as

$$R_0 = \frac{f_0}{K_p},$$

$$R_1 = \frac{f_1}{K_p},$$

and

$$R_2 = \frac{f_2}{K_p}.$$

Disease genetic models imply a specific relationship between genotype and phenotype. It includes dominant, additive, recessive, and multiplicative models. Assuming a genetic penetrance parameter $\gamma(\gamma > 1)$ and baseline penetrance parameter f_0, four disease models are defined in Table 3.3. If either one or two copies of the risk allele D increases γ-fold for disease risk, the model is referred to as a dominant model; if the risk of disease increases γ-fold for the genotype DD and $\gamma/2$-fold for genotype Dd, the model is called an additive model; if the risk of disease increases γ-fold only for the genotype DD, the genetic model is referred to as a recessive model; and if the risk of the disease increases γ-fold with each additional disease allele D, the genetic model is referred to as a multiplicative model. The genetic model with complete penetrance is summarized in Table 3.4.

Penetrance and risk estimation can only be derived directly from prospective cohort studies. In association studies, we often use a case–control study design. Cases and controls are retrospectively sampled from disease and normal populations. In a case–control study, the genotype frequencies are calculated as follows. Let $P(G_j)$, $P^A(G_j)$, and $P^N(G_j)$, $j = 0, 1, 2$, be the frequencies of the genotype G_j in the general population, disease population (cases),

TABLE 3.3 Genetic Disease Models

	Penetrance		
Disease Model	***DD***	***Dd***	***dd***
Dominant	$f_0\gamma$	$f_0\gamma$	f_0
Additive	$f_0\gamma$	$1/2f_0\gamma$	f_0
Recessive	$f_0\gamma$	f_0	f_0
Multiplicative	$f_0\gamma^2$	$f_0\gamma$	f_0

TABLE 3.4 Genetic Disease Models with Complete Penetrance

	Penetrance		
Disease Model	***DD***	***Dd***	***dd***
Dominant	1	1	0
Additive	1	1/2	0
Recessive	1	0	0
Multiplicative	γ^2	γ	1

and normal population (controls), respectively. The frequencies of the genotypes in cases and controls are

$$\begin{aligned} P^A(G_j) &= P(G_j|\text{Cases}) = \frac{P(G_j, \text{Cases})}{K_p} = \frac{P(G_j)f_j}{K_p}, \\ P^N(G_j) &= P(G_j|\text{Controls}) = \frac{P(G_j, \text{Controls})}{(1-K_p)} = \frac{P(G_j)(1-f_j)}{(1-K_p)}. \end{aligned} \tag{3.20}$$

Example 3.1

Assume the HWE at the disease locus in the general population and that the disease allele frequency is $p=0.05$ and the penetrance is $f_0=0.50$, $f_1=0.12$, and $f_2=0.03$. The frequencies of the three genotypes in the general population are

$$\begin{aligned} P(G_0) &= p^2 = 0.0025, \\ P(G_1) &= 2pq = 0.095, \\ P(G_2) &= q^2 = 0.9025. \end{aligned}$$

Thus, the prevalence of the disease is

$$K_p = 0.50*0.0025 + 0.12*0.095 + 0.03*0.9025 = 0.0397.$$

The genotype frequencies in cases are

$$P^A(G_0) = \frac{0.0025*0.5}{0.0397} = 0.03147,$$

$$P^A(G_1) = \frac{0.095*0.12}{0.0397} = 0.2870,$$

and

$$P^A(G_2) = \frac{0.9025*0.03}{0.0397} = 0.6815.$$

The genotype frequencies in controls are

$$P^N(G_0) = \frac{0.0025 * 0.50}{1 - 0.0397} = 0.001302,$$

$$P^N(G_1) = \frac{0.095 * (1 - 0.12)}{1 - 0.0397} = 0.08028,$$

and

$$P^N(G_2) = \frac{0.9025 * (1 - 0.03)}{1 - 0.0397} = 0.9116.$$

This shows that genotype frequencies in the controls are close but not equal to that in the general population.

3.1.4 Odds Ratio

An odds ratio (OR) is a measure of association between a genetic variant and a disease in a case–control study (Pagano and Gauvreau 1993). In individuals with a risk factor, the odds of disease are defined as the conditional probability of being affected on the risk factor compared with the probability of being unaffected on the risk factor. In other words, the odds of disease are defined as

$$\frac{P(\text{disease} \mid \text{risk factor})}{P(\text{control} \mid \text{risk factor})}. \tag{3.21}$$

In individuals with no risk factor, the odds of disease are defined as the conditional probability of being affected compared with the probability of unaffected, given no risk factor:

$$\frac{P(\text{disease} \mid \text{no risk factor})}{P(\text{control} \mid \text{no risk factor})}. \tag{3.22}$$

The OR of interest is thus defined as

$$\text{OR} = \frac{\dfrac{P(\text{disease} \mid \text{risk factor})}{P(\text{control} \mid \text{risk factor})}}{\dfrac{P(\text{disease} \mid \text{no risk factor})}{P(\text{control} \mid \text{no risk factor})}}. \tag{3.23}$$

TABLE 3.5 A Contingency Table with Disease Status and a Risk Factor

	R (Risk Factor)	**$\bar{R}$ (No Risk Factor)**	**Total**
$D=1$ (cases)	a	b	$a+b$
$D=0$ (controls)	c	d	$c+d$
Total	$a+c$	$b+d$	n

Suppose that D represents a disease status. $D=1$ indicates a case and $D=0$ indicates a control. Let R be the presence of a risk factor and $\bar{R}$ be the absence of the risk factor. The OR in Equation 3.23 can be expressed as (Table 3.5)

$$\mathrm{OR}=\frac{\dfrac{P(D=1|R)}{P(D=0|R)}}{\dfrac{P(D=1|\bar{R})}{P(D=0|\bar{R})}}. \tag{3.24}$$

Suppose that the data are arranged in the form of a 2×2 contingency table. Then, it is clear that

$$\begin{aligned}\pi_1&=P(D=1|R)=\frac{a}{a+c},\\ 1-\pi_1&=P(D=0|R)=\frac{c}{a+c},\\ \pi_2&=P(D=1|\bar{R})=\frac{b}{b+d},\\ 1-\pi_2&=P(D=0|\bar{R})=\frac{d}{b+d}.\end{aligned} \tag{3.25}$$

Therefore, the OR is

$$\mathrm{OR}=\frac{\pi_1(1-\pi_2)}{\pi_2(1-\pi_1)}=\frac{ad}{bc}. \tag{3.26}$$

The OR can be estimated from the observed data. The inference problem for the OR is that the distribution of the OR is skewed to the right. To overcome this problem, the natural logarithm is used to transform the OR. Fortunately, the natural logarithm of the OR is asymptotically distributed as a normal. Let $\mu_{OR}=E[\log(\mathrm{OR})]$. Then, we have the following Theorem 3.1.

Theorem 3.1

The natural logarithm of the estimate $\hat{\mathrm{OR}}$ is asymptotically distributed as a normal distribution, $N\left(\mu_{OR},\frac{1}{a}+\frac{1}{b}+\frac{1}{c}+\frac{1}{d}\right)$.

Intuitively, we can see that (a, c) follows a binomial distribution with variance

$$\begin{aligned} \operatorname{var}(\hat{\pi}_1) &= \frac{\pi_1(1-\pi_1)}{a+c}, \\ \operatorname{var}(\hat{\pi}_2) &= \frac{\pi_2(1-\pi_2)}{b+d}. \end{aligned} \tag{3.27}$$

Let $h_1 = \log\frac{\pi_1}{1-\pi_1}$, $h_2 = \log\frac{\pi_2}{1-\pi_2}$, $h = h_1 - h_2 = \log(\mathrm{OR})$.

Since h is a nonlinear function of π, we need to use an approximate formula to calculate its variance. Recall from standard statistical theory and the Taylor expansion (Lehmann 1983) that

$$\operatorname{var}(f(\hat{x})) = [f'(\mu_x)]^2 \operatorname{var}(\hat{x}), \quad \text{where } \mu_x = E[\hat{x}]. \tag{3.28}$$

Applying formula (3.28) to function $\log(\hat{\pi})$, we have

$$\begin{aligned} \operatorname{var}(h_1) &= \operatorname{var}(\log(\hat{\pi}_1)) + (\log(1-\hat{\pi}_1)) \\ &= \left[\frac{1}{\pi_1} + \frac{1}{1-\pi_1}\right]^2 \operatorname{var}(\pi_1) \\ &= \frac{1}{[\pi_1(1-\pi_1)]^2} \frac{\pi_1(1-\pi_1)}{a+c} \\ &= \frac{1}{\pi_1(1-\pi_1)(a+c)} \\ &= \frac{1}{\frac{a}{a+c}\frac{c}{a+c}(a+c)} \\ &= \frac{1}{a} + \frac{1}{c}. \end{aligned} \tag{3.29}$$

Similarly, we have

$$\operatorname{var}(h_2) = \frac{1}{b} + \frac{1}{d}. \tag{3.30}$$

Since samples from cases and controls are independent, we have

$$\operatorname{var}(\log(\mathrm{OR})) = \operatorname{var}(h) = \operatorname{var}(h_1) + \operatorname{var}(h_2) = \frac{1}{a} + \frac{1}{b} + \frac{1}{c} + \frac{1}{d}. \tag{3.31}$$

Now it is easy to calculate $100(1-\alpha)\%$ confidence interval of the OR. It follows from Equation 3.31 that the standard deviation of log (OR) is

$$SE=\sqrt{\frac{1}{a}+\frac{1}{b}+\frac{1}{c}+\frac{1}{d}}.$$

Thus, the $100(1-\alpha)\%$ confidence interval of the log (OR) is

$$CI(\log(OR))=(\log(OR)-z_{1-\alpha/2}SE,\log(OR)+z_{1-\alpha/2}SE)$$

and

$$CI(OR)=OR\left(e^{-z_{1-\alpha/2}SE},e^{z_{1-\alpha/2}SE}\right). \tag{3.32}$$

Another quantity for measuring the effect of the risk factor on the disease is relative risk. Relative risk is defined as the ratio of the probability of disease among individuals exposed to the risk factor over the probability of disease among individuals unexposed to the risk factor:

$$RR=\frac{\pi_1}{\pi_2}=\frac{a(b+d)}{b(a+c)}. \tag{3.33}$$

Similarly to OR, we can also calculate the variance and confidence interval. The variance of the natural logarithm of relative risk RR is (Exercise 3.3)

$$\text{var}(\log RR)=\frac{c}{a(a+c)}+\frac{d}{b(b+d)}. \tag{3.34}$$

Define the standard deviation:

$$SE_{RR}=\sqrt{\frac{c}{a(a+c)}+\frac{d}{b(b+d)}}. \tag{3.35}$$

Then, the $100(1-\alpha)\%$ of relative risk RR is

$$CI(RR)=RR\left(e^{-z_{1-\alpha/2}SE_{RR}},e^{z_{1-\alpha/2}SE_{RR}}\right). \tag{3.36}$$

Example 3.2

Consider the data shown in Table 3.6. The OR is

$$OR=\frac{1021*612}{617*824}=1.229.$$

TABLE 3.6 A Contingency Table with Disease Status and Two Alleles

Allele	*A*	*a*	Total
Cases	1021	617	1638
Controls	824	612	1436
Total	1845	1229	3074

The variance of $\log(\hat{OR})$ is calculated using Equation 3.31:

$$\text{var}\left(\log(\hat{OR})\right) = \frac{1}{1021} + \frac{1}{671} + \frac{1}{824} + \frac{1}{612} = 0.005448.$$

Then, we obtain $SE = \sqrt{0.005448} = 0.0738$. A 95% confidence interval for the odds ratio is (1.0635, 1.4203).

3.1.5 Single Marker Association Analysis

Genome-wide association studies (GWAS) are emerging as a promising tool for genetic analysis of complex diseases (Korte and Farlow 2013). The traditional GWAS is a variant-by-variant analysis. We test the association of the markers with the disease one marker at a time. The primary assumption for association studies is that a mutation (a disease allele) increases disease susceptibility. Under this assumption, one expects that the disease allele will occur more frequently in the affected individuals (cases) than in the unaffected (controls). The standard χ^2 test for association studies is to identify the disease locus by comparing the differences in allele/haplotype frequencies between the affected and unaffected individuals. More precisely, the χ^2 statistic is a quadratic form of difference of allele/haplotype frequencies between the affected and unaffected individuals (Akey et al. 2001). Similar to the χ^2 test, a 2×2 contingency table is used to test for a pair of frequencies independent of disease status.

3.1.5.1 Contingency Tables

Genetic data in cases and controls can be organized into a contingency table (Table 3.7) to examine the association of the genetic variant with the disease. We first study a 2×2 table for the allele-based test. Then, a 2×2 table is extended to a 2×3 table for the genotype-based test.

TABLE 3.7 Contingency Table for Allele-Based Association Test in Case–Control Analysis

Allele	G	*g*	Total
Cases	a	b	$a + b$
Controls	c	d	$c + d$
Total	$a + c$	$b + d$	n

A two-dimensional contingency table generates counts for the joint distribution of two categorical variables: disease status and genetic variant. In our representation, the row variable indicates the disease status of individuals (case and control categories) and column variables indicates the alleles carried by individuals. The observed counts for the four mutually exclusive categories follow a multinomial distribution. The null hypothesis of no association of the allele with the disease is equivalent to the hypothesis of independence between the allele and disease, that is, H_0: Genetic allele and disease are independent.

Let $P_{11} = P(G, \text{cases})$, $P_{12} = P(g, \text{cases})$, $P_{21} = P(G, \text{controls})$, and $P_{22} = P(g, \text{controls})$.

The marginal probabilities are

$$P(\text{cases}) = \frac{a+b}{n}, \quad P(\text{controls}) = \frac{c+d}{n}, \quad P(G) = \frac{a+c}{n}, \quad \text{and} \quad P(g) = \frac{b+d}{n}.$$

Under the null hypothesis of independence between the allele and disease, the joint probabilities are reduced to

$$P_{11} = \frac{a+b}{n}\frac{a+c}{n}, \quad P_{12} = \frac{a+b}{n}\frac{b+d}{n}, \quad p_{21} = \frac{c+d}{n}\frac{a+c}{n}, \quad \text{and} \quad P_{22} = \frac{c+d}{n}\frac{b+d}{n}.$$

Since the joint distribution of the observed counts is a multinomial distribution, the expectation of counts is equal to

$$E[a] = \frac{(a+b)(a+c)}{n}, \quad E[b] = \frac{(a+b)(b+d)}{n},$$

$$E[c] = \frac{(c+d)(a+c)}{n}, \quad \text{and} \quad E[d] = \frac{(c+d)(b+d)}{n}.$$

The results are summarized in Table 3.8.

The test for association between alleles and disease in the contingency table is the χ^2 test. It compares the observed counts in each category, denoted by O, with the expected counts under the null hypothesis of no association, denoted by E. In other words, we calculate the sum

$$\chi^2 = \sum_{i=1}^{rc} \frac{(O_i - E_i)^2}{E_i}, \tag{3.37}$$

TABLE 3.8 Expected Counts

Allele	G	g	Total
Cases	$E_1 = (a+b)(a+c)/n$	$E_2 = (a+b)(b+d)/n$	$a+b$
Controls	$E_3 = (c+d)(a+c)/n$	$E_4 = (c+d)(b+d)/n$	$c+d$
Total	$a+c$	$b+d$	N

where r and c are the numbers of rows and columns, respectively. Under the null hypothesis of no association of alleles with the disease, the χ^2 test statistic is distributed as a central $\chi^2_{(r-1)(c-1)}$ distribution.

Example 3.3

Data are shown in Table 3.6. The expected counts are listed in Table 3.9. The value of χ^2 is 7.8149, and the P-value is less than 0.0052. A contingency table can also be used for testing the association of genotype with the disease. Consider a 2×3 contingency table for the genotype-based association test (Table 3.10). The data produced consist of six counts of the numbers of genotypes (GG, Gg, and gg) in cases and controls. The expected six counts are summarized in Table 3.11. The statistic defined in Equation 3.37 can still be used for testing the association of genotype with disease, but here $c=3$. The degrees of freedom of the test is now equal to $(r-1)(c-1)=(2-1)(3-1)=2$. Under the null hypothesis of no association, the statistic χ^2 is asymptotically distributed as a central $\chi^2_{(2)}$ distribution. Similar to using a contingency table for allele and genotype-based association tests, a $2\times c$ contingency table can be used to test the association of multiple alleles or haplotypes with the disease. Suppose that the number of alleles, genotypes, or haplotypes is c. The χ^2 statistic is then asymptotically distributed as a central $\chi^2_{(c-1)}$ distribution under the null hypothesis of no association.

TABLE 3.9 Expected Counts for the Data in Table 3.6

Allele	***G***	***g***
Cases	983.1197	654.8803
Controls	861.8803	574.1197

TABLE 3.10 Contingency Table for Genotype-Based Association Test in Case–Control Analysis

Allele	***GG***	***Gg***	***gg***	**Total**
Cases	25	20	15	60
Controls	10	20	35	65
Total	35	40	50	125

TABLE 3.11 Contingency Table for Genotype-Based Association Test in Case–Control Analysis

Allele	***GG***	***Gg***	***gg***	**Total**
Cases	$E_1=(a+b+c)(a+d)/n$	$E_2=(a+b+c)(b+e)/n$	$E_3=(a+b+c)(c+f)/n$	$a+b+c$
Controls	$E_4=(d+e+f)(a+d)/n$	$E_5=(d+e+f)(b+e)/n$	$E_6=(d+e+f)(c+f)/n$	$d+e+f$
Total	$a+d$	$b+e$	$c+f$	n

3.1.5.2 Fisher's Exact Test

Fisher's exact test is often used for sparse contingency tables where the minor allele frequency (MAF) or sample size is small. Consider Table 3.7 where we assume that the two margins $(a+b)$ and $(a+c)$ are fixed. Suppose that $a+c$ alleles of G are sampled. Among them, a samples are from cases. The total number of samples is n. The variables $(n, a, a+c)$ under the assumption of no association follow a hypergeometric distribution:

$$\frac{\binom{a+b}{a}\binom{c+d}{c}}{\binom{n}{a+c}} = \frac{(a+b)!(a+c)!(c+d)!(b+d)!}{n!a!b!c!d!}. \tag{3.38}$$

Define the indicator variable for the disease status:

$$X_i = \begin{cases} 1 & \text{cases} \\ 0 & \text{controls.} \end{cases}$$

Define the variable summarizing X_i:

$$Y = \sum_{i=1}^{(a+c)} X_i.$$

It is clear that $P(X_i = 1) = \frac{a+b}{n}$ and $P(X_i = 1, X_j = 1) = \frac{a+b}{n}\frac{a+b-1}{n-1}$.

Thus, we have

$$E(X_i) = \frac{a+b}{n}, \tag{3.39}$$

$$\operatorname{var}(X_i) = E[X_i] - (E[X_i])^2 = \frac{(a+b)(c+d)}{n^2}, \tag{3.40}$$

and

$$\begin{aligned} \operatorname{cov}(X_i, X_j) &= E[X_i X_j] - E[X_i]E[X_j] \\ &= \frac{(a+b)(a+b-1)}{n(n-1)} - \frac{(a+b)^2}{n^2} \\ &= -\frac{(a+b)(c+d)}{n^2(n-1)}. \end{aligned} \tag{3.41}$$

Combining Equations 3.39 through 3.41 yields the expectation and variance of the variable Y:

$$E[Y]=E[a]=\frac{(a+c)(a+b)}{n},$$
$$\operatorname{var}(Y)=\operatorname{var}(a)=\frac{(a+b)(c+d)(a+c)(b+d)}{n^2(n-1)}. \quad (3.42)$$

We observed that $E[a]$ calculated by the hypergeometric distribution is exactly the same as that calculated by the multinomial distribution. The P-value for the two-sided Fisher's exact test is calculated by summing all the probabilities that counts Y in the table are extreme or more extreme than the observed count y_0:

$$P=P(Y\le y_0|a+b,a+c,n)+P(Y>y_0|a+b,a+c,n)$$
$$=\sum_{Y\le y_0}P(Y=y|a+b,a+c,n)+\sum_{Y>y_0}P(Y=y|a+b,a+c,n). \quad (3.43)$$

Example 3.4

Table 3.12 shows the five possible subtables for the observed marginal totals, 6, 6, 4, and 8, and the observed data are subtable III in Table 3.12. For each subtable, we calculate its conditional probability, given the fixed marginal totals using Equation 3.38. The probability of each subtable is listed underneath. The subtables III and IV correspond to the extreme cases where in cases or controls the allele G is not observed (subtables with the red color in the first row). Therefore, the P-value of the test for association of the allele with the disease is the summation of the probability in III and IV:

$$P=0.0303+0.0303=0.0606.$$

3.1.5.3 *The Traditional χ^2 Test Statistic*

An alternative to the contingency table for association testing, the traditional χ^2 association test compares the differences in allele, or genotype, or haplotype frequencies between cases and controls. Some types of markers such as microsatellite markers at one locus have multiple alleles. Although haplotypes are spanned by markers at multiple loci, the statistical distributions of the haplotypes are the same as that of multiple alleles at one locus. Therefore, the statistical methods for testing the association of multiple alleles can be applied to testing the association of haplotypes. Let the number of alleles, or genotypes or haplotypes be m. Let $n^A=\left[n_1^A,\ldots,n_m^A\right]^T$ and $n^G=[n_1,\ldots,n_m]^T$ be vectors of the number of alleles, or genotypes, or haplotypes in affected and unaffected individuals, respectively. Let n_A and n_G be the number of sampled affected and unaffected individuals, respectively. Let $P^A=\left[P_1^A,\ldots,P_m^A\right]^T$ and

TABLE 3.12 Set of Five Tables with Marginal Totals 6, 6, 4, and 8

	Allele	*G*	*g*	Total
I	Cases	1	5	6
	Controls	3	3	6
	Total	4	8	12
		Pr = 0.2424		
	Allele	*G*	*g*	Total
II	Cases	2	4	6
	Controls	2	4	6
	Total	4	8	11
		Pr = 0.4545		
	Allele	*G*	*g*	Total
III	Cases	4	2	6
	Controls	0	6	6
	Total	4	8	12
		Pr = 0.0303		
	Allele	*G*	*g*	Total
IV	Cases	0	6	6
	Controls	4	2	6
	Total	4	8	12
		Pr = 0.0303		
	Allele	*G*	*g*	Total
V	Cases	3	3	6
	Controls	1	5	6
	Total	4	8	12
		Pr = 0.2424		

$P^G = [P_1, \ldots, P_m]^T$ be vectors of allele, or genotype, or haplotype frequencies in the affected and unaffected individuals, respectively. Define

$$\hat{P}_i^A = \frac{n_i^A}{n_A} \quad \text{and} \quad \hat{P}_i^G = \frac{n_i}{n_G}.$$

Let $P^A = \left[P_1^A, \ldots, P_m^A\right]^T$ and $\hat{P}^G = \left[\hat{P}_1, \ldots, \hat{P}_m\right]^T$. By the standard statistical theory (Lehmann 1983), we know that the vectors n^A and n^G follow multinomial distributions with the following variance–covariance matrix:

$$\Pi^A = n_A\left[\operatorname{diag}\left(P_1^A, \ldots, P_m^A\right) - P^A\left(P^A\right)^T\right]$$

and

$$\Pi = n_G\left[\operatorname{diag}\left(P_1, \ldots, P_m\right) - PP^T\right], \text{ respectively.}$$

Here $\mathrm{diag}\left(P_1^A,\ldots,P_m^A\right)$ and $\mathrm{diag}(P_1,\ldots,P_m)$ denote diagonal matrices with the diagonal elements $P_1^A,\ldots,P_m^A$ and $P_1,\ldots,P_m$, respectively. Let

$$\Sigma^A=\mathrm{diag}\left(P_1^A,\ldots,P_m^A\right)-P^A\left(P^A\right)^T \quad \text{and} \quad \Sigma=\mathrm{diag}\left(P_1,\ldots,P_m\right)-P^G\left(P^G\right)^T.$$

The allele or haplotype frequencies are asymptotically distributed as the following multivariate normal distribution:

$$N\left(P^A,\frac{1}{n_A}\Sigma^A\right) \quad \text{and} \quad N\left(P^G,\frac{1}{n_G}\Sigma\right), \text{ respectively.}$$

One form of the standard χ^2 statistic for case–control association studies is given by

$$T=\left(\hat{P}^A-\hat{P}^G\right)^T\Lambda^-\left(\hat{P}^A-\hat{P}^G\right), \tag{3.44}$$

where $\Lambda=\frac{1}{n_A}\Sigma^A+\frac{1}{n_G}\Sigma$ and Λ^- is a generalized inverse of the matrix Λ.

Under the null hypothesis of no association of the marker with the disease, T is asymptotically distributed as a central $\chi^2_{(m-1)}$ distribution.

If we ignore the terms $-P_i^2$ and $-P_iP_j(i,j=1,\ldots,m)$ in the elements of the matrix Σ, the variance–covariance matrix Σ is reduced to

$$\Sigma\approx\begin{bmatrix} P_1 & 0 & \cdots & 0 \\ 0 & P_2 & \cdots & 0 \\ \cdots & \cdots & \cdots & \cdots \\ 0 & 0 & \cdots & P_m \end{bmatrix}=\mathrm{diag}\left(P_i\right).$$

Similarly, we have $\Sigma^A\approx\mathrm{diag}\left(P_i^A\right)$ for the affected individuals. Thus, T can be reduced to

$$T=\sum_{i=1}^m\frac{\left(P_i-P_i^A\right)^2}{\frac{P_i}{2n_G}+\frac{P_i^A}{2n_A}}. \tag{3.45}$$

If we assume that the numbers of affected and unaffected individuals are equal, i.e., $n_A=n_G=n$, then the χ^2 test statistic T can be further reduced to

$$T=2n\sum_{i=1}^m\frac{\left(P_i-P_i^A\right)^2}{P_i+P_i^A}, \tag{3.46}$$

which is exactly the formula of the standard χ^2 test statistic in Chapman and Wijsman (1998).

Example 3.5

Data are shown in Table 3.6. The vector of allele frequencies in cases and controls are $\hat{P}^A = [0.6233, 0.3767]^T$ and $\hat{P}^G = [0.5738, 0.4262]^T$.

The variance–covariance matrices of the vector of allele frequencies in cases and controls are respectively given by

$$\frac{1}{n_A}\Sigma_A = \begin{bmatrix} 0.0001433 & -0.0001433 \\ -0.0001433 & 0.0001433 \end{bmatrix},$$

$$\frac{1}{n_G}\Sigma = \begin{bmatrix} 0.0001703 & -0.0001703 \\ -0.0001703 & 0.00017303 \end{bmatrix},$$

$$\Lambda = \begin{bmatrix} 0.0001433 & -0.0001433 \\ -0.0001433 & 0.0001433 \end{bmatrix} + \begin{bmatrix} 0.0001703 & -0.0001703 \\ -0.0001703 & 0.00017303 \end{bmatrix}$$

$$= \begin{bmatrix} 0.000336 & -0.0003136 \\ -0.0003136 & 0.0003136 \end{bmatrix}.$$

One can easily see that the covariance matrix Λ is singular. The test statistic T can be expressed as

$$T = \frac{(0.6233 - 0.5738)^2}{0.000336} = 7.8139.$$

Thus, the P-value of the association test statistic T is

$$P - \text{value} = 0.0052,$$

which is the same as that calculated by the contingency table in Example 3.3.

3.1.6 Multimarker Association Analysis

Although most researchers acknowledge that genetic variation provides valuable information for the diagnosis, prevention, and treatment of complex diseases, there is no universally accepted consensus on how genetic variation contributes to the cause of complex disease. Two different basic views on the genetic architecture of complex diseases lead to two different strategies for analyzing complex diseases.

The popular view on the mechanisms of the common diseases is to assume that a single marker acts independently and can explain the pathogenesis of the disease. A widely used strategy for unraveling the genetic structure of a common disease is single-locus analysis, testing the association of a single variant one by one. The strategies focusing on the single marker with large marginal effects on disease risk have resulted in only limited success in genetic studies of complex diseases (Hartwell 2004).

The single marker paradigm for genetic analysis, which has proven successful in dissecting genetic structures of Mendelian diseases, may not lead to success in genetic studies of complex diseases.

An alternative approach is joint association analysis of multiple markers. An increasing number of genetic association studies demonstrate the limitations of the attempt to explain phenotype variation by a single variant. Accumulated evidence suggests that complex diseases are due to multiple correlated genetic variants. There are an increasing number of researchers who advocate taking a systems-level approach to complex diseases. The new concept concerning complex disease is to assume that the development of disease should be considered as a dynamic process with joint effects of multiple loci. Consequently, the genetic effects on the phenotype can be observed only when multiple mutations hit the biological processes. Uncovering association of multiple loci and their nonadditive relationships with disease susceptibility require developing novel statistics, which aggregate information across multiple loci and jointly test their association with the disease. We first introduce the Hotelling T^2 statistic to test the association of multiple common variants with the disease. The methods for testing the association of multiple rare variants will be studied in the next sections.

3.1.6.1 Generalized T^2 Test Statistic

Consider a design in which n_A cases from an affected population and n_G control subjects from a comparable unaffected population are sampled. Suppose that there are k markers typed in the samples. The jth marker has alleles B_j and b_j with population frequencies P_{B_j} and P_{b_j}, respectively. Define an indicator variable for the genotype of the jth marker for the ith individual from the affected population:

$$X_{ij} = \begin{cases} 1 & B_jB_j \\ 0 & B_jb_j \\ -1 & b_jb_j \end{cases}.$$

Similarly, we define an indicator variable, Y_{ij}, for an individual from the unaffected population. For each individual, we define a vector of genotype profiles X_i and Y_i, respectively, in cases and controls: $X_i = (X_{i1},\ldots,X_{ik})^T$, $Y_i = (Y_{i1},\ldots,Y_{ik})^T$.

For each marker, in cases and controls, we calculate the mean values $\bar{X}_j$ and $\bar{Y}_j$ of the genotype indicator variables in cases and controls:

$$\bar{X}_j = \frac{1}{n_A}\sum_{i=1}^{n_A} X_{ij}, \quad \bar{Y}_j = \frac{1}{n_G}\sum_{i=1}^{n_G} Y_{ij}.$$

Assembling all the mean values $\bar{X}_j$ and $\bar{Y}_j, j = 1,\ldots,k$ into vectors $\bar{X}$ and $\bar{Y}$, $\bar{X} = \left(\bar{X}_1,\ldots,\bar{X}_k\right)^T$, $\bar{Y} = \left(\bar{Y}_1,\ldots,\bar{Y}_k\right)^T$.

We assume that the covariance matrices of the genotype profiles in cases and controls are equal. The pooled sample variance–covariance matrix of the indicator variables for the marker genotypes is defined as

$$S = \frac{1}{n_A + n_G - 2}\left[\sum_{i=1}^{n_A}(X_i - \bar{X})(X_i - \bar{X})^T + \sum_{i=1}^{n_G}(Y_i - \bar{Y})(Y_i - \bar{Y})^T\right].$$

Hotelling's (1931) T^2 statistic is then defined as

$$T^2 = \frac{1}{\frac{1}{n_A} + \frac{1}{n_G}}(\bar{X} - \bar{Y})^T S^{-1}(\bar{X} - \bar{Y}). \tag{3.47}$$

Under the null hypothesis of no association of markers with the disease, the statistic T^2 is asymptotically distributed as a central $\chi^2_{(k)}$ distribution.

Example 3.6

Consider two SNPs: rs3094315 and rs2073813. Suppose that 101 individuals from cases and 111 individuals from controls are sampled. The mean vectors of the genotype profiles in cases and controls are

$$\bar{X} = \begin{bmatrix} 0.4455 \\ 0.3960 \end{bmatrix}, \quad \bar{Y} = \begin{bmatrix} 0.3514 \\ 0.2883 \end{bmatrix}.$$

The pooled sample covariance matrix is

$$S = \begin{bmatrix} 0.3249 & 0.2425 \\ 0.2425 & 0.2902 \end{bmatrix}.$$

Its inverse is

$$S^{-1} = \begin{bmatrix} 8.1793 & -6.8349 \\ -6.8349 & 9.1573 \end{bmatrix}.$$

The value of the T^2 statistic is $T^2 = 2.1234$ and the P-value is 0.3459.

Two SNPs are not jointly associated with the disease.

3.1.6.2 The Relationship between the Generalized T^2 Test and Fisher's Discriminant Analysis

The generalized T^2 test can be derived from Fisher's linear discriminant analysis. Consider two populations (cases and controls), which are to be separated. Let x be a vector of observations (observed genotype profiles) and $y = a^T x$ be a linear combination of the

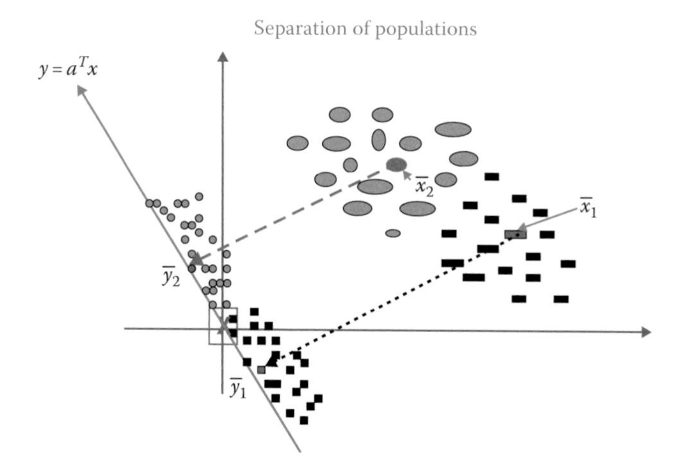

FIGURE 3.2 Scheme for linear discriminant analysis.

observations (Figure 3.2). Let $\bar{x}_1$ and $\bar{x}_2$ be means of the observations in populations 1 and 2, respectively. Denote the mean difference between cases and controls, $d = \bar{x}_1 - \bar{x}_2$. Let $\bar{y}_1 = a^T\bar{x}_1$, $\bar{y}_2 = a^T\bar{x}_2$.

The square of the distance between two points in the lines $y = a^T x$ projected by the original means of observations in cases and controls is

$$\begin{aligned}(\bar{y}_1 - \bar{y}_2)^2 &= a^T(\bar{x}_1 - \bar{x}_2)(\bar{x}_1 - \bar{x}_2)^T a \\ &= a^T dd^T a.\end{aligned}$$

We assume that the covariance matrices of observations in populations 1 and 2 are equal and denoted by Σ. Then, the variance of $\bar{y}_1 - \bar{y}_2$ is given by

$$\begin{aligned}\operatorname{var}(\bar{y}_1 - \bar{y}_2) &= a^T \operatorname{cov}(\bar{x}_1 - \bar{x}_2)a \\ &= a^T\left(\frac{1}{n_A}\Sigma + \frac{1}{n_G}\Sigma\right)a = \frac{n_A + n_G}{n_A n_G}a^T\Sigma a.\end{aligned}$$

The estimate of variance of $\bar{y}_1 - \bar{y}_2$ is given by $\frac{n_A + n_G}{n_A n_G}a^T Sa$, where S is the pooled estimate of the covariance matrix Σ. Our goal is to select the vector a achieving separation of the sample means with $\bar{y}_1$ and $\bar{y}_2$ as large as possible. To reach this goal, we maximize the following ratio:

$$\max \frac{|\bar{y}_1 - \bar{y}_2|^2}{\operatorname{var}(\bar{y}_1 - \bar{y}_2)} = \frac{a^T dd^T a}{a^T Sa}. \tag{3.48}$$

Since a constant will not affect the results of maximization in Equation 3.48, $\frac{n_1 + n_2}{n_1 n_2}a^T Sa$ is replaced by $a^T Sa$.

To solve the optimization problem (3.48), we pose the following constraint to normalize the data:

$$a^T Sa = 1.$$

Therefore, the optimization (3.48) can be reduced to the following optimization problem:

$$\begin{aligned} \max \quad & a^T dd^T a \\ \text{s.t.} \quad & a^T Sa = 1. \end{aligned} \tag{3.49}$$

The constrained optimization problem (3.49) can be reduced to the nonconstrained optimization problem by the Lagrange multiplier method:

$$\max \quad F = a^T dd^T a + \lambda\left(1 - a^T Sa\right).$$

Taking the derivative of F with respect to a yields

$$dd^T a - \lambda Sa = 0. \tag{3.50}$$

Multiplying a^T on both sides of the equation, we have

$$\lambda = \left(a^T d\right)^2.$$

Thus, $a = \frac{1}{a^T d} S^{-1} d$, which implies that $a^T dd^T a = d^T S^{-1} d$. Therefore, $y = \left(\bar{x}_1 - \bar{x}_2\right)^T S^{-1} x$.

The distance between two populations is defined as

$$D^2 = \frac{1}{\frac{1}{n_A} + \frac{1}{n_G}} \left(\bar{x}_1 - \bar{x}_2\right)^T S^{-1} \left(\bar{x}_1 - \bar{x}_2\right). \tag{3.51}$$

Replacing $\bar{x}_1$ by $\bar{X}$ and $\bar{x}_2$ by $\bar{Y}$ in Equation 3.51, the square of distance between cases and controls is exactly the same as the T^2 statistic.

3.2 POPULATION-BASED MULTIVARIATE ASSOCIATION ANALYSIS FOR NEXT-GENERATION SEQUENCING

Resequencing of exomes and whole genomes will generate unprecedentedly massive, high-dimensional genetic variation data that allows nearly a complete evaluation of the genetic variation including tens of millions of common (>5% population frequency), low-frequency (1%< the variance and <5% population frequency), and rare variants (<1% population frequency) in the typical human genome and provides a powerful tool to comprehensively catalogue human genetic variation. Despite their promise, next-generation sequencing (NGS) technologies suffer from three remarkable limitations: high error rates, enrichment of rare variants, and large proportion of missing values. Since an individual rare variant would have a relatively small impact on the common disease and the rare variants have very low frequencies in the populations, the power of the traditional variant-by-variant analytical tools that are mainly designed for the purpose of detecting common variants for testing the association of rare variants with disease will be limited. Developing new analytical tools for the analysis of the massive sequencing data poses a novel and great challenge to statistical analysis (Bacanu et al. 2011).

Genetic studies of complex diseases are undergoing a paradigm shift from the single market analysis to the joint analysis of multiple variants in a genomic region that can be genes or other functional units, which are often referred to as gene or genome region-based association analysis. The principle behind gene-based association analysis is to aggregate information across multiple variants in an analysis unit in which all rare variants are collapsed and treated as a single variable for analysis (Luo et al. 2011; Pan et al. 2014). Two types of statistical methods based on multivariate and functional analysis for joint testing the association of a group of variants with disease are developed. In the past several years, various versions of group association analysis for NGS data have been developed. It is difficult to cover all these methods. We will mainly focus on the basic ideas underlying gene-based association tests with NGS data.

3.2.1 Multivariate Group Tests

3.2.1.1 Collapsing Method

An essential problem for rare variant association analysis is that frequencies of rare variants are low. The low frequencies of the variants will decrease the power and inflate the type 1 errors. To overcome the limitation due to the low frequency of the rare variants, we collapse all the rare variants across the gene or genomic region into a single variable (Li and Leal 2008). Assume that n_A and n_G individuals from cases and controls, respectively, are sampled. We define an indicator variable for the individual in cases as

$$x_i = \begin{cases} 1 & \text{presence of rare variants} \\ 0 & \text{otherwise} \end{cases}, \quad i = 1, \ldots, n_A. \tag{3.52}$$

The indicator variable y_i can be similarly defined for the controls. The data can be assembled into $2 \times t$ tables. The contingency table analysis, exact test, and χ^2 test discussed in Sections 3.1.5.1 through 3.1.5.3 can be used to define the collapsed variable. For example, in the χ^2 test, we calculate the average number of individuals carrying at least one rare variant in the gene in cases and controls:

$$\bar{X} = \frac{1}{n_A}\sum_{i=1}^{n_A} x_i \quad \text{and} \quad \bar{Y} = \frac{1}{n_G}\sum_{i=1}^{n_G} y_i.$$

Define the test statistic:

$$T_{col} = \frac{\left(\bar{X} - \bar{Y}\right)^2}{\frac{S}{n_A} + \frac{S}{n_G}}, \tag{3.53}$$

$$S = \frac{1}{n_A + n_G - 2}\left[\sum_{i=1}^{n_A}\left(x_i - \bar{X}\right)^2 + \sum_{i=1}^{n_G}\left(y_i - \bar{Y}\right)^2\right].$$

Under the null hypothesis of no association of a set of rare variants with the diseases, the statistic T_{col} is asymptotically distributed as a central $\chi^2_{(1)}$ distribution.

Example 3.7

As an illustration, the collapse method is applied to the published resequencing dataset of ANGPTL4 in the Dallas Heart Study (Romeo et al. 2007). A total of 93 sequence variations were identified from 3553 samples. The individuals whose plasma triglyceride levels were less than or equal to the 25th percentile are classified as the low triglyceride group, and the individuals whose plasma triglyceride were greater than or equal to the 75th percentile are grouped as the high triglyceride group. We can also similarly select groups of high levels and low levels of high-density lipoprotein cholesterol (HDL), cholesterol, low-density lipoprotein (LDL), and very low–density lipoprotein (VLDL). The variants with frequencies less than or equal to 3% were defined as a rare variant for the easier comparison with the results in Romeo et al. (2007). *P*-values for testing association of rare variants in the gene ANGPTL4 with triglyceride, cholesterol, HDL, LDL, and VLDL in the Dallas Heart Study are 3.82×10^{-6}, 0.9640, 3.59×10^{-5}, and $0.3210, 2.01 \times 10^{-5}$, respectively.

3.2.1.2 Combined Multivariate and Collapsing Method

The combined multivariate and collapsing (CMC) method (Li and Leal 2008) partitions the variants into common and rare variants in a gene or genomic region. Then, all the rare variants are clustered into several groups. The rare variants in each group are collapsed into a single variable that is taken as a single pseudoindependent site. Finally, the generalized T^2 statistic is applied to all common variants and all pseudoindependent sites (rare variant sites). Suppose that there are m pseudoindependent sites and k common variants. In cases, indicator variables for each pseudoindependent site are defined in Section 3.2.1.1 and denoted by x_{ij}, $j = 1, \ldots, m$, and indicator variables for the genotypes at the common variant loci are denoted by x_{ij}, $j = m+1, \ldots, m+k$. Similarly, indicator variables y_{ij}, $j = 1, \ldots, m+k$ are defined for controls. Define two vectors of indicator variables, respectively, for cases and controls:

$$X_i = \left[x_{i1}, \ldots, x_{im}, x_{im+1}, \ldots, x_{im+k}\right]^T$$

and

$$Y_i = \left[y_{i1}, \ldots, y_{im}, y_{im+1}, \ldots, y_{im+k}\right]^T.$$

Define

$$\bar{X}_j = \frac{1}{n_A}\sum_{i=1}^{n_A} X_{ij}, \quad \bar{Y}_j = \frac{1}{n_G}\sum_{i=1}^{n_G} Y_{ij},$$

$$\bar{X} = \left(\bar{X}_1, \ldots, \bar{X}_{m+k}\right)^T, \quad \bar{Y} = \left(\bar{Y}_1, \ldots, \bar{Y}_{m+k}\right)^T,$$

and

$$S = \frac{1}{n_A + n_G - 2}\left[\sum_{i=1}^{n_A}\left(X_i - \bar{X}\right)\left(X_i - \bar{X}\right)^T + \sum_{i=1}^{n_G}\left(Y_i - \bar{Y}\right)\left(Y_i - \bar{Y}\right)^T\right].$$

Hotelling's (1931) T^2 statistic defined in (3.47)

$$T^2 = \frac{1}{\frac{1}{n_A} + \frac{1}{n_G}}\left(\bar{X} - \bar{Y}\right)^T S^{-1}\left(\bar{X} - \bar{Y}\right) \tag{3.54}$$

can be used to test the association of gene with disease. Under the null hypothesis of no association of any variant in the gene (genomic region) with the disease, T^2 is asymptotically distributed as a central $\chi^2_{(m+k)}$ distribution.

3.2.1.3 Weighted Sum Method

A serious drawback of the CMC method is that the outcome of the test depends on the selection of threshold. How to determine a threshold on the frequencies for selection of rare variants is a difficult problem. Consider k loci. Let n be the total number of sampled individuals (affected and unaffected). We often observe that the frequency of mutation is low, and hence the mutation is a minor allele. Therefore, for the convenience of discussion, we assume that the mutant allele is a minor allele. For each variant i, the frequency of the minor allele is estimated as

$$q_i = \frac{m_i^U + 1}{2\left(n^U + 1\right)},$$

where

m_i^U is the number of minor alleles observed for variant i in controls
n^U is the number of sampled individuals in the controls

To avoid zero in the estimate of frequency of the minor allele, we add one to the numerator and two to the denominator of the frequency estimate. The standard deviation of the estimate of the frequency of the minor allele is

$$w_i = \sqrt{nq_i(1-q_i)}. \tag{3.55}$$

Let I_{ij} be the number of minor alleles in the variant i for individual j. Summation of I_{ij} weighted by the standard deviation w_i of the frequency of minor allele is defined as the risk score of individual j, which contributes to the risk of disease:

$$r_j = \sum_{i=1}^{k} \frac{I_{ij}}{w_i}, \quad j = 1, \ldots, n. \tag{3.56}$$

All r_j are ranked. Similar to the Wilcoxon test, we summarize all ranks of the individuals in cases:

$$x = \sum_{j \in A} \text{rank}(r_j). \tag{3.57}$$

To find the mean and standard variation of x, the affected/unaffected (case/control) is permutated among the individuals, and calculations in Equations 3.54 through 3.57 are repeated l times to produce $x_1^*, \ldots, x_l^*$ under the null hypothesis of no association. Calculate mean and standard deviation of the rank summation x in cases:

$$\hat{\mu} = \sum_{j=1}^{l} x_j^* \quad \text{and} \quad \hat{\sigma} = \sqrt{\sum_{j=1}^{l} (x_j^* - \mu)^2}.$$

The z score is defined as

$$z = \frac{x - \hat{\mu}}{\hat{\sigma}}. \tag{3.58}$$

Under the null hypothesis of no association of any variant in the gene with the disease, the z score is asymptotically distributed as a standard normal distribution, $N(0, 1)$.

3.2.2 Score Tests and Logistic Regression

3.2.2.1 Score Function

Various rare variant association tests are based on the logistic regression model. Test statistics are the score tests or the modified score tests (Lee et al. 2014). Logistic regression models assume that n individuals are sampled. Consider p genetic variant sites in a genomic region and m covariates including intercept, age, environment variables, race, and top principal components of genetic variation for controlling population structure. Let G_{ij} be an indicator variable for the genotype at the jth variant site of the ith individual, defined in the previous sections, and X_{ij} be the jth covariate of the ith individual. Define the covariate matrix and genotype matrix:

$$X = \begin{bmatrix} X_{11} & \cdots & X_{1m} \\ \vdots & \ddots & \vdots \\ X_{n1} & \cdots & X_{nm} \end{bmatrix} \text{ and } G = \begin{bmatrix} G_{11} & \cdots & G_{1p} \\ \vdots & \ddots & \vdots \\ G_{n1} & \cdots & G_{np} \end{bmatrix}.$$

Consider a dichotomous phenotype (binary trait); let y_i denote the phenotype of the ith individual and $Y = [y_1, \ldots, y_n]^T$ be a vector of phenotypes. The logistic model for association analysis is

$$\log \frac{P(y_i = 1)}{1 - P(y_i = 1)} = \log \frac{\pi_i}{1 - \pi_i} = \alpha_0 + X_i \alpha + G_i \beta, \tag{3.59a}$$

where $\pi_i = P(y_i = 1)$.

Let

$$H_i = [1 \quad X_i], \quad Z_i = [H_i \quad G_i],$$

$$H = \begin{bmatrix} H_1 \\ \vdots \\ H_n \end{bmatrix}, \quad Z = \begin{bmatrix} Z_1 \\ \vdots \\ Z_n \end{bmatrix}, \quad \xi = \begin{bmatrix} \alpha_0 \\ \alpha \end{bmatrix}, \text{ and } \gamma = \begin{bmatrix} \xi \\ \beta \end{bmatrix} = \begin{bmatrix} \alpha_0 \\ \alpha \\ \beta \end{bmatrix}.$$

Equation 3.59a can also be written as

$$\log \frac{P(y_i = 1)}{1 - P(y_i = 1)} = \log \frac{\pi_i}{1 - \pi_i} = H_i \xi + G_i \beta = Z_i \gamma. \tag{3.59b}$$

Suppose that the responses $y_1, \ldots, y_n$ are observed values of independent random variables $Y_1, \ldots, Y_n$ following a binomial distribution with index 1 and parameter π_i. The likelihood function is $L(\pi, y) = \prod_{i=1}^{n} \pi_i^{y_i} (1 - \pi_i)^{1 - y_i}$, where the constant function of y not involving π has been omitted because it does not contradict to the estimation and inference. The log-likelihood is

$$l(\pi, y) = \sum_{i=1}^{n} \left[y_i \log \frac{\pi_i}{1 - \pi_i} + \log(1 - \pi_i) \right]. \tag{3.60}$$

Define the score function

$$U = \begin{bmatrix} U_1 \\ U_2 \end{bmatrix} = \begin{bmatrix} \dfrac{\partial l}{\partial \xi} \\ \dfrac{\partial l}{\partial \beta} \end{bmatrix}$$

and Fisher information matrix for γ

$$I=-E\left(\frac{\partial^2 l}{\partial\gamma\partial\gamma^T}\right)=\begin{bmatrix}-E\left(\frac{\partial^2 l}{\partial\xi\partial\xi^T}\right) & -E\left(\frac{\partial^2 l}{\partial\xi\partial\beta^T}\right)\\ -E\left(\frac{\partial^2 l}{\partial\beta\partial\xi^T}\right) & -E\left(\frac{\partial^2 l}{\partial\beta\partial\beta^T}\right)\end{bmatrix}=\begin{bmatrix}I_{\xi\xi} & I_{\xi\beta}\\ I_{\beta\xi} & I_{\beta\beta}\end{bmatrix}.$$

Under the null hypothesis $H_0:\gamma=0$, π_i is given by $\pi_i=\frac{1}{1+1}=\frac{1}{2}$.

Therefore, the weigh matrix is

$$W=\text{diag}\left(\frac{1}{2},\ldots,\frac{1}{2}\right). \tag{3.61}$$

We can show that the score function for β is given by

$$U_2=\sum_{i=1}^{n}(y_i-\pi_i)G_i^T \tag{3.62}$$

and the Fisher information matrix for γ is (Appendix 3A)

$$\begin{aligned}I&=\begin{bmatrix}H^TWH & H^TWG\\ G^TWH & G^TWG\end{bmatrix}\\ &=Z^TWZ.\end{aligned} \tag{3.63}$$

3.2.2.2 Score Tests

Since under the null hypothesis $H_0:\gamma=0$, we have $E(U)=0$. The vector score U is a sum of terms corresponding to individual observations and hence is asymptotically normal with mean zero and covariance matrix I. We are interested in testing the association of genetic variation with the disease. Therefore, the null hypothesis of interest is

$$H_0:\beta=0,$$

and the alternative hypothesis is

$$H_a:\beta\neq 0.$$

We consider two scenarios for defining test statistics:

1. *Scenario 1*: If the nuisance vector ξ is known, then the score test statistics of H_0 is defined as

$$T_{score1}=U_2^TI_{\beta\beta}^{-1}U_2. \tag{3.64}$$

Under the null hypothesis $H_0:\beta=0$, the score test statistic T_{score1} is asymptotically distributed as a central $\chi^2_{(p)}$ distribution.

2. *Scenario 2*: If the nuisance vector ξ is unknown and should be estimated, setting is equivalent to setting the score function $U_1=0$. Given $U_1=0$, the conditional distribution of U_2 is normal with mean zero and the covariance matrix is (Exercise 3.12)

$$I_{2.1}=\operatorname{cov}\left(U_2 \mid U_1=0\right)=I_{\beta\beta}-I_{\beta\xi}I_{\xi\xi}^{-1}I_{\xi\beta}. \tag{3.65}$$

The score statistic for testing association of the genetic variants, given covariates are estimated from data, is defined as

$$T_{score2}=U_2^T I_{2.1}^{-1}U_2. \tag{3.66}$$

Again, under the null hypothesis $H_0:\beta=0$, the score test statistic T_{score2} is asymptotically distributed as a central $\chi^2_{(p)}$ distribution.

3.2.3 Application of Score Tests for Association of Rare Variants

Two methods, weighted function methods and the adaptive association test, can be considered as applications of the logistic regression model and score tests to association analysis of rare variants. We first introduce weighted function methods.

3.2.3.1 Weighted Function Method

Lin and Tang (2011) used the score test as a general framework for developing methods to detect the association of rare variants with disease. Since degrees of freedom of the score test statistics that depend on the number of genetic variants in the genomic region are often large, to improve the power of the tests, we need to reduce the degree of freedom of the test. One way to do this is to project high-dimensional data to one-dimensional space by a linear transformation, $\beta=\tau\phi$, where τ is a scalar constant and ϕ is a weight function. Now the null hypothesis $H_0:\beta=0$ is transformed to $H_0:\tau=0$. Equation 3.59a is reduced to

$$\begin{aligned}\eta_i&=\log\frac{P(y_i=1)}{1-P(y_i=1)}=\log\frac{\pi_i}{1-\pi_i}=\alpha_0+X_i\alpha+G_i\phi\tau\\&=\alpha_0+X_i\alpha+S_i\tau,\end{aligned} \tag{3.67}$$

where $S_i=G_i\phi$ is a $p\times 1$ vector of weights.

The score function is then reduced to

$$U_2=\sum_{i=1}^{n}(y_i-\pi_i)S_i. \tag{3.68}$$

Under the null hypothesis $H_0 : \tau = 0$, π_i is

$$\pi_i = \frac{e^{H_i \xi}}{1 + e^{H_i \xi}}, \tag{3.69}$$

which implies

$$v_i = \pi_i \left(1 - \pi_i\right) = \frac{e^{H_i \xi}}{\left(1 + e^{H_i \xi}\right)^2}. \tag{3.70}$$

Recall that the Fisher information matrix is given by

$$\begin{aligned} I_{\xi\xi} &= H^T W H \\ &= \begin{bmatrix} H_1^T & \cdots & H_n^T \end{bmatrix} \begin{bmatrix} v_1 & \cdots & 0 \\ \vdots & \ddots & \vdots \\ 0 & \cdots & v_n \end{bmatrix} \begin{bmatrix} H_1 \\ \vdots \\ H_n \end{bmatrix} \\ &= \sum_{i=1}^{n} v_i H_i^T H_i, \end{aligned} \tag{3.71}$$

$$\begin{aligned} I_{\xi\beta} &= H^T W S \\ &= \left(\sum_{i=1}^{n} v_i S_i H_i \right)^T, \end{aligned} \tag{3.72}$$

$$\begin{aligned} I_{\beta\xi} &= I_{\xi\beta}^T \\ &= \sum_{i=1}^{n} v_i S_i H_i, \end{aligned} \tag{3.73}$$

and

$$\begin{aligned} I_{\beta\beta} &= S^T W S \\ &= \sum_{i=1}^{n} v_i S_i^2. \end{aligned} \tag{3.74}$$

Using Equation 3.65, we obtain the conditional information matrix I, given ξ is estimated:

$$\begin{aligned} I_{2.1} &= I_{\beta\beta} - I_{\beta\xi} I_{\xi\xi}^{-1} I_{\xi\beta} \\ &= \sum_{i=1}^{n} v_i S_i^2 - \sum_{i=1}^{n} v_i S_i H_i \left(\sum_{i=1}^{n} v_i H_i^T H_i \right)^{-1} \left(\sum_{i=1}^{n} v_i S_i H_i \right)^T. \end{aligned} \tag{3.75}$$

The score test statistic

$$T_{score2} = U_2^T I_{2.1}^{-1} U_2$$

is asymptotically distributed as a central $\chi^2_{(1)}$ under the null hypothesis $H_0 : \tau = 0$.

In the absence of covariates, i.e., $\alpha = 0$, then the score function is reduced to

$$U_2 = \sum_{i=1}^{n} (y_i - \pi) S_i, \tag{3.76}$$

where

$$\pi = \frac{e^{\alpha_0}}{1 + e^{\alpha_0}}$$

and the maximum likelihood estimate $\hat{\pi}$ is

$$\bar{Y} = \frac{1}{n} \sum_{i=1}^{n} y_i .$$

Equation 3.76 can be rewritten as

$$U_2 = \sum_{i=1}^{n} (y_i - \bar{Y}) S_i . \tag{3.77}$$

It is easy to see under $\alpha = 0$ that

$$v_i = \bar{Y}(1 - \bar{Y}). \tag{3.78}$$

After some algebra, we obtain

$$I_{2.1} = \bar{Y}(1 - \bar{Y}) \sum_{i=1}^{n} S_i^2 - \frac{1}{n} \bar{Y}(1 - \bar{Y}) \left(\sum_{i=1}^{n} S_i \right)^2 . \tag{3.79}$$

Again, the test statistic is defined as

$$T_{score2} = U_2^T I_{2.1}^{-1} U_2, \tag{3.80}$$

where U_2 and $I_{2.1}$ are defined in Equations 3.77 and 3.79, respectively. The results were reported in Lin and Tang (2011).

3.2.3.2 Sum Test and Adaptive Association Test

The Sum tests and adaptive tests that are robust in the presence of no-risk variants and allow for both risk variants and protective variants are based on score functions (Pan et al. 2014). They assume that the probability of being affected is estimated and the covariates are absent. Under these assumptions, the data matrices H and G are respectively reduced to

$$H = \begin{bmatrix} 1 \\ \vdots \\ 1 \end{bmatrix} \text{ and } G = \begin{bmatrix} G_1 \\ \vdots \\ G_n \end{bmatrix}.$$

The estimate of the probability π_i is $\hat{\pi}_i = \bar{Y}$. The Fisher information matrix is

$$I = \begin{bmatrix} n\bar{Y}(1-\bar{Y}) & \bar{Y}(1-\bar{Y})\sum_{i=1}^{n} G_i \\ \bar{Y}(1-\bar{Y})\sum_{i=1}^{n} G_i^T & \bar{Y}(1-\bar{Y})\sum_{i=1}^{n} G_i^T G_i \end{bmatrix},$$

which implies that the conditional information matrix, given the parameter α_0 is estimated by

$$\begin{aligned} I_{2.1} &= \bar{Y}(1-\bar{Y})\sum_{i=1}^{n} G_i^T G_i - \bar{Y}(1-\bar{Y})\sum_{i=1}^{n} G_i^T \bar{Y}(1-\bar{Y})\sum_{i=1}^{n} G_i / \left(n\bar{Y}(1-\bar{Y})\right) \\ &= \bar{Y}(1-\bar{Y})\sum_{i=1}^{n} (G_i - \bar{G})^T (G_i - \bar{G}). \end{aligned} \tag{3.81}$$

The score function is

$$\begin{aligned} U_2 &= \sum_{i=1}^{n} (y_i - \bar{Y}) G_i^T \\ &= \sum_{i=1}^{n} (y_i - \bar{Y}) \begin{bmatrix} G_{i1} \\ \vdots \\ G_{ip} \end{bmatrix}. \end{aligned} \tag{3.82}$$

It is interesting to note that the score function for an individual variant j is

$$U_j = \sum_{i=1}^{n} (y_i - \bar{Y}) G_{ij} = U_{2j}. \tag{3.83}$$

The score test statistic for the joint p variant analysis is

$$T_{joint} = U_2^T I_{2.1}^{-1} U_2. \tag{3.84}$$

The score test statistic for the marginal single-variant analysis is

$$T_j = \frac{U_j^2}{V_{jj}}, \tag{3.85}$$

where V_{jj} is the jth diagonal element of the conditional information matrix $I_{2.1}$ in Equation 3.81 and the minimum P-value test for the genomic region combining marginal score tests is

$$T_{U\min p} = \max_{1 \le j \le p} T_j. \tag{3.86}$$

3.2.3.3 The Sum Test

The Sum test assumes that the p variants share a common association parameter:

$$\beta_1 = \cdots = \beta_p = \beta_c.$$

The null hypothesis is then defined as $H_0 : \beta_c = 0$.

The score function and conditional information matrix are respectively given by

$$U_2 = \sum_{i=1}^{n} \left(y_i - \bar{Y}\right) \sum_{j=1}^{p} G_{ij}$$
$$= \sum_{j=1}^{p} U_j, \tag{3.87}$$

and

$$I_{2.1} = \bar{Y}\left(1 - \bar{Y}\right) \sum_{i=1}^{n} \left(\bar{G}_{i.} - \bar{G}_{..}\right)^2, \tag{3.88}$$

where $\bar{G}_{i.} = \frac{1}{p} \sum_{j=1}^{p} G_{ij}$ and $\bar{G}_{..} = \frac{1}{n} \sum_{i=1}^{n} \bar{G}_{i.}$.

Equation 3.87 defines **the Sum test** (Pan et al. 2014) as follows:

$$T_{Sum} = \sum_{j=1}^{p} U_j. \tag{3.89}$$

If we assume that the variants are in linkage equilibrium, then asymptotically

$$\sum_{i=1}^{n} G_i^T G_i \approx \begin{bmatrix} \sum_{i=1}^{n} G_{i1}^2 & \cdots & 0 \\ \vdots & \ddots & \vdots \\ 0 & \cdots & \sum_{i=1}^{n} G_{ip}^2 \end{bmatrix}. \tag{3.90}$$

Using Equation 3.90, we can show that the score test under linkage equilibrium assumption is

$$T_{SE} = \sum_{j=1}^{p} \frac{U_j^2}{\sum_{i=1}^{n} \left(G_{ij} - \bar{G}_{.j}\right)^2}. \tag{3.91}$$

If the variance for each variant in Equation 3.91 is ignored, then T_{SE} is reduced to the **SSU test** (Pan et al. 2014):

$$T_{SSU} = \sum_{j=1}^{p} U_j^2. \tag{3.92}$$

The Sum test can be extended to a more general case:

$$T_G = \sum_{j=1}^{p} \xi_j U_j, \tag{3.93}$$

where ξ_j, $j = 1, \ldots, p$ are weights. Depending on the genetic variant data to assign various weights, a number of adaptive tests have been developed (Pan et al. 2014). Since the score provides rich information on the strength of genetic effect, we can use weight $\xi_j = U_j^{r-1}$ for an integer $\gamma \geq 1$ to develop a new class of **adaptive tests (Sum of Powered Score)**:

$$T_{SPU(r)} = \sum_{j=1}^{p} U_j^r. \tag{3.94}$$

Define $\|U\|_r = \left(\sum_{j=1}^{p} |U_j|^r \right)^{1/r}$. When $r \to \infty$, we can show that $T_{SPU(r)} \propto \|U\|_r \to \max_{1 \leq j \leq p} |U_j|$.

Although the adaptive test in some scenarios can improve the power of detecting the association, the major drawback of the adaptive test is that, in general, the distributions of the adaptive tests are unknown and usually employ permutations to calculate their P-values. Permutations for genome-wide association tests require heavy computations.

3.2.4 Variance-Component Score Statistics and Logistic Mixed Effects Models

In the previous section, logistic fixed effects models and score tests are used to test the association of rare variants with disease. In this section, logistic fixed effects models are extended to logistic mixed effects models and variance-component score statistics will be explored for rare variant association analysis. For logistic mixed effects models, more mathematics is involved. For the self-sufficiency of this book, we will introduce necessary mathematics to help readers understand the principles underlying variance-component score tests for association analysis with NGS data.

3.2.4.1 Logistic Mixed Effects Models for Association Analysis

3.2.4.1.1 Model To change the logistic fixed effects models (3.59a) to logistic mixed effects models, we simply assume that the genetic effects are random effects in the model (3.59a). In other words, we start with the conditional distribution of y, given random genetic effects β (McCulloch and Searle 2001):

$$y|\beta \sim f_{Y|\beta}(y|\beta)$$

$$\begin{aligned} f_{Y|\beta}(y|\beta) &\propto \pi^{y}(1-\pi)^{1-y} \\ &\propto \exp\left\{y\log\frac{\pi}{1-\pi}-(-\log(1-\pi))\right\}. \end{aligned}$$

$f_{Y|\beta}(y|\beta)$ can also be denoted by

$$f_{Y|\beta}(y|\beta)=\exp\{y\theta-b(\theta)\}, \tag{3.95}$$

where

$$\theta=\log\frac{\pi}{1-\pi} \tag{3.96}$$

and

$$b(\theta)=\log\left(1+e^{\theta}\right). \tag{3.97}$$

This gives

$$\frac{\partial b(\theta)}{\partial\theta}=\frac{e^{\theta}}{1+e^{\theta}}=\pi. \tag{3.98}$$

π is the mean of y, given β, i.e.,

$$E\left[y|\beta\right]=\mu=\pi.$$

It is important to model this mean as a linear model in both fixed and random factors:

$$\begin{aligned} g(\mu) = g(\pi) = \log\frac{\pi}{1-\pi} = \theta = \alpha_0 + X\alpha + G\beta \\ = Z\gamma + G\beta, \end{aligned} \tag{3.99}$$

where $Z = [1, X]$, $\gamma = \begin{bmatrix} \alpha_0 \\ \alpha \end{bmatrix}$, where $g(.)$ is a logit function and is usually referred to as the link function that links the conditional mean of y and the linear form of genetic and non-genetic predictors, β is a p dimensional vector of random variables with mean of zero and variance matrix $\tau^2 B$, and other variables are defined as before.

To complete the model, we assign a distribution to the random genetic effects:

$$\beta \sim f_\beta(\beta) = (2\pi)^{-p/2} |D(\tau)|^{-1/2} \exp\left\{-\frac{1}{2}\beta^T D^{-1}\beta\right\}. \tag{3.100}$$

It is well known (Exercise 3.10) that

$$\mathrm{var}\left(\frac{\partial \log f_{y|\beta}(y)}{\partial\theta}\right) = -E\left[\frac{\partial^2 \log f_{Y|\beta}(y)}{\partial\beta^2}\right]. \tag{3.101}$$

This equality implies that the variance function $v(\mu)$ is (Appendix 3B)

$$v(\mu) = v(\pi) = \pi(1-\pi) = \mathrm{var}(y). \tag{3.102}$$

3.2.4.1.2 Likelihood and Its Score Functions Suppose that n individuals with observations $y_1, \ldots, y_n, X_1, \ldots, X_n, G_1, \ldots, G_n$ are sampled. We can easily write down a likelihood function for the model (3.95) as

$$L = \int \prod_{i=1}^{n} f_{Y|\beta}(y_i|\beta) f_\beta(\beta) d\beta. \tag{3.103}$$

Its log-likelihood function is then given by

$$\begin{aligned} l = \log \int \prod_{i=1}^{n} f_{Y|\beta}(y_i|\beta) f_\beta(\beta) d\beta \\ = \log f_Y(y). \end{aligned} \tag{3.104}$$

Similar to the logistic fixed effects model, the score function for the logistic mixed effects model is defined as

$$U = \begin{bmatrix} U_\gamma \\ U_\tau \end{bmatrix} = \begin{bmatrix} \frac{\partial l}{\partial r} \\ \frac{\partial l}{\partial \tau} \end{bmatrix}.$$

Now we first calculate the score function U_γ for fixed effects.

It follows from Equation 3.104 that

$$\frac{\partial l}{\partial \gamma} = \frac{\frac{\partial}{\partial \gamma} \int \prod_{i=1}^{n} f_{Y|\beta}(y_i|\beta) f_\beta(\beta) d\beta}{f_Y(y)}. \tag{3.105}$$

Since θ involves the random genetic effects β, to remove the impact of the random variables on the derivative, we consider the following transformation:

$$\frac{f_\beta(\beta)}{f_Y(y)} = \frac{1}{f_{Y|\beta}(y|\beta)} \frac{f_\beta(\beta) f_{Y|\beta}(y|\beta)}{f_Y(y)} = \frac{1}{f_{Y|\beta}(y|\beta)} f_{\beta|Y}(\beta|Y). \tag{3.106}$$

Substituting Equation 3.106 into Equation 3.105, we obtain

$$\begin{aligned} \frac{\partial l}{\partial \gamma} &= \frac{\frac{\partial}{\partial \gamma} \int \prod_{i=1}^{n} f_{Y|\beta}(y_i|\beta) f_{\beta|Y}(\beta|Y) d\beta}{f_{Y|\beta}(y|\beta)} \\ &= \int \frac{\partial}{\partial \gamma} \log(f_{Y|\beta}(y|\beta) f_{\beta|Y}(\beta|Y) d\beta. \end{aligned} \tag{3.107}$$

Using Equation 3.95, we obtain

$$\begin{aligned} \frac{\partial}{\partial \gamma} \log\left(f_{Y|\beta}(y|\beta)\right) &= \frac{\partial}{\partial \gamma} \sum_{i=1}^{n} \left[y_i \theta_i - b(\theta_i)\right] \\ &= \sum_{i=1}^{n} \left[y_i - \frac{\partial b(\theta_i)}{\partial \theta_i}\right] \frac{\partial \theta_i}{\partial \gamma}. \end{aligned}$$

Using Equation 3.62 gives

$$\frac{\partial}{\partial \gamma} \log\left(f_{Y|\beta}(y|\beta)\right) = \sum_{i=1}^{n} (y_i - \pi_i) Z_i^T. \tag{3.108}$$

Substituting Equation 3.108 into Equation 3.107 yields

$$U_\gamma = \frac{\partial l}{\partial \gamma} = \int \sum_{i=1}^{n} \left[(y_i - \pi_i) Z_i^T \right] f_{\beta|Y}(\beta|Y) d\beta$$
$$= E\left[\sum_{i=1}^{n} (y_i - \pi_i) Z_i^T | Y \right] = Z^T (Y - E[\pi|Y]), \tag{3.109}$$

where $Z = \begin{bmatrix} Z_1 \\ \vdots \\ Z_n \end{bmatrix}$, $Y = \begin{bmatrix} y_1 \\ \vdots \\ y_n \end{bmatrix}$, and $\pi = \begin{bmatrix} \pi_1 \\ \vdots \\ \pi_n \end{bmatrix}$.

We can similarly show that (Appendix 3C)

$$U_\tau = -\frac{1}{2} E\left[\text{Trace}\left(D^{-1} \frac{\partial D^T}{\partial \tau} \right) - \beta^T D^{-1} \frac{\partial D}{\partial \tau} D^{-1} \beta | Y \right]. \tag{3.110}$$

3.2.4.1.3 Penalized Quasi-Likelihood Quasi-likelihood for the logistic model is defined as

$$Q(\mu; y) = \int_y^\mu \frac{y - t}{v(t)} dt, \tag{3.111}$$

where $v(t)$ is a variant function. The quasi-likelihood model behaves like a log-likelihood function. Indeed, using the derivative of integral gives

$$\frac{\partial Q}{\partial \mu} = \frac{y - \mu}{v(\mu)}. \tag{3.112}$$

It is clear that

$$E\left[\frac{\partial Q}{\partial \mu} \right] = \frac{E[y] - \mu}{v(\mu)} = 0,$$

$$\text{var}\left(\frac{\partial Q}{\partial \mu} \right) = \frac{\text{var}(y - \mu)}{(v(\mu))^2} = \frac{1}{v(\mu)},$$

and

$$-E\left[\frac{\partial^2 Q}{\partial \mu^2}\right] = -E\left[\frac{-v(\mu) - \frac{\partial v(\mu)}{\partial \mu}(y-\mu)}{(v(\mu))^2}\right]$$

$$= \frac{v(\mu) + \frac{\partial v(\mu)}{\partial \mu}(E[y]-\mu)}{(v(\mu))^2}$$

$$= \frac{v(\mu)}{(v(\mu))^2} = \frac{1}{v(\mu)}.$$

This shows that quasi-likelihood behaves like a log-likelihood function and can be used as a log-likelihood. The quasi-likelihood for the logistic regression is

$$Q(\mu;y) = \int_y^\mu \frac{y-t}{v(t)}dt$$

$$= \int_y^\mu \frac{y-t}{t(1-t)}dt$$

$$= y\log\frac{\mu}{1-\mu} + \log(1-\mu) + y\log\frac{1-y}{y} - \log(1-y)$$

$$\approx y\log\frac{\mu}{1-\mu} + \log(1-\mu), \tag{3.113}$$

where $\mu = \pi = \frac{e^\theta}{1+e^\theta}$.

The quasi-likelihood can also be written as

$$Q(\mu;y) = y\theta - \log(1+e^\theta). \tag{3.114}$$

Equation 3.114 is exactly the same as Equation 3.95. This demonstrates that the quasi-likelihood is the conditional distribution of y on the given β. This also shows that the quasi-likelihood $Q(\mu;y)$ does not include information on the distribution of the random vector β. To include the variance structure information into the quasi-likelihood, we incorporate a penalty function defined as a quadratic form of the random vector (3.100) into the quasi-likelihood:

$$\text{PQL} = \sum_{i=1}^{n} Q(\mu;y_i) - \frac{1}{2}\beta^T D^{-1}\beta$$

$$= \sum_{i=1}^{n}\left[y_i\theta - \log(1+e^\theta)\right] - \frac{1}{2}\beta^T D^{-1}\beta. \tag{3.115}$$

Substituting Equation 3.99 into Equation 3.115 gives

$$\mathrm{PQL}=\sum_{i=1}^{n}[y_i\left(Z_i\gamma+G_i\beta\right)-\log(1+\exp(Z_i\gamma+G_i\beta]-\frac{1}{2}\beta^T D^{-1}\beta, \tag{3.116}$$

where Z_i and G_i are the ith row vectors of the matrices Z and G, respectively.

A score function of the PQL is defined as

$$U_{PQL}=\begin{bmatrix}\frac{\partial \mathrm{PQL}}{\partial\gamma}\\ \frac{\partial \mathrm{PQL}}{\partial\beta}\end{bmatrix}=\begin{bmatrix}\sum_{i=1}^{n}\left[y_iZ_i-\frac{\exp(Z_i\gamma+G_i\beta)Z_i}{1+\exp(Z_i\gamma+G_i\beta)}\right]\\ \sum_{i=1}^{n}\left[y_iG_i-\frac{\exp(Z_i\gamma+G_i\beta)G_i}{1+\exp(Z_i\gamma+G_i\beta)}\right]\end{bmatrix}. \tag{3.117}$$

U_{PQL} is often called the quasi-score function.

To find the score function of the PQL, we use Equation 3.116 to obtain

$$\frac{\partial \mathrm{PQL}}{\partial\gamma}=Z^T\left(y-\mu\right) \tag{3.118}$$

and

$$\frac{\partial \mathrm{PQL}}{\partial\beta}=G^T\left(y-\mu\right)-D^{-1}\beta, \tag{3.119}$$

where

$$y=\begin{bmatrix}y_1\\ \vdots\\ y_n\end{bmatrix},\quad \mu=\begin{bmatrix}\pi_1\\ \vdots\\ \pi_n\end{bmatrix}.$$

The Fisher information matrix is (Appendix 3D)

$$I=\begin{bmatrix}Z^TWZ & Z^TWG\\ G^TWZ & G^TWG+D^{-1}\end{bmatrix}, \tag{3.220}$$

where

$$W=\begin{bmatrix}\pi_1\left(1-\pi_1\right) & 0 & \cdots & 0\\ 0 & \pi_2\left(1-\pi_2\right) & \cdots & 0\\ \vdots & \ddots & \ddots & \vdots\\ 0 & 0 & \cdots & \pi_n\left(1-\pi_n\right)\end{bmatrix}.$$

The fixed effects γ and random effects β should satisfy the equation

$$U_{PQL} = 0. \tag{3.221}$$

Equation 3.221 is a system of nonlinear equations. The standard Newton–Raphson algorithm or Fisher scoring algorithm is widely used for solving the system of nonlinear equations (McCulloch and Searle 2001). The algorithm is summarized as follows (Appendix 3E).

Scoring Algorithm 3.1

Step 1. Select initial values $\gamma^{(0)}$ and $\beta^{(0)}$.

Step 2. Iteratively solve the following normal equations using Gaussian or other algebraic methods until convergence:

$$\begin{bmatrix} Z^T W^{(k)} Z & Z^T W^{(k)} G \\ G^T W^{(k)} Z & D^{-1} + G^T W^{(k)} G \end{bmatrix} \begin{bmatrix} \gamma^{(k+1)} \\ \beta^{(k+1)} \end{bmatrix} = \begin{bmatrix} Z^T W t^{(k)} \\ G^T W t^{(k)} \end{bmatrix}, \tag{3.222}$$

where

$$t^{(k)} = Z\gamma^{(k)} + G\beta^{(k)} + \left(W^{(k)}\right)^{-1}\left(y - \mu^{(k)}\right), \tag{3.223}$$

$$W^{(k)} = \text{diag}\left(\pi_1^{(k)}\left(1 - \pi_1^{(k)}\right), \ldots, \pi_n^{(k)}\left(1 - \pi_n^{(k)}\right)\right),$$

$$\mu_i^{(k)} = \pi_i^{(k)} = \frac{e^{\theta^{(k)}}}{1 + e^{\theta^{(k)}}},$$

and

$$\theta^{(k)} = Z\gamma^{(k)} + G\beta^{(k)}.$$

Scoring Algorithm 3.1 only estimates the fixed effects and random effects; it does not estimate the covariance component τ. Therefore, this will not provide sufficient information for rare-variant association tests under the logistic mixed effects models. Next we introduce a working variable and transform the quasi-likelihood solution to a linear mixed model. By iteratively solving the linear mixed models, we estimate the variance component.

3.2.4.1.4 Working Variate and Linear Mixed Models The linking function can be expanded around the mean of y_i using a Taylor expansion:

$$g(y_i) = g(\mu_i) + g_\mu^{'}(\mu_i)(y_i - \mu_i), \tag{3.224}$$

where $g_\mu^{'}(\mu_i)$ denotes $\frac{\partial g(\mu_i)}{\partial \mu}$ evaluated at μ_i.

Substituting Equation 3.99 into Equation 3.224 gives

$$g(y_i) = Z_i\gamma + G_i\beta + g_\mu^{'}(\mu_i)(y_i - \mu_i). \tag{3.225}$$

Let $t_i = g(y_i)$ and $t = [t_1, \ldots, t_n]^T$. Equation 3.225 can be rewritten as

$$t = Z\gamma + G\beta + \Delta(y - \mu), \tag{3.226}$$

where $\Delta = \text{diag}(g_\mu(\mu_1), \ldots, g_\mu(\mu_n))$ and $g_\mu(\mu_i)$ is evaluated at $\beta = 0$.

We can show that iteratively solving the linear mixed model (3.226) is equivalent to iteratively solving the normal equation (3.222) (Appendix 3F). Therefore, we can use the linear mixed models to estimate the variance components. We assume that var(t) is given by

$$V = \text{var}(t) = \tau^2 G D_0 G^T + W^{-1} \tag{3.227}$$

and t is distributed as $N(Z\gamma, V)$.

The log-likelihood for t is

$$l(\gamma, V | Z, G, y) = -\frac{n}{2}\log(2\pi) - \frac{1}{2}\log|V| - \frac{1}{2}(t - Z\gamma)^T V^{-1}(t - Z\gamma). \tag{3.228}$$

Since our goal is to infer variance components, we want to remove the fixed effects from the model. To achieve this, it is necessary to transform the data such that

$$KZ = 0, \tag{3.229}$$

where the rank of K is equal to $N - r(Z)$.

Applying this transformation to the model (3.226) gives

$$t^* = Kt = KG\beta + K\Delta(y - \mu). \tag{3.230}$$

The log-likelihood for t^*, which is often called restricted maximum likelihood (REML) is

$$l(t^*) = -\frac{1}{2}(n - r(Z))\log(2\pi) - \frac{1}{2}\log(KVK^T) - \frac{1}{2}(t^*)^T (KVK^T)^{-1} t^*. \tag{3.231}$$

Since $-\frac{1}{2}(n-r(Z))\log(2\pi)$ is a constant and is not involved in the variance components, it can be ignored. Equation 3.231 can be further reduced to (Appendix 3G)

$$l(t)=-\frac{1}{2}\log|V|-\frac{1}{2}\log\left|Z^TV^{-1}Z\right|-\frac{1}{2}(t-Z\hat{\gamma})^TV^{-1}(t-Z\hat{\gamma}), \tag{3.232}$$

where $\hat{\gamma}=\left(Z^TV^{-1}Z\right)^{-1}Z^TV^{-1}t$.

3.2.4.1.5 Score Functions and Fisher Information Matrix We start with the results of several derivatives. Recall that

$$V=\tau^2GD_0G^T+W^{-1}.$$

Therefore, we have

$$V_\tau=\frac{\partial V}{\partial\tau^2}=GD_0G^T. \tag{3.233}$$

Applying chain rule (Equation 1.158) and Equations 1.160 and 3.233, we obtain

$$\frac{\partial\log|V|}{\partial\tau^2}=\text{Trace}\left(V^{-1}V_\tau\right). \tag{3.234}$$

Applying Equations 1.156 and 3.233 gives

$$\frac{\partial V^{-1}}{\partial\tau^2}=-V^{-1}V_\tau V^{-1}. \tag{3.235}$$

By similar arguments, we have

$$\frac{\partial\log\left|Z^TV^{-1}Z\right|}{\partial\tau^2}=-\text{Trace}\left(\left(Z^TV^{-1}Z\right)^{-1}Z^TV^{-1}V_\tau V^{-1}Z\right). \tag{3.236}$$

Using Equations 3.234, 3.235, and 3.236, we can derive the score function:

$$\begin{aligned}U_\tau=\frac{\partial l(t)}{\partial\tau^2}&=-\text{Trace}\left(V^{-1}V_\tau\right)+\text{Trace}\left(\left(Z^TV^{-1}Z\right)^{-1}Z^TV^{-1}V_\tau V^{-1}Z\right)\\&\quad+\frac{1}{2}(t-Z\hat{\gamma})^TV^{-1}V_\tau V^{-1}(t-Z\hat{\gamma}).\end{aligned} \tag{3.237}$$

The score function can be further reduced by transforming the estimator $\hat{\gamma}$ to the original data. Note that

$$V^{-1}(t - Z\hat{\gamma}) = V^{-1}\left(t - Z\left(Z^T V^{-1} Z\right)^{-1} Z^T V^{-1} t\right)$$
$$= \left(V^{-1} - V^{-1} Z\left(Z^T V^{-1} Z\right)^{-1} Z^T V^{-1}\right) t$$
$$= Pt, \tag{3.238}$$

where

$$P = V^{-1} - V^{-1} Z\left(Z^T V^{-1} Z\right)^{-1} Z^T V^{-1}. \tag{3.239}$$

Substituting Equations 3.238 and 3.239 into Equation 3.237, we obtain the reduced score function

$$U_\tau = -\frac{1}{2}\text{Trace}\left(PV_\tau\right) + \frac{1}{2} t^T PV_\tau Pt. \tag{3.240}$$

Now we calculate the second derivative of log-likelihood with respect to the variance components. We can show that the formula of the derivative of P with respect to the variance component is given by (Exercise 3.15)

$$\frac{\partial P}{\partial \tau^2} = -PV_\tau P. \tag{3.241}$$

Using chain rule and Equations 1.164 and 3.241, we obtain

$$\frac{\partial \text{Trace}\left(PV_\tau\right)}{\partial \tau^2} = -\text{Trace}\left(V_\tau PV_\tau P\right)$$
$$= -\text{Trace}\left(PV_\tau PV_\tau\right). \tag{3.242}$$

Similarly, we have

$$\frac{\partial}{\partial \tau^2}\left(t^T PV_\tau Pt\right) = -2t^T PV_\tau PV_\tau Pt. \tag{3.243}$$

Combining Equations 3.242 and 3.243, we obtain the second derivatives of log-likelihood with respect to the variance component:

$$\frac{\partial U_\tau}{\partial \tau^2} = \frac{1}{2}\text{Trace}\left(PV_\tau PV_\tau\right) - t^T PV_\tau PV_\tau Pt. \tag{3.244}$$

The Fisher information is then given by

$$I_{\tau\tau} = -E\left[\frac{\partial U_\tau}{\partial \tau^2}\right] = \frac{1}{2}\text{Trace}\left(PV_\tau PV_\tau\right). \tag{3.245}$$

Fisher Scoring Algorithm 3.2

Step 1. Select initial value τ_0^2.
Step 2. Iteratively update $\tau_{(k)}^2$ until convergence to a final solution by using

$$\tau_{(k+1)}^2 = \tau_{(k)}^2 + \left(I_{\tau\tau}^{(k)}\right)^{-1} U_\tau^{(k)}, \tag{3.246}$$

where the score and Fisher information are evaluated using the current estimates.

3.2.4.2 Sequencing Kernel Association Test

Sequencing kernel association test (SKAT) (Lee et al. 2012, 2014; Wu et al. 2011) that is based on a logistic mixed effects model can be used for rare variant association analysis. The SKAT assumes the following logistic model:

$$\text{logit}\, P\left(y_i = 1\right) = \alpha_0 + X_i\alpha + G_i\beta, \tag{3.247}$$

where variables are defined as before. One way to test the association of variants within a region with disease is to test the null hypothesis:

$$H_0 : \beta = 0.$$

Since the number of rare variants in a gene is often large and each variant makes small risk of the disease, the power of the standard likelihood ratio test is often low. To increase the power of the test, the SKAT aggregates the genetic variation across the gene and tests the variance component under the null hypothesis:

$$H_0 : \tau = 0.$$

To heuristically derive the SKAT, we consider only the quadratic term in the score function (3.237):

$$U_\tau \approx \left(t - Z\hat{\gamma}\right)^T V^{-1} V_\tau V^{-1}\left(t - Z\hat{\gamma}\right). \tag{3.248}$$

Recall that we have

$$W\Delta = 1$$

when we evaluate the matrices W and Δ under the null hypothesis.

We can show that

$$V^{-1}(t - Z\hat{\gamma}) = (y - \mu). \tag{3.249}$$

Thus, Equation 3.248 can be reduced to

$$U_\tau \approx (y - \hat{\mu})^T GD_0G^T (y - \hat{\mu}), \tag{3.250}$$

where $\hat{\mu}_i = \frac{e^{\hat{\alpha}_0 + X_i\hat{\alpha}}}{1 + e^{\hat{\alpha}_0 + X_i\hat{\alpha}}}$ and D_0 are a constant matrix involved in defining the variance–covariance of the random genetic effects. Replacing D_0 by W gives the SKAT statistic (Equation 3 in Wu et al. 2011)

$$Q = (y - \hat{\mu})K(y - \hat{\mu}), \tag{3.251}$$

where $K = GWG^T$ and $W = \text{diag}\,(w_1, \ldots, w_p)$ with each weight w_i prespecified.

Wu et al. (2011) showed that under the null hypothesis,

$$Q \sim \sum_{i=1}^{n} \lambda_i \chi_{1,i}^2, \tag{3.252}$$

where λ_i are the eigenvalues of the matrix $P_0^{1/2} K P_0^{1/2}$, $P_0 = V - V\tilde{X}\left(\tilde{X}^T V \tilde{X}\right)^{-1} \tilde{X}^T V$, $\tilde{X} = [1, X]$, $V = \text{diag}\left(\hat{\mu}_1(1-\hat{\mu}_1), \ldots, \hat{\mu}_n(1-\hat{\mu}_n)\right)$, and $\chi_{1,i}^2$ represents independent $\chi_{(1)}^2$ random variables. Wu et al. (2011) used the Davies (1980) exact method to obtain the distribution of Q.

3.3 POPULATION-BASED FUNCTIONAL ASSOCIATION ANALYSIS FOR NEXT-GENERATION SEQUENCING

Genetic association analysis of complex diseases is undergoing a paradigm shift from the single marker analysis to the joint analysis of multiple variants in a genomic region (Neale and Sham 2004). In Section 3.2, we introduced a multivariate approach to association analysis of groups of variants. Although multivariate group association tests have higher power than the individual marker tests, they also suffer limitations. First, group association tests do not leverage linkage disequilibrium (LD) in the data. And second, since sequence errors are cumulative when rare variants are grouped, group tests are sensitive to the genotyping errors and missing data. All multivariate group association tests assume discrete genomic models. To utilize the advantages of both individual variant analysis and group tests and address the limitations inherent to individual variant analysis and group tests, in this section, we introduce functional data analysis (FDA) as an alternative approach to association analysis with NGS data, which utilize a genome continuum model as a general principle, and stochastic calculus and functional data analysis as tools for developing statistical methods.

3.3.1 Introduction

It is increasingly recognized that the genome is transmitted not in points but rather in segments. Instead of modeling the genome as a few separated individual loci, modeling the genome as a whole will enrich information on genetic variation across the genome. It has been shown that the number of genetic variants in large samples is approximately distributed as a Poisson process with its intensity depending on the total mutation rate (Joyce and Tavare 1995). The intensity of the Poisson process within a genomic region can be interpreted as a function of the genomic location. A collection of genetic variants for each individual can be viewed as a realization of the Poisson process.

The traditional variant-by-variant association tests compare differences in allele frequencies between cases and controls, and the multilocus association tests collectively compare differences in allele frequencies or haplotype frequencies between cases and controls across the gene (Lin et al. 2002). The FDA is an alternative approach to multivariate association analysis. Figure 3.3 presents the resequencing data for the gene ANGPTL4 in the Dallas Heart Study. As shown in Figure 3.3, in the FDA-based association analysis, the problem of collectively testing the association of multiple variants with the diseases can be transformed to test the difference of the two underlying random functions or stochastic processes between cases and controls.

There are several remarkable features of employing FDA techniques in developing statistics to test the association of a genomic region of multiple variants with the diseases. First, the low frequency of rare variants can only be found in a small number of individuals at a specific locus, and FDA techniques can effectively pool the information of multiple rare variants at a region. Second, FDA approaches viewing the genotype profiles as a function of the genomic location can achieve an overall significance level in testing the association of multiple variants. Third, functional principal component analysis (FPCA) will compress

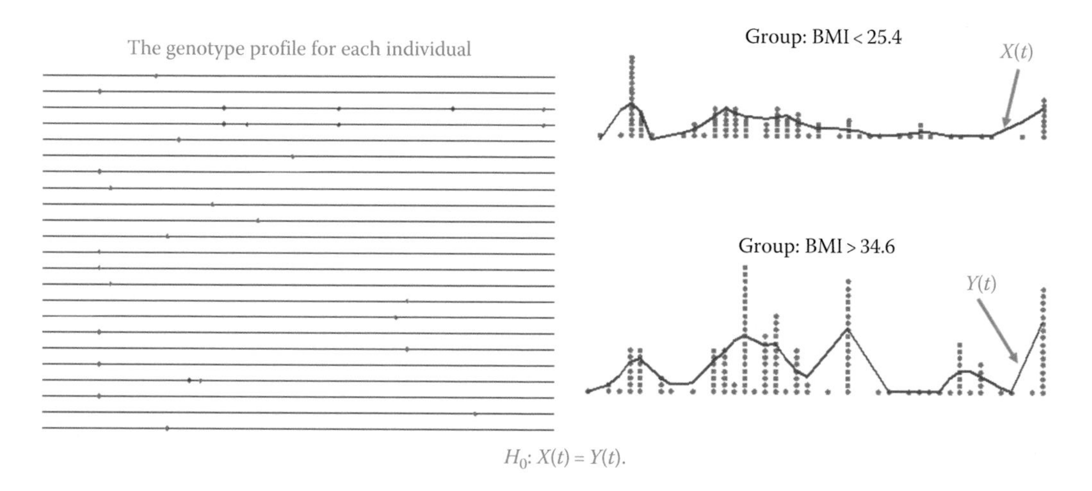

FIGURE 3.3 Resequencing Data: ANGPTL4 in the Dallas Heart Study. Left: a red point represents a rare allele at a genomic position for an individual. Right: a red point represents the number of rare alleles at a genomic position and blue curves summarize the genotype files in cases and controls fitting the data.

the data into a few top components, which will effectively remove noisy data, largely reduce the degree of freedom of the test, and eventually improve power for the test. Fourth, the FPCA approach can transform the original highly correlated genotype data into independent principal component scores. Fifth, the FDA approach can take into account the linkage disequilibrium information among the set of variants being tested, especially when we test haplotype differences between cases and controls. Recently developed functional data analyses techniques (Ramsay and Silverman 2005) are ideally suited for association tests using NGS data.

3.3.2 Functional Principal Component Analysis for Association Test

3.3.2.1 Model and Principal Component Functions

Consider two types of genetic variant profiles: genotype profiles and haplotype profiles. Let t be the position of a genetic variant along a chromosome or within a genomic region and T be the length of the genomic region being considered. For convenience, we rescale the region $[0, T]$ to $[0, 1]$. Because the density of genetic variants is high, we can view t as a continuous variable in the interval $[0, 1]$. Assume that n_A cases and n_G controls are sampled and sequenced.

We first define the genotype of the ith case as

$$Y_i(t) = \begin{cases} 2 & \text{MM} \\ 1 & \text{Mm}, \quad i = 1, \ldots, n_A \\ 0 & \text{mm} \end{cases} \tag{3.253}$$

where M is an allele at the genomic position t. Similarly, we can define a similar function $X_i(t)$, $(i = 1, \ldots, n_G)$ for the ith control. Next we define a haplotype profile. Assume that haplotypes of an individual in the genomic region are available. We define a haplotype function $Y_i(t)$ of the ith case as

$$Y_i(t) = \begin{cases} 1 & \text{M} \\ 0 & \text{m} \end{cases} \quad i = 1, \ldots, 2n_A \tag{3.254}$$

We can similarly define a haplotype function $X_i(t)$ for the ith control.

Similar to principal component analysis (PCA) for multivariate data where we consider a linear combination of variables to capture the variations contained in the entire data, we can consider a linear combination of functional values:

$$f = \int_0^1 \beta(t) X(t) dt, \tag{3.255}$$

where $\beta(t)$ is a weight function and $X(t)$ is a centered genotype or haplotype function defined in Equation 3.253 or Equation 3.254. To capture the genetic variations in the genotype or haplotype functions, we chose weight function $\beta(t)$ to maximize the variance of f. By the formula for the variance of stochastic integral (1.187), we have

$$\text{var}(f) = \int_0^1 \int_0^1 \beta(s) R(s,t) \beta(t) ds dt, \tag{3.256}$$

where $R(s,t)$ is the covariance function of the process $X(t)$. Since multiplying $\beta(t)$ by a constant will not change the maximizer of the variance var(f), we impose a constraint to make the solution unique:

$$\int_0^1 \beta^2(t) dt = 1. \tag{3.257}$$

Therefore, to find the weight function, we seek to solve the following optimization problem:

$$\begin{aligned} &\max \quad \int_0^1 \int_0^1 \beta(s) R(s,t) \beta(t) ds dt \\ &\text{s.t.} \quad \int_0^1 \beta^2(t) dt = 1. \end{aligned} \tag{3.258}$$

In Section 1.5.3, we show that the weight function $\beta(t)$ that solves the problem (3.258) should satisfy the following integral equation

$$\int_0^1 R(s,t) \beta(t) dt = \lambda \beta(s). \tag{3.259}$$

for an appropriate eigenvalue λ. The left side of the integral equation (3.259) defines an integral transform R of the weight function β. Therefore, the integral transform of the covariance function $R(s,t)$ is referred to as the covariance operator R. The integral equation (3.259) can be rewritten as

$$R\beta = \lambda\beta, \tag{3.260}$$

where $\beta(t)$ is an eigenfunction and referred to as a principal component function. Equation 3.260 is also referred to as an eigenequation. Clearly, the eigenequation (3.260) looks the same as the eigenequation for the multivariate PCA if the covariance operator and eigenfunction are replaced by a covariance matrix and an eigenvector.

Provided the functions X_i and Y_i are not linearly dependent, there will be only $N-1$ nonzero eigenvalues, where N is the total number of sampled individuals ($N = n_A + n_G$).

Eigenfunctions satisfying the eigenequation are orthonormal (Ramsay and Silverman 2005). In other words, Equation 3.260 generates a set of principal component functions:

$$R\beta_k = \lambda_k \beta_k, \quad \text{with } \lambda_1 \geq \lambda_2 \geq \cdots$$

These principal component functions satisfy

1. $\int_0^1 \beta_k^2(t)dt = 1$ and
2. $\int_0^1 \beta_k(t)\beta_m(t)dt = 0 \quad \text{for all } k < m.$

The principal component function β_1 with the largest eigenvalue is referred to as the first principal component function, and the principal component function β_2 with the second largest eigenvalue is referred to as the second principal component function, etc.

3.3.2.2 *Computations for the Principal Component Function and the Principal Component Score*

The eigenfunction is an integral function and difficult to solve in closed form. A general strategy for solving the eigenfunction problem in (3.260) is to convert the continuous eigen-analysis problem to an appropriate discrete eigen-analysis task (Luo et al. 2011; Ramsay and Silverman 2005). In this section, we introduce two methods: discretization and basis function expansion methods to achieve this conversion. As will be discussed later, these two methods are not the same, and one or the other may be more appropriate in specific situations.

3.3.2.2.1 Discretization Method In practice, the available genetic variant profiles are a function of discrete genomic positions. Assume that in a genomic region there are K variable loci, which are indexed as $t_1, t_2, \ldots, t_K$. For the ith individual, the observed genetic variant profile can be expressed as $X_i(t_1), \ldots, X_i(t_K)$. The covariance function $R(s, t)$ at these loci can be written as a matrix:

$$R = \begin{bmatrix} R(s_1, t_1), & \cdots & R(s_1, t_K) \\ \cdots & \cdots & \cdots \\ R(s_K, t_1) & \cdots & R(s_K, t_K) \end{bmatrix}. \tag{3.261}$$

Let $w_k = \frac{t_{k+1} - t_{k-1}}{2}$, $k = 1, \ldots, K$. The principal component function $\beta(t)$ at K loci is a vector and is written as $\beta = [\beta(t_1), \ldots, \beta(t_K)]^T$. By methods for numerical integration, the integral equation (3.260) can be converted to an ordinary matrix eigenequation. For each s_k, we have

$$R\beta(s_k) = \int R(s_k, t)\beta(t)dt \approx \sum_l R(s_k, t_l)\beta(t_l)w_l. \tag{3.262}$$

Then, Equation 3.260 has the approximate discrete form

$$RW\beta = \lambda\beta, \tag{3.263}$$

where $W = \text{diag}(w_1, \ldots, w_K)$.

Let $u = W^{\frac{1}{2}}\beta$. Then, Equation 3.261 can be reduced to

$$W^{\frac{1}{2}}RW^{\frac{1}{2}}u = \lambda u. \tag{3.264}$$

Equation 3.262 is the usual eigenequation from multivariate analysis. Compute the eigenvalues λ_k and eigenvectors u_k of $W^{\frac{1}{2}}RW^{\frac{1}{2}}$. Then, $\beta_k = W^{-\frac{1}{2}}u_k$ and λ_k are a pair of discrete eigenfunctions and eigenvalues, respectively, of the original functional eigenequation (3.260).

3.3.2.2.2 Basis Function Expansion Method Another method for solving the functional eigenequation (3.260) is to expand each genetic variant profile $X_i(t)$ as a linear combination of the basis function ϕ_j:

$$X_i(t) = \sum_{j=1}^{T} C_{ij}\phi_j(t) \tag{3.265}$$

Define the vector-valued function $X(t) = [X_1(t), \ldots, X_N(t)]^T$ and the vector-valued function $\phi(t) = [\phi_1(t), \ldots, \phi_T(t)]^T$. The joint expansion of all N genetic variant profiles can be expressed as

$$X(t) = C\phi(t) \tag{3.266}$$

where the matrix C is given by

$$C = \begin{bmatrix} C_{11}, & \cdots & C_{1T} \\ \cdots & \cdots & \cdots \\ C_{N1} & \cdots & C_{NT} \end{bmatrix}.$$

In matrix form, we can express the variance–covariance function of the genetic variant profiles as

$$\begin{aligned} R(s,t) &= \frac{1}{N} X^T(s) X(t) \\ &= \frac{1}{N} \phi^T(s) C^T C \phi(t). \end{aligned} \tag{3.267}$$

Similarly, the eigenfunction $\beta(t)$ can be expanded as

$$\beta(t)=\sum_{j=1}^{T} b_j\phi_j(t)$$

or

$$\beta(t)=\phi(t)^T b, \tag{3.268}$$

where $b=[b_1,\ldots,b_T]^T$. Substituting expansions (3.267) and (3.268) of variance–covariance $R(s, t)$ and eigenfunction $\beta(t)$ into the functional eigenequation (3.260), we obtain

$$\frac{1}{N}C^T CWb=\lambda b, \tag{3.269}$$

where

$$W=\int_T \phi(t)\phi^T(t)dt.$$

Normalization condition $\int_T \beta^2(t)dt=1$ implies that

$$b^T Wb=1. \tag{3.270}$$

Let $u=W^{\frac{1}{2}}b$. Then, the eigenequation (3.269) and normalization condition (3.270) can be reduced to

$$\frac{1}{N}W^{\frac{1}{2}}C^T CW^{\frac{1}{2}}u=\lambda u, \quad u^T u=1. \tag{3.271}$$

Solving the multivariate eigenvalue and eigenvector problems in Equation 3.271 will yield the eigenvalue λ and eigenvector u. Then, the eigenfunction $\beta(t)$ is finally given by

$$\beta(t)=\phi^T(t)W^{-\frac{1}{2}}u. \tag{3.272}$$

If the basis functions $\phi_j(t)$ are orthonormal, then $W=I$, an identity matrix.

3.3.2.3 Test Statistic

We use the pooled genetic variant profiles $X_i(t)$ of cases and $Y_i(t)$ of controls to estimate the principal component function $\phi_j(t)$ using both discretizing and basis expansion methods.

The original genetic variant functions $X_i(t)$ and $Y_i(t)$ can be expressed as a linear combination of eigenfunctions

$$X_i(t) = \sum_j \xi_{ij}\phi_j(t)$$

and

$$Y_i(t) = \sum_j \eta_{ij}\phi_j(t), \tag{3.273}$$

where $\phi_j(t), j = 1, 2, \ldots$ are orthonormal functions. To find coefficients of expansions, multiplying eigenfunctions $\phi_j(t)$ on both sides of Equation 3.273 and using orthonormal properties $\int_T \phi_j(t)\phi_k(t)dt = 0$ and $\int_T \phi_j^2(t)dt = 1$, we obtain

$$\xi_{ij} = \int_T X_i(t)\phi_j(t)dt, \tag{3.274}$$

$$\eta_{ij} = \int_T Y_i(t)\phi_j(t)dt. \tag{3.275}$$

Since the genetic variant functions $X_i(t)$ and $Y_i(t)$ are centered, the expectation of the expansion coefficients is

$$E[\xi_{ij}] = \int_T E[X_i(t)]\phi_j(t)dt = 0 \quad \text{and} \quad E[\eta_{ij}] = \int_T E[Y_i(t)]\phi_j(t)dt = 0.$$

Using Equations 1.187 and 3.259, we obtain

$$\begin{aligned} \operatorname{cov}(\xi_{ij}, \xi_{ik}) &= \int_T\int_T \phi_j(s)R(s,t)\phi_k(t)dsdt \\ &= \lambda_k \int_T \phi_j(s)\phi_k(s)ds = 0, \quad \forall j \neq k \end{aligned} \tag{3.276}$$

and

$$\begin{aligned} \operatorname{var}(\xi_{ij}) &= \int_T\int_T \phi_j(s)R(s,t)\phi_j(t)dsdt \\ &= \lambda_j \int_T \phi_j^2(s)ds \\ &= \lambda_j, \quad j = 1, 2, \ldots \end{aligned} \tag{3.277}$$

Similarly, we can show $\text{cov}(\eta_{ij}, \eta_{ik}) = 0, \forall j \neq k$ and $\text{var}(\eta_{ij}) = \lambda_j, j = 1, 2, \ldots$. In other words, ξ_{ij} and η_{ij} are uncorrelated random variables with zero mean and variances λ_j with $\sum_j \lambda_j < \infty$.

Define the averages $\bar{\xi}_j$ and $\bar{\eta}_j$ of the principal component scores ξ_{ij} and η_{ij} in the cases and controls:

$$\bar{\xi}_j = \frac{1}{n_A} \sum_{i=1}^{n_A} \xi_{ij} \quad \text{and} \quad \bar{\eta}_j = \frac{1}{n_G} \sum_{i=1}^{n_G} \eta_{ij}.$$

Define vectors of average of the principal component scores in cases and controls, respectively, by

$$\bar{\xi} = \left[\bar{\xi}_1, \ldots, \bar{\xi}_k\right]^T \quad \text{and} \quad \bar{\eta} = \left[\bar{\eta}_1, \ldots, \bar{\eta}_k\right]^T.$$

When the genetic variant profiles are defined by a genotype function, then **the FPCA-based statistic** for testing the association of a genomic region with the disease is defined as

$$T_{FPC} = \left(\bar{\xi} - \bar{\eta}\right)^T \Lambda^{-1} \left(\bar{\xi} - \bar{\eta}\right), \tag{3.278}$$

where

$$S = \frac{1}{n_A + n_G - 1} \left[\sum_{i=1}^{n_A} \left(\xi_i - \bar{\xi}\right)\left(\xi_i - \bar{\xi}\right)^T + \left(\eta_i - \bar{\eta}\right)\left(\eta_i - \bar{\eta}\right)^T \right],$$

$$\Lambda = \left(\frac{1}{n_A} + \frac{1}{n_G} \right) S.$$

If the genetic variant profiles are defined by a haplotype function, then the number of sampled individuals n_A and n_G in cases and controls in Equation 3.278 should be replaced by the number of sampled chromosomes $2n_A$ and $2n_G$ in cases and controls. Under the null hypothesis of no association of the genomic region, the test statistic T_{FPC} is asymptotically distributed as a central $\chi^2_{(k)}$ distribution.

3.3.3 Smoothed Functional Principal Component Analysis for Association Test

FPCA can greatly enhance the power to detect association of variants. However, when the genetic variant functions in FPCA rapidly change within the genomic region, the basis expansion in the FPCA cannot approximate the genetic variation data well, which will decrease the power of FPCA. Figure 3.4 shows the original genotype data and fitted genotype curves by the FPCA and smoothed FPCA method. Figure 3.4 shows the genotype curve fitted by the FPCA rapidly varies, but the genotype curve fitted by the smoothed FPCA changes smoothly.

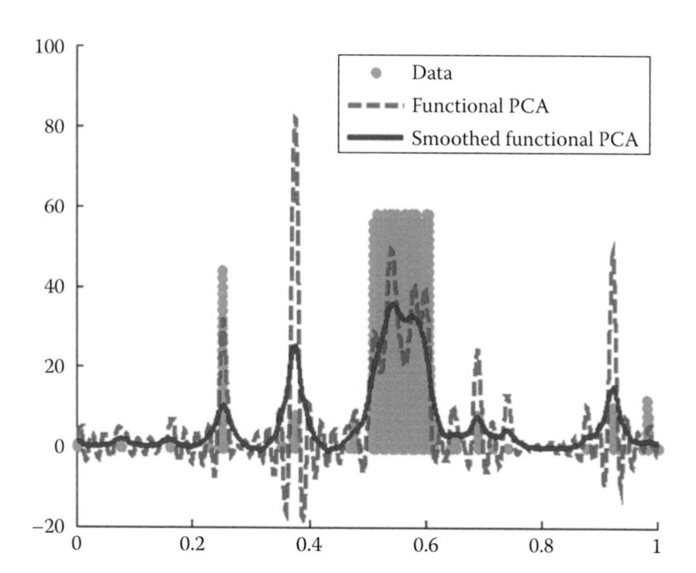

FIGURE 3.4 Resequencing data and fitted curves: MUC6 in Myocardial Infarction Studies. The green point represents the count of alleles in the mapped genomic position. The red dotted line represents the fitted genetic variation curve by the FPCA, and the blue solid line represents the fitted genetic variation curve by the smooth FPCA.

The smoothed FPCA for testing the association of rare variants combines a measure of goodness-of-fit with a roughness penalty to retain the advantages of basis expansion but circumvent its limitation (Luo et al. 2013). Smoothing of the estimated principal component curves can substantially reduce their variability and improve the approximation of genetic variation data by the smoothed functional principal components, which will finally lead to increasing the power of the tests.

3.3.3.1 A General Framework for the Smoothed Functional Principal Component Analysis

The observed genetic variant profiles are often not smooth, which leads to substantial variability in the estimated functional principal component curves. To improve the smoothness of the estimated functional principal component curves, we impose the roughness penalty on the functional principal component weight functions. We balance the goodness-of-fit and the roughness of the estimated functional principal component curves.

We often penalize the roughness of the functional principal component curve by its integrated squared second derivative. The balance between the goodness-of-fit and the roughness of the function is controlled by a smoothing parameter μ. We define an extended inner product as

$$(f,g)_\mu = \int f(t)g(t)dt + \mu\int D^2 f(t)D^2 g(t)dt, \tag{3.279}$$

where $D^2 f(t) = \frac{d^2 f(t)}{dt^2}$. The penalized sample variance is defined as

$$F = \frac{\operatorname{var}\left(\int_i^1 x(t)\beta(t)dt\right)}{\|\beta(t)\|_\mu^2}, \tag{3.280}$$

where $\|\beta(t)\|_\mu^2 = \int_0^1 \beta^2(t)dt + \mu\int_0^1 \left[D^2\beta(t)\right]^2 dt$.

The smoothed functional principal components can be obtained by solving the following integral equation (for derivation, please read Section 1.5.4):

$$\int_0^1 R(s,t)\beta(s)ds = \lambda\left[\beta(t) + \mu D^4\beta(t)\right]. \tag{3.281}$$

Note that when $\mu = 0$, the smoothed functional principal components analysis (SFPCA) is reduced to an unsmoothed functional principal component analysis.

3.3.3.2 Computations for the Smoothed Principal Component Function

The eigenfunction is an integral function and difficult to solve in closed form. A general strategy for solving the eigenfunction problem in (3.281) is to convert the continuous eigen-analysis problem to an appropriate discrete eigen-analysis task. We use basis function expansion methods to achieve this conversion.

Let $\{\phi_j(t)\}$ be the series of Fourier functions. For each j, define $\omega_{2j-1} = \omega_{2j} = 2\pi j$. We expand each genetic variant profile $X_i(t)$ as a linear combination of the basis function ϕ_j:

$$X_i(t) = \sum_{j=1}^{T} C_{ij}\phi_j(t). \tag{3.282}$$

Define the vector-valued function $X(t) = [X_1(t), \ldots, X_N(t)]^T$ and the vector-valued function $\phi(t) = [\phi_1(t), \ldots, \phi_T(t)]^T$. The joint expansion of all N genetic variant profiles can be expressed as

$$X(t) = C\phi(t), \tag{3.283}$$

where the matrix C is given by

$$C = \begin{bmatrix} C_{11} & \cdots & C_{1T} \\ \cdots & \cdots & \cdots \\ C_{N1} & \cdots & C_{NT} \end{bmatrix}.$$

In matrix form, we can express the variance–covariance function of the genetic variant profiles as

$$R(s,t)=\frac{1}{N}X^T(s)X(t)$$
$$=\frac{1}{N}\phi^T(s)C^TC\phi(t). \tag{3.284}$$

Similarly, the eigenfunction $\beta(t)$ can be expanded as

$$\beta(t)=\sum_{j=1}^{T}b_j\phi_j(t) \quad \text{and} \quad D^4\beta(t)=\sum_{j=1}^{T}\omega_j^4 b_j\phi_j(t)$$

or

$$\beta(t)=\phi(t)^T b \quad \text{and} \quad D^4\beta(t)=\phi(t)^T S_0 b, \tag{3.285}$$

where $b=[b_1,\ldots,b_T]^T$ and $S_0=\operatorname{diag}\left(\omega_1^4,\ldots,\omega_T^4\right)$. Let $S=\operatorname{diag}\left(\left(1+\mu\omega_1^4\right)^{-\frac{1}{2}},\ldots,\left(1+\mu\omega_T^4\right)^{-\frac{1}{2}}\right)$. Then, we have

$$\beta(t)+\mu D^4\beta(t)=\phi(t)^T S^{-2}b. \tag{3.286}$$

Substituting expansions (3.284) and (3.286) of variance–covariance $R(s, t)$ and eigenfunction $\beta(t)$ into the functional eigenequation (3.281), we obtain

$$\phi(t)^T\frac{1}{N}C^TCb=\lambda\phi^T(t)S^{-2}b. \tag{3.287}$$

Since Equation 3.287 must hold for all t, we obtain the following eigenequation:

$$\frac{1}{N}C^TCb=\lambda S^{-2}b, \tag{3.288}$$

which can be rewritten as

$$\left[S\left(\frac{1}{N}C^TC\right)S\right]\left[S^{-1}b\right]=\lambda\left[S^{-1}b\right],$$

or

$$S\left(\frac{1}{N}C^TC\right)Su=\lambda u, \tag{3.289}$$

where $u=S^{-1}b$. Thus, $b=Su$ and $\beta(t)=\phi(t)^Tb$ are solutions to eigenequation (3.281).

Note that $<u_j, u_j> = 1$ and $<u_j, u_k> = 0$, for $k<j$. Therefore, we obtain a set of orthonormal eigenfunctions with an inner product of two functions defined in Equation 3.279, as shown in Equation 3.290:

$$\|\beta_j\|_\mu^2 = b_j^T S^{-2} b_j = u_j^T S S^{-2} S u_j = 1 \quad \text{and} \quad <\beta_j, \beta_k>_\mu = b_j^T S^{-2} b_k = u_j^T u_k = 0. \quad (3.290)$$

3.3.3.3 Test Statistic

We use the pooled genetic variant profiles $X_i(t)$ of cases and $Y_i(t)$ of controls to estimate the set of orthonormal principal component function $\beta_j(t), j = 1, 2, \ldots, k$ (eigenfunctions) using the basis expansion methods. Similar to the previous section, by the K-L decomposition, the smoothed functional principal component score can be obtained by

$$\xi_{ij} = < x_i(t), \beta_j(t) >_\mu \quad \text{and} \quad \eta_{ij} = < y_i(t), \beta_j(t) >_\mu, \quad j = 1, 2, \ldots, k.$$

We denote vectors of averages of functional principal component scores in cases and controls by $\bar{\xi} = \left[\bar{\xi}_1, \ldots, \bar{\xi}_k\right]^T$ and $\bar{\eta} = \left[\bar{\eta}_1, \ldots, \bar{\eta}_k\right]^T$, where $\bar{\xi}_j = \sum_{i=1}^{n_A} \xi_{ij}$ and $\bar{\eta}_j = \sum_{i=1}^{n_G} \eta_{ij}$, $j = 1, 2, \ldots, k$, and define the pooled covariance matrix

$$S = \frac{1}{n_A + n_G - 2}\left[\sum_{i=1}^{n_A} (\xi_i - \bar{\xi})(\xi_i - \bar{\xi})^T + \sum_{i=1}^{n_G} (\eta_i - \bar{\eta}(\eta_i - \bar{\eta})^T\right],$$

where $\xi_i = [\xi_{i1}, \ldots, \xi_{ik}]^T, \eta_i = [\eta_{i1}, \ldots, \eta_{ik}]^T$.

Let $\Lambda = \left(\frac{1}{n_A} + \frac{1}{n_G}\right) S$.

Then, the statistic is defined as

$$T_{SFPC} = \left(\bar{\xi} - \bar{\eta}\right)^T \Lambda^{-1} \left(\bar{\xi} - \bar{\eta}\right). \quad (3.291)$$

Under the null hypothesis of no association of the genomic region with a disease, the statistic T_{SFPC} is asymptotically distributed as a central $\chi^2_{(k)}$ distribution.

3.3.3.4 Power Comparisons

To evaluate the performance of the FPCA-based statistics and other association tests introduced in the previous sections for testing the association of a set of variants with disease, we used simulated data to estimate their power to detect a true association. We considered four disease models: additive, dominant, recessive, and multiplicative. We used MS software (Hudson 2002) to simulate 1,000,000 individuals with 240 variants (60 common and 180 rare variants in a 30 kb region).

An individual's disease status was determined based on the individual's genotype and the penetrance for each locus. Let A_i be a rare risk allele at the ith locus. Let $G_{k_i}(k = 0, 1, 2)$ be the genotypes $a_i a_i$, $A_i a_i$, and $A_i A_i$, respectively, and f_{ki} be the penetrance of genotypes G_{k_i} at the

ith locus. The relative risk (RR) at the ith locus is defined as $R_{1i} = \frac{f_{1i}}{f_{0i}}$ and $R_{2i} = \frac{f_{2i}}{f_{0i}}$, where f_{0i} is the baseline penetrance of the wild-type genotype at the ith variant site. We assume that for the additive disease model, $R_{2i} = 2R_{1i} - 1$; for the dominant disease model, $R_{2i} = R_{1i}$; for the recessive disease model, $R_{1i} = 1$; and for the multiplicative disease model, $R_{2i} = R_{1i}^2$. The genotype relative risk was assumed to be inversely proportional to the MAF where the population attributable risk (PAR) of each group was assumed to be 0.005. We assumed that the relative risks across all variant sites are equal and that the variants influence disease susceptibility independently (i.e., no epistasis). Each individual was assigned to the group of cases or controls depending on their disease status. The process for sampling individuals from the population of 2,000,000 haplotypes was repeated until the desired samples were reached for each disease model.

Figures 3.5 and 3.6 plot the power curves of 12 statistics: smoothed FPCA, discretization (Luo et al. 2013); smoothed FPCA, Fourier expansion (Luo et al. 2013); FPCA, discretization

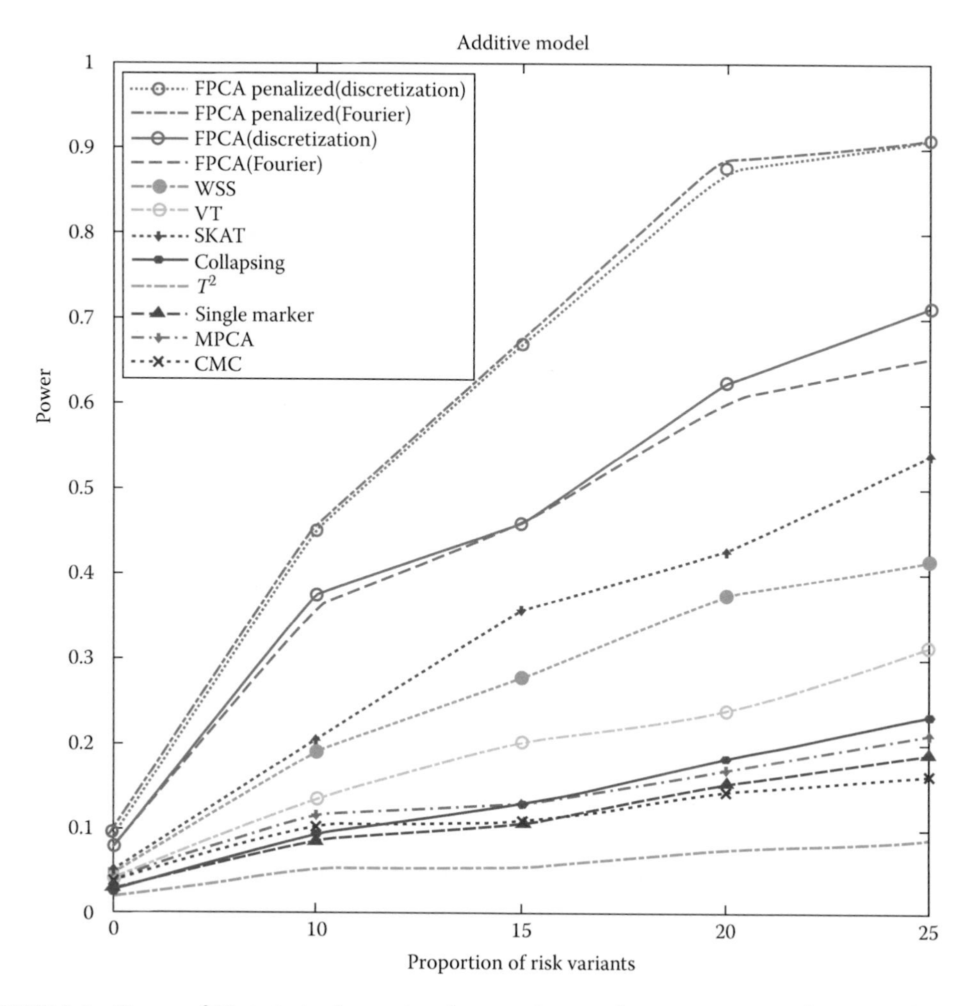

FIGURE 3.5 Power of 12 statistics for testing the association of rare variants as a function of proportion of risk variants under the additive models.

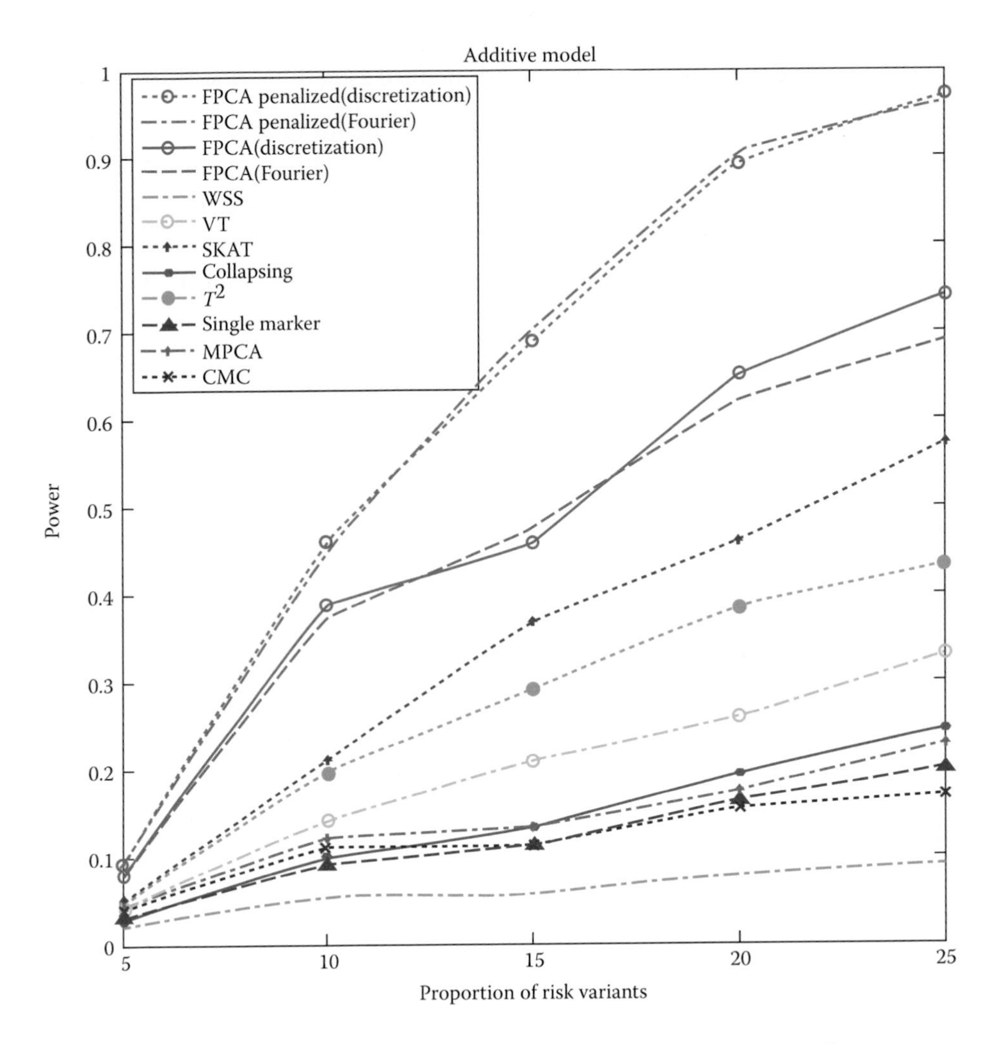

FIGURE 3.6 Power of 12 statistics for testing the association of both common and rare variants as a function of proportion of risk variants under the additive models.

(Luo et al. 2011); FPCA, Fourier expansion (Luo et al. 2011); sequence kernel association test (SKAT) (Wu et al. 2011); weighted sum statistic (WSS) (Madsen and Browning 2009); variable threshold (VT) (Price et al. 2010b); multivariate principal component (MPC)–based statistic; collapsing method (Li and Leal 2008); generalized T^2 statistic (Xiong et al. 2002); single marker χ^2 test where permutation was used to adjust for multiple testing; and the CMC method (variants with frequencies ≤0.005 were collapsed) as a function of the proportion of risk-increasing variants for testing the association of rare and both common and rare variants with disease at the significance level of 0.05 under additive disease models, assuming a baseline penetrance of 0.01 and that 2000 cases and 2000 controls were sampled. Figure 3.7 shows the power curves of 12 statistics as a function of sample sizes at the significance level of 0.05 under the additive model, assuming 7.5% of risk variants and 7.5% of protective variants. We can observe several remarkable features from these figures. First, the power of

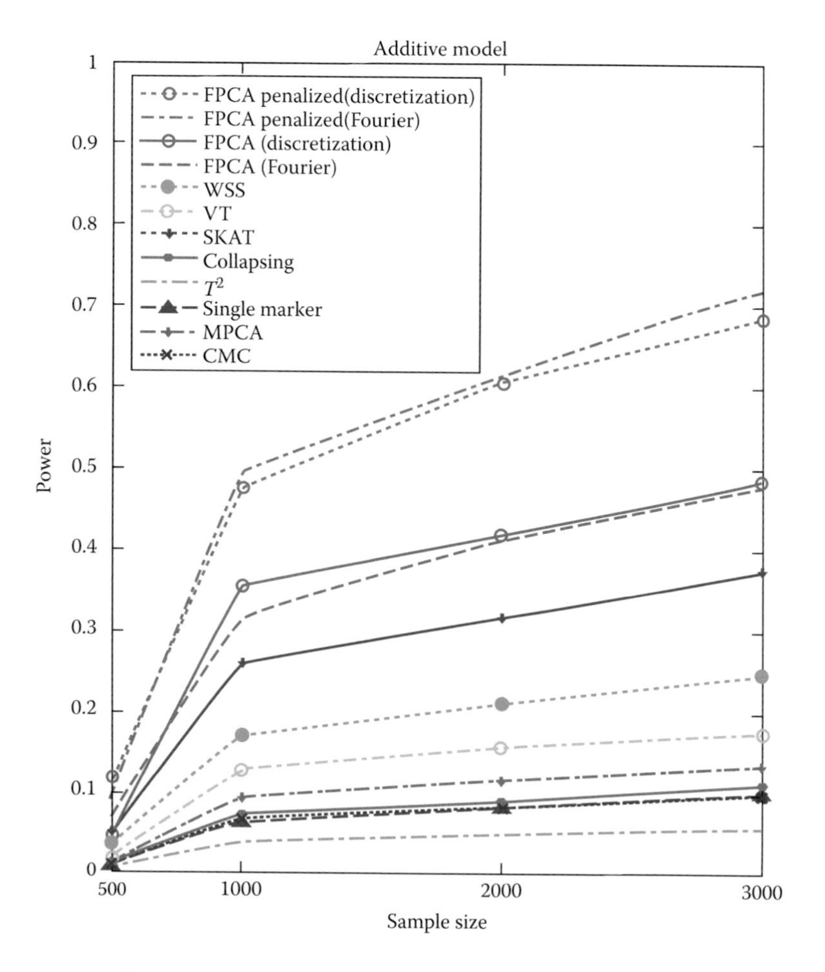

FIGURE 3.7 Power of 12 statistics for testing the association of 7.5% of risk rare variants and 7.5% of protection rare variants as a function of sample sizes.

the FDA-based statistics is much higher than that of all multivariate statistics. Second, the power of the SFPCA-based statistics is higher than that of the standard FPCA-based statistics. Third, the methods for genotype function expansion do not have much impact on the power of the tests. Fourth, among the multivariate rare variant association tests, the SKAT is the most powerful statistic.

3.3.3.5 Application to Real Data Examples

To evaluate their performance, 12 statistics are applied to the *ANGPTL3* sequence and phenotype data from the Dallas Heart Study with 3553 individuals (Romeo et al. 2007). Since the smoothed FPCA method requires that each individual should have at least two rare variants in the genomic region being tested, we excluded 98 individuals with only one rare variant. To examine the phenotypic effects of rare variants in *ANGPTL3*, we selected two groups of individuals with the lowest and highest quartiles of five traits

TABLE 3.13 *P*-Values of 11 Statistics for Testing the Association of Rare Variants in *ANGPTL3* with Five Traits in the Dallas Heart Study

	Phenotype				
Statistical Method	**BMI**	**Cholesterol**	**Triglyceride**	**VLDL**	**HDL**
Smoothed FPCA (discrete)	1.72E−02	2.45E−02	4.39E−08	1.46E−08	4.54E−03
Smoothed FPCA (Fourier)	1.68E−02	2.42E−02	4.35E−08	1.23E−08	4.23E−03
FPCA (discretization)	2.96E−02	2.68E−02	9.82E−08	4.91E−08	5.86E−03
FPCA (Fourier)	2.48E−02	2.50E−02	9.55E−08	4.86E−08	5.70E−03
SKAT	4.96E−02	3.83E−02	7.68E−06	2.40E−07	2.60E−03
T^2	2.16E−01	3.07E−01	1.40E−03	1.60E−03	1.85E−01
Collapsing	2.90E−01	9.57E−02	4.50E−01	1.90E−01	3.20E−01
Chi square (permutation)	1.28E−01	1.69E−01	4.80E−03	5.60E−03	2.16E−02
CMC	1.96E−01	5.40E−03	1.10E−01	1.13E−01	9.40E−01
WSS	1.34E−01	4.05E−01	4.00E−04	5.00E−04	2.37E−01
VT	2.12E−01	2.76E−01	2.00E−05	5.00E−05	9.95E−02
MPCA	3.98E−02	2.13E−01	1.56E−05	5.76E−04	1.38E−02

related to lipid metabolism. The individuals whose plasma triglyceride levels are less than or equal to the 25th percentile are classified as the lowest quartiles of the triglyceride, and the individuals whose plasma triglyceride levels are greater than or equal to the 75th percentile are grouped as the highest quartile of the triglyceride. We can similarly classify the individuals as the lowest and highest quartiles of high-density lipoprotein cholesterol (HDL), total cholesterol, very low-density lipoprotein cholesterol (VLDL), and body mass index (BMI). *P*-values from previous 12 statistics in power evaluations for testing the association of rare variants in *ANGPTL3* with the five traits are summarized in Table 3.13. For the CMC method, variants with an allele frequency below 0.005 were collapsed. We observe that only FPCA-based statistics and SKAT detect significant association of *ANGPTL3* with triglyceride and VLDL and *P*-values calculated using the FPCA-based statistics are smaller than that using SKAT.

To illustrate that many rare variant association tests can also be applied to variants of all frequencies, eight statistics, SFPCA, FPCA, SKAT, CMC, χ^2 statistic, WSS, VT, and MPCA, are applied to a GWAS of schizophrenia data, which are downloaded from dbGaP to test the association of variants within a genomic region with schizophrenia. The samples are of European origin and included 1,135 individuals with schizophrenia and 1,362 controls with 727,479 typed SNPs. The total number of genes being tested is 13,804. The threshold for declaring genome-wide significance after the Bonferroni correction is 3.6×10^{-6}. The results are summarized in Table 3.14. We observe that the SFPCA identifies 10 genes that are significantly associated with schizophrenia, and the *P*-values using SFPCA are much smaller than that using other statistics. Some results can be confirmed by the literature. For example, PDLIM5 was reported to be associated with schizophrenia and bipolar disorder (Zhao et al. 2009), CERKL was associated with narcolepsy (Shimada et al. 2010), and HAAO was associated with Parkinson's disease (Kim et al. 2006).

TABLE 3.14 *P*-Values of the Top 10 Genes Significantly Associated with Schizophrenia Identified by SFPCA

Gene	SFPCA	FPCA	SKAT	CMC	T^2	χ^2	WSS	VT	MPCA
LRSAM1	1.70E−19	3.00E−06	5.70E−04	2.10E−03	2.10E−03	2.30E−04	2.30E−01	3.00E−02	4.20E−03
RUNDC3B	3.10E−18	2.40E−08	8.80E−08	5.60E−07	5.60E−07	9.00E−07	2.60E−01	2.50E−01	8.30E−01
DTL	2.90E−12	5.80E−04	3.20E−06	3.50E−04	3.50E−04	8.10E−07	4.30E−01	5.10E−01	7.70E−01
MRPS17	4.20E−12	3.40E−05	1.90E−04	2.10E−03	2.10E−03	3.70E−03	5.00E−02	1.30E−01	4.10E−02
PDLIM5	1.60E−11	1.10E−06	2.50E−05	5.50E−05	5.50E−05	5.60E−05	5.80E−05	1.70E−03	3.30E−05
CECR1	1.70E−11	4.30E−05	6.30E−06	5.30E−05	5.30E−05	3.60E−03	7.00E−02	1.00E−02	2.60E−02
CERKL	2.20E−11	8.30E−07	4.40E−06	5.20E−05	5.20E−05	1.40E−03	3.10E−01	3.20E−01	2.60E−01
EVI5	3.30E−11	6.80E−09	3.80E−03	5.20E−02	5.20E−02	2.00E−05	1.30E−04	5.00E−05	6.30E−06
HAAO	5.70E−11	5.40E−01	5.20E−01	3.00E−08	3.00E−08	4.60E−04	5.00E−02	4.20E−01	6.00E−01
MTA3	7.20E−11	6.30E−01	9.40E−07	6.70E−07	6.70E−07	4.60E−04	5.80E−01	3.00E−02	9.40E−01

SOFTWARE PACKAGE

Plink (http://pngu.mgh.harvard.edu/~purcell/plink/) is designed for genome-wide association studies with common variants. R-Package "AssotesteR" (https://cran.r-project.org/web/packages/AssotesteR/AssotesteR.pdf) and R-package "aSPU" (https://cran.r-project.org/web/packages/aSPU/aSPU.pdf) are software packages that perform score-based association tests with rare variants. Fitting the logistic mixed effects models using PQL can use R-package "glmmPQL" (http://www.inside-r.org/r-doc/mass/glmmPQL). SKAT (http://www.hsph.harvard.edu/skat/) is a program for testing the association of a set of SNPs (gene or genomic region) with a continuous or binary trait. A program for implementing the FPCA and smoothed FPCA for association tests can be downloaded from our website, http://www.sph.uth.tmc.edu/hgc/faculty/xiong/index.htm.

APPENDIX 3A: FISHER INFORMATION MATRIX FOR γ

First we calculate U_2. By definition of score function, we have

$$\begin{aligned} U_2 = \frac{\partial l}{\partial \beta} &= \sum_{i=1}^{n}\left[y_i\left(\frac{1}{\pi_i}\frac{\partial \pi_i}{\partial \beta} + \frac{1}{1-\pi_i}\frac{\partial \pi_i}{\partial \beta}\right) - \frac{1}{1-\pi_i}\frac{\partial \pi_i}{\partial \beta}\right] \\ &= \sum_{i=1}^{n}\left[\frac{y_i}{\pi_i(1-\pi_i)} - \frac{1}{1-\pi_i}\right]\frac{\partial \pi_i}{\partial \beta} \\ &= \sum_{i=1}^{n}\left[\frac{y_i-\pi_i}{\pi_i(1-\pi_i)}\frac{\partial \pi_i}{\partial \beta}\right]. \end{aligned} \tag{3A.1}$$

Now we calculate $\frac{\partial \pi_i}{\partial \beta}$. Let

$$\eta_i = \log\frac{\pi_i}{1-\pi_i} = \log \pi_i - \log(1-\pi_i). \tag{3A.2}$$

Taking derivative with respect to η_i on both sides of Equation 3A.2, we obtain

$$1 = \frac{1}{\pi_i}\frac{\partial \pi_i}{\partial \eta_i} + \frac{1}{(1-\pi_i)}\frac{\partial \pi_i}{\partial \eta_i}. \tag{3A.3}$$

Solving Equation 3A.3 for $\frac{\partial \pi_i}{\partial \eta_i}$ yields

$$\frac{\partial \pi_i}{\partial \eta_i} = \pi_i(1-\pi_i). \tag{3A.4}$$

Combining Equations 3.59a and 3A.2, we have

$$\frac{\partial \eta_i}{\partial \beta} = G_i^T. \tag{3A.5}$$

Using chain rule and Equations 3A.4 and 3A.5, we obtain

$$\frac{\partial \pi_i}{\partial \beta} = \frac{\partial \pi_i}{\partial \eta_i}\frac{\partial \eta_i}{\partial \beta} = \pi_i\left(1-\pi_i\right)G_i^T. \tag{3A.6}$$

Substituting Equation 3A.6 into Equation 3A.1, we obtain the score function

$$\begin{aligned} U_2 &= \sum_{i=1}^{n}\left[\frac{y_i-\pi_i}{\pi_i\left(1-\pi_i\right)}\right]\pi_i\left(1-\pi_i\right)G_i^T \\ &= \sum_{i=1}^{n}\left(y_i-\pi_i\right)G_i^T. \end{aligned} \tag{3A.7}$$

Similar to Equation 3A.6, we can obtain from Equation 3A.2 that

$$\frac{\partial \pi_i}{\partial \beta^T} = \frac{\partial \pi_i}{\partial \eta_i}\frac{\partial \eta_i}{\partial \beta^T} = \pi_i\left(1-\pi_i\right)G_i. \tag{3A.8}$$

The Fisher information for β is

$$\begin{aligned} I_{\beta\beta} &= -E\left(\frac{\partial^2 l}{\partial\beta\partial\beta^T}\right) \\ &= -E\left(\frac{\partial U_2}{\partial \beta^T}\right) \\ &= -E\left(\sum_{i=1}^{n}\left(-G_i^T\frac{\partial \pi_i}{\partial \beta^T}\right)\right) \\ &= \sum_{i=1}^{n}\pi_i\left(1-\pi_i\right)G_i^T G_i \\ &= G^T W G, \end{aligned} \tag{3A.9}$$

where

$$W = \text{diag}\left(\pi_i\left(1-\pi_i\right)\right). \tag{3A.10}$$

Similarly, we have

$$U_1 = \sum_{i=1}^{n}\left(y_i-\pi_i\right)H_i^T \tag{3A.11}$$

and

$$I_{\xi\xi} = H^T W H. \tag{3A.12}$$

Now we calculate the Fisher information matrix $I_{\xi\beta}$. By definition of information matrix, we have

$$\begin{aligned} I_{\xi\beta} = I_{\beta\xi} &= -E\left(\frac{\partial^2 l}{\partial\beta\partial\xi^T}\right) \\ &= -E\left(\frac{\partial U_2}{\partial\xi^T}\right) \\ &= -E\left(\frac{\partial\left(\sum_{i=1}^n (y_i-\pi_i)G_i^T\right)}{\partial\xi^T}\right) \\ &= \sum_{i=1}^n G_i^T \frac{\partial\pi_i}{\partial\xi^T}. \end{aligned} \tag{3A.13}$$

Now we calculate $\frac{\partial\pi_i}{\partial\xi^T}$. It follows from Equation 3.59b that

$$\frac{\partial\eta_i}{\partial\xi} = H_i. \tag{3A.14}$$

Combining Equations 3A.3 and 3A.14 and using the chain rule, we obtain

$$\begin{aligned} \frac{\partial\pi_i}{\partial\xi^T} &= \frac{\partial\pi_i}{\partial\eta}\frac{\partial\eta}{\partial\xi^T} \\ &= \pi_i(1-\pi_i)H_i. \end{aligned} \tag{3A.15}$$

Substituting Equation 3A.15 into Equation 3A.13, we obtain

$$\begin{aligned} I_{\xi\beta} &= \sum_{i=1}^n G_i^T\pi_i(1-\pi_i)H_i \\ &= G^T WH. \end{aligned} \tag{3A.16}$$

APPENDIX 3B: VARIANCE FUNCTION $v(\mu)$

Using Equation 3.95, we obtain

$$\log f_{Y|\beta}(y) = y\theta - b(\theta). \tag{3B.1}$$

Thus, we have

$$\frac{\partial \log f_{Y|\beta}(y)}{\partial\theta} = y - \frac{\partial b(\theta)}{\partial\theta}, \tag{3B.2}$$

which implies that

$$\frac{\partial^2 \log f_{Y|\beta}(y)}{\partial\theta^2} = -\frac{\partial^2 b(\theta)}{\partial\theta^2}. \tag{3B.3}$$

Using Equations 3.98 and 3B.3 gives

$$1-\pi = \frac{1}{1+e^{\theta}} \tag{3B.4}$$

and

$$\frac{\partial^2 \log f_{Y|\beta}(y)}{\partial\theta^2} = -\frac{e^{\theta}}{\left(1+e^{\theta}\right)^2} = -\pi(1-\pi). \tag{3B.5}$$

It follows from Equation 3B.2 that

$$\operatorname{var}\left(\frac{\partial \log f_{Y|\beta}(y)}{\partial\theta}\right) = \operatorname{var}\left(y - \frac{\partial b(\theta)}{\partial\theta}\right) = \operatorname{var}(y). \tag{3B.6}$$

Using Equations 3.98, 3B.3, 3B.5, and 3B.6 gives

$$\operatorname{var}(y) = -E\left[\frac{\partial^2 \log f_{Y|\beta}(y)}{\partial\theta^2}\right] = \pi(1-\pi). \tag{3B.7}$$

We denote

$$\frac{\partial^2 b(\theta)}{\partial\theta^2} = v(\mu) = v(\pi) = \pi(1-\pi), \tag{3B.8}$$

where $v(\mu)$ is called the variance function.

APPENDIX 3C: DERIVATION OF SCORE FUNCTION FOR U_τ

By definition of U_τ, we have

$$\begin{aligned}
U_\tau &= \frac{\partial l}{\partial\tau} \\
&= \frac{1}{f_Y(y)}\frac{\partial f_Y(y)}{\partial\tau} \\
&= \frac{1}{f_Y(y)}\frac{\partial}{\partial\tau}\int f_\beta(\beta) f_{\beta|Y}(y|\beta)\,d\beta \\
&= \frac{1}{f_Y(y)}\int \frac{\partial f_\beta(\beta)}{\partial\tau} f_{\beta|Y}(y|\beta)\,d\beta \\
&= \frac{1}{f_Y(y)}\int \frac{\partial \log f_\beta(\beta)}{\partial\tau} f_\beta(\beta) f_{\beta|Y}(y|\beta)\,d\beta.
\end{aligned} \tag{3C.1}$$

Recall that the conditional density function $f_{\beta|Y}(\beta|Y)$ can be expressed as

$$f_{\beta|Y}(\beta|Y) = \frac{1}{f_Y(y)} f_\beta(\beta) f_{Y|\beta}(y|\beta). \tag{3C.2}$$

Substituting Equation 3C.2 into Equation 3C.1 gives

$$\begin{aligned} U_\tau &= \int \frac{\partial \log f_\beta(\beta)}{\partial \tau} f_{\beta|Y}(\beta|Y) d\beta \\ &= E\left[\frac{\partial \log f_\beta(\beta)}{\partial \tau} \middle| Y \right]. \end{aligned} \tag{3C.3}$$

When β is distributed as $N(0, D(\tau))$ (3.100), we have

$$\log f_\beta(\beta) = -\frac{p}{2}\log(2\pi) - \frac{1}{2}\log|D(\tau)| - \frac{1}{2}\beta^T D^{-1}\beta. \tag{3C.4}$$

Using Equations 1.158 and 1.160 gives

$$\frac{\partial \log|D(\tau)|}{\partial \tau} = \text{Trace}\left(D^{-1} \frac{\partial D^T}{\partial \tau} \right). \tag{3C.5}$$

Using formula for derivative of inverse matrix (Equation 1.156) and Equation 3C.5, we obtain

$$\frac{\partial \log f_\beta(\beta)}{\partial \tau} = -\frac{1}{2}\text{Trace}\left(D^{-1} \frac{\partial D^T}{\partial \tau} \right) + \frac{1}{2}\beta^T D^{-1} \frac{\partial D}{\partial \tau} D^{-1}\beta. \tag{3C.6}$$

Substituting Equation 3C.6 into Equation 3C.3 leads to

$$U_\tau = -\frac{1}{2} E\left[\text{Trace}\left(D^{-1} \frac{\partial D^T}{\partial \tau} \right) - \beta^T D^{-1} \frac{\partial D}{\partial \tau} D^{-1}\beta \middle| Y \right].$$

APPENDIX 3D: FISHER INFORMATION MATRIX OF PQL

We first derive the score function. By definition, we have

$$\begin{aligned} \frac{\partial \text{PQL}}{\partial \gamma} &= \sum_{i=1}^{n} [y_i - \pi_i] Z_i^T \\ &= Z^T(y - \mu) \end{aligned} \tag{3D.1}$$

and

$$\frac{\partial \mathrm{PQL}}{\partial \beta} = \sum_{i=1}^{n} (y_i - \pi_i) G_i^T - D^{-1}\beta$$
$$= G^T (y - \mu) - D^{-1}\beta. \tag{3D.2}$$

To find the Fisher information matrix, we first calculate the second derivative of the PQL. Using Equation 3D.1 gives

$$\frac{\partial^2 \mathrm{PQL}}{\partial \gamma \partial \gamma^T} = \sum_{i=1}^{n} \left[-Z_i^T \frac{\partial \pi_i}{\partial \gamma^T} \right]. \tag{3D.3}$$

Using Equation 3.99 and the chain rule, we obtain

$$\frac{\partial \pi_i}{\partial \gamma^T} = \frac{\partial \pi_i}{\partial \theta} \frac{\partial \theta}{\partial \gamma^T}$$
$$= \frac{1}{\partial \theta / \partial \pi_i} \frac{\partial \theta}{\partial \gamma^T}$$
$$= \frac{1}{1/\pi_i (1 - \pi_i)} Z_i$$
$$= \pi_i (1 - \pi_i) Z_i. \tag{3D.4}$$

Substituting Equation 3D.4 into Equation 3D.3 gives

$$\frac{\partial^2 \mathrm{PQL}}{\partial \gamma \partial \gamma^T} = \sum_{i=1}^{n} \left[-Z_i^T \pi_i (1 - \pi_i) Z_i \right]$$
$$= -Z^T W Z, \tag{3D.5}$$

where

$$W = \begin{bmatrix} \pi_1(1-\pi_1) & 0 & \cdots & 0 \\ 0 & \pi_2(1-\pi_2) & \cdots & 0 \\ \vdots & \ddots & \iddots & \vdots \\ 0 & 0 & \cdots & \pi_n(1-\pi_n) \end{bmatrix}.$$

Thus, we have

$$I_{\gamma\gamma} = E\left[-\frac{\partial^2 \mathrm{PQL}}{\partial \gamma \partial \gamma^T} \right] = Z^T W Z.$$

Similarly, we can prove

$$I_{\gamma\beta} = Z^T WG,$$
$$I_{\beta\gamma} = G^T WZ,$$
$$I_{\beta\beta} = G^T WG + D^{-1}.$$

APPENDIX 3E: SCORING ALGORITHM

The Fisher scoring algorithm for solving Equation 3.221 takes the form

$$\begin{bmatrix} \gamma^{(k+1)} \\ \beta^{(k+1)} \end{bmatrix} = \begin{bmatrix} \gamma^{(k)} \\ \beta^{(k)} \end{bmatrix} + \left(I^{(k)}\right)^{-1} \begin{bmatrix} Z^T\left(y-\mu^{(k)}\right) \\ G^T\left(y-\mu^{(k)}\right) - D^{-1}\beta^{(k)} \end{bmatrix}. \tag{3E.1}$$

Multiplying $I^{(k)}$ on both sides of Equation 3E.1, we obtain

$$I^{(k)} \begin{bmatrix} \gamma^{(k+1)} \\ \beta^{(k+1)} \end{bmatrix} = I^{(k)} \begin{bmatrix} \gamma^{(k)} \\ \beta^{(k)} \end{bmatrix} + \begin{bmatrix} Z^T\left(y-\mu^{(k)}\right) \\ G^T\left(y-\mu^{(k)}\right) - D^{-1}\beta^{(k)} \end{bmatrix}. \tag{3E.2}$$

We expand the first term on the right side of Equation 3E.2 using the Fisher score (Equation 3.220)

$$I^{(k)} \begin{bmatrix} \gamma^{(k)} \\ \beta^{(k)} \end{bmatrix} = \begin{bmatrix} Z^T W^{(k)}\left(Z\gamma^{(k)} + G\beta^{(k)}\right) \\ G^T W^{(k)}\left(Z\gamma^{(k)} + G\beta^{(k)}\right) \end{bmatrix}. \tag{3E.3}$$

Combining Equation 3E.3 with the second term on the right side of Equation 3E.2 gives

$$I^{(k)} \begin{bmatrix} \gamma^{(k)} \\ \beta^{(k)} \end{bmatrix} + \begin{bmatrix} Z^T\left(y-\mu^{(k)}\right) \\ G^T\left(y-\mu^{(k)}\right) - D^{-1}\beta^{(k)} \end{bmatrix} = \begin{bmatrix} Z^T W^{(k)} \\ G^T W^{(k)}\left(Z\gamma^{(k)} + G\beta^{(k)} + \left(W^{(k)}\right)^{-1}\left(y-\mu^{(k)}\right)\right) \end{bmatrix}. \tag{3E.4}$$

Now we expand the term on the left side of Equation 3E.2 using Equation 3.220:

$$I^{(k)} \begin{bmatrix} \gamma^{(k+1)} \\ \beta^{(k+1)} \end{bmatrix} = \begin{bmatrix} Z^T W^{(k)} Z & Z^T W^{(k)} G \\ G^T W^{(k)} Z & D^{-1} + G^T W^{(k)} G \end{bmatrix} \begin{bmatrix} \gamma^{(k+1)} \\ \beta^{(k+1)} \end{bmatrix}. \tag{3E.5}$$

Combing Equations 3E.4 and 3E.5, we obtain (Liu et al. 2008)

$$\begin{bmatrix} Z^T W^{(k)} Z & Z^T W^{(k)} G \\ G^T W^{(k)} Z & D^{-1} + G^T W^{(k)} G \end{bmatrix} \begin{bmatrix} \gamma^{(k+1)} \\ \beta^{(k+1)} \end{bmatrix} = \begin{bmatrix} Z^T W \left(Z\gamma^{(k)} + G\beta^{(k)} + \left(W^{(k)}\right)^{-1} \left(y - \mu^{(k)}\right)\right) \\ G^T W \left(Z\gamma^{(k)} + G\beta^{(k)} + \left(W^{(k)}\right)^{-1} \left(y - \mu^{(k)}\right)\right) \end{bmatrix}.$$

APPENDIX 3F: EQUIVALENCE BETWEEN ITERATIVELY SOLVING LINEAR MIXED MODEL AND ITERATIVELY SOLVING THE NORMAL EQUATION

The variance–covariance matrix of the working variate t is

$$\operatorname{var}(t) = GDG^T + \Delta^* \operatorname{var}(y - \mu)\Delta^*. \tag{3F.1}$$

Recall that

$$\Delta^* = W^{-1}. \tag{3F.2}$$

Substituting Equation 3F.2 into Equation 3F.1 gives

$$\psi = \operatorname{var}(t) = GDG^T + W^{-1} \operatorname{var}(y - \mu) W^{-1}. \tag{3F.3}$$

The least square estimates of the fixed and random effects are

$$\begin{aligned} \begin{bmatrix} \gamma^{(k+1)} \\ \beta^{(k+1)} \end{bmatrix} &= \left\{ \begin{bmatrix} Z^T \\ G^T \end{bmatrix} \psi^{-1} [Z \quad G] \right\}^{-1} \begin{bmatrix} Z^T \\ G^T \end{bmatrix} \psi^{-1} \left\{ [Z \quad G] \begin{bmatrix} \gamma^{(k)} \\ \beta^{(k)} \end{bmatrix} + \Delta^* \left(y - \mu^{(k)}\right) \right\} \\ &= \begin{bmatrix} \gamma^{(k)} \\ \beta^{(k)} \end{bmatrix} + \left\{ \begin{bmatrix} Z^T \\ G^T \end{bmatrix} \psi^{-1} [Z \quad G] \right\}^{-1} \begin{bmatrix} Z^T \\ G^T \end{bmatrix} \psi^{-1} \Delta^* \left(y - \mu^{(k)}\right). \end{aligned} \tag{3F.4}$$

To make Equation 3F.4 equivalent to Equation 3E.1, it requires

$$\begin{bmatrix} Z^T \psi^{-1} Z & Z^T \psi^{-1} Z \\ G^T \psi^{-1} Z & G^T \psi^{-1} G \end{bmatrix} = \begin{bmatrix} Z^T W^{(k)} Z & Z^T W^{(k)} G \\ G^T W^{(k)} Z & D^{-1} + G^T W^{(k)} G \end{bmatrix}, \tag{3F.5}$$

$$Z^T \psi^{-1} \Delta^* = Z^T, \tag{3F.6}$$

and

$$G^T \psi^{-1} \Delta^* \left(y - \mu^{(k)}\right) = G^T \left(y - \mu^{(k)}\right) - D^{-1} \beta^{(k)}. \tag{3F.7}$$

When $\psi^{-1} = W$ and $\mu^{(k)}$ is evaluated at $\beta = 0$, Equations 3F.5 and 3F.6 hold. If we make an additional assumption of absence of penalization term, Equation 3F.7 will also hold. From Equation 3F.3, we can see that these assumptions require that the variance components of the random effects are weak. Under the null hypothesis of no association of the genetic variants with the disease, Equations 3F.5 and 3F.6 will hold and maximization of quasi-likelihood problem can be transformed to iteratively solving a linear mixed effect problem using working variates.

APPENDIX 3G: EQUATION REDUCTION

We start derivation with selection of K (Wakefield 2008). Recall that the vector of residuals of least square estimate is

$$\begin{aligned} R &= t - Z\hat{\gamma} \\ &= t - Z\left(Z^T Z\right)^{-1} Z^T t \\ &= \left(I - Z\left(Z^T Z\right)^{-1} Z^T\right)t \\ &= (I - H)t. \end{aligned} \quad (3G.1)$$

Applying the model to the residuals, we obtain

$$\begin{aligned} R &= (I - H)t \\ &= (I - H)(Z\gamma + G\beta + \varepsilon) \\ &= \left[\left(Z - Z\left(Z^T Z\right)^{-1} Z^T Z\right)\right]\gamma + (I - H)(G\beta + \varepsilon) \\ &= (I - H)(G\beta + \varepsilon). \end{aligned} \quad (3G.2)$$

Although R is not dependent on γ, its rank is $n - m - 1$, and hence, its distribution will be degenerate. We need to augment R to full rank.

Consider the transformation

$$U = K^T t,$$

where K is an $n \times (n - m - 1)$ dimension matrix with

$$KK^T = I - H \quad (3G.3)$$

and

$$K^T K = I. \quad (3G.4)$$

Then, using Equations 3G.3 and 3G.4 gives

$$\begin{aligned} U &= K^T t \\ &= K^T K K^T t \\ &= K^T (I - H)t. \end{aligned} \quad (3G.5)$$

Substituting Equation 3G.2 into Equation 3G.5, we obtain

$$U = K^T R. \tag{3G.6}$$

Note that

$$\begin{aligned} K^T Z &= K^T K K^T Z \\ &= K^T (I - H) Z \\ &= K^T (Z - Z) \\ &= 0, \end{aligned} \tag{3G.7}$$

which implies

$$\begin{aligned} U &= K^T t \\ &= K^T G\beta + B^T \varepsilon \end{aligned} \tag{3G.8}$$

and

$$E[U] = 0.$$

We make the following transformation:

$$\begin{bmatrix} U \\ F \end{bmatrix} = \begin{bmatrix} K^T \\ B^T \end{bmatrix} t, \tag{3G.9}$$

where $B^T = (Z^T V^{-1} Z)^{-1} Z^T V^{-1}$ and $F = \hat{\gamma}$.

We make changes of variables and derive distribution of the changed variables. The Jacobian matrix of the transformation (3G.9) is

$$\begin{aligned} |J| &= \left| \frac{\partial (U, F)}{\partial Y} \right| \\ &= [K \quad B] \\ &= \left| \begin{bmatrix} K^T \\ B^T \end{bmatrix} [K \quad B] \right|^{1/2} \\ &= \begin{vmatrix} K^T K & K^T B \\ B^T K & B^T B \end{vmatrix}^{1/2}. \end{aligned} \tag{3G.10}$$

Applying the matrix partition formula

$$|A| = |A_{11}| \left| A_{22} - A_{21} A_{11}^{-1} A_{12} \right|$$

to Equation 3G.10, we obtain

$$|J| = |K^TK|^{1/2} \left|B^TB - B^TK\left(K^TK\right)^{-1}K^TB\right|^{1/2}. \tag{3G.11}$$

Substituting Equations 3G.3 and 3G.4 into Equation 3G.11 gives

$$\begin{aligned} |J| &= |I|^{1/2} \left|B^TB - B^T(I-H)B\right|^{1/2} \\ &= \left|B^T(I - I + H)B\right|^{1/2} \\ &= \left|B^THB\right|^{1/2}. \end{aligned} \tag{3G.12}$$

Note that

$$\begin{aligned} B^TH &= \left(Z^TV^{-1}Z\right)^{-1} Z^TV^{-1}Z\left(Z^TZ\right)^{-1}Z^T \\ &= \left(Z^TZ\right)^{-1}Z^T, \end{aligned} \tag{3G.13}$$

which implies

$$\begin{aligned} B^THB &= \left(Z^TZ\right)^{-1} Z^TV^{-1}Z\left(Z^TV^{-1}Z\right)^{-1} \\ &= \left(Z^TZ\right)^{-1}. \end{aligned} \tag{3G.14}$$

Substituting Equation 3G.14 into Equation 3G.11 gives

$$|J| = \left|Z^TZ\right|^{-1/2}. \tag{3G.15}$$

By distribution theory for transformation of random vectors, we obtain

$$\begin{aligned} f(U,F) &= \frac{1}{|J|} f(Y) \\ &= \frac{1}{|J|}(2\pi)^{-n/2}|V|^{-1/2} \exp\left\{-\frac{1}{2}(t - Z\gamma)^T V^{-1}(t - Z\gamma)\right\}. \end{aligned} \tag{3G.16}$$

We can show that

$$\operatorname{cov}(U,F) = 0. \tag{3G.17}$$

Since both random variables U and F are normal, U and F are independent. Recall that

$$F = \left(Z^TV^{-1}Z\right)^{-1} Z^TV^{-1}t.$$

The variance of U is

$$\operatorname{var}(F)=\left(Z^TV^{-1}Z\right)^{-1}Z^TV^{-1}VV^{-1}Z\left(Z^TV^{-1}Z\right)^{-1}$$
$$=\left(Z^TV^{-1}Z\right)^{-1}.$$

Thus, the density function of U is

$$f(F)=(2\pi)^{-\frac{m+1}{2}}\left|Z^TV^{-1}Z\right|^{1/2}\exp\left\{-\frac{1}{2}(F-\gamma)^T\left(Z^TV^{-1}Z\right)(F-\gamma)\right\}. \tag{3G.18}$$

Note that

$$Z^T(I-H)=Z^T-Z^TZ\left(Z^TZ\right)^{-1}Z^T=0 \tag{3G.19}$$

and

$$(t-Z\gamma)^TV^{-1}(t-Z\gamma)=(t-Z\hat{\gamma})^TV^{-1}(t-Z\hat{\gamma})+(\hat{\gamma}-\gamma)^T\left(Z^TV^{-1}Z\right)(\hat{\gamma}-\gamma). \tag{3G.20}$$

Therefore, using Equations 3G.15, 3G.16, 3G.18, and 3G.19, we can obtain the density function $f(U)$:

$$f(U)=\frac{f(U,F)}{f(F)}$$
$$=\frac{\frac{1}{|J|}(2\pi)^{-n/2}|V|^{-1/2}\exp\left\{-\frac{1}{2}(t-Z\gamma)^TV^{-1}(t-Z\gamma)\right\}}{(2\pi)^{-\frac{m+1}{2}}\left|Z^TV^{-1}Z\right|^{1/2}\exp\left\{-\frac{1}{2}(F-\gamma)^T\left(Z^TV^{-1}Z\right)(F-\gamma)\right\}}$$
$$=(2\pi)^{-\frac{n-m-1}{2}}\left|Z^TZ\right|^{-1/2}|V|^{-1/2}\left|Z^TV^{-1}Z\right|^{-1/2}\exp\left\{-\frac{1}{2}(t-Z\hat{\gamma})^TV^{-1}(t-Z\hat{\gamma})\right\}.$$

EXERCISES

Exercise 3.1 Table E.1 gives results of MN blood genotyping of 6129 American Caucasians. Calculate the frequencies of two alleles.

Exercise 3.2 Show the following equalities:

$$P(Aa)-2pq=-2D_A$$
$$P(aa)-q^2=D_A.$$

TABLE E.1

MM	MN	NN	Total
1787	3037	1305	6129

Exercise 3.3 The variance of natural logarithm of relative risk RR is

$$\operatorname{var}(\log RR) = \frac{c}{a(a+c)} + \frac{d}{b(b+d)}.$$

Exercise 3.4 Assume that the data are given in Table 3.6. Calculate the relative risk and its 95% confidence interval.

Exercise 3.5 Consider Table E.2. Calculate the P-value for testing the association of genotype with disease using the χ^2 test defined in Equation 3.37.

Exercise 3.6 Suppose that the observed number of alleles G and g are summarized in Table E.3. Calculate the P-value of exact test for the association of allele with disease.

Exercise 3.7 Show

$$\operatorname{var}\left(n^A, n^A\right) = \Pi^A = n_A\left[\operatorname{diag}\left(P_1^A, \ldots, P_m^A\right) - P^A\left(P^A\right)^T\right]$$

and

$$\operatorname{var}\left(n^G, n^G\right) = \Pi = n_G\left[\operatorname{diag}\left(P_1, \ldots, P_m\right) - PP^T\right].$$

Exercise 3.8 Calculate the variance–covariance matrix of $\hat{P}^A - \hat{P}^G$.

TABLE E.2 Contingency Table for Genotype-Based Association Test in Case–Control Analysis

Allele	GG	Gg	gg	Total
Cases	25	20	15	60
Controls	10	20	35	65
Total	35	40	50	125

TABLE E.3 Contingency Table with Marginal Totals 9, 9, 5, and 13

Allele	G	g	Total
Cases	1	8	9
Controls	4	5	9
Total	5	13	18

Exercise 3.9 Suppose that both the numbers of sampled individuals in cases and controls are equal to 1000. Haplotype frequencies in cases and controls are

$$P^A = [0.1425, 0.7512, 0.0000, 0.1063]^T \quad \text{and}$$

$$P^G = [0.1240, 0.7834, 0.0011, 0.0915]^T, \text{ respectively.}$$

Calculate the P-value of the χ^2 test for the association of haplotypes with the disease.

Exercise 3.10 Show that covariance matrix of score function is equal to the Fisher information matrix, i.e.,

$$\text{cov}(U, U) = -E\left(\frac{\partial^2 l}{\partial \beta \partial \beta^T}\right).$$

Exercise 3.11 Show Equation 3.74:

$$\frac{\partial \eta_i}{\partial \xi} = H_i.$$

Exercise 3.12 Show that the conditional covariance matrix of U_2, given U_1, is

$$I_{2.1} = \text{cov}(U_2 | U_1 = 0) = I_{\beta\beta} - I_{\beta\xi} I_{\xi\xi}^{-1} I_{\xi\beta}.$$

Exercise 3.13 We assume that the p variants share a common association parameter:

$$\beta_1 = \cdots = \beta_p = \beta_c.$$

Show that the score function and conditional information matrix are respectively given by

$$U_2 = \sum_{i=1}^{n} (y_i - \bar{Y}) \sum_{j=1}^{p} G_{ij}$$

$$= \sum_{j=1}^{p} U_j$$

and

$$I_{2.1} = \bar{Y}(1 - \bar{Y}) \sum_{i=1}^{n} (\bar{G}_{i.} - \bar{G}_{..})^2,$$

where

$$\bar{G}_{i.} = \frac{1}{p} \sum_{j=1}^{p} G_{ij}$$

$$\bar{G}_{..} = \frac{1}{n} \sum_{i=1}^{n} \bar{G}_{i}.$$

Exercise 3.14 Show that the score test under linkage equilibrium assumption is

$$T_{SE} = \sum_{j=1}^{p} \frac{U_j^2}{\sum_{i=1}^{n} \left(G_{ij} - \bar{G}_{.j}\right)^2}.$$

Exercise 3.15 Prove the formula of the derivative of P with respect to the variance component:

$$\frac{\partial P}{\partial \tau^2} = -PV_\tau P.$$

Exercise 3.16 Show that under the null hypothesis $H_0 : \tau = 0$, we have

$$V^{-1}\left(t - Z\hat{\gamma}\right) = \left(y - \mu\right).$$

Exercise 3.17 Show the correlation between U and F in Equation 3G.17 is zero, i.e.,

$$\operatorname{cov}\left(U, F\right) = 0.$$

Exercise 3.18 Prove the following equation:

$$\left(t - Z\gamma\right)^T V^{-1}\left(t - Z\gamma\right) = \left(t - Z\hat{\gamma}\right)^T V^{-1}\left(t - Z\hat{\gamma}\right) + \left(\hat{\gamma} - \gamma\right)^T \left(Z^T V^{-1} Z\right)\left(\hat{\gamma} - \gamma\right).$$

Exercise 3.19 Calculate the variance of summation of the functional principal components

$$\operatorname{var}\left(\sum_{j=1}^{K} \xi_{ij}\right).$$

Exercise 3.20 Show

$$\left\|\beta_j\right\|_\mu^2 = 1 \quad \text{and} \quad <\beta_j, \beta_k>_\mu = 0.$$

CHAPTER 4

Association Studies for Quantitative Traits

4.1 FIXED EFFECT MODEL FOR A SINGLE TRAIT

4.1.1 Introduction

Multifactorial traits that vary greatly among individuals may be considered as resulting from the combined effects of many genetic and environmental quantities. For this reason, they are often called quantitative traits. For example, crop yields, blood pressure, and body weight and height, among others, exhibit continuous variation (Figure 4.1). It is well documented that many traits that vary continuously are determined by a number of loci, each with small effects and working in concert with environmental factors. Quantitative genetics approaches have broad applications. Quantitative traits may be risk factors for diseases. Therefore, quantitative genetics can serve as a tool to unravel mechanisms of diseases. Quantitative genetics can also be used for animal and plant improvement. The purpose of quantitative genetics is to study how the quantitative traits are determined by the genetic factors and their interaction with environmental factors.

4.1.2 Genetic Effects

4.1.2.1 Variation Partition

Analysis of variance will show that phenotype variation is due to genetics and environments (Falconer and Mackay 1996). The phenotypic value can be divided into two components attributable to the influence of genotype and environment. A quantitative trait is influenced by genetic factors. It is important to study the relationship between phenotype and genotype. Consider a locus with k genotypes, denoted by $G_1, \dots, G_k$. Assume that the genotypes $G_1, \dots, G_k$ have population frequencies $f_1, \dots, f_k$. We assume random mating and Hardy–Weinberg equilibrium. Let y denote the measured phenotype values and $f(y|G_i)$ be the conditional density function of the phenotype y, given the individual with the genotype G_i. The mean phenotype value of the individuals with the genotype G_i is then

$$g_i = E[Y|G_i] = \int yf(y|G_i)dy. \quad (4.1)$$

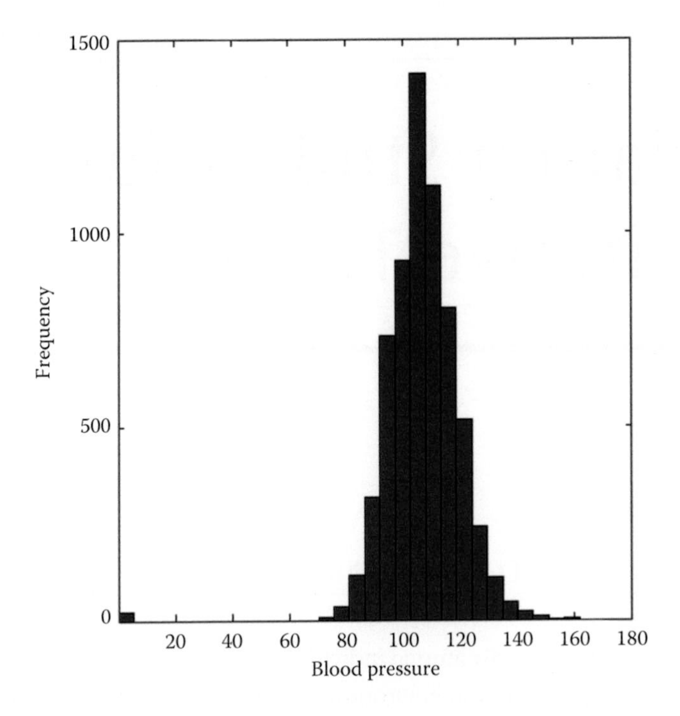

FIGURE 4.1 Continuous distribution of blood pressure.

Define the genetic model:

$$Y = E[Y|G] + \varepsilon,$$

where

$E[Y|G]$ is the average of the phenotypic values due to genetics
ε is the variation due to environment and noise

The total variance σ_p^2 is defined as

$$\begin{aligned}\sigma_P^2 &= E\left\{\left(Y - E[Y]\right)^2\right\} \\ &= E\left\{\left(Y - E[Y|G] + E[Y|G] - E[Y]\right)^2\right\} \\ &= E\left\{\left(Y - E[Y|G]\right)^2\right\} + E\left\{\left(E[Y|G] - E[Y]\right)^2\right\} \\ &= \sigma_E^2 + \sigma_G^2,\end{aligned} \tag{4.2}$$

where

$\sigma_G^2 = E\left\{\left(E[Y|G] - E[Y]\right)^2\right\}$ is the genotypic variance
$\sigma_E^2 = E\left\{\left(Y - E[Y|G]\right)^2\right\}$ is the environmental variance

The latter expression indicates that σ_P^2 is the sum of the between-genotype variance and the average within-genotype variance, which is entirely due to environmental effects. We have thus partitioned, for a given trait, the total phenotypic variability into two components: (1) genetic variance σ_G^2 and (2) environmental variance σ_E^2. The average phenotypic values of the traits in the population are given by

$$\bar{g} = E[Y] = E\left[E[Y|G]\right] = \sum_{i=1}^{k} f_i g_i. \tag{4.3}$$

The sampling formula for the genetic variance σ_G^2 is

$$\sigma_G^2 = E\left\{\left(E[Y|G] - E[Y]\right)^2\right\} = \sum_{i=1}^{k} f_i\left(g_i - \bar{g}\right)^2. \tag{4.4}$$

Similarly, the sampling formula for the total variance is

$$\sigma_P^2 = E\left\{\left(Y - E[Y]\right)^2\right\} = \frac{1}{n}\sum_{j=1}^{n}\left(y_j - \bar{y}\right)^2, \tag{4.5}$$

where $\bar{y} = \frac{1}{n}\sum_{j=1}^{n} y_j$.

The environment variance σ_E^2 is then given by

$$\sigma_E^2 = \sigma_P^2 - \sigma_G^2. \tag{4.6}$$

4.1.2.2 Genetic Additive and Dominance Effects

Consider a single locus with two alleles, A_1 and A_2, with allele frequencies p and q, respectively. We denote the genotypic value of one homozygote A_1A_1 by G_{11}, the other homozygote A_2A_2 by G_{22}, and the heterozygote A_1A_2 by G_{12}. We can now see how the gene frequencies influence the mean of the trait in the population as a whole. Assuming the Hardy–Weinberg equilibrium, it follows from Equation 4.3 that the population mean μ is given by

$$\begin{aligned} \mu &= E[Y] \\ &= p^2 G_{11} + 2pqG_{12} + q^2 G_{22}. \end{aligned} \tag{4.7}$$

In order to deduce the properties of a population connected with its family structure, we have to deal with the transmission of values from parent to offspring, and this cannot be done by means of genotype values alone, because parents pass on their alleles and not their genotypes to the next generation, genotype being created afresh in each generation. A new measure of value is therefore needed, which will refer to alleles and not genotype. The new value associated with the allele as distinct from genotypes is known as the average allelic effect.

The average effect of a particular allele is the mean deviation from the population mean of individuals, which received an allele from one parent and another allele from the other parent randomly chosen from the population.

Let α_1 and α_2 be the respective genic effects with overall population mean μ; the statistical model for the three genotypic values can be expressed as

$$\begin{aligned} G_{11} &= \mu + 2\alpha_1 + e_1 \\ G_{12} &= \mu + \alpha_1 + \alpha_2 + e_2 \\ G_{22} &= \mu + 2\alpha_2 + e_3, \end{aligned} \tag{4.8}$$

where e_1, e_2, and e_3 are the respective deviations of the genotypic values from their expectations on the basis of a perfect fit of the model. We also assume that

$$p\alpha_1 + q\alpha_2 = 0. \tag{4.9}$$

We then obtain μ, α_1, and α_2 by minimizing $Q = \sum_{i=1}^{3} f_i e_i^2$, i.e.,

$$Q = p^2 (G_{11} - \mu - 2\alpha_1)^2 + 2pq(G_{12} - \mu - \alpha_1 - \alpha_2)^2 + q^2 (G_{22} - \mu - 2\alpha_2)^2. \tag{4.10}$$

Setting $\frac{\partial Q}{\partial \mu} = 0, \frac{\partial Q}{\partial \alpha_1} = 0$, and $\frac{\partial Q}{\partial \alpha_2} = 0$ and solving these equations, we obtain

$$\begin{aligned} \hat{\mu} &= p^2 G_{11} + 2pqG_{12} + q^2 G_{22} \\ \hat{\alpha}_1 &= q\left[pG_{11} + (q - p)G_{12} - qG_{22}\right] \\ \hat{\alpha}_2 &= -p\left[pG_{11} + (q - p)G_{12} - qG_{22}\right]. \end{aligned} \tag{4.11}$$

The substitution effect is defined as

$$\alpha = \alpha_1 - \alpha_2 = pG_{11} + (q - p)G_{12} - qG_{22}. \tag{4.12}$$

The deviations of genotypic values from the fitted values, known as dominance deviations or effects, are then

$$\begin{aligned} \hat{e}_1 &= q^2 (G_{11} - 2G_{12} + G_{22}) = q^2 \delta \\ \hat{e}_2 &= -pq(G_{11} - 2G_{12} + G_{22}) = -pq\delta \\ \hat{e}_3 &= p^2 (G_{11} - 2G_{12} + G_{22}) = p^2 \delta, \end{aligned} \tag{4.13}$$

where $\delta = G_{11} - 2G_{12} + G_{22}$ is referred to as the genetic dominant effect. This gives $\sum f_i e_i = 0$, as expected.

There is another way to derive the genetic effects. Suppose that the allele A_1 is transmitted from the father to the child. Another allele of the child is randomly transmitted from mother.

The allele A_1 is transmitted from the mother with the probability p and the allele A_2 transmitted from the mother with the probability q. Then, the genic effect of the allele A_1 is given by

$$\hat{\alpha}_1 = pG_{11} + qG_{12} - \mu$$
$$= q\left[pG_{11} + (q-p)G_{12} - qG_{22}\right].$$

$\hat{\alpha}_2$ can be similarly derived.

4.1.2.3 Genetic Variance

The amount of variation is often measured and expressed as the variance. The variance is simply the mean of the squared deviations of genotypic values from the population mean. Mathematically, the genetic variance is defined as

$$\sigma_G^2 = p^2(G_{11}-\mu)^2 + 2pq(G_{12}-\mu)^2 + q^2(G_{22}-\mu)^2$$
$$= p^2(2\alpha_1 + e_1)^2 + 2pq(\alpha_1 + \alpha_2 + e_2)^2 + q^2(2\alpha_2 + e_3)^2. \quad (4.14)$$

Now we decompose the genetic variance into the genetic additive variance σ_a^2 and dominance variance σ_d^2. Some algebra gives

$$4p^2\alpha_1 e_1 + 4pq(\alpha_1 + \alpha_2)e_2 + 4q^2\alpha_2 e_3$$
$$= [q - q + p - p]4p^2q^2\alpha\delta$$
$$= 0. \quad (4.15)$$

Substituting Equation 4.15 into Equation 4.14 yields

$$\sigma_G^2 = 4p^2\alpha_1^2 + 2pq(\alpha_1 + \alpha_2)^2 + 4q^2\alpha_2^2 + p^2e_1^2 + 2pqe_2^2 + q^2e_3^2$$
$$= \sigma_A^2 + \sigma_D^2,$$

where

$$\sigma_A^2 = 4p^2\alpha_1^2 + 2pq(\alpha_1 + \alpha_2)^2 + 4q^2\alpha_2^2 \quad (4.16)$$

and

$$\sigma_D^2 = p^2e_1^2 + 2pqe_2^2 + q^2e_3^2. \quad (4.17)$$

Substituting Equations 4.11 and 4.12 into Equation 4.16 gives

$$\sigma_A^2 = 4p^2q^2\alpha^2 + 2pq(q-p)^2\alpha^2 + 4q^2p^2\alpha^2$$
$$= 2pq\alpha^2. \quad (4.18)$$

Similarly, we can show that

$$\sigma_D^2 = p^2 q^2 \delta^2. \tag{4.19}$$

σ_A^2 involves only a substitution effect and hence is referred to as additive variance, and σ_D^2 involves only the dominance effect and hence is referred to as dominance variance.

Example 4.1

We assume that 756 individuals were typed at the marker rs10838371 in gene HBG2. The frequency of the major allele Q was 0.92. The mean glucoses of three genotypes QQ, Qq, and qq were 4.98, 5.24, and 8.0, respectively. The population mean was 5.04. The substitution effect and dominance effect were, respectively,

$$\alpha = 0.92 * 4.98 + (0.08 - 0.92) * 5.24 - 0.08 * 8.0 = -0.46,$$

$$\delta = 4.98 - 2 * 5.24 + 8.00 = 2.50.$$

The genetic additive and dominance variances were, respectively,

$$\sigma_A^2 = 2 * 0.92 * 0.08 * (-0.46)^2 = 0.031$$

and

$$\sigma_D^2 = 0.92^2 * 0.082 * 2.5^2 = 0.034.$$

4.1.3 Linear Regression for a Quantitative Trait

Assume that n individuals are sampled. Let Y_i be a phenotypic value of the ith individual. A simple linear regression model for a quantitative trait is given by

$$Y_i = \mu + X_i \alpha + Z_i \delta + \varepsilon_i, \tag{4.20}$$

where μ is an overall mean and ε_i are independent and identically distributed normal variables with zero mean and variance σ_e^2:

$$X_i = \begin{cases} 2q, & A_1A_1 \\ q - p, & A_1A_2 \\ -2p, & A_2A_2 \end{cases} \quad \text{and} \quad Z_i = \begin{cases} q^2, & A_1A_1 \\ -qp, & A_1A_2 \\ p^2, & A_2A_2 \end{cases}. \tag{4.21}$$

We add a constant $-q + p$ into X_i. Then, the indicator variable will be transformed to

$$X_i = \begin{cases} 1, & A_1A_1 \\ 0, & A_1A_2 \\ -1, & A_2A_2 \end{cases},$$

which is a widely used indicator variable for additive effect in quantitative genetic analysis. Consider a marker locus with two alleles M and m having frequencies P_M and P_m, respectively. Let D be the disequilibrium coefficient of the linkage disequilibrium (LD) between the marker locus and the quantitative trait locus. The relationship between the phenotypic value Y_i and the genotype at the marker locus is given by

$$Y_i = \mu_m + X_i^m \alpha_m + Z_i^m \delta_m + \varepsilon_i, \tag{4.22}$$

where indicator variables X_i and Z_i are given by

$$X_i^m = \begin{cases} 2P_m, & MM \\ P_m - P_M, & Mm \\ -2P_M, & mm \end{cases} \quad \text{and} \quad Z_i^m = \begin{cases} P_m^2, & MM \\ -P_M P_m, & Mm \\ P_M^2, & mm \end{cases}. \tag{4.23}$$

Assuming the genetic model for the quantitative trait in Equation 4.20, it can be shown that

$$\hat{\alpha}_m \xrightarrow{a.s.} \frac{D}{P_M P_m}\alpha \quad \text{and} \quad \hat{\delta}_m \xrightarrow{a.s.} \frac{D}{P_M^2 P_m^2}\delta,$$

which implies that the estimators $\hat{\alpha}_m$ and $\hat{\delta}_m$ are consistent (Appendix 4A).

Assume that the marker is located at the genomic position t and the trait locus is at the genomic position s. Let $D(t, s)$ be the disequilibrium coefficient between the marker and trait loci. Let $\hat{\alpha}(t)$ and $\hat{\delta}(t)$ be the estimated genetic additive and dominant effects at the marker locus, and $\alpha(t)$ and $\delta(t)$ be the true additive and dominant effects at the marker locus, respectively. Let $P_M(t)$ and $P_m(t)$ be the frequencies of the marker alleles M and m at the genomic position t, respectively. Then, $\hat{\alpha}(t)$ and $\hat{\delta}(t)$ will be almost surely convergent to

$$\hat{\alpha}(t) \xrightarrow{a.s.} \frac{D(s,t)}{P_M(t)P_m(t)}\alpha(s) \quad \text{and} \quad \hat{\delta}(t) \xrightarrow{a.s.} \frac{D^2(s,t)}{P_M^2(t)P_m^2(t)}\delta(s). \tag{4.24}$$

Suppose that there are K trait loci, which are located at the genomic positions $s_1, \ldots, s_K$. The jth trait locus has the additive effect $\alpha(s_j)$ and dominance effect $\delta(s_j)$, respectively. Then, we have

$$\begin{aligned} \hat{\alpha}(t) &\xrightarrow{a.s.} \frac{1}{P_M(t)P_m(t)} \sum_{j=1}^{K} D(t, s_j)\alpha(s_j) \\ \hat{\delta}(t) &\xrightarrow{a.s.} \frac{1}{P_M^2(t)P_m^2(t)} \sum_{j=1}^{K} D^2(t, s_j)\delta(s_j), \end{aligned} \tag{4.25}$$

where $D(t, s_j)$ is the disequilibrium coefficient between the marker at the genomic position t and the trait locus at the genomic position s_j.

Equation 4.20 or 4.22 can be written in a matrix form:

$$Y = W\beta + \varepsilon. \tag{4.26}$$

The least square estimator of the regression coefficients β is given by

$$\hat{\beta} = \left(W^T W\right)^{-1} W^T Y. \tag{4.27}$$

The variance of the estimators $\hat{\beta}$ is

$$\mathrm{var}\left(\hat{\beta}\right) = \sigma_e^2 \left(W^T W\right)^{-1}, \tag{4.28}$$

where

$$\hat{\sigma}_e^2 = \frac{\left(Y - X\hat{\beta}\right)^T \left(Y - X\hat{\beta}\right)}{n-3}.$$

Asymptotically, we have

$$\mathrm{var}\left(\hat{\alpha}\right) = \frac{\sigma_e^2}{2nP_m P_M} \quad \text{and} \quad \mathrm{var}\left(\hat{\delta}\right) = \frac{\sigma_e^2}{nP_M^2 P_m^2}.$$

Statistics for testing the presence of genetic additive and dominance effects are

$$T_a = \frac{2nP_M P_m \hat{\alpha}^2}{\hat{\sigma}_e^2} \quad \text{and} \quad T_d = \frac{nP_M^2 P_m^2 \hat{\delta}^2}{\hat{\sigma}_e^2}. \tag{4.29}$$

Under the null hypothesis of no genetic additive or dominance effect, T_a or T_d is asymptotically distributed as a central $\chi^2_{(1)}$ distribution. We can also use them to test the presence of the quantitative trait locus (QTL). Under the null hypothesis of no QTL, $T_a + T_d$ is asymptotically distributed as a central $\chi^2_{(2)}$ distribution.

In general, suppose that the hypothesis being tested is given by

$$H\beta = h, \tag{4.30}$$

where

H is a $q \times p$ matrix
h is a given q-dimensional vector

Then, we have

$$\mathrm{var}\left(H\hat{\beta}\right) = H\mathrm{Var}\left(\hat{\beta}\right)H^T = \hat{\sigma}^2 H\left(W^T W\right)^{-1} H^T.$$

Define the test statistic as

$$T=\frac{\left(H\hat{\beta}-h\right)^{T}\left[H\left(W^{T}W\right)^{-1}H^{T}\right]^{-1}\left(H\hat{\beta}-h\right)}{\hat{\sigma}^{2}}. \tag{4.31}$$

Under the null hypothesis (4.30), the test statistic T is asymptotically distributed as a central $\chi^2_{(q)}$ distribution. Let $\beta=[\mu,\alpha,\delta]^T$ and Λ be the matrix obtained by removing the first row and the first column of the covariance matrix $\text{Var}\left(\hat{\beta}\right)$. The statistic for testing $\alpha=\delta=0$ is then reduced to

$$T=\left[\hat{\alpha},\hat{\delta}\right]\Lambda^{-1}\begin{bmatrix}\hat{\alpha}\\ \hat{\delta}\end{bmatrix}. \tag{4.32}$$

Then, under the null hypothesis $H_0:\alpha=0,\ \delta=0$, T is asymptotically distributed as a central $\chi^2_{(2)}$ distribution. We can show (Appendix 4A) that the noncentrality parameter of the test asymptotically converges to

$$T\xrightarrow{a.s.}\frac{n}{\sigma_e^2}\left[\frac{D^2}{P_M P_m P_Q P_q}\sigma_A^2+\frac{D^4}{P_M^2 P_m^2 P_Q^2 P_q^2}\sigma_D^2\right].$$

Next we study the squared multiple correlation coefficient, denoted as R^2. The squared multiple correlation coefficient is given by

$$R^2=\frac{\left[\text{cov}\left(w\beta,y\right)\right]^2}{\text{var}\left(w\beta\right)\sigma_y^2}=\frac{\Sigma_{yw}\Sigma_{ww}^{-1}\Sigma_{wy}}{\sigma_y^2}. \tag{4.33}$$

The squared multiple correlation coefficient R^2 quantifies the proportion of the phenotypic variation explained by the genetic variation.

If we only consider the genetic additive effect model

$$y=\mu+x\alpha+\varepsilon,$$

then at the trait locus, we have

$$\begin{aligned}R^2&=\frac{\alpha^2\,\text{var}\left(x\right)}{\sigma_y^2}\\&=\frac{2pq\alpha^2}{\sigma_p^2}\\&=\frac{\sigma_a^2}{\sigma_p^2}.\end{aligned} \tag{4.34}$$

Similarly, we can show that at the marker locus, R^2 is

$$R^2 = \frac{D^2}{pqP_M P_m} \frac{\sigma_a^2}{\sigma_p^2}. \tag{4.35}$$

In other words, R^2 is a function of the LD, frequencies at both marker and trait loci, and the ratio of the genetic additive variance over the total phenotypic variance.

4.1.4 Multiple Linear Regression for a Quantitative Trait

Consider L marker loci, which are located at the genomic positions $t_1, \ldots, t_L$. The multiple linear regression model for a quantitative trait is given by

$$Y_i = \mu_m + \sum_{l=1}^{L} X_{il}\alpha_l + \sum_{l=1}^{L} Z_{il}\delta_l + \varepsilon_i, \tag{4.36}$$

where indicator variables X_{il} and Z_{il} are similarly defined as that in Equation 4.23. Let P_{Ml} and P_{ml} be the frequencies of the alleles M_l and m_l of the marker located at the genomic positions t_l, respectively, and $D(t_l, t_j)$ be the disequilibrium coefficient of the LD between the marker at the genomic position t_l and the marker at the genomic position t_j. Assume that there are K trait loci defined as before. Let $D(t_j, s_k)$ be the disequilibrium coefficient of the LD between the marker at the genomic position t_j and the trait locus at the genomic position s_k. By a similar argument as that in Appendix 4A, we have

$$\hat{\alpha}_m \xrightarrow{a.s.} P_A^{-1}\alpha \quad \text{and} \quad \hat{\delta}_m \xrightarrow{a.s.} P_D^{-1}\delta, \tag{4.37}$$

where $\hat{\alpha}_m = \left[\hat{\alpha}_1, \ldots, \hat{\alpha}_L\right]^T, \hat{\delta}_m = \left[\hat{\delta}_1, \ldots, \hat{\delta}_L\right]^T,$

$$P_A = \begin{bmatrix} 2P_{M_1}P_{m1} & D(t_1, t_2) & \cdots & D(t_1, t_L) \\ D(t_2, t_1) & 2P_{M_2}P_{m_2} & \cdots & D(t_1, t_L) \\ \cdots & \cdots & \cdots & \cdots \\ D(t_L, t_1) & D(t_L, t_2) & \cdots & 2P_{M_L}P_{m_L} \end{bmatrix},$$

$$P_D = \begin{bmatrix} P_{M_1}^2 P_{m_1}^2 & D^2(t_1, t_2) & \cdots & D^2(t_1, t_L) \\ D^2(t_2, t_1) & P_{M_2}^2 P_{m_2}^2 & \cdots & D^2(t_1, t_L) \\ \cdots & \cdots & \cdots & \cdots \\ D^2(t_L, t_1) & D^2(t_L, t_2) & \cdots & P_{M_L}^2 P_{m_L}^2 \end{bmatrix},$$

$$\alpha = \left[\sum_{j=1}^{K} D(t_1, s_j)\alpha(s_j), \ldots, \sum_{j=1}^{K} D(t_L, s_j)\alpha(s_j)\right]^T,$$

and

$$\delta = \left[\sum_{j=1}^{K} D^2(t_1, s_j)\delta(s_j), \ldots, \sum_{j=1}^{K} D^2(t_L, s_j)\delta(s_j) \right]^T.$$

If we assume that all markers are in linkage equilibrium, then Equation 4.37 is reduced to

$$\hat{\alpha}(t) = \frac{1}{P_{M_t} P_{m_t}} \sum_{j=1}^{k} D(t, s_j)\alpha(s_j), \tag{4.38}$$

$$\hat{\delta}(t) = \frac{1}{P_{M_t}^2 P_{m_t}^2} \sum_{j=1}^{K} D^2(t, s_j)\delta(s_j), \tag{4.39}$$

which is exactly the same as Equation 4.25. In other words, under the assumption of linkage equilibrium among markers, multiple linear regression can be decomposed into a number of simple regressions. Under this assumption and the genetic additive effect model without interactions, the squared multiple correlation coefficient between the phenotype and marker located at the genomic position t is given by

$$\begin{aligned} R^2 &= \frac{2P_{M_t} P_{m_t} \alpha^2(t)}{\sigma_P^2} \\ &\approx \frac{1}{\sigma_P^2} \sum_{j=1}^{K} \frac{D^2(t, s_j)}{P_Q(s_j) P_q(s_j) P_{M_t} P_{m_t}} \sigma_a^2(s_j), \end{aligned} \tag{4.40}$$

where

$P_Q(s_j)$ and $P_q(s_j)$ are the frequencies of alleles Q and q of the trait locus at the genomic position s_j, respectively

P_{M_t} and P_{m_t} are the frequencies of the alleles M and m of the marker at the genomic position t, respectively

$\sigma_a^2(s_j)$ is the genetic additive variance of the trait locus at the genomic position s_j

We can show that (Exercise 3.5)

$$R^2 \le \sum_{j=1}^{K} \frac{\sigma_a^2(s_j)}{\sigma_P^2}.$$

To test the association of multiple markers with the quantitative trait, we rewrite Equation 4.36 in a matrix form:

$$Y = W\beta + \varepsilon,$$

where $Y = \begin{bmatrix} y_1 \\ \vdots \\ y_n \end{bmatrix}, W = \begin{bmatrix} 1 & x_{11} & \cdots & x_{1L} & z_{11} & \cdots & z_{1L} \\ \vdots & \ddots & \iddots & \vdots & \ddots & \iddots & \vdots \\ 1 & x_{n1} & \cdots & x_{nL} & z_{n1} & \cdots & z_{nL} \end{bmatrix}$, and $\beta = [\mu, \alpha_1, \ldots, \alpha_L, \delta_1, \ldots, \delta_L]^T$.

The least square estimator of the regression coefficients β is given by

$$\hat{\beta}=\left(W^TW\right)^{-1}W^TY.$$

Let $\beta_g=[\alpha_1,\ldots,\alpha_L,\delta_1,\ldots,\delta_L]^T$, $\text{var}\left(\hat{\beta}\right)=\sigma_e^2\left(W^TW\right)^{-1}$, $\hat{\sigma}_e^2=\frac{\left(Y-X\hat{\beta}\right)^T\left(Y-X\hat{\beta}\right)}{n-2L-1}$, and Λ be the matrix obtained by removing the first row and the first column of the covariance matrix $\text{Var}\left(\hat{\beta}\right)$. Similar to the previous section, we can define the test statistic:

$$T_Q=\left(\hat{\beta}_g\right)^T\Lambda^{-1}\hat{\beta}_g. \tag{4.41}$$

Under the null hypothesis $H_0:\alpha_1=\ldots=\alpha_L=0,\ \delta_1=\ldots,\delta_L=0$, T_Q is asymptotically distributed as a central $\chi^2_{(2L)}$ distribution. To investigate what factors affect the power of the test for the detection of QTLs, we derive the formula for the calculation of the noncentral parameter. Asymptotically, the noncentrality parameter of the test in Equation 4.41 is (Appendix 4A)

$$T_Q\to\frac{n}{\sigma_e^2}\left[\sum_{j=1}^{L}\sum_{k=1}^{L}2D_{jk}\alpha_j\alpha_k+\sum_{j=1}^{L}\sum_{k=1}^{L}D_{jk}^2\delta_j\delta_k\right],$$

where

D_{ij} is the coefficient of the LD between the ith marker and jth marker
P_{M_j} and P_{m_j} are the frequencies of alleles M_j and m_j at the jth marker, respectively

We consider three special scenarios:

1. All markers are trait loci.
   We define the genetic additive covariance and dominance covariance as follows:

$$\sigma^2_{A(j,k)}=2D_{jk}\alpha_j\alpha_k,$$
$$\sigma^2_{D(j,k)}=D_{jk}^2\delta_j\delta_k.$$

   The noncentrality parameter of the test under this scenario asymptotically converges to

$$T_Q\to\frac{n}{\sigma_e^2}\left[\sum_{j=1}^{L}\sum_{k=1}^{L}\sigma^2_{A(i,j)}+\sum_{j=1}^{L}\sum_{k=1}^{L}\sigma^2_{D(j,k)}\right].$$

2. All markers are trait loci and are in linkage equilibrium.

Under this scenario, the above equation is reduced to

$$T_Q \to \frac{n}{\sigma_e^2}\left[\sum_{j=1}^{L}\sigma_{A(j)}^2 + \sum_{j=1}^{L}\sigma_{D(j)}^2\right].$$

This formula shows that the noncentrality parameter or power of the test depends on the genetic additive variance, dominance variance, and the strength of the LD between the marker and trait loci, between the trait loci, and between markers. In other words, the power of the test depends on the genomic structure of the genes. Some genes are easier to detect associations than other genes.

4.2 GENE-BASED QUANTITATIVE TRAIT ANALYSIS

Next-generation sequencing techniques will generate unprecedented massive, high-dimensional genetic variation data and provide a powerful tool for detecting the entire allelic spectrum of the causal genetic variations. Despite their promise, next-generation sequencing (NGS) platforms also have three specific disadvantages: high error rates, enrichment of rare variants, and large proportion of missing values (Bansal et al. 2010). To meet challenges in QTL analysis raised by NGS, a simple and natural idea is group tests that combine multiple rare variants into a single variable to predict quantitative phenotype variation (Bacanu et al. 2011). The regression-type group tests that aggregate information across variants for prediction have a higher power than the individual variant tests.

However, regression-type group tests ignore differences in genetic effects among SNPs at different genomic locations and do not leverage LD in the data. Analysis from low-dimensional data to high-dimensional genomic data demands changes in statistical methods from multivariate data analysis to functional data analysis. An alternative to group regression tests, the functional linear model (FLM) with scale response is a natural extension of the multivariate regression for quantitative genetic analysis. The FLM can collectively test for the association of genetic variants with quantitative traits, but it can also allow for heterogeneity of genetic effects. It can utilize the merit of both individual and group tests. In Chapter 3, we introduce SKAT as a statistical method for testing association of a set of variants with a dichotomous phenotype. In this chapter, we will also briefly investigate SKAT that can be used for testing the association of a set of rare variants with a quantitative trait.

4.2.1 Functional Linear Model for a Quantitative Trait

4.2.1.1 Model

The multiple linear regression model, which jointly uses multiple marker information, in general, might have a higher power to detect a QTL than the simple linear regression model. However, as the number of markers increase, the degree of freedom of the test statistics will also increase. As a typical example, in UK10K dataset, we observe that more than 1400 genes have more than 500 SNPs. This will compromise the power of multiple regressions for identifying QTLs. To reduce the degrees of freedom in the model due to the presence of a large number of rare variants in the model, we consider a functional linear model with scalar response for a quantitative trait where the genetic effects are defined as a function of genomic position, and the quantitative trait is predicted by the genotype score function (Fan et al. 2013; Luo et al. 2012).

Let t be a genomic position. Define a genotype function $X_i(t)$ for an additive effect and genotype function $Z_i(t)$ for a dominance effect of the ith individual as

$$X_i(t)=\begin{cases}2P_m(t), & MM\\ P_m(t)-P_M(t), & Mm\\ -2P_M(t), & mm\end{cases} \tag{4.42}$$

$$Z_i(t)=\begin{cases}P_m^2(t), & MM\\ -P_M(t)P_m(t), & Mm\\ P_M^2(t), & mm\end{cases} \tag{4.43}$$

where

M and m are two alleles of the marker at the genomic position t

$P_M(t)$ and $P_m(t)$ are the frequencies of the alleles M and m, respectively

Let Y_i be a phenotype value of the ith individual. A functional linear model for a quantitative trait with genetic additive effect is defined as

$$Y_i=\mu+\int_0^T X_i(t)\alpha(t)dt+\int_0^T Z_i(t)\delta(t)dt+\varepsilon_i, \tag{4.44}$$

where

ε_i are independent and identically distributed normal variables with mean of zero and variance σ_e^2

T is the length of the genome region being considered

$\alpha(t)$ and $\delta(t)$ are the putative genetic additive effect and dominance functions of the marker at the genomic position t, respectively

For convenience, the genome region $[0, T]$ is rescaled to $[0, 1]$. If the integrals in Equation 4.43 are discretized, the functional linear model will be reduced to multiple linear regression models (4.36).

4.2.1.2 Parameter Estimation

Three types of approaches have been developed for estimating regression coefficient functions in the functional linear models (Febrero-Bande et al. 2010). The first approach uses basis function expansion and penalized methods to estimate regression coefficient functions (Ramsay and Silverman 2005). The second approach uses functional principal component expansions and least square to estimate regression coefficient functions (Cardot et al. 1999). The third approach uses nonparametric estimates based on kernels (Frank and Friedman 1993).

We explore to use restricted basis functions and functional principal component expansions to estimate the additive and dominant effect functions. Since the genotype functions are nonperiodic functions, we use *B*-spline basis functions to expand the genotype functions and additive and dominant effect functions.

Let the domain $[0,1]$ be subdivided into knot spans by a set of nondecreasing numbers, $0 = u_0 \le u_1 \le u_2 \le \cdots \le u_m = 1$. The $u_i^{'}$ s are called knots.

The *i*th *B*-spline basis function of degree *p*, written as $B_{i,p}(t)$, is defined recursively as follows:

$$B_{i,0}(t) = \begin{cases} 1 & \text{if } u_i \le t \le u_{i+1} \\ 0 & \text{otherwise} \end{cases}$$

$$B_{i,p}(t) = \frac{t - u_i}{u_{i+p} - u_i} B_{i,p-1}(t) + \frac{u_{i+p+1} - t}{u_{i+p+1} - u_{i+1}} B_{i+1,p-1}(t).$$

B-spline basis functions have two important features:

1. Basis function $B_{i,p}(t)$ is nonzero only on $p+1$ knot spans: $[u_i, u_{i+1}), [u_{i+1}, u_{i+2}), \ldots, [u_{i+p}, u_{i+p+1})$.
2. Given any knot span $[u_i, u_{i+1})$, there are at most $p+1$ degree p basis functions that are nonzero, namely,

$$B_{i-p,p}(t), B_{i-p+1,p}(t), B_{i-p+2,p}(t), \ldots, B_{i-1,p}(t) \text{ and } B_{i,p}(t).$$

Let $B_K(t)$ be a *B*-spline basis function if we set *p* to be a specific integer (i.e., $p = 3$ yields cubic *B*-spline basis series). We expand the genotype functions $X_i(t)$ and $Z_i(t)$ in terms of *B*-spline basis functions. Let $B(t)$ be a vector of *B*-spline basis functions of length K_G. Then, we have

$$X_i(t) = \sum_{k=1}^{K_G} u_{ik} B_k(t) \quad \text{and} \quad Z_i(t) = \sum_{k=1}^{K_G} v_{ik} B_k(t). \tag{4.45}$$

The coefficients of the expansion u_{ik} and v_{ik} can be obtained by minimizing the least square criterion:

$$\sum_{j=1}^{T} \left[X_i(t_j) - \sum_{k=1}^{K_G} B_k(t_j) u_{ik} \right]^2 \quad \text{and} \quad \sum_{j=1}^{T} \left[Z_i(t_j) - \sum_{k=1}^{K_G} B_k(t_j) v_{ik} \right]^2. \tag{4.46}$$

Let $X_i = [X_i(t_1), \ldots, X_i(t_T)]^T$, $Z_i = [Z_i(t_1), \ldots, Z_i(t_T)]^T$, $u_i = [u_{i1}, \ldots, u_{iK_G}]^T$, $v_i = [v_{i1}, \ldots, v_{iK_G}]^T$,

$$B = \begin{bmatrix} B_1(t_1) & \cdots & B_1(t_T) \\ \cdots & \cdots & \cdots \\ B_{K_G}(t_1) & \cdots & B_{K_G}(t_T) \end{bmatrix}.$$

The least square estimators of the expansion coefficients are then given by

$$\hat{u}_i = \left(B^T B\right)^{-1} B^T X_i \quad \text{and} \quad \hat{v}_i = \left(B^T B\right)^{-1} B^T Z_i. \tag{4.47}$$

The expansion of the genetic effect functions can be similarly written as

$$\alpha(t) = \sum_{k=1}^{K_\beta} \alpha_k \theta_k(t) = \theta^T \alpha \quad \text{and} \quad \delta(t) = \sum_{k=1}^{K_\beta} \delta_k \theta_k(t) = \theta^T \delta, \tag{4.48}$$

where $\theta = [\theta_1(t), \ldots, \theta_{K_\beta}(t)]^T$, $\alpha = [\alpha_1, \ldots, \alpha_{K_\beta}]^T$, and $\delta = [\delta_1, \ldots, \delta_{K_\beta}]^T$.

Let $Y = [Y_1, \ldots, Y_n]^T$, $X = [X_1, \ldots, X_n]^T$, $Z = [Z_1, \ldots, Z_n]^T$, $U = [u_1, \ldots, u_n]^T$, $V = [v_1, \ldots, v_n]^T$, $B(t) = [B_1(t), \ldots, B_{K_G}(t)]^T$. Then, we have

$$\begin{aligned} X &= UB(t), \\ Z &= VB(t). \end{aligned}$$

The functional linear model (4.44) can then be written as

$$\begin{aligned} Y &= \mu I + \int_0^T UB(t)\theta^T(t)\alpha dt + \int_0^T VB(t)\theta^T(t)\delta dt + \varepsilon \\ &= \mu I + U\left[\int_0^T B(t)\theta^T(t)dt\right]\alpha + V\left[\int_0^T B(t)\theta^T(t)dt\right]\delta + \varepsilon. \end{aligned} \tag{4.49}$$

Let

$$J_{\beta\theta} = \begin{bmatrix} \int_0^T B_1(t)\theta_1(t)dt & \cdots & \int_0^T B_1(t)\theta_{K_\beta}(t)dt \\ \cdots & \cdots & \cdots \\ \int_0^T B_{K_\beta}(t)\theta_1(t)dt & \cdots & \int_0^T B_{K_\beta}(t)\theta_{K_\beta}(t)dt \end{bmatrix},$$

$W = [1, UJ_{B\theta}, VJ_{B\theta}]$, and $\beta = [\mu, \alpha^T, \delta^T]^T$. Then, Equation 4.49 can be rewritten as

$$Y = W\beta. \tag{4.50}$$

The least square estimate of the parameter vector β is given by

$$\hat{\beta} = \left(W^T W\right)^{-1} W^T Y.$$

If we consider only the genetic additive effect function, Equation 4.49 is reduced to

$$Y = \mu I + U\left[\int_0^T B(t)\theta^T(t)dt\right]\alpha + \varepsilon. \tag{4.51}$$

We can also expand genotype functions in terms of functional principal component scores discussed in Chapter 3:

$$x_i(t) = \sum_{k=1}^{L} \xi_{ik}\varphi_k(t),$$

$$z_i(t) = \sum_{k=1}^{L} \eta_{ik}\varphi_k(t), \tag{4.52}$$

where

$$\xi_{ik} = \int_0^T x_i(t)\varphi_k(t) \quad \text{and} \quad \eta_{ik} = \int_0^T z_i(t)\varphi_k(t)dt. \tag{4.53}$$

Then, model (4.44) can be reduced to

$$\begin{aligned} Y_i &= \mu + \int_0^T \sum_{k=1}^{L} \xi_{ik}\varphi_k(t)\alpha(t)dt + \int_0^T \sum_{k=1}^{L} \eta_{ik}\varphi_k\delta(t)dt + \varepsilon_i \\ &= \mu + \sum_{k=1}^{L} \xi_{ik}\int_0^T \varphi_k(t)\alpha(t)dt + \sum_{k=1}^{L} \eta_{ik}\int_0^T \varphi_k(t)\delta(t)dt + \varepsilon_i \\ &= \mu + \sum_{k=1}^{L} \xi_{ik}\alpha_k + \sum_{k=1}^{L} \eta_{ik}\delta_k + \varepsilon_i, \end{aligned} \tag{4.54}$$

where $\alpha_k = \int_0^T \varphi_k(t)\alpha(t)dt$ and $\delta_k = \int_0^T \varphi_k(t)\delta(t)dt$.

Let

$$Y=\begin{bmatrix} Y_1 \\ \vdots \\ Y_n \end{bmatrix},\quad W=\begin{bmatrix} 1 & \xi_{11} & \cdots & \xi_{1L} & \eta_{11} & \cdots & \eta_{1L} \\ \vdots & \ddots & \cdot^{\cdot^{\cdot}} & \vdots & \ddots & \cdot^{\cdot^{\cdot}} & \vdots \\ 1 & \xi_{n1} & \cdots & \xi_{nL} & \eta_{n1} & \cdots & \eta_{nL} \end{bmatrix},\quad \varepsilon=\begin{bmatrix} \varepsilon_1 \\ \vdots \\ \varepsilon_n \end{bmatrix},$$

and $\beta=\left[\mu,\alpha_1,\ldots,\alpha_L,\delta_1,\ldots,\delta_L\right]^T$.

Model (4.54) can be written in a matrix form:

$$Y=W\beta+\varepsilon. \tag{4.55}$$

Equation 4.54 indicates that after the functional principal component expansion, the functional linear model (4.44) is reduced to the multiple linear model (4.36) where the expansion coefficients of genotype functions replace the genotype indicator variables. Again, the least square estimate of the parameter vector β is given by

$$\hat{\beta}=\left(W^TW\right)^{-1}W^TY. \tag{4.56}$$

The estimators of the genetic additive and dominance effect functions are

$$\alpha(t)=\sum_{k=1}^{L}\alpha_k\varphi_k(t) \quad\text{and}\quad \delta(t)=\sum_{k=1}^{L}\delta_k\varphi_k(t). \tag{4.57}$$

We can show that the estimators of the expansion coefficients of the genetic additive and dominance effects asymptotically converge to (Appendix 4B)

$$\hat{\alpha}_k\xrightarrow{a.s.}\frac{2}{\lambda_k}\left(\int_0^T D(t,s)\varphi_k(t)dt\right)\alpha, \tag{4.58}$$

$$\hat{\delta}_k\xrightarrow{a.s.}\frac{1}{\gamma_k}\left(\int_0^T D^2(t,s)\varphi_k(t)\delta\right),\quad k=1,\ldots,L, \tag{4.59}$$

where the true QTL is located at the genomic position s and $D(t,s)$ is the LD coefficient between the marker located at the genomic position t and the true QTL, λ_k and γ_k are eigenvalues of functional principal component expansions of the genotype functions for the additive effect and dominance effects, respectively.

We can also show that the variance of the estimators asymptotically converges to (Appendix 4B)

$$\begin{aligned} \operatorname{var}\left(\hat{\alpha}_k\right)&\xrightarrow{a.s.}\frac{\sigma_e^2}{n\lambda_k}, \\ \operatorname{var}\left(\hat{\delta}_k\right)&\xrightarrow{a.s.}\frac{\sigma_e^2}{n\gamma_k}. \end{aligned} \tag{4.60}$$

Let

$$W_r = \left[1, UJ_{B\theta}\right] \text{ and } \beta_r = \left[\mu \quad \alpha^T\right]^T.$$

Then, Equation 4.50 is reduced to

$$Y = W_r \beta_r. \tag{4.61}$$

Therefore, the least square estimator of the parameter vector β_r is

$$\hat{\beta}_r = \left(W_r^T W_r\right)^{-1} W_r^T Y. \tag{4.62}$$

4.2.1.3 Test Statistics

An essential problem in genetic studies of the quantitative trait is to test the association of a genomic region with the quantitative trait. Formally, we investigate the problem of testing the following hypothesis:

$$H_0 : \alpha(t) = 0 \text{ and } \delta(t) = 0, \ \forall t \in [0, T] \tag{4.63}$$

against

$$H_a : \alpha(t) \neq 0 \text{ or } \delta(t) \neq 0 \text{ or } \alpha(t) \neq 0 \text{ and } \delta(t) \neq 0.$$

If the genetic effect functions $\alpha(t)$ and $\beta(t)$ are expanded in terms of the basis functions

$$\alpha(t) = \theta(t)^T \alpha \text{ and } \delta(t) = \theta(t)^T \delta,$$

then testing the null hypothesis H_0 in Equation 4.63 is equivalent to testing the hypothesis

$$H_0 : \alpha = 0 \text{ and } \delta = 0. \tag{4.64}$$

Let $\hat{b} = \left[\alpha^T, \delta^T\right]^T$ and A be the matrix obtained by removing the first row of the matrix $(W^T W)^{-1} W^T$. Then, we obtain from Equation 4.50 that

$$\hat{b} = AY.$$

The covariance matrix of the estimator $\hat{b}$ is then given by

$$\Lambda = \text{var}\left(\hat{b}\right) = \left(AA^T\right)\hat{\sigma}_e^2,$$

where

$$\hat{\sigma}_e^2 = \frac{1}{N-2K_\beta-1} Y^T \left[I - W\left(W^T W\right)^{-1} W^T \right] Y. \tag{4.65}$$

Statistics for testing the association of the genomic region with the quantitative trait can be constructed by estimators of the expansion coefficients of the genetic effect functions. Define the test statistic:

$$T_Q = \hat{b}^T \Lambda^{-1} \hat{b}. \tag{4.66}$$

Then, under the null hypothesis $H_0:\alpha=0$ and $\delta=0$, T_Q is asymptotically distributed as a central $\chi^2_{(2K_\beta)}$ distribution.

The noncentrality parameter of the test T_Q is

$$T_Q \to \frac{n}{\sigma_e^2}\left[\sum_{k=1}^{L} \lambda_k \alpha_k^2 + \sum_{k=1}^{L} \gamma_k \delta_k^2\right]$$

or

$$T_Q \to \frac{n}{\sigma_e^2}\left\{\left[\frac{2}{P_Q P_q} \sum_{k=1}^{L} \lambda_k \alpha_k^2 \frac{1}{\lambda_k}\left(\int_0^T D(t,s)\varphi_k(t)dt\right)^2\right]\sigma_A^2 + \left[\frac{1}{P_Q^2 P_q^2} \sum_{k=1}^{L} \frac{1}{\gamma_k}\left(\int_0^T D^2(t,s)\varphi_k(t)\right)^2\right]\sigma_D^2\right\}.$$

The noncentrality parameter and power of the test depends on the sample size, variance of errors, frequencies of trait alleles, genetic additive and dominance variance, variance of the coefficients of the genotype function expansions, and coefficients of the measure of linkage disequilibrium FPCA expansion.

We also can use the F test for linear regression to test for association. Let the full model be given by

$$Y = W\beta \tag{4.67}$$

and the reduced model be given by

$$Y = \mu I. \tag{4.68}$$

Then, the F statistic is

$$T_F = \frac{\left\{\left(Y-\bar{Y}\right)^T\left(Y-\bar{Y}\right)-Y^T\left[I-W\left(W^TW\right)^{-1}W^T\right]Y\right\}\left(N-2K_\beta-1\right)}{Y^T\left[I-W\left(W^TW\right)^{-1}W^T\right]Y\cdot 2K_\beta}. \tag{4.69}$$

Under the null hypothesis of no association, the statistic T_F is distributed as $F_{2K_\beta, N-2K_\beta-1}$ distribution.

Next we consider a special case where only the genetic additive function $\alpha(t)$ is studied. Let A_r be the matrix that is obtained by removing the first row of the matrix $\left(W_r^TW_r\right)^{-1}W_r^T$ and $b_r=[\mu,\alpha^T]^T$. Then, it follows from Equation 4.62 that

$$\hat{b}_r = A_rY.$$

The covariance matrix of the estimator $\hat{b}_r$ is

$$\Lambda_r = \left(A_rA_r^T\right)\sigma_{er}^2,$$

where

$$\sigma_{er}^2 = \frac{1}{N-K_\beta-1}Y^T\left[I-W_r\left(W_r^TW_r\right)^{-1}W_r^T\right]Y.$$

Then, two statistics for testing the association of the genomic region are

$$T_{rQ} = \hat{b}_r^T\Lambda_r^{-1}\hat{b}_r \tag{4.70}$$

and

$$T_{rF} = \frac{N-K_\beta-1\left(Y-\bar{Y}\right)^T\left(Y-\bar{Y}\right)-Y^T\left[I-W_r\left(W_r^TW_r\right)^{-1}W_r^T\right]Y}{Y^T\left[I-W_r\left(W_r^TW_r\right)^{-1}W_r^T\right]Y}. \tag{4.71}$$

Under the null hypothesis of no association of the genomic region with quantitative trait $H_0:\alpha=0$, T_{rQ} is asymptotically distributed as a central $\chi^2_{(K_\beta)}$ distribution and T_{rF} is asymptotically distributed as a $F_{K_\beta, N-K_\beta-1}$ distribution.

4.2.2 Canonical Correlation Analysis for Gene-Based Quantitative Trait Analysis

4.2.2.1 Multivariate Canonical Correlation Analysis

Alternative to FLM, canonical correlation analysis (CCA) provides another statistical framework for testing the association of a gene or genomic region with a quantitative trait (Press 2011). The goal of CCA is to seek optimal correlation between a quantitative trait and a linear combination of SNPs within a gene. The CCA measures the strength of association between the multiple SNPs and the trait.

Consider a quantitative trait y and L SNPs with indicator variables for the genetic additive and dominance effects $x_1, \ldots, x_L, z_1, \ldots, z_L$ that are similarly defined in Equation 4.23. Let $w = [x_1, \ldots, x_L, z_1, \ldots, z_L]$. Define the variance and covariance matrices:

$$\Sigma_{yy} = \sigma_y^2, \quad \Sigma_{yg} = \begin{bmatrix} \Sigma_{yx} & \Sigma_{yz} \end{bmatrix}, \text{ and } \Sigma_{gg} = \begin{bmatrix} \Sigma_{xx} & \Sigma_{xz} \\ \Sigma_{zx} & \Sigma_{zz} \end{bmatrix},$$

where

$$\Sigma_{yx} = \begin{bmatrix} \operatorname{cov}(y, x_1) & \cdots & \operatorname{cov}(y, x_L) \end{bmatrix} = \Sigma_{xy}^T,$$

$$\Sigma_{xx} = \begin{bmatrix} \sigma_{x_1x_1} & \cdots & \sigma_{x_1x_L} \\ \vdots & \ddots & \vdots \\ \sigma_{x_Lx_1} & \cdots & \sigma_{x_Lx_L} \end{bmatrix}, \quad \Sigma_{xz} = \begin{bmatrix} \sigma_{x_1z_1} & \cdots & \sigma_{x_1z_L} \\ \vdots & \ddots & \vdots \\ \sigma_{x_Lz_1} & \cdots & \sigma_{x_Lz_L} \end{bmatrix}, \text{ and } \Sigma_{zz} = \begin{bmatrix} \sigma_{z_1z_1} & \cdots & \sigma_{z_1z_L} \\ \vdots & \ddots & \vdots \\ \sigma_{z_Lz_1} & \cdots & \sigma_{z_Lz_L} \end{bmatrix}.$$

Recall Equation 1.220 defined in Chapter 1:

$$R = \Sigma_{yy}^{-1/2} \Sigma_{yg} \Sigma_{gg}^{-1} \Sigma_{gy} \Sigma_{yy}^{-1/2}.$$

Since $\Sigma_{yy} = \sigma_y^2$ is a number, the matrix R is reduced to

$$R = \frac{\Sigma_{yg} \Sigma_{gg}^{-1} \Sigma_{gy}}{\sigma_y^2}.$$

Its eigenvalue is

$$\lambda^2 = \frac{\Sigma_{yg} \Sigma_{gg}^{-1} \Sigma_{gy}}{\sigma_y^2}. \tag{4.72}$$

Let $\hat{\sigma}_y^2, S_{yg}, S_{gg}$, and S_{gy} be sampling versions of $\sigma_y^2, \Sigma_{yg}, \Sigma_{gg}$, and Σ_{gy}. Then,

$$\hat{\lambda}^2 = \frac{S_{yg} S_{gg}^{-1} S_{gy}}{\hat{\sigma}_y^2}. \tag{4.73}$$

The statistic for testing the association of the gene with the trait is defined as

$$T_{CCA} = -N \log\left(1 - \hat{\lambda}^2\right). \tag{4.74}$$

Under the null hypothesis of no association of the gene with the trait, T_{CCA} is a central $\chi^2_{(2L)}$ distribution.

The noncentrality parameter of the test T_{CCA} is

$$\bar{T}_{CCA} = -N \log\left(1-\lambda^2\right). \tag{4.75}$$

We can show that (Appendix 4C) asymptotically

$$\bar{T}_{CCA} > T_Q,$$

where T_Q is defined in Equation 4.41. This indicates that the CCA has a higher power to detect trait locus than the test statistic defined in Equation 4.41 for multiple linear regression analysis. However, as we show in Chapter 5, in general, the tests defined in CCA are almost equivalent to the tests defined in linear regression analysis.

4.2.2.2 Functional Canonical Correlation Analysis

Sequenced genes or genomic regions may contain more than thousands of SNPs. The degree of freedom of the test T_{CCA} is the number of SNPs in the gene being tested and hence is very large. To reduce the degrees of freedom of the test, we can use functional data analysis techniques to reduce the dimensions of the data and extend the multivariate CCA to functional CCA (FCCA) for gene-based QTL analysis. Since functional CCA for a single trait genetic analysis is a special case of the functional CCA for genetic analysis of multiple traits, detailed discussion will be provided in Section 5.4.3.

4.3 KERNEL APPROACH TO GENE-BASED QUANTITATIVE TRAIT ANALYSIS

Next-generation sequencing produces high-dimensional genomic data. Dimension reduction is a key to the success of the sequence-based association analysis. If the model for sequencing data is too highly parameterized, it will react too strongly to the data, which will come overfitting the sequencing data, and will fail to learn the underlying data generating process. The concept of "kernels" will provide a flexible and efficient method for data reduction. However, kernel algorithms and reproducing kernel Hilbert space (RKHS) are difficult to understand. We first review RKHS and some simple kernel algorithms (Shawe-Taylor and Cristianini 2004) and then introduce kernel regression and kernel canonical correlation analysis. Finally, we will discuss the relationship between kernel algorithms and functional data analysis.

4.3.1 Kernel and RKHS

4.3.1.1 Kernel and Nonlinear Feature Mapping

Consider an example. We have two variables in one dimension arranged in an interval $[-3, 3]$ as shown in Figure 4.2. We want to separate the red point + from the green point * by a linear classifier. In the original data space, it is clear that we are unable to separate two types of points by a linear classifier. However, if we make a nonlinear feature map,

$$\phi(x) = x^2.$$

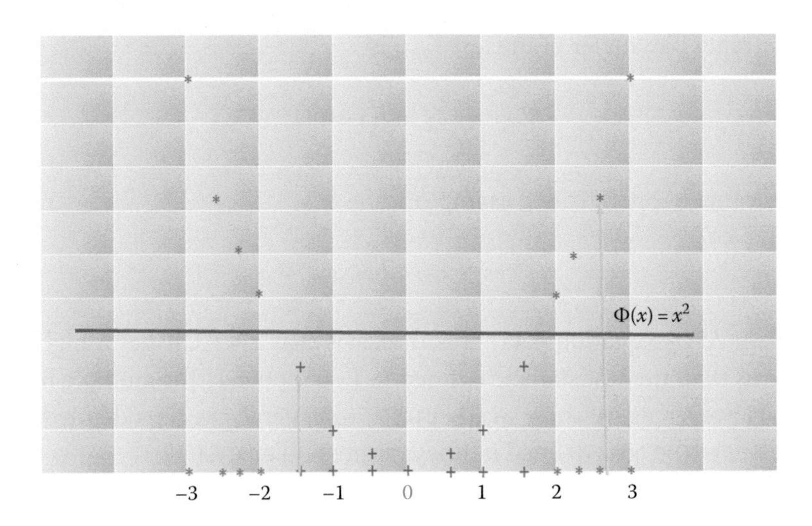

FIGURE 4.2 Pattern of feature map $\Phi(x)=x^2$.

As Figure 4.2 shows, in the mapped feature space, a line can easily separate the red points from the green points.

This example shows that if we choose the map appropriately, complex relations in the original data space can be simplified in the mapped feature space. We start defining a feature map with a vector space. Let $x \in R^m$. Consider a feature map

$$\phi(x): R^m \to R^N.$$

For example,

$$x = \begin{bmatrix} x_1 \\ x_2 \end{bmatrix} \text{ and } \phi(x) = \begin{bmatrix} x_1 \\ x_2 \\ x_1 x_2 \end{bmatrix}.$$

Although a feature map $\phi(x)$ can efficiently transform the nonlinear relations in the original feature space into linear relations in the mapped new feature space, it is, in general, difficult to explicitly find such a nonlinear map. To avoid finding complicated nonlinear feature maps, we introduce the concept of a kernel. A kernel is a similarity measure. We use an inner product to measure similarity between two vectors x and y. In a vector space, an inner product is defined as

$$\langle x, y \rangle = x^T y = \sum_{i=1}^{m} x_i y_i.$$

Recall that in Chapter 1, a Euclidean norm of x is defined as

$$\|x\|_2 = \sqrt{< x, x >} = \sqrt{x^T x}.$$

A linear space in which an inner product is defined is referred to as a Hilbert space that is denoted by H. Suppose that x and y are mapped to a new feature space via the map $\phi(x)$. Then, the inner product of new points, $\phi(x)$ and $\phi(y)$, in the new feature space is defined as

$$\langle \phi(x), \phi(y) \rangle = \phi^T(x)\phi(y) = \sum_{i=1}^{m} \phi_i(x)\phi_i(y). \tag{4.76}$$

Now we are ready to define a kernel.

Definition (Kernel).

A function $k: \chi \times \chi \rightarrow R$ is called a kernel if there exists a map $\phi: \chi \rightarrow \mathrm{H}$ such that for all $x, y \in \chi$,

$$k(x, y) = \langle \phi(x), \phi(y) \rangle_H. \tag{4.77}$$

Example 4.2

Consider two points $x = [x_1, x_2]^T$ and $y = [y_1, y_2]^T$ in R^2. Define a kernel function:

$$\begin{aligned} k(x,y) &= (x^T y)^2 \\ &= (x_1 y_1 + x_2 y_2)^2 \\ &= x_1^2 y_1^2 + x_2^2 y_2^2 + 2x_1 y_1 x_2 y_2 \\ &= \left[x_1, \sqrt{2}x_1 x_2, x_2\right]^T \left[y_1, \sqrt{2}y_1 y_2, y_2\right] \\ &= \phi^T(x)\phi(y), \end{aligned}$$

where $\phi(x) = \left[x_1, \sqrt{2}x_1 x_2, x_2\right]^T$ and $\phi(y) = \left[y_1, \sqrt{2}y_1 y_2, y_2\right]^T$ are feature mapping from input space R^2 to higher feature space R^3. It is clear that we can directly compute the kernel function $k(x, y)$ without knowing the feature map ϕ.

Example 4.3

Consider a linear kernel: $k(x, y) = x^T y$.

Both feature maps

$$\phi(x) = x \quad \text{and} \quad \phi(x) = \begin{bmatrix} x/\sqrt{2} \\ x/\sqrt{2} \end{bmatrix} \text{ define the same linear kernel } k(x,y).$$

Example 4.4 The Gaussian Kernel

The Gaussian kernel is defined as

$$k(x, y) = \exp\left(-\frac{\|x - y\|_2^2}{2\sigma^2}\right). \tag{4.78}$$

Recall that $\|x - y\|_2^2 = x^T x - 2x^T y + y^T y$.

The Gaussian kernel can then be reduced to

$$k(x, y) = \exp\left(\frac{x^T y}{\sigma^2}\right)\exp\left(-\frac{\|x\|_2^2}{2\sigma^2}\right)\exp\left(-\frac{\|y\|_2^2}{2\sigma^2}\right). \tag{4.79}$$

Define the feature map

$$\phi(x) = \begin{bmatrix} 1 \\ \frac{x}{\sigma} \\ \vdots \\ \frac{x^j}{\sigma\sqrt{j!}} \\ \vdots \end{bmatrix} \exp-\left(\frac{\|x\|_2^2}{2\sigma^2}\right).$$

It is clear that $k(x, y) = \exp\left(-\frac{\|x - y\|_2^2}{2\sigma^2}\right) = \phi^T(x)\phi(y)$.

Therefore, the Gaussian function defines a kernel.

Example 4.5

Consider a nonempty set χ and a feature map:

$$\phi(x) = \begin{bmatrix} \phi_1(x) \\ \vdots \\ \phi_i(x) \\ \vdots \end{bmatrix},$$

where $\phi_i(x)$ is in L_2.

Define a kernel as $K(x, y) = \phi^T(x)\phi(y) = \sum_{i=1}^{\infty} \phi_i(x)\phi_i(y)$.

4.3.1.2 The Reproducing Kernel Hilbert Space

Now we study how to use kernels to define functions on a space. The space of such functions is called a reproducing kernel Hilbert space (RKHS) (Gretton 2015a). As a motivating example, we consider a function:

$$f(x) = a_1 x_1 + a_2 x_2 + a_3 x_1 x_2. \tag{4.80}$$

We define two feature maps:

$$f(.) = \begin{bmatrix} a_1 \\ a_2 \\ a_3 \end{bmatrix} \tag{4.81}$$

and

$$\phi(x) = \begin{bmatrix} x_1 \\ x_2 \\ x_3 \end{bmatrix}.$$

The function $f(x)$ that is often referred to as the function evaluated at a particular point x can be expressed as an inner product in feature space:

$$\begin{aligned} f(x) &= f^T(.)\phi(x) \\ &= \langle f(.), \phi(x) \rangle_H \end{aligned} \tag{4.82}$$

We can use the feature map $\phi(x)$ to define a kernel:

$$\begin{aligned} k(x,y) &= x_1 y_1 + x_2 y_2 + x_3 y_3 \\ &= \phi^T(x)\phi(y). \end{aligned} \tag{4.83}$$

We can denote

$$\phi(x) = k(.,x) \quad \text{and} \quad \phi(y) = k(.,y).$$

Then, function $f(x)$ can be defined as

$$f(x) = \langle f(.), k(.,x) \rangle_H, \tag{4.84}$$

and the kernel $k(x, y)$ can be written as

$$k(x, y) = \langle k(.,x), k(.,y) \rangle_H. \tag{4.85}$$

Equation 4.84 shows that the evaluation of function $f(.)$ at point x is an inner product of a kernel in a feature space. We can view $\phi(x)$ or $k(.,x)$ as a feature map from input space R^2 to feature space R^3 or as a function mapping from R^2 to R. H is a space of function from R^2 to R.

Now we consider an infinite-dimensional functional space. We define an inner product in the functional space as

$$\langle f(x), g(x)\rangle = \int_T f(x)g(x)dx. \tag{4.86}$$

Consider a set of orthonormal basis functions: $\{e_i(x), i=1,2,\ldots\}$ with

$$\langle e_i, e_j\rangle = \int_T e_i(x)e_j(x)dx = \begin{cases} 1 & i=j \\ 0 & i\neq j \end{cases}. \tag{4.87}$$

Define an infinite-dimensional feature map $\phi(x)$:

$$\phi(x) = \begin{bmatrix} \sqrt{\lambda_1}e_1(x) \\ \vdots \\ \sqrt{\lambda_i}e_i(x) \\ \vdots \end{bmatrix}$$

and a kernel $k(x,y)$:

$$k(x,y) = \phi^T(x)\phi(y) = \infty\sum_{i=1}\lambda_i e_i(x)e_i(y). \tag{4.88}$$

Define a function $f(.)$:

$$f(.) = \begin{bmatrix} f_1 \\ \vdots \\ f_i \\ \vdots \end{bmatrix}.$$

Evaluation of function f at x is

$$f(x) = \langle f(.), \phi(x)\rangle = \sum_{i=1}^{\infty} f_i\sqrt{\lambda_i}e_i(x). \tag{4.89}$$

If we take $k(.,x) = \phi(x)$, then Equation 4.89 is reduced to

$$f(x) = \langle f(.), k(.,x)\rangle.$$

In the above example, the question is how to find f_i. Next we study a general class of functions and investigate a general procedure to define kernel functions. We consider a real function defined on the interval $[-\pi, \pi]$ with periodic boundary conditions. It is well known that the function can be expanded in terms of the Fourier series:

$$f(x)=\sum_{l=-\infty}^{\infty} f_j e^{ilx}, \tag{4.90}$$

where $e^{ilx}=\cos(lx)+i\sin(lx)$. When the function is real, its Fourier expansion coefficients are conjugate symmetric:

$$f_{-l}=\overline{f_l}.$$

Suppose that a kernel is a function of a single argument:

$$k(x,y)=k(x-y),$$

where $k(x)$ can be any function that can be expanded by the Fourier series.

Its Fourier series representation is

$$k(x)=\sum_{l=-\infty}^{\infty} k_l e^{ilx}.$$

Now we define the feature map

$$\phi(x)=k(.,x)=\begin{bmatrix} \vdots \\ \sqrt{k_l}\, e^{-ilx} \end{bmatrix}.$$

Its inner product is

$$\begin{aligned}\langle k(.,x),k(.,y)\rangle &= \phi^T(x)\overline{\phi(y)} \\ &= \sum_{l=-\infty}^{\infty} k_l e^{-ilx} e^{ily} \\ &= \sum_{l=-\infty}^{\infty} k_l e^{il(y-x)} = k(x-y).\end{aligned}$$

Define function $f(.)$ as

$$f(.)=\begin{bmatrix} \vdots \\ \frac{f_l}{\sqrt{k_l}} \\ \vdots \end{bmatrix}.$$

Then, evaluation of function $f(.)$ at x is

$$\begin{aligned} f(x) &= \langle f(.), \overline{k(.,x)} \rangle \\ &= \sum_{k=-\infty}^{\infty} \frac{f_l}{\sqrt{k_l}} \sqrt{k_l} e^{ilx} \\ &= \sum_{-\infty}^{\infty} f_l e^{ilx}. \end{aligned}$$

Definition 4.1

(Reproducing kernel Hilbert space)

Let H be a Hilbert space of real value functions defined on a nonempty set, χ. A function $k(.,.): \chi \times \chi \to R$ is a reproducing kernel of a Hilbert space H, and H is a reproducing kernel Hilbert space, if for any function in the Hilbert space, we have

(1) For all $x \in \chi$, kernel function $k(.,x) \in H$
(2) As to the reproducing property, for all $x \in \chi$ and function $f \in H$, $\langle f(.), k(.,x) \rangle_H = f(x)$

In particular, for all $x, y \in \chi$,

$$k(x,y) = \langle k(.,x), k(.,y) \rangle_H . \tag{4.91}$$

Equation 4.91 indicates that $\phi(x) = k(.,x)$ is a valid feature map.

The operator of the function evaluation can give an equivalent definition of reproducing kernel Hilbert space.

Definition 4.2

(Evaluation operator)

Let H be a Hilbert space of functions, $f: \chi \to R$. For a fixed $x \in R$, the map $\delta_x: H \to R$ is called the function evaluation operator and denoted by

$$\delta_x f = f(x).$$

It is easy to show that the evaluation operator is linear. By definition, we have

$$\delta_x(\alpha f + \beta g) = \alpha f(x) + \beta g(x) = \alpha \delta_x f + \beta \delta_x g.$$

Now we give another RKHS definition based on the evaluation operator.

Definition 4.3

(Reproducing kernel Hilbert space based on the evaluation operator)

H is an RKHS if for all $x \in \chi$, the evaluation operator δ_x is bounded. In other words, there exists a corresponding $\lambda_x \geq 0$ such that for all $f \in H$,

$$|\delta_x f| = |f(x)| \leq \lambda_x \|f\|_H . \tag{4.92}$$

The definition implies that when two functions have the same RKHS norms, two functions agree at every point:

$$|f(x) - g(x)| = |\delta_x (f - g)| \leq \lambda_x \|f - g\|_H .$$

Now we show that two definitions are equivalent.

Theorem 4.1

RKHS equivalence theorem.

Evaluation operator δ_x is a bounded linear operator on a Hilbert space H if and only if H has a reproducing kernel.

Proof.

We first show that if H has a reproducing kernel, then the evaluation operator is bounded. By definition,

$$\begin{aligned}
|\delta_x f| &= |f(x)| \\
&= |\langle K(.,x), f(.)\rangle|_H \\
&=\leq \|k(.x)\|_H \|f\|_H \quad \text{(by Cauchy–Schwarz inequality)} \\
&= \left(\langle K(.,x), k(.,x)\rangle_H\right)^{1/2} \|f\|_H \\
&= \lambda_x \|f\|_H ,
\end{aligned}$$

where $\lambda_x = (\langle k(.,x), k(.,x)\rangle_H)^{1/2}$.

This shows that the evaluation operator is bounded.

Another direction can be shown by the Riesz theorem. Recall that the Riesz theorem (Pavone 1994) stated that if δ is a bounded linear operator on a Hilbert space H, then there exists some $g \in H$ such that for every $f \in H$, we have

$$\delta(f) = \langle f, g\rangle_H . \tag{4.93}$$

Consider bounded linear evaluation operator δ_x. It follows from Equation 4.93 that

$$\delta_x(f) = \langle f, g_x\rangle_H . \tag{4.94}$$

Let $g_x(y) = k(y, x)$ for all $y \in \chi$. It is clear that

$$k(.,x) = g_x \in H. \tag{4.95}$$

Recall that

$$\delta_x(f) = f(x). \tag{4.96}$$

Substituting Equations 4.95 and 4.96 into Equation 4.94 gives

$$f(x) = \langle f, k(.,x) \rangle_H,$$

which states that k is the reproducing kernel.

Assume that observations $x_1, x_2, \ldots$ are given. Feature maps $k(x_1,.), k(x_2,.), \ldots$ can be taken as a set of bases in the RKHS. Functions $f(x)$ in RKHS can then be expressed as linear combinations of the bases $k(x_1,.), k(x_2,\ldots), \ldots$:

$$f(.) = \sum_{i=1}^{m} \alpha_i k(x_i,.). \tag{4.97}$$

Assume that the function $g(.)$ can be expressed as

$$g(.) = \sum_{j=1}^{n} \beta_j k(x_j,.).$$

Now the inner product $\langle f(.), g(.) \rangle_H$ is given by

$$\begin{aligned} \langle f(.), g(.) \rangle_H &= \sum_{i=1}^{m} \sum_{j=1}^{n} \alpha_i \beta_j \langle k(x_i,.), k(x_j,.) \rangle_H \\ &= \sum_{i=1}^{m} \sum_{j=1}^{n} \alpha_i \beta_j k(x_i, x_j) \\ &= \alpha^T K \beta, \end{aligned} \tag{4.98}$$

where

$$\alpha = \begin{bmatrix} \alpha_1 \\ \vdots \\ \alpha_m \end{bmatrix}, \quad \beta = \begin{bmatrix} \beta_1 \\ \vdots \\ \beta_n \end{bmatrix}, \quad K = \begin{bmatrix} k(x_1, x_1) & \cdots & k(x_1, x_n) \\ \vdots & \vdots & \vdots \\ k(x_m, x_1) & \cdots & k(x_m, x_n) \end{bmatrix}.$$

The matrix is referred to as a Gram matrix.

Example 4.6 FPCA and Karhunen–Loeve Expansion

Recall that in functional principal component analysis (Section 1.5.3), covariant function is defined as $R(s,t)=\text{cov}(X(s),X(t))$, and functional principal components can be obtained by solving integral eigenfucntion (1.193a):

$$\int_T R(s,t)\beta(s)ds = \lambda\beta(t), \tag{4.99}$$

where
- $\beta(s)$ is a functional principal component
- λ is an eigenevalue

The covariance function induces the kernel operator:

$$(R\beta)(t) = \int_T R(s,t)\beta(s)ds. \tag{4.100}$$

In Equations 1.193a and 1.193b, we show that

$$(R\beta_j)(t) = \lambda_j\beta_j(t), \quad j = 1,2,\ldots. \tag{4.101}$$

We can show that $R(s,t)$ can be expressed as

$$R(s,t) = \sum_{j=1}^{\infty}\lambda_j\beta_j(s)\beta_j(t). \tag{4.102}$$

In the functional RKHS, we can define a kernel by functional principal components:

$$k(s,t) = \sum_{j=1}^{\infty}\beta_j(s)\beta_j(t). \tag{4.103}$$

In FPCA of genomic analysis, we take a genotypic profile as a genotype function, $x(t)$, and expand it in terms of functional principal component (Equation 1.208):

$$x(t) = \sum_{j=1}^{\infty}\xi_j\beta_j(t), \tag{4.104}$$

where

$$\xi_j = \langle x(t),\beta_j(t)\rangle = \int_T x(t)\beta_j(t)dt.$$

It is clear that

$$\langle x(.), k(.,t)\rangle_H = \int_T x(s)\sum_{j=1}^{\infty}\beta_j(s)\beta_j(t)ds$$
$$= \sum_{j=1}^{\infty}\int_T x(s)\beta_j(s)ds\beta_j(t)$$
$$= \sum_{j=1}^{\infty}\xi_j\beta_j(t) = x(t).$$

This shows that $k(s,t)$ is a reproducing kernel.

4.3.2 Covariance Operator and Dependence Measure

The covariance operator is an extension of the covariance matrix. It is a useful tool for assessing dependence between variables and hence forms a foundation for association analysis that is based on either multivariate analysis or functional data analysis.

However, the covariance operator involves an abstract concept and is difficult to understand. Few statistical and genetic literatures cover covariance operator materials. To lay a foundation for modern association analysis, we introduce the basic theory of the covariance operator and dependence measure.

4.3.2.1 Hilbert–Schmidt Operator and Norm

Consider two separable Hilbert spaces, Γ and H. Let T be an operator: $T: H \to \Gamma$. If there exists an orthonormal basis, $e_i, i = 1, 2, \ldots$, in H such that $\sum_{i=1}^{n}\left\|Te_j\right\|^2 < \infty$, then T is called a **Hilbert–Schmidt** operator.

Let $(e_i)_{i\in I}$ be an orthonormal basis in Γ and $(h_j)_{j\in J}$ be the orthonormal basis in H. The **Hilbert–Schmidt norm** of the operator T is defined as

$$\|T\|_{HS}^2 = \sum_{j\in J}\|Th_j\|_\Gamma^2. \tag{4.105}$$

However, $Th_j \in \Gamma$ and can be expressed as

$$Th_j = \sum_{i\in I}\langle Th_j, e_i\rangle_\Gamma e_i \tag{4.106}$$

and

$$\|Th_j\|_\Gamma^2 = \sum_{i\in I}\left|\langle Th_j, e_i\rangle_\Gamma\right|^2. \tag{4.107}$$

Substituting Equations 4.106 and 4.107 into Equation 4.105 gives

$$\|T\|_{HS}^2 = \sum_{j\in J}\sum_{j\in J}\left|\langle Th_j, e_i\rangle_\Gamma\right|^2. \qquad (4.108)$$

Assume that $i=1,2,\ldots$ and $j=1,2,\ldots,$. Denote a_{ij} $\langle Th_j, e_i\rangle_\Gamma$. Then, $\|T\|_{HS}^2$ can be written as

$$\|T\|_{HS}^2 = \sum_{i=1}^{\infty}\sum_{j=1}^{\infty}\left|a_{ij}\right|^2.$$

We can show that the Hilbert–Schmidt norm of the operator is independent of the choice of orthonormal basis (Exercise 4.11).

Example 4.7 Hilbert–Schmidt Integral Kernels

Let $L^2(\Omega)$ be the space of square-integrable functions and $k(x,y)\in L^2(\Omega\times\Omega)$ be a kernel. Define integral operators with the Hilbert–Schmidt kernel:

$$T_K u(x) = \int_\Omega k(x,y)u(y)dy.$$

The space $L^2(\Omega)$ has a countable orthonormal basis, $\{\phi_i(x), i=1,2,\ldots\}$, and hence $\{\phi_i(x)\phi_j(y), i=1,2,\ldots,j=1,2,\ldots\}$ is an orthonormal basis of $L^2(\Omega\times\Omega)$. The kernel function $k(x,y)$ can be expanded as

$$k(x,y) = \sum_{i=1}^{\infty}\sum_{j=1}^{\infty} k_{ij}\phi_i(x)\phi_j(y),$$

where $k_{ij} = \int_\Omega\int_\Omega k(x,y)\phi_i(x)\phi_j(y)dxdy$.

By definition of the Hilbert–Schmidt norm of the operator T_K, we have

$$\begin{aligned}
\|T_K\|_{HS}^2 &= \sum_{i=1}^{\infty}\sum_{j=1}^{\infty}\left|\langle T_K\phi_j, \phi_i\rangle\right|^2 \\
&= \sum_{i=1}^{\infty}\sum_{j=1}^{\infty}\left|\int_\Omega\int_\Omega k(x,y)\phi_i(x)\phi_j(y)dxdy\right|^2 \\
&= \sum_{i=1}^{\infty}\sum_{j=1}^{\infty} k_{ij}^2.
\end{aligned}$$

The Hilbert–Schmidt norm of the operator forms a Hilbert space, written as $HS(H,\Gamma)$. Let $S: H \to \Gamma$ and $T: H \to \Gamma$ be two compact linear operators. Define the inner product

$$\langle S,T \rangle_{HS} = \sum_{j \in J} \langle Sh_j, Th_j \rangle_\Gamma . \tag{4.109}$$

Recall that $\{e_i, i=1,2,\ldots\}$ is an orthonormal basis in Γ. The elements Sh_j and Th_j in Γ can be expanded as

$$Sh_j = \sum_{i \in I} \langle Sh_j, e_i \rangle_\Gamma e_i \tag{4.110}$$

and

$$Th_j = \sum_{i \in I} \langle Th_j, e_i \rangle_\Gamma e_i . \tag{4.111}$$

Therefore,

$$\langle Sh_j, Th_j \rangle_\Gamma = \sum_{i \in I} \langle Sh_j, e_i \rangle_\Gamma \langle Th_j, e_i \rangle_\Gamma . \tag{4.112}$$

Substituting Equation 4.112 into Equation 4.109 gives

$$\langle S,T \rangle_{HS} = \sum_{i \in I} \sum_{j \in J} \langle Sh_j, e_i \rangle_\Gamma \langle Th_j, e_i \rangle_\Gamma .$$

Result 4.1

The inner product $\langle S, T \rangle_{HS}$ can be calculated by

$$\langle S,T \rangle_{HS} = \sum_{i \in I} \sum_{j \in J} \langle Sh_j, e_i \rangle_\Gamma \langle Th_j, e_i \rangle_\Gamma . \tag{4.113}$$

4.3.2.2 Tensor Product Space and Rank-One Operator

The tensor product is a useful concept for functional analysis but less present in statistical and genetic literature. Here, we intuitively and briefly introduce a tensor product and rank-one operator in a Hilbert space (Gretton 2015).

4.3.2.2.1 Tensor Product for Vectors and Matrices Tensor product is a generalization of the ordinary product of scalar numbers. Let a and b be two numbers. Their product is simply ab. Now suppose that a and b are two vectors. What does a product of two vectors mean?

In Chapter 1, we define an inner product of two vectors as a linear map of two vectors to a real number. Can we define a linear map: $T:(u,v)\to w$? The tensor product of two vectors, denoted by $\otimes$, is such a product. The tensor product is also called a Kronecker product or direct product.

Consider an example. Let

$$u=\begin{bmatrix} u_1 \\ \vdots \\ u_n \end{bmatrix} \text{ and } v=\begin{bmatrix} v_1 \\ \vdots \\ v_m \end{bmatrix}.$$

The tensor product $u\otimes v$ is defined as

$$u\otimes v = uv^T = \begin{bmatrix} u_1 \\ \vdots \\ u_n \end{bmatrix}\begin{bmatrix} v_1 & \cdots & v_m \end{bmatrix} = \begin{bmatrix} u_1v_1 & \cdots & u_1v_n \\ \vdots & \vdots & \vdots \\ u_nv_1 & \cdots & u_nv_m \end{bmatrix}. \quad (4.114)$$

Now we define the tensor product of two matrices. Let

$$A=\begin{bmatrix} a_{11} & \cdots & a_{1m} \\ \vdots & \vdots & \vdots \\ a_{n1} & \cdots & a_{nm} \end{bmatrix} \text{ and } B=\begin{bmatrix} b_{11} & \cdots & b_{1l} \\ \vdots & \vdots & \vdots \\ b_{k1} & \cdots & b_{kl} \end{bmatrix}.$$

The tensor product of two matrices, denoted as $A\otimes B$, is defined as

$$A\otimes B=\begin{bmatrix} a_{11}B & \cdots & a_{1m}B \\ \vdots & \vdots & \\ a_{n1}B & \cdots & a_{nm}B \end{bmatrix}. \quad (4.115)$$

4.3.2.2.2 Rank-One Operator and Tensor Product in Hilbert Space There are two ways to define a tensor product in a Hilbert space. One way is to define the tensor product via axioms. Another way is to define the tensor product via rank-one operator on a Hilbert space. We take a second approach to define the tensor product in a Hilbert space.

Recall that one-rank matrix T can be formed by two vectors:

$$T = uv^T = u\otimes v. \quad (4.116)$$

Viewing the one-rank matrix T as an operator on a linear space for any vector f in the linear space, we have

$$Tf = (u\otimes v)f = uv^T f = (v^T f)u. \quad (4.117)$$

Using Equation 4.113, we can define the rank-one operator and tensor product in a Hilbert space. Let H_1 and H_2 be two Hilbert spaces and $u \in H_1, v \in H_2$. An operator $T: H_1 \to H_2$ is of rank one if and only if

$$Tf = \langle v, f \rangle_{H_2} u. \tag{4.118}$$

The tensor product $u \otimes v$ in the Hilbert space is then defined as a rank-one operator:

$$(u \otimes v) f = \langle v, f \rangle_{H_2} u. \tag{4.119}$$

Now we show the result.

Result 4.2

The tensor product $u \otimes v$ is a Hilbert–Schmidt operator.

Proof.

Let $\{e_j\}_{j \in J}$ be the orthonormal bases in the Hilbert space H_1. From Equation 4.101, we can compute the norm:

$$\begin{aligned} \|v \otimes u\|_{HS}^2 &= \sum_{j \in J} \|(v \otimes u) e_j\|_{H_2}^2 \\ &= \sum_{j \in J} \left\| \langle u, e_j \rangle_{H_1} v \right\|_{H_2}^2 \\ &= \|v\|_{H_2}^2 \sum_{j \in J} \left| \langle u, e_j \rangle_{H_1} \right|^2 \\ &= \|v\|_{H_2}^2 \|u\|_{H_1}^2 . \end{aligned}$$

We assume that u and v are bounded. Therefore, the Hilbert–Schmidt norm of the tensor product $u \otimes v$ is bounded, which indicates that the tensor product or rank-one operator $u \otimes v$ is a Hilbert–Schmidt operator.

Next we study how to define the inner product in a tensor product space. Let $L \in HS(H_1, H_2)$. We first compute

$$\langle L, v \otimes u \rangle_{HS} .$$

From Equation 4.105, we have

$$\begin{aligned} \langle L, v \otimes u \rangle_{HS} &= \sum_{j \in J} \langle Le_i, (v \otimes u) e_j \rangle_{H_2} \\ &= \sum_{j \in J} \left\langle Le_j, \langle u, e_j \rangle_{H_1} v \right\rangle_{H_2} \\ &= \sum_{j \in J} \langle u, e_j \rangle_{H_1} \langle Le_j, v \rangle_{H_2} . \end{aligned} \tag{4.120}$$

Next we compute

$$\langle v, Lu \rangle_{H_2}.$$

Suppose that the expansion of u in terms of the orthonormal basis is

$$u = \sum_{j \in J} \langle u, e_j \rangle_{H_1} e_j, \quad (4.121)$$

which gives

$$\begin{aligned} \langle v, Lu \rangle_{H_2} &= \left\langle v, L \sum_{j \in J} \langle u, e_j \rangle_{H_1} e_j \right\rangle_{H_2} \\ &= \sum_{j \in J} \langle u, e_j \rangle_{H_1} \langle v, Le_j \rangle_{H_2}. \end{aligned} \quad (4.122)$$

Comparing Equations 4.116 and 4.118 gives

$$\langle L, v \otimes u \rangle_{HS} = \langle v, Lu \rangle_{H_2}. \quad (4.123)$$

Equation 4.119 allows defining the inner product between two rank-one operators. Let $L = b \otimes a$. Using Equation 4.119, we obtain

$$\begin{aligned} \langle b \otimes a, v \otimes u \rangle_{HS} &= \langle L, v \otimes u \rangle_{HS} \\ &= \langle v, Lu \rangle_{H_2}. \end{aligned} \quad (4.124)$$

It follows from Equation 4.115 that

$$\begin{aligned} Lu &= (b \otimes a) u \\ &= \langle a, u \rangle_{H_1} b. \end{aligned} \quad (4.125)$$

Substituting Equation 4.121 into Equation 4.120 gives

$$\langle v, Lu \rangle_{H_2} = \langle b, v \rangle_{H_2} \langle a, u \rangle_{H_1}.$$

Therefore, we obtain Result 4.3.

Result 4.3

The inner product of two tensor products in the Hilbert–Schmidt operator space can be computed by

$$\langle b \otimes a, v \otimes u \rangle_{HS} = \langle b, v \rangle_{H_2} \langle a, u \rangle_{H_1}. \quad (4.126)$$

4.3.2.3 Cross-Covariance Operator

The covariance matrix between two random vectors in the multivariate analysis investigates the linear relationships between random variables. However, in many cases, we need to consider the nonlinear relationships between the variables via the kernel approach. The cross-covariance operator is a generalization of the covariance matrix to infinite-dimensional feature space.

For the convenience of presentation, we assume that all random variables are centered. Recall the covariance matrix $\tilde{\Sigma}_{XY} = E\left[XY^T\right]$ and $f^T\tilde{\Sigma}_{XY}g = E\left[\left(f^Tx\right)\left(g^Ty\right)^T\right]$, where

$$X = \begin{bmatrix} x_1 \\ \vdots \\ x_n \end{bmatrix}, \quad Y = \begin{bmatrix} y_1 \\ \vdots \\ y_m \end{bmatrix}, \quad f = \begin{bmatrix} f_1 \\ \vdots \\ f_n \end{bmatrix}, \quad \text{and} \quad g = \begin{bmatrix} g_1 \\ \vdots \\ g_m \end{bmatrix}.$$

Similar to Equation 4.116, we have

$$\tilde{\Sigma}_{XY} = E\left[X \otimes Y\right]. \tag{4.127}$$

We often use the covariance matrix to measure linear dependence. If two variables or two vectors of variables are nonlinearly related, the nonlinear dependence measure should be developed. Let $f(X)$ and $g(Y)$ be nonlinear functions of random variables; we extend linear covariance to nonlinear covariance to measure nonlinear dependence between two variables:

$$\operatorname{cov}\left(f(X), g(Y)\right) = E\left[f(X)g(Y)\right] - E\left[f(X)\right]E\left[g(Y)\right]. \tag{4.128}$$

Taking $f(X) = I_A(X)$ and $g(Y) = I_B(Y)$ as indicator functions for the sets A and B gives

$$\begin{aligned} \operatorname{cov}\left(f(X), g(Y)\right) &= E\left[I_A(X)I_B(Y)\right] - E\left[I_A(X)\right]E\left[I_B(Y)\right] \\ &= P\left(X \in A, Y \in B\right) - P\left(X \in A\right)P\left(Y \in B\right). \end{aligned} \tag{4.129}$$

Equation 2.33 indicates that nonlinear covariance $\operatorname{cov}(I_A(X), I_B(Y))$ is equal to zero if and only if the random variables X and Y are independent. We do not need to make normal distribution assumptions of the random variables X and Y to ensure that $\operatorname{cov}(f(X), g(Y)) = 0$ implies the independence between the random variables X and Y. Therefore, the maximum of nonlinear covariance over all possible nonlinear functions

$$\sup_{f,g} \operatorname{cov}\left(f(X), g(Y)\right) \tag{4.130}$$

can be used to measure the dependence between random variables.

RKHS is a powerful tool for functional analysis. We restrict functions in the RKHS. Assume that H_1 and H_2 are reproducing kernel Hilbert spaces with respective reproducing kernels k and l. We consider two feature maps, ϕ and ψ. After nonlinear mapping from input spaces to feature spaces nonlinear relationships between the original variables become linear in the feature space. In the feature space, Equation 4.123 is transformed to

$$\tilde{\Sigma}_{XY} = E_{X,Y}\left[\phi(X)\otimes\phi(Y)\right], \tag{4.131}$$

where $\phi(X)$ and $\phi(Y)$ are the feature maps from the original input space to the feature space. The cross product or tensor product $\phi(X)\otimes\phi(Y)$ is a random variable in $HS(H_1, H_2)$. First we show that the expectation operator of the cross product $E_{X,Y}[\phi(X)\otimes\phi(Y)]$ in the Hilbert space exists. In other words, we show that for any operator or element, $A \in HS(H_1, H_2)$, the linear functional $E_{X,Y}\langle\phi(X)\otimes\phi(Y), A\rangle_{HS}$ can be written uniquely as the inner product:

$$\langle g, A\rangle_{HS} = E_{X,Y}\langle\phi(X)\otimes\phi(Y), A\rangle_{HS}, g \in HS(H_1, H_2). \tag{4.132}$$

We denote g by $\tilde{\Sigma}_{XY}$. Therefore, Equation 4.128 can be written as

$$\langle E_{X,Y}[\phi(X)\otimes\phi(Y), A\rangle_{HS} = E_{X,Y}\langle\phi(X)\otimes\phi(Y), A\rangle_{HS}. \tag{4.133}$$

Now we show the results of the existence of the cross-covariance operator.

Results 4.4

If we assume $E_{XY}[\|\phi(X)\otimes\phi(Y)\|_{HS}] < \infty$, then there exists a unique element in the HS space, denoted by $\tilde{\Sigma}_{XY}$, such that

$$\langle\tilde{\Sigma}_{XY}, A\rangle_{HS} = E_{XY}\left[\langle\phi(X)\otimes\phi(Y), A\rangle_{HS}\right]. \tag{4.134}$$

We can write $\tilde{\Sigma}_{XY} = E_{X,Y}\left[\phi(X)\phi(Y)\right]$.

Proof.

Assume $A \in HS$. The inner product $\langle\phi(X)\otimes\phi(Y), A\rangle_{HS}$ is a functional of A. The expectation $E_{XY}[\langle\phi(X)\otimes\phi(Y), A\rangle_{HS}]$ is a function of $\langle\phi(X)\otimes\phi(Y), A\rangle_{HS}$. Therefore, we can view $E_{XY}[\langle\phi(X)\otimes\phi(Y), A\rangle_{HS}]$ as a linear functional of $A \in HS$. If we can show that the linear functional $E_{XY}[\langle\phi(X)\otimes\phi(Y), A\rangle >_{HS}]$ is bounded, then by the Riesz representation theorem, there is a unique element, $g \in HS$, such that

$$\langle g, A\rangle_{HS} = E_{XY}\left[\langle<\phi(X)\otimes\phi(Y), A\rangle\right]_{HS}.$$

We denote g by $\tilde{\Sigma}_{XY}$. Equation 4.130 is then proved.

Now we show that $E_{X,Y}$ is the bounded linear functional. Using Jensen's inequality gives

$$\left|E_{X,Y}\left[\left\langle \phi(X)\otimes\phi(Y),A\right\rangle_{HS}\right]\right|\leq E_{X,Y}\left[\left|\left\langle \phi(X)\otimes\phi(Y),A\right\rangle >_{HS}\right|\right]. \quad (4.135)$$

Applying the Cauchy–Schwartz inequality, the inner product should be less than the product of the norms of their components, i.e.,

$$\left|\left\langle \phi(X)\otimes\phi(Y),A\right\rangle\right|_{HS}\leq\|A\|_{HS}\|\phi(X)\otimes\phi(Y)\|_{HS},$$

which implies

$$E_{X,Y}\left[\left|\left\langle \phi(X)\otimes\phi(Y),A\right\rangle\right|\right]\leq E_{X,Y}\left[\|A\|_{HS}\|\phi(X)\otimes\phi(Y)\|_{HS}\right]$$
$$<\infty. \quad \text{(by assumption)} \quad (4.136)$$

This shows that the linear functional $E_{X,Y}[\langle\phi(X)\otimes\phi(Y),A\rangle_{HS}]$ is bounded.

Before we solve the optimization problem (4.130) for assessing independence, we need to compute $\text{cov}(f(X),g(X))$.

Recall from Equation 4.78 that

$$f(X)=\left\langle f,\phi(X)\right\rangle_{H_1} \quad \text{and} \quad g(Y)=\left\langle g,\phi(Y)\right\rangle_{H_2}. \quad (4.137)$$

The cross-covariance operator is a useful tool for achieving this. For the noncentered nonlinear covariance, we have

$$\text{cov}\left(f(X),g(X)\right)=E_{X,Y}\left[f(X)g(Y)\right]. \quad (4.138)$$

Substituting Equation 4.133 into Equation 4.134 gives

$$\text{cov}\left(f(X),g(X)\right)=E_{X,Y}\left[\left\langle f,\varphi(X)\right\rangle_{H_1}\left\langle g,\varphi(Y)\right\rangle_{H_2}\right]. \quad (4.139)$$

It follows from Equation 4.122 that

$$\left\langle f,\phi(X)\right\rangle_{H_1}\left\langle g,\phi(Y)\right\rangle_{H_2}=\left\langle f\otimes g,\phi(X)\otimes\phi(Y)\right\rangle_{HS}. \quad (4.140)$$

Substituting Equation 4.136 into Equation 4.135, we obtain

$$\text{cov}\left(f(X),g(X)\right)=E_{X,Y}\left[\left\langle f\otimes g,\phi(X)\otimes\phi(Y)\right\rangle_{HS}\right]. \quad (4.141)$$

Let $A=f\otimes g$. Using Equation 4.130 gives

$$E_{X,Y}\left[\left\langle f\otimes g,\phi(X)\otimes\phi(Y)\right\rangle_{HS}\right]=\left\langle\tilde{\Sigma}_{XY},f\otimes g\right\rangle_{HS}. \quad (4.142)$$

It follows from Equation 4.119 that

$$\left\langle \tilde{\Sigma}_{XY}, f \otimes g \right\rangle_{HS} = \left\langle f, \tilde{\Sigma} g \right\rangle_{HS}.$$

This proves Result 4.5.

Result 4.5

$$\operatorname{cov}\left(f(X), g(X)\right) = \left\langle f, \tilde{\Sigma} g \right\rangle_{HS}. \tag{4.143}$$

This result indicates that the nonlinear covariance can be calculated by the cross-covariance operator. Now we define the centered cross-covariance operator as

$$\begin{aligned} \Sigma_{XY} &= \tilde{\Sigma}_{XY} - \left(E_X\left[\phi(X)\right]\right) \otimes \left(E_Y\left[\phi(Y)\right]\right) \\ &= E_{X,Y}\left[\phi(X) \otimes \phi(Y)\right] - \left(E_X\left[\phi(X)\right]\right) \otimes \left(E_Y\left[\phi(Y)\right]\right). \end{aligned} \tag{4.144}$$

Applying the traditional sampling techniques gives the following sampling formula for the centered cross-covariance operator:

$$\hat{\Sigma}_{XY} = \frac{1}{n}\sum_{i=1}^{n} \phi(x_i) \otimes \phi(y_i) - \left(\frac{1}{n}\sum_{i=1}^{n} \phi(x_i)\right) \otimes \left(\frac{1}{n}\sum_{i=1}^{n} \phi(y_i)\right). \tag{4.145}$$

Using Equation 4.112, we can write $\phi(x_i) \otimes \phi(y_i)$ as $\phi(x_i)\phi^T(y_i)$. Then, Equation 4.111 can be reduced to

$$\begin{aligned} \hat{\Sigma}_{XY} &= \frac{1}{n}\sum_{i=1}^{n} \phi(x_i)\phi^T(y_i) - \left(\frac{1}{n}\sum_{i=1}^{n} \phi(x_i)\right)\left(\frac{1}{n}\sum_{i=1}^{n} \phi(y_i)\right)^T \\ &= \frac{1}{n}\Phi(x)\Phi^T(x) - \left(\frac{1}{n}\Phi(x)\mathbf{1}_n\right)\left(\frac{1}{n}\Phi(y)\mathbf{1}_n\right)^T \\ &= \frac{1}{n}\Phi(x)\left[I_n - \frac{1}{n}\mathbf{1}_n\mathbf{1}_n^T\right]\Phi^T(y) \\ &= \frac{1}{n}\Phi(x)H\Phi^T(y), \end{aligned} \tag{4.146}$$

where $\Phi(x) = [\phi(x_1), \ldots, \phi(x_n)]$, $\Phi(y) = [\phi(y_1), \ldots, \phi(y_n)]$, $\mathbf{1}_n = [1, 1, \ldots, 1]^T$, $H = I_n - \frac{1}{n}\mathbf{1}_n\mathbf{1}_n^T$, and I_n is an identity matrix.

It is clear that

$$\hat{\mu}_x = \frac{1}{n}\Phi(x)\mathbf{1}_n \quad \text{and} \quad \hat{\mu}_y = \frac{1}{n}\Phi(y)\mathbf{1}_n. \tag{4.147}$$

4.3.2.4 Dependence Measure and Covariance Operator

Magnitude of dependence can be quantified by the covariance operator. We study two dependence measures. One measure is derived from kernel canonical correlation analysis. Another measure is the norm of the Hilbert–Schmidt norm of the cross-covariance operator.

4.3.2.4.1 Dependence Measure and Kernel Canonical Correlation Analysis Recall from Equation 4.126 that assessing independence is equivalent to solving the following optimization problem:

$$\begin{aligned} \max_{f,g} \quad & \langle f, \hat{\Sigma}_{XY} g \rangle_{H_2} \\ \text{s.t.} \quad & \|f\|_{H_2} = 1 \quad . \\ & \|g\|_{H_1} = 1 \end{aligned} \tag{4.148}$$

To solve it, the optimization problem is first rewritten in terms of kernel functions.

Consider the function $f(x) \in RKHS$. The function $f(x)$ can be expressed as

$$\begin{aligned} f(x) &= \sum_{i=1}^{n} \alpha_i \left[\phi(x_i) - \frac{1}{n} \Phi(x) 1_n \right] \\ &= \left[\phi(x_1) - \frac{1}{n} \Phi(x) 1_n, \ldots, \phi(x_n) - \frac{1}{n} \Phi(x) 1_n \right] \begin{bmatrix} \alpha_1 \\ \vdots \\ \alpha_n \end{bmatrix} \\ &= \left(\Phi(x) - \frac{1}{n} \Phi(x) 1_n 1_n^T \right) \alpha \\ &= \Phi(x) H \alpha. \end{aligned} \tag{4.149}$$

Similarly, we have

$$g(y) = \Phi(y) H \beta. \tag{4.150}$$

Using Equations 4.142, 4.145, and 4.146, we write the objective function in the optimization problem (4.148) as

$$\begin{aligned} \langle g, \hat{\Sigma}_{XY} f \rangle_{H_2} &= \frac{1}{n} \alpha^T H \Phi^T(x) \Phi(x) H \Phi^T(y) \Phi(y) H \beta \\ &= \frac{1}{n} \alpha^T \hat{K}_{xx} \hat{K}_{yy} \beta, \end{aligned} \tag{4.151}$$

where

$\hat{K}_{xx} = H \Phi^T(x) \Phi(x) H$ is a centered Gram matrix for variable X

$\hat{K}_{yy} = H \Phi^T(y) \Phi(y) H$ is a centered Gram matrix for variable Y

Similarly, we have

$$\|f\|_{H_2}^2 == \alpha^T \hat{K}_{xx}\alpha \tag{4.152}$$

and

$$\|g\|_{H_1}^2 = \beta^T \hat{K}_{yy}\beta. \tag{4.153}$$

Equation 4.144 is then reduced to

$$\begin{aligned} \max_{\alpha,\beta} \quad & \frac{1}{n}\alpha^T \hat{K}_{xx}\hat{K}_{yy}\beta \\ \text{s.t.} \quad & \alpha^T \hat{K}_{xx}\alpha = 1 \\ & \beta^T \hat{K}_{yy}\beta = 1 \end{aligned} .$$

The optimization problem (4.148) can be solved by the Lagrange multiplier method:

$$L(\alpha,\beta,\lambda,\gamma) = \frac{1}{n}\alpha^T \hat{K}_{xx}\hat{K}_{yy}\beta + \frac{\lambda}{2}\left(1-\alpha^T \hat{K}_{xx}\alpha\right) + \frac{\gamma}{2}\left(1-\beta^T \hat{K}_{yy}\beta\right).$$

Maximizing the Lagrangian $L(\alpha,\beta,\lambda,\gamma)$ with respect to α and β gives the generalized eigen-equation (Appendix 4D):

$$\begin{bmatrix} 0 & \frac{1}{n}\hat{K}_{xx}\hat{K}_{yy} \\ \frac{1}{n}\hat{K}_{yy}\hat{K}_{xx} & 0 \end{bmatrix} \begin{bmatrix} \alpha \\ \beta \end{bmatrix} = \lambda \begin{bmatrix} \hat{K}_{xx} & 0 \\ 0 & \hat{K}_{yy} \end{bmatrix} \begin{bmatrix} \alpha \\ \beta \end{bmatrix}. \tag{4.154}$$

Solving eigenequation (4.150), we can obtain the following dependence measure (Appendix 4D).

Result 4.6

Let $R_{xy} = \hat{K}_{xx}^{1/2}\hat{K}_{yy}^{1/2}$ and ρ be the largest singular value of the matrix R_{xy}. Dependence measure is defined as either ρ/n or

$$\frac{1}{n^2}\text{Trace}\left(\hat{K}_{xx}\hat{K}_{yy}\right). \tag{4.155}$$

4.3.2.5 Dependence Measure and Hilbert–Schmidt Norm of Covariance Operator

The covariance operator characterizes nonlinear dependence between two sets of variables. We can expect that the magnitude of the covariance operator can quantify the degrees of dependence. Therefore, we use the Hilbert–Schmidt norm of covariance operator as a dependence measure between two sets of variables.

Recall from Equation 4.142 that the centered sampling covariance operator is given by

$$\hat{\Sigma}_{XY} = \frac{1}{n}\Phi(x)H\Phi^T(y).$$

The covariance operator is expressed as a matrix. The Hilbert–Schmidt norm of the operator can be computed as an inner product of its representation matrix. The inner product between two matrices, A and B, denoted by $\langle A, B\rangle$ is defined as (Liu et al. 2013)

$$\langle A,B\rangle = \text{Trace}\left(A^T B\right).$$

Therefore, the HS norm of the covariance operator $\hat{\Sigma}_{XY}$ is

$$\begin{aligned}\left\|\hat{\Sigma}_{XY}\right\|_{HS}^2 &= \left\langle \hat{\Sigma}_{XY}, \hat{\Sigma}_{XY}\right\rangle_{HS} \\ &= \text{Trace}((\hat{\Sigma}_{XY})^T\hat{\Sigma}_{XY}) \\ &= \frac{1}{n^2}\text{Trace } (\Phi(y)H\Phi^T(x)\Phi(x)H\Phi^T(y)) \\ &= \frac{1}{n^2}\text{Trace } (H\Phi^T(x)\Phi(x)HH\Phi^T(y)\Phi(y)H). \end{aligned} \tag{4.156}$$

Recall that the centered Gram matrices are

$$\hat{K}_{xx} = H\Phi^T(x)\Phi(x)H, \tag{4.157}$$

$$\hat{K}_{yy} = H\Phi^T(y)\Phi(y)H. \tag{4.158}$$

Substituting Equations 4.153 and 4.154 into Equation 4.152 gives the estimate of the dependence measure.

Result 4.7

$$\left\|\hat{\Sigma}_{XY}\right\|_{HS}^2 = \frac{1}{n^2}\text{Trace } (\hat{K}_{xx}\hat{K}_{yy}), \tag{4.159}$$

which is called a Hilbert–Schmidt dependence measure.

We can see that Equation 4.155 is exactly the same as that in Equation 4.151.

As a special case, we consider a linear final dimensional case. Consider a sampled dataset $\{(x_1 y_1), \ldots, (x_n, y_n)\}$. Let $\Sigma_{xx} = \text{cov}(X, X)$, $\Sigma_{xy} = \text{cov}(X, Y)$ and $\Sigma_{yy} = \text{cov}(Y, Y)$. We want to find α and β such that

$$\begin{aligned} &\max_{\alpha,\beta} \quad \alpha^T\Sigma_{xy}\beta \\ &\text{s.t.} \quad \alpha^T\Sigma_{xx}\alpha = 1. \\ &\qquad\ \ \beta^T\Sigma_{yy}\beta = 1 \end{aligned} \tag{4.160}$$

Let $u = \Sigma_{xx}^{1/2}\alpha$ and $v = \Sigma_{yy}^{1/2}\beta$, and Equation 4.156 can be transformed to

$$\begin{aligned} \max_{u,v} \quad & u^T \Sigma_{xx}^{-1/2} \Sigma_{xy} \Sigma_{yy}^{-1/2} v \\ \text{s.t.} \quad & u^T u = 1 \\ & v^T v = 1 \end{aligned} \quad . \tag{4.161}$$

Comparing Equation 4.157 with Equation 4D.7, we obtain

$$\Sigma_{xx}^{-1/2}\Sigma_{xy}\Sigma_{yy}^{-1/2} = \frac{1}{n}\hat{K}_{xx}^{1/2}\hat{K}_{yy}^{1/2}. \tag{4.162}$$

It follows from Equation 4.158 that

$$\text{Trace}\left(\Sigma_{yy}^{-1/2}\Sigma_{yx}\Sigma_{xx}^{-1}\Sigma_{xy}\Sigma_{yy}^{-1/2}\right) = \frac{1}{n^2}\text{Trace}\left(\hat{K}_{xx}\hat{K}_{yy}\right). \tag{4.163}$$

Therefore, we have the result.

Result 4.8

The linear dependence measure is given by

$$\text{Trace}\left(\Sigma_{yy}^{-1/2}\Sigma_{yx}\Sigma_{xx}^{-1}\Sigma_{xy}\Sigma_{yy}^{-1/2}\right). \tag{4.164}$$

It follows from Equations 1.220 and 4.160 that the linear dependence measure can be computed by linear canonical correlation analysis.

4.3.2.6 Kernel-Based Association Tests

4.3.2.6.1 Kernel Measure of Independence as a General Tool for Developing Association Tests Many statistical methods such as regressions, correlation analysis, and canonical correlation analysis are developed for testing the association of genetic variants with phenotypes. Although these test statistics are different, the principal for various association tests is to assess the independence between two sets of variables. In other words, we assess that knowing information on a set of variables provides no additional information about another set of variables. The traditional genetic association tests often assume the linear relations between genetic variants and phenotypes and normal error distribution. However, in practice, the assumptions of linear relations and normal errors are often violated. Dependence measures studied in the previous section do not make such assumptions. Therefore, the dependence measure provides a general framework for developing general association tests without linear and normal distribution assumptions.

Independence can be characterized by the cross-covariance operator $\hat{\Sigma}_{XY}$ on RKHS. The statistics for testing independence between X and Y are based on the Hilbert–Schmidt norm of the cross-covariance operator $\left\|\hat{\Sigma}_{XY}\right\|_{HS}^2$ (Equation 4.155). Let $\lambda_{x,1} \geq \lambda_{x,2} \geq \ldots \geq \lambda_{x,n}$ be the eigenvalues of the centered kernel matrix $\hat{K}_{xx}$ and $\lambda_{y,1} \geq \lambda_{y,2} \geq \ldots \geq \lambda_{y,n}$ be the eigenvalues of the centered kernel matrix $\hat{K}_{yy}$. The statistics for testing independence between X and Y are defined as

$$T_I = \frac{1}{n}\text{Trace}\left(\hat{K}_{xx}\hat{K}_{yy}\right). \tag{4.165}$$

Zhang et al. (2012) derived the asymptotic distribution of the test statistic T_I under the null hypothesis.

Theorem 4.2

(Independence test).

Under the null hypothesis that two sets of random variables X and Y are independent, the statistic T_I defined in Equation 4.161 is asymptotically distributed as

$$\frac{1}{n^2}\sum_{i=1}^{n}\sum_{j=1}^{n}\lambda_{x,i}\lambda_{y,j}z_{ij}^2, \tag{4.166}$$

where z_{ij}^2 is independently and identically distributed as a central $\chi_{(1)}^2$ distribution.

Using statistics T_I to test association of genetic variants with the phenotype, we first select kernel functions for the genetic variants and phenotype. Then, we compute the centered kernel matrices $\hat{K}_{xx}$ and $\hat{K}_{yy}$. We use a single value decomposition method to calculate the eigenvalues of the kernel matrices. Distribution is computed by simulations. We draw i.i.d. random samples from each z_{ij}^2 according to a central $\chi_{(1)}^2$ distribution. We compute values using Equation 4.162 and repeat simulations multiple times. Finally, the distribution of the test T_I will be obtained.

Using simulations to find the null distribution of the test is time consuming. The null distribution of the test can also be approximated. Below, we introduce approximating the null distribution by a two-parameter Gamma distribution (Zhang et al 2012).

Theorem 4.3

Gamma distribution approximation to independence measure–based test.

The null distribution of the independence test statistic T_I can be approximated by the $\Gamma(\tau,\theta)$ distribution:

$$P(t) = t^{\tau-1}\frac{e^{-t/\theta}}{\theta^{\tau}\Gamma(\tau)}, \tag{4.167}$$

where

$$\tau = \frac{\left(E[T_I]\right)^2}{\text{var}(T_I)}, \quad \theta = \frac{\text{var}(T_I)}{E(T_I)},$$

$$E[T_I] = \frac{1}{n^2}\text{Trace}\left(\hat{K}_{xx}\right)\text{Trace}\left(\hat{K}_{yy}\right),$$

and

$$\text{var}(T_I) = \frac{2}{n^4}\text{Trace}\left(K_{xx}^2\right)\text{Trace}\left(K_{yy}^2\right).$$

4.3.2.6.2 Sequencing Kernel Association Test (SKAT) SKAT (Wu et al. 2011) can be viewed as a special case of the modified kernel independence test. Consider a gene or a genomic region and a phenotype. We want to test the association of a gene or genomic region with the phenotype. For the ith individual, the phenotype variable is denoted by y_i, a vector of covariates such as age, sex, and environments, among others, is denoted by $X_i = [x_{i1}, \ldots, x_{iq}]$, and a vector of genotype indicator variables at p loci is denoted by $G_i = [G_{i1}, \ldots, G_{ip}]$. Let

$$y = \begin{bmatrix} y_1 \\ \vdots \\ y_n \end{bmatrix}, \quad X = \begin{bmatrix} 1 & X_1 \\ \vdots & \vdots \\ 1 & X_n \end{bmatrix}, \quad G = \begin{bmatrix} G_1 \\ \vdots \\ G_n \end{bmatrix}, \quad \text{and} \quad \hat{\mu} = X\left(X^T X\right)^{-1} X^T Y.$$

Define a kernel matrix for the genotypes as $K = GWG^T$, where $W = \text{diag}(w_1, \ldots, w_p)$ is a weight matrix. The SKAT test is defined as

$$Q = (y - \hat{\mu})^T K (y - \hat{\mu}). \tag{4.168}$$

To see the relationship between the SKAT and the dependence measure–based independence test, the SKAT statistic Q can be transformed to

$$Q = \text{Trace}\left(K(y - \hat{\mu})(y - \hat{\mu})^T\right). \tag{4.169}$$

If the outer product, $(y - \hat{\mu})(y - \hat{\mu})^T$, is replaced by a distance measure, $D = (d_{ij})_{n \times n}$, $d_{ij} = \|y_i - y_j\|_2^2$, the SKAT statistic is equivalent to the dependence measure–based independence test.

The kernel matrix measures the genetic similarity between individuals. Weight is often set to $\sqrt{w_j} = \text{Beta}(MAF_j, \alpha, \beta)$, where *Beta* represents the beta distribution density function and MAF is the frequency of the minor allele at the jth locus. When $\alpha = \beta = 0.5$, $\sqrt{w_j} = \frac{1}{\sqrt{MAF_j(1 - MAF_j)}}$.

Under the null hypothesis of no association, the SKAT statistic Q follows a mixture of chi-square distributions (Wu et al. 2011). Let $P_0 = V - VX(X^TVX)^{-1}X^TV$ and $V = \hat{\sigma}_0^2 I$, where $\hat{\sigma}_0^2$ is the estimator of σ^2 under the null model, $y = X\alpha + \varepsilon$. Then, Q is asymptotically distributed as

$$Q \sim \sum_{i=1}^{n} \lambda_i \chi_{1,i}^2, \tag{4.170}$$

where

$\lambda_i, i = 1, 2, \ldots, n$ are eigenvalues of the matrix $P_0^{1/2} K P_0^{1/2}$

$\chi_{1,i}^2$ are independent χ_1^2 random variables

4.4 SIMULATIONS AND REAL DATA ANALYSIS

4.4.1 Power Evaluation

To illustrate the simulations for evaluating the performance of statistics for testing the association of the genetic variants with a single trait, we introduce the procedures for power simulations with next-generation sequencing data.

A true quantitative genetic model is given as follows. Consider L trait loci, which are located at the genomic positions $t_1, \ldots, t_L$. Let A_l be a risk allele at the lth trait locus. The following multiple linear regression is used as an additive genetic model for a quantitative trait:

$$Y_i = \mu_m + \sum_{l=1}^{L} X_{il}\alpha_l + \varepsilon_i,$$

where

$$X_{il} = \begin{cases} 2(1-P_l), & A_l A_l \\ 1-2P_l & A_l a_l \\ -2P & a_l a_l, \end{cases}$$

$\alpha_l = P_l G_{11}^l + (1-2P_l) G_{12}^l - (1-P_l) G_{22}^l$, G_{11}^l, G_{12}^l and G_{22}^l are genotypic values of the genotypes A_1A_1, A_1a_1 and a_1a_1, respectively, and ε_i is distributed as a standard normal distribution $N(0,1)$.

We considered four disease models: additive, dominant, recessive, and multiplicative. The relative risks across all variant sites are assumed to be equal, and the variants are assumed to influence the trait independently (i.e., no epistasis). Let $f_0 = 1$ be a baseline penetrance that is defined as the contribution of the wild genotype to the trait variation and r be a risk parameter. For the dominant model, we assume $G_{11}^l = rf_0, G_{12}^l = rf_0$ and $G_{22}^l = f_0$. Thus, the genetic additive effect is defined as $\alpha_l = (1 - P_l)(r - 1)f_0$. Similarly, the genetic additive effects are defined as $\alpha_l = (r-1)f_0$, $\alpha_l = (rP_l + 1 - P_l)(r-1)f_0$, and $\alpha_l = (r-1)P_l f_0$ for additive, multiplicative, and recessive disease models, respectively.

In our sequenced 1000 individuals with European origin (data have not been published), the average number of SNPs per kb is 8 SNPs. However, the average number of SNPs per kb will rapidly increase as the number of sequenced individuals increases. It was recently reported that the average number of SNPs per kb in 202 drug target genes sequenced in 12,514 European subjects is about 48 SNPs (Nelson et al. 2012). Due to the expense of whole-genome sequencing, in most genetic studies, only thousands of individuals are often sequenced. In the simulations for power studies, 30 SNPs per kb is assumed. For simplicity, 25%, 15%, and 60%, respectively, of the SNPs are often taken as common, low-frequency, and rare SNPs. Consider a 30 kb region (average length of a gene). Since the average number of SNPs per kb is 30 SNPs, the number of SNPs in a gene is assumed to be 900 SNPs in the simulations.

The MS software (Hudson 2002) can be used to generate a population of 2,000,000 chromosomes with the above variants. Two haplotypes were randomly sampled from the population and assigned to an individual, and 10% of the rare variants are randomly selected as causal variants. A total of 5000 simulations were repeated for power calculation.

Figures 4.3 and 4.4 plot the power curves of ten statistics: the FLM; the smoothed FLM; the SKAT; two collapsing-based regression tests, RVT1 and RVT2 (Morris and Zeggini 2010); simple regression where permutation was used to adjust for multiple testing; multiple regression; regression on principal components (PCA); and WSS and VT for testing association of rare variants in the genomic region under additive model with 2000 individuals in the presence of 18 risk variants and 9 risk variants and 9 protective rare variants, respectively. These power curves are a function of the risk parameter at the significance level $\alpha = 0.05$. These two figures show that the functional linear model for quantitative trait analysis had the highest power.

4.4.2 Application to Real Data Examples

To illustrate how to apply the introduced statistical methods to real data analysis, we present three examples. The first example is an application of nine statistics to the ANGPTL4 sequence and phenotype data from the Dallas Heart Study (Romeo et al. 2007). A total of 93 variants were identified from 3553 individuals. The total number of rare variants with a minor allele frequency below 0.03 in the dataset was 71. The study included six quantitative traits: plasma triglyceride levels (Trig), high-density lipoprotein cholesterol (HDL), total cholesterol, very low–density lipoprotein cholesterol (VLDL), and body mass index (BMI). *P*-values from the FLM, SKAT, two collapsing-based regression tests, simple regression, multiple regression, regression on principal components, WSS, and VT for testing the association of rare variants in ANGPTL4 with the six quantitative traits are summarized in Table 4.1 where *P*-values for WSS, VT, and simple regression were obtained by permutations.

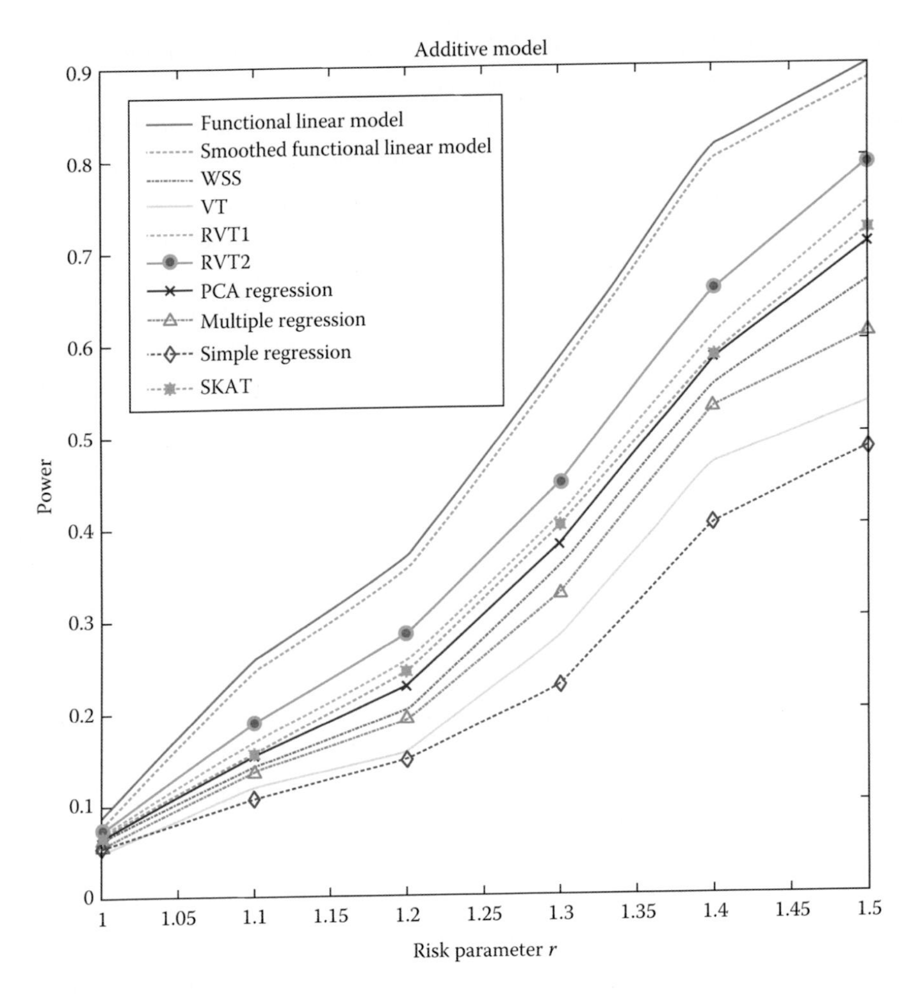

FIGURE 4.3 Power of the 10 statistics for testing association with a quantitative trait under the additive model with rare risk variants only.

The second example is the application of nine statistics to rare expression quantitative trait loci (eQTLs) analysis. Genetic variation in the low-coverage resequencing data of the 1000 Genomes Project (released March 2010) and the 15 gene expressions acquired by RNA sequencing (RNA-seq) in the lymphoblastoid cell lines (LCLs) from 60 individuals of European origin (CEU) are analyzed (Montgomery et al. 2010). The expression of a gene is measured by a normalized overall expression level of the gene. We take a gene as a unit of rare eQTL association analysis. A total of 2533 genes that consisted of SNPs with MAF < 5% were included in the analysis. A P-value for declaring significant association after the Bonferroni correction for multiple tests was 1.97×10^{-5}. Nine statistics, the FLM, SKAT, two collapsing-based regression tests, simple regression (SRG), multiple

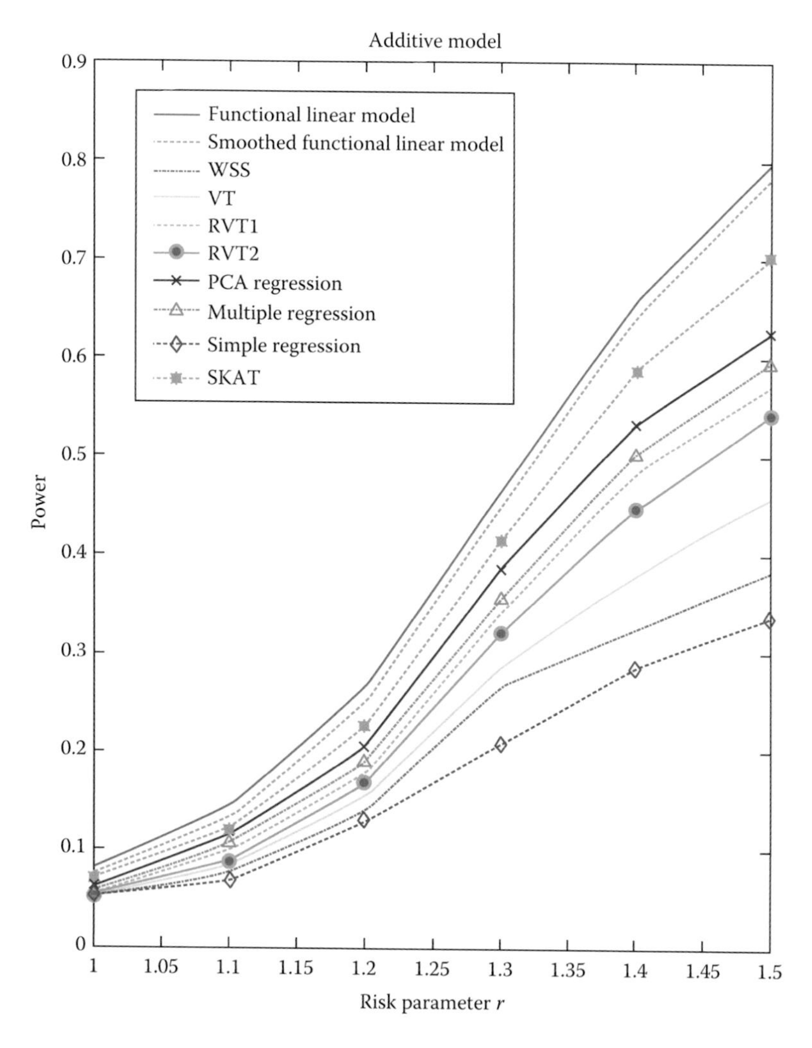

FIGURE 4.4 Power of the 10 statistics for testing association with a quantitative trait under the additive model with both rare risk variants and protective variants.

regression (MRG), regression on principal components (PCA), WSS, and VT, are used to test for the association of rare variants in 2533 genes with 15 RNA-seq expressions. The FLM identified 13 genes; the SKAT identified three genes; and SRG, MRG, PCA, WSS, and VT identified one gene with rare SNPs, which were significantly associated with the expressions of 13 genes after the Bonferroni correction for multiple tests. It was reported that one SNP (rs7639979) in the gene PRSS50 was significantly associated with the expression of PRSS46 with P-value $< 1.76 \times 10^{-6}$. The P-values of 14 genes calculated by the FLM and other 8 statistics for testing their association with the whole-gene expressions were summarized in Table 4.2.

TABLE 4.1 *P*-Values of Statistics for Testing Association of Rare Variants in ANGPTL4 with Six Traits in the Dallas Heart Study

	Phenotypes					
Statistical Methods	**BMI**	**Cholesterol**	**Triglycerides**	**VLDL**	**LDL**	**HDL**
FLM	3.72E−06	4.81E−02	5.48E−07	3.67E−06	1.77E−07	4.27E−06
RVT1	4.18E−06	7.74E−01	2.21E−03	2.13E−03	7.29E−01	2.13E−03
RVT2	6.93E−05	7.18E−01	1.82E−04	1.89E−04	8.93E−01	6.84E−04
PCA	3.93E−05	7.55E−01	2.58E−06	4.59E−06	5.86E−01	4.49E−03
Multiple regression	9.25E−03	1.93E−01	3.19E−03	3.36E−03	2.54E−07	1.02E−02
Simple regression	1.10E−02	7.10E−01	5.50E−02	1.60E−01	1.60E−01	5.00E−03
VT	1.20E−05	2.65E−01	4.95E−01	7.25E−01	3.10E−01	3.80E−01
WSS	5.39E−02	3.55E−01	3.00E−01	8.00E−02	4.60E−01	2.50E−01
SKAT	1.03E−03	7.49E−01	1.57E−01	1.88E−01	2.34E−01	1.05E−01

The third example is the application of five statistics, FLM, SLM (SRG), MLM (MRG), PCA, and SKAT, to the CHARGE-S studies, which generated low-coverage whole-genome sequencing data of 955 individuals from the ARIC (Atherosclerosis Risk in Communities) study, Framingham Heart Study, and Cardiovascular Health Study (CHS) longitudinal cohorts after quality control with rich phenotypes including HDL and LDL levels (Zhang et al. 2014). Figure 4.5 presents a Manhattan plot showing *P*-values of five statistics for testing the association of genes in Chromosome 2 with HDL. Figure 4.6 presents a Manhattan plot showing *P*-values of five statistics for testing the association of genes in Chromosome 10 with LDL. The results again show that the FLM continues to outperform the other four statistics.

SOFTWARE PACKAGE

The code with MATLAB® implementation of a kernel-based statistical hypothesis test for independence can be downloaded from the website http://www.kyb.mpg.de/bs/people/arthur/indep.htm. SKAT (http://www.hsph.harvard.edu/skat/) is a software for testing the association of a set of SNPs (gene or genomic region) with continuous or binary traits. A program for implementing the functional linear model for testing the association of a gene or a genomic region can be downloaded from the website http://www.sph.uth.tmc.edu/hgc/faculty/xiong/index.htm. The R package "candisc" for canonical correlation analysis can be downloaded from the website https://cran.r-project.org/web/packages/candisc/index.html.

TABLE 4.2 *P*-Values of 14 Genes Associated with the Expressions of 13 Genes

Gene	FLM	RVT1	RVT2	SRG	MRG	PCA	WSS	VT	SKAT
PRR19	1.93E−05	4.86E−01	7.95E−01	2.97E−02	1.68E−03	3.66E−04	1.09E−01	6.93E−02	5.73E−05
OR4Q3	5.83E−01	6.82E−01	6.82E−01	8.01E−02	4.61E−01	5.84E−01	4.46E−01	6.44E−01	1.27E−01
C12orf53	9.14E−08	1.28E−01	2.23E−01	5.94E−02	9.33E−05	5.63E−03	3.96E−02	6.93E−02	1.22E−05
PLAC2	6.43E−06	3.55E−01	6.62E−01	9.90E−02	4.20E−03	1.92E−01	4.16E−01	4.95E−01	4.62E−04
KLK11	5.33E−05	4.19E−01	3.69E−01	8.91E−02	4.27E−02	5.46E−02	1.39E−01	3.17E−01	5.97E−04
NDUFB7	6.69E−07	5.64E−01	2.14E−01	4.65E−01	3.94E−04	5.75E−02	2.57E−01	2.67E−01	1.94E−04
C22orf27	1.18E−05	9.03E−03	6.70E−03	1.98E−02	6.68E−04	2.56E−02	9.90E−03	3.96E−02	4.04E−04
CHRNE	3.26E−06	9.08E−01	6.04E−01	1.98E−02	2.66E−03	3.31E−03	9.90E−03	9.90E−02	4.59E−05
FUT1	1.64E−05	3.04E−01	6.98E−01	2.97E−02	5.55E−03	8.93E−04	9.90E−03	5.94E−02	1.10E−04
FFAR3	3.72E−06	5.71E−01	3.79E−01	3.27E−01	4.44E−03	2.64E−01	2.08E−01	1.68E−01	8.42E−04
CCL11	1.52E−05	5.33E−05	5.85E−05	9.90E−03	1.43E−03	3.69E−03	9.90E−03	9.90E−03	1.24E−04
GGT6	1.13E−05	3.94E−01	5.74E−01	8.91E−02	9.88E−04	6.27E−03	5.94E−02	1.39E−01	2.64E−04
C19orf46	6.79E−11	7.62E−02	7.62E−02	2.97E−02	2.55E−04	1.99E−02	2.77E−01	9.90E−03	6.15E−07
PRSS50	8.65E−09	5.55E−05	4.26E−04	1.00E−05	5.68E−06	1.49E−07	1.00E−05	1.00E−05	3.25E−07

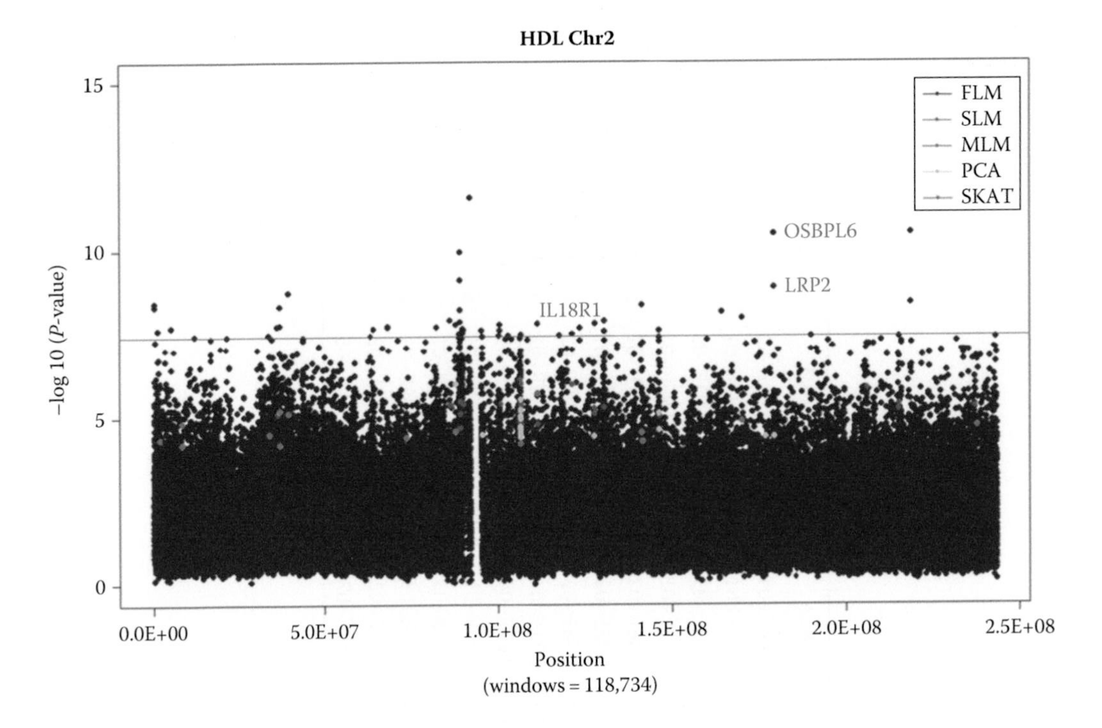

FIGURE 4.5 Manhattan plot showing *P*-values of five statistics for testing association of genes in Chromosome 2 with HDL.

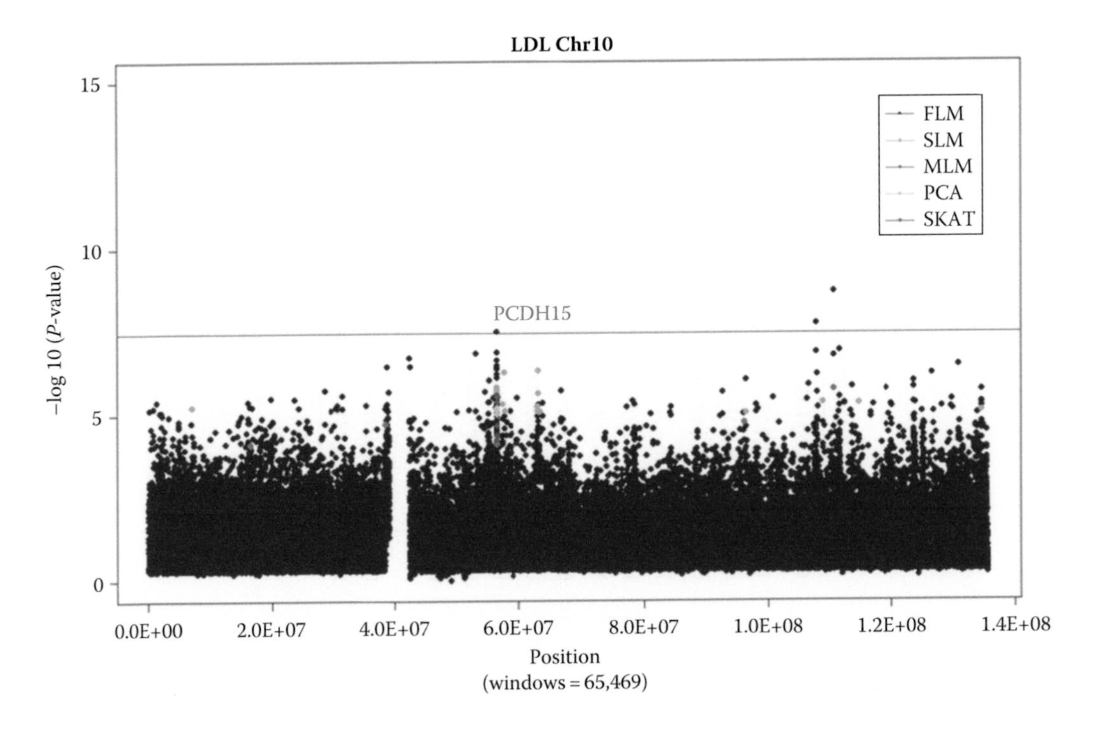

FIGURE 4.6 Manhattan plot showing *P*-values of five statistics for testing association of genes in Chromosome 10 with LDL.

APPENDIX 4A: CONVERGENCE OF THE LEAST SQUARE ESTIMATOR OF THE REGRESSION COEFFICIENTS

Assume that n individuals are sampled. Let $Y=[Y_1,\ldots,Y_n]^T$, $X^m=\left[X_1^m,\ldots,X_n^m\right]^T$, $Z^m\left[Z_1^m,\ldots,Z_n^m\right]^T$, $\beta=[\mu_m,\alpha_m,\delta_m]^T$, $\varepsilon=[\varepsilon_1,\varepsilon_2,\ldots\varepsilon_n]^T$, $W=[1,X^m,Z^m]^T$, $\mathbf{1}=[1,1,\ldots,1]^T$. Then, Equation 4.22 can be written in a matrix form:

$$Y=W\beta+\varepsilon. \tag{4A.1}$$

The least square estimator of the regression coefficients β is given by

$$\hat{\beta}=\left(W^TW\right)^{-1}W^TY. \tag{4A.2}$$

The true genetic model in Equation 4.20 is

$$Y_i=\mu+X_i\alpha+Z_i\delta+\varepsilon_i. \tag{4A.3}$$

It follows from Equation 4.23 that

$$\begin{aligned}E\left[X_1^m\right]&=2P_mP_M^2+2\left(P_m-P_M\right)\left(P_mP_M\right)-2P_MP_m^2\\&=2P_mP_M\left(P_M+P_m-P_M-P_m\right)=0,\end{aligned}$$

$$E\left[Z_1^m\right]=P_m^2P_M^2-2\left(P_mP_M\right)^2+P_M^2P_m^2=0,$$

$$\begin{aligned}E\left[\left(X_1^m\right)^2\right]&=\left(2P_m\right)^2P_M^2+2\left(P_m-P_M\right)^2\left(P_mP_M\right)+\left(-2P_M\right)^2P_m^2\\&=2P_mP_M,\end{aligned}$$

$$\begin{aligned}E\left[\left(Z_1^m\right)^2\right]&=\left(P_m^2\right)^2P_M^2+2\left(-P_MP_m\right)^2\left(P_mP_M\right)+\left(P_M^2\right)^2P_m^2\\&=P_m^2P_M^2,\end{aligned}$$

$$\begin{aligned}E\left[X_1^mZ_1^m\right]&=\left(2P_m\right)P_m^2P_M^2-2P_mP_M\left(P_m-P_M\right)P_mP_M-2P_M^3P_m^2\\&=0.\end{aligned}$$

Since

$$\frac{1}{n}W^TW=\begin{bmatrix}1 & \frac{1}{n}\sum_i X_i^m & \frac{1}{n}\sum_i Z_i^m\\ \frac{1}{n}\sum_i X_i^m & \frac{1}{n}\sum_i\left(X_i^m\right)^2 & \frac{1}{n}\sum_i\left(X_i^mZ_i^m\right)\\ \frac{1}{n}\sum_i Z_i^m & \frac{1}{n}\sum_i X_i^mZ_i^m & \frac{1}{n}\left(\sum_i Z_i^m\right)^2\end{bmatrix},$$

by large number theory, we have

$$\frac{1}{n}W^TW \xrightarrow[a.s.]{} \begin{bmatrix} 1 & 0 & 0 \\ 0 & 2P_mP_M & 0 \\ 0 & 0 & P_m^2P_M^2 \end{bmatrix}. \tag{4A.4}$$

It follows from Equations 4.20 and 4.21 that

$$E[Y_1] = \mu,$$

$$\begin{aligned} E\left[X_1^m X_1\right] &= 2P_m\left[2qP_{MA_1}^2 + 2(q-p)P_{MA_1}P_{MA_2} - 2pP_{MA_2}^2\right] \\ &\quad + (P_m - P_M)\left[(2q)^2 2P_{MA_1}P_{mA_1} + (q-p)(P_{MA_1}P_{mA_2} + P_{MA_2}P_{mA_1}) + 2(-2p)P_{MA_2}P_{mA_2}\right] \\ &\quad + (-2P_M)\left[(2q)P_{mA_1}^2 + 2(q-p)P_{mA_1}P_{mA_2} + (-2q)P_{mA_2}^2\right] \\ &= 4P_mP_MD + 2(P_m - P_M)^2 D + 4P_mP_MD \\ &= 2D. \end{aligned}$$

We can similarly show that

$$E\left[X_1^m Z_1\right] = 0,$$

$$E\left[Z_1^m X_1\right] = 0,$$

and

$$\begin{aligned} E\left[Z_1^m Z_1\right] &= P_m^2 D^2 + 2P_M P_m D^2 + P_M^2 D^2 \\ &= D^2. \end{aligned}$$

By the large number theory and Equation 4A.3,

$$\frac{1}{n}W^TY = \begin{bmatrix} \frac{1}{n}\sum_i Y_i \\ \frac{1}{n}\sum_i X_i^m Y_i \\ \frac{1}{n}\sum_i Z_i^m Y_i \end{bmatrix} \xrightarrow[a.s.]{} \begin{bmatrix} \mu \\ 2D\alpha \\ D^2\delta \end{bmatrix}. \tag{4A.5}$$

Combining Equations 4A.4 and 4A.5, we obtain

$$\alpha_m \xrightarrow{a.s.} \frac{D}{P_M P_m}\alpha \quad \text{and} \quad \delta_m \xrightarrow{a.s.} \frac{D^2}{P_M^2 P_m^2}\delta. \tag{4A.6}$$

When the marker is at the trait locus, we have

$$D = P_M - P_M^2 = P_M P_m,$$

which leads to

$$\alpha_m \xrightarrow{a.s.} \alpha \quad \text{and} \quad \delta_m \xrightarrow{a.s.} \delta.$$

The estimators of the genetic additive and dominance variance at the marker locus are

$$\begin{aligned} \sigma_{A(m)}^2 &= 2P_M P_m \alpha_m^2 \xrightarrow{a.s.} \frac{D^2}{P_M P_m P_Q P_q}\sigma_A^2, \\ \sigma_{D(m)}^2 &= P_M^2 P_m^2 \delta_m^2 \xrightarrow{a.s.} \frac{D^4}{P_M^2 P_m^2 P_Q^2 P_q^2}\sigma_D^2. \end{aligned} \tag{4A.7}$$

It follows from Equations 4A.2 that the variance of the estimators of the parameters in the model is

$$\Lambda = \text{var}\left(\hat{\beta}\right) = \sigma_e^2 \left(W^T W\right)^{-1}. \tag{4A.8}$$

The noncentrality parameter of the test in Equation 4.32 is

$$\begin{aligned} T &= \hat{\beta}_g^T \Lambda^{-1} \hat{\beta}_g \\ &\xrightarrow{a.s.} \frac{n\left(2P_M P_m \alpha_m^2 + P_M^2 P_m^2 \delta_m^2\right)}{\sigma_e^2} \\ &\xrightarrow{a.s.} \frac{n}{\sigma_e^2}\left[\frac{D^2}{P_M P_m P_Q P_q}\sigma_A^2 + \frac{D^4}{P_M^2 P_m^2 P_Q^2 P_q^2}\sigma_D^2\right]. \end{aligned} \tag{4A.9}$$

Next we study the noncentrality parameter of the test in Equation 4.41 for multiple linear regression models. Let

$$X = \begin{bmatrix} x_{11} & \cdots & x_{1L} \\ \vdots & \ddots & \vdots \\ x_{n1} & \cdots & x_{nL} \end{bmatrix}, \quad Z = \begin{bmatrix} z_{11} & \cdots & z_{1L} \\ \vdots & \ddots & \vdots \\ z_{n1} & \cdots & z_{nL} \end{bmatrix}, \quad W = \begin{bmatrix} 1 & X & Z \end{bmatrix}.$$

By large number theory, we have

$$\frac{1}{n}W^TW \xrightarrow{a.s.} \begin{bmatrix} 1 & 0 & 0 \\ 0 & \Lambda_1 & 0 \\ 0 & 0 & \Lambda_2 \end{bmatrix}, \tag{4A.10}$$

where

$$\Lambda_1 = \begin{bmatrix} 2P_{M_1}P_{m_1} & \cdots & 2D_{1L} \\ \vdots & \ddots & \vdots \\ 2D_{L1} & \cdots & 2P_{M_L}P_{m_L} \end{bmatrix}, \Lambda_2 = \begin{bmatrix} P_{M_1}^2 P_{m_1}^2 & \cdots & D_{1L}^2 \\ \vdots & \ddots & \vdots \\ D_{L1}^2 & \cdots & P_{M_L}^2 P_{m_L}^2 \end{bmatrix}$$

D_{ij} is the coefficient of the LD between the ith marker and jth marker
P_{Mj} and P_{mj} are the frequencies of alleles M_j and m_j at the jth marker, respectively

Let

$$\Lambda = \begin{bmatrix} \Lambda_1 & 0 \\ 0 & \Lambda_2 \end{bmatrix}.$$

Then, the noncentrality parameter of the test defined in Equation 4.41 almost surely converges to

$$T \xrightarrow{a.s.} \frac{n}{\sigma_e^2}\left[\alpha^T\Lambda_1\alpha + \delta^T\Lambda_2\delta\right] = \frac{n}{\sigma_e^2}\left[\sum_{j=1}^{L}\sum_{k=1}^{L} 2D_{jk}\alpha_j\alpha_k + \sum_{j=1}^{L}\sum_{k=1}^{L} D_{jk}^2\delta_j\delta_k\right], \tag{4A.11}$$

where $D_{jj} = P_{Mj}P_{mj}$.

We consider three special scenarios:

1. All markers are trait loci.
   We define the genetic additive covariance and dominance covariance as follows:

$$\begin{aligned} \sigma_{A(j,k)}^2 &= 2D_{jk}\alpha_j\alpha_k, \\ \sigma_{D(j,k)}^2 &= D_{jk}^2\delta_j\delta_k. \end{aligned} \tag{4A.12}$$

   When $j = k$, we have

$$\sigma_A^2 = 2P_{Q_j}P_{q_j}\alpha_j^2 \quad \text{and} \quad \sigma_D^2 = P_{Q_j}^2 P_{q_j}^2 \delta_j^2.$$

   Under this scenario, Equation 4A.11 is reduced to

$$T \xrightarrow{a.s.} \frac{n}{\sigma_e^2}\left[\sum_{j=1}^{L}\sum_{k=1}^{L}\sigma_{A(i,j)}^2 + \sum_{j=1}^{L}\sum_{k=1}^{L}\sigma_{D(j,k)}^2\right]. \tag{4A.13}$$

2. All markers are trait loci and are in linkage equilibrium.
   Under this scenario, $D_{jk}=0$, for all $j \neq k$ and $D_{jj}=P_{Qj}P_{qj}$. Equation 4A.11 is reduced to

$$T \xrightarrow{a.s.} \frac{n}{\sigma_e^2}\left[\sum_{j=1}^{L}\sigma_{A(j)}^2 + \sum_{j=1}^{L}\sigma_{D(j)}^2\right]. \tag{4A.14}$$

3. There is the presence of a trait locus that is in linkage disequilibrium with the markers. We denote the coefficient of LD between the jth marker and the trait locus by D_j and genetic additive variance and dominance variance, respectively, by

$$\sigma_A^2 = 2P_Q P_q \alpha^2 \quad \text{and} \quad \sigma_D^2 = P_Q^2 P_q^2 \delta^2.$$

Using Equation 4A.6, we obtain

$$\alpha_j \to \frac{D_j}{P_{M_j}P_{m_j}}\alpha, \quad 2\alpha_j\alpha_k \to \frac{2D_jD_k}{P_{M_j}P_{m_j}P_{M_k}P_{m_k}}\alpha^2 = \frac{D_jD_k 2P_QP_q\alpha^2}{P_{M_j}P_{m_j}P_{M_k}P_{m_k}P_QP_q} = \frac{D_jD_k}{P_{M_j}P_{m_j}P_{M_k}P_{m_k}P_QP_q}\sigma_A^2,$$

$$\delta_j \to \frac{D_j^2}{P_{M_j}^2P_{m_j}^2}\delta, \quad \delta_j\delta_k = \frac{D_j^2D_k^2}{P_{M_j}^2P_{m_j}^2P_{M_k}^2P_{m_k}^2P_Q^2P_q^2}\sigma_D^2. \tag{4A.15}$$

Substituting Equation 4A.15 into Equation 4A.11, we obtain

$$T \xrightarrow{a.s.} \frac{n}{\sigma_e^2}\left[\left(\sum_{j=1}^{L}\sum_{k=1}^{L}\frac{D_jD_k}{P_{M_j}P_{m_j}P_{M_k}P_{m_k}P_QP_q}\right)\sigma_A^2 + \left(\sum_{j=1}^{L}\sum_{k=1}^{L}\sum\frac{D_j^2D_k^2}{P_{M_j}^2P_{m_j}^2P_{M_k}^2P_{m_k}^2P_Q^2P_q^2}\sigma_D^2\right)\right]$$

$$= \frac{n}{\sigma_e^2}\left(f_A\sigma_A^2 + f_D\sigma_D^2\right),$$

where

$$f_A = \sum_{j=1}^{L}\sum_{k=1}^{L}\frac{D_jD_k}{P_{M_j}P_{m_j}P_{M_k}P_{m_k}P_QP_q}$$

and

$$f_D = \sum_{j=1}^{L}\sum_{k=1}^{L}\frac{D_j^2D_k^2}{P_{M_j}^2P_{m_j}^2P_{M_k}^2P_{m_k}^2P_Q^2P_q^2}. \tag{4A.16}$$

APPENDIX 4B: CONVERGENCE OF REGRESSION COEFFICIENTS IN THE FUNCTIONAL LINEAR MODEL

Recall that

$$\xi = \begin{bmatrix} \xi_{11} & \cdots & \xi_{1L} \\ \vdots & \ddots & \vdots \\ \xi_{n1} & \cdots & \xi_{nL} \end{bmatrix} \text{ and } \eta = \begin{bmatrix} \eta_{11} & \cdots & \eta_{1L} \\ \vdots & \ddots & \vdots \\ \eta_{n1} & \cdots & \eta_{nL} \end{bmatrix}.$$

By large number theory and Exercise 3.6, we have

$$\begin{gathered} \frac{1}{n}\sum_{i=1}^{n}\xi_{ik}^2 \xrightarrow{a.s.} E\left[\xi_{ik}^2\right] = \lambda_k, \quad \frac{1}{n}\sum_{i=1}^{n}\xi_{ik}\xi_{il} \xrightarrow{a.s.} 0, \quad \frac{1}{n}\sum_{i=1}^{n}\xi_{ik}\eta_{il} \xrightarrow{a.s.} 0, \\ \frac{1}{n}\sum_{i=1}^{n}\eta_{ik}^2 \xrightarrow{a.s.} E\left[\eta_{ik}^2\right] = \gamma_k, \quad \frac{1}{n}\sum_{i=1}^{n}\eta_{ik}\eta_{il} = 0, \quad \frac{1}{n}\sum_{i=1}^{n}\xi_{ik} \xrightarrow{a.s.} 0, \quad \frac{1}{n}\sum_{i=1}^{n}\eta_{ik} \xrightarrow{a.s.} 0, \end{gathered} \tag{4B.1}$$

which implies

$$\begin{aligned} &\frac{1}{n}\xi^T\xi \xrightarrow{a.s.} \Lambda = \operatorname{diag}(\lambda_1,\ldots,\lambda_L), \\ &\frac{1}{n}\eta^T\eta \xrightarrow{a.s.} \Sigma = \operatorname{diag}(\gamma_1,\ldots,\gamma_L), \\ &\frac{1}{n}\xi^T\eta \xrightarrow{a.s.} 0. \end{aligned} \tag{4B.2}$$

Thus, we have

$$\frac{1}{n}W^TW \xrightarrow{a.s.} \begin{bmatrix} 1 & 0 & 0 \\ 0 & \Lambda & 0 \\ 0 & 0 & \Sigma \end{bmatrix}. \tag{4B.3}$$

Recall that

$$\operatorname{var}(\hat{\beta}) = \sigma_e^2\left(W^TW\right)^{-1}. \tag{4B.4}$$

Combining Equations 4B.3 and 4B.4 gives

$$\mathrm{var}(\hat{\beta}) \approx \frac{\sigma_e^2}{n}\begin{bmatrix} 1 & 0 & 0 \\ 0 & \Lambda^{-1} & 0 \\ 0 & 0 & \Sigma^{-1} \end{bmatrix}.$$

Therefore, we have

$$\mathrm{var}(\hat{\alpha}_k) \approx \frac{\sigma_e^2}{n\lambda_k} \quad \text{and} \quad \mathrm{var}(\hat{\delta}_k) \approx \frac{\sigma_e^2}{n\gamma_k}, \quad k = 1,\ldots,L. \tag{4B.5}$$

Assume that the true genetic model is

$$Y_i = \mu + x_i(s)\alpha + z_i(s)\delta + \varepsilon_i, \tag{4B.6}$$

where $x_i(s)$ and $z_i(s)$ are genotype functions for the genetic additive and dominance effect at the trait locus s.

Similarly, by Equations 4.53 and 4.57 and large number theory, we obtain

$$\begin{aligned}
\frac{1}{n}\sum_{i=1}^{n}\xi_{ik}Y_i &= \frac{1}{n}\sum_{i=1}^{n}\left[\int_0^T x_i(t)\varphi_k(t)dt\right]Y_i \\
&= \frac{1}{n}\sum_{i=1}^{n}\left[\int_0^T x_i(t)\varphi_k(t)dt\right]\left[\mu + x_i(s)\alpha + z_i(s)\delta\right] \\
&\xrightarrow{a.s.} \frac{1}{n}\sum_{i=1}^{n}\int_0^T x_i(t)x_i(s)\varphi_k(t)dt\alpha \\
&\xrightarrow{a.s.} \left[\int_0^T 2D(t,s)\varphi_k(t)dt\right]\alpha, \quad k = 1,\ldots,L,
\end{aligned} \tag{4B.7}$$

where $D(t,s)$ is the coefficient of LD between the marker located in the genomic position t and trait locus at the genomic position s.

Similarly, we have

$$\frac{1}{n}\sum_{i=1}^{n}\eta_{ik}Y_i \xrightarrow{a.s.} \left[\int_0^T D^2(t,s)\varphi_k(t)dt\right]\delta, \quad k = 1,\ldots,L. \tag{4B.8}$$

Combining Equations 4B.3, 4B.7, and 4B.8 gives

$$\hat{\beta}=\left(\frac{1}{n}W^TW\right)^{-1}\left(\frac{1}{n}W^TY\right)\xrightarrow{a.s.}\begin{bmatrix}\mu\\ \left[\int_0^T 2D(t,s)\varphi_1(t)dt\right]\alpha/\lambda_1\\ \vdots\\ \left[\int_0^T 2D(t,s)\varphi_L(t)dt\right]\alpha/\lambda_L\\ \left[\int_0^T D^2(t,s)\varphi_1(t)dt\right]\delta/\gamma_1\\ \vdots\\ \left[\int_0^T D^2(t,s)\varphi_L(t)dt\right]\delta/\gamma_L\end{bmatrix}. \tag{4B.9}$$

In other words, we obtain

$$\hat{\alpha}_k\xrightarrow{a.s.}\frac{2}{\lambda_k}\left(\int_0^T D(t,s)\varphi_k(t)dt\right)\alpha, \tag{4B.10}$$

$$\hat{\delta}_k\xrightarrow{a.s.}\frac{1}{\gamma_k}\left(\int_0^T D^2(t,s)\varphi_k(t)\right)\delta,\quad k=1,\ldots,L. \tag{4B.11}$$

Under this model, it follows from Equations 4.66 and 4B.2 that the noncentrality parameter of the test is

$$T\xrightarrow{a.s.}\frac{n}{\sigma_e^2}\left[\sum_{k=1}^{L}\lambda_k\alpha_k^2+\sum_{k=1}^{L}\gamma_k\delta_k^2\right]. \tag{4B.12}$$

It follows from Equations 4B.10 and 4B.11 that

$$\begin{aligned}\hat{\alpha}_k^2&\to\frac{4}{\lambda_k^2}\left[\int_0^T D(t,s)\varphi_k(t)dt\right]^2\alpha^2=\frac{2}{\lambda_k^2P_QP_q}\left[\int_0^T D(t,s)\varphi_k(t)dt\right]^2\sigma_A^2,\\ \hat{\delta}_k^2&\to\frac{1}{\gamma_k^2}\left[\int_0^T D^2(t,s)\phi_k(t)\right]^2\delta^2=\frac{1}{\gamma_k^2P_Q^2P_q^2}\left[\int_0^T D^2(t,s)\phi_k(t)\right]^2\sigma_D^2.\end{aligned} \tag{4B.13}$$

Substituting Equation 4B.13 into Equation 4B.12 gives

$$T \to \frac{n}{\sigma_e^2}\left\{\left[\frac{2}{P_Q P_q}\sum_{k=1}^{L}\frac{1}{\lambda_k}\left(\int_0^T D(t,s)\varphi_k(t)dt^2\right)\right]\sigma_A^2+\left[\frac{1}{P_Q^2 P_q^2}\sum_{k=1}^{L}\frac{1}{\gamma_k}\left(\int_0^T D^2(t,s)\varphi_k(t)^2\right)\right]\sigma_D^2\right\}. \tag{4B.14}$$

APPENDIX 4C: NONCENTRALITY PARAMETER OF THE CCA TEST

Let $G=\begin{bmatrix} X & Z \end{bmatrix}$.

Then, the least square estimators and their variance in the multiple regression model (4.36) are

$$\hat{\beta}_g=\left(G^TG\right)^{-1}G^TY \quad \text{and} \quad \Lambda=\text{var}\left(\hat{\beta}_g\right)=\sigma_Y^2\left(G^TG\right)^{-1}.$$

$$\begin{aligned} T_Q &= \left(\hat{\beta}_g\right)^T\Lambda^{-1}\hat{\beta}_g \\ &= Y^TG\left(G^TG\right)^{-1}\Lambda^{-1}\left(G^TG\right)^{-1}G^TY \\ &= \frac{1}{\sigma_Y^2}Y^TG\left(G^TG\right)^{-1}G^TY \\ &\to \frac{n}{\sigma_Y^2}\Sigma_{yg}\left(\Sigma_{gg}\right)^{-1}\Sigma_{gy}=n\lambda^2. \end{aligned} \tag{4C.1}$$

We can show that (Exercise 4.7)

$$-\log(1-x)>x. \tag{4C.2}$$

Equation 4C.1 gives

$$\bar{T}_{CCA}=-n\log\left(1-\frac{T_Q}{n}\right).$$

It follows from Equation 4C.2 that

$$\bar{T}_{CCA}=-n\log\left(1-\frac{T_Q}{n}\right)>n\frac{T_Q}{n}=T_Q.$$

APPENDIX 4D: SOLUTION TO THE CONSTRAINED NONLINEAR COVARIANCE OPTIMIZATION PROBLEM AND DEPENDENCE MEASURE

Taking the derivative of the Lagrangian with respect to the vectors α and β gives

$$\frac{\partial L(\alpha,\beta,\lambda,\gamma)}{\partial\alpha}=\frac{1}{n}\hat{K}_{xx}\hat{K}_{yy}\beta-\lambda\hat{K}_{xx}\alpha=0, \tag{4D.1}$$

$$\frac{\partial L(\alpha,\beta,\lambda,\gamma)}{\partial \beta}=\frac{1}{n}\hat{K}_{yy}\hat{K}_{xx}\alpha-\gamma\hat{K}_{yy}\beta=0. \tag{4D.2}$$

Recall that

$$\alpha^T\hat{K}_{xx}\alpha=1 \quad \text{and} \quad \beta^T\hat{K}_{yy}\beta=1. \tag{4D.3}$$

Multiplying by α^T on both sides of Equation 4D.1 and by β^T on both sides of Equation 4D.2 and applying Equation 4D.3, we obtain

$$\frac{1}{n}\alpha^T\hat{K}_{xx}\hat{K}_{yy}\beta=\lambda, \tag{4D.4}$$

$$\frac{1}{n}\beta^T\hat{K}_{yy}\hat{K}_{xx}\alpha=\gamma. \tag{4D.5}$$

It follows from Equations 4D.4 and 4D.5 that

$$\lambda=\gamma.$$

Then, the eigenequations (4D.1) and (4D.2) can be written as

$$\begin{bmatrix} 0 & \frac{1}{n}\hat{K}_{xx}\hat{K}_{yy} \\ \frac{1}{n}\hat{K}_{yy}\hat{K}_{xx} & 0 \end{bmatrix}\begin{bmatrix}\alpha\\ \beta\end{bmatrix}=\lambda\begin{bmatrix}\hat{K}_{xx} & 0\\ 0 & \hat{K}_{yy}\end{bmatrix}\begin{bmatrix}\alpha\\ \beta\end{bmatrix}. \tag{4D.6}$$

Let $u=\hat{K}_{xx}^{1/2}\alpha$ and $v=\hat{K}_{yy}^{1/2}\beta$. The optimization problem (4.148) is transformed to

$$\begin{aligned} \max_{u,v} \quad & \frac{1}{\text{n}}u^T\hat{K}_{xx}^{1/2}\hat{K}_{yy}^{1/2}v \\ \text{s.t.} \quad & u^Tu=1 \\ & v^Tv=1 \end{aligned} \quad . \tag{4D.7}$$

Similarly, using the Lagrange multiplier approach to solve the optimization problem, we can obtain the eigenequation

$$\frac{1}{n}\hat{K}_{xx}^{1/2}\hat{K}_{yy}^{1/2}v=\lambda u, \tag{4D.8}$$

$$\frac{1}{n}\hat{K}_{yy}^{1/2}\hat{K}_{xx}^{1/2}u=\lambda v. \tag{4D.9}$$

Substituting Equation 4D.9 into Equation 4D.8 gives the eigenequation

$$\frac{1}{n^2}\hat{K}_{xx}^{1/2}\hat{K}_{yy}\hat{K}_{xx}^{1/2}u=\lambda^2 u. \tag{4D.10}$$

Assume that the single value decomposition of $\hat{K}_{xx}^{1/2}\hat{K}_{yy}^{1/2}$ is

$$\hat{K}_{xx}^{1/2}\hat{K}_{yy}^{1/2}=U\Lambda V^T, \tag{4D.11}$$

where

$$\Lambda=\begin{bmatrix}\rho_1 & \cdots & 0\\ \vdots & \vdots & \vdots\\ 0 & \cdots & \rho_n\end{bmatrix}$$

with $\rho_1\geq\rho_2\geq\cdots\geq\rho_n$.

Then, it follows from Equation 4D.11 that

$$\hat{K}_{xx}^{1/2}\hat{K}_{yy}^{1/2}\left(\hat{K}_{xx}^{1/2}\hat{K}_{yy}^{1/2}\right)^T=\hat{K}_{xx}^{1/2}\hat{K}_{yy}\hat{K}_{xx}^{1/2}=U\Lambda^2U^T. \tag{4D.12}$$

It is clear from Equations 4D.10 and 4D.12 that

$$\frac{\rho_1^2}{n^2}=\lambda^2,$$

which implies that a dependence measure, λ, is equal to $\frac{\rho_1}{n}$.

It follows from Equation 4D.12 that

$$\frac{1}{n^2}\Lambda^2=\frac{1}{n^2}U^T\hat{K}_{xx}^{1/2}\hat{K}_{yy}\hat{K}_{xx}^{1/2}U$$

or

$$\begin{aligned}\frac{1}{n^2}\sum_{i=1}^{n}\rho_i^2&=\frac{1}{n^2}\text{Trace}\left(U^T\hat{K}_{xx}^{1/2}\hat{K}_{yy}\hat{K}_{xx}^{1/2}U\right)\\&=\frac{1}{n^2}\text{Trace}\left(\hat{K}_{xx}\hat{K}_{yy}\right).\end{aligned} \tag{4D.13}$$

Therefore, $\frac{1}{n^2}\text{Trace}\left(\hat{K}_{xx}\hat{K}_{yy}\right)$ can also be taken as a dependence measure.

EXERCISES

Exercise 4.1 Show that under which conditions, we obtain

$$\sigma_P^2 = E\left\{\left(Y - E[Y]\right)^2\right\}$$
$$= E\left\{\left(Y - E[Y|G]\right)^2\right\} + E\left\{\left(E[Y|G] - E[Y]\right)^2\right\}.$$

Exercise 4.2 Considering a single locus with two alleles, A_1 and A_2, we call the genotypic value of one homozygote $+a$, that of the other homozygote $-a$, and that of the heterozygote d. Show that

(1) The substitution effect is $\alpha = \alpha_1 - \alpha_2 = a + (q - p)d$

(2) The genetic additive variance and dominance variance are, respectively, given by

$$\sigma_A^2 = 2pq\alpha^2$$
$$\sigma_D^2 = 4p^2q^2d^2.$$

Exercise 4.3 Show

$$E\left[X_1^m Z_1\right] = 0,$$

$$E\left[Z_1^m X_1\right] = 0,$$

and

$$E\left[Z_1^m Z_1\right] = D^2.$$

Exercise 4.4 Prove

$$R^2 = \frac{\left[\operatorname{cov}(w\beta, y)\right]^2}{\operatorname{var}(w\beta)\sigma_y^2} = \frac{\Sigma_{yw}\Sigma_{ww}^{-1}\Sigma_{wy}}{\sigma_y^2}.$$

Exercise 4.5 Show

$$D^2(t, s_j) \le P_{M_t} P_{m_t} P_Q(s_j) P_q(s_j).$$

Exercise 4.6 Suppose that the functional principal component expansions of the genotype functions for the genetic additive and dominance effects, respectively, are

$$x_i(t) = \sum_{k=1}^{L} \xi_{ik} \varphi_k(t),$$

$$z_i(t) = \sum_{k=1}^{L} \eta_{ik} \varphi_k(t).$$

Show that

(i) $E[\xi_{ik}] = 0$, $E[\eta_{ik}] = 0$;

(ii) $\text{var}(\xi_{ik}) = \lambda_k$, $\text{var}(\eta_{ik}) = \gamma_k$; and

(iii) $E[\xi_{ik}\xi_{il}] = 0$, $E[\eta_{ik}\eta_{il}] = 0$, $l \neq k$, $E[\xi_{ik}\eta_{il}] = 0$.

Exercise 4.7 Show for $x > 0$ we have $-\log(1-x) > x$.

Exercise 4.8 Show that the squared multiple correlation coefficient in the multiple regression model is given by

$$R^2 = \frac{\left[\text{cov}\left(\beta^T x, y\right)\right]^2}{\text{var}\left(\beta^T x\right)\sigma_y^2} = \frac{\Sigma_{yx}\Sigma_{xx}^{-1}\Sigma_{xy}}{\sigma_y^2} = \frac{SS_{reg}}{S_{YY}}.$$

Exercise 4.9 Consider a function

$$f(x) = \begin{cases} 1 & |x| \leq T \\ 0 & T \leq T < \pi \end{cases}.$$

Calculate the coefficient of its Fourier series expansion.

Exercise 4.10 Show that if $\|f\|_H = \|g\|_H$, then $\|f - g\|_H = 0$.

Exercise 4.11 Show that the Hilbert–Schmidt norm of the operator is independent of the choice of orthonormal basis.

Exercise 4.12 Let

$$A = \begin{bmatrix} 1 & 5 \\ 2 & 4 \end{bmatrix} \text{ and } B = \begin{bmatrix} 2 & 3 & 6 \end{bmatrix}.$$

Calculate the tensor product $A \otimes B$.

Exercise 4.13 Let $f(X) \in H_x$ and $g(Y) \in H_y$, where H_x and H_y are RKHS with kernel functions $k_x(X, X)$ and $k_y(Y, Y)$ for random variables X and Y, respectively. Σ_{yx} is a cross-covariance operator between X and Y. Show that

$$\langle g, \Sigma_{yx} f \rangle_{H_y} = \text{cov}\left(f(X), g(y)\right).$$

CHAPTER 5

Multiple Phenotype Association Studies

On December 18, 2014, a *Catalog of Published Genome-Wide Association Studies* (GWAS) reported significant association of 15,177 SNPs with more than 700 traits in 2087 publications (National Human Genome Research Institute 2015). It is reported that more than 4.6% of the SNPs and 16.9% of the genes were significantly associated with more than one trait (Solovieff et al. 2013). These results demonstrate that genetic pleiotropic effects, which refer to the effects of a genetic variant affecting multiple traits, play a crucial role in uncovering genetic structures of correlated phenotypes (Chen et al. 2015; Hill and Zhang 2012; Wagner and Zhang 2011). Most genetic analyses of quantitative traits have focused on a single trait association analysis, analyzing each phenotype independently (Stephens 2013). However, multiple phenotypes are correlated. The integrative analysis of correlated phenotypes, which test for the association of a genetic variant with multiple traits, often increases statistical power to identify genetic associations and increases the precision of genetic effect estimation (Aschard et al. 2014).

Three major approaches are commonly used to explore the association of genetic variants with multiple correlated phenotypes: multiple regression methods, integration of *P*-values of univariate analysis, and dimension reduction methods (Ray et al. 2016). The methods for multiple phenotype association studies can be classified into three categories: designed for only common variants, designed for only rare variants, and designed for both common and rare variants. This chapter will cover all three categories of multiple phenotype association analysis methods.

As the number of phenotypes increases, the proportion of subjects missing at least one observation increases exponentially (Dahl et al. 2016). Imputing missing phenotypes is an essential for multiple phenotype association analysis.

5.1 PLEIOTROPIC ADDITIVE AND DOMINANCE EFFECTS

To rigorously investigate multiple phenotype association analysis, we need to formally define pleiotropic genetic additive and dominance effects. Consider a single locus with two alleles: A1 and A2 with allele frequencies, respectively, and K traits. For the jth trait, we denote its genotypic values for the homozygote A_1A_1, the other homozygote A_2A_2, and the heterozygote A_1A_2 by $G_{11}^{(j)}$, $G_{22}^{(j)}$, and $G_{12}^{(j)}$, respectively. Similar to Section 4.1.2.2, we can derive the genetic additive and dominance effects for the jth trait:

$$\alpha^{(j)} = pG_{11}^{(j)} + (q-p)G_{12}^{(j)} - qG_{22}^{(j)}, \tag{5.1}$$

$$\delta^{(j)} = G_{11}^{(j)} - 2G_{12}^{(j)} + G_{22}^{(j)}. \tag{5.2}$$

Thus, the genotypic values for the jth trait can be expressed in terms of genetic additive and dominance effects:

$$\begin{aligned} G_{11}^{(j)} &= \mu^{(j)} + 2q\alpha^{(j)} + q^2\delta^{(j)} \\ G_{12}^{(j)} &= \mu^{(j)} + (q-p)\alpha^{(j)} - pq\delta^{(j)} \\ G_{22}^{(j)} &= \mu^{(j)} - 2p\alpha^{(j)} + p^2\delta^{(j)}, \quad j = 1,2,\ldots,K. \end{aligned} \tag{5.3}$$

Let

$$G = \begin{bmatrix} G_{11}^{(1)} & \cdots & G_{11}^{(K)} \\ G_{12}^{(1)} & \cdots & G_{12}^{(K)} \\ G_{22}^{(1)} & \cdots & G_{22}^{(K)} \end{bmatrix}, \quad \mu = \begin{bmatrix} \mu^{(1)} & \cdots & \mu^{(K)} \end{bmatrix}, \quad \alpha = \begin{bmatrix} \alpha^{(1)} & \cdots & \alpha^{(K)} \end{bmatrix},$$

$$\delta = \begin{bmatrix} \delta^{(1)} & \cdots & \delta^{(K)} \end{bmatrix}, \quad 1 = \begin{bmatrix} 1 \\ 1 \\ 1 \end{bmatrix}, \quad X = \begin{bmatrix} 2q \\ (q-p) \\ -2p \end{bmatrix}, \quad \text{and} \quad Z = \begin{bmatrix} q^2 \\ -pq \\ p^2 \end{bmatrix}.$$

Then, we have

$$G = 1\mu + X\alpha + Z\delta. \tag{5.4}$$

Let $G = \begin{bmatrix} X & Z \end{bmatrix}$ and $B = \begin{bmatrix} \alpha \\ \delta \end{bmatrix}$. Then, Equation 5.4 can be further written as

$$G = 1\mu + GB. \tag{5.5}$$

Next we define the genetic additive and dominance covariance. Similar to Section 4.1.2.3, we define a genetic covariance between two traits:

$$\begin{aligned} \sigma_G^{j,k} &= p^2\left(G_{11}^{(j)} - \mu^{(j)}\right)\left(G_{11}^{(k)} - \mu^{(k)}\right) + 2pq\left(G_{12}^{(j)} - \mu^{(j)}\right)\left(G_{12}^{(j)} - \mu^{(j)}\right) + q^2\left(G_{22}^{(j)} - \mu^{(j)}\right)\left(G_{22}^{(k)} - \mu^{(k)}\right) \\ &= p^2\left(2\alpha_1^{(j)} + e_1^{(j)}\right)\left(2\alpha_1^{(k)} + e_1^{(k)}\right) + 2pq\left(\alpha_1^{(j)} + \alpha_2^{(j)} + e_2^{(j)}\right)\left(\alpha_1^{(j)} + \alpha_2^{(j)} + e_2^{(j)}\right) \\ &\quad + q^2\left(2\alpha_2^{(j)} + e_3^{(j)}\right)\left(2\alpha_2^{(k)} + e_3^{(k)}\right). \end{aligned} \tag{5.6}$$

Substituting Equation 5.3 into Equation 5.6 gives

$$\sigma_G^{j,k} = p^2\left(2q\alpha^{(j)} + q^2\delta^{(j)}\right)\left(2q\alpha^{(k)} + q^2\delta^{(k)}\right) + 2pq\left((q-p)\alpha^{(j)} - pq\delta^{(j)}\right)\left((q-p)\alpha^{(k)} - pq\delta^{(k)}\right)$$
$$+ q^2\left(-2p\alpha^{(j)} + p^2\delta^{(j)}\right)\left(-2p\alpha^{(k)} + p^2\delta^{(k)}\right)$$
$$= \sigma_a^{j,k} + \sigma_d^{j,k}, \tag{5.7}$$

where

$$\sigma_a^{j,k} = p^2\left(2q\alpha^{(j)}\right)\left(2q\alpha^{(k)}\right) + 2pq\left((q-p)\alpha^{(j)}\right)\left((q-p)\alpha^{(k)}\right) + q^2\left(-2p\alpha^{(j)}\right)\left(-2p\alpha^{(k)}\right)$$
$$= 2pq\alpha^{(j)}\alpha^{(k)} \tag{5.8}$$

and

$$\sigma_d^{j,k} = p^2\left(q^2\delta^{(j)}\right)\left(q^2\delta^{(k)}\right) + 2pq\left(-pq\delta^{(j)}\right)\left(-pq\delta^{(k)}\right) + q^2\left(p^2\delta^{(j)}\right)\left(p^2\delta^{(k)}\right)$$
$$= p^2q^2\delta^{(j)}\delta^{(k)}. \tag{5.9}$$

Define

$$\alpha = \begin{bmatrix} \alpha^{(1)} \\ \vdots \\ \alpha^{(K)} \end{bmatrix} \quad \text{and} \quad \delta = \begin{bmatrix} \delta^{(1)} \\ \vdots \\ \delta^{(K)} \end{bmatrix}.$$

The genetic additive covariance and dominance covariance matrices are defined as

$$\Sigma_a = 2pq\alpha\alpha^T \quad \text{and} \quad \Sigma_d = p^2q^2\delta\delta^T, \tag{5.10}$$

respectively.

Formula (5.10) is similar to Equations 4.18 and 4.19. Replacing scalars α and δ in Equations 4.18 and 4.19 by vectors α and δ gives formula (5.10).

5.2 MULTIVARIATE MARGINAL REGRESSION

5.2.1 Models

Assume that n individuals with K correlated traits, $Y_1, Y_2, \ldots, Y_K$, are sampled. Let X_{il} and Z_{il} be the indicator variables for the genotypes associated with the genetic additive and dominance effects of the ith individual at the lth SNP. Let μ_j be an overall mean of the jth trait, α_{lj} and δ_{lj} be the genetic additive and dominance effect of the lth SNP associated with the jth trait, $\varepsilon_i = [\varepsilon_{i1}, \ldots, \varepsilon_{ik}]^T$ be a vector of errors distributed as a normal distribution with a mean vector of zeros and a $K \times K$ variance–covariance matrix Σ, and $\varepsilon_1, \ldots, \varepsilon_n$ assumed to be independent.

Consider the multivariate regression model for association analysis of multiple phenotypes:

$$\begin{bmatrix} y_{11} & \cdots & y_{1K} \\ \vdots & \ddots & \vdots \\ y_{n1} & \cdots & y_{nK} \end{bmatrix} = \begin{bmatrix} 1 & h_{11} & \cdots & h_{1l} \\ \vdots & \vdots & \ddots & \vdots \\ 1 & h_{n1} & \cdots & h_{nl} \end{bmatrix} \begin{bmatrix} \mu_1 & \cdots & \mu_K \\ \eta_{11} & \cdots & \eta_{1K} \\ \vdots & \ddots & \vdots \\ \eta_{l1} & \cdots & \eta_{lK} \end{bmatrix}$$

$$+ \begin{bmatrix} x_1 \\ \vdots \\ x_n \end{bmatrix} \begin{bmatrix} \alpha_1 & \cdots & \alpha_K \end{bmatrix} + \begin{bmatrix} z_1 \\ \vdots \\ z_n \end{bmatrix} \begin{bmatrix} \delta_1 & \cdots & \delta_K \end{bmatrix} + \begin{bmatrix} \varepsilon_{11} & \cdots & \varepsilon_{1K} \\ \vdots & \ddots & \vdots \\ \varepsilon_{n1} & \cdots & \varepsilon_{nK} \end{bmatrix}. \quad (5.11)$$

Equation 5.11 can be written in a compact form:

$$Y = W\eta + X\alpha + Z\delta + \varepsilon, \quad (5.12)$$

where

Y is a phenotype matrix
W is a covariate matrix
X is a genotype indicator vector for the genetic additive effect
Z is a genotype indicator vector for the genetic dominance effect
η is a coefficient matrix associated with W
α is a row vector of genetic additive effects
δ is a row vector of genetic dominance effects
ε is an error matrix

We assume that $\varepsilon_i = [\varepsilon_{i1}, \ldots, \varepsilon_{iK}]$ is normally distributed with mean zero and covariance matrix Σ and $\varepsilon_1, \ldots, \varepsilon_n$ are independent. Covariates include age, sex, race, and PCA for population structure correction.

For the convenience of discussion, the model (5.12) can be further rewritten as

$$Y = HB + \varepsilon, \quad (5.13)$$

where $H = \begin{bmatrix} W & X & Z \end{bmatrix}$ and $B = \begin{bmatrix} \eta \\ \alpha \\ \delta \end{bmatrix}$.

5.2.2 Estimation of Genetic Effects

5.2.2.1 Least Square Estimation

The widely used method for estimation of the parameters in the model (5.12) is the least square estimation. Let

$$Y = \begin{bmatrix} Y_1 & \cdots & Y_K \end{bmatrix}, \; B = \begin{bmatrix} B_1 & \cdots & B_K \end{bmatrix}, \text{ and } \varepsilon = \begin{bmatrix} \varepsilon_1 & \cdots & \varepsilon_K \end{bmatrix}.$$

Then, the model (5.13) can be reduced to

$$Y_j = HB_j + \varepsilon_j, \quad j = 1,\ldots,K. \tag{5.14}$$

The least square estimate for B is to minimize the residual sum of squares:

$$F = \sum_{j=1}^{K} (Y_j - HB_j)^T (Y_j - HB_j). \tag{5.15}$$

Define the residual sum of squares and cross products matrix:

$$(Y - HB)^T (Y - HB) = \begin{bmatrix} (Y_1 - HB_1)^T (Y_1 - HB_1) & \cdots & (Y_1 - HB_1)^T (Y_K - HB_K) \\ \vdots & \ddots & \vdots \\ (Y_K - HB_K)^T (Y_1 - HB_1) & \cdots & (Y_K - HB_K)^T (Y_K - HB_K) \end{bmatrix}. \tag{5.16}$$

It is clear that F in Equation 5.15 is equal to the trace of the residual sum of squares and cross products matrix:

$$F = \text{Trace}\left((Y - HB)^T (Y - HB)\right). \tag{5.17}$$

Using the formula for the derivative of the trace function with respect to the matrix in Equation 1.166 gives

$$\frac{\partial F}{\partial B} = -2H^T (Y - HB).$$

Setting $\frac{\partial F}{\partial B} = 0$ and solving the resulting equation, we obtain

$$\hat{B} = \left(H^T H\right)^{-1} H^T Y, \tag{5.18}$$

or

$$\left[\hat{B}_1,\ldots,\hat{B}_K\right] = \left(H^T H\right)^{-1} H^T \left[Y_1,\ldots,Y_K\right]. \tag{5.19}$$

In conformity with the univariate least square estimate for the single phenotype, we take

$$\hat{B}_j = \left(H^T H\right)^{-1} H^T Y_j. \tag{5.20}$$

After the estimated regression coefficients are available, we can use them to calculate the matrix of predicted values:

$$\hat{Y} = H\hat{B} = H\left(H^T H\right)^{-1} H^T Y. \tag{5.21}$$

The residual matrix is given by

$$\hat{\varepsilon} = Y - \hat{Y} = \left[I - H\left(H^T H\right)^{-1} H^T\right]Y, \tag{5.22}$$

which implies the following residual sum of squares and cross products matrix:

$$\hat{\varepsilon}^T\hat{\varepsilon} = Y^T\left(I - H\left(H^T H\right)^{-1} H^T\right)Y. \tag{5.23}$$

Next we study the properties of the estimators. First, we show that the estimator is unbiased. It follows from Equations 5.13 and 5.18 that

$$E\left[B\right] = E\left[\left(H^T H\right)^{-1} H^T\left(HB + \varepsilon\right)\right] = B.$$

To calculate the covariance matrix of the estimator $\hat{B}$, we use a vector operation. In distribution theory, it is generally easier to work with a vector than a matrix. The vector of a matrix A, denoted by $A = [a_1, \ldots, a_k]$, is defined as

$$\text{vec}\left(A\right) = \begin{bmatrix} a_1 \\ \vdots \\ a_k \end{bmatrix}.$$

It is well known that (Graybill 1983)

$$\text{vec}\left(ABC^T\right) = \left(C \otimes A\right)\text{vec}\left(B\right), \tag{5.24}$$

where $\otimes$ denotes the Kronecker product.

Using Equations 5.18 and 5.24 gives

$$\text{vec}\left(\hat{B}\right) = (I \otimes\left(\left(H^T H\right)^{-1} H^T\right)\text{vec}\left(Y\right). \tag{5.25}$$

Recall that

$$\text{vec}\left(Y\right) = \begin{bmatrix} Y_1 \\ \vdots \\ Y_K \end{bmatrix}$$

and

$$\text{cov}\left(\text{vec}(Y)\right) = \Sigma \otimes I. \tag{5.26}$$

Using Equations 5.25 and 5.26, we obtain the covariance matrix of the estimators $\text{vec}\left(\hat{B}\right)$:

$$\begin{aligned}\text{cov}\left(\text{vec}(\hat{B})\right) &= \left[I \otimes \left(\left(H^T H\right)^{-1} H^T\right)\right][\Sigma \otimes I][I \otimes \left(H^T \left(H^T H\right)^{-1}\right) \\ &= \Sigma \otimes \left(H^T H\right)^{-1}.\end{aligned} \tag{5.27}$$

It follows from Equation 5.27 that

$$\text{cov}\left(\hat{B}_j, \hat{B}_k\right) = \sigma_{jk}\left(H^T H\right)^{-1}. \tag{5.28}$$

Next we study the covariance matrix of the residuals $\hat{\varepsilon}$. Equation 5.22 gives

$$\hat{\varepsilon}_j = \left[I - H\left(H^T H\right)^{-1} H^T\right] Y_j \tag{5.29}$$

and

$$\begin{aligned}E\left[\hat{\varepsilon}_j\right] &= \left[I - H\left(H^T H\right)^{-1} H^T\right] E\left[Y_j\right] \\ &= \left[I - H\left(H^T H\right)^{-1} H^T\right] H B_j = 0.\end{aligned} \tag{5.30}$$

This implies that

$$\begin{aligned}E\left[\hat{\varepsilon}_j^T \hat{\varepsilon}_k\right] &= E\left[\text{Trace}\left(\hat{\varepsilon}_j^T \hat{\varepsilon}_k\right)\right] \\ &= E\left[\text{Trace}\left(Y_j^T \left[I - H\left(H^T H\right)^{-1} H^T\right]\left[I - H\left(H^T H\right)^{-1} H^T\right] Y_k\right)\right] \\ &= E\left[\text{Trace}\left(\left[I - H\left(H^T H\right)^{-1} H^T\right] Y_k Y_j^T\right)\right] \\ &= \text{Trace}\left(\left[I - H\left(H^T H\right)^{-1} H^T\right] E\left[Y_k Y_j^T\right]\right) \\ &= \text{Trace}\left(\left[I - H\left(H^T H\right)^{-1} H^T\right] \sigma_{jk} I\right) \\ &= (n - r)\sigma_{jk},\end{aligned} \tag{5.31}$$

where $r = \text{rank}(H)$.

Equation 5.31 indicates that the unbiased estimator of σ_{jk} is

$$\hat{\sigma}_{jk} = \frac{1}{n-r}\hat{\varepsilon}_j^T\hat{\varepsilon}_k. \quad (5.32)$$

Recall that

$$\hat{\varepsilon}_j^T\hat{\varepsilon}_k = \sum_{i=1}^{n}\hat{\varepsilon}_{ij}\hat{\varepsilon}_{ik},$$

which implies that

$$\hat{\sigma}_{jk} = \frac{1}{n-r}\sum_{i=1}^{n}\hat{\varepsilon}_{ij}\hat{\varepsilon}_{ik}. \quad (5.33)$$

Equation 5.32 gives

$$\hat{\Sigma} = \frac{1}{n-r}\hat{\varepsilon}^T\hat{\varepsilon}. \quad (5.34)$$

In summary, we obtain the following result.

Result 5.1

The least square estimator for the regression coefficients is

$$\hat{B} = \left(H^T H\right)^{-1} H^T Y,$$

and the covariance matrix of the estimators is

$$\operatorname{cov}\left(\operatorname{vec}\left(\hat{B}\right)\right) = \Sigma \otimes \left(H^T H\right)^{-1},$$

where Σ is estimated by

$$\hat{\Sigma} = \frac{1}{n-r}\hat{\varepsilon}^T\hat{\varepsilon}, \quad \hat{\varepsilon} = \left[I - H\left(H^T H\right)^{-1} H^T\right]Y.$$

Similar to Appendix 4A, we can obtain the asymptotic results.

Along the line in Appendix 4A, we can easily show that

$$\frac{1}{n}H^T H = \frac{1}{n}\begin{bmatrix} W^T \\ X^T \\ Z^T \end{bmatrix}\begin{bmatrix} W & X & Z \end{bmatrix} \xrightarrow{a.s.} \begin{bmatrix} \frac{1}{n}W^T W & 0 & 0 \\ 0 & 2P_M P_m & 0 \\ 0 & 0 & P_M^2 P_m^2 \end{bmatrix}$$

and

$$\frac{1}{n}H^TY = \frac{1}{n}\begin{bmatrix} W^TY \\ X^TY \\ Z^TY \end{bmatrix} \xrightarrow{a.s.} \begin{bmatrix} \frac{1}{n}W^TY \\ 2D\alpha_m \\ D^2\delta_m \end{bmatrix},$$

where $\alpha_m = \begin{bmatrix} \alpha_m^{(1)} & \cdots & \alpha_m^{(K)} \end{bmatrix}$ and $\delta_m = \begin{bmatrix} \delta_m^{(1)} & \cdots & \delta_m^{(K)} \end{bmatrix}$.

Therefore, we obtain the following result.

Result 5.2

Asymptotically, the vectors of the estimated genetic additive and dominance effects at the marker converge to

$$\alpha_m \xrightarrow{a.s.} \frac{D}{P_M P_m}\alpha \quad \text{and} \quad \delta_m \xrightarrow{a.s.} \frac{D^2}{P_M^2 P_m^2}\delta. \tag{5.35}$$

When the marker locus is a trait locus, Equation 5.35 is reduced to

$$\alpha_m \xrightarrow{a.s.} \alpha \quad \text{and} \quad \delta_m \xrightarrow{a.s.} \delta.$$

The estimated genetic additive covariance and dominance covariance matrices converge to

$$\hat{\Sigma}_q = 2P_M P_m \alpha_m \alpha_m^T \xrightarrow{a.s.} \frac{D^2}{P_M P_m P_Q P_q}\Sigma_a, \tag{5.36}$$

$$\hat{\Sigma}_d = P_M^2 P_m^2 \delta_m \delta_m^T \xrightarrow{a.s.} \frac{D^4}{P_M^2 P_m^2 P_Q^2 P_q^2}\Sigma_d. \tag{5.37}$$

5.2.2.2 Maximum Likelihood Estimator

We begin with a log-likelihood function (Johnson and Wichern 2002). Let the phenotype data matrix be organized in terms of individual samples:

$$Y = \begin{bmatrix} Y_1^T \\ \vdots \\ Y_n^T \end{bmatrix},$$

where Y_j is assumed to be normally distributed as $N(B^T H_j, \Sigma)$. We can write

$$H^T = \begin{bmatrix} H_1^T \\ \vdots \\ H_n^T \end{bmatrix}.$$

Thus, we have

$$\begin{aligned} Y - HB &= \begin{bmatrix} Y_1^T \\ \vdots \\ Y_n^T \end{bmatrix} - \begin{bmatrix} H_1^T \\ \vdots \\ H_n^T \end{bmatrix} B = \begin{bmatrix} Y_1^T - H_1^T B \\ \vdots \\ Y_n^T - H_n^T B \end{bmatrix} \\ &= \begin{bmatrix} Y_1 - B^T H_1 & \cdots & Y_n - B^T H_n \end{bmatrix}^T. \end{aligned} \tag{5.38}$$

Equation 5.38 implies

$$(Y - HB)^T (Y - HB) = \sum_{j=1}^{n} (Y_j - B^T H_j)(Y_j - B^T H_j)^T$$

and

$$\begin{aligned} \sum_{j=1}^{n} (Y_j - B^T H_j)^T \Sigma^{-1} (Y_j - B^T H_j) &= \sum_{j=1}^{n} \text{Trace}\left[(Y_j - B^T H_j)^T \Sigma^{-1} (Y_j - B^T H_j)\right] \\ &= \sum_{j=1}^{n} \text{Trace}\left[\Sigma^{-1} (Y_j - B^T H_j)(Y_j - B^T H_j)^T\right] \\ &= \text{Trace}\left[\sum_{j=1}^{n} \left[\Sigma^{-1} (Y_j - B^T H_j)(Y_j - B^T H_j)^T\right]\right] \\ &= \text{Trace}\left[\Sigma^{-1} (Y - HB)^T (Y - HB)\right]. \end{aligned} \tag{5.39}$$

The log-likelihood function can be written as

$$l(B, \Sigma) = -\frac{nK}{2} \log(2\pi) - \frac{n}{2} \log|\Sigma| - \frac{1}{2} \text{Trace}\left[\Sigma^{-1} (Y - HB)^T (Y - HB)\right]. \tag{5.40}$$

To find the matrix B that maximizes the log-likelihood function, we set

$$\frac{\partial l(B, \Sigma)}{\partial B} = 0 \tag{5.41}$$

and

$$\frac{\partial l(B,\Sigma)}{\partial \Sigma}=0. \tag{5.42}$$

Using the formula for the derivative of the trace function with respect to the matrix (1.164), Equation 5.41 is reduced to

$$\frac{\partial l(B,\Sigma)}{\partial B}=H^T(Y-HB)\Sigma^{-1}=0. \tag{5.43}$$

Multiplying Σ from the right side of Equation 5.43 gives

$$H^T(Y-HB)=0. \tag{5.44}$$

Solving Equation 5.44 for the matrix B, we obtain the maximum likelihood estimator of the matrix B:

$$\hat{B}=\left(H^TH\right)^{-1}H^TY, \tag{5.45}$$

if the matrix H is of full rank.

Equation 5.45 is exactly the same as Equation 5.18. This indicates that both the least square and maximum likelihood estimators are equivalent.

Next we solve Equation 5.42. Using Equations 1.160 and 1.168, we obtain

$$\frac{\partial l(B,\Sigma)}{\partial \Sigma}=-\frac{n}{2}\Sigma^{-1}+\frac{1}{2}\Sigma^{-1}(Y-HB)^T(Y-HB)\Sigma^{-1}=0. \tag{5.46}$$

Solving Equation 5.46 gives

$$\hat{\Sigma}=\frac{1}{n}\left(Y-H\hat{B}\right)^T\left(Y-H\hat{B}\right). \tag{5.47}$$

Recall that

$$\begin{aligned}\hat{\varepsilon}&=Y-H\hat{B}=Y-H\left(H^TH\right)^{-1}H^TY\\&=\left[I-H\left(H^TH\right)^{-1}H^T\right]Y\\&=\left[I-H\left(H^TH\right)^{-1}H^T\right](HB+\varepsilon)\\&=\left[I-H\left(H^TH\right)^{-1}H^T\right]\varepsilon.\end{aligned} \tag{5.48}$$

Note that the trace of $[I - H(H^T H)^{-1} H^T]$ is

$$\text{Trace}\left[I - H\left(H^T H\right)^{-1} H^T\right] = n - r, \tag{5.49}$$

where r is the rank of the matrix H. Therefore, the eigenvalue decomposition of $[I - H(H^T H)^{-1} H^T]$ is

$$\left[I - H\left(H^T H\right)^{-1} H^T\right] = U \Lambda U^T = \sum_{j=1}^{n-r} \lambda_j u_j u_j^T. \tag{5.50}$$

Since

$$\begin{aligned}\left[I - H\left(H^T H\right)^{-1} H^T\right] &= \left[I - H\left(H^T H\right)^{-1} H^T\right]^2 \\ &= U \Lambda^2 U^T,\end{aligned}$$

we have

$$\Lambda^2 = \Lambda,$$

which implies

$$\Lambda = \text{diag}\left(1, 1, \ldots, 1, 0, \ldots, 0\right).$$

Therefore, Equation 5.50 is reduced to

$$\left[I - H\left(H^T H\right)^{-1} H^T\right] = \sum_{j=1}^{n-r} u_j u_j^T. \tag{5.51}$$

It follows from Equations 5.47 and 5.48 that

$$\begin{aligned}n\hat{\Sigma} &= \left(Y - H\hat{B}\right)^T \left(Y - H\hat{B}\right) \\ &= \varepsilon^T \left[I - H\left(H^T H\right)^{-1} H^T\right] \varepsilon.\end{aligned} \tag{5.52}$$

Using Equations 5.51 and 5.52 gives

$$n\hat{\Sigma} = \sum_{j=1}^{n-r} \varepsilon^T u_j u_j^T \varepsilon. \tag{5.53}$$

Now we derive the distribution of the linear combination $\varepsilon^T u_j$. Recall that

$$\varepsilon^T u_j = \begin{bmatrix} \varepsilon_1 & \cdots & \varepsilon_n \end{bmatrix} \begin{bmatrix} u_{1j} \\ \vdots \\ u_{nj} \end{bmatrix}$$

$$= \sum_{i=1}^{n} u_{ij} \varepsilon_i . \tag{5.54}$$

It is clear that

$$E\left[\varepsilon^T u_j\right] = E\left[\varepsilon^T\right] u_j = 0, \tag{5.55}$$

$$\operatorname{var}\left(\varepsilon^T u_j\right) = \left(\sum_{i=1}^{n} u_j^2\right) \Sigma$$

$$= \Sigma, \tag{5.56}$$

and

$$\operatorname{cov}\left(\varepsilon^T u_j, \varepsilon^T u_k\right) = \left(u_j^T u_k\right) \Sigma = 0. \tag{5.57}$$

Let $V_j = \varepsilon^T u_j$. Equations 5.55 through 5.57 indicate that the variables $V_j, j = 1, \ldots, n-r$ are independent and normally distributed variables, $N(0, \Sigma)$. Therefore, by definition of Wishart distribution, $n\hat{\Sigma} = \sum_{j=1}^{n-r} V_j V_j^T$ has the Wishart distribution $W_{K,n-r}(\Sigma)$.

In summary, we have the following result.

Result 5.3

Let r be the rank of the matrix H and K be the number of traits. Assume that the errors ε_i, $i = 1, \ldots, n$ are independently and normally distributed as $N(0, \Sigma)$. The maximum likelihood estimators of the regression coefficient matrix B and covariance matrix Σ are given, respectively, by

$$\hat{B} = \left(H^T H\right)^{-1} H^T Y$$

and

$$\hat{\Sigma} = \frac{1}{n}\left(Y - H\hat{B}\right)^T \left(Y - H\hat{B}\right).$$

$n\hat{\Sigma}$ is distributed as $W_{K,n-r}(\Sigma)$.

5.2.3 Test Statistics

5.2.3.1 *Classical Null Hypothesis*

Assume

$$B = \begin{bmatrix} \eta \\ \alpha \\ \delta \end{bmatrix} = \begin{bmatrix} B_{(1)} \\ B_{(2)} \end{bmatrix},$$

where

$$B_{(2)} = \begin{bmatrix} \alpha \\ \delta \end{bmatrix}.$$

The null hypothesis for no association is

$$H_0 : \alpha = 0 \quad \text{and} \quad \delta = 0$$

or

$$B_{(2)} = 0.$$

The regression model can be written as

$$Y = H_1 B_1 + H_2 B_2 + \varepsilon,$$

where

$$H = \begin{bmatrix} H_{1,n\times(r-2)} & H_{2,n\times 2} \end{bmatrix}.$$

5.2.3.1.1 Likelihood Ratio Test Likelihood ratio statistics can be used to test the presence of genetic effects. Under the null hypothesis $H_0 : B_2 = 0$, the regression model is

$$Y = H_1 B_1 + \varepsilon_1. \tag{5.58}$$

The likelihood ratio is

$$\Lambda = \frac{\max_{B_1} L(B_1, \Sigma_1)}{\max_B L(B, \Sigma)} = \frac{L(\hat{B}_1, \hat{\Sigma}_1)}{L(\hat{B}, \hat{\Sigma})} = \left(\frac{|\hat{\Sigma}|}{|\hat{\Sigma}_1|} \right)^{\frac{n}{2}}$$

or

$$-2\log\Lambda = -n\log\frac{|\hat{\Sigma}|}{|\hat{\Sigma}_1|} = -n\log\frac{|n\hat{\Sigma}|}{|n\hat{\Sigma}+n(\hat{\Sigma}_1-\hat{\Sigma})|} \tag{5.59}$$

$$-n\log\frac{|\hat{\Sigma}|}{|\hat{\Sigma}_1|}$$

or

$$-\left(n-r-\frac{K+1}{2}\right)\log\frac{|\hat{\Sigma}|}{|\hat{\Sigma}_1|}. \tag{5.60}$$

The likelihood ratio statistic is asymptotically distributed as a central $\chi^2_{(2K)}$ distribution.

5.2.3.2 *The Multivariate General Linear Hypothesis*

The single variate general linear hypothesis has the form

$$C\beta-\theta=\mathbf{0},$$

where

- C is a $q\times r$-dimensional matrix
- β is an r-dimensional vector
- θ and $\mathbf{0}$ are q-dimensional vectors

The hypothesis assesses the contribution of the genetic variants to a trait. The parameter β is a column of the vector. However, for the multivariate regression, the B matrix has multiple columns involving multiple traits. We need to consider differences in the contributions of the genetic variants to the multiple traits and test linear hypothesis about the multiple columns of the matrix B. Therefore, we should consider both the rows and columns of the matrix B. The multivariate general linear hypothesis has the form

$$CBM-\theta=\mathbf{0}, \tag{5.61}$$

where

- C is a $q\times r$-dimensional matrix
- B is an $r\times K$-dimensional matrix
- M is a $K\times l$-dimensional matrix
- θ is a $q\times l$-dimensional matrix
- $\mathbf{0}$ is a $q\times l$-dimensional matrix

We assume that the rank of the matrix $C=q\leq r$ and the rank of the matrix $M=l\leq K$.

Example 5.1

Consider a sex and a race variable as covariates in the model and two traits. We want to test the association of a genetic variant with two traits. The general linear hypothesis for this problem has the form

$$H_0:$$

$$\begin{bmatrix} 0 & 0 & 0 & 1 & 0 \\ 0 & 0 & 0 & 0 & 1 \end{bmatrix} \begin{bmatrix} \eta_{01} & \eta_{02} \\ \eta_{11} & \eta_{12} \\ \eta_{21} & \eta_{22} \\ \alpha_1 & \alpha_2 \\ \delta_1 & \delta_2 \end{bmatrix} \begin{bmatrix} 1 & 0 \\ 0 & 1 \end{bmatrix} = \begin{bmatrix} 0 & 0 \\ 0 & 0 \end{bmatrix}$$

or

$$H_0:$$

$$\begin{bmatrix} \alpha_1 & \alpha_2 \\ \delta_1 & \delta_2 \end{bmatrix} = \begin{bmatrix} 0 & 0 \\ 0 & 0 \end{bmatrix}.$$

5.2.3.3 Estimation of the Parameter Matrix under Constraints

Before we introduce additional statistics for testing association, we first estimate the parameters in the model under the constraint of Equation 5.61. We attempt to find the matrix B that minimizes (Izenman 2008)

$$\text{Trace}\left[(Y - HB)^T (Y - HB)\right] \tag{5.62}$$

$$\text{Subject to} \quad CBM - \theta = \mathbf{0}.$$

Using Lagrange multipliers, the constrained optimization problem (5.62) can be transformed into the following unconstrained optimization problem:

$$F = \text{Trace}\left[(Y - HB)^T (Y - HB)\right] + \text{Trace}\left[\Lambda(CBM - \theta)\right], \tag{5.63}$$

where Λ is a matrix of the Langrage multipliers.

Using formula (1.164), we obtain

$$\frac{\partial F}{\partial B} = -H^T (Y - HB) + C^T \Lambda M^T.$$

Setting $\frac{\partial F}{\partial B} = 0$ gives

$$H^T HB + C^T \Lambda M^T = H^T Y. \tag{5.64}$$

Solving Equation 5.64 for the matrix B, we obtain

$$B^* = \left(H^T H\right)^{-1} H^T Y - \left(H^T H\right)^{-1} C^T \Lambda M^T$$
$$= \hat{B} - \left(H^T H\right)^{-1} C^T \Lambda M^T. \tag{5.65}$$

where $\hat{B} = \left(H^T H\right)^{-1} H^T Y$ is the estimator of the parameter matrix B in the multivariate linear regression without constraints. Equation 5.65 implies that the estimator of the parameter matrix B consists of two parts. The first part in Equation 5.65 is the estimator of the parameter matrix B without constraints and the second part in Equation 5.65 is due to constraints.

Next we estimate the matrix Λ. Substituting Equation 5.65 into Equation 5.61 gives

$$C\left(H^T H\right)^{-1} C^T \Lambda M^T M = C\hat{B}M - \theta. \tag{5.66}$$

Solving Equation 5.66 for the matrix Λ, we obtain

$$\Lambda = \left[C\left(H^T H\right)^{-1} C^T\right]^{-1} \left(C\hat{B}M - \theta\right)\left(M^T M\right)^{-1}. \tag{5.67}$$

Substituting Equation 5.67 into Equation 5.65 gives

$$B^* = \hat{B} - \left(H^T H\right)^{-1} C^T \left[C\left(H^T H\right)^{-1} C^T\right]^{-1} \left(C\hat{B}M - \theta\right)\left(M^T M\right)^{-1} M^T. \tag{5.68}$$

We can easily check that the matrix B^* satisfies the constraint Equation 5.61.

5.2.3.4 Multivariate Analysis of Variance (MANOVA)

Analysis of variance is an important tool for the analysis in the single variate linear regression. Next we extend the single analysis of variance to multivariate analysis of variance. Consider the residual under the constrained model

$$\varepsilon^* = Y - HB^*$$
$$= Y - H\hat{B} + H\left(\hat{B} - B^*\right)$$
$$= \hat{\varepsilon} + H\left(\hat{B} - B^*\right), \tag{5.69}$$

where $\hat{\varepsilon}$ is the residual for the original unconstrained model. Recall that

$$H^T \hat{\varepsilon} = H^T \left[I - H\left(H^T H\right)^{-1} H^T\right] Y = 0. \tag{5.70}$$

It follows from Equations 5.69 and 5.70 that the residual sum of squares Σ^* under the constrained model is

$$\begin{aligned} n\Sigma^* &= \left(\varepsilon^*\right)^T \varepsilon^* \\ &= \hat{\varepsilon}^T\hat{\varepsilon} + \left(\hat{B}-B^*\right)^T H^T H\left(\hat{B}-B^*\right) \\ &= n\hat{\Sigma} + \left(\hat{B}-B^*\right)^T H^T H\left(\hat{B}-B^*\right) \\ &= n\hat{\Sigma} + n\hat{\Sigma}_H, \end{aligned} \tag{5.71}$$

where $n\hat{\Sigma}_H = \left(\hat{B}-B^*\right)^T H^T H\left(\hat{B}-B^*\right)$ is the sum of squares due to constraints.

Using Equation 5.68, the matrix $\hat{\Sigma}_H$ can be reduced to

$$n\hat{\Sigma}_H = M\left(M^T M\right)^{-1}\left(C\hat{B}M-\theta\right)^T\left[C\left(H^T H\right)^{-1}C^T\right]^{-1}\left(C\hat{B}M-\theta\right)\left(M^T M\right)^{-1}M^T. \tag{5.72}$$

Under the null hypothesis $H_0 : CB-\theta=0$, where $M=I$, as a special case of Equation 5.72, we have

$$n\hat{\Sigma}_H = \left(C\hat{B}-\theta\right)^T\left[C\left(H^T H\right)^{-1}C^T\right]^{-1}\left(C\hat{B}-\theta\right). \tag{5.73}$$

Result 5.4

The matrix version of the residual sum of squares, $n\hat{\Sigma}$, for the unconstrained model is $n\hat{\Sigma} = \hat{\varepsilon}^T\hat{\varepsilon} = Y^T\left[I-H\left(H^T H\right)^{-1}H^T\right]Y$, and the extra sum of squares and cross products due to the hypothesis, $n\hat{\Sigma}_H$, is

$$n\hat{\Sigma}_H = n\Sigma^* - n\hat{\Sigma} = M\left(M^T M\right)^{-1}\left(C\hat{B}M-\theta\right)^T\left[C\left(H^T H\right)^{-1}C^T\right]^{-1}\left(C\hat{B}M-\theta\right)\left(M^T M\right)^{-1}M^T.$$

5.2.3.5 Other Multivariate Test Statistics

In the univariate linear regression analysis, we often use F statistics to test the null hypothesis. Let SSE be the sum of squares for the error and SSH be the sum of squares for the hypothesis. The F statistics is defined as

$$F = \frac{\mathrm{SSH}/q}{\mathrm{SSE}/(n-r)} = \mathrm{SSE}^{-1}\mathrm{SSH}\frac{n-r}{q}.$$

To extend the F statistics to the multivariate case, we let the degrees of freedom be absorbed into the corresponding multivariate sum of squares. Define the $l \times l$ residual sum of squares and cross products matrix as

$$E = M^T Y^T \left[I - H\left(H^T H\right)^{-1} H^T \right] YM, \tag{5.74}$$

and $l \times l$ hypothesis sum of squares and cross products matrix as

$$G = \left(C\hat{B}M - \theta\right)^T \left[C\left(H^T H\right)^{-1} C^T \right]^{-1} \left(C\hat{B}M - \theta\right). \tag{5.75}$$

The extension of F statistics to the multivariate case is realized through a function (determinant, trace, or largest eigenvalue) of the quantity $E^{-1}G$ (Izenman 2008).

Let $s = \min(q, l)$ be the rank of the matrix $E^{-1}G$, $\lambda_1 \geq \lambda_2 \geq \cdots \geq \lambda_s$ be the eigenvalues of the matrix $E^{-1}G$, and $\rho_1 \geq \rho_2 \geq \cdots \geq \rho_s$ be the eigenvalues of the matrix $G(G+E)^{-1}$. We can show that

$$\rho_i = \frac{\lambda_i}{1+\lambda_i}. \tag{5.76}$$

In fact, by definition of eigenvalues of the matrix $G(G+E)^{-1}$, ρ_i satisfies the equation

$$\left|\rho I - G\left(G+E\right)^{-1}\right| = 0. \tag{5.77}$$

Equation 5.77 can be reduced to

$$\left|\rho\left(G+E\right) - G\right| \left|\left(G+E\right)^{-1}\right| = 0,$$

which implies

$$\left|\rho\left(G+E\right) - G\right| = 0$$

or

$$\left|\left(\rho - 1\right)G + \rho E\right| = \left|\frac{\rho}{1-\rho} I - GE^{-1}\right| \left|E^{-1}\right| = 0$$

or

$$\left|\frac{\rho}{1-\rho} I - GE^{-1}\right| = 0. \tag{5.78}$$

If follows from Equation 5.78 that

$$\frac{\rho_i}{1-\rho_i}=\lambda_i$$

or

$$\rho_i=\frac{\lambda_i}{1+\lambda_i}. \tag{5.79}$$

We can also show that

$$\frac{|E|}{|E+G|}=\prod_{i=1}^{s}\frac{1}{1+\lambda_i}. \tag{5.80}$$

It is easy to see that

$$\frac{|E|}{|E+G|}=\frac{1}{|I+E^{-1}G|}=\prod_{i=1}^{s}\frac{1}{\gamma_i}, \tag{5.81}$$

where γ_i are eigenvalues of the matrix $I+E^{-1}G$. By definition of eigenvalues of the matrix $I+E^{-1}G$, we have

$$\left|\gamma I-\left(I+E^{-1}G\right)\right|=\left|(\gamma-1)I-E^{-1}G\right|=0,$$

which implies

$$\gamma_i-1=\lambda_i$$

or

$$\gamma_i=1+\lambda_i. \tag{5.82}$$

Substituting Equation 5.82 into Equation 5.81 gives

$$\frac{|E|}{|E+G|}=\prod_{i=1}^{s}\frac{1}{1+\lambda_i}. \tag{5.83}$$

Now we give four related test statistics.

Result 5.5

Four other tests are defined as

$$\text{Wilks' lambda} = \frac{|E|}{|E+G|} = \prod_{i=1}^{s} \frac{1}{1+\lambda_i},$$

$$\text{Pillai's trace} = \text{Trace}\left[G(G+E)^{-1}\right] = \sum_{i=1}^{s} \frac{\lambda_i}{1+\lambda_i},$$

$$\text{Hotelling} - \text{Lawley trace} = \text{Trace}\left[GE^{-1}\right] = \sum_{i=1}^{s} \lambda_i,$$

and

$$\text{Roy's largest root} = \frac{\lambda_1}{1+\lambda_1}.$$

Next we study the distributions of the four statistics. In general, exact formulas for easy distribution calculations are not available for any of the four test statistics (Muller and Peterson 1984). Efficient approximate distributions for the four test statistics are usually used for testing the association.

5.2.3.5.1 Wilks' Lambda We first consider Wilks' lambda Λ. The statistic is defined as

$$\Lambda = \frac{|E|}{|E+G|}, \tag{5.84}$$

which can be reduced to

$$\begin{aligned} \Lambda &= \frac{|\hat{\Sigma}|}{|\hat{\Sigma} + \hat{\Sigma}_H|} \\ &= \frac{|\hat{\Sigma}^* - \hat{\Sigma}_H|}{|\hat{\Sigma}^*|}. \end{aligned} \tag{5.85}$$

If we assume that the matrix M is nonsingular, let $\hat{\Sigma}_{\varepsilon^* H} = \varepsilon^* H^T, \hat{\Sigma}_{H\varepsilon^*} = H(\varepsilon^*)^T$ and $\hat{\Sigma}_{HH} = H^T H$. Then, $\hat{\Sigma}_H = \hat{\Sigma}_{\varepsilon^* H} \hat{\Sigma}_{HH}^{-1} \hat{\Sigma}_{H\varepsilon^*}$. Equation 5.85 can be further reduced to

$$\Lambda = \left| I - (\hat{\Sigma}^*)^{-1/2} \hat{\Sigma}_{\varepsilon^* H} \hat{\Sigma}_{HH}^{-1} \hat{\Sigma}_{H\varepsilon^*} (\hat{\Sigma}^*)^{-1/2} \right|.$$

Consider the canonical correlation between two sets of variables: ε^* and H. Let $\rho_1^2 \geq \rho_2^2 \geq \cdots \geq \rho_s^2$ be the eigenvalues of the matrix $\left(\hat{\Sigma}^*\right)^{-1/2} \hat{\Sigma}_{\varepsilon^* H} \hat{\Sigma}_{HH}^{-1} \hat{\Sigma}_{H\varepsilon^*} \left(\hat{\Sigma}^*\right)^{-1/2}$. Then, eigenvalues of the matrix $I - \left(\hat{\Sigma}^*\right)^{-1/2} \hat{\Sigma}_{\varepsilon^* H} \hat{\Sigma}_{HH}^{-1} \hat{\Sigma}_{H\varepsilon^*} \left(\hat{\Sigma}^*\right)^{-1/2}$ can be found by solving the equation

$$\left| \mu I - \left(I - \left(\hat{\Sigma}^*\right)^{-1/2} \hat{\Sigma}_{\varepsilon^* H} \hat{\Sigma}_{HH}^{-1} \hat{\Sigma}_{H\varepsilon^*} \left(\hat{\Sigma}^*\right)^{-1/2} \right) \right| = 0,$$

which implies

$$\left| (1-\mu) I - \left(\hat{\Sigma}^*\right)^{-1/2} \hat{\Sigma}_{\varepsilon^* H} \hat{\Sigma}_{HH}^{-1} \hat{\Sigma}_{H\varepsilon^*} \left(\hat{\Sigma}^*\right)^{-1/2} \right| = 0. \tag{5.86}$$

Solving Equation 5.86 for μ yields

$$\mu_i = 1 - \rho_i^2, \quad i = 1, \ldots, s.$$

Therefore, we have

$$\Lambda = \prod_{i=1}^{s} \left(1 - \rho_i^2\right). \tag{5.87}$$

If we define $E = \Sigma = Y^T[I - H(H^T H)^{-1} H^T] Y$ and $G = \Sigma_{YY} - \Sigma$, then, ρ_i^2 are the eigenvalues of the matrix $\Sigma_{YY}^{-1/2} \Sigma_{YH} \Sigma_{HH}^{-1} \Sigma_{HY} \Sigma_{YY}^{-1/2}$, which are canonical correlations between the variables.

It is well known that $\Lambda^{1/s}$ is a geometric mean of the $\left(1-\rho_i^2\right)$'s. The approximation uses $\Lambda^{1/s}$ as a key quantity for distribution calculations.

Define the F approximation as

$$F_\Lambda = \frac{1 - \Lambda^{1/t}}{\Lambda^{1/t}} \frac{mt - 2u}{lq}, \tag{5.88}$$

where

$$u = \frac{lq-2}{4}, m = n - r - \frac{l-q+1}{2}$$

$$t = \begin{cases} \dfrac{l^2 q^2 - 4}{l^2 + q^2 - 5} & l^2 + q^2 - 5 > 0 \\ 1 & l^2 + q^2 - 5 \leq 0 \end{cases}$$

n is the sample size
r is the number of columns of the matrix H
q is the number of rows of the matrix C
l is the number of columns of M

For the univariate model, $l=1$, $t=1$, F_Λ is reduced to the classical F statistic:

$$F = \frac{\rho^2/\text{d.f. model}}{\left(1-\rho^2\right)/\text{d.f. error}}.$$

Comparing Equation 5.87 with Wilks' lambda, we obtain

$$\lambda_i = \frac{\rho_i^2}{1-\rho_i^2}. \tag{5.89}$$

Using Equation 5.89, we can express the four statistics in terms of canonical correlations.

Result 5.6

Four other tests are defined as

$$\text{Wilks' lambda} = \frac{|E|}{|E+G|} = \prod_{i=1}^{s}\left(1-\rho_i^2\right), \tag{5.90}$$

$$\text{Pillai's trace} = \text{PB} = \text{Trace}\left[G\left(G+E\right)^{-1}\right] = \sum_{i=1}^{s}\rho_i^2, \tag{5.91}$$

$$\text{Hotelling-Lawley trace} = \text{HLT} = \text{Trace}\left[GE^{-1}\right] = \sum_{i=1}^{s}\frac{\rho_i^1}{1-\rho_i^2}, \tag{5.92}$$

and

$$\text{Roy's largest root} = \text{RLR} = \rho_1^2. \tag{5.93}$$

We can also have an F approximation to the distribution of the other three statistics.

Result 5.7

F Approximation to the Distribution of Three Test Statistics

$$F_{HLT} = \frac{\dfrac{\text{HLT}/s}{ql}}{\dfrac{1}{s\left(n-l-1\right)+2}}$$

with $F_{ql,s(n-l-1)+2}$ distribution, where s is the rank of G;

$$F_{PB} = \frac{\dfrac{\text{PB}}{ql}}{\dfrac{(s-\text{PB})}{s(n+s-l)}}$$

with $F_{ql,s(n+s-l)}$ distribution; and

$$F_{RLR} = \frac{\dfrac{\text{RLR}}{q}}{\dfrac{1-\text{RLR}}{n-\text{rank}(H)}}$$

with $F_{q,n-\text{rank}(H)}$ distribution.

5.3 LINEAR MODELS FOR MULTIPLE PHENOTYPES AND MULTIPLE MARKERS

5.3.1 Multivariate Multiple Linear Regression Models

A multiple marginal regression quantitative trait model for a single marker can be easily extended to multiple markers. Consider p marker loci, which are located at the genomic positions $t_1, \dots, t_p$. The genotype matrices X and Z and genetic effects α and δ in Equations 5.11 and 5.12 are changed to

$$X = \begin{bmatrix} x_{11} & \cdots & x_{1p} \\ \vdots & \vdots & \vdots \\ x_{n1} & \cdots & x_{np} \end{bmatrix}, \quad Z = \begin{bmatrix} z_{11} & \cdots & z_{1p} \\ \vdots & \vdots & \vdots \\ z_{n1} & \cdots & z_{np} \end{bmatrix}, \quad \alpha = \begin{bmatrix} \alpha_{11} & \cdots & \alpha_{1K} \\ \vdots & \vdots & \vdots \\ \alpha_{p1} & \cdots & \alpha_{pK} \end{bmatrix},$$

$$\text{and} \quad \delta = \begin{bmatrix} \delta_{11} & \cdots & \delta_{1K} \\ \vdots & \vdots & \vdots \\ \delta_{p1} & \cdots & \delta_{pK} \end{bmatrix}.$$

After the genotype matrices X and Z and genetic effects α and δ are expanded, Equation 5.12 can be used for a multivariate multiple linear regression genetic model of the multiple phenotypes:

$$Y = W\eta + X\alpha + Z\delta + \varepsilon$$

or

$$Y = HB + \varepsilon. \tag{5.94}$$

The formula for the estimation of the genetic effects and test statistics studied in Section 5.2 can be extended to multiple markers without changes. What we need to change are the theoretic asymptotic results. Let P_{Ml} and P_{ml} be the frequencies of the alleles M_l and m_l of the marker located at the genomic position t_l, respectively, and $D(t_l, t_j)$ be the disequilibrium coefficient of the LD between the marker at the genomic position t_l and the marker at the genomic position t_j. Assume that there are L trait loci defined as before. Let $D(t_j, s_l)$ be the disequilibrium coefficient of the linkage disequilibrium (LD) between the marker at the genomic position t_j and the trait locus at the genomic position s_l. By a similar argument as that in Appendix 4A, we have the following result.

Result 5.8

$$\hat{\alpha}^{(m)} \xrightarrow{a.s.} P_A^{-1}\alpha_0 \quad \text{and} \quad \hat{\delta}^{(m)} \xrightarrow{a.s.} P_D^{-1}\delta_0, \tag{5.95}$$

where

$$\alpha^{(m)} = \begin{bmatrix} \alpha_{11}^{(m)} & \cdots & \alpha_{1K}^{(m)} \\ \vdots & \vdots & \vdots \\ \alpha_{p1}^{(m)} & \cdots & \alpha_{pK}^{(m)} \end{bmatrix} \quad \text{and} \quad \delta^{(m)} = \begin{bmatrix} \delta_{11}^{(m)} & \cdots & \delta_{1K}^{(m)} \\ \vdots & \vdots & \vdots \\ \delta_{p1}^{(m)} & \cdots & \delta_{pK}^{(m)} \end{bmatrix}$$

are the potential genetic additive and dominance effect matrices at the marker loci, respectively;

$$\alpha = \begin{bmatrix} \alpha_{1s_1} & \cdots & \alpha_{Ks_1} \\ \vdots & \vdots & \vdots \\ \alpha_{1s_p} & \cdots & \alpha_{Ks_p} \end{bmatrix} \quad \text{and} \quad \delta = \begin{bmatrix} \delta_{1s_1} & \cdots & \delta_{Ks_1} \\ \vdots & \vdots & \vdots \\ \delta_{1s_p} & \cdots & \delta_{Ks_p} \end{bmatrix}$$

are the true genetic additive and dominance effect matrices, respectively; and

$$P_A = \begin{bmatrix} 2P_{M_1}P_{m_1} & D(t_1, t_2) & \cdots & D(t_1, t_L) \\ D(t_2, t_1) & 2P_{M_2}P_{m_2} & \cdots & D(t_1, t_L) \\ \cdots & \cdots & \cdots & \cdots \\ D(t_L, t_1) & D(t_L, t_2) & \cdots & 2P_{M_L}P_{m_L} \end{bmatrix}, \quad P_D = \begin{bmatrix} P_{M_1}^2 P_{m_1}^2 & D^2(t_1, t_2) & \cdots & D^2(t_1, t_L) \\ D^2(t_2, t_1) & 2P_{M_2}^2 P_{m_2}^2 & \cdots & D^2(t_1, t_L) \\ \cdots & \cdots & \cdots & \cdots \\ D^2(t_L, t_1) & D^2(t_L, t_2) & \cdots & 2P_{M_L}^2 P_{m_L}^2 \end{bmatrix}$$

$$D_1 = \begin{bmatrix} D(t_1, s_1) & \cdots & D(t_1, s_p) \\ \vdots & \vdots & \vdots \\ D(t_L, s_1) & \cdots & D(t_L, s_p) \end{bmatrix} \quad \text{and} \quad D_2 = \begin{bmatrix} D^2(t_1, s_1) & \cdots & D^2(t_1, s_p) \\ \vdots & \vdots & \vdots \\ D^2(t_L, s_1) & \cdots & D^2(t_L, s_p) \end{bmatrix},$$

and

$$\alpha_0 = 2D_1\alpha \quad \text{and} \quad \delta_0 = D_2\delta.$$

Other asymptotic results can be similarly changed.

5.3.2 Multivariate Functional Linear Models for Gene-Based Genetic Analysis of Multiple Phenotypes

When data are generated by next-generation sequencing, gene-based association analysis of rare variants should be developed. In this section, we extend the functional linear model for association analysis of a single trait, studied in Chapter 4, to multiple traits. Again, we consider K correlated traits. Let y_{ik} be the kth phenotype value of the ith individual. A functional linear model for multiple quantitative traits with a genetic additive effect is defined as

$$y_{ik} = \mu_k + h_i^T \eta_k + \int_T X_i(t)\alpha_k(t)dt + \int_T Z_i(t)\delta_k(t)dt + \varepsilon_{ik}, \tag{5.96}$$

where $h_i = [h_{i1}, \ldots, h_{im}]^T$ and $\eta_k = [\eta_{1k}, \ldots, \eta_{mk}]^T$, $X_i(t)$ and $Z_i(t)$ are as defined in Equations 4.42 and 4.43, $\alpha_k(t)$ and $\delta_k(t)$ are the kth putative genetic additive effect and dominance functions of the marker at the genomic position t affecting the kth trait, respectively, $\varepsilon_i = [\varepsilon_{i1}, \ldots, \varepsilon_{ik}]^T$ is a vector of errors distributed as a normal distribution with a mean vector of zeros and a $K \times K$ variance–covariance matrix Σ, and $\varepsilon_1, \ldots, \varepsilon_n$ is assumed to be independent.

We expand the genotype functions $X_i(t)$ and $Z_i(t)$ in terms of eigenfunctions (functional principal components):

$$X_i(t) = \sum_{j=1}^{J} \xi_{ij}\varphi_j(t),$$

$$Z_i(t) = \sum_{j=1}^{J} \gamma_{ij}\varphi_j(t), \tag{5.97}$$

where $\xi_{ij} = \int_T X_i(t)\varphi_j(t)dt$ and $\gamma_{ij} = \int_T Z_i(t)\varphi_j(t)dt$.

Substituting Equation 5.97 into Equation 5.96 gives

$$\begin{aligned} y_{ik} &= \mu_k + h_i^T \eta_k + \int_T \sum_{j=1}^{J} \xi_{ij}\varphi_j(t)\alpha_k(t)dt + \int_T \sum_{j=1}^{J} \gamma_{ij}\varphi_j(t)\delta_k(t)dt + \varepsilon_{ik} \\ &= \mu_k + h_i^T \eta_k + \sum_{j=1}^{J} \xi_{ij} \int_T \varphi_j(t)\alpha_k(t)dt + \sum_{j=1}^{J} \gamma_{ij} \int_T \varphi_j(t)\delta_k(t)dt + \varepsilon_{ik} \\ &= \mu_k + h_i^T \eta_k + \sum_{j=1}^{J} \xi_{ij}\alpha_{jk} + \sum_{j=1}^{J} \gamma_{ij}\delta_{jk} + \varepsilon_{ik}. \end{aligned} \tag{5.98}$$

Let

$$Y=\begin{bmatrix} y_{11} & \cdots & y_{1K} \\ \vdots & \ddots & \vdots \\ y_{n1} & \cdots & y_{nK} \end{bmatrix},\quad \xi=\begin{bmatrix} \xi_{11} & \cdots & \xi_{1J} \\ \vdots & \vdots & \vdots \\ \xi_{n1} & \cdots & \xi_{nJ} \end{bmatrix},\quad \alpha=\begin{bmatrix} \alpha_{11} & \cdots & \alpha_{1K} \\ \vdots & \vdots & \vdots \\ \alpha_{J1} & \cdots & \alpha_{JK} \end{bmatrix},$$

$$\gamma=\begin{bmatrix} \gamma_{11} & \cdots & \gamma_{1J} \\ \vdots & \vdots & \vdots \\ \gamma_{n1} & \cdots & \gamma_{nJ} \end{bmatrix},\quad \delta=\begin{bmatrix} \delta_{11} & \cdots & \delta_{1K} \\ \vdots & \vdots & \vdots \\ \delta_{J1} & \cdots & \delta_{JK} \end{bmatrix},\quad W=\begin{bmatrix} 1 & h_{11} & \cdots & h_{1l} \\ \vdots & \vdots & \ddots & \vdots \\ 1 & h_{n1} & \cdots & h_{nl} \end{bmatrix},$$

$$\eta=\begin{bmatrix} \mu_1 & \cdots & \mu_K \\ \eta_{11} & \cdots & \eta_{1K} \\ \vdots & \ddots & \vdots \\ \eta_{l1} & \cdots & \eta_{lK} \end{bmatrix},\quad \text{and}\quad \varepsilon=\begin{bmatrix} \varepsilon_{11} & \cdots & \varepsilon_{1K} \\ \vdots & \ddots & \vdots \\ \varepsilon_{n1} & \cdots & \varepsilon_{nK} \end{bmatrix}.$$

Then, Equation 5.98 can be written in a matrix form:

$$Y=W\eta+\xi\alpha+\gamma\delta+\varepsilon, \tag{5.99}$$

or

$$Y=HB+\varepsilon, \tag{5.100}$$

where $H=\begin{bmatrix} W & \xi & \gamma \end{bmatrix}$ and $B=\begin{bmatrix} \eta \\ \alpha \\ \delta \end{bmatrix}$.

Then, the multivariate functional linear model is reduced to the traditional multivariate multiple linear model. All estimation methods and statistics studied in Section 5.2 can be used for solving the problem (5.99) or (5.100).

5.3.2.1 Parameter Estimation

The least square estimator for the regression coefficients is

$$\hat{B}=\left(H^T H\right)^{-1} H^T Y,$$

and the covariance matrix of the estimators is

$$\operatorname{cov}\left(\operatorname{vec}\left(\hat{B}\right)\right)=\Sigma\otimes\left(H^T H\right)^{-1},$$

where Σ is estimated by

$$\hat{\Sigma}=\frac{1}{n-r}\hat{\varepsilon}^T\hat{\varepsilon},\quad \hat{\varepsilon}=\left[I-H\left(H^T H\right)^{-1} H^T\right]Y.$$

5.3.2.2 *Null Hypothesis and Test Statistics*

Assume

$$B = \begin{bmatrix} \eta \\ \alpha \\ \delta \end{bmatrix} = \begin{bmatrix} B_{(1)} \\ B_{(2)} \end{bmatrix},$$

where

$$B_{(2)} = \begin{bmatrix} \alpha \\ \delta \end{bmatrix}.$$

The null hypothesis for no association is

$$H_0 : \alpha = 0 \quad \text{and} \quad \delta = 0$$

or

$$B_{(2)} = 0.$$

The regression model can be written as

$$Y = H^{(1)} B_1 + H^{(2)} B_2 + \varepsilon,$$

where

$$H = \begin{bmatrix} H^{(1)}_{n\times(l+1)} & H^{(2)}_{n\times(2J)} \end{bmatrix}.$$

Let

$$\hat{\Sigma} = \frac{1}{n} Y^T \left[I - H\left(H^T H\right)^{-1} H^T \right] Y$$

and

$$\hat{\Sigma}_1 = \frac{1}{n} Y^T \left[I - H_1\left(H_1^T H_1\right)^{-1} H_1^T \right] Y.$$

Then, the test statistic is defined as

$$T_F = -n \log \frac{|\hat{\Sigma}|}{|\hat{\Sigma}_1|}, \tag{5.101}$$

or

$$T_F = -\left(n - 2J - l - 1 - \frac{K - 2J + 1}{2}\right)\log\frac{|\hat{\Sigma}|}{|\hat{\Sigma}_1|}. \tag{5.102}$$

Under the null hypothesis of no association, the statistic T_F is asymptotically distributed as a central $\chi^2_{(2KJ)}$ distribution.

5.3.2.3 *Other Multivariate Test Statistics*

The results in Section 5.2.3.5 can also be extended to the multivariate functional linear model.

Define

$$E = n\hat{\Sigma} \quad \text{and} \quad G = n\left(\hat{\Sigma}_1 - \hat{\Sigma}\right).$$

Other ways to define the four other tests are

$$\text{Wilks' lambda} = \frac{|E|}{|E+G|} = \prod_{i=1}^{s}\frac{1}{1+\lambda_i},$$

$$\text{Pillai's trace} = \text{Trace}\left[G\left(G+E\right)^{-1}\right] = \sum_{i=1}^{s}\frac{\lambda_i}{1+\lambda_i},$$

$$\text{Hotelling-Lawley trace} = \text{Trace}\left[GE^{-1}\right] = \sum_{i=1}^{s}\lambda_i,$$

and

$$\text{Roy's largest root} = \frac{\lambda_1}{1+\lambda_1},$$

where $\lambda_1 \geq \lambda_2 \geq \cdots \geq \lambda_s$ are the eigenvalues of the matrix $E^{-1}G$.

Let $\rho_1 \geq \rho_2 \geq \cdots \geq \rho_s$ be the eigenvalues of the matrix $G(G+E)^{-1}$. The F approximation to the null distributions of the four statistics are given in Equation 5.103, and Results 5.6 and 5.7 can be adopted here.

5.3.2.4 Wilks' Lambda

Let $\Lambda = \prod_{i=1}^{s}\left(1-\rho_i^2\right)$.

Define F approximation as

$$F_\Lambda = \frac{1-\Lambda^{1/t}}{\Lambda^{1/t}}\frac{mt-2u}{lq}, \tag{5.103}$$

where

$$u = \frac{lq-2}{4},\ m = n-r-\frac{l-q+1}{2}$$

$$t = \begin{cases} \frac{l^2q^2-4}{l^2+q^2-5} & l^2+q^2-5>0 \\ 1 & l^2+q^2-5\le 0 \end{cases}$$

n is the sample size

$r = l+2J+1$

$q = l = 2J$

5.3.2.5 F Approximation to the Distribution of Three Test Statistics

$$F_{HLT} = \frac{\frac{\text{HLT}/s}{ql}}{\frac{1}{s(n-l-1)+2}}$$

with $F_{ql,s(n-l-1)+2}$ distribution, where s is the rank of G;

$$F_{PB} = \frac{\frac{\text{PB}}{ql}}{\frac{(s-\text{PB})}{s(n+s-l)}}$$

with $F_{ql,s(n+s-l)}$ distribution; and

$$F_{RLR} = \frac{\frac{\text{RLR}}{q}}{\frac{1-\text{RLR}}{n-\text{rank}(H)}}$$

with $F_{q,n-\text{rank}(H)}$ distribution, where

$$\text{HLT} = \text{Trace}[GE^{-1}]$$

$$\text{PB} = \text{Trace}[G(G+E)^{-1}]$$

$$\text{RLR} = \rho_1^2$$

5.4 CANONICAL CORRELATION ANALYSIS FOR GENE-BASED GENETIC PLEIOTROPIC ANALYSIS

5.4.1 Multivariate Canonical Correlation Analysis (CCA)

Multivariate CCA for quantitative trait analysis of a single trait studied in Section 4.2.2 can be directly extended to multiple traits. We only need to extend a single trait, y, to multiple traits, Y. In other words, we consider multiple trait $Y=[y_1,\ldots,y_K]$ and L SNPs with indicator variables for the genetic additive and dominance effects $x_1,\ldots,x_L$, $z_1,\ldots,z_L$ that are similarly defined in Equation 4.23. Let $w=[x_1,\ldots,x_L,z_1,\ldots,z_L]$ and define the variance and covariance matrices:

$$\Sigma_{yy}=\begin{bmatrix}\sigma_{y_1y_2} & \cdots & \sigma_{y_1y_K}\\ \vdots & \vdots & \vdots\\ \sigma_{y_Ky_1} & \cdots & \sigma_{y_Ky_K}\end{bmatrix},\quad \Sigma_{yg}=\begin{bmatrix}\Sigma_{yx} & \Sigma_{yz}\end{bmatrix},\quad \text{and}\quad \Sigma_{gg}=\begin{bmatrix}\Sigma_{xx} & \Sigma_{xz}\\ \Sigma_{zx} & \Sigma_{zz}\end{bmatrix},$$

where

$$\Sigma_{yx}=\begin{bmatrix}\sigma_{y_1x_1} & \cdots & \sigma_{y_1x_L}\\ \vdots & \vdots & \vdots\\ \sigma_{y_Kx_1} & \cdots & \sigma_{y_Kx_L}\end{bmatrix}=\Sigma_{yx}^T,\ \Sigma_{yz}=\begin{bmatrix}\sigma_{y_1z_1} & \cdots & \sigma_{y_1z_L}\\ \vdots & \vdots & \vdots\\ \sigma_{y_Kz_1} & \cdots & \sigma_{y_Kz_L}\end{bmatrix}=\Sigma_{zy}^T,$$

$$\Sigma_{xx}=\begin{bmatrix}\sigma_{x_1x_1} & \cdots & \sigma_{x_1x_L}\\ \vdots & \ddots & \vdots\\ \sigma_{x_Lx_1} & \cdots & \sigma_{x_Lx_L}\end{bmatrix},\ \Sigma_{xz}=\begin{bmatrix}\sigma_{x_1z_1} & \cdots & \sigma_{x_1z_L}\\ \vdots & \ddots & \vdots\\ \sigma_{x_Lz_1} & \cdots & \sigma_{x_Lz_L}\end{bmatrix},\ \text{and}\ \Sigma_{zz}=\begin{bmatrix}\sigma_{z_1z_1} & \cdots & \sigma_{z_1z_L}\\ \vdots & \ddots & \vdots\\ \sigma_{z_Lz_1} & \cdots & \sigma_{z_Lz_L}\end{bmatrix}.$$

Define

$$R=\Sigma_{yy}^{-1/2}\Sigma_{yg}\Sigma_{gg}^{-1}\Sigma_{gy}\Sigma_{yy}^{-1/2}. \tag{5.104}$$

Assume that the eigendecomposition of the matrix R is

$$R=U\Lambda U^T, \tag{5.105}$$

where $\Lambda=\text{diag}(\lambda_1,\ldots,\lambda_K)$.

From Equation 1.248, we can define the test statistic for testing the association of the gene or genomic region with the multiple traits as

$$T_{CCA}=-n\sum_{i=1}^{P}\log\left(1-\lambda_i^2\right), \tag{5.106}$$

or

$$T_{CCA} = -\left[n-1-\frac{(2L+K+1)}{2}\right]\sum_{i=1}^{p}\log\left(1-\lambda_i^2\right), \tag{5.107}$$

where $p=\min(2L,K)$.

Under the null hypothesis of no association of the gene with the trait, T_{CCA} is a central $\chi^2_{(2KL)}$ distribution.

5.4.2 Kernel CCA

Kernel CCA is a nonlinear extension of canonical correlation analysis with positive definite kernels (Fukumizu et al. 2007; Larson et al. 2014). Let $f(X)\in H_x$ and $g(Y)\in H_y$, where H_x and H_y are RKHS with kernel functions $k_x(X,X)$ and $k_y(Y,Y)$ for random variables X and Y, respectively. Similar to the linear CCA, the goal of the kernel CCA is to seek nonlinear functions $f(X)$ and $g(Y)$ such that they have a maximum correlation. In other words, we want to solve the following optimization problem:

$$\max_{f,g} \frac{\operatorname{cov}\left(f(X),g(Y)\right)}{\sqrt{\operatorname{var}\left(f(X)\right)\operatorname{var}\left(g(Y)\right)}}. \tag{5.108}$$

Since $f(X)\in H_x$ and $g(Y)\in H_y$ and H_x and H_y are RKHS, Equation 4.84 gives a general representation of functions $f(X)$ and $g(Y)$:

$$f(X)=<f(.),k(.,X)>_{H_x} \tag{5.109}$$

and

$$g(Y)=<g(.),k(.,Y)>_{H_y}. \tag{5.110}$$

By kernel theory we can show that the optimization problem can be transformed to (Appendix 5A)

$$\begin{aligned} &\max_{\alpha,\beta} \ \alpha^T K_x H K_y \beta \\ &\text{s.t.} \ \alpha^T K_x H K_x \alpha = 1 \\ &\qquad \beta^T K_y H K_y \beta = 1, \end{aligned} \tag{5.111}$$

where

$$K_x = \begin{bmatrix} k_x(x_1, x_1) & \cdots & k_x(x_1, x_n) \\ \vdots & \vdots & \vdots \\ k_x(x_n, x_1) & \vdots & k_x(x_n, x_n) \end{bmatrix}, \; K_y = \begin{bmatrix} k_y(y_1, y_1) & \cdots & k_y(y_1, y_n) \\ \vdots & \vdots & \vdots \\ k_y(y_n, y_1) & \cdots & k_y(y_n, y_n) \end{bmatrix}$$

$H = I_n - \frac{1}{n}\mathbf{1}\mathbf{1}^T$, $\mathbf{1} = [1, 1, \ldots, 1]^T$

I_n is an $n \times n$-dimensional identity matrix

$\alpha = [\alpha_1, \alpha_2, \ldots, \alpha_n]^T$

$\beta = [\beta_1, \beta_2, \ldots, \beta_n]^T$

Using the Lagrangian multiplier method to solve the optimization problem, we obtain the following eigenequation:

$$\begin{bmatrix} 0 & K_x H K_y \\ K_y H K_x & 0 \end{bmatrix} \begin{bmatrix} \alpha \\ \beta \end{bmatrix} = \lambda \begin{bmatrix} K_x H K_x & 0 \\ 0 & K_y H K_y \end{bmatrix} \begin{bmatrix} \alpha \\ \beta \end{bmatrix}. \tag{5.112}$$

However, the eigenequation problem (5.112) is ill-defined. The optimization problem should be regularized. The Lagrangian can be written as

$$L(\alpha, \beta, \lambda, \mu) = \alpha^T K_x H K_y \beta + \frac{\lambda}{2}\left(N - \alpha^T K_x H K_x \alpha - \eta \alpha^T \alpha\right) + \frac{\mu}{2}\left(N - \beta^T K_y H K_y \beta - \eta \beta^T \beta\right). \tag{5.113}$$

Solving the optimization problem, we set the derivative of the Lagrangian with respect to α and β equal to zero:

$$\begin{aligned} \frac{\partial L}{\partial \alpha} &= K_x H K_y \beta - \lambda\left(K_x H K_x \alpha + \eta \alpha\right) = 0 \\ \frac{\partial L}{\partial \beta} &= K_y H K_x \alpha - \mu\left(K_y H K_y \beta + \eta \beta\right) = 0 \end{aligned} \tag{5.114}$$

or

$$\begin{bmatrix} 0 & K_x H K_y \\ K_y H K_x & 0 \end{bmatrix} \begin{bmatrix} \alpha \\ \beta \end{bmatrix} = \lambda \begin{bmatrix} K_x H K_x + \eta I & 0 \\ 0 & K_y H K_y + \eta I \end{bmatrix} \begin{bmatrix} \alpha \\ \beta \end{bmatrix}. \tag{5.115a}$$

Equation 5.115a can be further reduced to the eigenequation

$$Ru = \lambda^2 u, \tag{5.115b}$$

where

$R = (K_x H K_x + \eta I)^{-1/2} K_x H K_y (K_y H K_y + \eta I)^{-1} K_y H K_x (K_x H K_x + \eta I)^{-1/2}$

$u = (K_x H K_x + \eta I)^{1/2} \alpha$

$\beta = (K_y H K_y + \eta I)^{-1} (K_y H K_x)(K_x H K_x + \eta I)^{-1/2} u$

5.4.3 Functional CCA

In the traditional functional canonical correlation analysis (CCA), we study correlations between pairs of observed random curves. However, in the genetic analysis of multiple traits, the genotype files in a gene or a genomic region with NGS data can be taken as an observed curve, but multiple phenotypes cannot be taken as a curve. Multiple phenotypes are multiple variables. Therefore, in this section, we investigate correlations between the observed genotype profile curve and multiple variables.

Let t be a genomic position. A genotype function, $X(t)$, for an additive effect and a genotype function, $Z(t)$, for a dominance effect of the ith individual are defined as in Equations 4.42 and 4.43. Consider K multiple traits: $Y=[Y_1, Y_2, \ldots, Y_K]$.

A general tool for functional CCA is functional principal component analysis (FPCA). The observed genetic variant profiles are often not smooth, which leads to substantial variability in the estimated functional principal component curves. To improve the smoothness of the estimated functional principal component curves, we impose the roughness penalty on the functional principal component weight functions. We balance the goodness of fit and the roughness of the estimated functional principal component curves.

We often penalize the roughness of the functional principal component curve by its integrated squared second derivative. The balance between the goodness of fit and the roughness of the function is controlled by a smoothing parameter, μ (Luo et al. 2013). We define an extended inner product as

$$(f,g)_\mu \int f(t)g(t)dt+\mu\int D^2 f(t)D^2 g(t)dt, \tag{5.116}$$

where $D^2 f(t)=\dfrac{d^2 f(t)}{dt^2}$.

The penalized sample variance is defined as

$$F=\frac{\mathrm{Var}\left(\int_0^1 x(t)\beta(t)dt\right)}{\beta(t)_\mu^2}, \tag{5.117}$$

where $\beta(t)_\mu^2=\int_0^1 \beta^2(t)dt+\mu\int_0^1\left[D^2\beta(t)\right]^2 dt$.

The smoothed functional principal components can be obtained by solving the following integral equation:

$$\int_0^1 R(s,t)\beta(s)ds=\lambda\left[\beta(t)+\mu D^4\beta(t)\right]. \tag{5.118}$$

Note that when $\mu=0$, the smoothed functional principal component analysis is reduced to an unsmoothed functional principal component analysis.

The eigenfunction is an integral function and difficult to solve in closed form. A general strategy for solving the eigenfunction problem in Equation 5.118 is to convert the continuous eigen-analysis problem to an appropriate discrete eigen-analysis task. Details of this process can be found in Luo et al. (2013).

We expand the genotype profiles $X(t)$ and $Z(t)$ in terms of orthonormal eigenfunctions (functional principal components) $\{\varphi_j(t), j=1,2,\ldots,J\}$ and $\{\psi_l(t), l=1,2,\ldots,L\}$:

$$X(t)=\sum_{j=1}^{J}\xi_j\varphi_j(t), \tag{5.119}$$

$$Z(t)=\sum_{l=1}^{L}\eta_j(t)\psi_l(t), \tag{5.120}$$

where

$\xi_j = <X(t), \varphi_j(t)>_\mu$
$\eta_j = <Z(t), \psi_j t)>_\mu$

The concept of canonical variables as a linear combination of the variables $a_1^T X + a_2^T Z$ (inner product between two vectors) can be extended as an inner product between two functions:

$$G=\int_T \alpha_1(t)X(t)dt+\int_T \alpha_2(t)Z(t)dt. \tag{5.121}$$

Substituting Equations 5.119 and 5.120 into Equation 5.121 gives

$$\begin{aligned} G &= \sum_{j=1}^{J}\xi_j\int_T \alpha_1(t)\phi_j(t)dt+\sum_{l=1}^{L}\eta_l\int_T \alpha_2(t)\psi_j \\ &= \sum_{j=1}^{J}\xi_j a_{1j}+\sum_{l=1}^{L}\eta_j a_{2j} \\ &= \begin{bmatrix} a_{11} & \cdots & a_{1J}\end{bmatrix}\begin{bmatrix}\xi_1\\ \vdots \\ \xi_J\end{bmatrix}+\begin{bmatrix} a_{21} & \cdots & a_{2L}\end{bmatrix}\begin{bmatrix}\eta_1\\ \vdots \\ \eta_L\end{bmatrix} \\ &= a_1^T\xi+a_2^T\eta \\ &= a^T\theta, \end{aligned} \tag{5.122}$$

where $a=\begin{bmatrix}a_1\\ a_2\end{bmatrix}$ and $\theta=\begin{bmatrix}\xi\\ \eta\end{bmatrix}$.

It is well known that the expansion coefficients are random variables (Tran 2008). Therefore, the covariance between the canonical variables of X, Z, and Y is

$$\begin{aligned} &\operatorname{cov}\left(\int_T \alpha_1(t)X(t)dt + \int_T \alpha_2(t)Z(t)dt, Yb\right) \\ &= \operatorname{cov}\left(a^T\theta, Yb\right) \\ &= a^T \operatorname{cov}(\theta, Y)b, \end{aligned} \tag{5.123}$$

where

$$\Sigma_{GY} = \operatorname{cov}(\theta, Y) = \begin{bmatrix} \operatorname{cov}(\xi, Y) \\ \operatorname{cov}(\eta, Y) \end{bmatrix},$$

$$\operatorname{cov}(\xi, Y) = \begin{bmatrix} \operatorname{cov}(\xi_1, Y_1) & \cdots & \operatorname{cov}(\xi_1, Y_J) \\ \vdots & \vdots & \vdots \\ \operatorname{cov}(\xi_J, Y_1) & \cdots & \operatorname{cov}(\xi_J, Y_J) \end{bmatrix} \text{and} \operatorname{cov}(\eta, Y) = \begin{bmatrix} \operatorname{cov}(\eta_1, Y_1) & \cdots & \operatorname{cov}(\eta_1, Y_K) \\ \vdots & \vdots & \vdots \\ \operatorname{cov}(\eta_L, Y_1) & \cdots & \operatorname{cov}(\eta_L, Y_K) \end{bmatrix}.$$

Similarly, we have $\operatorname{var}(G) = a^T \operatorname{cov}(\theta, \theta)a$, where

$$\begin{aligned} \Sigma_{GG} &= \operatorname{cov}(\theta, \theta) \\ &= \begin{bmatrix} \operatorname{cov}(\xi, \xi) & \operatorname{cov}(\xi, \eta) \\ \operatorname{cov}(\eta, \xi) & \operatorname{cov}(\eta, \eta) \end{bmatrix}. \end{aligned}$$

Let $\Sigma_{\theta\theta} = \operatorname{cov}(\theta, \theta)$, $\Sigma_{\theta Y} = \operatorname{cov}(\theta, Y)$, and $\Sigma_{YY} = \operatorname{cov}(Y, Y)$.

The functional CCA seeks coefficient vectors a and b such that

$$\max_{a,b} \frac{a^T \Sigma_{\theta Y} b}{\sqrt{(a^T \Sigma_{\theta\theta} a)} \sqrt{b^T \Sigma_{YY} b}} \tag{5.124}$$

or

$$\begin{aligned} &\max_{a,b} \quad a^T \Sigma_{\theta y} b \\ &\text{s.t.} \quad a^T \Sigma_{\theta\theta} a = 1 \\ &\qquad\quad b^T \Sigma_{YY} b = 1. \end{aligned} \tag{5.125}$$

For the convenience of presentation, we assume that $J + L \leq K$. Otherwise, switch the order of the data between the FPC score matrices θ and phenotypes Y.

Standard CCA studied in Section 1.62 can be used to solve the problem (5.124). Similar to Equation 1.220, we obtain

$$R = \Sigma_{YY}^{-1/2}\Sigma_{Y\theta}\Sigma_{\theta\theta}^{-1}\Sigma_{\theta Y}\Sigma_{YY}^{-1/2}. \quad (5.126)$$

The matrix R can be eigendecomposed into

$$R = U\Lambda U^T, \quad (5.127)$$

where $\Lambda = \text{diag}(\lambda_1, \ldots, \lambda_p)$ and $p = \min(J+L, K)$. Let $q = \max(J+L, K)$. The statistic for testing the association of the gene or genomic region with the K phenotypes is

$$T_{FCCA} = -n\sum_{i=1}^{P}\log\left(1-\lambda_i^2\right), \quad (5.128)$$

or

$$T_{FCCA} = -\left[n-1-\frac{(p+q+1)}{2}\right]\sum_{i=1}^{p}\log\left(1-\lambda_i^2\right), \quad (5.129)$$

where n is a sampled number of individuals.

Under the null hypothesis of no association of the gene with the trait, T_{CCA} is a central $\chi^2_{((J+L)K)}$ distribution.

5.4.4 Quadratically Regularized Functional CCA

To further reduce the dimension of the data, we present a novel powerful quadratically regularized functional CCA for pleiotropic genetic analysis. Assume that n individuals are sampled. We expand the genotype profiles $x_i(t)$ and $z_i(t)$ of each individual in terms of orthonormal eigenfunctions:

$$x_i(t) = \sum_{j=1}^{J}\xi_{ij}\varphi_j(t), \quad (5.130)$$

$$z_i(t) = \sum_{l=1}^{L}\eta_{ij}(t)\psi_l(t), \quad (5.131)$$

where

$$\xi_{ij} = < x_i(t), \varphi_j(t) >_\mu \quad \text{and} \quad \eta_{ij} = < Z(t), \psi_j t) >_\mu.$$

The quantities ξ_{ij} and η_{ij} are called functional principal component scores of the genotype profiles of the ith individual.

Let y_{ij} be the jth trait value of the ith individual. Define the data matrices

$$\xi = \begin{bmatrix} \xi_{11} & \cdots & \xi_{1J} \\ \vdots & \vdots & \vdots \\ \xi_{n1} & \cdots & \xi_{nJ} \end{bmatrix}, \quad \eta = \begin{bmatrix} \eta_{11} & \cdots & \eta_{1L} \\ \vdots & \vdots & \vdots \\ \eta_{n1} & \cdots & \eta_{nL} \end{bmatrix}, \quad Y = \begin{bmatrix} y_{11} & \cdots & y_{1K} \\ \vdots & \vdots & \vdots \\ y_{n1} & \cdots & y_{nK} \end{bmatrix}, \quad \text{and} \quad G = [\xi \quad \eta].$$

We seek the following quadratically recognized matrix factorization (Udell et al. 2016):

$$\min_{A,B} \ \|G - AB\|_F^2 + \gamma\|A\|_F^2 + \gamma\|B\|_F^2, \tag{5.132}$$

where

$A \in R^{n\times l}$ and $B \in R^{l\times(J+L)}$ are two factor matrices

$\gamma \geq 0$ is the regularization parameter

Since we restrict $\|A\|_F^2 \leq \alpha$ and $\|B\|_F^2 \leq \alpha$, we can expect that the single values of the product matrix AB will be reduced.

Assume that the single value decomposition (SVD) of G is given by

$$G = U\Lambda V^T = [u_1, \ldots, u_r] \begin{bmatrix} \lambda_1 & \cdots & 0 \\ \vdots & \ddots & \vdots \\ 0 & \cdots & \lambda_r \end{bmatrix} \begin{bmatrix} v_1^T \\ \vdots \\ v_r^T \end{bmatrix}, \tag{5.133}$$

where $U \in R^{m\times r}$ and $V \in R^{n\times r}$ have orthonormal columns, $\lambda_1 \geq \lambda_2 \geq \cdots \geq \lambda_r > 0$, $r = \text{Rank}(G)$, U and V are referred to as the left and right singular vectors of the matrix G, respectively, and $\lambda_1, \ldots, \lambda_r$ are the singular values of G. Using matrix calculus, we can find the solution to the optimization problem (5.132) (Udell et al. 2016) (this will also be shown in Appendix 8D, Chapter 8):

$$A = U_\Omega(\Lambda_\Omega - \gamma I)^{1/2} \quad \text{and} \quad B = (\Lambda_\Omega - \gamma I)^{1/2} V_\Omega^T, \tag{5.134}$$

where U_Ω and V_Ω denote the submatrix of U, V with columns indexed by Ω, respectively; $|\Omega| \leq l$ with $\lambda_i \geq \gamma$ for $i \in \Omega$; and similarly, we denote Λ_Ω. For example, if $\Omega = \{i: 1, 2, 3\}$, then we have $U_\Omega = [u_1, u_2, u_3]$, $\Lambda_{(3)} = \text{diag}\ (\lambda_1 - \gamma, \lambda_2 - \gamma, \lambda_3 - \gamma)$, and $V_\Omega = [v_1, v_2, v_3]$. In general, Equation 5.134 can also be written as

$$A = U_l \tilde{\Lambda}_l^{1/2}, \quad B = \tilde{\Lambda}_l^{1/2} V_l^T, \tag{5.135}$$

where $U_l = [u_1, \ldots, u_l]$, $V_l = [v_1, \ldots, v_l]$, $\tilde{\Lambda}_l = \text{diag}\left((\lambda_1 - \gamma)_+, \ldots, (\lambda_l - \gamma)_+\right)$, $(a)_+ = \max(a, 0)$. Finally, the matrix G will be approximated by a low-rank model:

$$\tilde{G} = AB. \tag{5.136}$$

Similarly, we can obtain the low-rank model approximation $\tilde{Y}$ of the matrix Y. We assume that the ranks of the matrices $\tilde{G}$ and $\tilde{Y}$ are p_1 and p_2, respectively. We assume that $p_1 \le p_2$. Denote $p = \min(p_1, p_2)$.

We use the low-rank model approximation of the original functional principal component scores of the genotype profiles and the phenotype data to define the sampling covariance matrices. Define

$$\hat{\Sigma}_{GG} = \tilde{G}^T \tilde{G}, \quad \tilde{\Sigma}_{GY} = \tilde{G}^T Y = \tilde{\Sigma}_{YG}^T, \quad \tilde{\Sigma}_{YY} = \tilde{Y}^T \tilde{Y},$$

and matrix

$$\tilde{R} = \tilde{\Sigma}_{YY}^{-1/2} \tilde{\Sigma}_{YG} \tilde{\Sigma}_{GG}^{-1} \tilde{\Sigma}_{GY} \tilde{\Sigma}_{YY}^{-1/2}. \tag{5.137}$$

Again, using Equation 5.127 gives the eigenvector decomposition of the matrix R with eigenvalues $\lambda_1 \ge \lambda_2 \ge \cdots \ge \lambda_p$.

Similar to test statistics in Equations 5.128 and 5.129, the statistic for testing the association of the gene or genomic region with the K phenotypes using quadratically recognized functional CCA is

$$T_{QFCCA} = -n \sum_{i=1}^{P} \log\left(1 - \lambda_i^2\right), \tag{5.138}$$

or

$$T_{QFCCA} = -\left[n - 1 - \frac{(p_1 + p_2 + 1)}{2}\right] \sum_{i=1}^{p} \log\left(1 - \lambda_i^2\right). \tag{5.139}$$

Under the null hypothesis of no association of the gene with the trait, T_{QFCCA} is a central $\chi^2_{(p_1 p_2)}$ distribution. In general, $p_1 p_2$ is much less than $(J+L)K$. Therefore, the statistic T_{QFCCA} has a higher power to detect association than the FCCA-based statistic T_{FCCA}.

5.5 DEPENDENCE MEASURE AND ASSOCIATION TESTS OF MULTIPLE TRAITS

Kernel-based statistics for single trait association tests studied in Section 4.3.2.6 can be directly extended to the association test of multiple traits. We only need to replace the single phenotype by multiple phenotypes in calculating the kernel matrix K_{yy}. Such a typical test is called Gene Association with Multiple Traits (GAMuT) (Broadaway et al. 2016). The GAMuT uses the dependence measure–based test for independence between two sets of multivariate variables: multiple phenotypes and multilocus genotypes. Recall in

Equations 4.157 and 4.158 that we define the centered kernel matrix $\hat{K}_{yy}$ for the phenotypes and centered matrix for the genotypes $\hat{K}_{xx}$. Consider two vectors, X and X'. Typical kernels are

1. Polynomial kernel: $K(X,X') = (<X,X'> + c)^d$, where $d \geq 1$ and $c \geq 0$.
2. Taylor series kernel: Assume $a_n \geq 0$ for all $n \geq 0$ and define kernel

$$K(X, X') = \sum_{n=1}^{\infty} a_n <X, X'>^n.$$

3. Exponential kernel: $K(X,X') = \exp(<X,X'>)$.
4. Gaussian kernel: $K(X, X') = \exp\left(-\frac{\|X - X'\|^2}{\sigma^2}\right)$.

For the genotypes, we can define additional kernels:

1. IBS kernel: $K(G_i, G_j) = \frac{\sum_{li=1}^{L}\left(2 - |G_{il} - G_{jl}|\right)}{2L}$.
2. $K_{xx} = GWG^T$, where

$$G = [X \quad Z] = \begin{bmatrix} x_{11} & \cdots & x_{1p} & z_{11} & \cdots & z_{1p} \\ \vdots & \vdots & & \vdots & \vdots & \vdots \\ x_{n1} & \cdots & x_{np} & z_{n1} & \cdots & z_{np} \end{bmatrix}, \quad W = \text{diag}(w_{11}, \ldots, w_{1p}, w_{21}, \ldots, w_{2p}).$$

If the genetic dominance effects are ignored, then

$$G = \begin{bmatrix} x_{11} & \cdots & x_{1p} \\ \vdots & \vdots & \vdots \\ x_{n1} & \cdots & x_{np} \end{bmatrix} \quad \text{and} \quad W = \text{diag}(w_1, \ldots, w_p).$$

The statistic T_I in Equation 4.165 is

$$T_I = \frac{1}{n}\text{Trace}\left(\hat{K}_{xx}\hat{K}_{yy}\right). \tag{5.140}$$

Theorem 4.2 stated that under the null hypothesis that two sets of random variables X and Y are independent, the statistic T_I is asymptotically distributed as

$$\frac{1}{n^2}\sum_{i=1}^{n}\sum_{j=1}^{n}\lambda_{x,i}\lambda_{y,j}z_{ij}^2, \tag{5.141}$$

where z_{ij}^2 are independently and identically distributed as a central $\chi_{(1)}^2$ distribution.

5.6 PRINCIPAL COMPONENT FOR PHENOTYPE DIMENSION REDUCTION

5.6.1 Principal Component Analysis

Principal component analysis (PCA) is one of the oldest and most widely used tools for dimension reduction and visualization and is applied to pleiotropic genetic analysis of multiple traits (Aschard et al. 2014; Gao et al. 2015; Zhou et al. 2015). PCA seeks a linear combination of multiple traits such that it projects high-dimensional phenotype data to a low-dimensional space with a few directions (axes) along which data variation is maximized.

PCA studied in Section 1.5.1 can be applied to dimension reduction of multiple phenotypes. Recall that the original phenotype matrix is defined as

$$Y = \begin{bmatrix} Y_1 & \cdots & Y_K \end{bmatrix}.$$

The K new variables that are linear combinations of the original phenotypes Y can be expressed as

$$Y^* = YE, \tag{5.142}$$

where

$$Y^* = \begin{bmatrix} Y_1^* & \cdots & Y_K^* \end{bmatrix}$$

and

$$E = \begin{bmatrix} e_{11} & \cdots & e_{1K} \\ \vdots & \ddots & \vdots \\ e_{K1} & \cdots & e_{KK} \end{bmatrix} = \begin{bmatrix} e_1 & \cdots & e_K \end{bmatrix}.$$

The transformed variables Y_k^* are called pseudophenotypes. PCA is to find $e_1, \ldots, e_K$ such that the variance of independent Y_k^* is maximized. In other words, PCA seeks to solve the following optimization problem:

$$\begin{aligned} &\max_E \quad \text{Trace}\left(E^T \Sigma E\right) \\ &\text{Subject to} \quad E^T E = I_K, \end{aligned} \tag{5.143}$$

where $\Sigma = \text{var}(Y)$.

The sampling estimation of the covariance matrix $\Sigma = \frac{1}{n} Y^T HY$, where $H = I_n - \frac{1}{n} \mathbf{1}\mathbf{1}^T$; $\mathbf{1}$ is an n-dimensional vector of ones.

By the Lagrange multiplier method, the constrained optimization problem (5.143) can be transformed into the following unconstrained optimization problem:

$$\max_E \quad \text{Trace}\left(E^T \Sigma E\right) + \text{Trace}\left(\Lambda\left(I - E^T E\right)\right). \tag{5.144}$$

Using formula (1.164), the optimal conditions for solving the optimization problem (5.144) are to solve the following eigenequation problem:

$$\Sigma E = E\Lambda, \tag{5.145}$$

where $\Lambda = \text{diag}(\lambda_1, \lambda_2, \ldots, \lambda_K)$.

After the eigenequation problem is solved, we can use Equation 5.142 to obtain the pseudophenotypes Y^*. All or partial pseudophenotypes can be used as phenotypes for pleiotropic analysis using multivariate linear models, multivariate functional linear models, CCA, FCCA, and quadratically regularized FCCA.

5.6.2 Kernel Principal Component Analysis

As Figure 5.1 shows, two classes are linearly inseparable in the input space. However, if we use nonlinear mapping to map the data from the input space to feature space, it is clear that two classes can be linearly separated in the feature space. PCA attempts to efficiently represent the data by finding orthonormal axes, which maximally capture the data variation. However, when the data are linearly inseparable in the original input space, principal components cannot maximally capture the data variation. To overcome the limitation of linear PCA, a nonlinear PCA is introduced. As Figure 5.2 shows, the kernel PCA can effectively account for data variation in the feature space.

Suppose that the phenotypes Y are mapped to a feature space via a nonlinear function (Schölkopf et al. 1997), $\Phi(Y_i)$, $i = 1, 2, \ldots, n$, where $\Phi(Y_i)$ is a K-dimensional vector in the feature space.

Let

$$\tilde{\Phi}(Y) = \begin{bmatrix} \tilde{\Phi}(Y_1)^T \\ \vdots \\ \tilde{\Phi}(Y_n)^T \end{bmatrix} = H\Phi(Y), \quad \Phi(Y) = \begin{bmatrix} \Phi(Y_1) \\ \vdots \\ \Phi(Y_n) \end{bmatrix}$$

and

$$Y^* = \tilde{\Phi}(Y)V, \tag{5.146}$$

where V is a K-dimensional vector in the feature space.

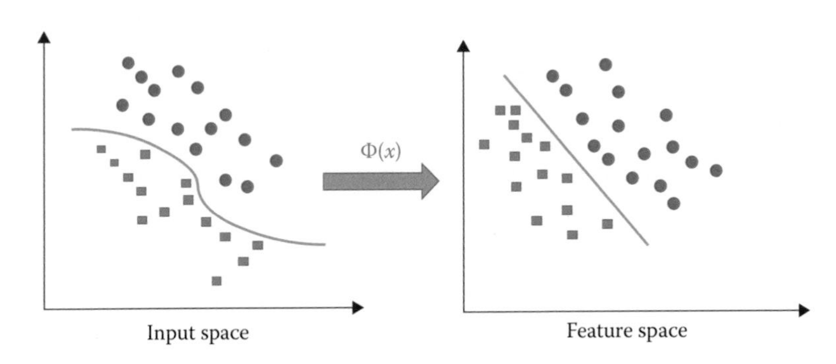

FIGURE 5.1 Power of seven statistics for testing the association of a gene with eight phenotypes.

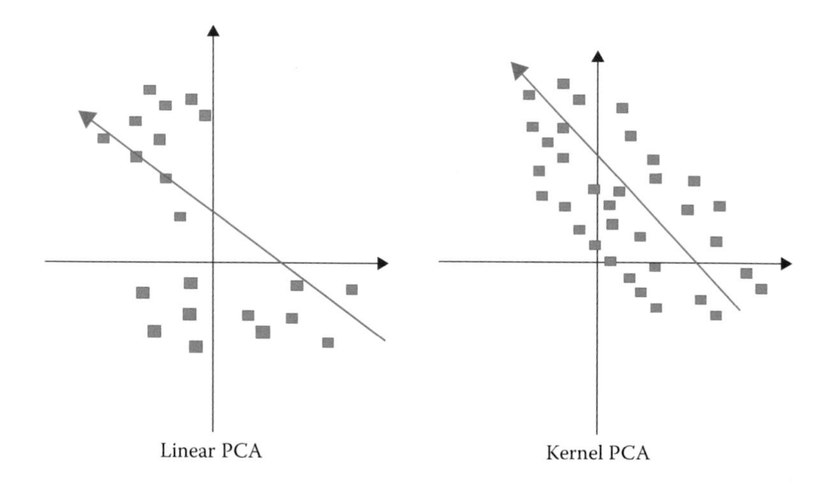

FIGURE 5.2 Scheme of feature mapping.

Similar to Equation 5.145, in the feature space, PCA seeks to solve the following eigenequation:

$$\Sigma V = \lambda V, \tag{5.147}$$

where

$$\Sigma = \frac{1}{n}\sum_{i=1}^{n}\tilde{\Phi}(Y_i)\tilde{\Phi}(Y_i)^T, \tag{5.148}$$

$$V = \sum_{j=1}^{n}\alpha_j\tilde{\Phi}(Y_j), \tag{5.149}$$

and $\tilde{\Phi}(Y_i) = H\Phi(Y_i)$ is the centered feature mapping. Let $\tilde{K}(Y_i, Y_j) = \tilde{\Phi}(Y_i)^T\tilde{\Phi}(Y_j)$.

Substituting Equations 5.148 and 5.149 into Equation 5.147 gives

$$\begin{aligned}\Sigma V &= \frac{1}{n}\sum_{i=1}^{n}\tilde{\Phi}(Y_i)\tilde{\Phi}(Y_i)^T\sum_{j=1}^{n}\alpha_j\tilde{\Phi}(Y_j)\\ &= \frac{1}{n}\sum_{i=1}^{n}\tilde{\Phi}(Y_i)\sum_{j=1}^{n}\alpha_j\tilde{\Phi}(Y_i)^T\tilde{\Phi}(Y_j)\\ &= \frac{1}{n}\sum_{i=1}^{n}\tilde{\Phi}(Y_i)\sum_{j=1}^{n}\alpha_j\tilde{K}(Y_i, Y_j),\end{aligned}$$

which implies

$$\frac{1}{n}\sum_{i=1}^{n}\tilde{\Phi}(Y_i)\sum_{j=1}^{n}\alpha_j\tilde{K}(Y_i, Y_j) = \lambda\sum_{j=1}^{n}\alpha_j\tilde{\Phi}(Y_j). \tag{5.150}$$

Multiplying by $\tilde{\Phi}(Y_l)^T$ on both sides of Equation 5.150, we obtain

$$\frac{1}{n}\sum_{i=1}^{n}\tilde{\Phi}(Y_l)^T\tilde{\Phi}(Y_i)\sum_{j=1}^{n}\alpha_j\tilde{K}(Y_i,Y_j)=\lambda\sum_{j=1}^{n}\alpha_j\tilde{\Phi}(Y_l)^T\tilde{\Phi}(Y_j),\quad l=1,\ldots,n. \tag{5.151}$$

Let

$$\tilde{K}(Y_{i,\cdot})=\left[\tilde{K}(Y_i,Y_1)\quad\cdots\quad\tilde{K}(Y_i,Y_n)\right]\ \text{and}\ \ \alpha=\left[\alpha_1,\ldots,\alpha_n\right]^T.$$

Then, Equation 5.151 can be reduced to

$$\frac{1}{n}\sum_{i=1}^{n}\tilde{K}(Y_l,Y_i)\tilde{K}(Y_{i,\cdot})\alpha=\lambda\sum_{j=1}^{n}\alpha_j\tilde{K}(Y_l,Y_j),\quad l=1,\ldots,n. \tag{5.152}$$

Equation 5.152 can be written in a matrix form:

$$\tilde{K}^2\alpha=n\lambda\tilde{K}\alpha,$$

or equivalently

$$\tilde{K}\alpha=n\lambda\alpha. \tag{5.153}$$

From Equations 5.146 and 5.149, we obtain the mapped phenotype values or pseudophenotype values:

$$\begin{aligned}
Y^* &= \tilde{\Phi}(Y)V \\
&= \tilde{\Phi}(Y)\tilde{\Phi}(Y)^T\alpha \\
&= \begin{bmatrix}\tilde{\Phi}(Y_1)^T \\ \vdots \\ \tilde{\Phi}(Y_n)^T\end{bmatrix}\left[\tilde{\Phi}(Y_1)\quad\cdots\quad\tilde{\Phi}(Y_n)\right]\alpha \\
&= \begin{bmatrix}\tilde{\Phi}(Y_1)^T\tilde{\Phi}(Y_1) & \cdots & \tilde{\Phi}(Y_1)^T\tilde{\Phi}(Y_n) \\ \vdots & \vdots & \vdots \\ \tilde{\Phi}(Y_n)^T\tilde{\Phi}(Y_1) & \cdots & \tilde{\Phi}(Y_n)^T\tilde{\Phi}(Y_n)\end{bmatrix}\alpha \\
&= \begin{bmatrix}\tilde{K}(Y_1,Y_n) & \cdots & \tilde{K}(Y_1,Y_n) \\ \vdots & \vdots & \vdots \\ \tilde{K}(Y_n,Y_1) & \cdots & \tilde{K}(Y_n,Y_n)\end{bmatrix}\alpha \\
&= \tilde{K}\alpha=n\lambda\alpha.
\end{aligned} \tag{5.154}$$

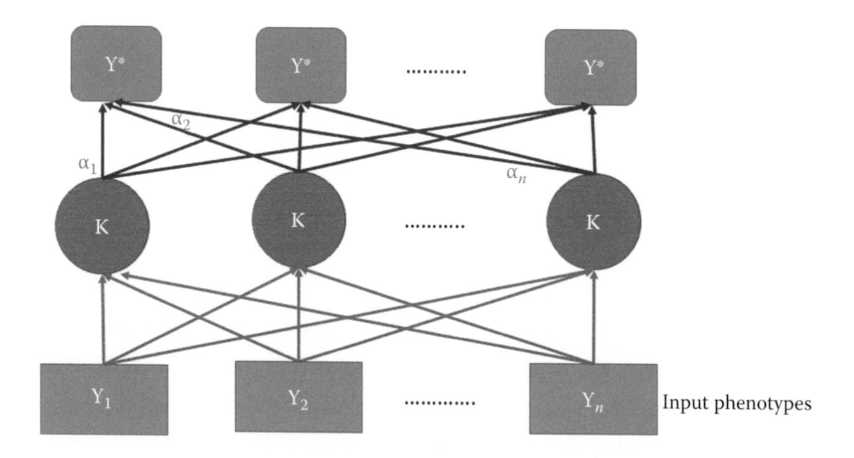

FIGURE 5.3 Linear PCA and kernel PCA.

The procedure of kernel PCA for generating pseudophenotypes is shown in Figure 5.3. Specifically,

1. Select a kernel function.
2. Calculate kernel Gram matrix.
3. Perform eigendecomposition of the kernel matrix $\tilde{K}$:

$$\tilde{K} = U\Lambda U^T.$$

4. Select $q \leq \min(K, n)$ eigenvectors from the matrix $U = [U_q, U_{n-q}]$ and calculate the pseudophenotype matrix: $Y^* = \tilde{K}U_q$ or $Y^* = YU_q$.

Output of the kernel PCA is the pseudophenotype values. We use Y^* as phenotypes for further genetic analysis of phenotypes.

5.6.3 Quadratically Regularized PCA or Kernel PCA

In Section 1.5.1.2, we showed that the variance of the first principal component is the largest eigenvalue value of the covariance matrix Σ of X and the proportion of the total variance due to the first principal component is

$$\frac{\lambda_1}{\lambda_1 + \lambda_2 + \cdots + \lambda_k}.$$

To increase the proportion of the first principal component in total variation of the low-dimensional space, we reduce every eigenvalue by a constant γ. We can show that

$$\frac{\lambda_1 - \gamma}{\lambda_1 - \gamma + \lambda_2 - \gamma + \cdots + \lambda_k - \gamma} > \frac{\lambda_1}{\lambda_1 + \lambda_2 + \cdots + \lambda_k}.$$

To achieve this goal, we extend the PCA to a quadratically regularized PCA.

Let $Z = Y$ or $Z = Y^*$. Assume $Z \in R^{n \times K}$. Similar to the quadratically regularized FCCA, we factorize the matrix Z to the product of two factor matrices, $A \in R^{n \times l}$ and $B \in R^{l \times K}$. The quadratically regularized PCA problem is mathematically formulated as

$$\min_{A,B} \ \|Z - AB\|_F^2 + \gamma \|A\|_F^2 + \gamma \|B\|_F^2, \tag{5.155}$$

where $\gamma \geq 0$ is the regularization parameter and $\|.\|_F$ is the Frobenius norm of a matrix, i.e., the square root of the sum of the squares of the entries. Since we restrict $\|A\|_F^2 \leq \alpha$ and $\|B\|_F^2 \leq \alpha$, we can expect that the single values of the product matrix AB will be reduced. When $\gamma = 0$, the quadratically regularized PCA problem is reduced to the traditional PCA.

Assume that the single value decomposition (SVD) of Z is given by

$$Z = U \Lambda V^T = \left[u_1, \ldots, u_r\right] \begin{bmatrix} \lambda_1 & \cdots & 0 \\ \vdots & \ddots & \vdots \\ 0 & \cdots & \lambda_r \end{bmatrix} \begin{bmatrix} v_1^T \\ \vdots \\ v_r^T \end{bmatrix}, \tag{5.156}$$

where $U \in R^{n \times r}$ and $V \in R^{K \times r}$ have orthonormal columns, $\lambda_1 \geq \lambda_2 \geq \cdots \geq \lambda_r > 0$, $r = \text{Rank}(Z)$, U and V are referred to as the left and right singular vectors of the matrix Z, respectively, and $\lambda_1, \ldots, \lambda_r$ are the singular values of Z.

Using matrix calculus, we can find the solution to the optimization problem (5.155) (Udell et al. 2016):

$$A = U_\Omega \left(\Lambda_\Omega - \gamma I\right)^{1/2} \quad \text{and} \quad B = \left(\Lambda_\Omega - \gamma I\right)^{1/2} V_\Omega^T, \tag{5.157}$$

where U_Ω, V_Ω denote the submatrix of U, V with columns indexed by Ω, respectively, $|\Omega| \leq l$ with $\lambda_i \geq \gamma$ for $i \in \Omega$, and similarly, we denote Λ_Ω. For example, if $\Omega = \{i : 1, 2, 3\}$, then we have $U_\Omega = [u_1, u_2, u_3]$, $\Lambda_{(3)} = \text{diag}(\lambda_1 - \gamma, \lambda_2 - \gamma, \lambda_3 - \gamma)$, and $V_\Omega = [v_1, v_2, v_3]$. In general, Equation 5.157 can also be written as

$$A = U_l \tilde{\Lambda}_l^{1/2}, \quad B = \tilde{\Lambda}_l^{1/2} V_l^T, \tag{5.158}$$

where $U_l = [u_1, \ldots, u_l]$, $V_l = [v_1, \ldots, v_l]$, $\tilde{\Lambda}_l = \text{diag}\left(\left(\lambda_1 - \gamma\right)_+, \ldots, \left(\lambda_l - \gamma\right)_+\right)$, $(a)_+ = \max(a, 0)$.

Finally, we take $Z^* = AB$ as the matrix of pseudophenotype values for pleiotropic genetic analysis.

5.7 OTHER STATISTICS FOR PLEIOTROPIC GENETIC ANALYSIS

5.7.1 Sum of Squared Score Test

The sum of squared score test (SSU) originally developed for testing the association of multiple SNPs with a single trait (Pan et al. 2014) can be extended to test multiple traits (Ray et al. 2016). It follows from Equations 5.40 and 1.164 that under the global null hypothesis of no association, the derivative of the log-likelihood function with respect to the coefficient matrix B is

$$\frac{\partial l}{\partial B} = H^T Y \Sigma^{-1}, \tag{5.159}$$

where $H = \begin{bmatrix} X & Z \end{bmatrix}$.

Under the null hypothesis of no association, $B = 0$, the vector of marginal scores is

$$\begin{aligned} U_M &= \text{vec}\left(\frac{\partial l}{\partial B}\right) \\ &= \left(\Sigma^{-1} \otimes H^T\right) \text{vec}(Y). \end{aligned} \tag{5.160}$$

If we do not consider the correlation structure of the multiple phenotypes, then the covariance matrix is given by

$$\Sigma = \sigma_0^2 I_K, \tag{5.161}$$

where $\sigma_0^2 = \frac{1}{K(n-1)} \sum_{i=1}^{n} \sum_{j=1}^{K} y_{ij}^2$.

Substituting Equation 5.161 into Equation 5.160 gives the $2K$-dimensional vector of marginal scores:

$$\begin{aligned} U_M &= \frac{1}{\sigma_0^2} \begin{bmatrix} H^T & \cdots & 0 \\ \vdots & \vdots & \vdots \\ 0 & \cdots & H^T \end{bmatrix} \begin{bmatrix} Y_1 \\ \vdots \\ Y_K \end{bmatrix} \\ &= \frac{1}{\sigma_0^2} \begin{bmatrix} H^T Y_1 \\ \vdots \\ H^T Y_K \end{bmatrix} \\ &= \frac{1}{\sigma_0^2} \begin{bmatrix} \left(I_2 \otimes Y_1^T\right) H \\ \vdots \\ I_2 \otimes Y_K^T) H \end{bmatrix} \\ &= \frac{1}{\sigma_0^2} \left(I_2 \otimes Y^T\right) H. \end{aligned} \tag{5.162}$$

The SSU test statistic for the association of multiple SNPs with the multiple traits is defined as

$$T_S = U_M^T U_M. \tag{5.163}$$

Under H_0, the distribution of the SSU test statistic T_s can be approximated by a mixed χ^2 distribution, $a\chi_{(d)}^2 + b$ (Ray et al. 2016; Zhang 2005). Let $\lambda_1 \geq \lambda_2 \geq \cdots \geq \lambda_{2K}$ be eigenvalues of the matrix $\frac{1}{n\sigma_0^2}H^T H\left(I_2 \otimes Y^T Y\right)$. Let $m=2K$. The parameters a, b, and d in the mixed χ^2 distribution can be calculated by

$$a=\frac{\sum_{j=1}^{M}\lambda_j^3}{\sum_{j=1}^{M}\lambda_j^2},\quad b=\sum_{j=1}^{M}\lambda_j-\frac{\left(\sum_{j=1}^{M}\lambda_j^2\right)^2}{\sum_{j=1}^{M}\lambda_j^3},\quad d=\frac{\left(\sum_{j=1}^{M}\lambda_j^2\right)^3}{\left(\sum_{j=1}^{M}\lambda_j^3\right)^2}. \tag{5.164}$$

When we only consider genetic additive effects, the vector of marginal scores is reduced to

$$U_M=\frac{1}{\sigma_0^2}Y^T X. \tag{5.165}$$

The eigenvalues $\lambda_1 \geq \lambda_2 \geq \cdots \geq \lambda_K$ of the matrix $\frac{1}{n\sigma_0^2}H^T H Y^T Y$ will be used to calculate the parameters a, b, and d in Equation 5.165.

5.7.2 Unified Score-Based Association Test (USAT)

A limitation for the SSU test is that the SSU does not explicitly explore correlation structures of the multiple traits in the test statistic. Although MANOVA can incorporate the correlation structure of the traits, it also suffers from lack of power when the genetic effects for each trait are similar and are in the same direction as the correlation. To utilize the merit of both the SSU test and MANOVA and overcome their limitations, a score-based association test (USAT) that combines the USAT and MANOVA for a single marker was proposed (Ray et al. 2016).

Let T_M be a test statistic, for example, likelihood ration test statistic or Wilks' lambda in MANOVA. Define a weighted statistic:

$$T_\omega=\omega T_M+(1-\omega)T_S, \tag{5.166}$$

where $\omega \in [0,1]$ is the weight as a USAT statistic. When $\omega=0$, the statistic T_ω is reduced to the SSU statistic T_S, and $\omega=1$, the statistic T_ω is reduced the MANOVA test T_M. The power of the USAT statistic depends on the weight we chose. Since the weight is unknown, the optimal weight can be determined by

$$T_{USAT}=\min_{0\leq\omega\leq1} P_\omega, \tag{5.167}$$

where P_ω is the P-value of the test statistic T_ω.

The distribution of the USAT can be found in Ray et al. (2016). Since the USAT can only consider one SNP at a time and is difficult to apply to the next-generation sequencing data, its application to pleiotropic genetic analysis is limited.

5.7.3 Combining Marginal Tests

Assume that the traits are uncorrelated. Then, we can test a trait at a time. Let P_k be a P-value for testing the association of genetic variants with the kth trait. Then, Fisher's method can be used to combine K P-values into a global P-value for testing the association with the K traits. Define the statistic:

$$T_a = -2\log \sum_{k=1}^{K} P_k.$$

Then, under the null hypothesis of no association and assumption of independence tests, asymptotically, T_a is distributed as a central $\chi^2_{(2K)}$ distribution. If the correlation is present, Fisher's method of combining P-values will inflate the type 1 error rates.

We may also use the Bonferroni correction to combine P-values. Define the statistic:

$$T_{\min} = \min\left(KP_k\right).$$

Under the assumption of no association and independence tests, $T_{\min}$ is distributed as the minimum of independent $U(0, 1)$ variables (Ray et al. 2016).

5.7.4 FPCA-Based Kernel Measure Test of Independence

All SKAT, GAMuT, and other kernel measure tests of independence use the original genotype data. However, in many scenarios, particularly for next-generation sequencing data, the FPCA provides a powerful tool for data dimension reduction. Using FPC scores will improve the power of the tests. For all kernel measure–based tests, we can use FPC scores to replace the original genotype data.

Let ξ_{ij} and η_{ij} be functional principal component scores of the genotype profiles of the ith individual calculated in Equations 5.130 and 5.131. Define the data matrices:

$$\xi = \begin{bmatrix} \xi_{11} & \cdots & \xi_{1J} \\ \vdots & \vdots & \vdots \\ \xi_{n1} & \cdots & \xi_{nJ} \end{bmatrix}, \quad \eta = \begin{bmatrix} \eta_{11} & \cdots & \eta_{1L} \\ \vdots & \vdots & \vdots \\ \eta_{n1} & \cdots & \eta_{nL} \end{bmatrix},$$

$G = \left[\xi \quad \eta\right]$, and $\lambda_1, \ldots, \lambda_J$ and $\gamma_1, \ldots, \gamma_L$ are the eigenvalues associated with the FPC scores ξ of the genotype indicators for the additive effects and the FPC score η of the genotype indicators for the dominance effects, respectively.

Let $W_\xi = \operatorname{diag}\left(\frac{1}{\lambda_1(1-\lambda_1)}, \ldots, \frac{1}{\lambda_J(1-\lambda_J)}\right)$, $W_\eta = \operatorname{diag}\left(\frac{1}{\gamma_1(1-\gamma_1)}, \ldots, \frac{1}{\gamma_L(1-\gamma_L)}\right)$, and $W = \begin{bmatrix} W_\xi & 0 \\ 0 & W_\eta \end{bmatrix}$. Define the kernel matrix for the FPC scores: $K_{xx} = GWG^T$ and $\hat{K}_{xx} = HK_{xx}H$. Let $\hat{K}_{yy}$ be defined as before.

Then, the statistic defined in Equation 5.140

$$T_I = \frac{1}{n}\text{Trace}\left(\hat{K}_{xx}\hat{K}_{yy}\right) \tag{5.168}$$

can be used to test association of the gene or genomic region with the multiple traits.

5.8 CONNECTION BETWEEN STATISTICS

For the simplicity of discussion, here we only consider genetic additive effects. The conclusions can be easily extended to genetic dominance effects. In the multivariate linear model, we can show that the matrix of regression coefficients can be estimated by

$$\hat{B} = \Sigma_{XX}^{-1}\Sigma_{XY}. \tag{5.169}$$

The contribution of the genetic additive effects to the phenotypic variation is

$$S_{reg} = \Sigma_{YX}\Sigma_{XX}^{-1}\Sigma_{XY}. \tag{5.170}$$

Similar to a single trait, the matrix of the proportions of genetic additive effect to the variation of the multiple traits is defined as

$$H_{her} = \Sigma_{YY}^{-1/2}\Sigma_{YX}\Sigma_{XX}^{-1}\Sigma_{XY}\Sigma_{YY}^{-1/2}. \tag{5.171}$$

This is exactly the same as the R matrix in Equation 5.104. This shows the equivalence between linear models and CCA.

Next we study the connection among PCA, linear models, and CCA. Suppose that the singular value decomposition (SVD) of the phenotype matrix Y is

$$Y = U\Lambda V^T, \tag{5.172}$$

which implies that

$$Y^TY = V\Lambda^2V^T. \tag{5.173}$$

Assume that the covariance matrices are approximated by their sampling formulas:

$$\hat{\Sigma}_{YY} = Y^TY, \quad \hat{\Sigma}_{YX} = Y^TX = \hat{\Sigma}_{XY}^T, \quad \text{and} \quad \hat{\Sigma}_{XX} = X^TX.$$

We rewrite Equation 5.145 as

$$\hat{\Sigma}_{YY}E = E\Lambda_{\Sigma}$$

or

$$Y^T Y = E \Lambda_\Sigma E^T. \tag{5.174}$$

Comparing Equations 5.173 and 5.62 gives

$$E = V, \quad \Lambda_\Sigma = \Lambda^2. \tag{5.175}$$

Thus, the pseudophenotypes are

$$Y^* = YV. \tag{5.176}$$

Using Equation 5.176, we obtain the variance matrix of the pseudophenotypes and covariance matrices between pseudophenotypes and genotypes:

$$\hat{\Sigma}_{Y^*Y^*} = \left(Y^*\right)^T Y^* = V^T Y^T Y V = V^T \hat{\Sigma}_{YY} V, \tag{5.177}$$

$$\hat{\Sigma}_{Y^*X} = \left(Y^*\right)^T X = V^T Y^T X = V^T \hat{\Sigma}_{YX} = \left(\hat{\Sigma}_{XY^*}\right)^T. \tag{5.178}$$

The coefficient matrix B^* regressing the pseudophenotypes on the genotypes in the multivariate linear model is

$$B^* = \hat{\Sigma}_{XX}^{-1} \hat{\Sigma}_{XY^*} = \hat{\Sigma}_{XX}^{-1} \hat{\Sigma}_{XY} V = BV. \tag{5.179}$$

Therefore, the sum of squares due to regression in the pseudophenotype linear models is

$$\begin{aligned} S_{reg}^* &= \hat{\Sigma}_{Y^*X} \hat{\Sigma}_{XX}^{-1} \hat{\Sigma}_{XY^*} \\ &= V^T \hat{\Sigma}_{YX} \hat{\Sigma}_{XX}^{-1} \hat{\Sigma}_{XY} V \\ &= V^T S_{req} V. \end{aligned} \tag{5.180}$$

Next we study how the PCA for phenotypes affects the CCA. It is well known that the eigenvalues of the matrix

$$R_Y = \hat{\Sigma}_{YY}^{-1} \hat{\Sigma}_{YX} \hat{\Sigma}_{XX}^{-1} \hat{\Sigma}_{XY} \tag{5.181}$$

equal to the eigenvalues of the matrix R in Equation 5.89.

Now we calculate the R_Y^* matrix for the pseudophenotypes. Substituting Equations 5.177 and 5.178 into Equation 5.181 gives

$$\begin{aligned} R_{Y^*} &= \left(V^T \hat{\Sigma}_{YY} V\right)^{-1} V^T \hat{\Sigma}_{YX} \hat{\Sigma}_{XX}^{-1} \hat{\Sigma}_{XY} V \\ &= V^{-1} R_Y V. \end{aligned} \tag{5.182}$$

We define the total heritability of the multiple phenotypes as

$$\begin{aligned} H_Y &= \text{Trace}\left(R_{Y^*}\right) \\ &= \text{Trace}\left(R_Y\right) \\ &= \sum_{k=1}^{K} \lambda_k^2, \end{aligned} \tag{5.183}$$

which indicates that if we use a full-rank matrix, V, to transform the phenotypes, then such transformation will not change the total heritability of the multiple phenotypes and the sum of squares of all the canonical correlations.

Now assume that

$$Y^* = YV_r, \tag{5.184}$$

where $r < K$.

Then, it is easy to see that

$$B^* = BV_r, \tag{5.185}$$

and

$$S_{reg}^* = V_r^T S_{reg} V_r. \tag{5.186}$$

Expressing R_Y^2 in terms of singular values of the SVD of the phenotype matrix Y, it follows from Equation 5.172 that

$$\hat{\Sigma}_{YY} = V \Lambda^2 V^T, \tag{5.187}$$

and

$$\hat{\Sigma}_{YX} = V \Lambda U^T X = \left(\hat{\Sigma}_{XY}\right)^T. \tag{5.188}$$

Thus, R_Y^2 can be written as

$$R_Y^2 = V\Lambda^{-1}U^T X\left(X^T X\right)^{-1} X^T U\Lambda V^T. \tag{5.189}$$

From Equation 4.168, it follows that

$$R_{Y^*} = \left(V_r^T V\Lambda^2 V^T V_r\right)^{-1} V_r^T V\Lambda U^T X\left(X^T X\right)^{-1} X^T U\Lambda V^T V_r. \tag{5.190}$$

Suppose that

$$Y = \begin{bmatrix} U_r & U_{K-r} \end{bmatrix} \begin{bmatrix} \Lambda_r & 0 \\ 0 & \Lambda_{K-r} \end{bmatrix} \begin{bmatrix} V_r^T \\ V_{K-r}^T \end{bmatrix}. \tag{5.191}$$

Using Equation 5.191, we can show

$$V_r^T V\Lambda^2 V^T V_r = \Lambda_r^2, \tag{5.192}$$

$$\Lambda_r^{-2} V_r^T V\Lambda U^T = \Lambda_r^{-1} U_r^T, \tag{5.193}$$

and

$$U\Lambda V^T V_r = U_r \Lambda_r. \tag{5.194}$$

Substituting Equations 5.192, 5.193, and 5.194 into Equation 5.190, we obtain

$$R_{Y^*} = \Lambda_r^{-1} U_r X\left(X^T X\right)^{-1} X^T U_r \Lambda_r. \tag{5.195}$$

It follows from Equation 5.189 that

$$X\left(X^T X\right)^{-1} X^T = U\Lambda V^T R_Y^2 V\Lambda^{-1} U^T. \tag{5.196}$$

Substituting Equation 5.196 into Equation 5.195 gives

$$R_{Y^*} = \Lambda_r^{-1} U_r U\Lambda V^T R_Y^2 V\Lambda^{-1} U^T U_r \Lambda_r. \tag{5.197}$$

We can show that

$$\Lambda_r^{-1} U_r U\Lambda V^T = V_r^T \tag{5.198}$$

and

$$V\Lambda^{-1}U^TU_r\Lambda_r = V_r. \tag{5.199}$$

Substituting Equations 5.198 and 5.199 into Equation 5.197, we obtain the following interesting relations between R_{Y*} and R_Y:

$$R_{Y^*} = V_r^T R_Y^2 V_r. \tag{5.200}$$

Similarly, for kernel PCA, we have

$$B_{KPCA} = BU_q, \tag{5.201}$$

$$S_{reg}^{KPCA} = U_q^T S_{reg} U_q, \tag{5.202}$$

$$R_{KPCA} = U_q^T R_Y^2 U_q.$$

Now we investigate replacing the original genotype data by their FPC scores in the functional linear models and FCCA. For simplicity, we only consider genetic additive effects. First, we consider regression coefficients in the functional linear model. We can show that the regression coefficient matrix in the functional linear mode, sum of squares due to regression, and R_{FCCA} matrix are, respectively, given by (Appendix 5B)

$$\hat{B}_{FLM} = \left(\Phi^T X^T X \Phi\right)^{-1} \Phi^T X^T Y, \tag{5.203}$$

$$S_{reg}^{FLM} = \left(Y^T Y\right)^{-1} Y^T X \Phi \left(\Phi^T X^T X \Phi\right)^{-1} \Phi^T X^T Y, \tag{5.204}$$

and

$$\hat{R}_{FCCA} = \left(Y^T Y\right)^{-1} Y^T X \Phi \left(\Phi^T X^T X \Phi\right)^{-1} \Phi^T X^T Y. \tag{5.205}$$

Four approaches can be used for testing the association of genetic variation with the multiple phenotypes: *P*-value combination of single trait analysis, regression, CCA, and dependence measure–based tests (kernel methods). Since the power of single trait analysis is limited, we mainly focus on the remaining three methods. The methods can be classified into two categories: original data and dimension-reduced data. Figure 5.4 lists the current and near-future methods for genetic pleiotropic analysis. Many of them can be applied to next-generation sequencing data. These methods have their merits and limitations. Large-scale simulations and real data analysis are needed to evaluate their performance.

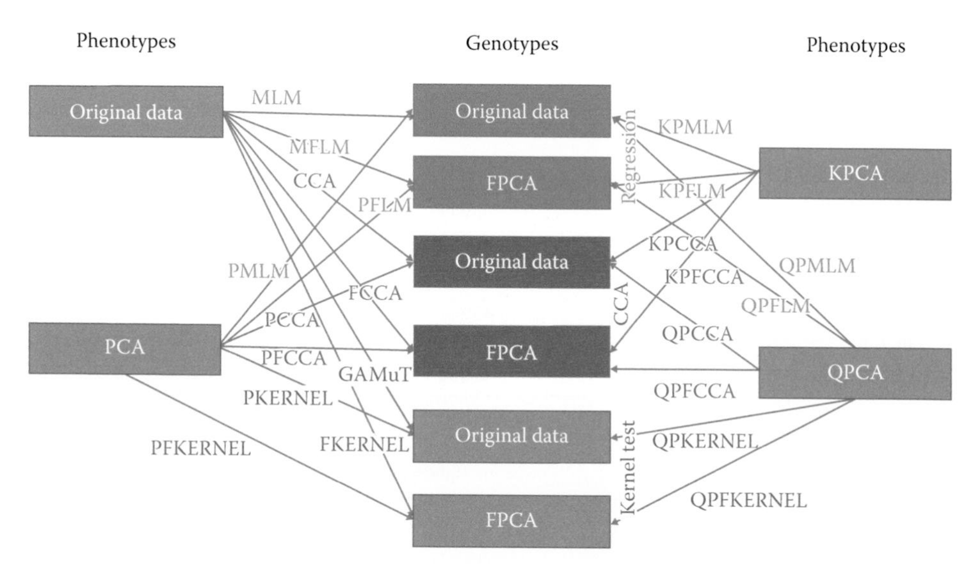

FIGURE 5.4 Scheme of kernel PCA.

5.9 SIMULATIONS AND REAL DATA ANALYSIS

5.9.1 Type 1 Error Rate and Power Evaluation

To verify the feasibility of the several statistics for testing associations of genes with multiple phenotypes, we will perform large-scale simulations to validate the null distribution of the test statistics and calculate their type 1 error rates. We used the coalescent software MS (Hudson 2002) to generate a population of 2,000,000 chromosomes with 400 loci under the neutrality model. Correlations ranging from 0.01 to 0.5 were randomly assigned to the eight phenotypes, and 5000 simulations were conducted. Table 5.1 shows type 1 error rates for the nominal level of seven statistics where QRFCCA denotes a quadratically regularized functional canonical analysis test; FCCA is a functional canonical correlation test; PCA is a multivariate principal component analysis test; "Min *P*-value" denotes the minimum of *P*-values calculated by MANOVA(multivariate ANOVA), a SNP at a time, adjusted by false discovery rate (FDR); USAT is a Unified Score-Based Association Test (Ray et al. 2016); and GAMuT is a Gene Association with Multiple Traits. Table 5.1 shows that type 1 error rates of the statistics QRFCCA, FCCA, PCA, and USAT were not appreciably different from the nominal levels, while Min *P*-value and MANOVA were conservative.

To further evaluate the performance of the statistics for testing the association of genetic variants with multiple traits, a large simulation to make power comparisons should be conducted. The power evaluation of various methods for detecting association with multiple phenotypes is complicated. It depends on the correlation structure of the multiple phenotypes and pattern of linkage disequilibrium and allelic spectrum. Here we present a simple simulation for illustration purpose. We use data as in type 1 error simulations, but we assumed that each phenotype had five associating SNPs, i.e., SNP1–5 were associated with phenotype 1, SNP6–10 associated with phenotype 2, etc. SNPs explain 20% of total

TABLE 5.1 Type 1 Error Rates of Six Statistics for Testing the Association of a Gene with Eight Phenotypes

Sample Size	500	1000	1500	2000
Min *P*-value	0.031	0.031	0.030	0.031
MANOVA	0.041	0.041	0.040	0.040
USAT	0.055	0.053	0.050	0.050
GAMuT	0.056	0.051	0.042	0.047
PCA	0.053	0.053	0.050	0.053
FCCA	0.054	0.054	0.052	0.055
QRFCCA	0.051	0.054	0.055	0.051

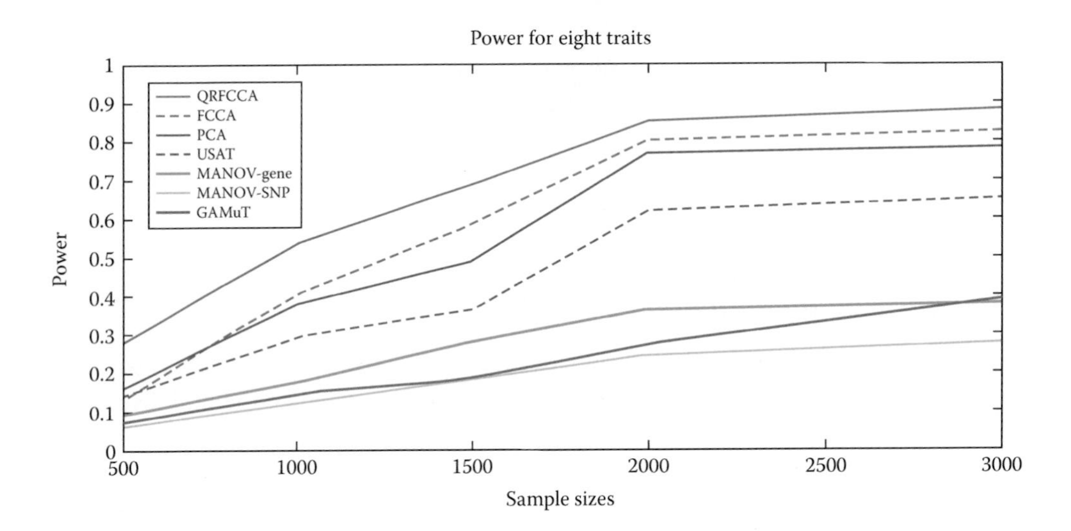

FIGURE 5.5 Methods for genetic analysis of multiple phenotypes.

phenotype variance. As shown in Figure 5.5, the power of the QRFCCA was the highest, followed by the FCCA. The power of the MANOVA was the lowest.

5.9.2 Application to Real Data Example

Genetic pleiotropic analysis of multiple traits has recently been intensively investigated. Six recently developed statistics, QRFCCA, CCA, PCA, MANOVA, GAMuT, and USAT, are applied to testing association with 20 traits including seven cognitive functions, seven lipid metabolisms, and five environmental traits in the Rush Alzheimer Disease dataset. It included 1,707 individuals and 53,295 genes. The *P*-value for declaring significance is 9.38E−07. The total number of significant genes discovered using QRFCCA, CCA, PCA, MANOVA, GAMuT, and USAT are 114, 31, 27, 13, 4, and 0, respectively. APOC1, TOMM40, and CSMD1 are included in the set of 114 genes. They are associated with AD or AD markers. Manhattan plots of the results are shown in Figure 5.6. The results demonstrated that QRFCCA substantially outperforms other statistics and has a high power to identify genes associated with multiple traits.

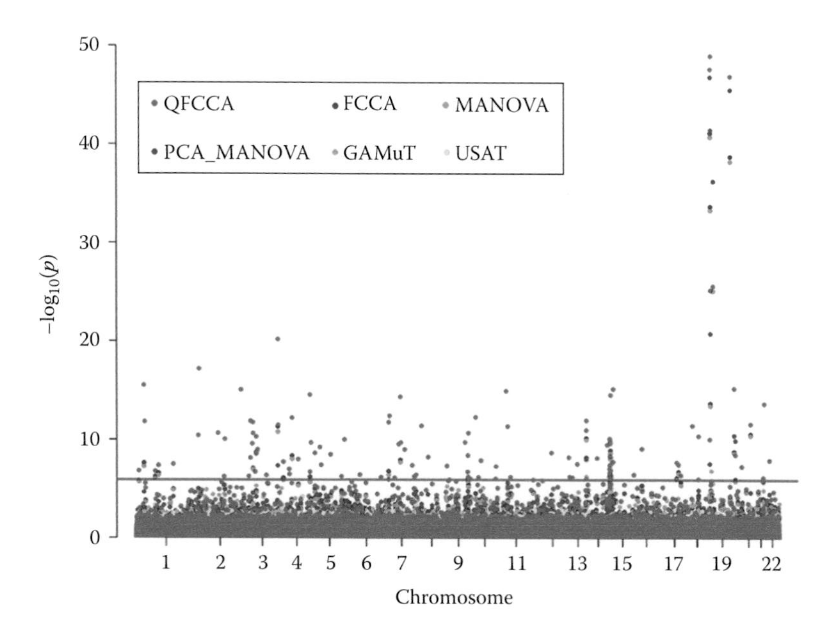

FIGURE 5.6 Manhattan plot showing P-values of six statistics for testing the association of genes across the genome with 20 traits.

SOFTWARE PACKAGE

The code with MATLAB® implementation of a kernel-based statistical hypothesis test for independence can be downloaded from http://www.kyb.mpg.de/bs/people/arthur/indep.htm. An R package "candisc" for canonical correlation analysis can be downloaded from the website https://cran.r-project.org/web/packages/candisc/index.html. Alternatively, we can use the package from the website (https://cran.r-project.org/web/packages/CCA/index.html) for CCA. The R package "MultiPhen" (https://cran.r-project.org/web/packages/MultiPhen/index.html) uses a proportional odds regression model for pleiotropic genetic analysis. An R package for MANOVA can be downloaded from the website https://stat.ethz.ch/R-manual/R-patched/library/stats/html/summary.manova.html. GAMuT-based kernel regression can be downloaded from the website http://www.genetics.emory.edu/labs/epstein/software. A software for implementing QRFCCA and FCCA is in the website http://www.sph.uth.tmc.edu/hgc/faculty/xiong/index.htm.

APPENDIX 5A OPTIMIZATION FORMULATION OF KERNEL CCA

Equation 5.109 gives the sampling formula for the expectation $E[f(X)]$:

$$E\left[f(X)\right] \approx \frac{1}{n}\sum_{j=1}^{n} f(x_j) = \frac{1}{n}\sum_{j=1}^{n} < f(.), k_x(., x_j) >_{H_x}. \tag{5A.1}$$

Function $f(.)$ can be expressed as

$$f(.) = \sum_{i=1}^{n} \alpha_i k_x(., x_i). \tag{5A.2}$$

Substituting Equation 5A.2 into Equation 5A.1 gives

$$E\left[f(X)\right]=m_x\approx\frac{1}{n}\sum_{j=1}^{n}<\sum_{i=1}^{n}\alpha_i k_x(.,x_i),k_x(.,x_j)>_{H_x}$$

$$=\frac{1}{n}\sum_{i=1}^{n}\alpha_i\sum_{j=1}^{n}k_x(x_i,x_j). \tag{5A.3}$$

Similarly, we have

$$g(.)=\sum_{i=1}^{n}\beta_i k_y(.,y_i), \tag{5A.4}$$

and

$$E\left[g(Y)\right]=m_x\approx\frac{1}{n}\sum_{i=1}^{n}\beta_i\sum_{j=1}^{n}k_y(y_i,y_j). \tag{5A.5}$$

It follows from Equations 5A.2 and 5A.3 that

$$f(X)-E[f(X)\}=\sum_{i=1}^{n}\alpha_i\left[k_x(X,x_i)-\frac{1}{n}1^T K_x(.,x_i)\right], \tag{5A.6}$$

where $1^T K_x(.,x_i)=\sum_{u=1}^{n}K_x(x_u,x_i)$.

Similarly, we have

$$g(Y)-E[g(Y)\}=\sum_{i=1}^{n}\beta_i\left[k_y(Y,y_i)-\frac{1}{n}1^T K_y(.,y_i)\right], \tag{5A.7}$$

where $1^T K_y(.,y_i)=\sum_{u=1}^{n}K_y(x_u,y_i)$.

Thus, the sampling covariance between $f(X)$ and $f(Y)$ is

$$\operatorname{cov}(f(X),g(Y))\approx\frac{1}{n}\sum_{l=1}^{n}[f(x_l)-m_x][g(y_l)-m_y]$$

$$=\frac{1}{n}\sum_{l=1}^{n}\left[\sum_{i=1}^{n}\alpha_i(k_x(x_l,x_i)-m_x)\right]\left[\sum_{j=1}^{n}\beta_j(k_y(y_l,y_j)-m_y)\right]$$

$$=\frac{1}{n}\sum_{l=1}^{n}\sum_{i=1}^{n}\sum_{j=1}^{n}\alpha_i\beta_j\left(k_x(x_l,x_i)-\frac{1}{n}1^T K_x(.,x_i)\right)\left(k_y(y_l,y_j)-1^T K_y(.,y_i)\right)$$

$$=\frac{1}{n}\alpha^T K_x H K_x. \tag{5A.8}$$

Similarly, we can show that

$$\operatorname{var}\left(f(X)\right)=\alpha^T K_x H K_x \alpha \quad \text{and} \quad \operatorname{var}\left(g(y)\right)=\beta^T K_y H K_y. \tag{5A.9}$$

The optimization problem (5.108) can be transformed to the following optimization problem:

$$\begin{aligned} \max_{f,g} \quad & \operatorname{cov}\left(f(X), g(X)\right) \\ \text{s.t.} \quad & \operatorname{var}\left(f(X)\right)=1 \\ & \operatorname{var}\left(g(Y)\right)=1 \end{aligned} \tag{5A.10}$$

Using Equations 5A.8, 5A.9, and 5A.10 gives

$$\begin{aligned} \max_{\alpha,\beta} \quad & \alpha^T K_x H K_y \beta \\ \text{s.t.} \quad & \alpha^T K_x H K_x \alpha = 1 \\ & \beta^T K_y H K_y \beta = 1. \end{aligned}$$

APPENDIX 5B DERIVATION OF THE REGRESSION COEFFICIENT MATRIX IN THE FUNCTIONAL LINEAR MODE, SUM OF SQUARES DUE TO REGRESSION, AND R_{FCCA} MATRIX

Assume that the genotype function $x_i(t)$ is expanded in terms of eigenfunctions:

$$x_i(t)=\sum_{j=1}^{J} \xi_{ij}\phi_j(t), \tag{5B.1}$$

where $\phi_j(t)$ is the jth eigenfunction and

$$\xi_{ij}=\int_T x_i(t)\phi_j(t). \tag{5B.2}$$

Denote $x_i(t_1)$ by x_{i1}. Then, we have

$$\xi_{ij}=\sum_{l=1}^{P} x_i(t_l)\phi_j(t_l), \tag{5B.3}$$

where t_j is the genomic position of the jth SNP.

Equation 5B.3 can be written in a matrix form:

$$\begin{bmatrix} \xi_{11} & \cdots & \xi_{1J} \\ \vdots & \vdots & \vdots \\ \xi_{n1} & \cdots & \xi_{nJ} \end{bmatrix} = \begin{bmatrix} x_{11} & \cdots & x_{1P} \\ \vdots & \vdots & \vdots \\ x_{n1} & \cdots & x_{nP} \end{bmatrix} \begin{bmatrix} \phi_1(t_1) & \cdots & \phi_J(t_1) \\ \vdots & \vdots & \vdots \\ \phi_1(t_P) & \cdots & \phi_J(t_P) \end{bmatrix}$$

or

$$\xi = X\Phi. \tag{5B.4}$$

Using Equation 5B.4, we can obtain the regression coefficient matrix in the functional linear model

$$\begin{aligned}\hat{B}_{FLM} &= \left(\xi^T\xi\right)^{-1}\xi^T Y \\ &= \left(\Phi^T X^T X\Phi\right)^{-1}\Phi^T X^T Y.\end{aligned} \tag{5B.5}$$

The sum of squares due to regression S_{reg}^{FLM} is

$$S_{reg}^{FLM} = \left(Y^TY\right)^{-1} Y^T X\Phi\left(\Phi^T X^T X\Phi\right)^{-1}\Phi^T X^T Y. \tag{5B.6}$$

By the similar argument, R_{FCCA} in the FCCA is

$$\begin{aligned}\hat{R}_{FCCA} &= \left(Y^TY\right)^{-1} Y^T\xi\left(\xi^T\xi\right)^{-1}\xi^T Y \\ &= \left(Y^TY\right)^{-1} Y^T X\Phi\left(\Phi^T X^T X\Phi\right)^{-1}\Phi^T X^T Y.\end{aligned} \tag{5B.7}$$

EXERCISES

Exercise 5.1 Prove Equation 5.7:

$$\sigma_G^{j,k} = \sigma_a^{j,k} + \sigma_d^{j,k}.$$

Exercise 5.2 Show that the least square estimator $\hat{B} = \left(H^T H\right)^{-1} H^T Y$ minimizes the generalized variance $|(Y - HB)^T(Y - HB)|$.

Exercise 5.3 We assume that $\varepsilon_i = [\varepsilon_{i1}, \ldots, \varepsilon_{iK}]$ is normally distributed with mean zero and covariance matrix Σ and $\varepsilon_1, \ldots, \varepsilon_K$ are independent. Show that

$$\operatorname{cov}\left(\operatorname{vec}\left(Y\right)\right) = \Sigma \otimes I.$$

Exercise 5.4 Show Result 5.8.

Exercise 5.5 Extend the multivariate linear model

$$Y = H_1B_1 + H_2B_2 + \varepsilon$$

for the one marker to the p markers.

Exercise 5.6 Extend the test statistic in Equation 5.60 and its null distribution to the p markers.

Exercise 5.7 Verify

$$K_x H K_y = \sum_{l=1}^{n}\sum_{i=1}^{n}\sum_{j=1}^{n} \alpha_i \beta_j \left(k_x(x_l, x_i) - \frac{1}{n} 1^T K_x(., x_i) \right)\left(k_y(y_l, y_j) - 1^T K_y(., y_i)\right).$$

Exercise 5.8 Show that using the Lagrangian multiplier method to solve the optimization problem (5.111), we obtain the following eigenequation:

$$\begin{bmatrix} 0 & K_x H K_y \\ K_y H K_x & 0 \end{bmatrix}\begin{bmatrix} \alpha \\ \beta \end{bmatrix} = \lambda \begin{bmatrix} K_x H K_x & 0 \\ 0 & K_y H K_y \end{bmatrix}\begin{bmatrix} \alpha \\ \beta \end{bmatrix}.$$

Exercise 5.9 Show in Equation 5.114 that $\lambda = \mu$.

Exercise 5.10 Show that eigenequation (5.115b) can be further reduced to

$$(K_x H K_x + \eta I)^{-1/2} K_x H K_y (K_y H K_y + \eta I)^{-1} K_y H K_x (K_x H K_x + \eta I)^{-1/2} u = \lambda^2 u,$$

where

$$u = (K_x H K_x + \eta I)^{1/2} \alpha,$$

$$\beta = (K_y H K_y + \eta I)^{-1} (K_y H K_x)(K_x H K_x + \eta I)^{-1/2} u.$$

Exercise 5.11 It is known that the proportion of total variance due to the first principal component is

$$\frac{\lambda_1}{\lambda_1 + \lambda_2 + \cdots + \lambda_k}.$$

Please show that to increase the proportion of the first principal component in the total variation of the low-dimensional space, we can reduce every eigenvalue by a constant, γ. In other words, we have

$$\frac{\lambda_1 - \gamma}{\lambda_1 - \gamma + \lambda_2 - \gamma + \cdots + \lambda_k - \gamma} > \frac{\lambda_1}{\lambda_1 + \lambda_2 + \cdots + \lambda_k}.$$

Exercise 5.12 Compare the summation of the canonical correlations of the FCCA and QFCCA: $\text{Trace}(R)$ and $\text{Trace}(\tilde{R})$.

Exercise 5.13 Assume $a_n \geq 0$ for all $n \geq 0$. Show

$$K\left(X, X^{'}\right) = \sum_{n=1}^{\infty} a_n <X, X^{'}>^n$$

defines a kernel.

Exercise 5.14 Let $Y^* = YV_r$. Show $S_{reg}^* = V_r^T S_{reg} V_r$.

Exercise 5.15 Let $Y = U\Lambda V^T$. Show that

$$R_Y^2 = V\Lambda^{-1}U^T X\left(X^T X\right)^{-1} X^T U\Lambda V^T.$$

Exercise 5.16 Suppose that

$$Y = \begin{bmatrix} U_r & U_{K-r} \end{bmatrix} \begin{bmatrix} \Lambda_r & 0 \\ 0 & \Lambda_{K-r} \end{bmatrix} \begin{bmatrix} V_r^T \\ V_{K-r}^T \end{bmatrix}.$$

Show

$$V_r^T V \Lambda^2 V^T V_r = \Lambda_r^2,$$

$$\Lambda_r^{-2} V_r^T V \Lambda U^T = \Lambda_r^{-1} U_r^T,$$

and

$$U\Lambda V^T V_r = U_r \Lambda_r.$$

Exercise 5.17 Show

$$\Lambda_r^{-1} U_r U \Lambda V^T = V_r^T$$

and

$$V\Lambda^{-1} U^T U_r \Lambda_r = V_r.$$

CHAPTER 6

Family-Based Association Analysis

Population-based sample design is the current major study design for association studies. However, many rare variants are from recent mutations in pedigrees (Chakravarti 2011; Lupski et al. 2011; Najmabadi et al. 2011). The inability of common variants to account for most of the supposed heritability and the low power of population-based analysis tests for the association of rare variants have led to a renewed interest in family-based design with enrichment for risk alleles to detect the association of rare variants. It is hypothesized that an individual's disease risk is likely to come from the collected action of common variants segregating in the population and rare variants recently arising in extended pedigrees.

Family-based designs have several remarkable features over the population-based association studies (Wijsman 2016). Family data convey more information than case–control data. Family data not only include genetic information across the genome but also contain correlation between individuals. The segregation of rare variants in families offers information on multiple copies of the segregated rare variants. Family data provide rich information on the transmission of genetic variants from generation to generation, which will improve accuracy for imputation of rare variants.

The classical method for genetic analysis in the pedigree-based designs is linkage analysis that analyzes cosegregation of the traits and genetic variants within pedigrees. Linkage analysis utilizes linkage segregation information on the pedigree but ignores linkage disequilibrium information in the populations. There is increasing consensus that for common variants, association analyses are more powerful than linkage analysis to detect risk alleles (Ott et al. 2015). However, both cosegregation of traits and markers in pedigrees and association between trait and genetic variants in populations carry complementary but useful information (Wijsman 2016). It is increasingly recognized that analyzing samples from populations and pedigrees separately is highly inefficient. It is natural to unify population and family study designs for association studies. The unified approach can correct for unknown population stratification, family structure, and cryptic relatedness while maintaining high power in the sequence-based association studies. In this chapter, we will mainly focus on the statistical methods for a unified approach to the genetic analysis, which can utilize both linkage and linkage disequilibrium information and can be applied to both population-based and family-based designs.

6.1 GENETIC SIMILARITY AND KINSHIP COEFFICIENTS

6.1.1 Kinship Coefficients

An individual receives at any given locus a copy of a randomly chosen one of the two alleles from his father and (independently) a copy of a randomly chosen one of the two alleles from his mother. Therefore, two individuals may receive alleles that are from the same ancestral origin. Sharing ancestral genetic materials is characterized via a concept of identity by descent (IBD). The concept of IBD is originally designed for measuring the shared genetic materials. The shared genetic materials can be an allele in an SNP or can be a segment of a genome. Two alleles are IBD if one is a copy of the other or if they are both copies of the same allele present in some remote ancestor. In Figure 6.1, father transmits allele 1 to both sons, but mother transmits allele 3 to one son and allele 4 to another son. Two sons share allele 1 transmitted from father, and hence they have one allele IBD. Alleles that are IBD will have the same base sequences; however, the alleles with the same base sequences may not be IBD.

The first quantity for measuring IBD is the inbreeding coefficient. The inbreeding coefficient studies IBD of two alleles of a single individual. The individuals that contain pairs of IBD alleles are called inbred. The **inbreeding coefficient** is defined as the probability that his or her two alleles at any autosomal locus are IBD. The inbreeding coefficient of an individual, i, is denoted by f_i. Let P_A and P_a be frequencies of alleles A and a, respectively, and P_{AA} be the frequency of the genotype AA. Then, the genotype frequency P_{AA} can be written as

$$P_{AA} = (1-f)P_A^2 + fP_A = P_A^2 + fP_A(1-P_A). \tag{6.1}$$

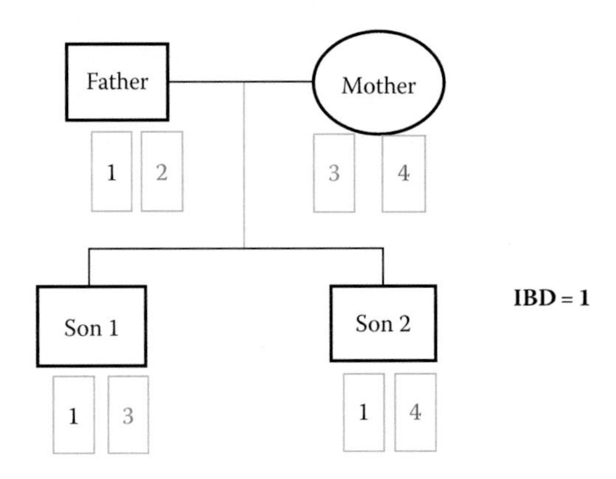

FIGURE 6.1 IBD sharing.

The second quantity for measuring IBD is a kinship coefficient. The kinship coefficient studies IBD of two alleles of two individuals. The kinship coefficient Φ_{ij} between two individuals, i and j, is defined as the probability that an allele selected randomly from individual i and an allele selected randomly from the same autosomal locus of individual j are IBD.

The inbreeding coefficient and kinship coefficient have the following relation:

$$\Phi_{ii} = \frac{1}{2}(1 + f_i) \tag{6.2}$$

and

$$f_i = \Phi_{kl}, \tag{6.3}$$

where k and l are parents of individual i.

Example 6.1 Parent and Offspring

The kinship coefficient between the parent and offspring is $\Phi_{ij} = \frac{1}{4}$.

Example 6.2 Full Sibs

Consider parents, 1 and 2, and their children, 3 and 4, as shown in Figure 6.2. Suppose that allele A is randomly selected from child 4 with probability $\frac{1}{2}$ and the probability of randomly selecting allele A from child 3 and allele A from the father is the kinship coefficient Φ_{31}. Therefore, the probability of sharing allele A between two sibs is $\frac{1}{2}\Phi_{31}$.

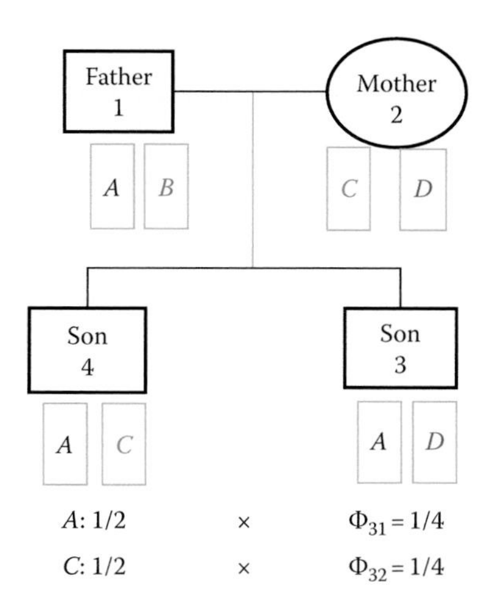

FIGURE 6.2 Scheme for calculation of kinship coefficient.

Similarly, the probability of sharing allele C between two sibs is $\frac{1}{2}\Phi_{32}$. The kinship coefficient between two sibs is

$$\Phi_{34} = \frac{1}{2}\Phi_{31} + \frac{1}{2}\Phi_{32} = \frac{1}{4}.$$

Example 6.3 Half Sibs

Consider half sibs (Figure 6.3). Take paternal allele A in child 4. The probability of sharing allele A between two children, 4 and 5, should be equal to $\frac{1}{2}\Phi_{51}$. Similarly, consider maternal allele B in child 4. The probability of sharing allele B between two children, 4 and 5, should be equal to $\frac{1}{2}\Phi_{52}$. The kinship coefficient between two half sibs is

$$\Phi_{45} = \frac{1}{2}\Phi_{51} + \frac{1}{2}\Phi_{52} = \frac{1}{2}\times 0 + \frac{1}{2}\times\frac{1}{4} = \frac{1}{8}.$$

Example 6.4 First Cousins

We can show that the kinship coefficient of the first cousins is $\frac{1}{16}$ (Exercise 6.2).

The general procedure for calculating kinship coefficients between members of a pedigree is given below:

1. Order members in the pedigree such that every parent precedes his or her children.
2. The kinship coefficients are calculated from the left top downwards recursively.
3. The recursive calculation formulas are given by

$$\Phi_{ij} = \begin{cases} 0 & \text{if } i \text{ and } j \text{ are founders} \\ \frac{1}{2} & i = j, i \text{ is a founder} \\ \frac{1}{2}\left(\Phi_{jk} + \Phi_{jl}\right) & k \text{ and } l \text{ are parents of } i \\ \frac{1}{2}\left(1 + \Phi_{kl}\right) & i = j, k, \text{and } l \text{ are parents of } i. \end{cases}$$

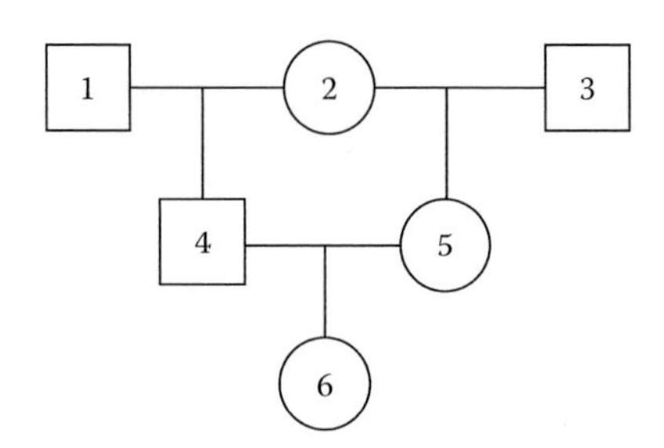

FIGURE 6.3 A three-generation pedigree.

Example 6.5 A Three-Generation Pedigree (Figure 6.3)

Applying the general procedure for computing kinship coefficients to pedigree in Figure 6.3, we obtain the following kinship coefficient matrix:

$$\Phi = \begin{bmatrix} \frac{1}{2} & 0 & 0 & \frac{1}{4} & 0 & \frac{1}{8} \\ 0 & \frac{1}{2} & 0 & \frac{1}{4} & \frac{1}{4} & \frac{1}{4} \\ 0 & 0 & \frac{1}{2} & 0 & \frac{1}{4} & \frac{1}{8} \\ \frac{1}{4} & \frac{1}{4} & 0 & \frac{1}{2} & \frac{1}{8} & \frac{5}{16} \\ 0 & \frac{1}{4} & \frac{1}{4} & \frac{1}{8} & \frac{1}{2} & \frac{5}{16} \\ \frac{1}{8} & \frac{1}{4} & \frac{1}{8} & \frac{5}{16} & \frac{5}{16} & \frac{9}{16} \end{bmatrix}.$$

6.1.2 Identity Coefficients

Consider a single locus with two alleles in two individuals, i and j. There are $C_4^4 + C_4^3 + C_4^2 + 3 + C_4^0 = 15$ possible configurations of identity by descent for four alleles as shown in Figure 6.4. The identity of two alleles can exist within an individual or between individuals. For example, in S_1, the identity of two alleles exists within an individual and between individuals. Figure 6.4 describes the relationships between the identity states and the condensed identity states. Now we define the condensed coefficients of identity and the probability of the condensed identity states:

$$\Delta_1 = P(S_1),\ \ \Delta_2 = P(S_6),\ \ \Delta_3 = P(S_2 \cup S_3),\ \ \Delta_4 = P(S_7),\ \ \Delta_5 = P(S_4 \cup S_5),\ \ \Delta_6 = P(S_8),$$
$$\Delta_7 = P(S_9 \cup S_{12}),\ \ \Delta_8 = P(S_{10} \cup S_{11} \cup S_{13} \cup S_{14}),\ \text{ and }\ \Delta_9 = P(S_{15}).$$

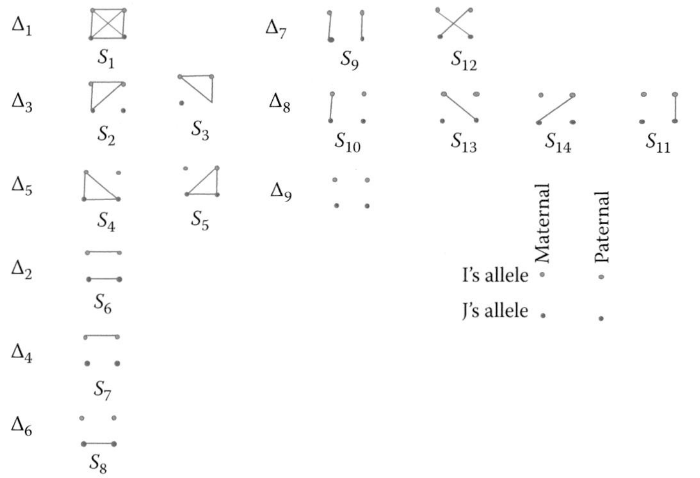

FIGURE 6.4 Fifteen possible configurations of identity by descent for four alleles in two individuals.

The condensed coefficient of identity depends on the genetic relationships between two individuals. For example, consider noninbred full sibs. Since the probability of inheriting the same paternal allele is 0.5, the probability of inheriting the same maternal allele is 0.5, and they are independent, then we have $P(S_9)=0.5*0.5=0.25$. Also, it is impossible for the noninbred full sibs to inherit the same paternal and maternal allele. Thus, $P(S_{12})=0$ and $\Delta_7=P(S_9)+P(S_{12})=0.25$. By the same arguments, we have $P(S_{10})=P(S_{11})=0.25$ and $P(S_{13})=P(S_{14})=0$. Thus, we have $\Delta_8=P(S_{10}\bigcup S_{11}\bigcup S_{13}\bigcup S_{14})=0.25+0.25+0+0=0.50$. We note that the probabilities of not inheriting paternal alleles and not inheriting maternal alleles are 0.5 and 0.5, respectively. Two events are independent. Therefore, we have $\Delta_9=P(S_{15})=0.5*0.5=0.25$.

6.1.3 Relation between Identity Coefficients and Kinship Coefficients

It is easy to see that $\Delta_2=\Delta_4=\Delta_6=\Delta_9=0$. Kinship coefficient can be calculated by conditioning on the identity states. For example, conditioning on $S_2\bigcup S_3$, the probability that two alleles that are drawn randomly from individuals i and j are IBD is $\frac{1}{2}$. Similar arguments can be applied to $S_4\bigcup S_5$ and $S_9\bigcup S_{12}$. Conditioning on $S_{10}\bigcup S_{11}\bigcup S_{13}\bigcup S_{14}$ and S_1, the probabilities that two alleles that are drawn randomly from individuals i and j are IBD are $\frac{1}{4}$ and 1, respectively. Therefore, the kinship coefficient is given by (Lynch and Walsh 1998)

$$\Phi_{ij}=\Delta_1+\frac{1}{2}\left(\Delta_3+\Delta_5+\Delta_7\right)+\frac{1}{4}\Delta_8. \tag{6.4}$$

Next we calculate Φ_{ii} (Figure 6.5). The kinship coefficient Φ_{ii} is the probability that two randomly drawn alleles from individual i are IBD. We assume that individual i has two alleles, A_1 and A_2, at the locus. We randomly draw an allele from the locus, replace it, and then randomly draw another allele. The four possibilities are shown in Figure 6.5. Suppose that we first draw allele A_1 with the probability $\frac{1}{2}$. Then, we replace A_1 and randomly draw the second allele. There are two possibilities. We may still draw allele A_1 with probability $\frac{1}{2}$, or we draw allele A_2 with probability $\frac{1}{2}$. The probability that two alleles A_1 and A_2 are IBD is equal to 1. Let f_i be the inbreeding coefficient of individual i. The probability that alleles A_1 and A_2 is f_i.

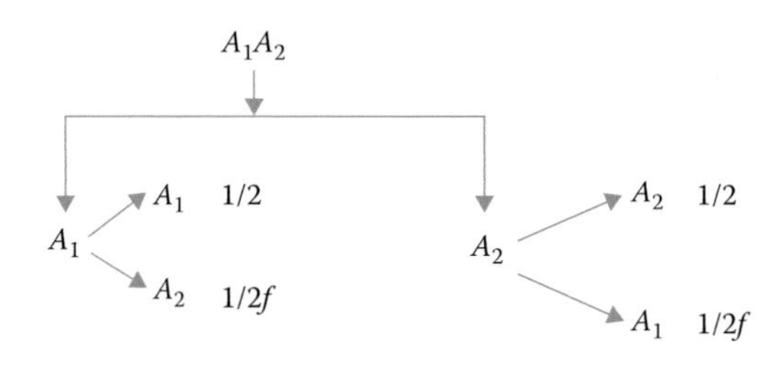

FIGURE 6.5 Scheme for calculation of substitution effect.

Therefore, the probability along the left paths in Figure 6.5 is $\frac{1}{4}+\frac{1}{4}f_i$. By similar arguments, the probability along the right path in Figure 6.5 is also $\frac{1}{4}+\frac{1}{4}f_i$. The kinship Φ_{ii} is

$$\Phi_{ii}=\frac{1}{4}+\frac{1}{4}f_i+\frac{1}{4}+\frac{1}{4}f_i=\frac{1}{2}+\frac{1}{2}f_i. \tag{6.5}$$

Next we calculate Φ_{ij}. In the following calculations, we assume no inbreeding. Therefore, we have $\Delta_i=0,\ i=1,\ldots,6$. We first consider the kinship Φ_{po} between a parent and its offspring. Their identity states are characterized by $S_{10}\bigcup S_{11}\bigcup S_{13}\bigcup S_{14}$. If we assume that the parent is the mother, the identity states are S_{10} and S_{14}. To form the identity state S_{10}, we need to draw a maternal allele from the mother with probability $\frac{1}{2}$ and randomly draw a maternal allele from its offspring with probability $\frac{1}{2}$, and thus the probability $P(S_{10})=\frac{1}{2}\times\frac{1}{2}=\frac{1}{4}$. Similarly, we obtain $P(S_{14})=\frac{1}{4}$. Next we assume that the parent is the father. By a similar argument, $P(S_{11})=\frac{1}{4}$ and $P(S_{13})=\frac{1}{4}$. Therefore, we obtain $\Delta_8=P(S_{10}\bigcup S_{11}\bigcup S_{13}\bigcup S_{14})=1$ and $\Delta_1=\Delta_3=\Delta_5=\Delta_7=0$. Substituting these quantities into Equation 6.4 gives $\Phi_{p0}=\frac{1}{8}$.

We then calculate the kinship Φ_{hs} between half sibs. The identity states of the half sibs are S_{10} and S_{14}. Suppose that the mother is shared between two half sibs. We randomly draw a maternal allele from one half sib with probability $\frac{1}{2}$. The probability that the allele randomly drawn from another half sib is IBD with the first drawn allele is $\frac{1}{2}$. Thus, $P(S_{10})=\frac{1}{2}\times\frac{1}{2}=\frac{1}{4}$. If we assume that the father is shared by two half sibs, then by a similar argument, we obtain $P(S_{14})=\frac{1}{4}$. As a consequence, we have $\Delta_8=P(S_{10})+P(S_{14})=\frac{1}{2}$. Other identity coefficients are equal to zero. Therefore, using Equation 6.4, we obtain the kinship coefficient between half sibs: $\Phi_{hs}=\frac{1}{4}\Delta_8=\frac{1}{8}$.

By the similar arguments, we calculate the kinship coefficients for other simple pedigrees, which are summarized in Table 6.1.

TABLE 6.1 Identity Coefficients for Simple Pedigrees under the Assumption of No Inbreeding

Relationship	Δ_7	Δ_8	Δ_9	Φ	Δ
Parent–offspring	0	1	0	1/4	0
Full siblings	1/4	1/2	1/4	1/4	1/4
Half siblings	0	1/2	1/2	1/8	0
First cousins	0	1/4	3/4	1/16	0
Double first cousins	1/16	6/16	9/16	1/8	1/16
Second cousins	0	1/16	15/16	1/64	0
Uncle–nephew	0	1/2	1/2	1/8	0

6.1.4 Estimation of Genetic Relations from the Data

Genetic relation information includes both recent genetic relation information such as pedigree information and distant genetic relations such as population structure (Conomos et al. 2016). Theoretic estimation of kinship coefficients in the presence of population structure has a limitation. Therefore, we study the direct estimation of recent genetic relatedness from the data.

6.1.4.1 A General Framework for Identity by Descent

To reveal the genetic relations, we first introduce indicator variables for measuring genetic information (Weir and Hill 2002; Zheng and Weir 2016). Consider two alleles at each locus. Define x_{ijkl} as an allele indicator variable for the kth allele, $k=1,2$, at the lth locus in the jth individual sampled from the ith population, $j=1,\ldots,n_i$; $i=1,\ldots,n$, i.e.,

$$x_{ijkl} = \begin{cases} 1 & k\text{th allele, } l\text{th locus, } j\text{th individual, } i\text{th population} \\ 0 & \text{otherwise.} \end{cases}$$

Let g_{ijl} be the genotype of the jth individual from the ith population at the lth locus. We have $g_{ijl}=x_{ij1l}+x_{ij2l}$. The expectations of the indicator variables are given by

$$E\left[x_{ijkl}\right] = p_l, \tag{6.6}$$

where p_l is the overall or ancestral frequency of the reference allele k in the single reference population.

Similarly, we obtain

$$E\left[x_{ijkl}^2\right] = E\left[x_{ijkl}\right] = p_l \tag{6.7}$$

and a variance of x_{ijkl}:

$$\text{var}\left(x_{ijkl}\right) = E\left[x_{ijkl}^2\right] - \left(E\left[x_{ijkl}\right]\right)^2 = p_l - p_l^2 = p_l\left(1-p_l\right). \tag{6.8}$$

Now we calculate the covariance between indicator variables of the alleles. Let F_{ij} be the total inbreeding coefficient of the jth individual in the ith population, Φ_i or Φ_{ii} be the within kinship coefficient, and $\Phi_{ii'}$ be the kinship coefficient between the ith population and i'th population. Then, we have

$$E\left[x_{ijkl}x_{ijk'l}\right] = \left(1-F_{ij}\right)p_l^2 + F_{ij}p_l = p_l^2 + p_l\left(1-p_l\right)F_{ij} \tag{6.9}$$

and

$$\begin{aligned} \text{cov}\left(x_{ijkl}, x_{ijk'l}\right) &= E\left[x_{ijkl}x_{ijk'l}\right] - E\left[x_{ijkl}\right]E\left[x_{ijk'l}\right] \\ &= p_l^2 + p_l\left(1-p_l\right)F_{ij} - p_l^2 = p_l\left(1-p_l\right)F_{ij}. \end{aligned} \tag{6.10}$$

Next we consider the covariance of the indicator variables for the alleles from the different populations. By similar arguments, we have

$$E\left[x_{ijkl}x_{ij'k'l}\right]=(1-\Phi_i)p_l^2+\Phi_i p_l=p_l^2+p_l(1-p_l)\Phi_i, \tag{6.11}$$

and

$$\begin{aligned}\operatorname{cov}\left(x_{ijkl},x_{ij'k'l}\right)&=E\left[x_{ijkl}x_{ij'k'l}\right]-E\left[x_{ijkl}\right]E\left[x_{ij'k'l}\right]\\&=p_l^2+p_l(1-p_l)\Phi_i-p_l^2=p_l(1-p_l)\Phi_i.\end{aligned} \tag{6.12}$$

$$E\left[x_{ijkl}x_{i'j'k'l}\right]=p_l^2+p_l(1-p_l)\Phi_{ii'}, \tag{6.13}$$

and

$$\operatorname{cov}\left(x_{ijkl},x_{i'j'k'l}\right)=p_l(1-p_l)\Phi_{ii'}. \tag{6.14}$$

Next we investigate the expectation and variance of allele frequencies in the population. For the simplicity of discussion, we assume that each population has only one sampled individual. In this case, we denote $F_{ij}=\Phi_i$. Define the average frequency the alleles at the lth locus in the ith population as

$$\bar{p}_{il}=\frac{1}{2}\left(x_{ij1l}+x_{ij2l}\right) \tag{6.15}$$

and the average allele frequency at the lth locus as

$$\bar{p}_l=\frac{1}{n}\sum_{i=1}^{n}\bar{p}_{il}. \tag{6.16}$$

Then, it is clear that

$$E[\bar{p}_l]=\frac{1}{n}\sum_{i=1}^{n}E[\bar{p}_{il}]=p_l. \tag{6.17}$$

Using Equations 6.8, 6.10, and 6.15 gives

$$\begin{aligned}\operatorname{var}(\bar{p}_{il})&=\frac{1}{4}\left[\operatorname{var}(x_{ij1l})+\operatorname{var}(x_{ij2l})+2\operatorname{cov}(x_{ijj1l},x_{ij2l})\right]\\&=\frac{1}{4}\left[2p_l(1-p_l)+2p_l(1-p_l)\Phi_i\right]=\frac{1}{2}p_l(1-p_l)(1+\Phi_i).\end{aligned} \tag{6.18}$$

$$\begin{aligned}\operatorname{cov}\left(\bar{p}_{il}, \bar{p}_{i'l}\right) &= \frac{1}{4}\operatorname{cov}\left(x_{ij1l} + x_{ij2l}, x_{i'j1l} + x_{i'j2l}\right) \\ &= \frac{1}{4} 4 p_l\left(1-p_l\right)\Phi_{ii'} = p_l\left(1-p_l\right)\Phi_{ii'}.\end{aligned} \tag{6.19}$$

Now we are ready to calculate the variance of the average frequency $\bar{p}_l$. By definition, we have

$$\operatorname{var}\left(\bar{p}_l\right) = \frac{1}{n^2}\left[\sum_{i=1}^{n}\operatorname{var}\left(\bar{p}_{il}\right) + 2\sum_{i=1}^{n}\sum_{i'=1, i'\neq i}^{n}\operatorname{cov}(\bar{p}_{il}, \bar{p}_{i'l})\right]. \tag{6.20}$$

Substituting Equations 6.18 and 6.19 into Equation 6.20, we obtain

$$\begin{aligned}\operatorname{var}\left(\bar{p}_l\right) &= \frac{1}{n^2}\left[\sum_{i=1}^{n}\frac{1}{2}p_l\left(1-p_l\right)\left(1+\Phi_i\right) + 2\sum_{i=1}^{n}\sum_{i'=1, i'\neq i}^{n} p_l\left(1-p_l\right)\Phi_{ii'}\right] \\ &= \frac{1}{2n}p_l\left(1-p_l\right)\left(1+\Phi_I\right) + \frac{n-1}{n}p_l\left(1-p_l\right)\Phi_T,\end{aligned} \tag{6.21}$$

where $\Phi_I = \frac{1}{n}\sum_{i=1}^{n}\Phi_i$ is the average inbreeding coefficient and $\Phi_T = \frac{1}{n(n-1)}\sum_{i=1}^{n}\sum_{i'=1,i\neq i'}^{n}\Phi_{ii'}$ the average kinship coefficient.

Using Equation 6.21, we can easily calculate $E\left[\bar{p}_l\left(1-\bar{p}_l\right)\right]$:

$$\begin{aligned}E\left[\bar{p}_l\left(1-\bar{p}_l\right)\right] &= E\left[\bar{p}_l\right] - E\left[\bar{p}_l^2\right] \\ &= p_l - \left(\operatorname{var}\left(\bar{p}_l\right) + \left(E\left[\bar{p}_l\right]\right)^2\right) \\ &= \frac{n-1}{n}p_l\left(1-p_l\right)\left(1-\Phi_T\right) + \frac{1}{2n}p_l\left(1-p_l\right)\left(1-\Phi_I\right).\end{aligned} \tag{6.22}$$

6.1.4.2 Kinship Matrix or Genetic Relationship Matrix in the Homogeneous Population

Although the genealogy relationship between individuals in the same pedigrees can be directly specified, the relationships between individuals in different pedigrees are usually unknown. In the presence of hidden population substructures and cryptic relatedness in the samples, the genealogy relationships between individuals in the different pedigrees cannot be ignored. The kinship matrix or genetic relationship matrix includes both the pedigree relationships of the related individuals and population structures. In general, the kinship matrix or genetic relationship matrix is unknown and can be estimated by the genetic variants in the data.

Let G be an empirical genetic relationship matrix (GRM):

$$\hat{G}_{ik} = \frac{1}{L}\frac{1}{n_i n_k}\sum_{l=1}^{L}\sum_{u=1}^{n_i}\sum_{v=1}^{n_k}\frac{(g_{iul}-2\bar{p}_l)(g_{kvl}-2\bar{p}_l)}{2\bar{p}_l(1-\bar{p}_l)}, \tag{6.23}$$

where n_i and n_k are the respective numbers of individuals sampled from the ith and kth populations.

The expectation of $\hat{\Phi}_{ik}$ is given by

$$E\left[\hat{G}_{ik}\right] = \frac{1}{L}\sum_{l=1}^{L}\frac{E\left[(g_{i1l}-2\bar{p}_l)(g_{k1l}-2\bar{p}_l)\right]}{2\bar{p}_l(1-\bar{p}_l)}. \tag{6.24}$$

We can show that (Appendix 6A)

$$E\left[(g_{i1l}-2\bar{p}_l)^2\right] = 2p_l(1-p_l)\left(1+\Phi_i+2\frac{n-1}{n}\Phi_T-4\psi_i\right)+\frac{2}{n}p_l(1-p_l)(\Phi_I+2\Phi_i-1), \tag{6.25}$$

$$E\left[(g_{i1l}-2\bar{p}_l)(g_{k1l}-2\bar{p}_l)\right] = 4p_l(1-p_l)\left(\Phi_{ik}+\frac{n-1}{n}\Phi_T-\psi_i-\psi_k\right) + \frac{2}{n}p_l(1-p_l)(\Phi_I+\Phi_i+\Phi_k-1), \tag{6.26}$$

where $\psi_i = \sum_{k=1}^{n}\frac{\Phi_{ik}}{n}$, $\Phi_{ii} = \Phi_i$.

When the number of sampled individuals is large, Equations 6.25 and 6.26 can be reduced to

$$\frac{E\left[(g_{i1l}-2\bar{p}_l)^2\right]}{\bar{p}_l(1-\bar{p}_l)} \approx \frac{2(1+\Phi_i)}{1-\Phi_T}+4\frac{\Phi_T-2\psi_i}{1-\Phi_T}, \tag{6.27}$$

$$\frac{E\left[(g_{i1l}-2\bar{p}_l)(g_{k1l}-2\bar{p}_l)\right]}{\bar{p}_l(1-\bar{p}_l)} \approx \frac{4\Phi_{ik}}{1-\Phi_T}+4\frac{\Phi_T-\psi_i-\psi_k}{1-\Phi_T}. \tag{6.28}$$

6.1.4.3 Kinship Matrix or Genetic Relationship Matrix in the General Population

6.1.4.3.1 Genetic Models and Estimation of Recent and Distant Genetic Relatedness To study the variance and covariance of the genotype frequencies in the general population, we first need to specify genetic models in the general population (Conomos et al. 2016). The general population consists of the current population, which the individuals are sampled from; the recent ancestral population; and the common ancestral population (Figure 6.6). Assume that n individuals are sampled at time t_{now} from a current structured population descended

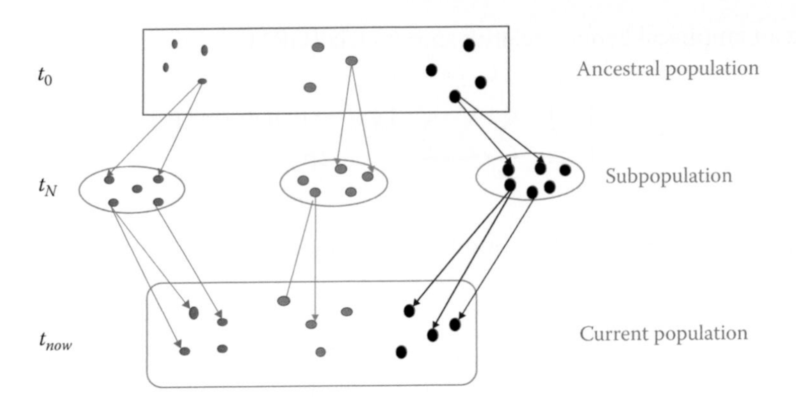

FIGURE 6.6 Population structure.

from N distinct subpopulations at t_N. We further assume that these N subpopulations came from a common ancestral population at t_0. The genetic background for n sampled individuals is that the genome of the ith individual in the current population was inherited from the N subpopulations with proportions $a_i = \left[a_i^1, \ldots, a_i^N\right]^T$, $a_i^k \ge 0$, and $\sum_{k=1}^{N} a_i^k = 1$. Consider L loci. Let $p_l^k, l = 1, \ldots, L, k = 1, \ldots, N$ be the reference allele frequency at the l locus in the kth population, and $\mathrm{P}_l = \left[p_l^1, \ldots, p_l^N\right]^T$ be the vector of subpopulation-specific allele frequencies. Suppose that p_l^k are random variables. Since the N subpopulations descended from a common ancestral population, using Equations 6.6 and 6.14, we obtain the mean and covariance of the population-specific allele frequencies:

$$E\left[\mathrm{P}_l\right] = p_l \mathbf{1}$$

and

$$\operatorname{var}\left(\mathrm{P}_l\right) = p_l\left(1 - p_l\right)\Phi_N, \tag{6.29}$$

where

1 is a N dimensional column vector of 1

$\mathbf{\Phi}_N$ is a $N \times N$ matrix that defines the within and between subpopulation correlations of alleles

The GRM can be used to estimate the population structure (distant genetic relatedness) and the kinship coefficients (recent genetic relatedness). Now we study its asymptotic properties under the general genetic models. The entries $\hat{G}_{ik}$ in the GRM measure the genotype correlations for a pair of individuals, i, j.

We can show that (Appendix 6B)

$$E\left[g_{il} g_{jl}\right] = 4p_l^2 + 4p_l\left(1 - p_l\right)\left(\phi_{ij} + \theta_{ij} - \sum_{m \in A_{ij}} \phi_{ij|m} \theta_{mm}\right). \tag{6.30}$$

$$E\left[\hat{G}_{ik}\right] \approx 2\left(\phi_{ij} + \theta_{ij} - \sum_{m \in A_{ij}} \phi_{ij|m}\theta_{mm}\right), \tag{6.31}$$

where

ϕ_{ij} is the kinship coefficient between individuals i and j

$\theta_{ij} = a_i^T \Phi_N a_j$ is the coancestry coefficient between individuals i,j due to population structure

A_{ij} is the set of most recent common ancestors of individuals i and j

$\phi_{ij|m} = \left(\frac{1}{2}\right)^{n_{mi}+n_{mj}-1}(1+f_m)$ is the kinship coefficient of the allele of the individuals i and j tracing back to the common ancestor m

$\theta_{mm} = a_m^T \Phi_N a_m$ is the coancestry coefficient between individual m with itself due to population structure

As Equation 6.31 indicates, in the general genetic models, the elements of the GRM measure both kinship coefficients ϕ_{ij} due to recent genetic similarity and the coancestry coefficients θ_{ij} and θ_{mm} due to the distant genetic similarity in the population. If we consider a homogeneous population, then $\hat{G}_{ik} \to 2\phi_{ij}$, the GRM will only estimate the kinship coefficients. In general, recent and distant genetic correlations are confounded, the estimators of the kinships coefficients and coancestry coefficient due to population structure cannot be separately estimated. Below we will introduce new statistics to estimate the kinship coefficients and coancestry coefficients in the presence of population structure separately.

6.1.4.3.2 Estimation of Recent Kinships To remove the impact of population structure on the estimation of the recent kinship coefficients, we can use individual allele frequencies p_{il} and p_{jl} to replace population allele frequency p_s in calculation of the GRM. In other words, the kinship coefficient ϕ_{ij} will be estimated by

$$\hat{\phi}_{ij} = \frac{\sum_{l=1}^{L}(g_{il} - 2p_{il})(g_{jl} - 2p_{jl})}{4\sum_{l=1}^{L}\sqrt{p_{il}(1-p_{il})p_{jl}(1-p_{jl})}}. \tag{6.32}$$

In Appendix 6C, we show that

$$\hat{\phi}_{ij} \to \phi_{ij} - \sum_{m \in A_{ij}} \phi_{ij|m} \frac{\theta_{mm} - d_\phi(i,j)}{1 - d_\phi(i,j)}, \tag{6.33}$$

where $d_\phi(i,j)$ is defined in Equation 6C.8.

The estimator of the kinship coefficients, which uses genotype values centered and scaled by individual-specific allele frequencies, measures the genetic correlation due to alleles shared IBD between individuals i,j from recent common ancestors and removes the population structure from the kinship estimation.

To use Equation 6.32 for kinship estimation, we need to estimate the individual-specific allele frequencies. Regression of the genotypes on the principal components (PCs) can be used to estimate the individual-specific allele frequencies (Conomos et al. 2016). The PCs contain information on population structure. Let $V=[V^1, \ldots, V^D]$ be an $n \times D$ matrix with column vectors being the top D PCs. Let g_l be a vector of the genotype values of n individuals at SNP: l. Define the linear regression model

$$g_l = \mathbf{1}\beta_0 + V\beta, \tag{6.34}$$

where

$\beta=[\beta_1, \ldots, \beta_D]^T$ is a vector of regression coefficients for each of the PCs
$\mathbf{1}$ is a n-dimensional vector of 1s

Regression of g_l on the PCs is equivalent to the regression of g_l on the true ancestries of the sampled individuals. Therefore, the regression model (6.34) can be used to estimate the individual-specific allele frequencies:

$$\hat{p}_{il} = \frac{1}{2}\left(\hat{\beta}_0 + \sum_{d=1}^{D} \hat{\beta}_d V_i^d\right), \tag{6.35}$$

where V_i^d is the PC score of individual i. Since for each d, the average of PC score V^d is zero, $\frac{1}{2}\hat{\beta}_0$ is equal to the sample average allele frequency or population allele frequency p_l. The regression coefficients $\hat{\beta}_d$ measures deviation in allele frequency from the population allele frequency due to the ancestry components. We substitute the estimation of individual-specific allele frequencies $\hat{p}_{il}$ and $\hat{p}_{jl}$ into Equation 6.32 to give the PC-based estimation of kinship coefficient:

$$\hat{\phi}_{ij} = \frac{\sum_{l=1}^{L}\left(g_{il} - 2\hat{p}_{il}\right)\left(g_{jl} - 2\hat{p}_{jl}\right)}{4\sum_{l=1}^{L}\sqrt{\hat{p}_{il}\left(1-\hat{p}_{il}\right)\hat{p}_{jl}\left(1-\hat{p}_{jl}\right)}}. \tag{6.36}$$

From the analysis in Appendix 6C, we can see that for unrelated pairs of individuals, the estimated kinship coefficients will be close to zero, no matter whether the population structure is presented in the samples or not. In the presence of population structure but with no admixture among the N subpopulations, the estimated kinship coefficients are still close to the true kinship coefficients: $\hat{\phi}_{ij} \to \phi_{ij}$.

6.1.4.3.3 Estimation of Inbreeding Coefficient in the Presence of Population Structure The GRM defined in Equation 6.23 can also be used to estimate the inbreeding coefficient when we take $j=i$:

$$\hat{f}_i = \hat{G}_{ii} = \frac{1}{L}\sum_{l=1}^{L}\frac{E\left[\left(g_{il} - 2\bar{p}_l\right)^2\right]}{2\bar{p}_l\left(1-\bar{p}_l\right)}. \tag{6.37}$$

If the true P_l is known when the number of SNPs is large, then we have (Exercise 6.6)

$$\hat{G}_{ii} - 1 \rightarrow f_i\left[1 - \theta_{M(i)P(i)}\right] + \theta_{M(i)P(i)}, \tag{6.38}$$

where $M(i)$ and $P(i)$ denote the mother and father of individual i, respectively. Equation 6.38 shows that the estimator of the inbreeding coefficient converges to the true inbreeding coefficient only when the population is homogeneous. Otherwise, the estimator is not consistent then the population structure is present.

Similar to the estimation of the kinship coefficient, to remove the impact of the population structure on the estimation of inbreeding coefficient, we can use the individual-specific allele frequency to replace the population allele frequency in Equation 6.37. We can use the PC score–based linear regression to estimate the individual-specific allele frequency. Therefore, the PC-based estimator of the inbreeding coefficient f_i of individual i is given by

$$\hat{f}_i = \frac{1}{L}\sum_{l=1}^{L} \frac{E\left[\left(g_{il} - 2\hat{p}_{il}\right)^2\right]}{2\hat{p}_{il}\left(1 - \hat{p}_{il}\right)}, \tag{6.39}$$

where the individual-specific allele frequency $\hat{p}_{il}$ is estimated by Equation 6.35.

We can show that (Exercise 6.7)

$$\hat{f}_i \rightarrow f_i\left[1 - \frac{\theta_{M(i)P(i)} - \theta_{ii}}{1 - \theta_{ii}}\right] + \frac{\theta_{M(i)P(i)} - \theta_{ii}}{1 - \theta_{ii}}, \tag{6.40}$$

where $\theta_{ii} = a_i^T \Phi_N a_i$.

In the presence of discrete population substructure, $\theta_{M(i)P(i)} = \theta_{ii}$ implies that $\hat{f}_i \rightarrow f_i$. Even in the presence of general population structure, the bias, asymptotically, is small.

6.1.4.4 Coefficient of Fraternity

We have studied the identity of a single allele by descent. Now we extend the identity of a single allele by descent to the identity of genotype by descent. There are scenarios that lead to the identity of genotype by descent (Figure 6.7). Let m_x and f_x be the maternal and paternal allele of individual x, respectively. The first scenario is that both maternal allele of two individuals are identical by descent and both paternal allele of two individuals are identical by descent with probability $\phi_{m_x m_y}\phi_{f_x f_y}$. The second scenario is that the maternal allele of individual x and paternal allele of individual y and the paternal allele of individual x and maternal allele of individual y are identical by descent with probability $\phi_{m_x f_y}\phi_{f_x m_y}$. Let Δ_{xy} denote the coefficient of fraternity. The coefficient of fraternity, which is defined as the probability of the genotypes (two alleles) of individuals x and y are identical by descent, is given by

$$\Delta_{xy} = \phi_{m_x m_y}\phi_{f_x f_y} + \phi_{m_x f_y}\phi_{f_x m_y}. \tag{6.41}$$

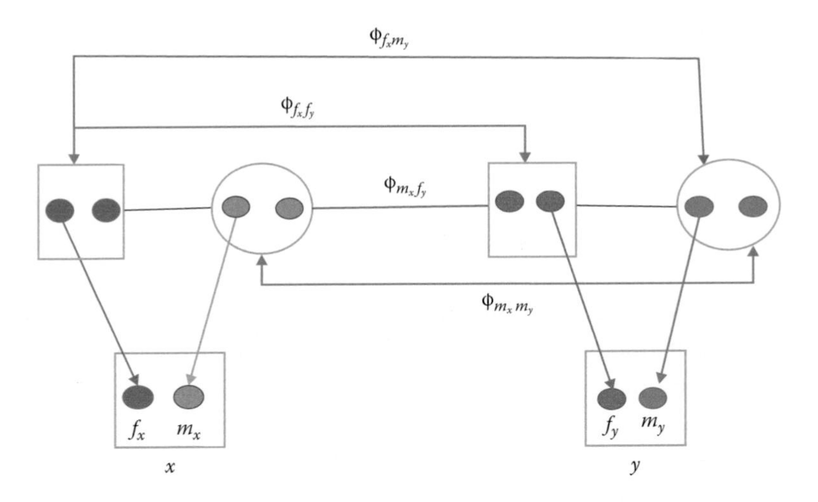

FIGURE 6.7 Genotypes of two individuals are identical by descent.

If inbreeding is not present, Equation 6.41 can be reduced to

$$\Delta_{xy} = \Delta_7 \tag{6.42}$$

Example 6.6 Full Sibs

For full sibs, both sibs share a mother ($m_x = m_y = m$) and father ($f_x = f_y = f$). Therefore, Equation 6.41 is reduced to

$$\Delta_{xy} = \phi_{mm}\phi_{ff} + \phi_{mf}^2. \tag{6.43}$$

If the parents of two individuals are unrelated and parents are not inbred, then $\phi_{mm} = \phi_{ff} = \frac{1}{2}$ and $\phi_{mf} = 0$. Thus, we have

$$\Delta_{xy} = \frac{1}{2} \times \frac{1}{2} = \frac{1}{4}.$$

6.2 GENETIC COVARIANCE BETWEEN RELATIVES

Genetic inheritance occurs between relatives and have a crucial impact on the trait covariance between relatives. In this section, we study genetic additive covariance, dominance covariance, and gene–gene interaction covariance between relatives and effects of lineage and linkage disequilibrium on the genetic covariance between relatives.

6.2.1 Assumptions and Genetic Models

We first make the following assumptions for the genetic models (Lynch and Walsh 1998):

1. All of the genetic variation is caused in autosomal loci.
2. Mating is random.
3. Without linkage between loci.

4. No maternal effects.
5. Ignore gene–environment interaction.
6. No selection.

Before we study the genetic covariance between relatives, we introduce the genetic models that are the basis for investigating the genetic covariance. Let x and y be a pair of individuals. Consider two loci, A with alleles A_i and A_j and B with alleles A_k and A_l. Let p_1, q_1, p_2, and q_2 be frequencies of alleles A_i, A_j, A_k, and A_l, respectively. Let $G_{ijkl}(x)$ and $G_{ijkl}(y)$ be the genotypic values of the genotypes A_iA_j and A_kA_l of the individuals x and y, respectively. Consider the following genetic model:

$$G_{ijkl}(x) = \mu_G + \left[\alpha_i^x + \alpha_j^x + \alpha_k^x + \alpha_l^x\right] + \left[\delta_{ij}^x + \delta_{kl}^x\right] + \left[(\alpha\alpha)_{ik}^x + (\alpha\alpha)_{il}^x + (\alpha\alpha)_{jk}^x + (\alpha\alpha)_{jl}^x\right],$$
$$G_{ijkl}(y) = \mu_G + \left[\alpha_i^y + \alpha_j^y + \alpha_k^y + \alpha_l^y\right] + \left[\delta_{ij}^y + \delta_{kl}^y\right] + \left[(\alpha\alpha)_{ik}^y + (\alpha\alpha)_{il}^y + (\alpha\alpha)_{jk}^y + (\alpha\alpha)_{jl}^y\right], \quad (6.44)$$

where

μ_G represents an overall mean
α represents additive effect
δ represents dominance effect
$(\alpha\alpha)$ represents additive × additive interaction effect

For simplicity, in the genetic model, we do not include additive × dominance interaction effect, dominance × additive interaction effect, and dominance × dominance interaction effect. Extension to including three additional interaction effects is straightforward.

6.2.2 Analysis for Genetic Covariance between Relatives

After the genetic models are assumed, we can calculate the genetic covariance between relatives using identical by descent and quantitative genetic theory. Based on the genetic model (6.44), the general formula for calculation of the genetic covariance between two individuals, x and y, is given by

$$V_G = \mathrm{cov}\left(G_{ijkl}(x), G_{ijkl}(y)\right) = V_A(x, y) + V_D(x, y) + V_{AA}(x, y), \quad (6.45)$$

where $V_A(x,y)$, $V_D(x,y)$, and $V_{AA}(x,y)$ are the genetic additive, dominance, and dominance × dominance interaction covariances between individuals x and y, respectively.

We start with the calculation of $V_A(x,y)$. Since the means of all genetic effects are zero, it follows from Equation 6.45 that

$$\begin{aligned} V_A(x, y) &= E\left[\left(\alpha_i^x + \alpha_j^x\right)\left(\alpha_i^y + \alpha_j^y\right)\right] \\ &= E\left[\alpha_i^x\alpha_i^y\right] + E\left[\alpha_i^x\alpha_j^y\right] + E\left[\alpha_j^x\alpha_i^y\right] + E\left[\alpha_j^x\alpha_j^y\right] \\ &= 4\phi_{xy}E\left[\alpha_i^2\right]. \end{aligned} \quad (6.46)$$

Next we calculate $E\left[\alpha_1^2\right]$. It follows from Equations 4.11 and 4.12 that α_1 can take either $q_1\alpha$ with probability p_1 or $p_1\alpha$ with probability q_1 where α is a substitution effect. Therefore, we have

$$E\left[\alpha_i^2\right] = p_1 q_1^2 \alpha^2 + q_1 p_1^2 \alpha^2 = p_1 q_1 \alpha^2 = \frac{1}{2} V_A, \tag{6.47}$$

where V_A is the genetic additive variance. Substituting Equation 6.47 into Equation 6.46 gives

$$V_A(x, y) = 2\phi_{xy} V_A. \tag{6.48}$$

Now we consider the dominant genetic covariance between relatives. Considering locus 1 and using the genetic model (6.44), we obtain the dominant covariance between individuals x and y:

$$V_D(x, y) = E\left[\delta_{ij}^x \delta_{ij}^y\right]. \tag{6.49}$$

Since the probability that the genotypes (two alleles) of two individuals are identical by descent is equal to the coefficient of fraternity Δ_{xy}, the dominant covariance between two individuals are given by

$$V_D(x, y) = \Delta_{xy} E\left[\delta_{ij}^2\right]. \tag{6.50}$$

Using Equation 4.13, we obtain

$$\begin{aligned} E\left[\delta_{ij}^2\right] &= p_1^2 \left(q_1^2 \delta\right)^2 + 2 p_1 q_1 \left(-p_1 q_1 \delta\right)^2 + q_1^2 \left(p_1^2 \delta\right)^2 \\ &= p_1^2 q_1^2 \delta^2 \left(q_1^2 + 2 p_1 q_1 + p_1^2\right) \\ &= p_1^2 q_1^2 \delta^2 \\ &= V_{D1}. \end{aligned} \tag{6.51}$$

Substituting Equation 6.51 into Equation 6.50 gives the dominant covariance at locus 1 between individuals x and y:

$$V_D(x, y) = \Delta_{xy} V_{D_1}. \tag{6.52}$$

Since Equation 6.52 holds for every locus, summing over all loci, we obtain the dominance genetic covariance between individuals x and y:

$$V_D(x, y) = \Delta_{xy} V_D. \tag{6.53}$$

Finally, we calculate the genetic interaction covariance between individuals x and y. Similarly, from the genetic model (6.44), the genetic interaction covariance is given by

$$V_{AA}(x,y)=E\left[\left((\alpha\alpha)_{ik}^{x}+(\alpha\alpha)_{il}^{x}+(\alpha\alpha)_{jk}^{x}+(\alpha\alpha)_{jl}^{x}\right)\left((\alpha\alpha)_{ik}^{y}+(\alpha\alpha)_{il}^{y}+(\alpha\alpha)_{jk}^{y}+(\alpha\alpha)_{jl}^{y}\right)\right]$$
$$=16E\left[(\alpha\alpha)_{ik}^{x}(\alpha\alpha)_{ik}^{y}\right]. \quad (6.54)$$

If a random allele drawn from the first locus in individual x is identical by descent with one drawn from individual y and a random allele drawn from the second locus in individual x is identical by descent with one drawn from individual y, then we have

$$E\left[(\alpha\alpha)_{ik}^{x}(\alpha\alpha)_{ik}^{y}\right]=\phi_{xy}^{2}E\left[(\alpha\alpha)^{2}\right]. \quad (6.55)$$

But we can show (Exercise 6.8) that

$$E\left[(\alpha\alpha)^{2}\right]=\frac{1}{4}V_{AA}, \quad (6.56)$$

where V_{AA} is additive × additive interaction variance.

Substituting Equations 6.55 and 6.56 into Equation 6.54 gives

$$V_{AA}(x,y)=4\phi_{xy}^{2}V_{AA}. \quad (6.57)$$

We observe that the additive, dominant, and additive × additive covariances between relatives consist of (1) identity coefficients and (2) a component of genetic variance. This rule can be extended to higher-order interaction covariance (Lynch and Walsh 1998). Since we assume that allele frequencies at different loci are independent, the high-order interaction covariance is equal to the product of the probability of identity for each component additive effect, the probability of identity for each component dominance effect, and the corresponding variance component. For example, the additive × dominance interaction covariance $V_{AD}(x,y)$ is equal to $2\phi_{xy}\Delta_{xy}V_{AD}$, and the dominance × dominance interaction covariance $V_{DD}(x,y)$ is equal to $\Delta_{xy}^{2}V_{DD}$. In general, the genetic covariance between relatives can be computed by

$$V_G(x,y)=2\phi_{xy}V_A+\Delta_{xy}V_D+(2\phi_{xy})^2V_{AA}+2\phi_{xy}\Delta_{xy}V_{AD}+\Delta_{xy}^2V_{DD}+(2\phi_{xy})^3V_{AAA}$$
$$+(2\phi_{xy})^2\Delta_{xy}V_{AAD}+\cdots \quad (6.58)$$

Genetic covariance for a few relative pairs, which is calculated using identity coefficients in Table 6.1, is summarized in Table 6.2.

TABLE 6.2 Coefficients for the Components of Genetic Covariance between Relatives

Relationship	V_A	V_D	V_{AA}	V_{AD}	V_{DD}
Parent–offspring	1/2		1/4		
Full siblings	1/2	1/4	1/4	1/8	1/16
Half siblings	1/4		1/16		
First cousins	1/8		1/64		
Double first cousins	1/4	1/16	1/16	1/64	1/256
Second cousins	1/32		1/1024		
Uncle–nephew	1/4		1/16		

6.3 MIXED LINEAR MODEL FOR A SINGLE TRAIT

The presence of population structure often causes spurious association. Mixed linear models incorporating genetic correlation information among the sampled individuals into them are often used for genetic association analysis of quantitative traits to avoid spurious association (Yang et al. 2010, 2014). Mixed linear models serve (1) association analysis of quantitative traits, (2) estimation of heritability, and (3) complex phenotype prediction.

Genetic random effect models that are based on the genetic relationship matrix (GRM) discussed in Section 6.2 is a key component for mixed linear models. Therefore, we start with introducing single genetic random effect model that is accurate when kinship coefficients from pedigrees are available. However, when whole-genome sequencing or exome sequencing are used to measure the GRM, the single random effect models assume that all SNPs including both common and rare variants have the same effect size distribution are not appropriate. We then extend the single random effect models to multiple random effect models with different effect size random variables, which divide the whole genome into several regions and then group SNPs in close proximity together.

6.3.1 Genetic Random Effect

6.3.1.1 Single Random Variable

We begin to study the genetic random effect model with a single random variable (Lee et al. 2011; Speed and Balding 2014). Assume that n individuals are sampled. Let y_i be the phenotypic value of individual i. Consider the genetic random effect model

$$Y = g + e, \tag{6.59}$$

where $Y = [y_1, \ldots, y_n]^T$, $g = [g_1, \ldots, g_n]^T$, and $e = [e_1, \ldots, e_n]^T$ with $e \sim N\left(0, I\sigma_e^2\right)$. We assume that $g_i, i = 1, \ldots, n$ are random variables. The correlation coefficient between g_i and g_k is denoted by

$$r_{ik} = \frac{\operatorname{cov}(g_i, g_k)}{\sqrt{\operatorname{var}(g_i)\operatorname{var}(g_k)}}. \tag{6.60}$$

There are two ways to define the genetic additive effect for the individual. Consider a locus with two alleles, D and d. The frequencies of alleles D and d are denoted by p and q, respectively. The genotypic values of the genotypes DD, Dd, and dd are denoted by G_{11}, G_{12}, and G_{22}, respectively. Let α_1 and α_2 be the respective genic effects of the alleles D and d, respectively. The substitution effect is defined as $\alpha = \alpha_1 - \alpha_2$.

Using Equations 4.11 and 4.12, we define the genetic additive effect g_i as

$$g_i = \begin{cases} 2\alpha & DD \\ \alpha & Dd \\ 0 & dd. \end{cases} \tag{6.61}$$

The genetic additive effect g_i can also be defined as

$$g_i = \begin{cases} 2q\alpha & DD \\ (q-p)\alpha & Dd \\ -2p\alpha & dd. \end{cases} \tag{6.62}$$

Define an indicator variable for the genotypes as

$$x_i = \begin{cases} 2 & DD \\ 1 & Dd \\ 0 & dd, \end{cases} \tag{6.63}$$

or

$$x_i = \begin{cases} 2q & DD \\ (q-p) & Dd \\ -2p & dd. \end{cases} \tag{6.64}$$

We then can denote $g_i = x_i\alpha$. It is easy to see that if x_i is defined in Equation 6.63, then (Exercise 6.9)

$$E[x_i] = 2p \quad \text{and} \quad \text{var}(x_i) = 2pq. \tag{6.65}$$

Similarly, we can show (Exercise 6.10) that if x_i is defined in Equation 6.64, then we have

$$E[x_i] = 0 \quad \text{and} \quad \text{var}(x_i) = 2pq. \tag{6.66}$$

Covariance between g_i and g_k is given by

$$\operatorname{cov}(g_i, g_k) = E\left[(x_i - E(x_i))(x_k - E(x_k))\alpha^2\right], \tag{6.67}$$

where $E[x_i] = E[x_k] = \mu$ for a particular SNP.

Similarly, we can calculate the correlation between g_i and g_k by

$$\operatorname{corr}(g_i, g_k) = \frac{E\left[(x_i - \varpi)(x_k - \mu)\alpha^2\right]}{2pq}. \tag{6.68}$$

Since individuals x_i and x_k cannot be replicated, we use L SNPs to approximate the correlation $\operatorname{corr}(g_i, g_k)$ as follows:

$$\operatorname{corr}(g_i, g_k) = \frac{1}{L}\sum_{j=1}^{L} \frac{(x_{ij} - \mu_j)(x_{kj} - \mu_j)E\left[\alpha_j^2\right]}{2p_jq_j}, \tag{6.69}$$

where p_j is the frequency of the reference allele at the SNP j, $q_j = 1 - p_j$ and $\mu_j = 2p_j$ if the indicator variable for the genotype at the SNP j is defined by Equation 6.63 or $\mu_j = 0$ if the indicator variable for the genotype at the SNP j is defined by Equation 6.64.

If we make **two assumptions**, (1) that each SNP independently affects the phenotype variation and (2) that the distribution of the random contribution of all SNPs is identical (Krishna Kumar et al. 2016), then $E\left[\alpha_j^2\right] = V_A$, where V_A is a genetic additive variance. Thus, Equation 6.69 can be reduced to

$$\operatorname{corr}(g_i, g_k) = \frac{1}{L}\sum_{j=1}^{L} \frac{(x_{ij} - \mu_j)(x_{kj} - \mu_j)}{2p_jq_j} V_A. \tag{6.70}$$

Inspired by Equation 6.70, we can further define the genetic random effect model as (Yang et al. 2010)

$$y_i = g_i + e_i = \sum_{j=1}^{L} z_{ij}u_j + e_j, \tag{6.71}$$

where

- g_i is the total genetic random effect of individual i, $g_i = \sum_{j=1}^{L} z_{ij}u_j$, $z_{ij} = \frac{x_{ij} - \mu_{ju}}{\sqrt{2p_jq_j}}$
- u_j is an independent random genetic effect with mean zero and variance σ_u^2 for SNP j
- e_i is the residual effect and distributed as a normal distribution, $e_i \sim N(0, \sigma_e^2)$ with σ_e^2 being the residual variance

Equation 6.71 can be written in a matrix form:

$$Y = Zu + e, \tag{6.72}$$

where $z_i = [z_{i1}, \ldots, z_{iL}]^T, Z = [z_1, \ldots, z_n]^T$, and $u = [u_1, \ldots, u_L]^T$. We assume that

$$\Sigma_u = \mathrm{cov}(u, u) = \sigma_u^2 I_L, \tag{6.73}$$

where I_L is an L-dimensional identity matrix. Then, the covariance matrix of the vector of phenotypes Y is

$$\begin{aligned} \mathrm{var}(Y) &= ZZ^T \sigma_u^2 + I_n \sigma_e^2 \\ &= \frac{ZZ^T}{L} L\sigma_u^2 + I_n \sigma_e^2 \\ &= GV_A + I_n \sigma_e^2, \end{aligned} \tag{6.74}$$

where

$G = \frac{ZZ^T}{L} = \frac{1}{L}\sum_{j=1}^{L} \frac{(x_{ij} - \mu_j)(x_{kj} - \mu_j)}{2p_j q_j}$ is the GRM

$V_A = L\sigma_u^2$ is the total genetic additive variance

I_n is an n-dimensional identity matrix

As we discussed in Section 6.2, the GRM includes both recent genetic relatedness such as family relationships and distant genetic relatedness such as population substructure. If the mixed model is used for association analysis, both recent and distant genetic relatedness should be considered in the analysis. Originally, in deriving genetic random effect models, we assume that all variants are causal. However, the causal variants are unknown; we use all SNPs across the autosome chromosomes in the calculation of the GRM.

6.3.1.2 Multiple Genetic Random Effects

In the single genetic random effect model, we assume that all SNPs have the homogeneous distribution of variant effect size across the genome (Speed and Balding 2014; Weissbrod et al. 2016). However, different SNP classes may have different effect size distributions. We can divide SNPs into different groups. The SNPs in each group have similar effect sizes. We assign a random effect for each group.

Consider M groups. We define M random effects $g^1, \ldots, g^M$ with the genetic relationship matrix specified by $G^1, \ldots, G^M$ and the corresponding genetic additive variances $V_{A_1}, \ldots, V_{A_M}$. Consider multiple genetic random effect model

$$Y = \sum_{m=1}^{M} g^m + e, \tag{6.75}$$

where $g^m \sim N(0, G_m V_{A_m})$ and $e \sim N\left(0, \mathbf{I}_n \sigma_2^2\right)$.

For each group of SNPs, similar to Equation 6.72, we define the normalized genotype indicator variable matrix Z^m. Let the number of SNPs in group m be L_m. The multiple genetic random effect model (6.75) can be further specified as

$$Y = \sum_{m=1}^{M} Z^m u^m + e, \tag{6.76}$$

where u^m is a vector of genetic random effects with distribution $N\left(0, \frac{V_{Am}}{L_m} I_{L_m}\right)$ and the genetic relationship matrix is defined as $G_m = \frac{Z^m \left(Z^m\right)^T}{L_m}$.

6.3.2 Mixed Linear Model for Quantitative Trait Association Analysis

6.3.2.1 Mixed Linear Model

The random effect model can be extended to a mixed linear model including covariates such as gender, race, age, principal components, biomarkers, and clinical variables, among others. The effects of these covariates on the phenotype variation are fixed. Let $w_i = [w_{i_1}, \ldots, w_{iq}]^T$ be a vector of covariates of individual i and $\alpha = [\alpha_1, \ldots, \alpha_q]^T$ be a vector of fixed effects associated with covariates. A mixed linear model for relating genotypes to phenotypes is defined as

$$y_i = w_i \alpha + \sum_{m=1}^{M} Z_i^m u^m + e_i, \tag{6.77}$$

where $Z_i^m = \left[z_{i_1}^m, \ldots, z_{iL_m}^m\right]^T$.

If we take the genetic additive effects of the SNPs as fixed effects and the genetic background and population structure components as random effects, the model (6.77) can be extended to the more general mixed linear model

$$y_i = w_i \alpha + \pi_i \beta + \sum_{m=1}^{M} Z_i^m u^m + e_i, \tag{6.78}$$

where

$\pi_i = [\pi_{i1}, \ldots, \pi_{ip}]^T$ is a vector of the indicator variables for the genotypes of p SNPs
$\beta_1 = [\beta_1, \ldots, \beta_p]^T$ is a vector of the genetic additive effects

The mixed linear model (6.78) can be written in a matrix form:

$$Y = W\alpha + \Pi\beta + Zu + e, \tag{6.79}$$

where

$$Y=\begin{bmatrix} y_1 \\ \vdots \\ y_n \end{bmatrix}, W=\begin{bmatrix} w_{11} & \cdots & w_{1q} \\ \vdots & \vdots & \vdots \\ w_{n1} & \cdots & w_{nq} \end{bmatrix}, \Pi=\begin{bmatrix} \pi_{i1} & \cdots & \pi_{ip} \\ \vdots & \vdots & \vdots \\ \pi_{n1} & \cdots & \rho_{np} \end{bmatrix}, Z^m=\begin{bmatrix} z_{11}^m & \cdots & z_{1L_1}^m \\ \vdots & \vdots & \vdots \\ z_{n1}^m & \cdots & z_{nL_n}^m \end{bmatrix}, u=\begin{bmatrix} u^1 \\ \vdots \\ u^M \end{bmatrix},$$

$$e=\begin{bmatrix} e_1 \\ \vdots \\ e_n \end{bmatrix}, e \sim N\left(0,\sigma_e^2 I_n\right), \text{ and } Z=\left[Z^1 \dots Z^m\right].$$

The conditional mean of Y, given u, is

$$E\left[Y|u\right]=W\alpha+\Pi\beta+Zu. \tag{6.80}$$

We assume that the vector of random effects follows distribution

$$u \sim N\left(0, D\right), \tag{6.81}$$

where

$$D=\begin{bmatrix} \frac{V_{A_1}}{L_1} I_{L_1} & \cdots & 0 \\ \vdots & \vdots & \vdots \\ 0 & \cdots & \frac{V_{AM}}{L_M} I_{LM} \end{bmatrix}.$$

Therefore, the vector of the phenotype Y is distributed as

$$Y \sim N\left(W\alpha+\Pi\beta, V\right), \tag{6.82}$$

where $V = ZDZ^T + I_n\sigma_e^2$.

Equation 6.82 shows that the fixed effects enter only the mean, whereas the variance of the random effects enter only the variance of Y.

6.3.2.2 Estimating Fixed and Random Effects

6.3.2.2.1 Estimating Fixed Effects for Known Variance   Let

$$H=\left[W \quad \Pi\right] \text{ and } \gamma=\begin{bmatrix} \alpha \\ \beta \end{bmatrix}.$$

Then, distribution (6.82) can be written as

$$Y \sim N\left(H\gamma, V\right). \tag{6.83}$$

The log-likelihood is

$$l = -\frac{n}{2}\log(2\pi) - \frac{1}{2}\log|V| - \frac{1}{2}(Y - H\gamma)^T V^{-1}(Y - H\gamma). \tag{6.84}$$

To estimate the fixed effects, taking derivative of the log-likelihood with respect to the fixed effect γ and setting it equal to zero gives

$$H^T V^{-1}(Y - H\gamma) = 0,$$

which leads to

$$\hat{\gamma} = \left(H^T V^{-1} H\right)^{-1} H^T V^{-1} Y. \tag{6.85}$$

The variance of the estimator $\hat{\gamma}$ is

$$\text{var}(\hat{\gamma}) = \left(H^T V^{-1} H\right)^{-1} H^T V^{-1} V V^{-1} H \left(H^T V^{-1} H\right)^{-1} = \left(H^T V^{-1} H\right)^{-1}. \tag{6.86}$$

The last p components in $\hat{\gamma}$ are the estimators $\hat{\beta}$ of the genetic additive effects of p SNPs, and the last p rows and p columns of the variance matrix var($\hat{\gamma}$) form the variance matrix var($\hat{\beta}$).

The null hypothesis for testing the association of the p SNPs with a trait is

$$H_0 : \beta = 0.$$

A widely used χ^2 statistic is defined as

$$T = \hat{\beta}^T \Lambda^{-1} \hat{\beta}, \tag{6.87}$$

where $\Lambda = \text{var}(\hat{\beta})$.

Under the null hypothesis of no association of p SNPs with the trait, the statistic T will asymptotically be distributed as a central $\chi^2_{(p)}$ distribution.

6.3.2.2.2 Estimating Fixed Effects for Unknown Variance When variance V is unknown, we need to simultaneously estimate the fixed effects and variance. If we do not use GRM information and directly estimate the variance from the samples, we set both derivatives of the log-likelihood with respect to the fixed effects and variance matrix equal to zero (Exercise 6.11):

$$H^T V^{-1}(Y - H\gamma) = 0, \tag{6.88}$$

$$-\frac{1}{2}V^{-1} + \frac{1}{2}V^{-1}(Y - H\gamma)(Y - H\gamma)^T V^{-1} = 0. \tag{6.89}$$

Equations 6.88 and 6.89 can be further reduced to

$$\gamma = \left(H^T V^{-1} H\right)^{-1} H^T V^{-1} Y, \tag{6.90}$$

$$V = (Y - H\gamma)(Y - H\gamma)^T. \tag{6.91}$$

Iteratively solving Equations 6.90 and 6.91 gives the estimators of γ and V.

6.3.2.2.3 Predicting Random Effects Genetic effects can also be modeled as random effects. A fixed effect is assumed to be a constant. However, a random effect is assumed to come from a population of effects. The observed random effects are a random selection from population of effects and are a realization of population effects. They are determined by the observed phenotypes Y. Therefore, the random effects can be predicted by $E[u|\,Y]$. To calculate $E[u|\,Y]$, we start a joint distribution of u and Y. We assume that u and Y are jointly distributed as

$$\begin{bmatrix} u \\ Y \end{bmatrix} \sim N\left(\begin{pmatrix} 0 \\ H\gamma \end{pmatrix} \quad \Sigma\right),$$

where

$$\Sigma = \begin{bmatrix} D & DZ^T \\ ZD & V \end{bmatrix}. \tag{6.92}$$

The conditional mean $E[u|\,Y]$ is given by

$$\begin{aligned} E[u|Y] &= E[u] + \operatorname{cov}(u, Y)\left[\operatorname{var}(Y)\right]^{-1}\left(Y - E[Y]\right) \\ &= DZ^T V^{-1}(Y - H\gamma). \end{aligned} \tag{6.93}$$

Therefore, $\hat{u} = DZ^T V^{-1}(Y - H\gamma)$ can be taken as a prediction of a random effect. The variance of the predictor of the random effects is

$$\operatorname{var}(\hat{u}) = DZ^T V^{-1} ZD. \tag{6.94}$$

It is well known that

$$\begin{aligned} \operatorname{var}(u) &= \operatorname{var}\left(E[u|Y]\right) + E\left[\operatorname{var}(u|Y)\right] \\ &= \operatorname{var}(\hat{u}) + E\left[\operatorname{var}(u|Y)\right] \\ &\geq \operatorname{var}(\hat{u}). \end{aligned} \tag{6.95}$$

Equation 6.95 shows that the variance of the predictor of the genetic random effects is less than the variance of the estimator of taking random effects as fixed effects.

6.3.3 Estimating Variance Components

In the estimation of both fixed and random effects, we assume that the variance matrix V is known. However, in general, the variance matrix is unknown and needs to be estimated. Consider a general mixed linear model:

$$Y = H\gamma + g + e, \tag{6.96}$$

where $g = \sum_{i=1}^{m} g_i$, $g_i = Z_i^m u_m$, $V_A = \sum_{m=1}^{M} G_m V_{Am}$, $V_p = V = V_A + \mathbf{I}_n \sigma_e^2$, V_A denotes the total genetic additive variance, and V_p denotes the total phenotypic variance. The total phenotypic variance is partitioned into the genetic additive variance explained by M groups of SNPs and residual variance due to environments and other nongenetic factors.

Maximum likelihood (ML) and residual maximum likelihood (REML), also known as restricted maximum likelihood, are widely used classical methods for estimating variance parameters (Gumedze and Dunne 2011; Lynch and Walsh 1998). Since the ML estimators do not consider the degrees of freedom lost in the estimation of the fixed and random effects, they are biased. REML for variance component estimation can overcome this limitation. Therefore, in general, the REML estimators of the variance component may be better than the ML estimators.

6.3.3.1 ML Estimation of Variance Components

Due to its simplicity, we start with ML estimation of variance components. The marginal distribution of the phenotypes Y in the mixed linear model (6.96) is given by $N(H\gamma, V)$. The log-likelihood function of Y is

$$l(\gamma, V | H, Y) = -\frac{n}{2}\log(2\pi) - \frac{1}{2}\log|V| - \frac{1}{2}(Y - H\gamma)^T V^{-1}(Y - H\gamma). \tag{6.97}$$

The variance components V_{Am} and σ_e^2 are embedded within the matrix V. Since both the fixed effects γ and variance components are unknown, we will use the ML to simultaneously estimate them.

We can show (Appendix 6D) that

$$\frac{\partial l}{\partial \sigma_j^2} = -\frac{1}{2}\operatorname{Tr}\left(V^{-1}V_j\right) + \frac{1}{2}(Y - H\gamma)^T V^{-1}V_j V^{-1}(Y - H\gamma). \tag{6.98}$$

Setting the derivative equal to zero gives

$$\operatorname{Tr}\left(V^{-1}V_j\right) = (Y - H\gamma)^T V^{-1}V_j V^{-1}(Y - H\gamma), \tag{6.99}$$

where

$$\frac{\partial V}{\partial \sigma_j^2} = V_j = \begin{cases} I_n & \sigma_j^2 = \sigma_e^2 \\ G_m & \sigma_j^2 = V_{Am}. \end{cases}$$

Let

$$P = V^{-1} - V^{-1}H\left(H^T V^{-1} H\right)^{-1} H^T V^{-1}. \tag{6.100}$$

Equation 6.99 will be reduced to

$$\mathrm{Tr}\left(V^{-1}V_j\right) = Y^T P V_j P Y. \tag{6.101}$$

In the previous section, we show that the ML estimator of the fixed effects is given by

$$\hat{\gamma} = \left(H^T V^{-1} H\right)^{-1} H^T V^{-1} Y. \tag{6.102}$$

Iteratively solving Equations 6.101 and 6.102, finally, we can obtain the ML estimators of both fixed effects and variance components:

$$\hat{\gamma} = \left(H^T \hat{V}^{-1} H\right)^{-1} H^T \hat{V}^{-1} Y, \tag{6.103}$$

$$\mathrm{Tr}\left(\hat{V}^{-1}\right) = Y^T \hat{P}\hat{P} Y, \quad \text{when } \sigma_j^2 = \sigma_e^2, \tag{6.104}$$

$$\mathrm{Tr}\left(\hat{V}^{-1} G_m\right) = Y^T \hat{P} G_m \hat{P} Y, \quad m = 1, \ldots, M, \quad \text{when } \sigma_j^2 = V_{Am}, \tag{6.105}$$

where $\hat{P}$ is estimated by using

$$\hat{V} = \sum_{m=1}^{M} G_m \hat{V}_{Am} + \mathbf{I_n} \hat{\sigma}_e^2$$

and

$$\hat{P} = \hat{V}^{-1} - \hat{V}^{-1} H\left(H^T \hat{V}^{-1} H\right)^{-1} H^T \hat{V}^{-1}. \tag{6.106}$$

6.3.3.1.1 Covariance Matrix of the ML Estimators Let $\sigma = \left[\sigma_1^2, \ldots, \sigma_M^2, \sigma_e^2\right]$. The Fisher information matrix is defined as

$$F = -E\left(\begin{bmatrix} \dfrac{\partial^2 l}{\partial \gamma \partial \gamma^T} & \dfrac{\partial^2 l}{\partial \gamma \partial \sigma^T} \\ \dfrac{\partial^2 l}{\partial \sigma \partial \gamma^T} & \dfrac{\partial^2 l}{\partial \sigma \partial \sigma^T} \end{bmatrix}\right). \tag{6.107}$$

Define the matrix:

$$\Lambda = -E\left[\frac{\partial^2 l}{\partial\sigma\partial\sigma^T}\right] = \begin{bmatrix} \frac{1}{2}\mathrm{Tr}\left(V^{-1}G_1V^{-1}G_1\right) & \cdots & \frac{1}{2}\mathrm{Tr}\left(V^{-1}G_1V^{-1}G_M\right) & \frac{1}{2}\mathrm{Tr}\left(V^{-1}G_1V^{-1}\right) \\ \vdots & \vdots & \vdots & \\ \frac{1}{2}\mathrm{Tr}\left(V^{-1}G_1V^{-1}\right) & \cdots & \frac{1}{2}\mathrm{Tr}\left(V^{-1}G_MV^{-1}\right) & \frac{1}{2}\mathrm{Tr}\left(V^{-1}V^{-1}\right) \end{bmatrix}.$$

In Appendix 6E, we show

$$F = \begin{bmatrix} H^TV^{-1}H & 0 \\ 0 & \Lambda \end{bmatrix}. \tag{6.108}$$

Equation 6.108 demonstrates that the ML estimators of the fixed effects are uncorrelated with the ML estimators of the variance components.

Let $\hat{\theta} = \left[\hat{\gamma}^T, \hat{\sigma}^T\right]^T$. The covariance matrix of the ML estimators $\hat{\theta}$ is the inverse of the Fisher matrix:

$$\mathrm{cov}\left(\hat{\theta}\right) = F^{-1} = \begin{bmatrix} \left(H^TV^{-1}H\right)^{-1} & 0 \\ 0 & \Lambda^{-1} \end{bmatrix}. \tag{6.109}$$

Example 6.7 Consider the model

$$y_i = z_i u + e_i.$$

Assume that u is distributed as a normal variable, $u \sim N\left(0,\sigma_u^2\right)$, and e_i is the residual effect and distributed as a normal distribution, $e_i \sim N\left(0,\sigma_e^2\right)$, with σ_e^2 being the residual variance.

Then, $H=0$, which implies

$$\hat{\gamma} = 0 \quad \text{and} \quad \hat{P} = \hat{V}^{-1}.$$

Applying Equations 6.104 and 6.105 gives

$$\mathrm{Tr}\left(\hat{V}^{-1}\right) = Y^T\hat{V}^{-2}Y$$

and

$$\mathrm{Tr}\left(\hat{V}^{-1}ZZ^T\right) = Y^T\hat{V}^{-1}ZZ^T\hat{V}^{-1}Y,$$

where $Z=[z_1, \ldots, z_n]^T$ and $\hat{V} = ZZ^T\hat{\sigma}_u^2 + \mathbf{I}_n\hat{\sigma}_e^2$.

The Fisher information matrix is

$$F=\begin{bmatrix}0 & 0\\ 0 & \Lambda\end{bmatrix},$$

where

$$\Lambda=\frac{1}{2}\begin{bmatrix}\mathrm{Tr}(V^{-1}ZZ^TV^{-1}ZZ^T & \mathrm{Tr}\left(V^{-1}ZZ^TV^{-1}\right)\\ \mathrm{Tr}\left(V^{-1}ZZ^TV^{-1}\right) & \mathrm{Tr}\left(V^{-1}V^{-1}\right)\end{bmatrix}.$$

6.3.3.2 Restricted Maximum Likelihood Estimation

In spite of its very useful properties, the ML estimate of the variance components has its limitations (Hartville 1974). First, the ML estimation requires numerical solution to the constrained nonlinear optimization problem. Second, the ML estimation ignores the loss of degrees of freedom resulting from the estimation of fixed effects in the model. Third, the ML estimation requires the assumption of normal distribution of the phenotypes. Restricted maximum likelihood (REML) method for estimation of variance components has been developed to overcome these limitations.

REML considers a linear transformation of the observation vector Y to remove the fixed effects from the model (Gumedze and Dunne 2011; Lynch and Walsh 1998). Specifically, REML selects a linear combination of K^TY so that K^TY is of maximal rank but contains no information on the fixed effects. To achieve this, the matrix K should satisfy

$$K^TH=0. \tag{6.110}$$

It can be shown that the solution to Equation 6.110 is (Appendix 6F)

$$K=\left[I-\left(H^T\right)^{-}H^T\right]c, \tag{6.111}$$

where c is any vector.

From the distribution theory, K^TY is distributed as $N(0,K^TVK)$. The log-likelihood of K^TY is given by

$$l\left(\gamma,V\middle|H,K^TY\right)=-\frac{n}{2}\log(2\pi)-\frac{1}{2}\log\left|K^TVK\right|-\frac{1}{2}Y^TK\left(K^TVK\right)^{-1}K^TY. \tag{6.112}$$

The REML equation can be obtained from Equation 6.99 by replacing Y with K^TY, replacing V with K^TVK, and replacing V_j with K^TV_jK. The transformed Equation 6.99 becomes

$$\mathrm{Tr}\left(\left(K^TVK\right)^{-1}K^TV_jK\right)=Y^TK\left(K^TVK\right)^{-1}K^TV_jK\left(K^TVK\right)^{-1}K^TY. \tag{6.113}$$

In Appendix 6F, we show that

$$K\left(K^T V K\right)^{-1} K^T = P. \tag{6.114}$$

Substituting Equation 6.114 into Equation 6.113 gives the REML equation

$$\mathrm{Tr}\left(PV_j\right) = Y^T P V_j P Y, \quad j = 1, \ldots, M+1. \tag{6.115}$$

6.3.3.3 Numerical Solutions to the ML/REML Equations

The ML/REML equations are high-dimensional nonlinear equations. In general, there are no analytic solutions to the ML/REML equations generated by the ML/REML estimations. In this section, we review the numerical solutions to the ML/REML equations.

The Newton–Raphson (NR) and Fisher scoring algorithms are widely used methods for solving the ML/REML equations (Hartville 1977; Searle et al. 1992).

6.3.3.3.1 Newton–Raphson (NR) Methods The Newton–Raphson algorithm is a powerful method for solving a system of nonlinear equations (Beck 2014). It is based on the simple idea of linear approximation. Let θ be parameters in the mixed linear model. For the ML estimators, the parameters are defined as $\theta = [\gamma^T, \sigma^T]^T$, while for the REML estimators, the parameters are defined as $\theta = \gamma$, which include only variance components. Let $F(\theta) = \frac{\partial l}{\partial \theta}$, a column vector of the partial derivatives of the log-likelihood function with respect to the parameters. Both ML and REML estimators should satisfy the following equation:

$$F(\theta) = 0,$$

which is a system of high-dimensional nonlinear equations. To solve the system of nonlinear equations, we expand the function $F(\theta)$ in terms of a Taylor series:

$$F(\theta) = F\left(\theta^{(l)}\right) + \left.\frac{\partial F}{\partial \theta^T}\right|_{\theta^{(l)}} \left(\theta - \theta^{(l)}\right) = 0, \tag{6.116}$$

where $\frac{\partial F}{\partial \theta^T}$ is called the Hessian matrix and denoted by $H(\theta)$.

Solving Equation 6.116 gives the estimators of the parameters for the next iteration:

$$\theta^{(l+1)} = \theta^{(l)} - \left(H\left(\theta^{(l)}\right)\right)^{-1} F\left(\theta^{(l)}\right). \tag{6.117}$$

Equation 6.117 can be used iteratively to refine the estimators of the maximum on the $(l+1)$th iteration. To implement the iterative procedure (6.117), we need to calculate the score function $F(\theta)$ and the Hessian matrix $H(\theta)$. Since the score function and the Hessian matrix for the ML and REML estimators are different, below we will separately derive them.

6.3.3.3.2 Iteration Procedures for the ML Recall that the score function for the ML is given by

$$F(\theta)=\begin{bmatrix}\frac{\partial l}{\partial \gamma}\\ \frac{\partial l}{\partial \sigma}\end{bmatrix}=\begin{bmatrix}H^T V^{-1}(Y-H\gamma)\\ -\frac{1}{2}\mathrm{Tr}\left(V^{-1}V_1\right)+\frac{1}{2}(Y-H\gamma)^T V^{-1}V_1V^{-1}(Y-H\gamma)\\ \vdots\\ -\frac{1}{2}\mathrm{Tr}\left(V^{-1}V_{M+1}\right)+\frac{1}{2}(Y-H\gamma)^T V^{-1}V_{M+1}V^{-1}(Y-H\gamma)\end{bmatrix}. \quad (6.118)$$

Now we calculate the Hessian matrix for the ML.

From Equation 6.118, we obtain the second partial derivatives of the log-likelihood function with the parameters of the fixed effects γ:

$$\frac{\partial^2 l}{\partial\gamma\partial\gamma^T}=-H^T V^{-1}H. \quad (6.119)$$

Using Equation 6H.4, we obtain the second partial derivatives of the log-likelihood function with the parameters γ and σ:

$$\frac{\partial^2 l}{\partial\gamma\partial\sigma^T}=\left[-H^T V^{-1}V_1V^{-1}(Y-H\gamma)\quad\cdots\quad -H^T V^{-1}V_{M+1}V^{-1}(Y-H\gamma)\right] \quad (6.120)$$

and

$$\frac{\partial^2 l}{\partial\sigma\partial\gamma^T}=\left(\frac{\partial^2 l}{\partial\gamma\partial\sigma^T}\right)^T. \quad (6.121)$$

Again, using Equation 6H.4, we obtain

$$\frac{\partial^2 l}{\partial\sigma_i\partial\sigma_j}=\frac{1}{2}\mathrm{Tr}\left(V^{-1}V_iV^{-1}V_j\right)-(Y-H\gamma)^T V^{-1}V_iV^{-1}V_jV^{-1}(Y-H\gamma) \quad (6.122)$$

and

$$\frac{\partial^2 l}{\partial\sigma\partial\sigma^T}=\begin{bmatrix}\frac{\partial^2 l}{\partial\sigma_1\partial\sigma_1} & \cdots & \frac{\partial^2 l}{\partial\sigma_1\partial\sigma_{M+1}}\\ \vdots & \vdots & \vdots\\ \frac{\partial^2 l}{\partial\sigma_{M+1}\partial\sigma_1} & \cdots & \frac{\partial^2 l}{\partial\sigma_{M+1}\partial\sigma_{M+1}}\end{bmatrix}. \quad (6.123)$$

Combining Equations 6.119 through 6.123, we obtain the Hessian matrix for the ML:

$$H\left(\theta^{(l)}\right)=\begin{bmatrix}\frac{\partial^2 l}{\partial\gamma\partial\gamma^T} & \frac{\partial^2 l}{\partial\gamma\partial\sigma^T}\\ \frac{\partial^2 l}{\partial\sigma\partial\gamma^T} & \frac{\partial^2 l}{\partial\sigma\partial\sigma^T}\end{bmatrix}_{\theta^{(l)}}. \tag{6.124}$$

6.3.3.3.3 Iteration Procedures for REML Equation 6.115 implies that the score function for REML is

$$F\left(\theta_1,\ldots,\theta_{M+1}\right)=\begin{bmatrix}\frac{\partial l}{\partial\sigma_1}\\ \vdots\\ \frac{\partial l}{\partial\sigma_{M+1}}\end{bmatrix}=\begin{bmatrix}-\frac{1}{2}\mathrm{Tr}\left(PV_1\right)+\frac{1}{2}Y^TPV_1PY\\ \vdots\\ -\frac{1}{2}\mathrm{Tr}\left(PV_{M+1}\right)+\frac{1}{2}Y^TPV_{M+1}PY\end{bmatrix}. \tag{6.125}$$

To calculate the Hessian matrix for REML, we start to compute $\frac{\partial P}{\partial\sigma_j}$. It follows from Equations 6.114 and 6D.4 that

$$\frac{\partial P}{\partial\sigma_j}=-K\left(K^TVK\right)^{-1}K^TV_jK\left(K^TVK\right)^{-1}K^T. \tag{6.126}$$

Substituting Equation 6.114 into Equation 6.125 gives

$$\frac{\partial P}{\partial\sigma_j}=-PV_jP. \tag{6.127}$$

Recall in Equation 6.126 that

$$\frac{\partial l}{\partial\sigma_i}=-\frac{1}{2}\mathrm{Tr}\left(PV_i\right)+\frac{1}{2}Y^TPV_iPY. \tag{6.128a}$$

From Equations 6.127 and 6.128a, it follows that

$$\frac{\partial^2 l}{\partial\sigma_i\partial\sigma_j}=\frac{1}{2}\mathrm{Tr}\left(PV_iPV_j\right)-Y^TPV_iPV_jPY, \tag{6.128b}$$

which implies

$$H\left(\theta^{(l)}\right)=\begin{bmatrix} \frac{1}{2}\operatorname{Tr}\left(P^{(l)}V_1P^{(l)}V_1\right)-Y^TP^{(l)}V_1P^{(l)}V_1P^{(l)}Y & \cdots & \frac{1}{2}\operatorname{Tr}\left(P^{(l)}V_1P^{(l)}V_{M+1}\right)-Y^TP^{(l)}V_1P^{(l)}V_{M+1}P^{(l)}Y \\ \vdots & \vdots & \vdots \\ \frac{1}{2}\operatorname{Tr}\left(P^{(l)}V_{M+1}P^{(l)}V_1\right)-Y^TP^{(l)}V_{M+1}P^{(l)}V_1P^{(l)}Y & \cdots & \frac{1}{2}\operatorname{Tr}\left(P^{(l)}V_1P^{(l)}V_{M+1}\right)-Y^TP^{(l)}V_{M+1}P^{(l)}V_{M+1}P^{(l)}Y \end{bmatrix}. \tag{6.129}$$

6.3.3.3.4 Fisher Scoring Algorithm To reduce the computation of the NR method, we replace the Hessian matrix by its expectation in the NR algorithm. In the previous section, we define the Fisher information matrix as

$$F(\theta)=E\left[-\frac{\partial^2 l}{\partial\theta\partial\theta^T}\right]. \tag{6.130}$$

Equation 6.117 is then replaced by

$$\theta^{(l+1)}=\theta^{(l)}-\left(F\left(\theta^{(l)}\right)\right)^{-1}F\left(\theta^{(l)}\right). \tag{6.131}$$

Now we calculate the Fisher information matrix for the ML and REML estimators.

6.3.3.4 Fisher Information Matrix for the ML Estimators

The Fisher information matrix for the ML estimators is given in Equation 6.108. We can rewrite it as

$$F=\begin{bmatrix} H^TV^{-1}H & 0 & \cdots & 0 \\ 0 & \frac{1}{2}\operatorname{Tr}\left(V^{-1}V_1V^{-1}V_1\right) & \cdots & \frac{1}{2}\operatorname{Tr}\left(V^{-1}V_1V^{-1}V_{M+1}\right) \\ \vdots & \vdots & \vdots & \vdots \\ 0 & \frac{1}{2}\operatorname{Tr}\left(V^{-1}V_{M+1}V^{-1}V_1\right) & \cdots & \frac{1}{2}\operatorname{Tr}\left(V^{-1}V_{M+1}V^{-1}V_{M+11}\right) \end{bmatrix}. \tag{6.132}$$

6.3.3.4.1 Fisher Information Matrix for the REML Estimators In Equation 6.112 we showed that K^TY is distributed as $N(0, K^TVK)$ distribution. Therefore,

$$PY=K\left(K^TVK\right)^{-1}K^TY\sim N\left(0,P\right). \tag{6.133}$$

Using Theorem 4.6.1 in the book (Graybill 1976), we obtain

$$E\left[Y^TPV_iPV_jPY\right]=\operatorname{Tr}\left(V_iPV_jP\right)=\operatorname{Tr}\left(PV_iPV_j\right). \tag{6.134}$$

Using Equations 6.128b and 6.134, we obtain the element of the Fisher information matrix for the REML estimators:

$$E\left[-\frac{\partial^2 l}{\partial\sigma_i\partial\sigma_j}\right] = -\frac{1}{2}\operatorname{Tr}\left(PV_iPV_j\right) + E\left[Y^T PV_iPV_jPY\right]$$
$$= -\frac{1}{2}\operatorname{Tr}\left(PV_iPV_j\right) + \operatorname{Tr}\left(PV_iPV_j\right)$$
$$= \frac{1}{2}\operatorname{Tr}\left(PV_iPV_j\right). \quad (6.135)$$

Therefore, the Fisher information matrix for the REML estimators is given by

$$F(\theta) = \frac{1}{2}\begin{bmatrix} \operatorname{Tr}\left(PV_1PV_1\right) & \cdots & \operatorname{Tr}\left(PV_1PV_{M+1}\right) \\ \vdots & \vdots & \vdots \\ \operatorname{Tr}\left(PV_{M+1}PV_1\right) & \cdots & \operatorname{Tr}\left(PV_{M+1}PV_{M+1}\right) \end{bmatrix}. \quad (6.136)$$

6.3.3.5 *Expectation/Maximization (EM) Algorithm for ML Estimation*

The ML/REML methods require solving a system of high-dimensional nonlinear equations, which are due to unobserved random effects. If the random effects are known, it is much easier to find the ML/REML estimators. The EM algorithm attempts to augment the data in which we treat unobserved random effects as the known variables. The augmented dataset including Y and random variables u is referred to as the complete data. The dataset including only the observed Y is referred to as incomplete data. After data are augmented into complete data, the maximization procedures are then applied to the complete data.

We use the conditional expectation of the random variable u, given the observed Y to augment the data. Therefore, we first need to derive the joint distribution of the variables Y and u. Again, consider the linear mixed model

$$Y = H\gamma + Zu + e. \quad (6.137)$$

Recall that Equation 6.92 specifies the joint distribution of Y and U:

$$\begin{bmatrix} u \\ Y \end{bmatrix} \sim N\left(\begin{pmatrix} 0 \\ H\gamma \end{pmatrix} \quad \Sigma\right),$$

where

$$\Sigma = \begin{bmatrix} D & DZ^T \\ ZD & V \end{bmatrix}. \quad (6.138)$$

From standard multivariate statistic theory (Anderson 1984), we can obtain the conditional mean of the variables u, given the observed phenotypic values Y

$$\begin{aligned} E[u|Y] &= E[u] + \operatorname{cov}(u, Y^T)[\operatorname{var}(Y)]^{-1}(Y - E[Y]) \\ &= DZ^T V^{-1}(Y - H\gamma) \\ &= DZ^T PY \end{aligned} \tag{6.139}$$

and conditional covariance matrix of variables u, given the observed phenotypic values Y

$$\begin{aligned} \Sigma_{u|Y} &= \operatorname{var}(u) - \operatorname{cov}(u, Y^T)[\operatorname{var}(Y)]^{-1}\operatorname{cov}(Y, u^T) \\ &= D - DZ^T V^{-1} ZD. \end{aligned} \tag{6.140}$$

Next we consider the joint distribution of Y and e:

$$\begin{bmatrix} e \\ Y \end{bmatrix} \sim N\left(\begin{bmatrix} 0 \\ H\gamma \end{bmatrix} \quad \Sigma_{eY} \right), \tag{6.141}$$

where

$$\Sigma_{eY} = \begin{bmatrix} \sigma_e^2 I_n & \sigma_e^2 I_n \\ \sigma_e^2 I_n & V \end{bmatrix},$$

which implies that the conditional mean of errors e, given the observed phenotypic values Y

$$\begin{aligned} E[e|Y) &= \sigma_e^2 V^{-1}(Y - H\gamma) \\ &= \sigma_e^2 PY, \end{aligned} \tag{6.142}$$

and the conditional covariance matrix errors e, given the observed phenotypic values Y

$$\Sigma_{e|Y} = \sigma_e^2 I_n - \sigma_e^4 V^{-1}. \tag{6.143}$$

If we assume that the random variables u are known, we can use quadratic forms of the variables u to estimate the variance components. Assume that a vector of variables, x, is distributed as a normal distribution with mean μ and covariance matrix Λ. The expectation of a quadratic form $x^T Ax$ can be calculated by

$$E\left[x^T Ax\right] = \operatorname{Tr}(A\Lambda) + \mu^T A\mu. \tag{6.144}$$

Recall that in Equation 6.78 we defined

$$Zu = \sum_{m=1}^{M} Z^m u^m \tag{6.145}$$

and

$$D_m = V_{Am} \frac{I_{Lm}}{L_m}, \tag{6.146}$$

where

V_{Am} is the genetic additive variance of the mth random effect group
L_m is the number of SNPs in the mth random effect group

Therefore, the conditional mean $E[u^m|Y]$ and conditional covariance matrix $\Sigma_{u^m|Y} = \text{cov}\left(u^m, u^m|Y\right)$ are given by

$$E\left[u^m|Y\right] = D_m\left(Z^m\right)^T V^{-1}\left(Y - H\gamma\right) \tag{6.147}$$

and

$$\Sigma_{u^m|Y} = \text{cov}\left(u^m, u^m|Y\right) = D_m - D_m\left(Z^m\right)^T V^{-1} Z^m D_m, \tag{6.148}$$

respectively.

Therefore, using Equations 6.144 and 6.146, we have

$$E\left[\left(u^m\right)^T u^m\right] = \text{Tr}\left(V_{Am}\frac{I_{Lm}}{L_m}\right) = V_{Am}. \tag{6.149}$$

Equation 6.149 showed that the genetic additive variance can be estimated by the estimated random effects: $E[(u^m)^T u^m|Y]$. Using Equations 6.144, 6.147, and 6.148, we obtain

$$\begin{aligned} E\left[\left(u^m\right)^T u^m\middle|Y\right] &= \text{Tr}\left[\left(D_m - D_m\left(Z^m\right)^T V^{-1}Z^m D_m\right)I_{Lm}\right] \\ &\quad + \left(Y - H\gamma\right)^T V^{-1}Z^m D_m D_m\left(Z^m\right)^T V^{-1}\left(Y - H\gamma\right) \\ &= V_{Am} - \frac{V_{Am}^2}{L_m^2}\text{Tr}\left(V^{-1}V_m\right) + \frac{V_{Am}^2}{L_m^2}\left(Y - H\gamma\right)^T V^{-1}Z^m\left(Z^m\right)^T V^{-1}\left(Y - H\gamma\right). \end{aligned} \tag{6.150}$$

Similarly, we have

$$E\left[e^T e|Y\right] = n\sigma_e^2 + \sigma_e^4\left(\left(Y - H\gamma\right)^T V^{-1}V^{-1}\left(Y - H\gamma\right) - \text{Tr}\left(V^{-1}\right)\right). \tag{6.151}$$

By similar arguments, the fixed effects can be estimated by

$$\hat{\gamma} = \left(H^T H\right)^{-1} H^T\left(Y - \sum_{m=1}^{M} Z^m u^m\right).$$

Therefore, we have

$$E[\gamma|Y] = \left(H^T H\right)^{-1} H^T \left(Y - E\left[\sum_{m=1}^{M} Z^m u^m \middle| Y\right]\right)$$

$$= \left(H^T H\right)^{-1} H^T \left(Y - \sum_{m=1}^{M} Z^m D_m \left(Z^m\right)^T V^{-1}\left(Y - H\hat{\gamma}\right)\right)$$

$$= \left(H^T H\right)^{-1} H^T (H\hat{\gamma} + \hat{\sigma}_e^2 V^{-1}\left(Y - H\hat{\gamma}\right)$$

$$= \hat{\gamma} + \hat{\sigma}_e^2 \left(H^T H\right)^{-1} H^T V^{-1}\left(Y - H\hat{\gamma}\right). \tag{6.152}$$

Now we summarize the EM algorithm for the ML methods:

Step 0: Initialization.

Select initial values $V_{Am} = \frac{V_p}{M+1}$ and $\sigma_e^2 = \frac{V_p}{M+1}$. Let $V = ZDZ^T + I_n\sigma_e^2$, where

$$D = \begin{bmatrix} \frac{V_{A_1}}{L_1} I_{L_1} & \cdots & 0 \\ \vdots & \vdots & \vdots \\ 0 & \cdots & \frac{V_{A_M}}{L_M} I_{L_M} \end{bmatrix}.$$

Step 1: E-step.

Compute expected values of quadratic forms for the estimation of variance components:

$$\hat{A}_m^{(l)} = V_{Am}^{(l)} - \frac{\left(\hat{V}_{Am}^2\right)^{(l)}}{L_m^2} \operatorname{Tr}\left(\left(\hat{V}^{(l)}\right)^{-1} V_{Am}^{(l)}\right)$$

$$+ \frac{\left(\hat{V}_{Am}^2\right)^{(l)}}{L_m^2} \left(Y - H\hat{\gamma}^{(l)}\right)^T \left(\hat{V}^{(l)}\right)^{-1} ZZ^T \left(\hat{V}^{(l)}\right)^{-1} \left(Y - H\hat{\gamma}^{(l)}\right), \quad m = 1, \ldots, M \tag{6.153}$$

and

$$\hat{B}^{(l)} = n\left(\hat{\sigma}_e^2\right)^{(l)} + \left(\hat{\sigma}_e^4\right)^{(l)} \left(\left(Y - H\hat{\gamma}^{(l)}\right)^T \left(\hat{V}^{(l)}\right)^{-1} \left(\hat{V}^{(l)}\right)^{-1} \left(Y - H\hat{\gamma}^{(l)}\right) - \operatorname{Tr}\left(\left(\hat{V}^{(l)}\right)^{-1}\right)\right). \tag{6.154}$$

Compute the expected fixed effects at the current iteration:

$$\hat{S}^{(l)} = \hat{\gamma}^{(l)} + \left(\hat{\sigma}_e^2\right)^{(l)} \left(H^T H\right)^{-1} H^T \left(\hat{V}^{(l)}\right)^{-1} \left(Y - H\hat{\gamma}^{(l)}\right). \tag{6.155}$$

Step 2 (M-step). ML estimate of the complete data

$$\hat{V}_{Am}^{(l+1)} = \hat{A}_m^{(l)}, \quad m = 1, \ldots, M, \tag{6.156}$$

$$\left(\hat{\sigma}_e^2\right)^{(l+1)} = \hat{B}^{(l)}, \tag{6.157}$$

$$H\hat{\gamma}^{(l+1)} = H\hat{S}^{(l)}. \tag{6.158}$$

Step 3. Convergence.

Let $\varepsilon_1, \varepsilon_2$, and ε_3 be prespecified errors. If $\left\|V_{Am}^{(l+1)} - V_{Am}^{(l)}\right\| < \varepsilon_1$, $\left|\left(\hat{\sigma}_e^2\right)^{(l+1)} - \left(\hat{\sigma}_e^2\right)^{(l)}\right| < \varepsilon_2$, and $\left\|\hat{\gamma}^{(l+1)} - \hat{\gamma}^{(l)}\right\| < \varepsilon_3$, then iterations are convergent; set

$$\hat{V}_{Am} = \hat{V}_{Am}^{(l+1)}, \quad \hat{\sigma}_e^2 = \left(\hat{\sigma}_e^2\right)^{(l+1)}, \quad \text{and} \quad \hat{\gamma} = \hat{\gamma}^{(l+1)}.$$

Otherwise, increase l by 1 and return to step 1.

6.3.3.6 Expectation/Maximization (EM) Algorithm for REML Estimation

In Equations 6.152 and 6.154, replacing y by $K^T y$, H by $K^T H = 0$, Z by $K^T Z$, and V by $K^T VK$, we obtain the EM algorithm for the REML estimation:

Step 0. Initialization.

Let σ_p^2 be the variance of the phenotypes. Define $V_{Am}^{(0)} = \dfrac{\sigma_p^2}{m+1}$ and $\left(\sigma_e^2\right)^{(0)} = \dfrac{\sigma_p^2}{m+1}$.

Step 1: E-step.

Computer expected values of quadratic forms for the estimation of variance components:

$$\hat{A}_m^{(l)} = V_{Am}^{(l)} + \frac{\left(\hat{V}_{Am}^2\right)^{(l)}}{L_m^2}\left[Y^T P V_m P Y - \mathrm{Tr}\left(P V_m\right)\right], \quad m = 1, \ldots, M, \tag{6.159}$$

and

$$\hat{B}^{(l)} = n\left(\hat{\sigma}_e^2\right)^{(l)} + \left(\hat{\sigma}_e^4\right)^{(l)}\left[Y^T P P Y - \mathrm{Tr}(P)\right]. \tag{6.160}$$

Step 2: M-step.

Set

$$\hat{V}_{Am}^{(l+1)} = \hat{A}_m^{(l)}, \quad m = 1, \ldots, M, \tag{6.161}$$

$$\left(\hat{\sigma}_e^2\right)^{(l+1)} = \hat{B}^{(l)}.$$

Step 3. Convergence

Let ε_1 and ε_2 be prespecified errors. If $\left\|V_{Am}^{(l+1)} - V_{Am}^{(l)}\right\| < \varepsilon_1$ and $\left|\left(\hat{\sigma}_e^2\right)^{(l+1)} - \left(\hat{\sigma}_e^2\right)^{(l)}\right| < \varepsilon_2$, then iterations are convergent; set

$$\hat{V}_{Am} = \hat{V}_{Am}^{(l+1)} \quad \text{and} \quad \hat{\sigma}_e^2 = \left(\hat{\sigma}_e^2\right)^{(l+1)}.$$

Otherwise, increase l by 1 and return to step 1.

6.3.3.7 *Average Information Algorithms*

The average information algorithm is to replace the Fisher information matrix in the Fisher scoring algorithm by the average of the observed and expected information matrices called the average information matrix (Gilmour et al. 1995). Using Equations 6.129 and 6.136, we obtain the average information matrix for REML:

$$I_A\left(\theta^{(l)}\right) = \begin{bmatrix} \frac{1}{2}Y^T P^{(l)} V_1 P^{(l)} V_1 P^{(l)} Y & \cdots & \frac{1}{2}Y^T P^{(l)} V_1 P^{(l)} V_{M+1} P^{(l)} Y \\ \vdots & \vdots & \vdots \\ \frac{1}{2}Y^T P^{(l)} V_{M+1} P^{(l)} V_1 P^{(l)} Y & \cdots & \frac{1}{2}Y^T P^{(l)} V_{M+1} P^{(l)} V_{M+1} P^{(l)} Y \end{bmatrix}. \tag{6.162}$$

The EM algorithm is often used to compute the initial values for the REML algorithm.

6.3.4 Hypothesis Test in Mixed Linear Models

We consider the problem of testing null hypotheses that include both fixed effects and random effects in the mixed linear model. The general null hypothesis for the fixed effects is

$$H_0: L\gamma = 0 \tag{6.163}$$

versus

$$H_a: L\gamma \neq 0,$$

where L is a $k \times (p+q)$ dimensional matrix.

The widely used statistic for testing the null hypothesis (6.163) is the Wald statistic. It is defined as

$$T_\gamma = \hat{\gamma}^T L^T \left(L\left(H^T V^{-1} H\right)^{-1} L^T\right)^{-1} L\hat{\gamma}. \tag{6.164}$$

Under the null hypothesis $H_0: L\gamma = 0$, the test statistic T_γ is asymptotically distributed as a central $\chi^2_{(rank(L))}$ with rank (L) degrees of freedom.

In general, we are interested in testing the null hypothesis involving fixed effects β of genetic variants. Define the matrix L as

$$L = \begin{bmatrix} 0 & L_\beta \end{bmatrix},$$

where L_β is a $k \times q$ dimensional matrix.

Let

$$E = \left(H^T V^{-1} H\right)^{-1} = \begin{bmatrix} E_{\alpha\alpha} & E_{\alpha\beta} \\ E_{\beta\alpha} & E_{\beta\beta} \end{bmatrix},$$

where $E_{\beta\beta}$ is a $q \times q$ dimensional matrix corresponding to the vector β.

To test the null hypothesis for the β

$$H_0 : L_\beta \beta = 0 \quad \text{versus} \quad H_a : L_\beta \beta \neq 0,$$

we define the statistic

$$T_\beta = \hat{\beta}^T L_\beta^T \left(L_\beta E_{\beta\beta} L_\beta^T\right)^{-1} L_\beta \hat{\beta}. \tag{6.165}$$

Under the null hypothesis $H_0 : L_\beta \beta = 0$, the test statistic T_β is asymptotically distributed as a central $\chi^2_{(rank(L_\beta))}$ distribution with rank (L_β) degrees of freedom.

Next we discuss the test for the variance components. We consider three statistics: likelihood ratio statistic, the score statistic, and the Wald statistic for testing the variance components (Molenberghs and Verbeke 2007). To illustrate the basic idea behind the distribution of the test statistics, we first assume $m = 1$. The null hypothesis is defined as

$$H_0 : V_A = 0 \quad \text{versus} \quad H_0 : V_A = 0. \tag{6.166}$$

Let $l(V_A)$ be the log-likelihood function for the mixed linear model. The likelihood ratio test statistic is defined as

$$T_{LR} = 2\log\left(\frac{\max_{H_a} l(V_A)}{\max_{H_0} l(V_A)}\right). \tag{6.167}$$

Let $\hat{V}_A$ be the maximum likelihood estimate of V_A under the unconstrained parameterization, i.e., $V_A < 0$ is allowed. As Figure 6.8 showed, there are two scenarios to consider: case 1 where $\hat{V}_A$ is located in the alternative hypothesis area and case 2 where $\hat{V}_A$ is located outside the alternative hypothesis area. In case 2, we have

$$\max_{H_a} l(V_A) = \max_{H_0} l(V_A), \quad \text{which implies } T_{LR} = 0.$$

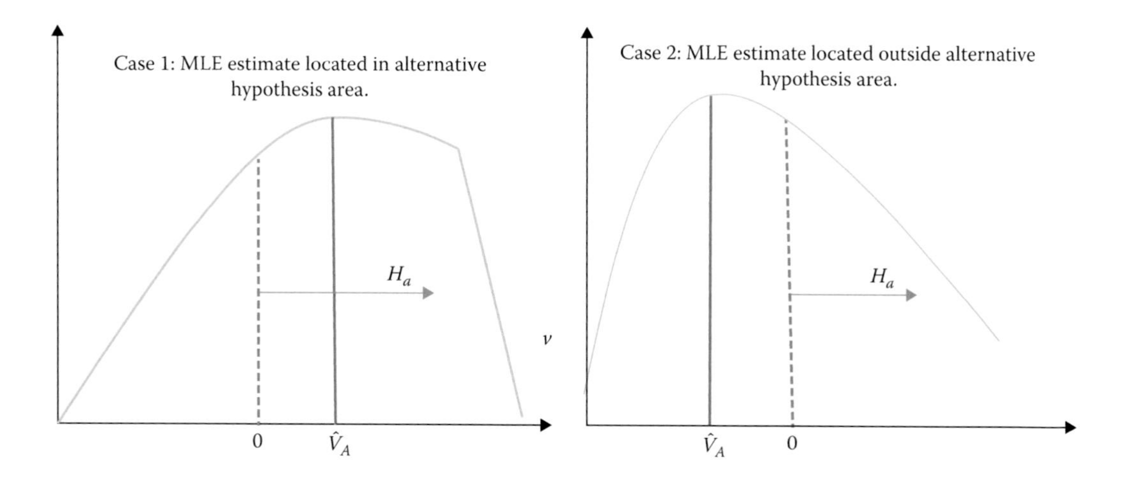

FIGURE 6.8 Maximum likelihood estimator and null hypothesis.

In case 1, $\max_{H_a} l(V_A) > \max_{H_0} l(V_A)$, the distribution theory of the likelihood ratio test statistic applies and T_{LR} is asymptotically distributed as a central $\chi^2_{(1)}$ distribution. Therefore, we have

$$\begin{aligned} P(T_{LR} > c \mid H_0) &= P(T_{LR} > c \mid H_0, \hat{V}_A \geq 0) P(\hat{V}_A \geq 0 \mid H_0) + P(T_{LR} > c \mid H_0, V_A < 0) P(V_A < 0 \mid H_0) \\ &= \frac{1}{2} P(\chi^2_{(1)} > c) + \frac{1}{2} P(\chi^2_0 > c), \end{aligned} \tag{6.168}$$

where $P(\hat{V}_A \geq 0 \mid H_0) = P(V_A < 0 \mid H_0) = \frac{1}{2}$ and χ^2_0 denotes the distribution with all probability mass at zero.

Now we consider the vector case. Suppose that all parameters γ and $\psi = [V_{A1}, \ldots, V_{Am}]^T$ are denoted by $\theta = [\gamma^T, \psi^T]^T$. The null hypothesis is defined as

$$H_0 : \psi = 0 \quad \text{versus} \quad H_a : \psi \in C, \tag{6.169}$$

where C is a closed and convex cone in the Euclidean space, with the vertex at the origin.

Set the score function

$$S(\theta) = \frac{\partial l}{\partial \theta} = \begin{bmatrix} \frac{\partial l}{\partial \gamma} \\ \frac{\partial l}{\partial \psi} \end{bmatrix} = \begin{bmatrix} S_\gamma \\ S_\psi \end{bmatrix}$$

and

$$H(\theta) = -\frac{\partial^2 l}{\partial \theta \partial \theta^T} = \begin{bmatrix} H_{\gamma\gamma} & H_{\gamma\psi} \\ H_{\psi\gamma} & H_{\psi\psi} \end{bmatrix}.$$

Let

$$\theta_H = \begin{bmatrix} \gamma \\ 0 \end{bmatrix} \text{ and } \hat{\theta}_H = \begin{bmatrix} \hat{\gamma} \\ 0 \end{bmatrix},$$

where $\hat{\gamma}$ is the maximum likelihood estimate of γ under the null hypothesis H_0. Define $S_N = N^{-1/2} S_\psi$. Then, three statistics, the likelihood ratio statistic, the score statistic, and the Wald statistic are respectively defined as

$$T_{LR} = 2\ln\left(\frac{\max\limits_{H_a} l(\psi)}{\max\limits_{H_0} l(\psi)}\right), \tag{6.170}$$

$$T_S = S_N^T H_{\psi\psi}^{-1}\left(\hat{\theta}_H\right) S_N - \inf\left\{\left(S_N - b\right)^T H_{\psi\psi}^{-1}\left(\hat{\theta}_H\right)\left(S_N - b\right) \middle| b \in C, \right. \tag{6.171}$$

and

$$T_w = \hat{\psi}^T V_{\psi\psi}^{-1} \hat{\psi} - \inf\left\{\left(\hat{\psi} - b\right)^T V_{\psi\psi}^{-1}\left(\hat{\psi} - b\right) \middle| b \in C, \right. \tag{6.172}$$

where

$$V = \text{cov}(\hat{\theta}) = \begin{bmatrix} V_{\gamma\gamma} & V_{\gamma\psi} \\ V_{\psi\gamma} & V_{\psi\psi} \end{bmatrix}.$$

The null hypothesis (6.166) and decision rule (6.168) can be taken as a special case. When $\hat{\theta}_u = \hat{\psi} = \hat{V}_A > 0$, the score test and Wald test at zero are nonnegative, and hence, the infimum in (6.171) and (6.172) becomes zero. Then, we have

$$T_S = S_N^T H_{\psi\psi}^{-1}\left(\hat{\theta}_H\right) S_N$$

and

$$T_W = \hat{\psi}^T V_{\psi\psi}^{-1} \hat{\psi}.$$

When $\hat{\theta}_u = \hat{\psi} = \hat{V}_A < 0$, the score test and Wald test at zero are negative. The infimum is reached for $b = 0$, which leads to $T_S = 0$ and $T_W = 0$.

The null distributions of three test statistics equal a weighted sum of chi-squared distributions. The weights are often difficult to determine. However, for some important cases, the null distributions can be analytically determined (Stram and Lee 1994).

Consider the null hypothesis $\psi = \begin{bmatrix} \psi_{11} & 0 \\ 0 & 0 \end{bmatrix}$ with $m \times m$ positive definite matrix ψ_{11} against the alternative hypothesis that matrix ψ is a $(m+1)\times(m+1)$ positive semidefinite matrix. Under this hypothesis, three statistics are asymptotically distributed as $\frac{1}{2}\chi^2_{(m)} + \frac{1}{2}\chi^2_{(m+1)}$. As special cases, when testing $V_A = 0$ versus $V_A > 0$, three statistics are asymptotically distributed as a mixture of $\frac{1}{2}\chi^2_{(0)} + \frac{1}{2}\chi^2_{(1)}$.

We can also consider the null hypothesis:

$$\psi_m = 0 \text{ against } \psi_m > 0 \quad (m = 1,\ldots,k).$$

In this case, the null distributions of three statistics are a mixture of

$$\sum_{m=0}^{k} 2^{-k} \binom{k}{m} \chi^2_{(m)}. \tag{6.173}$$

6.3.5 Mixed Linear Models for Quantitative Trait Analysis with Sequencing Data

Similar to association analysis for a quantitative trait in Chapter 4, two approaches, mixed multivariate linear models and mixed functional models, can also be applied to family-based quantitative trait analysis or quantitative trait analysis in the presence of population substructures with next-generation sequencing data. The sequence kernel association test (SKAT) is a typical mixed multivariate model that can be used for association testing for both qualitative and quantitative trait analyses. In Chapters 3 and 4, we have already introduced the SKAT for population-based association analysis. In this chapter, we briefly discuss that the SKAT can also be used for family-based or population-based but with population structure and association analysis.

To model family structures and population structures, the functional models with fixed effects can be extended to mixed functional models for association analysis where either fixed genetic effects or random effects can be modeled as functional curves. Two types of mixed functional models for quantitative trait association analysis in the presence of both family structures and population structures will be investigated.

6.3.5.1 Sequence Kernel Association Test (SKAT)

SKAT (Wu et al. 2011) comprises modified mixed models for testing the association of multiple variants in a genomic region or a gene with a phenotype in the presence of family or population structures. Assume that n individuals are sequenced in the genomic region with p SNPs. Covariates such as age, sex, race, and principal components of genetic variation for controlling population structures as fixed effects can be included in the model. These covariates for the ith individual is denoted as $X_i = [x_{i1}, \ldots, x_{iq}]$. The genotypes for the p SNPs are denoted as $Z_i = [z_{i1}, \ldots, z_{ip}]$. Let y_i be the phenotype of the ith individual.

Consider a linear model

$$y_i = \alpha_0 + X_i\alpha + Z_i\beta + \varepsilon_i, \tag{6.174}$$

where

α_0 is an intercept term
$\alpha = [\alpha_1, \ldots, \alpha_q]^T$ is the vector of regression coefficients for the covariates
$\beta = [\beta_1, \ldots, \beta_p]$ is the vector of regression coefficients for the SNPs
ε_i is an error with a mean of zero and a variance of σ_e^2

$$Y = \begin{bmatrix} y_1 \\ \vdots \\ y_n \end{bmatrix}, \quad 1 = \begin{bmatrix} 1 \\ \vdots \\ 1 \end{bmatrix}, \quad X = \begin{bmatrix} X_1 \\ \vdots \\ X_n \end{bmatrix}, \quad Z = \begin{bmatrix} Z_1 \\ \vdots \\ Z_n \end{bmatrix}, \quad \varepsilon = \begin{bmatrix} \varepsilon_1 \\ \vdots \\ \varepsilon_n \end{bmatrix}.$$

Equation 6.174 can be written in a matrix form:

$$Y = \alpha_0 1 + X\alpha + Z\beta + \varepsilon. \tag{6.175}$$

The score function for the genetic effects β is given by

$$S = \frac{\partial l}{\partial \beta} = Z^T\left(Y - \alpha_0 1 - X\alpha - Z\beta\right). \tag{6.176}$$

Taking a partial derivative with respect to β^T on both sides of Equation 6.176 gives

$$H_{\beta\beta} = -\frac{\partial^2 l}{\partial\beta\partial\beta^T} = Z^T Z. \tag{6.177}$$

The null hypothesis is $H_0: \beta = 0$ against the alternative hypothesis $H_a: \beta \neq 0$.
Under the null hypothesis, the score function is reduced to

$$S = \left.\frac{\partial l}{\partial \beta}\right|_{H_0} = Z^T\left(Y - \hat{\alpha}_0 1 - X\hat{\alpha}\right). \tag{6.178}$$

The score test is then defined as

$$T_S = S^T H_{\beta\beta}^{-1} S = Z\left(Z^T Z\right)^{-1} Z^T\left(Y - \hat{\alpha}_0 1 - X\hat{\alpha}\right). \tag{6.179}$$

For simple computation, $(Z^TZ)^{-1}$ is replaced by the weight matrix $W = \text{diag}(w_1, \ldots, w_p)$. Then, Equation 6.179 is reduced to

$$T_s = \left(Y - \hat{\alpha}_0 1 - X\hat{\alpha}\right)^T ZWZ^T\left(Y - \hat{\alpha}_0 1 - X\hat{\alpha}\right), \tag{6.180}$$

where

$$K = ZWZ^T = \begin{bmatrix} \sum_{l=1}^{p} w_l z_{1l}^2 & \cdots & \sum_{l=1}^{p} w_l z_{1l} z_{nl} \\ \vdots & \vdots & \vdots \\ \sum_{l=1}^{p} w_l z_{nl} z_{1l} & \cdots & \sum_{l=1}^{p} w_l z_{nl}^2 \end{bmatrix}.$$

The matrix K is referred to as a kernel matrix. Its (i,j)th element, $K(i,j) = \sum_{l=1}^{p} w_l z_{il} z_{kl}$, measures the genetic similarity between individuals i and j by p SNPs.

SKAT can also be derived from variance component analysis. To increase the power, each regression coefficient β_j is assumed a random variable that follows an arbitrary distribution with a mean of zero and a variance of $w_j\tau$ where w_j is a prespecified weight for the jth SNP and τ is a variance component. The null hypothesis is

$$H_0: \tau = 0 \quad \text{against} \quad H_a: \tau > 0. \tag{6.181}$$

The variance component score test can be defined as

$$T_S = \left(Y - \alpha_0 1 - X\hat{\alpha}\right)^T K \left(Y - \alpha_0 1 - X\hat{\alpha}\right), \tag{6.182}$$

where $K = ZWZ^T$. It is clear that the T_S defined in Equation 6.182 is exactly the same as that defined in Equation 6.180.

The power of the test depends on the choice of weights. Weights reflect the contribution of the SNPs to the phenotype variation. If the lth SNP makes a small contribution to the phenotype variation, a small weight, W_l, should be selected. In contrast, if the lth SNP makes a big contribution to the phenotype variation, a large weight, W_l, should be selected. However, in practice, it is unknown which variants make big contributions. It is often assumed that the genetic effect of the rare variant is inversely proportional to the frequency of the rare variants. Therefore, it is suggested to use beta function to model the distribution of the weights: $\sqrt{w_j} = \text{beta}\left(\text{MAF}_j, \alpha, \beta\right)$, where MAF denotes the minor allele frequency that are estimated using total samples and α and β are two specified parameters. Wu et al. (2011) recommended to set $\alpha = 1$ and $\beta = 25$.

The distribution of the test statistic T_S under the null hypothesis is a mixture of chi-squared

$$T_S \sim \sum_{i=1}^{n} \lambda_i \chi_{1,i}^2,$$

where $P_0 = V - V\tilde{X}\left(\tilde{X}^T V \tilde{X}\right)^{-1} \tilde{X}^T V$, $\tilde{X} = [1, X]$, $V = \hat{\sigma}_e^2 I$, $\lambda_j, j = 1, \ldots, n$ are the eigenvalues of $P_0^{1/2} K P_0^{1/2}$, and $\chi_{1,i}^2$ are independent $\chi_{(1)}^2$ random variable.

6.4 MIXED FUNCTIONAL LINEAR MODELS FOR SEQUENCE-BASED QUANTITATIVE TRAIT ANALYSIS

Mixed linear models are widely used for quantitative trait association studies (Kang et al. 2010; Lippert et al. 2011; Listgarten et al. 2010; Price et al. 2010a; Yu et al. 2006; Zhang et al. 2010; Zhou and Stephens 2012). Mixed linear models for quantitative trait association studies have two remarkable features. First, mixed linear models unify family and population study designs and can be applied to samples with arbitrary combinations of related and unrelated individuals. Second, mixed linear models are able to correct for confounding arising from population stratification, family structure, and cryptic relatedness. However, mixed linear models are designed for testing the association of common variants with quantitative traits. They are typically carried out by testing the association of single loci, one locus at a time.

It is now well documented that NGS can generate several millions or even dozens of millions of genetic variation data. As a consequence, these genetic variation data are so densely distributed across the genome that the genetic variation can be modeled as a function of genomic location. But, standard multivariate statistical analysis often fails with functional data. The emergence of NGS demands alternative approaches to the analytic methods for quantitative trait analysis. In this section, we introduce mixed functional linear models for sequence-based quantitative trait association studies to unify population and family study designs in which a continuous phenotype is taken as a scalar response, genetic variants across the genomic regions as functional predictors, and additive genetic background effects due to population and family structure and cryptic relatedness as random effects.

6.4.1 Mixed Functional Linear Models (Type 1)

Since it jointly uses multiple marker information, the multiple linear regression model, in general, might have more power to detect a quantitative trait locus (QTL) than the simple linear regression model. However, as the number of markers increases, the degree of freedom of the test statistics will also increase. This will compromise the power of multiple regressions for identifying QTLs. In addition, when the frequencies of the genetic variant become smaller and smaller, the variances of the estimators of the genetic effects will be larger and larger. To reduce the degrees of freedom in the model and variances of the estimators due to the presence of rare variants in the model, we consider a functional linear model for a quantitative trait.

Consider a genomic region, $[a, b]$. Let t be a genomic position in the genomic region $[a, b]$. Define a genotype profile, $X_i(t)$, of the ith individual as

$$X_i(t) = \begin{cases} 2P_m(t), & \text{MM} \\ P_m(t) - P_M(t), & \text{Mm} \\ -2P_M(t), & \text{mm}, \end{cases} \tag{6.183}$$

where

M and m are two alleles of the marker at the genomic position t

$P_M(t)$ and $P_m(t)$ are the frequencies of the alleles M and m, respectively

For the convenience of discussion, the function in the genomic region $[a, b]$ can be mapped to the region $[0, T]$; T is the length of the genome region being considered.

Let Y_i be a phenotype value of the ith individual. A functional linear model for a quantitative trait is defined as

$$Y_i = W_i\alpha + \int_0^T X_i(t)\beta(t)dt + u_i + \varepsilon_i, \tag{6.184}$$

where

$w_i = [w_{i_1}, \ldots, w_{iq}]^T$ is a vector of covariates of individual i including sex, age, race, environmental variables, and the top principal components of the genetic variants

$\alpha = [\alpha_1, \ldots, \alpha_q]^T$ is a vector of regression coefficients associated with covariates

ε_i are independent and identically distributed normal variables with a mean of zero and a variance of σ_e^2

u_i are random effects with a mean of zero that measures the genetic relationship among individuals

$\beta(t)$ is the genetic additive effect of the marker at the genomic position t

Let $u = [u_1, \ldots, u_n]^T$. Then, the variance–covariance matrix is given by $G = V_A A$, where the additive genetic relationship matrix A has elements $A_{ij} = 2\phi_{ij}$; ϕ_{ij} is the coefficient of coancestry (Lynch and Walsh 1998). The matrix A can also be taken as a function of the genealogy of the sampled individuals, including both the pedigree relationships and population structure. If the matrix A is unknown, it can be estimated by the genetic variants in the data. Consider m markers outside the genomic region being tested. Let z_{ik} be the indicator variable taking the values of the copy of the reference allele for the kth SNP of the ith individual and p_k be its frequency of the reference allele. The GRM matrix can be estimated by

$$A_{ij} = \frac{1}{m}\sum_{k=1}^{m} \frac{(z_{ik} - 2p_k)(z_{jk} - 2p_k)}{2p_k(1 - p_k)}. \tag{6.185}$$

For convenience, the genome region $[0,T]$ is rescaled to $[0,1]$. If the integrals in Equation 6.184 are discretized, the functional mixed linear model will be reduced to multiple linear regression models.

Similar to Equation 4.52, genotype function $X_i(t)$ can be expanded in terms of functional principal components (eigenfunctions):

$$x_i(t) = \sum_{k=1}^{p} \xi_{ik}\varphi_k(t), \tag{6.186}$$

where

$\varphi_k(t)$ are functional principal components (eigenfunctions)

$\xi_{ik} = \int_0^T x_i(t)\varphi_k(t)$ are functional principal component scores

Substituting Equation 6.186 into Equation 6.184 gives

$$
\begin{aligned}
Y_i &= W_i\alpha + \int_0^T \sum_{k=1}^{p} \xi_{ik}\phi_k(t)\beta(t)dt + u_i + \varepsilon_i \\
&= W_i\alpha + \sum_{k=1}^{p} \xi_{ik} \int_0^T \phi_k(t)\beta(t)dt + u_i + \varepsilon_i \\
&= W_i\alpha + \sum_{k=1}^{p} \xi_{ik} b_k + u_i + \varepsilon_i \\
&= W_i\alpha + \xi_i b + u_i + \varepsilon_i,
\end{aligned} \tag{6.187}
$$

where $\xi_i = [\xi_{i1}, \ldots, \xi_{ip}]$, $b = [b, \ldots, b_p]^T$.

Therefore, the mixed functional linear model (6.184) is transformed to a standard mixed linear model (6.78). We can use the ML to jointly estimate the genetic effects b and variances V_A and σ_e^2. An essential problem in genetic studies of the quantitative trait is to test the association of a genomic region with the quantitative trait. Formally, we investigate the problem of testing the following hypothesis:

$$H_0: \beta(t) = 0, \quad \forall t \in [0, T] \tag{6.188}$$

against

$$H_a: \beta(t) \neq 0.$$

If the genetic effect function $\beta(t)$ is expanded in terms of the eigenfunctions

$$\beta(t) = \varphi(t)^T b,$$

where $\varphi(t) = [\varphi_1(t), \ldots, \varphi_p(t)]^T$, then testing the null hypothesis H_0 in Equation 6.188 is equivalent to testing the hypothesis

$$H_0: b = 0. \tag{6.189}$$

Rewrite Equation 6.187 in a matrix form:

$$Y = H\gamma + u + \varepsilon, \tag{6.190}$$

where

$$Y = \begin{bmatrix} Y_1 \\ \vdots \\ Y_n \end{bmatrix}, \quad H = \begin{bmatrix} W_1 & \xi_1 \\ \vdots & \vdots \\ W_n & \xi_n \end{bmatrix}, \quad \gamma = \begin{bmatrix} \alpha \\ b \end{bmatrix}, \quad u = \begin{bmatrix} u_1 \\ \vdots \\ u_n \end{bmatrix}, \quad \text{and} \quad \varepsilon = \begin{bmatrix} \varepsilon_1 \\ \vdots \\ \varepsilon_n \end{bmatrix}.$$

The parameters γ can be estimated by

$$\hat{\gamma}=\left(H^T\hat{V}^{-1}H\right)^{-1}H^T\hat{V}^{-1}Y, \tag{6.191}$$

where $\hat{V}=\hat{V}_A A+\hat{\sigma}_e^2 I$.

Recall that the covariance matrix of the ML estimators $\hat{\gamma}$ is given by

$$\text{Var}(\hat{\gamma})=\left(H^T\hat{V}^{-1}H\right)^{-1}. \tag{6.192}$$

Let Λ be the matrix obtained by removing the first q row and the first q column of the covariance matrix $\text{Var}(\hat{\gamma})$. Define the test statistic

$$T_Q=\hat{b}^T\Lambda^{-1}\hat{b}. \tag{6.193}$$

Then, under the null hypothesis $H_0: b=0$, T_Q is asymptotically distributed as a central $\chi^2_{(P)}$ distribution.

6.4.2 Mixed Functional Linear Models (Type 2: Functional Variance Component Models)

Similar to the type 1 mixed functional linear model in Section 6.4.1 where the genetic effects are modeled as a function of genomic position, we can model the genetic random effects as a function of genomic position. In this section, we introduce type 2 mixed functional linear models: functional variance component models. Let $u(t)$ be a random effect function of genomic position t with a mean of zero and covariance of zero between two genomic positions. The variance of $u(t)$ is denoted by $\text{var}(u(t))=\sigma_A^2(t)$. A functional variance component model is defined as

$$Y_i=W_i\alpha+\int_0^T X_i(t)u(t)dt+g_{Ai}+\varepsilon_i, \tag{6.194}$$

where $g_{A_i}\sim N(0,V_A A)$ as u_i in Equation 6.184 models the genetic relationships among individuals, A is defined in Equation 6.185, and other parameters are defined in the model (6.184).

Again, using the genotype function expansion (6.186), the model (6.194) can be transformed to

$$\begin{aligned}Y_i&=W_i\alpha+\int_0^T\sum_{m=1}^{p}\xi_{im}\varphi_m(t)u(t)dt+g_{Ai}+\varepsilon_i\\&=W_i\alpha+\sum_{m=1}^{p}\xi_{im}\int_0^T\varphi_m(t)u(t)dt+g_{Ai}+\varepsilon_i\\&=W_i\alpha+\sum_{m=1}^{p}\xi_{im}u_m+g_{Ai}+\varepsilon_i,\end{aligned} \tag{6.195}$$

where $u_m = \int_0^T \varphi_m(t)u(t)dt$ is a random variable. By stochastic integral theory, we have

$$E[u_m] = \int_0^T \varphi_m(t)E[u(t)]dt = 0,$$

$$\begin{aligned} \operatorname{cov}(u_k, u_l) &= E\left[\int_0^T \varphi_k(t)u(t)dt \int_0^T \varphi_l(t)u(t)dt\right] \\ &= \int_0^T\int_0^T \varphi_k(t)E[u(t)u(s)]\varphi_l(s)dtds \\ &= \int_0^T \varphi_k(t)\varphi_l(t)\sigma_A^2(t)dt. \end{aligned} \tag{6.196}$$

Similarly, we can calculate the variance of u_m:

$$\operatorname{var}(u_m) = \int_0^T \varphi_m^2(t)\sigma_A^2(t)dt = V_{Am}. \tag{6.197}$$

If we assume that $\sigma_A^2(t) = \tau$, then we have

$$\operatorname{cov}(u_k, u_l) = 0, \quad k \neq l$$

and

$$\operatorname{var}(u_m) = \tau. \tag{6.198}$$

In general, we assume $\sigma_A^2(t) \neq \tau$, but

$$\operatorname{cov}(u_k, u_l) = 0, \quad k \neq l. \tag{6.199}$$

In this case, u_m are independent with a mean of zero and a variance of V_{Am}. If we assume $V_{Am} = \tau\psi_m$ and consider the model

$$Y_i = W_i\alpha + \int_0^T X_i(t)u(t)dt + \varepsilon_i, \tag{6.200}$$

then its eigenfunction expansion

$$Y_i = W_i\alpha + \sum_{m=1}^{p} \xi_{im}u_m + \varepsilon, \tag{6.201}$$

or its matrix form

$$Y = W\alpha + \xi u + \varepsilon$$

is reduced to SKAT with

$$\operatorname{cov}(\xi u, \xi u) = \tau \xi \Psi \xi^T, \tag{6.202}$$

where $\Psi = \text{diag}(\psi_1, \ldots, \psi_m)$.

In general, we consider the mixed linear model

$$Y_i = W_i\alpha + \sum_{m=1}^{p} \xi_{im} u_m + g_{Ai} + \varepsilon_i \tag{6.203}$$

or its matrix form

$$Y = W\alpha + Zu + e, \tag{6.204}$$

where

$$Y = \begin{bmatrix} y_1 \\ \vdots \\ y_n \end{bmatrix}, \quad W = \begin{bmatrix} w_{11} & \cdots & w_{1q} \\ \vdots & \vdots & \vdots \\ w_{n1} & \cdots & w_{nq} \end{bmatrix}, \quad Z = [\xi \quad I], \quad \xi = \begin{bmatrix} \xi_{11} & \cdots & \xi_{1p} \\ \vdots & \vdots & \vdots \\ \xi_{n1} & \cdots & \xi_{np} \end{bmatrix},$$

$$\alpha = \begin{bmatrix} \alpha_1 \\ \vdots \\ \alpha_q \end{bmatrix}, \quad u = \begin{bmatrix} u^m \\ g_A \end{bmatrix}, \quad u^m = \begin{bmatrix} u_1 \\ \vdots \\ u_p \end{bmatrix}, \quad g_A = \begin{bmatrix} g_{A1} \\ \vdots \\ g_{An} \end{bmatrix}, \quad \varepsilon = \begin{bmatrix} \varepsilon_1 \\ \vdots \\ \varepsilon_n \end{bmatrix},$$

$$V = \xi B \xi^T + V_A A + \sigma_e^2 I, \quad \text{and} \quad B = \text{diag}\left(V_{A_1}, \ldots, V_{A_p}\right).$$

If in Equation 6.79 we set $X = 0$, then the model (6.204) is reduced to the mixed linear model (6.79). All techniques in solving the mixed linear model (6.79) can be applied to solving the model (6.204).

6.5 MULTIVARIATE MIXED LINEAR MODEL FOR MULTIPLE TRAITS

The tools for association analysis of multiple traits in the presence of pedigree data and population structures are multivariate mixed linear models (Furlotte and Eskin 2015).

6.5.1 Multivariate Mixed Linear Model

Assume that n individuals with K correlated traits $Y_1, Y_2, \ldots, Y_K$ are sampled. Let w_{il} be the lth covariate of the ith individual; α_{lk} be the regression coefficient associated with the lth covariate and the kth trait; μ_k be an overall mean of the kth trait; x_{ij} be the indicator variable, taking values of copies of the reference allele for the jth SNP of the ith individual, $z_{ij} = \frac{x_{ij} - \mu_{ju}}{\sqrt{2p_j(1-p_j)}}$; u_{jk} be an independent random genetic effect with mean zero and variance σ_{jk}^2 for SNP j and trait k; and $\varepsilon_i = [\varepsilon_{i1}, \ldots, \varepsilon_{ik}]^T$ be a vector of errors distributed as a normal distribution with a mean vector of zeros and a $K \times K$ variance–covariance matrix Σ, and $\varepsilon_1, \ldots, \varepsilon_n$ are assumed to be independent.

Consider the multivariate regression model for association analysis of multiple phenotypes:

$$\begin{bmatrix} y_{11} & \cdots & y_{1K} \\ \vdots & \ddots & \vdots \\ y_{n1} & \cdots & y_{nK} \end{bmatrix} = \begin{bmatrix} 1 & w_{11} & \cdots & w_{1L} \\ \vdots & \vdots & \ddots & \vdots \\ 1 & w_{n1} & \cdots & w_{nL} \end{bmatrix} \begin{bmatrix} \mu_1 & \cdots & \mu_K \\ \alpha_{11} & \cdots & \alpha_{1K} \\ \vdots & \ddots & \vdots \\ \alpha_{L1} & \cdots & \varepsilon_{LK} \end{bmatrix} + \begin{bmatrix} z_{11} & \cdots & z_{1m} \\ \vdots & \vdots & \vdots \\ z_{n1} & \cdots & z_{nm} \end{bmatrix} \begin{bmatrix} u_{11} & \cdots & u_{1K} \\ \vdots & \vdots & \vdots \\ u_{m1} & \cdots & u_{mK} \end{bmatrix} + \begin{bmatrix} \varepsilon_{11} & \cdots & \varepsilon_{1K} \\ \vdots & \ddots & \vdots \\ \varepsilon_{n1} & \cdots & \varepsilon_{nK} \end{bmatrix}. \quad (6.205)$$

Equation 6.205 can be written in a compact form:

$$Y = W\alpha + Zu + \varepsilon, \quad (6.206)$$

where

Y is a phenotype matrix
W is an $n \times (L+1)$ dimensional covariate matrix
Z is an $n \times m$ dimensional indicator matrix for the genotypes
α is an $(L+1) \times K$ dimensional coefficient matrix associated with W
u is an $m \times K$ dimensional matrix of random effects
ε is an $n \times K$ dimensional error matrix

We assume that n rows of the error matrix ε are independent and identically distributed as $N(0, \Sigma)$ and random genetic effect matrix u has distribution $\text{vec}(u^T) \sim N(0, \Phi)$, where

$$\Phi_j = \text{cov}(u_j, u_j), \quad u_j = [u_{j1}, \ldots, u_{jK}]^T$$

and

$$\Phi = \begin{bmatrix} \Phi_1 & \cdots & 0 \\ \vdots & \vdots & \vdots \\ 0 & \cdots & \Phi_m \end{bmatrix}.$$

Applying a vector of a matrix operation to Equation 6.206 gives

$$\mathrm{vec}\left(Y^T\right)=\left(W\otimes I_K\right)\mathrm{vec}\left(\alpha^T\right)+\left(Z\otimes I_K\right)\mathrm{vec}\left(u^T\right)+\mathrm{vec}\left(\varepsilon^T\right). \tag{6.207}$$

Using Equation 6.207 and assumptions of distributions of the random genetic effects and errors, we obtain the distribution of the phenotypes Y:

$$\mathrm{vec}\left(Y^T\right)\sim N\left(\left(W\otimes I_K\right)\mathrm{vec}\left(\alpha^T\right),\left(Z\otimes I_K\right)\Phi\left(Z^T\otimes I_K\right)+I_n\otimes\Sigma\right). \tag{6.208}$$

Let $V=\mathrm{var}(\mathrm{vec}(Y^T))=(Z\otimes I_K)\Phi(Z^T\otimes I_K)+I_n\otimes\Sigma$. The log-likelihood function is

$$\begin{aligned}l(W,V)=-\frac{nK}{2}\log(2\pi)-\frac{1}{2}\log|V|-\frac{1}{2}\left(\mathrm{vec}\left(Y^T\right)-\left(W\otimes I_K\right)\mathrm{vec}\left(\alpha^T\right)\right)^T\\ *V^{-1}\left(\mathrm{vec}\left(Y^T\right)-\left(W\otimes I_K\right)\mathrm{vec}\left(\alpha^T\right)\right).\end{aligned}$$

To estimate the fixed effects, taking the derivative of the log-likelihood with respect to the fixed effect $\mathrm{vec}(\alpha^T)$ and setting it equal to zero gives the solution

$$\mathrm{vec}\left(\hat{\alpha}^T\right)=\left[\left(W^T\otimes I_K\right)V^{-1}\left(W\otimes I_K\right)\right]^{-1}\left(W^T\otimes I_K\right)V^{-1}\mathrm{vec}\left(Y^T\right). \tag{6.209}$$

Next we estimate the random effects. We assume that u and Y are jointly distributed as

$$\begin{bmatrix}\mathrm{vec}\left(u^T\right)\\ \mathrm{vec}\left(Y^T\right)\end{bmatrix}\sim N\left(\begin{pmatrix}0\\ \left(W\otimes I_K\right)\mathrm{vec}\left(\alpha^T\right)\end{pmatrix}\quad\Lambda\right), \tag{6.210}$$

where $\Lambda=\begin{bmatrix}\Phi & \Phi\left(Z^T\otimes I_K\right)\\ \left(Z\otimes I_K\right)\Phi & V\end{bmatrix}$.

The conditional mean $E[\mathrm{vec}(u^T)|\mathrm{vec}(Y^T)]$ is given by

$$E\left[\mathrm{vec}\left(u^T\right)\middle|\mathrm{vec}\left(Y^T\right)\right]=\Phi\left(Z^T\otimes I_K\right)V^{-1}\left(\mathrm{vec}\left(Y^T\right)-\left(W\otimes I_K\right)\mathrm{vec}\left(\alpha^T\right)\right). \tag{6.211}$$

Therefore, $\mathrm{vec}\left(\hat{u}^T\right)=\Phi\left(Z^T\otimes I_K\right)V^{-1}\left(\mathrm{vec}\left(Y^T\right)-\left(W\otimes I_K\right)\mathrm{vec}\left(\alpha^T\right)\right)$ can be taken as a prediction of a random effect. The variance of predictor of the random effects is

$$\mathrm{var}\left(\mathrm{vec}\left(\hat{u}^T\right)\right)=\Phi\left(Z^T\otimes I_K\right)V^{-1}\left(Z\otimes I_K\right)\Phi. \tag{6.212}$$

In Equation 6.209, we assume that the covariance matrix V is known. In practice, the covariance matrix V is, in general, unknown. It can also be estimated by maximizing the log-likelihood. Taking derivatives $\frac{\partial L(W,V)}{\partial V}$ and setting it to be zero gives (Exercise 6.13)

$$\hat{V} = \left(\text{vec}\left(Y^T\right) - \left(W \otimes I_K\right)\text{vec}\left(\alpha^T\right)\right)\left(\text{vec}\left(Y^T\right) - \left(W \otimes I_K\right)\text{vec}\left(\alpha^T\right)\right)^T. \quad (6.213)$$

The algorithms for estimation of the fixed effects and covariance matrix V are as follows:

Step 0: Initialization. Set

$$\text{vec}\left(\left(\hat{\alpha}^T\right)^{(0)}\right) = \left[\left(W^T \otimes I_K\right)\left(W \otimes I_K\right)\right]^{-1}\left(W^T \otimes I_K\right)\text{vec}\left(Y^T\right),$$

$$\hat{V}^{(0)} = \left(\text{vec}\left(Y^T\right) - \left(W \otimes I_K\right)\text{vec}\left(\left(\alpha^T\right)^{(0)}\right)\right)\left(\text{vec}\left(Y^T\right) - \left(W \otimes I_K\right)\text{vec}\left(\left(\alpha^T\right)^{(0)}\right)\right)^T.$$

Step 1: For $l = 1, 2, \ldots$

$$\text{vec}\left(\left(\hat{\alpha}^T\right)^{(l+1)}\right) = \left[\left(W^T \otimes I_K\right)\left(\hat{V}^{(l)}\right)^{-1}\left(W^T \otimes I_K\right)\right]^{-1}\left(W^T \otimes I_K\right)\left(\hat{V}^{(l)}\right)^{-1}\text{vec}\left(Y^T\right),$$

$$\hat{V}^{(l+1)} = \left(\text{vec}\left(Y^T\right) - \left(W \otimes I_K\right)\text{vec}\left(\left(\hat{\alpha}^T\right)^{(l+1)}\right)\right)\left(\text{vec}\left(Y^T\right) - \left(W \otimes I_K\right)\text{vec}\left(\left(\hat{\alpha}^T\right)^{(l+1)}\right)\right)^T.$$

Repeat step 1 until convergence.

6.5.2 Maximum Likelihood Estimate of Variance Components

Variance components are embedded within the matrix V. Similar to the single variate mixed linear model, we can show (Appendix 6H) that the estimators of the variance components can be obtained by solving the following equation:

$$\text{Tr}\left(V^{-1}V_{il}^{u}\right) = \left(\text{vec}\left(Y^T\right) - H\text{vec}\left(\alpha^T\right)\right)^T V^{-1}V_j V^{-1}\left(\text{vec}\left(Y^T\right) - H\text{vec}\left(\alpha^T\right)\right) \quad (6.214a)$$

or

$$\text{Tr}\left(V^{-1}V_{il}^{u}\right) = \left(\text{vec}\left(Y^T\right)\right)^T P V_{il}^{u} P\text{vec}\left(Y^T\right), \quad (6.214b)$$

where

$$P = V^{-1} - V^{-1}H\left(H^TV^{-1}H\right)^{-1}H^TV^{-1}, \quad H = W \otimes I_K$$

$$\frac{\partial V}{\partial \sigma_{il}^u} = V_{il}^u = \begin{cases} \dfrac{\partial V}{\partial \sigma_{ij}^l} & \sigma_{il}^u = \sigma_{il}^j \\ \dfrac{\partial V}{\partial \sigma_{il}^e} & \sigma_{il}^u = \sigma_{il}^e \end{cases}$$

$\frac{\partial V}{\partial \sigma_{ij}^l}$ and $\frac{\partial V}{\partial \sigma_{il}^e}$ are defined in Equations 6H.5 and 6H.6, respectively.

6.5.3 REML Estimate of Variance Components

Similar to the REML of variance components, we can find a matrix K_M such that

$$K_M^T H = 0, \tag{6.215}$$

where $H = W \otimes I_K$.

Therefore, $K_M^T \text{vec}\left(Y^T\right)$ is distributed as $N\left(0, K_M^T V K_M\right)$. Its log-likelihood is given by

$$l\left(W, V \middle| H, K_M^T Y\right) = -\frac{n}{2}\log(2\pi) - \frac{1}{2}\log\left|K_M^T V K_M\right| - \frac{1}{2}\left(\text{vec}\left(Y^T\right)\right)K_M\left(K_M^T V K_M\right)^{-1}K_M^T \text{vec}\left(Y^T\right). \tag{6.216}$$

The REML equation can be obtained from Equation 6.214a by replacing vec(Y^T) with $K_M^T \text{vec}\left(Y^T\right)$, replacing V with $K_M^T V K_M$, and replacing V_{il}^u with $K_M^T V_{il}^u K_M$ and using the following equality:

$$K_M\left(K_M^T V K_M\right)^{-1} K_M^T = P. \tag{6.217}$$

Indeed, the left side of Equation 6.214a is replaced by

$$\begin{aligned} \text{Tr}\left(\left(K_M^T V K_M\right)^{-1} K_M^T V_{il}^u K_M\right) &= \text{Tr}\left(K_M\left(K_M^T V K_M\right)^{-1} K_M^T V_{il}^u\right) \\ &= \text{Tr}\left(P V_{il}^u\right), \end{aligned} \tag{6.218}$$

and the right side of Equation 6.214a is replaced by

$$\left(\text{vec}\left(Y^T\right)\right)^T K_M\left(K_M^T V K_M\right)^{-1} K_M^T V_{il}^u K_M\left(K_M^T V K_M\right)^{-1} K_M^T \text{vec}\left(Y^T\right) = \left(\text{vec}\left(Y^T\right)\right)^T P V_{il}^u P \text{vec}\left(Y^T\right). \tag{6.219}$$

Combing Equations 6.214a, 6.218, and 6.219 gives

$$\mathrm{Tr}\left(PV_{il}^{u}\right)=\left(\mathrm{vec}\left(Y^{T}\right)\right)^{T}PV_{il}^{u}P\mathrm{vec}\left(Y^{T}\right). \tag{6.220}$$

Numerical techniques for solving Equation 6.101 in single variate mixed linear models can be extended to solving Equation 6.220 with more complex and intensive computations. The hypothesis testing for single variate mixed linear models can also be extended to multivariate mixed linear models. Because their extensions are straightforward, we do not repeat them here again.

6.6 HERITABILITY

Heritability quantifies how much of the phenotypic variation is due to genetic variation. Two types of heritability are explored to measure the proportion of the phenotype variance explained by the genetic variance: broad-sense heritability, which measure the contribution of the total genetic variation to the phenotype variation, and narrow-sense heritability, which quantifies the contribution of the genetic additive effect to the phenotype variation (Golan et al. 2014). Although heritability estimation is an essential issue for genetic studies of complex traits, there has been debate on how to estimate heritability in the past decades (Krishna Kumar et al. 2016). GWASs have identified tens of thousands of genetic variants significantly associated with hundreds of diseases. However, the significantly associated genetic variants explain only a small fraction of heritability. The heritability in the GWAS is missing. A number of approaches are proposed to solve heritability problems (Bonnet et al. 2015; Heckerman et al. 2016; Yang et al. 2010; Zaitlen and Kraft 2012).

In this section, we will introduce the definition of heritability. Since whole-genome sequencing data are widely available, we will mainly introduce heritability estimation methods in high-dimensional linear mixed models with GWAS and whole-genome sequencing data. The traditional approach to heritability estimation has focused on single trait heritability. Fewer methods for the estimation of multiple trait heritability have been developed. To fill this gap, we will investigate the estimation of the heritability of multiple traits. Genetic causal inference emerged as a novel powerful tool for genetic studies of complex diseases. Since it uses a mixed structural equation model as a general framework for causal heritability analysis, causal inference for the heritability of multiple traits will be introduced in *Big Data in Omics and Imaging: Integrated Analysis and Causal Inference*, Chapter 1.

6.6.1 Heritability Estimation for a Single Trait

6.6.1.1 Definition of Narrow-Sense Heritability

There are two ways to define the heritability: broad definition and narrow definition of heritability. The phenotype variance V_p is the sum of genetic variance V_G and environmental variance V_E (Zaitlen and Kraft 2012):

$$V_P = V_G + V_E.$$

Broad-sense heritability is defined as the ratio of total genetic variance to phenotypic variance:

$$H^2 = \frac{V_G}{V_P}. \tag{6.221}$$

The genetic variance includes the genetic additive variance V_A, dominance variance V_D, and genetic interaction variance V_I, i.e., $V_G = V_A + V_D + V_I$. The broad-sense heritability covers the contribution of the allele–allele interactions and gene–gene interactions to the phenotypic variation.

The narrow-sense heritability can be defined as the ratio of the genetic additive variance to phenotypic variance to capture the genetic "additive" contribution:

$$h^2 = \frac{V_A}{V_P}. \tag{6.222}$$

In this section, we focus on narrow-sense heritability.

6.6.1.2 Mixed Linear Model for Heritability Estimation

Consider m SNPs genotyped on n individuals. Let x_{ij} be the number of copies of the reference allele for the jth SNP of the ith individual and p_j be the frequency of the reference allele. The standardized genotype is defined as

$$z_{ij} = \frac{x_{ij} - 2p_j}{\sqrt{2p_j\left(1-2p_j\right)}}, \quad i = 1,\ldots,n, \; j = 1,\ldots,m. \tag{6.223}$$

Let y_i be the phenotype of the ith individual and $Y = [y_1, \ldots, y_n]^T$ be a vector of phenotypes. Consider a general mixed linear model (Yang et al. 2011)

$$Y = H\gamma + Zu + \varepsilon, \tag{6.224}$$

where

H is an $n \times q$ dimensional matrix of covariates
γ is a q-dimensional vector of regression coefficients associated with covariates
Z is an $n \times m$ dimensional matrix of standardized genotypes defined in Equation 6.223
u is an m-dimensional vector of random genetic effects with $u \sim N\left(0, \mathbf{I}_n\sigma_u^2\right)$
ε is a n-dimensional vector of residual effects with $\varepsilon \sim N\left(0, \mathbf{I}\sigma_e^2\right)$

The variance of the phenotypes is then given by

$$V = \operatorname{var}\left(Y\right) = \sigma_u^2 ZZ^T + \mathbf{I}_n\sigma_e^2. \tag{6.225}$$

Define $V_A = m\sigma_u^2$ and $G = \frac{ZZ^T}{m}$, where G is the genetic relationship matrix (GRM). Then, the phenotypic variance matrix can be expressed as

$$V = V_A G + \mathbf{I}_n \sigma_e^2. \tag{6.226}$$

The model in (6.224) is then equivalent to

$$Y = H\gamma + g + \varepsilon, \tag{6.227}$$

where g is an n-dimensional vector of the genetic random effects with $g \sim N(0, V_A G)$. Parameter estimation methods for the mixed linear models introduced in Section 6.3 can be used to estimate the genetic additive variance V_A. Then, using Equation 6.222, we can estimate the heritability.

In the model (6.227), we consider only one genetic factor. In general, we can extend the model (6.227) to l genetic factors:

$$Y = H\gamma + \sum_{j=1}^{l} g_j + \varepsilon, \tag{6.228}$$

where g_j is an n-dimensional vector of the random genetic effects with $g_j \sim N(0, V_{Aj} G_j)$, where V_{Aj} is the additive genetic variance of the jth genomic region and G_j is its corresponding GRM. The phenotypic variance is then equal to

$$V = \sum_{j=1}^{l} V_{A_j} G_j + I\sigma_e^2. \tag{6.229}$$

The heritability can then be estimated by

$$\hat{h}^2 = \frac{\sum_{j=1}^{l} \hat{V}_{A_j}}{\sum_{j=1}^{l} \hat{V}_{A_j} + \hat{\sigma}_e^2}. \tag{6.230}$$

The genetic additive variances and variance due to residuals are often estimated by the REML methods discussed in Section 6.3. Let θ be a vector of variance components $\left[V_{A_1}, \ldots, V_{A_l}, \sigma_e^2\right]^T$. For the completeness, we briefly describe the algorithms for estimation of variance components as follows (Yang et al. 2011).

Algorithms for Variance Component Estimation

Step 0: Initialization.

The EM algorithm will be used to find the initial values of the variance components.

(i) $\theta_j^{(0)} = \frac{V_p}{l+1}$, which is the initial values for he EM algorithm.

(ii) $\theta_j^{(1)} = \frac{1}{n}\left[\left(\theta_j^{(0)}\right)^2 Y^T PG_j PY + \text{Tr}(\theta_j^{(0)}\text{I} - \left(\theta_j^{(0)}\right)^2 PA_j\right]$.

Average information algorithm will be used to implement the REML algorithm for estimation of variance components.

Step 1: Newton–Raphson iteration is updated (6.117):

$$\theta^{(t+1)} = \theta^{(t)} - \left(H\left(\theta^{(t)}\right)\right)^{-1} F\left(\theta^{(t)}\right), \tag{6.231}$$

where

$$F\left(\theta^{(t)}\right) = \frac{\partial l}{\partial \theta} = \frac{1}{2}\begin{bmatrix} \text{Tr}\left(PG_1\right) - Y^T PG_1 PY \\ \vdots \\ r\left(PG_l\right) - Y^T PG_l PY \\ \text{Tr}\left(P\right) - Y^T PPY \end{bmatrix}, \tag{6.232}$$

$$H\left(\theta^{(t)}\right) = \frac{1}{2}\begin{bmatrix} Y^T PG_1 PG_1 PY & \cdots & Y^T PG_1 PG_l PY & Y^T PG_1 PPY \\ \vdots & \vdots & \vdots & \vdots \\ Y^T PG_l PG_1 PY & \cdots & Y^T PG_l PG_l PY & Y^T PG_l PPY \\ Y^T PPG_1 PY & \cdots & Y^T PPG_l PY & Y^T PPPY \end{bmatrix}, \tag{6.233}$$

$$P = V^{-1} - V^{-1}H\left(H^T V^{-1} H\right)^{-1} H^T V^{-1}. \tag{6.234}$$

Step 2: Calculate the log-likelihood function.

The log-likelihood function for REML can be rewritten as (Appendix 6G)

$$l = -\frac{1}{2}\log\left|V\right| - \frac{1}{2}\log\left|H^T V^{-1} H\right| - \frac{1}{2}\left(Y - H\hat{\lambda}\right)^T V^{-1}\left(Y - H\hat{\gamma}\right) \tag{6.235}$$

or

$$l = -\frac{1}{2}\log\left|V\right| - \frac{1}{2}\log\left|H^T V^{-1} H\right| - \frac{1}{2}Y^T PY. \tag{6.236}$$

Calculate

$$\hat{V}^{(t+1)} = \sum_{j=1}^{l}\left(\hat{V}_{A_j}\right)^{(t+1)} G_j + I\left(\hat{\sigma}_e^2\right)^{(t+1)}, \tag{6.237}$$

where $\left(\hat{V}_{A_j}\right)^{(t+1)} = \hat{\theta}_j^{(t+1)}$, $j = 1,\ldots,l$ and $\left(\hat{\sigma}_e^2\right)^{(t+1)} = \hat{\theta}_{(l+1)}^{(t+1)}$.

$$\hat{P}^{(t+1)} = \left(\hat{V}^{(t+1)}\right)^{-1} - \left(\hat{V}^{(t+1)}\right)^{-1} H\left(H^T\left(\hat{V}^{(t+1)}\right)^{-1} H\right)^{-1} H^T\left(\hat{V}^{(t+1)}\right)^{-1}. \quad (6.238)$$

Calculate

$$l^{(t+1)} = -\frac{1}{2}\log\left|\hat{V}^{(t+1)}\right| - \frac{1}{2}\log\left|H^T\left(\hat{V}^{(t+1)}\right)^{-1} H\right| - \frac{1}{2}Y^T\hat{P}^{(t+1)}Y. \quad (6.239)$$

Step 3: Convergence assessment.
If $|l^{(t+1)} - l^{(t)}| < 10^{-4}$, then stop. Otherwise if $t \leftarrow t+1$, go to step 1.

6.6.2 Heritability Estimation for Multiple Traits

6.6.2.1 Definition of Heritability Matrix for Multiple Traits

Let A denote what can be transmitted from a parent, i.e., breeding value, and P denote the phenotypic value of a parent. Then, for a single trait, an individual's estimated breeding value is the product of the phenotypic value and the heritability

$$A = h^2 P. \quad (6.240)$$

Equation 6.240 can be extended to multiple traits. Consider K traits. Let A be a K-dimensional vector of breeding values for the K traits, P be a K-dimensional vector of phenotypic values, and H be a $K \times K$ dimensional heritability matrix:

$$H = \begin{bmatrix} h_{11} & \cdots & h_{1K} \\ \vdots & \vdots & \vdots \\ h_{K1} & \cdots & h_{KK} \end{bmatrix}. \quad (6.241)$$

Then, Equation 6.240 can be extended to the prediction of breeding values for the K traits:

$$\begin{bmatrix} A_1 \\ \vdots \\ A_K \end{bmatrix} = \begin{bmatrix} h_{11} & \cdots & h_{1K} \\ \vdots & \vdots & \vdots \\ h_{K1} & \cdots & h_{KK} \end{bmatrix} \begin{bmatrix} P_1 \\ \vdots \\ P_K \end{bmatrix} \quad (6.242)$$

or

$$A = HP.$$

Denote the covariance matrix between breeding values and phenotypic values by Σ_{AP} and covariance matrix of the phenotypic values by Σ_p.

Recall that

$$P = A + D + E, \tag{6.243}$$

where

D is a K-dimensional vector of dominance deviation
E is a K-dimensional vector of environments

If we assume that the covariance between the breeding value and dominance deviation and the variance between the environment and genetic values are zero, then it follows from equation (6.243) that

$$\text{cov}(A,P) = \text{cov}(A,A) = \Sigma_A, \tag{6.244}$$

where

$$\Sigma_A = \begin{bmatrix} V_{11}^A & \cdots & V_{1K}^A \\ \vdots & \vdots & \vdots \\ V_{K1}^A & \cdots & V_{KK}^A \end{bmatrix}.$$

Using Equations 6.242 and 6.244 gives

$$\Sigma_A = H\Sigma_P. \tag{6.245}$$

Therefore, we define the heritability matrix as

$$H = \Sigma_A \Sigma_P^{-1}. \tag{6.246}$$

6.6.2.2 Connection between Heritability Matrix and Multivariate Mixed Linear Models

If we consider a SNP ($m = 1$), then the matrix Φ in Equation 6.208 is equal to

$$\Phi = \Sigma_A. \tag{6.247}$$

Recall that the multivariate mixed linear model (6.175) is

$$Y = W\alpha + Zu + \varepsilon. \tag{6.248}$$

We assume that

$$u^T \sim N(0, I_m \otimes \Sigma_A) \quad \text{and} \quad \varepsilon \sim N(0, I_n \otimes \Sigma_e). \tag{6.249}$$

Then, Equation 6.208 becomes

$$\operatorname{vec}\left(Y^T\right) \sim N\left(\left(W \otimes I_K\right)\operatorname{vec}\left(\alpha^T\right), \left(Z \otimes I_K\right)\left(I_m \otimes \Sigma_A\right)\left(Z^T \otimes I_K\right) + I_n \otimes \Sigma_e\right). \quad (6.250)$$

Let

$$V = \left(Z \otimes I_K\right)\left(I_m \otimes \Sigma_A\right)\left(Z^T \otimes I_K\right) + I_n \otimes \Sigma_e). \quad (6.251)$$

We can show that

$$\begin{aligned}\left(Z \otimes I_K\right)\left(I_m \otimes \Sigma_A\right)\left(Z^T \otimes I_K\right) &= \left(ZZ^T\right) \otimes \Sigma_A \\ &= G \otimes \Sigma_A,\end{aligned} \quad (6.252)$$

where $G = ZZ^T$ is a GRM. Equation 6.251 is then reduced to

$$V = G \otimes \Sigma_A + I_n \otimes \Sigma_e. \quad (6.253)$$

The heritability matrix can then be estimated by

$$\hat{H} = \hat{\Sigma}_A \hat{\Sigma}_P^{-1}, \quad (6.254)$$

where $\hat{\Sigma}_P = \hat{\Sigma}_A + \hat{\Sigma}_e$.

Estimation methods for the multivariate mixed linear models studied in Section 6.4 can be used to estimate variance component matrices $\hat{\Sigma}_A$ and $\hat{\Sigma}_e$.

6.6.2.3 *Another Interpretation of Heritability*

Heritability can also be interpreted proportions of phenotype variance explained by the genetic variance. Consider the fixed linear model for a single trait:

$$Y = X\gamma + \varepsilon, \quad (6.255)$$

where

- H is genotype matrix
- γ is the regression coefficients

The regression coefficient is estimated by

$$\hat{\gamma} = \left[\operatorname{var}(X)\right]^{-1} \operatorname{cov}(X,Y). \quad (6.256)$$

The variance due to regression contribution is $\operatorname{var}(X\hat{\gamma})$. The proportion of the phenotype variance explained by genetic variance is defined as

$$R^2 = \frac{\operatorname{var}(X\hat{\gamma})}{\operatorname{var}(Y)} = \frac{\hat{\gamma}^T \operatorname{var}(X)\hat{\gamma}}{\operatorname{var}(Y)} = \frac{\operatorname{cov}(X,Y)^T \left[\operatorname{var}(X)\right]^{-1} \operatorname{cov}(X,Y)}{\operatorname{var}(Y)}, \quad (6.257)$$

which is the squared multiple correlation coefficient.

Recall (Mitteroecker et al. 2016) that the genetic additive effect is

$$\gamma = \frac{\text{cov}(X, Y)}{\text{Var}(X)} \quad \text{and} \quad \text{Var}(X) = 2pq.$$

The genetic additive variance is

$$V_A = 2pq\gamma^2 = \text{var}(X)\frac{\left[\text{cov}(X, Y)\right]^2}{\left[\text{Var}(X)\right]^2} = \frac{\left[\text{cov}(X, Y)\right]^2}{\text{Var}(X)}. \tag{6.258}$$

Therefore, the heritability is

$$h^2 = \frac{V_A}{V_p} = \frac{\left[\text{cov}(X, Y)\right]^2}{\text{Var}(Y)\text{Var}(X)} = R^2. \tag{6.259}$$

Now we extend Equation 6.259 to multiple traits. Consider a multivariate linear model:

$$Y = XB + \varepsilon. \tag{6.260}$$

The matrix of regression coefficients is

$$\hat{B} = \Sigma_{xx}^{-1}\Sigma_{XY}. \tag{6.261}$$

The breeding values are given by

$$A = XB. \tag{6.262}$$

The covariance matrix of breeding values is

$$\begin{aligned} V_A &= \text{cov}(XB, XB) \\ &= B^T\Sigma_{xxx}B \\ &= \Sigma_{YX}\Sigma_{xx}^{-1}\Sigma_{XY}. \end{aligned} \tag{6.263}$$

Substituting Equation 6.263 to Equation 6.254 gives

$$H = \Sigma_{YX}\Sigma_{XX}^{-1}\Sigma_{XY}\Sigma_{YY}^{-1}, \tag{6.264}$$

which is equivalent to the R matrix in Equation 1.220 or extension of the squared multiple correlation coefficient.

6.6.2.4 Maximizing Heritability

Next we show that maximizing heritability (Mitteroecker et al. 2016) is equivalent to canonical correlation analysis. We consider a linear combination of multiple phenotypes, Yb, and a linear combination of genotypes at multiple loci, Xa, to transform the association analysis of multiple traits to the association analysis of a single trait. Define the linear genetic model for Yb and Xa:

$$Yb = (Xa)\alpha + \varepsilon. \tag{6.265}$$

The total genetic effect of the multiple genotypes on the multiple traits can be estimated by

$$\hat{\alpha} = \frac{\operatorname{cov}(Xa, Yb)}{\operatorname{var}(xa)} = \frac{a^T\Sigma_{xy}b}{a^T\Sigma_{xx}a}. \tag{6.266}$$

Using Equations 6.258, we obtain the genetic additive variance of Xa and heritability of Yb:

$$\sigma_{AI}^2 = \operatorname{var}(Xa)\alpha^2 = \frac{\left(a^T\Sigma_{xy}b\right)^2}{a^T\Sigma_{xx}a} \tag{6.267}$$

and

$$h_I^2 = \frac{\operatorname{var}(Xa)\alpha^2}{\operatorname{var}(Yb)} = \frac{\left(a^T\Sigma_{xy}b\right)^2}{b^T\Sigma_{yy}ba^T\Sigma_{xx}a}, \tag{6.268}$$

respectively.

It is clear that the squared multiple correlation coefficient is given by

$$R^2 = \frac{\left[\operatorname{cov}(Yb, (Xa)\alpha)\right]^2}{\operatorname{var}(Yb)\operatorname{var}((Xa)\alpha)} = \frac{\left[\operatorname{cov}(Yb, Xa)\right]^2\alpha^2}{\operatorname{var}(Yb)\operatorname{var}(Xa)\alpha^2} = \frac{\left(a^T\Sigma_{xy}b\right)^2}{b^T\Sigma_{yy}ba^T\Sigma_{xx}a} = h_I^2.$$

Next we seek the optimal combinations of the genotypes at multiple loci and the multiple traits to maximize the genetic additive effect, genetic additive variance, and heritability. We first find the maximum genetic additive effect.

Using Equation 6.266 and the Lagrangian multiplier method, we can solve the following optimization problem to obtain the maximum genetic additive effect:

$$L(a, b, \lambda) = a^T\Sigma_{xy}b + \frac{\lambda}{2}\left(1 - a^T\Sigma_{xx}a\right), \tag{6.269}$$

where λ is a multiplier.

Setting $\frac{\partial L(a, b, \lambda)}{\partial a} = \Sigma_{xy}b - \lambda\Sigma_{xx}a = 0$ gives

$$\Sigma_{xy}b = \lambda\Sigma_{xx}a$$

or

$$\Sigma_{xx}^{-1}\Sigma_{xy}b = \lambda a. \tag{6.270}$$

Suppose that the SVD of the matrix $\Sigma_{xx}^{-1}\Sigma_{xy}$ is given by

$$\Sigma_{xx}^{-1}\Sigma_{xy} = U_e\Lambda_e V_e^T. \tag{6.271}$$

Then, it follows from Equation 6.271 that a, b and the optimal genetic effect are the left and right singular vectors, and singular value of $\Sigma_{xx}^{-1}\Sigma_{xy}$, respectively. Similarly, we can show that the genetic additive variance σ_{Al}^2 is the square of the singular value of $\Sigma_{xx}^{-1}\Sigma_{xy}$. The maximum heritability can be obtained by setting the Lagrange function

$$L(a, b, \lambda, \mu) = \left(a^T\Sigma_{xy}b\right)^2 + \lambda\left(1 - a^T\Sigma_{xx}a\right) + \mu\left(1 - b^T\Sigma_{yy}b\right)$$

and

$$\begin{aligned} \frac{\partial L}{\partial a} &= 2\left(a^T\Sigma_{xy}b\right)\Sigma_{xy}b - 2\lambda\Sigma_{xx}a = 0, \\ \frac{\partial L}{\partial b} &= 2\left(a^T\Sigma_{xy}b\right)\Sigma_{yx}a - 3\mu\Sigma_{yy}b = 0. \end{aligned} \tag{6.272}$$

Let $K = \Sigma_{xx}^{-1/2}\Sigma_{xy}\Sigma_{yy}^{-1/2}$ and SVD of K be

$$\Sigma_{xx}^{-1/2}\Sigma_{xy}\Sigma_{yy}^{-1/2} = U_K\Lambda_K V_K^T. \tag{6.273}$$

Solving Equation 6.272, we obtain

$$\begin{aligned} a &= \Sigma_{xx}^{-1/2}u_k, \\ b &= \Sigma_{yy}^{-1/2}v_k, \\ h_l^2 &= \lambda = \tau_K^2, \end{aligned} \tag{6.274}$$

where u_k, v_k, and τ_K are the left and right singular vectors and singular value of the matrix K, respectively. Substituting Equation 6.274 into Equations 6.266 and 6.267 gives

$$\begin{aligned} \hat{\alpha} &= u_K^T K v_K, \\ \sigma_{Al}^2 &= \hat{\alpha}^2. \end{aligned} \tag{6.275}$$

Note that τ_K^2 is the eigenvalue of the matrix

$$K^T K = \Sigma_{yy}^{-1/2} \Sigma_{yx} \Sigma_{xx}^{-1} \Sigma_{xy} \Sigma_{yy}^{-1/2}. \quad (6.276)$$

It follows from Equation 6.276 and 1.220 that

$$K^T K = R.$$

This shows that the maximum heritability analysis is equivalent to CCA. This also shows that the maximum heritability is the largest eigenvalue of the heritability matrix.

6.7 FAMILY-BASED ASSOCIATION ANALYSIS FOR QUALITATIVE TRAIT

This section introduces a general strategy to analyze multiple common and rare variants across family- and population-based samples to unify family and population designs. Population-based sample design is the current major study design for association studies. However, many rare variants are from recent mutations in pedigrees (Lupski et al. 2011). The inability of common variants to account for most of the supposed heritability and the low power of population-based analysis tests for the association of rare variants have led to a renewed interest in family-based design with enrichment for risk alleles to detect the association of rare variants (Ott et al. 2011). It is hypothesized that an individual's disease risk is likely to come from the collected action of common variants segregating in the population and rare variants recently arising in extended pedigrees. It is increasingly recognized that analyzing samples from populations and pedigrees separately is highly inefficient (Liu and Thalamuthu 2011). It is natural to unify population and family study designs for association studies. The unified approach can correct for unknown population stratification, family structure, and cryptic relatedness while maintaining high power in the sequence-based association studies.

Functional data analysis methods combined with high-dimensional data reduction techniques will be taken as a general framework to unify population-based and family-based association analysis with the next-generation sequencing (NGS) data. Family-based functional principal component analysis (FPCA) with or without smoothing, generalized T^2, collapsing, weighted sum of square (WSS) and variable-threshold (VT) tests, and CMC and single marker association test statistics will be introduced. Most materials are based on our two papers (Shugart et al. 2012; Zhu and Xiong 2012).

6.7.1 The Generalized T^2 Test with Families and Additional Population Structures

Consider n sampled individuals from multiple families or unrelated individuals with unknown population structures. Assume that each individual has T genetic variants. Suppose that the genotypes of the ith individual at the tth genetic variant site are denoted by $a_t a_t$, $a_t A_t$, and $A_t A_t$, respectively. Assume that A_t is a risk allele. Define an indicator variable for the genotype as (Zhu and Xiong 2012)

$$Z_i^t = \begin{cases} 2 & A_t A_t \\ 1 & A_t a_t \\ 0 & a_t a_t \end{cases}, \quad i = 1, 2, \ldots, n, \quad t = 1, 2, \ldots, T.$$

Let

$$Z^t = \left[Z_1^t, \ldots, Z_n^t \right]^T \quad \text{and} \quad Z = \left[\left(Z^1\right)^T, \ldots, \left(Z^T\right)^T \right]^T.$$

Define $D_r = [u_1, \ldots, u_n]^T$ and $D_p = [1, 1, \ldots, 1]^T$, a column vector of 1s of length n, where

$$u_i = \begin{cases} 1 & \text{if } i \text{ is a case} \\ 0 & \text{if } i \text{ is a control.} \end{cases}$$

Define

$$H = \begin{bmatrix} \left(D_r - \frac{n_c}{n} D_p\right)^T & 0 & \cdots & 0 \\ 0 & \left(D_r - \frac{n_c}{n} D_p\right)^T & \cdots & 0 \\ \cdots & \cdots & \cdots & \cdots \\ 0 & 0 & \cdots & \left(D_r - \frac{n_c}{n} D_p\right)^T \end{bmatrix}$$

$$= I_{(T)} \otimes \left(D_r - \frac{n_c}{n} D_p \right)^T, \tag{6.277}$$

where

n_c is the number of affected individuals
$I_{(T)}$ is a T-dimensional identity matrix
$\otimes$ denotes the Kronecker product of two matrices

The generalized T^2 statistic with pedigree structures is defined as

$$T_F^2 = (HZ)^T \Gamma^{-1} HZ, \tag{6.278}$$

where $\Gamma = \text{cov}(HZ, HZ)$.

Let

$$\Sigma_z = \begin{bmatrix} \sigma_{11} & \sigma_{12} & \cdots & \sigma_{1T} \\ \sigma_{21} & \sigma_{22} & \cdots & \sigma_{2T} \\ \cdots & \cdots & \cdots & \cdots \\ \sigma_{T1} & \sigma_{T2} & \cdots & \sigma_{TT} \end{bmatrix}, \tag{6.279}$$

where $\sigma_{ij} = \text{cov}\left(Z_1^i, Z_1^j\right)$.

The generalized T^2 statistic was originally developed for unrelated individuals. A key issue for its extension to the family data is how to take account of the dependence relationships among individuals in the calculation of the covariance matrix of markers at multiple loci. The concept of IBD discussed in Section 6.1 is the basis for the calculation of the covariance matrix of the family genetic data.

It can be shown that (Appendix 6I)

$$\Lambda_z = \operatorname{cov}(Z, Z) = \Sigma_z \otimes \Phi, \tag{6.280}$$

where Φ is the kinship matrix and defined as

$$\Phi = \begin{bmatrix} 1+h_1 & 2\phi_{12} & \cdots & 2\phi_{1n} \\ 2\phi_{21} & 1+h_2 & \cdots & 2\phi_{2n} \\ \cdots & \cdots & \cdots & \cdots \\ 2\phi_{n1} & 2\phi_{n2} & \cdots & 1+h_n \end{bmatrix}, \tag{6.281}$$

where

h_i is the inbreeding coefficient of individual i

ϕ_{ij} is the kinship coefficient between individuals i and j

The matrix Σ_z can be estimated by

$$\hat{\Sigma}_z = \frac{1}{n-T}\sum_{i=1}^{n}\left(Z_i - \bar{Z}\right)\left(Z_i - \bar{Z}\right)^T, \tag{6.282}$$

where

$$Z_i = \left[Z_i^1, Z_i^2, \ldots, Z_i^T\right]^T, \quad \bar{Z} = \frac{1}{n}\sum_{i=1}^{n} Z_i.$$

Using Equations 6.277 and 6.280, we can calculate the covariance matrix Γ:

$$\begin{aligned} \Gamma &= \operatorname{cov}(HZ, HZ) \\ &= H\Lambda_z H^T \\ &= \left[I_{(T)} \otimes \left(D_r - \frac{n_c}{n} D_p\right)^T\right]\left[\Sigma_z \otimes \Phi\right]\left[I_{(T)} \otimes \left(D_r - \frac{n_c}{n} D_p\right)\right] \\ &= \left[\left(D_r - \frac{n_c}{n} D_p\right)^T \Phi \left(D_r - \frac{n_c}{n} D_p\right)\right]\Sigma_z. \end{aligned} \tag{6.283}$$

To establish the relationship between the test statistic T_F^2 for general pedigrees and the T^2 statistic for the population-based association test, we need to simplify HZ. It is easy to see that

$$\begin{aligned}\left(D_r-\frac{n_c}{n}D_p\right)^T Z^t &= \sum_{i\in \text{cases}} Z_i^t-\frac{n_c}{n}\sum_{i=1}^{n} Z_i^t\\ &= n_c\bar{Z}_A^t-\frac{n_c}{n}\left[n_c\bar{Z}_A^t+(n-n_c)\bar{Z}_G^t\right]\\ &= \frac{n_c(n-n_c)\left[\bar{Z}_A^t-\bar{Z}_G^t\right]}{n},\end{aligned} \tag{6.284}$$

where $\bar{Z}_A^t$ and $\bar{Z}_G^t$ are averages of the indicator variables for the genotypes at the tth variant site in cases and controls, respectively.

From Equation 6.284, it follows that

$$HZ=\frac{n_c(n-n_c)}{n}\begin{bmatrix}\bar{Z}_A^1-\bar{Z}_G^1\\ \vdots\\ \bar{Z}_A^T-\bar{Z}_G^T\end{bmatrix}. \tag{6.285}$$

Therefore, the test statistic T_F^2 can be simplified to

$$\begin{aligned}T_F^2 &= \frac{(HZ)^T\Sigma_z^{-1}HZ}{\left(D_r-\frac{n_c}{n}D_p\right)^T\Phi\left(D_r-\frac{n_c}{n}D_p\right)}\\ &= \frac{\left[\frac{n_c(n-n_c)}{n}\right]^2\left[\left(\bar{Z}_A-\bar{Z}_G\right)^T\Sigma_z^{-1}\left(\bar{Z}_A-\bar{Z}_G\right)\right]}{\left(D_r-\frac{n_c}{n}D_p\right)^T\Phi\left(D_r-\frac{n_c}{n}D_p\right)}\\ &= \frac{T^2}{\frac{n}{n_c(n-n_c)}\left(D_r-\frac{n_c}{n}D_p\right)^T\Phi\left(D_r-\frac{n_c}{n}D_p\right)}\\ &= \frac{T^2}{P_{corr}},\end{aligned} \tag{6.286}$$

where T^2 is the generalized T^2 statistic for the population-based association tests and $P_{corr}=\frac{n}{n_c(n-n_c)}\left(D_r-\frac{n_c}{n}D_p\right)^T\Phi\left(D_r-\frac{n_c}{n}D_p\right)$ is the correction factor to be applied to the generalized T^2 statistic to have a valid test in the presence of pedigree and population structures. The correction factor depends on kinship coefficients and the number of individuals in cases and controls. Under the null hypothesis of no association of the genomic region with the disease, T_F^2 is distributed as a central $\chi^2_{(T)}$ distribution with T degrees of freedom.

6.7.2 Collapsing Method

The population-based collapsing test can be extended to the families with known or unknown population structure. Consider n sampled individuals in the pedigrees. An indicator variable for the ith individual in the pedigrees is defined as

$$x_i = \begin{cases} 1 & \text{presence of rare variants in the region} \\ 0 & \text{otherwise.} \end{cases}$$

Let $X=[x_1,x_2,\ldots,x_n]^T$. Then, the expectation of the vector of indicator variables under the null hypothesis of no association of the genomic region with the disease is given by $E_0[X]=[p,p,\ldots,p]^T$, where $p=P(\text{presence of rare variants in the genomic region})$. Under the alternative model of association of the genomic region with the disease, we assume that

$$E[x_i]=\mu_i=p+u_ir,$$

where $0<p<1$, $0<p+r<1$, and $u_i = \begin{cases} 1 & \text{if } i \text{ is case} \\ 0 & \text{otherwise.} \end{cases}$

Define $\mu=[\mu_1,\mu_2,\ldots,\mu_n]^T$. The derivative of μ with respect to p is given by

$$D_p = \frac{\partial \mu}{\partial p} = [1, 1, \ldots, 1]^T.$$

Similarly, we have

$$D_r = \frac{\partial \mu}{\partial r} = u, \quad \text{where } u = [u_1, u_2, \ldots, u_n]^T.$$

We then calculate the covariance matrix of the vector X. Let h_i be the inbreeding coefficient of individual i and ϕ_{ij} be the kinship coefficient between individual i and j. Let $\sigma^2=p(1-p)$. Computing the expectation by conditioning, we have (Exercise 6.17)

$$\operatorname{Cov}(x_i,x_j)=2\phi_{ij}\sigma^2. \tag{6.287}$$

By the same token, we have

$$\operatorname{Var}(x_i)=(1+h_i)\sigma^2. \tag{6.288}$$

We assume that the kinship coefficient matrix is given by

$$\Phi = \begin{bmatrix} 1+h_1 & 2\phi_{12} & \cdots & 2\phi_{1n} \\ 2\phi_{21} & 1+h_2 & \cdots & 2\phi_{2n} \\ \cdots & \cdots & \cdots & \cdots \\ 2\phi_{n1} & 2\phi_{n2} & \cdots & 2\phi_{n2} \end{bmatrix}. \quad (6.289)$$

Combining Equations 6.287, 6.288, and 6.289, we can obtain the following covariance matrix of the vector X:

$$\Sigma = \text{Var}(X, X) = \sigma^2 \Phi. \quad (6.290)$$

Let

$$H_C = \left(D_r - \frac{n_c}{n} D_p\right)^T X,$$

where n_c is the number of cases.

The covariance matrix of H_C is given by

$$\begin{aligned} \Gamma &= \text{Cov}(H_C, H_C) \\ &= \left(D_r - \frac{n_c}{n} D_p\right)^T \Phi \left(D_r - \frac{n_c}{n} D_p\right) \sigma^2. \end{aligned}$$

The statistic for testing the association of a genomic region with the disease can be defined as

$$T_{CF} = \frac{H_C^2}{\Gamma}. \quad (6.291)$$

However,

$$\begin{aligned} H_C &= D_r^T X - \frac{n_c}{n} D_p^T X \\ &= \sum_{i \in \text{cases}} x_i - \frac{n_c}{n} \sum_{i=1}^{n} x_i \\ &= n_c \bar{X}_A - \frac{n_c}{n}\left(n_c \bar{X}_A + n_G \bar{X}_G\right) \\ &= \frac{n_c n_G}{n}\left(\bar{X}_A - \bar{X}_G\right), \end{aligned} \quad (6.292)$$

where

n_G is the number of controls

$\bar{X}_A$ and $\bar{X}_G$ are the averages of the indicator variables in cases and controls, respectively

The test statistic can then be rewritten as (Exercise 6.18)

$$T_{CF} = \frac{T_C}{P_{corr}}, \tag{6.293}$$

where

T_C is the population-based collapsing test statistic

$P_{corr} = \frac{n}{n_c n_G}\left(D_r - \frac{n_c}{n} D_p\right)^T \Phi \left(D_r - \frac{n_c}{n} D_p\right)$ is a correction factor

Under the null hypothesis of no association, T_{CF} is distributed as a central $\chi^2_{(1)}$ distribution.

6.7.3 CMC with Families

Now we extend the population-based CMC test to the families with known or unknown population structure. We previously extended the population-based generalized T^2 test and collapse test to the families. Combining the collapsing test and the generalized T^2 test with families, we can obtain the CMC test with families in the samples. Specifically, suppose that T variants can be classified as k groups of rare variants and m individual variant sites.

Define indicator variables for the k group of rare variants:

$$v_i^s = \begin{cases} 1 & \text{presence of rare variants in the } s\text{th group of the } i\text{th individual} \\ 0 & \text{otherwise} \end{cases}$$

and

$$P_s = P(\text{presence of the rare variants in the } s\text{th group}).$$

The variance of the indicator variable can be estimated by

$$\sigma_s^2 = P_s(1 - P_s), \quad s = 1, 2, \ldots, k.$$

Let

$$V^s = \begin{bmatrix} v_1^s \\ \vdots \\ v_n^s \end{bmatrix} \text{ and } V = \begin{bmatrix} V^1 \\ \vdots \\ V^k \end{bmatrix}.$$

Define

$$\eta = \begin{bmatrix} V \\ Z \end{bmatrix} \text{ and } H_{CMC} = I_{(k+m)} \otimes \left(D_r - \frac{n_c}{n} D_p\right)^T,$$

where the parameters in the above equations are defined as before. The vector η consists of two parts: one is for collapsed variants and the other is for uncollapsed variants.

We define a diagonal matrix:

$$\Sigma_v = \mathrm{diag}\left(\sigma_1^2, \sigma_2^2, \ldots, \sigma_k^2\right). \tag{6.294}$$

The covariance matrix is given by

$$\Lambda_v = \mathrm{cov}\left(V, V\right) = \Sigma_v \otimes \Phi. \tag{6.295}$$

Thus, the covariance matrix of η is given by

$$\Lambda = \begin{bmatrix} \Sigma_v & 0 \\ 0 & \Sigma_z \end{bmatrix} \otimes \Phi = \Sigma \otimes \Phi, \tag{6.296}$$

where

$$\Sigma = \begin{bmatrix} \Sigma_v & 0 \\ 0 & \Sigma_z \end{bmatrix}.$$

Then, by a similar argument as before, the covariance matrix of $H_{CMC}\eta$ is given by

$$\Gamma_{CMC} = \left(D_r - \frac{n_c}{n} D_p\right)^T \Phi \left(D_r - \frac{n_c}{n} D_p\right) \Sigma. \tag{6.297}$$

Thus, the family-based CMC statistic can be defined as

$$\begin{aligned} T_{CMCF} &= \left(H_{CMC}\eta\right)^T \Gamma_{CMC}^{-1} H_{CMC}\eta \\ &= \frac{\left(H_{CMC}\eta\right)^T \Sigma_{CMC}^{-1} H_{CMC}\eta}{\left(D_r - \frac{n_c}{n} D_p\right)^T \Phi \left(D_r - \frac{n_c}{n} D_p\right)} \\ &= \frac{\frac{n_c\left(n-n_c\right)}{n}\left[\left(\bar{V}_A - \bar{V}_G\right)^T \Sigma_v^{-1}\left(\bar{V}_A - \bar{V}_G\right) + \left(\bar{Z}_A - \bar{Z}_G\right)^T \Sigma_z^{-1}\left(\bar{Z}_A - \bar{Z}_G\right)\right]}{\frac{n_c\left(n-n_c\right)}{n}\left(D_r - \frac{n_c}{n} D_p\right)^T \Phi \left(D_r - \frac{n_c}{n} D_p\right)} \\ &= \frac{T_{CMC}}{P_{corr}}, \end{aligned} \tag{6.298}$$

where

$\bar{V}_A$ and $\bar{V}_G$ are the averages of the indicator variables in cases and controls, respectively
T_{CMC} is the CMC statistic for the population-based association test
P_{corr} is the correction factor defined as before

The test statistic T_{CMCF} follows a $\chi^2_{(k+m)}$ distribution with $(k+m)$ degrees of freedom, asymptotically, under the null hypothesis of no association of the genomic region being tested.

6.7.4 The Functional Principal Component Analysis and Smoothed Functional Principal Component Analysis with Families

The FPCA and smoothed FPCA can be applied to the population-based association studies. Now we extend them to a general case where multiple families and additional population structures are presented in the samples. Let $\beta_j(t), j=1,2,\ldots,k$ be a set of eigenfunctions that are formed from the genotype data of the sampled individuals under the SFPCA model. Let $x_i(t), i=1,2,\ldots,n$ be a genotypic function of the ith individual, where t is the genomic position, and defined as

$$x_i(t)=\begin{cases}2 & A_tA_t\\ 1 & A_ta_t\\ 0 & a_ta_t.\end{cases} \tag{6.299}$$

Suppose that the genotypic function $x_i(t)$ is expanded in terms of eigenfunctions that are formed by the SFPCA as

$$x_i(t)=\sum_{j=1}^{k}\xi_{ij}\beta_j(t), \tag{6.300}$$

where

$$<\beta_j,\beta_l>_\lambda=\int_T\beta_j(t)\beta_l(t)dt+\lambda\int_T\ddot{\beta}_j(t)\ddot{\beta}_l(t)dt=0$$

and the FPCA scores ξ_{ij} are

$$\xi_{ij}=<x_i,\beta_j>_\lambda=\int_T x_i(t)\beta_j(t)dt+\lambda\int_T\ddot{x}_i(t)\ddot{\beta}_j(t)dt,$$

where λ is a penalty parameter. When λ is equal to zero, expansion of Equation 6.300 will be reduced to the FPCA expansion.

Our purpose is to use the functional principal component scores to develop test statistics that can be applied to pedigrees. To achieve this, we first calculate the covariance matrix of the functional principal component scores. Let

$$\xi_{.j}=\left[\xi_{1j},\xi_{2j},\ldots,\xi_{nj}\right]^T,\quad \xi_{i.}=\left[\xi_{i1},\xi_{i2},\ldots,\xi_{ik}\right]^T,\quad \text{and}\quad \xi=\left[\xi_{.1},\xi_{.2},\ldots,\xi_{.k}\right]^T.$$

Define

$$\sigma_{jk}^{\xi}=\operatorname{cov}\left(\xi_{1j},\xi_{1k}\right)$$

and

$$\Sigma_{SFPCA} = \begin{bmatrix} \sigma_{11}^{\xi} & \sigma_{12}^{\xi} & \cdots & \sigma_{1k}^{\xi} \\ \sigma_{21}^{\xi} & \sigma_{22}^{\xi} & \cdots & \sigma_{2k}^{\xi} \\ \cdots & \cdots & \cdots & \cdots \\ \sigma_{k1}^{\xi} & \sigma_{k2}^{\xi} & \cdots & \sigma_{kk}^{\xi} \end{bmatrix}.$$

The matrix Σ_{ξ} can be estimated by

$$\hat{\Sigma}_{SFPCA} = \frac{1}{n-k}\sum_{i=1}^{n}\left(\xi_i - \bar{\xi}\right)\left(\xi_i - \bar{\xi}\right)^T. \tag{6.301}$$

Then, it can be shown that (Appendix 6J)

$$\begin{aligned} \Lambda_{SFPCAF} &= \operatorname{cov}\left(\xi, \xi\right) \\ &= \Sigma_{SFPCA} \otimes \Phi. \end{aligned} \tag{6.302}$$

Define

$$H_{FPCAF} = I_{(k)} \otimes \left(D_r - \frac{n_c}{n} D_p\right) \tag{6.303}$$

and

$$\Gamma_{SFPCA} = \operatorname{cov}\left(H_{FPCAF}\xi, H_{FPCAF}\xi\right).$$

It follows from Equation 6.302 that

$$\begin{aligned} \Gamma_{SFPCA} &= \left[I_{(k)} \otimes \left(D_r - \frac{n_c}{n} D_p\right)^T\right] \Lambda_{SFPCA}\left[I_{(k)} \otimes \left(D_r - \frac{n_c}{n} D_p\right)\right] \\ &= \left[\left(D_r - \frac{n_c}{n} D_p\right)^T \Phi \left(D_r - \frac{n_c}{n} D_p\right)\right] \Sigma_{SFPCA}. \end{aligned} \tag{6.304}$$

The family-based SFPCA statistic is then defined as

$$T_{SFPCAF} = \left(H_{FPCA}\xi\right)^T \Gamma_{SFPCA}^{-1} H_{FPCA}\xi. \tag{6.305}$$

When $\lambda = 0$, the family-based smoothed FPCA statistic T_{SFPCAF} in Equation 6.305 is reduced to the family-based FPCA statistic without smoothing.

Let $\bar{\xi}_A$ and $\bar{\xi}_G$ be the vector of averages of the functional principal component scores in cases and controls, respectively. It can be shown that the statistic T_{SFPCAF} can be simplified to (Appendix 6J)

$$T_{SFPCAF} = \frac{\left[\frac{n_c(n-n_c)}{n}\right]^2 \left(\bar{\xi}_A - \bar{\xi}_G\right)^T \Sigma_{SFPCA}^{-1}\left(\bar{\xi}_A - \bar{\xi}_G\right)}{\left(D_r - \frac{n_c}{n}D_p\right)^T \Phi\left(D_r - \frac{n_c}{n}D_p\right)}$$
$$= \frac{T_{SFPCA}}{P_{corr}}, \quad (6.306)$$

where

T_{SFPCA} is the population-based smoothed FPCA statistic
P_{corr} is the correction factor as defined previously

When penalty parameter λ is equal to zero, the family-based smoothed FPCA T_{SFPCAF} is reduced to the family-based FPCA statistic

$$T_{FPCAF} = \frac{T_{FPCA}}{P_{corr}}. \quad (6.307)$$

Under the null hypothesis of no association of the genomic region with disease, the test statistics T_{SFPCAF} and T_{FPCAF} will be asymptotically distributed as a central $\chi^2_{(k)}$ distribution where k is the number of functional principal components in the eigenequation expansion of genotypic functions.

SOFTWARE PACKAGE

R package "kinship2" for kinship coefficient and IBD computing can be found in http://r-forge.r-project.org. Software for heritability calculation can be found in http://cnsgenomics.com/software/gcta/. The software for the family-based association analysis with next-generation sequence data can be downloaded from https://sph.uth.edu/hgc/faculty/xiong/software-D.html.

APPENDIX 6A: GENETIC RELATIONSHIP MATRIX

We show that Equations 6.25 and 6.26 hold. We first calculate the variance of the indicator variable for the genotype. It is clear that

$$E\left[g_{ijl}\right] = E\left[x_{ij1l} + x_{ij2l}\right] = E\left[x_{ij1l}\right] + E\left[x_{ij2l}\right] = 2p_l \quad (6A.1)$$

and

$$\begin{aligned}\operatorname{var}\left(g_{ijl}\right) &= \operatorname{var}\left(x_{ij1l} + x_{ij2l}\right)\\ &= \operatorname{var}\left(x_{ij1l}\right) + \operatorname{var}\left(x_{ij2l}\right) + 2\operatorname{cov}\left(x_{ij1l}, x_{ij2l}\right)\\ &= 2p_l\left(1-p_l\right) + 2p_l\left(1-p_l\right)F_{ij} = 2p_l\left(1-p_l\right)\left(1+F_{ij}\right). \quad (6A.2)\end{aligned}$$

Next we calculate $E\left[\left(g_{i1l}-2\bar{p}_l\right)^2\right]$. We can expand it as follows:

$$E\left[\left(g_{i1l}-2\bar{p}_l\right)^2\right]=E\left[\left(g_{i1l}-2p_l+2\left(p_l-\bar{p}_l\right)\right)^2\right]$$
$$=\operatorname{var}\left(g_{i1l}\right)+4E\left[\left(g_{i1l}-2p_l\right)\left(p_l-\bar{p}_l\right)\right]+4\operatorname{var}(\bar{p}_l). \tag{6A.3}$$

Note that

$$E\left[\left(g_{i1l}-2p_l\right)\left(p_l-\bar{p}_l\right)\right]=-E\left[\left(x_{i11l}-p_l+x_{i12l}-p_l\right)\frac{1}{2n}\sum_{u=1}^{n}\left(x_{u11l}-p_l+x_{i12l}-p_l\right)\right]$$
$$=-\frac{1}{2n}\sum_{u=1}^{n}[\operatorname{cov}\left(x_{i11l},x_{u11l}\right)+\operatorname{cov}\left(x_{i11l},x_{u12l}\right)+\operatorname{cov}\left(x_{i12l},x_{u11l}\right)+\operatorname{cov}\left(x_{i12l},x_{u12l}\right)$$
$$=-\frac{1}{2n}\left\{\operatorname{cov}\left(x_{i11l},x_{i11l}\right)+\operatorname{cov}\left(x_{i12l},x_{i12l}\right)+\operatorname{cov}\left(x_{i11l},x_{i12l}\right)+\operatorname{cov}\left(x_{i12l},x_{i11l}\right)\right.$$
$$\left.+\sum_{u=1,u\neq i}^{n}\operatorname{cov}\left(x_{i11l},x_{u11l}\right)+\operatorname{cov}\left(x_{i11l},x_{u12l}\right)+\operatorname{cov}\left(x_{i12l},x_{u11l}\right)+\operatorname{cov}\left(x_{i12l},x_{u12l}\right)\right\}. \tag{6A.4}$$

Using Equation 6.8, we obtain

$$\begin{aligned}\operatorname{cov}\left(x_{i11l},x_{i11l}\right)&=\operatorname{var}\left(x_{i11l}\right)=p_l\left(1-p_l\right),\\ \operatorname{cov}\left(x_{i12l},x_{i12l}\right)&=\operatorname{var}\left(x_{i11l}\right)=p_l\left(1-p_l\right).\end{aligned} \tag{6A.5}$$

It follows from Equation 6.10 that

$$\begin{aligned}\operatorname{cov}\left(x_{i11l},x_{i12l}\right)&=p_l\left(1-p_l\right)\Phi_i,\\ \operatorname{cov}\left(x_{i12l},x_{i11l}\right)&=p_l\left(1-p_l\right)\Phi_i.\end{aligned} \tag{6A.6}$$

Similarly, using Equation 6.14 gives

$$\begin{aligned}\operatorname{cov}\left(x_{i11l},x_{u11l}\right)&=p_l\left(1-p_l\right)\Phi_{iu},\quad i\neq u,\\ \operatorname{cov}\left(x_{i11l},x_{u12l}\right)&=p_l\left(1-p_l\right)\Phi_{iu},\quad i\neq u,\\ \operatorname{cov}\left(x_{i12l},x_{u11l}\right)&=p_l\left(1-p_l\right)\Phi_{iu},\quad i\neq u,\\ \operatorname{cov}\left(x_{i12l},x_{u12l}\right)&=p_l\left(1-p_l\right)\Phi_{iu},\quad i\neq u.\end{aligned} \tag{6A.7}$$

Substituting Equations 6A.5, 6A.6, and 6A.7 into Equation 6A.4, we obtain

$$E\left[\left(g_{i1l}-2p_l\right)\left(p_l-\bar{p}_l\right)\right]=-\frac{1}{2n}\left[2p_l\left(1-p_l\right)+2p_l\left(1-p_l\right)\Phi_i\right]-\frac{1}{2n}\left[\sum_{u=1,u\neq i}^{n}4p_l\left(1-p_l\right)\Phi_{iu}\right]$$

$$=-\frac{1}{n}p_l\left(1-p_l\right)\left(1+\Phi_i+2\sum_{u=1,u\neq i}^{n}\Phi_{iu}\right)$$

$$=-p_l\left(1-p_l\right)\left[\frac{1}{n}\left(1-\Phi_i\right)+2\psi_i\right], \tag{6A.8}$$

where $\psi_i=\frac{1}{n}\sum_{u=1}^{n}\Phi_{ii}$.

Substituting Equations 6A.2, 6A.8, and 6.21 into Equation 6A.3 gives

$$E\left[\left(g_{i1l}-2\bar{p}_l\right)^2\right]=2p_l\left(1-p_l\right)\left(1+\Phi_i\right)-4p_l\left(1-p_l\right)\left[\frac{1}{n}\left(1-\Phi_i\right)+2\psi_i\right]$$

$$+4\left[\frac{1}{2n}p_l\left(1-p_l\right)\left(1+\Phi_I\right)+\frac{n-1}{n}p_l\left(1-p_l\right)\Phi_T\right]$$

$$=2p_l\left(1-p_l\right)\left(\left(1+\Phi_i\right)+2\frac{n-1}{n}\Phi_T-4\psi_i\right)+\frac{2}{n}p_l\left(1-p_l\right)\left(\Phi_I+2\Phi_i-1\right).$$

This shows that Equation 6.25 holds.

By similar arguments, we have

$$E\left[\left(g_{i1l}-2\bar{p}_l\right)\left(g_{k1l}-2\bar{p}_l\right)\right]=\operatorname{cov}\left(g_{i1l},g_{k1l}\right)+2E\left[\left(g_{i1l}-2p_l\right)\left(p_l-\bar{p}_l\right)\right]$$

$$+2E\left[\left(g_{k1l}-2p_l\right)\left(p_l-\bar{p}_l\right)\right]+4\operatorname{var}\left(\bar{p}_l\right). \tag{6A.9}$$

We can show that (Exercise 6.4)

$$\operatorname{cov}\left(g_{i1l},g_{k1l}\right)=4p_l\left(1-p_l\right)\Phi_{ik}. \tag{6A.10}$$

Substituting 6A.10, 6A.8, and 6.21 into Equation 6A.9 yields

$$E\left[\left(g_{i1l}-2\bar{p}_l\right)\left(g_{k1l}-2\bar{p}_l\right)\right]=4p_l\left(1-p_l\right)\left(\Phi_{ik}+\frac{n-1}{n}\Phi_T-\psi_i-\psi_k\right)$$

$$+\frac{2}{n}p_l\left(1-p_l\right)\left(\Phi_I+\Phi_i+\Phi_k-1\right).$$

APPENDIX 6B: DERIVATION OF EQUATION 6.30

For the self-contained, following the approach in Conomos et al. (2016), we summarize the proof as follows. We assume that the individual j descended from the N subpopulations. The average frequency of the allele at the l locus of the individual j is a linear combination of the N subpopulation-specific allele frequencies $p_{jl} = a_j^T \mathbf{P}_l$, where $\mathbf{P}_l$ is a vector of random population-specific allele frequencies. Since the N subpopulation descended from an ancestral population, we have $E\left[p_l^k\right] = p_l$ and $E[\mathbf{P}_l] = p_l\mathbf{1}$, where p_l is an ancestral frequency of the reference allele in the ancestral population. Using Equations 6.18 and 6.19, we obtain

$$\mathrm{var}\left(p_l^k\right) = \frac{1}{2} p_l\left(1-p_l\right)\left(1+F_k\right) = p_l\left(1-p_l\right)\Phi_k \tag{6B.1}$$

and

$$\mathrm{cov}\left(p_l^k, p_l^m\right) = p_l\left(1-p_l\right)\Phi_{km}, \tag{6B.2}$$

where $\Phi_k = \dfrac{1+F_k}{2}$.

Equations 6.26 and 6.1 can be written in a matrix form:

$$\mathrm{cov}\left(\mathbf{P}_l\right) = p_l\left(1-p_l\right)\Phi_N, \tag{6B.3}$$

where $\Phi_N = (\Phi_{km})_{N\times N}$ is an N-dimensional matrix.

The above discussions lead to

$$E\left[p_{jl}\right] = a_j^T E\left[\mathbf{P}_l\right] = p_l a_j^T \mathbf{1} = p_l, \tag{6B.4}$$

$$\begin{aligned}\mathrm{cov}\left(p_{il}, p_{jl}\right) &= a_i^T \,\mathrm{cov}\left(\mathbf{P}_l\right) a_j \\ &= p_l\left(1-p_l\right) a_i^T \Phi_N a_j = p_l\left(1-p_l\right)\theta_{ij},\end{aligned} \tag{6B.5}$$

where $\theta_{ij} = a_i^T \Phi_N a_j$, which can be viewed as the coancestry coefficient for a pair of individuals, i and j, in the presence of population structure.

Next we consider the impact of the pedigree on the covariance of the random genotype indicator variables. Let A_{ij} be the set of the most recent common ancestors of individuals i and j. The set A_{ij} may include individuals i or j. For example, if i directly descended from j then $A_{ij} = \{j\}$. For the sib pair i and j, the set A_{ij} is their two parents; for the half sib pair i and j, A_{ij} is their shared parent. The path diagram can be used to calculate the kinship coefficient ϕ_{ij} (Lynch and Walsh 1998). Suppose that the alleles of the individuals i and j can be traced the most recent common ancestor $m \in A_{ij}$. Let n_{mi} be the number of individuals in the path leading from ancestor m to individual i. If individuals m and i are the same individual, then $n_{mi} = 1$. If the individual m is the grandfather of the individual i, then $n_{mi} = 3$. Using the path diagram to trace back alleles of individuals i and j to their most recent common ancestors, we obtain the kinship coefficient between individuals i and j:

$$\phi_{ij} = \sum_{m \in A_{ij}} \left[\left(\frac{1}{2} \right)^{n_{mi}+n_{mj}-1} (1+f_m) \right] = \sum_{m \in A_{ij}} \phi_{ij|m}, \tag{6B.6}$$

where $\phi_{ij|m} = \left(\frac{1}{2} \right)^{n_{mi}+n_{mj}-1} (1+f_m)$ is referred to as the kinship coefficient of the allele of the individuals i and j tracing back to the common ancestor m.

We define $g_{il} = x_{i11l} + x_{i12l}$ as the indicator variable for the genotype at locus l in the ith population. Let p_{ml} be the average frequency of the allele at the l locus of individual m. Using Equation 6A.7, the conditional covariance between g_{il} and g_{jl}, given the common ancestor m and population-specific allele frequencies $\mathbf{P}_l$, is

$$\text{cov}\left(g_{il}, g_{jl} \middle| \mathbf{P}_l, m\right) = 4\phi_{ij|m} p_{ml} (1 - p_{ml}). \tag{6B.7}$$

Therefore, summarizing all the common ancestors in the set A_{ij}, we obtain the conditional covariance, given the population-specific allele frequencies $\mathbf{P}_l$:

$$\text{cov}\left(g_{il}, g_{jl} \middle| \mathbf{P}_l\right) = 4 \sum_{m \in A_{ij}} \phi_{ij|m} p_{ml} (1 - p_{ml}). \tag{6B.8}$$

It follows that

$$E\left[g_{il} g_{jl} \middle| \mathbf{P}_l\right] = \text{cov}\left(g_{il}, g_{jl} \middle| \mathbf{P}_l\right) + 4 p_{il} p_{jl}. \tag{6B.9}$$

Recall that $p_{il} = a_i^T \mathbf{P}_l$, $p_{jl} = a_j^T \mathbf{P}_l$, $p_{ml} = a_m^T \mathbf{P}_l$, and

$$\begin{aligned} \text{cov}(p_{il}, p_{jl}) &= p_l (1 - p_l) a_i^T \Phi_N a_j = p_l (1 - p_l) \theta_{ij}, \\ \text{var}(p_{ml}) &= p_l (1 - p_l) a_m^T \Phi_N a_m = p_l (1 - p_l) \theta_{mm}, \end{aligned} \tag{6B.10}$$

where θ_{mm} is referred to as the coancestry coefficient between individual m with itself due to population structure.

Note that

$$\begin{aligned} E\left[p_{ml}^2\right] &= \text{var}(p_{ml}) + \left(E[p_{ml}]\right)^2 \\ &= \text{var}(p_{ml}) + p_l^2, \\ E\left[p_{il} p_{jl}\right] &= \text{cov}(p_{il}, p_{jl}) + p_l^2. \end{aligned} \tag{6B.11}$$

Thus, using Equations 6.93 and 6.12, we obtain

$$\begin{aligned} E\left[p_{ml}(1-p_{ml})\right] &= E\left[p_{ml}\right] - E\left[p_{ml}^2\right] \\ &= p_l - p_l^2 - p_l(1-p_l)\theta_{mm} \\ &= p_l(1-p_l)(1-\theta_{mm}), \\ E\left[p_{il}p_{jl}\right] &= p_l^2 + p_l(1-p_l)\theta_{ij}. \end{aligned} \tag{6B.12}$$

Taking expectation over the distribution of population-specific allele frequencies $\mathbf{P}_l$ on both sides of Equation 6.74, we obtain

$$\begin{aligned} \operatorname{cov}(g_{il}, g_{jl}) &= 4\sum_{m\in A_{ij}} \phi_{ij|m} E\left[p_{ml}(1-p_{ml})\right], \\ E\left[g_{il}g_{jl}\right] &= 4\sum_{m\in A_{ij}} \phi_{ij|m} E\left[p_{ml}(1-p_{ml})\right] + 4p_{il}p_{jl}. \end{aligned} \tag{6B.13}$$

Substituting Equation 6.18 into Equation 6.19 gives

$$\begin{aligned} \operatorname{cov}(g_{il}, g_{jl}) &= 4p_l(1-p_l)\sum_{m\in A_{ij}} \phi_{ij|m}(1-\theta_{mm}) \\ &= 4p_l(1-p_l)\left(\phi_{ij} - \sum_{m\in A_{ij}} \phi_{ij|m}\theta_{mm}\right), \\ E\left[g_{il}g_{jl}\right] &= 4p_l^2 + 4p_l(1-p_l)\left(\phi_{ij} + \theta_{ij} - \sum_{m\in A_{ij}} \phi_{ij|m}\theta_{mm}\right). \end{aligned} \tag{6B.14}$$

Next we calculate var(g_{il}). By definition of genotype indicator variables, we have

$$\begin{aligned} \operatorname{var}(g_{il}|\mathbf{P}_1) &= \operatorname{var}(x_{i1l} + x_{i2l}|\mathbf{P}_1) \\ &= \operatorname{var}(x_{i1l}|\mathbf{P}_1) + \operatorname{var}(x_{i2l}|\mathbf{P}_1) + 2\operatorname{cov}(x_{i1l}, x_{i2l}|\mathbf{P}_1). \end{aligned} \tag{6B.15}$$

Similar to Equation 6.8, we can easily show that

$$\operatorname{var}(x_{i1l}|\mathbf{P}_1) = (1-f_i)p_{M(i)l} + f_i p_{il} - p_l^2.$$

Thus,

$$\operatorname{var}(x_{i1l}) = E\left[\operatorname{var}(x_{i1l}|\mathbf{P}_1)\right] = p_l(1-p_l). \tag{6B.16}$$

Similarly, we have

$$\operatorname{var}\left(x_{i2l}\right)=p_l\left(1-p_l\right). \tag{6B.17}$$

Let $M(i)$ and $P(i)$ be the mother allele and father allele of individual i, respectively. Then, we can show that (Exercise 6.5)

$$\operatorname{cov}\left(x_{i1l},x_{i2l}\right)=f_i p_l\left(1-p_l\right)+\left(1-f_i\right)p_l\left(1-p_l\right)\theta_{M(i)P(i)}, \tag{6B.18}$$

where $\theta_{M(i)P(I)}=a_{M(i)}^T\Phi_N a_{P(i)}$.

Therefore, using Equations 6.3, 6.115, and 6.120, we obtain

$$\begin{aligned}\operatorname{var}\left(g_{il}\right)&=\operatorname{var}\left(x_{i1l}\right)+\operatorname{var}\left(x_{i2l}\right)+2\operatorname{cov}\left(x_{i1l},x_{i2l}\right)\\&=2p_l\left(1-p_l\right)+2\left[f_i p_l\left(1-p_l\right)+\left(1-f_i\right)p_l\left(1-p_l\right)\theta_{M(i)P(i)}\right]\\&=2p_l\left(1-p_l\right)\left[1+\theta_{M(i)P(i)}+f_i\left(1-\theta_{M(i)P(i)}\right)\right]\\&=2p_l\left(1-p_l\right)\left(1+F_i\right),\end{aligned} \tag{6B.19}$$

where $F_i=f_i(1-\theta_{M(i)P(i)})+\theta_{M(i)P(i)}$.

APPENDIX 6C: DERIVATION OF EQUATION 6.33

Recall that

$$E\left[g_{il}g_{jl}\right]=4p_l^2+4p_l\left(1-p_l\right)\left(\phi_{ij}+\theta_{ij}-\sum_{m\in A_{ij}}\phi_{ij|m}\theta_{mm}\right) \tag{6C.1}$$

and

$$E\left[p_{il}p_{jl}\right]=p_l^2+p_l\left(1-p_l\right)\theta_{ij}. \tag{6C.2}$$

From the above equations, we can clearly observe that using the product of genotypes of individuals i,j to estimate the kinship coefficients will have two bias terms, θ_{ij} and $\sum_{m\in A_{ij}}\phi_{ij|m}\theta_{mm}$, due to population structure. Equation 6.19 shows that $E[p_{il}p_{jl}]$ involves θ_{ij}. To remove the bias term θ_{ij}, we can use individual allele frequencies p_{il} and p_{jl} to replace population allele frequency p_s in calculation of the GRM. In other words, the kinship coefficient ϕ_{ij} will be estimated by

$$\hat{\phi}_{ij}=\frac{\sum_{l=1}^{L}\left(g_{il}-2p_{il}\right)\left(g_{jl}-2p_{jl}\right)}{4\sum_{l=1}^{L}\sqrt{p_{il}\left(1-p_{il}\right)p_{jl}\left(1-p_{jl}\right)}}. \tag{6C.3}$$

Now we show that the estimator $\hat{\phi}_{ij}$ will asymptotically converge to ϕ_{ij} or it asymptotically has small bias.

Using the rule of computing the expectation by conditioning, we can show that

$$\begin{aligned} E\left[g_{il}p_{jl}\right] &= E\left[\left[Eg_{il}p_{jl} \middle| \mathbf{P}_l\right]\right] \\ &= E\left[p_{jl}E\left[g_{il} \middle| \mathbf{P}_l\right]\right] \\ &= 2E\left[p_{jl}p_{il}\right]. \end{aligned} \tag{6C.4}$$

Equation 6.12 implies that

$$\begin{aligned} E\left[\left(g_{il}-2p_{il}\right)\left(g_{jl}-2p_{jl}\right)\right] &= E\left[g_{il}g_{jl}\right]-2E\left[p_{il}g_{jl}\right]-2E\left[g_{il}p_{jl}\right]+4E\left[p_{il}p_{jl}\right] \\ &= E\left[g_{il}g_{jl}\right]-4E\left[p_{il}p_{jl}\right]. \end{aligned} \tag{6C.5}$$

Therefore, it follows from Equations 6.21 and 6.18 that asymptotically, we have

$$\hat{\phi}_{ij} \to \frac{E\left[g_{il}g_{jl}\right]-4E\left[p_{il}p_{jl}\right]}{4\sqrt{p_{il}\left(1-p_{il}\right)p_{jl}\left(1-p_{jl}\right)}}. \tag{6C.6}$$

Using Equation 6B.13 and 6.2 gives

$$E\left[g_{il}g_{jl}\right]-4E\left[p_{il}p_{jl}\right] = 4p_l\left(1-p_l\right)\left(\phi_{ij}-\sum_{m\in A_{ij}}\phi_{ij|m}\theta_{mm}\right). \tag{6C.7}$$

Define

$$E\left[\sqrt{p_{il}\left(1-p_{il}\right)p_{jl}\left(1-p_{jl}\right)}\right] = p_l\left(1-p_l\right)\left(1-d_{\varphi}\left(i,j\right)\right). \tag{6C.8}$$

Substituting Equation 6.1 into Equation 6.22 yields

$$\begin{aligned} \frac{E\left[g_{il}g_{jl}\right]-4E\left[p_{il}p_{jl}\right]}{4\sqrt{p_{il}\left(1-p_{il}\right)p_{jl}\left(1-p_{jl}\right)}} &= \frac{\phi_{ij}-\sum_{m\in A_{ij}}\phi_{ij|m}\theta_{mm}}{1-d_{\phi}\left(i,j\right)} \\ &= \frac{\phi_{ij}\left(1-d_{\phi}\left(i,j\right)\right)+\phi_{ij}d_{\phi}\left(i,j\right)-\sum_{m\in A_{ij}}\phi_{ij|m}\theta_{mm}}{1-d_{\phi}\left(i,j\right)} \\ &= \phi_{ij}-\sum_{m\in A_{ij}}\phi_{ij|m}\frac{\theta_{mm}-d_{\phi}\left(i,j\right)}{1-d_{\phi}\left(i,j\right)}. \end{aligned}$$

APPENDIX 6D: ML ESTIMATION OF VARIANCE COMPONENTS

Recall that the variance matrix V is

$$V = \sum_{m=1}^{M} G_m V_{Am} + \mathbf{I_n}\sigma_e^2. \tag{6D.1}$$

Then, the matrix derivative with respect to the variance component is given by

$$\frac{\partial V}{\partial \sigma_j^2} = V_j = \begin{cases} I_n & \sigma_j^2 = \sigma_e^2 \\ G_m & \sigma_j^2 = V_{Am}. \end{cases} \tag{6D.2}$$

Using Equations 1.158 and 1.160, we obtain

$$\frac{\partial \log(|V|)}{\partial \sigma_j^2} = \mathrm{Tr}\left[V^{-1}\left(\frac{\partial V}{\partial \sigma_j^2}\right)^T\right] = \mathrm{Tr}\left[V^{-1}V_j\right]. \tag{6D.3}$$

Using Equation 1.156 gives

$$\frac{\partial V^{-1}}{\partial \sigma_j^2} = -V^{-1}V_jV^{-1}. \tag{6D.4}$$

Using Equations 6D.3 and 6D.4, we can directly derive the derivative of the log-likelihood function with respect to the variance components:

$$\frac{\partial \log l(\gamma, V|H, Y)}{\partial \sigma_j^2} = -\frac{1}{2}\mathrm{Tr}\left(V^{-1}V_j\right) + \frac{1}{2}(Y - H\gamma)^T V^{-1}V_jV^{-1}(Y - H\gamma). \tag{6D.5}$$

Setting the derivative equal to zero gives

$$\mathrm{Tr}\left(V^{-1}V_j\right) = (Y - H\gamma)^T V^{-1}V_jV^{-1}(Y - H\gamma). \tag{6D.6}$$

Recall that the ML estimator of the fixed effects is given by

$$H^T V^{-1}(Y - H\gamma) = 0. \tag{6D.7}$$

Iteratively solving Equations 6D.7 and 6D.6, the solutions will converge the ML estimators of both fixed effects and variance components. Assume that $\hat{V}$ is the solution to Equation 6D.6 in the previous iteration. Then, in the current iteration, substituting $\hat{V}$ into Equation 6D.7 and solving it, we obtain the ML estimator of the fixed effect:

$$\hat{\gamma} = \left(H^T V^{-1} H\right)^{-1} H^T V^{-1} Y. \tag{6D.8}$$

Now we calculate

$$V^{-1}\left(Y-H\hat{\gamma}\right)=V^{-1}\left(Y-H\left(H^{T}V^{-1}H\right)^{-1}H^{T}V^{-1}Y\right)$$
$$=PY, \tag{6D.9}$$

where $P = V^{-1} - V^{-1}H(H^TV^{-1}H)^{-1}H^TV^{-1}$.

Substituting Equation 6D.9 into Equation 6D.6, we obtain

$$\operatorname{Tr}\left(V^{-1}V_j\right)=Y^TPV_jPY. \tag{6D.10}$$

Specifically, using Equation 6D.2 gives

$$\operatorname{Tr}\left(\hat{V}^{-1}\right)=Y^T\hat{P}\hat{P}Y, \quad \text{when } \sigma_j^2=\sigma_e^2, \tag{6D.11}$$

$$\operatorname{Tr}\left(\hat{V}^{-1}G_m\right)=Y^T\hat{P}G_m\hat{P}Y, \quad m=1,\ldots,M, \quad \text{when } \sigma_j^2=V_{Am}, \tag{6D.12}$$

where $\hat{P}$ is estimated by using

$$\hat{V}=\sum_{m=1}^{M}G_m\hat{V}_{Am}+\mathbf{I}_n\hat{\sigma}_e^2. \tag{6D.13}$$

APPENDIX 6E: COVARIANCE MATRIX OF THE ML ESTIMATORS

In Section 6.3.2.2.1, we showed

$$\frac{\partial l}{\partial \gamma}=H^TV^{-1}\left(Y-H\gamma\right). \tag{6E.1}$$

Therefore, we have

$$\frac{\partial^2 l}{\partial\gamma\partial\gamma^T}=-H^TV^{-1}H. \tag{6E.2}$$

Define the matrix

$$V_J=\left[G_1,\ldots,G_M,\mathbf{I}_n\right]. \tag{6E.3}$$

It follows from Equation 6D.4 that

$$\frac{\partial V^{-1}}{\partial\sigma}=-V^{-1}V_JV^{-1}. \tag{6E.4}$$

Using Equations 6E.1 and 6E.4, we obtain

$$\frac{\partial^2 l}{\partial\gamma\partial\sigma} = -H^T V^{-1} V_J V^{-1}(Y - H\gamma). \tag{6E.5}$$

In Equation 6E.5, we showed that

$$\frac{\partial l}{\partial\sigma_j^2} = -\frac{1}{2}\mathrm{Tr}\left(V^{-1}V_j\right) + \frac{1}{2}(Y - H\gamma)^T V^{-1}V_j V^{-1}(Y - H\gamma). \tag{6E.6}$$

Using Equation 6E.4 and taking a partial derivative of $\frac{\partial l}{\partial\sigma_j^2}$ with respect to σ_k^2, we obtain

$$\frac{\partial^2 l}{\partial\sigma_k^2\partial\sigma_j^2} = \frac{1}{2}\mathrm{Tr}\left(V^{-1}V_k V^{-1}V_j\right) - (Y - H\gamma)^T\left(V^{-1}V_k V^{-1}V_j V^{-1}\right)(Y - H\gamma). \tag{6E.7}$$

Since $E[Y] = H\gamma$, taking expectation on both sides of Equation 6E.5, we obtain

$$E\left[\frac{\partial^2 l}{\partial\gamma\partial\sigma^T}\right] = -H^T V^{-1} V_J V^{-1}\left(E[Y] - H\gamma\right) = 0. \tag{6E.8}$$

Recall that in Equation 6.83, we assume that

$$E\left[(Y - H\gamma)(Y - H\gamma)^T\right] = V,$$

which implies

$$\begin{aligned}
E\left[\frac{\partial^2 l}{\partial\sigma_k^2\partial\sigma_j^2}\right] &= \frac{1}{2}\mathrm{Tr}\left(V^{-1}V_k V^{-1}V_j\right) - E(Y - H\gamma)^T\left(V^{-1}V_k V^{-1}V_j V^{-1}(Y - H\gamma)\right] \\
&= \frac{1}{2}\mathrm{Tr}\left(V^{-1}V_k V^{-1}V_j\right) - E\left[\mathrm{Tr}\left(V^{-1}V_k V^{-1}V_j V^{-1}\right)(Y - H\gamma)(Y - H\gamma)^T\right] \\
&= \frac{1}{2}\mathrm{Tr}\left(V^{-1}V_k V^{-1}V_j\right) - \mathrm{Tr}\left(V^{-1}V_k V^{-1}V_j V^{-1}E\left[(Y - H\gamma)(Y - H\gamma)^T\right]\right) \\
&= \frac{1}{2}\mathrm{Tr}\left(V^{-1}V_k V^{-1}V_j\right) - \mathrm{Tr}\left(V^{-1}V_k V^{-1}V_j V^{-1}V\right) \\
&= -\frac{1}{2}\mathrm{Tr}\left(V^{-1}V_k V^{-1}V_j\right).
\end{aligned} \tag{6E.9}$$

Define the matrix

$$-E\left[\frac{\partial^2 l}{\partial\sigma\partial\sigma^T}\right]=\begin{bmatrix}\frac{1}{2}\mathrm{Tr}\left(V^{-1}G_1V^{-1}G_1\right) & \cdots & \frac{1}{2}\mathrm{Tr}\left(V^{-1}G_1V^{-1}G_M\right) & \frac{1}{2}\mathrm{Tr}\left(V^{-1}G_1V^{-1}\right)\\ \vdots & \vdots & & \vdots\\ \frac{1}{2}\mathrm{Tr}\left(V^{-1}G_1V^{-1}\right) & \cdots & \frac{1}{2}\mathrm{Tr}\left(V^{-1}G_MV^{-1}\right) & \frac{1}{2}\mathrm{Tr}\left(V^{-1}V^{-1}\right)\end{bmatrix}. \tag{6E.10}$$

Combining Equations 6E.2, 6E.8, and 6E.9, we obtain the Fisher matrix:

$$\begin{aligned}F&=-E\left(\begin{bmatrix}\frac{\partial^2 l}{\partial\gamma\partial\gamma^T} & \frac{\partial^2 l}{\partial\gamma\partial\sigma^T}\\ \frac{\partial^2 l}{\partial\sigma\partial\gamma^T} & \frac{\partial^2 l}{\partial\sigma\partial\sigma^T}\end{bmatrix}\right)\\ &=\begin{bmatrix}H^TV^{-1}H & 0\\ 0 & -E\left[\frac{\partial^2 l}{\partial\sigma\partial\sigma^T}\right]\end{bmatrix}.\end{aligned} \tag{6E.11}$$

APPENDIX 6F: SELECTION OF THE MATRIX K IN THE REML

For the self-contained, following the lines of Searle et al. (1992), we give a brief introduction for the selection of the matrix K. Assume that matrix K satisfies the equation

$$K^TH=0 \quad \text{or} \quad H^TK=0. \tag{6F.1}$$

Suppose that a general solution to Equation 6F.1 can be represented by

$$K=c-a, \tag{6F.2}$$

where c and a are any two vectors. Substituting Equation 6F.2 into Equation 6F.1 gives

$$H^Ta=H^Tc. \tag{6F.3}$$

Solving Equation 6F.3 for the vector a, we obtain

$$a=\left(H^T\right)^-H^Tc, \tag{6F.4}$$

where $(H^T)^-$ is a generalized inverse of the matrix H^T.

Substituting Equation 6F.4 into Equation 6F.2 leads to a solution matrix

$$K = \left[I - \left(H^T\right)^{-} H^T \right] c. \tag{6F.5}$$

Let K^+ be the Moore–Penrose inverse of matrix K. It is clear that $KK^+ = K(K^TK)^{-1}K^T$ and $HH^+ = H(H^TH)^{-}H^T$ are symmetric and idempotent. $K^TH = 0$ and $H^TK = 0$ imply $KK^+H = K(K^TK)^{-1}K^TH = 0$ and $HH^+K = H(H^TH)^{-}H^TK = 0$. Define $T = I - HH^+ - KK^+$.

After some algebra, we have

$$\begin{aligned} TT &= \left(I - HH^+ - KK^+\right)\left(I - HH^+ - KK^+\right) \\ &= I - HH^+ - KK^+ - HH^+ + HH^+HH^+ + HH^+KK^+ - KK^+ + KK^+HH^+ + KK^+KK^+ \\ &= I - HH^+ - KK^+ - HH^+ + HH^+ + 0 - KK^+ + 0 + KK^+ \\ &= I - HH^+ - KK^+ = T, \end{aligned} \tag{6F.6}$$

which shows that T is symmetric and idempotent.

It is well known that $r(HH^+) \le r(H)$. Note that $r(H) = r(HH^+H) \le r(HH^+)$. Thus, $r(HH^+) = r(H)$. $r(K) = \mathrm{Tr}(I - (H^T)^{-}H^T) = n - r(H)$. Next we show that $\mathrm{Tr}(TT^T) = 0$. In fact,

$$\begin{aligned} \mathrm{Tr}\left(TT^T\right) &= \mathrm{Tr}\left(I - HH^+ - KK^+\right) \\ &= n - \mathrm{Tr}\left(\mathrm{HH}^+\right) - \mathrm{Tr}\left(KK^+\right) \\ &= n - r(H) - r(K) \\ &= n - r(H) - \left(n - r(H)\right) = 0, \end{aligned} \tag{6F.7}$$

which implies $T = 0$. This shows

$$I - HH^+ = KK^+. \tag{6F.8}$$

Since covariance matrix V is positive definite, matrix V can be decomposed into $V = (V^{1/2})^2$. Thus, $K^TH = 0$ implies

$$\left(V^{1/2}K\right)^T V^{-1/2}H = K^T H = 0. \tag{6F.9}$$

Equation 6F.9 shows that we can replace K by $V^{1/2}K$ and replace H by $V^{-1/2}H$ in Equation 6F.8 and obtain

$$V^{1/2}K\left(K^TVK\right)^{-} K^TV^{1/2} = I - V^{-1/2}H\left(H^TV^{-1}H\right)^{-} H^TV^{-1/2}. \tag{6F.10}$$

Multiplying both sides of Equation 6F.10 by $V^{-/2}V^{-1/2}$, we obtain

$$K\left(K^TVK\right)^- K^T = V^{-1} - V^{-1}H\left(H^TV^{-1}H\right)^- H^TV^{-1}$$
$$= P. \tag{6F.11}$$

APPENDIX 6G: ALTERNATIVE FORM OF LOG-LIKELIHOOD FUNCTION FOR THE REML

Define the transformation of Y:

$$\begin{bmatrix} U \\ \Phi \end{bmatrix} = \begin{bmatrix} K^T \\ B^T \end{bmatrix} Y, \tag{6G.1}$$

where $B^T = (H^TV^{-1}H)^{-1}H^TV^{-1}$.

The Jacobian matrix of the transformation is

$$J = \begin{bmatrix} K^T \\ B^T \end{bmatrix}. \tag{6G.2}$$

Note that

$$|A| = \sqrt{\left|AA^T\right|}. \tag{6G.3}$$

Thus, the determinant of the Jacobian matrix is equal to

$$|J| = \sqrt{\left|\begin{bmatrix} K^T \\ B^T \end{bmatrix} \begin{bmatrix} K & B \end{bmatrix}\right|}$$
$$= \sqrt{\begin{vmatrix} K^TK & K^TB \\ B^TK & B^TB \end{vmatrix}}. \tag{6G.4}$$

Recall that

$$\begin{vmatrix} A_1 & A_2 \\ A_3 & A_4 \end{vmatrix} = |A_1|\left|A_4 - A_3A_1^{-1}A_2\right| \tag{6G.5}$$

and

$$K^TK = I. \tag{6G.6}$$

Substituting Equation 6G.5 into Equation 6G.4 gives

$$|J| = \left(|I|\left|B^T B - B^T KK^T B\right|\right)^{1/2}. \tag{6G.7}$$

Recall that

$$KK^T = I - H\left(H^T H\right)^{-1} H^T. \tag{6G.8}$$

Substituting Equation 6G.8 into Equation 6G.7 yields

$$\begin{aligned} |J| &= \left(\left|B^T B - B^T\left(I - H\left(H^T H\right)^{-1} H^T\right)B\right|\right)^{1/2} \\ &= \left(\left|B^T H\left(H^T H\right)^{-1} H^T B\right|\right)^{1/2}. \end{aligned} \tag{6G.9}$$

Using Equation 6G.1, we have

$$\begin{aligned} B^T H\left(H^T H\right)^{-1} H^T B &= \left(H^T V^{-1} H\right)^{-1} H^T V^{-1} H\left(H^T H\right)^{-1} H^T V^{-1} H\left(H^T V^{-1} H\right)^{-1} \\ &= \left(H^T H\right)^{-1}. \end{aligned} \tag{6G.10}$$

Substituting Equation 6G.10 into Equation 6G.9 gives

$$|J| = \left|H^T H\right|^{-1/2}. \tag{6G.11}$$

The density function $P(U, \Phi)$ of the transformed random variables U and Φ is given by

$$\begin{aligned} P(U,\Phi) &= \frac{1}{|J|} P(Y) \\ &= \frac{1}{\left|H^T H\right|^{-1/2}} (2\pi)^{-n/2} |V|^{-1/2} \exp\left\{-\frac{1}{2}(Y - H\gamma)^T V^{-1}(Y - H\gamma)\right\}. \end{aligned} \tag{6G.12}$$

Note that

$$E[\Phi] = \gamma$$

and

$$\begin{aligned} \operatorname{var}(\Phi) &= \left(H^T V^{-1} H\right)^{-1} H^T V^{-1} V V^{-1} H\left(H^T V^{-1} H\right)^{-1} \\ &= \left(H^T V^{-1} H\right)^{-1}. \end{aligned}$$

Thus, $\Phi \sim N(\gamma, (H^TV^{-1}V)^{-1})$. The density function $P(\Phi)$ is given by

$$P(\Phi)=(2\pi)^{-(p+q)/2}\left|H^TV^{-1}H\right|^{1/2}\exp\left\{-\frac{1}{2}(\hat{\Phi}-\gamma)^T\left(H^TV^{-1}H\right)(\hat{\Phi}-\gamma)\right\}. \quad (6G.13)$$

Now we show that random vectors U and Φ are independent where $U=K^T(Zu+e)$ and $\Phi=\gamma+B^T(Zu+e)$. First, we calculate the covariance matrix between U and Φ:

$$\begin{aligned}\operatorname{cov}(U,\Phi) &= E\left[U(\Phi-\gamma)^T\right]\\ &= E\left[K^T(Zu+e)(Zu+e)^T B\right]\\ &= K^T E[(Zu+e)(Zu+e)^T B\\ &= K^TVB\\ &= K^TVV^{-1}H\left(H^TV^{-1}H\right)^{-1}\\ &= K^TH\left(H^TV^{-1}H\right)^{-1}\\ &= 0.\end{aligned}$$

Since both random variables U and Φ are normal, uncorrelated variables U and Φ are independent.

Now we are ready to calculate the density function $P(U)$. Using Equations 6G.12 and 6G.13, we obtain

$$\begin{aligned}P(U) &= \frac{P(U,\Phi)}{P(\Phi)}\\ &= \frac{\dfrac{1}{\left|H^TH\right|^{-1/2}}(2\pi)^{-n/2}|V|^{-1/2}\exp\left\{-\frac{1}{2}(Y-H\gamma)^TV^{-1}(Y-H\gamma)\right\}}{(2\pi)^{-(p+q)/2}\left|H^TV^{-1}H\right|^{1/2}\exp\left\{-\frac{1}{2}(\hat{\gamma}-\gamma)^T\left(H^TV^{-1}H\right)(\hat{\gamma}-\gamma)\right\}}\\ &= \frac{1}{\left|H^TH\right|^{-1/2}}(2\pi)^{-(n-q-p)/2}|V|^{-1/2}\left|H^TV^{-1}H\right|^{-1/2}\exp\left\{-\frac{1}{2}(Y-H\hat{\gamma})^TV^{-1}(Y-H\hat{\gamma})\right\}.\end{aligned}$$

Therefore, the log-likelihood function of the transformed variables K^TY, ignoring constants, is given by

$$l=-\frac{1}{2}\log|V|-\frac{1}{2}\log\left|H^TV^{-1}H\right|-\frac{1}{2}(Y-H\hat{\gamma})^TV^{-1}(Y-H\hat{\gamma}). \quad (6G.14)$$

Using Equation 6.102, we can show that Equation 6G.14 can be rewritten as (Exercise 6.12)

$$l=-\frac{1}{2}\log\left|V\right|-\frac{1}{2}\log\left|H^TV^{-1}H\right|-\frac{1}{2}Y^TPY. \tag{6G.15}$$

APPENDIX 6H: ML ESTIMATE OF VARIANCE COMPONENTS IN THE MULTIVARIATE MIXED LINEAR MODELS

Let

$$\Phi_j=\operatorname{cov}\left(u_j,u_j\right)=\begin{bmatrix}\sigma_{11}^j & \cdots & \sigma_{1K}^j\\ \vdots & \vdots & \vdots\\ \sigma_{K1}^j & \cdots & \sigma_{KK}^j\end{bmatrix}.$$

Using Definition 1.5 (derivative of a matrix with respect to a scalar) gives

$$\frac{\partial\Phi_j}{\partial\sigma_{il}^j}=V_{il}^j=\begin{bmatrix}0 & \cdots & \cdots & \cdots & 0\\ \vdots & \vdots & \vdots & \vdots & \vdots\\ 0 & \cdots & 1 & \cdots & 0\\ \vdots & \vdots & \vdots & \vdots & \vdots\\ 0 & \cdots & \cdots & \cdots & 0\end{bmatrix}, \tag{6H.1}$$

i.e., the (i,l)th element of matrix V_{il}^j is 1 and all other elements in matrix V_{il}^j are zero.

Therefore, we have

$$\frac{\partial\Phi}{\partial\sigma_{il}^j}=\operatorname{diag}\left(0,\ldots,0,\ V_{il}^j,0,\ldots,0\right),\quad j=1,\ldots,m. \tag{6H.2}$$

Similarly, we have

$$V_{il}^e=\frac{\partial\Sigma}{\partial\sigma_{il}^e}=\begin{bmatrix}0 & \cdots & \cdots & \cdots & 0\\ \vdots & \vdots & \vdots & \vdots & \vdots\\ 0 & \cdots & 1 & \cdots & 0\\ \vdots & \vdots & \vdots & \vdots & \vdots\\ 0 & \cdots & \cdots & \cdots & 0\end{bmatrix}, \tag{6H.3}$$

where the (i,l)th element of matrix V_{il}^e is 1 and all other elements in the matrix V_{il}^e are zero.

Recall that

$$V=\left(Z\otimes I_K\right)\Phi\left(Z^T\otimes I_K\right)+I_n\otimes\Sigma. \tag{6H.4}$$

Then, using Equations 6H.2 and 6H.4, we obtain

$$\frac{\partial V}{\partial \sigma_{il}^{j}} = (Z \otimes I_K)\frac{\partial \Phi}{\partial \sigma_{il}^{j}}(Z^T \otimes I_K). \tag{6H.5}$$

For $K = 1$, we have $\frac{\partial V}{\partial \sigma_A^j} = Z_j Z_j^T$ (Exercise 6.14), which is the same as that in Equation 6H.2.

Again, using Equations 6H.3 and 6H.4 gives

$$\frac{\partial V}{\partial \sigma_{il}^{e}} = I_n \otimes V_{il}^{e}. \tag{6H.6}$$

The matrix derivative with respect to the variance component is then denoted by

$$\frac{\partial V}{\partial \sigma_{il}^{u}} = V_{il}^{u} = \begin{cases} \frac{\partial V}{\partial \sigma_{ij}^{l}} & \sigma_{il}^{u} = \sigma_{il}^{j} \\ \frac{\partial V}{\partial \sigma_{il}^{e}} & \sigma_{il}^{u} = \sigma_{il}^{e}. \end{cases} \tag{6H.7}$$

Using Equations 1.158 and 1.160, we obtain

$$\frac{\partial \log(|V|)}{\partial \sigma_{il}^{u}} = \mathrm{Tr}\left[V^{-1}\left(\frac{\partial V}{\partial \sigma_{il}^{u}}\right)^{T}\right] = \mathrm{Tr}\left[V^{-1}V_{il}^{u}\right]. \tag{6H.8}$$

Using Equation 1.156 gives

$$\frac{\partial V^{-1}}{\partial \sigma_{il}^{u}} = -V^{-1}V_{il}^{u}V^{-1}. \tag{6H.9}$$

Using Equations 6H.8 and 6H.9, we can directly derive the derivative of the log-likelihood function with respect to the variance components:

$$\frac{\partial l}{\partial \sigma_{il}^{u}} = -\frac{1}{2}\mathrm{Tr}\left[V^{-1}V_{il}^{u}\right] + \frac{1}{2}\left(\mathrm{vec}(Y^T) - (W \otimes I_K)\mathrm{vec}(\alpha^T)\right)^T V^{-1}V_{il}^{u}V^{-1}\left(\mathrm{vec}(Y^T) - (W \otimes I_K)\mathrm{vec}(\alpha^T)\right). \tag{6H.10}$$

Let

$$H = W \otimes I_K.$$

Using Equation 6.208, we obtain

$$\text{vec}\left(\hat{\alpha}^T\right) = \left(H^T V^{-1} H\right)^{-1} H^T V^{-1} \text{vec}\left(Y^T\right). \quad (6H.11)$$

Then, we have

$$V^{-1}\left[\text{vec}\left(Y^T\right) - (W \otimes I_K)\text{vec}\left(\alpha^T\right)\right] = V^{-1}\left[\text{vec}\left(Y^T\right) - H\left(H^T V^{-1} H\right)^{-1} H^T V^{-1}\text{vec}\left(Y^T\right)\right]$$
$$= P\text{vec}\left(Y^T\right), \quad (6H.12)$$

where

$$P = V^{-1} - V^{-1} H\left(H^T V^{-1} H\right)^{-1} H^T V^{-1}. \quad (6H.13)$$

Setting $\frac{\partial l}{\partial \sigma_{il}^u} = 0$ and using Equation 6H.12 leads to

$$\text{Tr}\left(V^{-1} V_{il}^u\right) = \left(\text{vec}\left(Y^T\right)\right)^T P V_{il}^u P\text{vec}\left(Y^T\right). \quad (6H.14)$$

APPENDIX 6I: COVARIANCE MATRIX FOR FAMILY-BASED T^2 STATISTIC

A key for deriving a covariance matrix for multiple markers of family members or population-sampled individuals in the presence of population structures is the concept of identical by descent (IBD).

Let

$$x_i^t = \begin{cases} 1 & A_t \\ 0 & a_t \end{cases}, \quad t = 1, 2, \ldots, m, \; i = 1, 2, \ldots, n.$$

Similarly, we define y_i^t. Then, we have

$$z_i^t = x_i^t + y_i^t, \quad t = 1, 2, \ldots, m, \; i = 1, 2, \ldots, n.$$

Define

$$\sigma_{tt} = 2P(A_t)(1 - P(A_t)), \quad t = 1, 2, \ldots, m.$$

Let h_i be the inbreeding coefficient of individual i and ϕ_{ij} be the kinship coefficient between individuals i and j. Using computing expectation by conditioning, we have

$$E\left[x_i^t x_j^t\right] = E\left[x_i^t E\left[x_j^t | x_i^t\right]\right] = 2\phi_{ij} E\left[\left(x_i^t\right)^2\right]. \tag{6I.1}$$

Similarly, we have

$$\begin{aligned} E\left[x_i^t y_j^t\right] &= 2\phi_{ij} E\left[x_i^t y_i^t\right], \\ E\left[y_i^t x_j^t\right] &= 2\phi_{ij} E\left[y_i^t x_i^t\right], \\ E\left[y_i^t y_j^t\right] &= 2\phi_{ij} E\left[\left(y_i^t\right)^2\right], \\ E\left[x_j^t + y_j^t\right] &= 2\phi_{ij} E\left[x_i^t + y_i^t\right]. \end{aligned} \tag{6I.2}$$

By definition of the covariance between variables z_i^t and z_j^t, we obtain

$$\operatorname{cov}\left(z_i^t, z_j^t\right) = E\left[x_i^t x_j^t + x_i^t y_j^t + y_i^t x_j^t + y_i^t y_j^t\right] - E\left[\left(x_i^t + y_i^t\right)\left(x_j^t + y_j^t\right)\right]. \tag{6I.3}$$

Substituting Equations 6I.1 and 6I.2 into Equation 6I.3, we obtain

$$\begin{aligned} \operatorname{cov}\left(z_i^t, z_j^t\right) &= 2\phi_{ij}\left\{E\left[\left(x_i^t + y_i^t\right)^2\right] - \left[E\left(x_i^t + y_i^t\right)\right]^2\right\} \\ &= 2\phi_{ij}\sigma_{tt}. \end{aligned} \tag{6I.4}$$

Similarly, we have

$$\operatorname{cov}\left(z_i^t, z_i^t\right) = \left(1 + h_{ii}\right)\sigma_{tt}. \tag{6I.5}$$

Combining Equations 6I.4 and 6I.5, we obtain

$$\operatorname{cov}\left(Z^t, Z^t\right) = \sigma_{tt}\Phi, \tag{6I.6}$$

where

$$\Phi = \begin{bmatrix} \left(1 + h_1\right) & 2\phi_{12} & \cdots & 2\phi_{1n} \\ 2\phi_{21} & 1 + h_2 & \cdots & 2\phi_{2n} \\ \cdots & \cdots & \cdots & \cdots \\ 2\phi_{n1} & 2\phi_{n2} & \cdots & 1 + h_n \end{bmatrix}. \tag{6I.7}$$

By a similar argument as that for Equation 6I.4, we have

$$\operatorname{cov}\left(z_i^{t_k}, z_j^{t_l}\right) = 2\phi_{ij}\sigma_{t_k t_l}. \tag{6I.8}$$

Combining Equations 6I.6 and 6I.8 leads to

$$\Lambda_z = \operatorname{cov}(Z, Z) = \begin{bmatrix} \operatorname{cov}(Z^1, Z^1) & \operatorname{cov}(Z^1, Z^2) & \cdots & \operatorname{cov}(Z^1, Z^m) \\ \operatorname{cov}(Z^2, Z^1) & \operatorname{cov}(Z^2, Z^2) & \cdots & \operatorname{cov}(Z^2, Z^m) \\ \cdots & \cdots & \cdots & \cdots \\ \operatorname{cov}(Z^m, Z^1) & \operatorname{cov}(Z^m, Z^1) & \cdots & \operatorname{cov}(Z^m, Z^m) \end{bmatrix}$$

$$= \Sigma_z \otimes \Phi, \tag{6I.9}$$

where

$$\Sigma_z = \begin{bmatrix} \sigma_{11} & \sigma_{12} & \cdots & \sigma_{1m} \\ \sigma_{21} & \sigma_{22} & \cdots & \sigma_{2m} \\ \cdots & \cdots & \cdots & \cdots \\ \sigma_{m1} & \sigma_{m2} & \cdots & \sigma_{mm} \end{bmatrix}.$$

APPENDIX 6J: FAMILY-BASED FUNCTIONAL PRINCIPAL COMPONENT ANALYSIS

First, we calculate $\operatorname{cov}(x_i(s), x_k(t))$. Using computing expectation by conditioning and IBD concept, we obtain

$$\begin{aligned} \operatorname{cov}(x_i(s), x_k(t)) &= E\left[E\left[x_i(s)x_k(t) \mid x_i(t)\right]\right] - E\left[x_i(s)\right]E\left[E\left[x_k(t) \mid x_i(t)\right]\right] \\ &= 2\phi_{ik}E\left[x_i(s)x_i(t)\right] - 2\phi_{ik}E\left[x_i(s)\right]E\left[x_i(t)\right] \\ &= 2\phi_{ik}R(s,t), \end{aligned} \tag{6J.1}$$

where $R(s,t)$ is a covariance function of the genotype indicator variables between genomic positions s and t.

Similarly, we have

$$\begin{aligned} \operatorname{cov}\left(x_i(s), \ddot{x}_k(t)\right) &= \frac{\partial^2}{\partial t^2}\operatorname{cov}\left(x_i(s), x_k(t)\right) \\ &= 2\phi_{ik}\frac{\partial^2 R(s,t)}{\partial t^2}, \end{aligned} \tag{6J.2}$$

$$\operatorname{cov}\left(\ddot{x}_i(s), x_k(t)\right) = 2\phi_{ik}\frac{\partial^2 R(s,t)}{\partial s^2}, \tag{6J.3}$$

and

$$\text{cov}\left(\ddot{x}_i(s), \ddot{x}_k(t)\right) = 2\phi_{ik}\frac{\partial^4 R(s,t)}{\partial s^2 \partial t^2}. \tag{6J.4}$$

From stochastic calculus (Henderson and Plaschko 2006), we can obtain

$$\text{cov}\left(\int_T x_1(t)\beta_j(t)dt, \int_T x_1(t)\beta_k(t)dt\right) = \int_T\int_T \beta_j(s)R(s,t)\beta_k(t)dsdt, \tag{6J.5a}$$

$$\text{cov}\left(\int_T x_1(t)\beta_j(t)dt, \int_T \ddot{x}_1(t)\beta_k(t)dt\right) = \int_T\int_T \beta_j(t)\frac{\partial^2 R(s,t)}{\partial t^2}\beta_k(t)dsdt, \tag{6J.5b}$$

$$\text{cov}\left(\int_T \ddot{x}_1(t)\ddot{\beta}_j(t)dt, \int_T x_1(t)\beta_k(t)dt\right) = \int_T\int_T \ddot{\beta}_j(s)\frac{\partial^2 R(s,t)}{\partial s^2}\beta_k(t)dsdt, \tag{6J.5c}$$

$$\text{cov}\left(\int_T \ddot{x}_1(t)\ddot{\beta}_j(t)dt, \int_T \ddot{x}_1(t)\ddot{\beta}_k(t)dt\right) = \int_T\int_T \ddot{\beta}_j(s)\frac{\partial^4 R(s,t)}{\partial s^2 \partial t^2}\ddot{\beta}_k(t)dsdt. \tag{6J.5d}$$

Combining Equations 6J.5a through 6J.5d, we obtain the covariance between functional principal component scores for the same individual without inbreeding:

$$\begin{aligned}
\sigma_{jk}^{\xi} &= \text{cov}\left(\xi_{1j}, \xi_{1k}\right) \\
&= \int_T\int_T \beta_j(s)R(s,t)\beta_k(t)dsdt + \lambda\int_T\int_T \beta_j(s)\frac{\partial^2 R(s,t)}{\partial t^2}\ddot{\beta}_k(t)dsdt \\
&\quad + \lambda\int_T\int_T \ddot{\beta}_j(s)\frac{\partial^2 R(s,t)}{\partial s^2}\beta_k(t)dsdt + \lambda^2\int_T\int_T \ddot{\beta}_j(s)\frac{\partial^4 R(s,t)}{\partial s^2 \partial t^2}\ddot{\beta}_j(t)dsdt.
\end{aligned} \tag{6J.6}$$

Using Equations 6J.1 through 6J.4 and 6J.6, we can obtain the covariance of the functional principal component scores between a pair of individuals:

$$\begin{aligned}
\text{cov}\left(\xi_{ij}, \xi_{lk}\right) &= \text{cov}\left(\int_T x_i(t)\beta_j(t)dt + \lambda\int_T \ddot{x}_i(t)\ddot{\beta}_j(t)dt, \int_T x_l(t)\beta_k(t)dt + \lambda\int_T \ddot{x}_l(t)\ddot{\beta}_k dt\right) \\
&= 2\phi_{il}\left\{\int_T\int_T \beta_j(s)R(s,t)\beta_k(t)dsdt + \lambda\int_T\int_T \beta_j(s)\frac{\partial^2 R(s,t)}{\partial t^2}\ddot{\beta}_k(t)dsdt\right. \\
&\quad \left. + \lambda\int_T\int_T \ddot{\beta}_j(s)\frac{\partial^2 R(s,t)}{\partial s^2}\beta_k(t)dsdt + \lambda^2\int_T\int_T \ddot{\beta}_j(s)\frac{\partial^4 R(s,t)}{\partial s^2 \partial t^2}\ddot{\beta}_j(t)dsdt\right\} \\
&= 2\phi_{il}\sigma_{jk}^{\xi}.
\end{aligned} \tag{6J.7}$$

Similarly, considering inbreeding, we can prove that

$$\operatorname{var}(\xi_{ij}) = (1+h_i)\sigma_{jj}^{\xi}. \tag{6J.8}$$

Define the covariance matrix of the vector of functional principal component score ξ as

$$\Lambda_\xi = \begin{bmatrix} \operatorname{var}(\xi_{.1}) & \operatorname{cov}(\xi_{.1}, \xi_{.2}) & \cdots & \operatorname{cov}(\xi_{.1}, \xi_{.k}) \\ \operatorname{cov}(\xi_{.2}, \xi_{.1}) & \operatorname{var}(\xi_{.2}) & \cdots & \operatorname{cov}(\xi_{.2}, \xi_{.k}) \\ \cdots & \cdots & \cdots & \cdots \\ \operatorname{cov}(\xi_{.k}, \xi_{.1}) & \operatorname{cov}(\xi_{.k}, \xi_{.1}) & \cdots & \operatorname{var}(\xi_{.k}) \end{bmatrix}.$$

But, we have

$$\operatorname{var}(\xi_{.j}) = \sigma_{jj}^{\xi}\Phi, \tag{6J.9}$$

$$\operatorname{cov}(\xi_{.j}, \xi_{.k}) = \sigma_{jk}^{\xi}\Phi. \tag{6J.10}$$

Let

$$\Sigma_{SFPCA} = \begin{bmatrix} \sigma_{11}^{\xi} & \sigma_{12}^{\xi} & \cdots & \sigma_{1k}^{\xi} \\ \sigma_{21}^{\xi} & \sigma_{22}^{\xi} & \cdots & \sigma_{2k}^{\xi} \\ \cdots & \cdots & \cdots & \cdots \\ \sigma_{k1}^{\xi} & \sigma_{k2}^{\xi} & \cdots & \sigma_{kk}^{\xi} \end{bmatrix}.$$

Then, by combining Equations 6J.9 and 6J.10, we obtain

$$\Lambda_{SFPCA} = \Sigma_{SFPCA} \otimes \Phi. \tag{6J.11}$$

Let

$$\bar{\xi}_{Aj} = \frac{1}{n_c}\sum_{i \in cases}\xi_{ij} \quad \text{and} \quad \bar{\xi}_{G_j} = \frac{1}{n-n_c}\sum_{i \in controls}\xi_{ij},$$

$$\bar{\xi}_A = \left[\bar{\xi}_{A1}, \ldots \bar{\xi}_{Ak}\right]^T \quad \text{and} \quad \bar{\xi}_G = \left[\bar{\xi}_{G1}, \ldots, \bar{\xi}_{Gk}\right]^T.$$

Note that

$$\begin{aligned} \left(D_r - \frac{n_c}{n}D_p\right)^T \xi_{.j} &= \sum_{i \in cases}\xi_{ij} - \frac{n_c}{n}\sum_{i=1}^{n}\xi_{ij} \\ &= n_c\bar{\xi}_{Aj} - \frac{n_c}{n}[n_c\bar{\xi}_{Aj} + (n-n_c)\bar{\xi}_{Gj} \\ &= \frac{n_c(n-n_c)}{n}\left(\bar{\xi}_{Aj} - \bar{\xi}_{Gj}\right). \end{aligned}$$

Therefore, we have

$$H_{SFPCAF}\xi = \left[I_{(k)} \otimes \left(D_r - \frac{n_c}{n} D_p \right)^T \right] \xi$$

$$= \begin{bmatrix} \left(D_r - \frac{n_c}{n} D_p \right)^T \xi_{.1} \\ \vdots \\ \left(D_r - \frac{n_c}{n} D_p \right)^T \xi_{.k} \end{bmatrix}$$

$$= \frac{n_c (n - n_c)}{n} \left(\bar{\xi}_A - \bar{\xi}_G \right). \tag{6J.12}$$

Using Equation 6J.12, we obtain the simple family-based FPCA test statistic:

$$T_{SFPCAF} = \left[\frac{n_c (n - n_c)}{n} \right]^2 \frac{\left(\bar{\xi}_A - \bar{\xi}_G \right)^T \Sigma_{SFPCA}^{-1} \left(\bar{\xi}_A - \bar{\xi}_G \right)}{\left(D_r - \frac{n_c}{n} D_P \right)^T \Phi \left(D_r - \frac{n_c}{n} D_p \right)}$$

$$= \frac{\frac{n_c (n - n_c)}{n} \left(\bar{\xi}_A - \bar{\xi}_G \right)^T \Sigma_{SFPCA}^{-1} \left(\bar{\xi}_A - \bar{\xi}_G \right)}{\frac{n}{n_c (n - n_c)} \left(D_r - \frac{n_c}{n} D_p \right)^T \left(D_r - \frac{n_c}{n} D_p \right)}$$

$$= \frac{T_{SFPCA}}{P_{corr}}. \tag{6J.13}$$

EXERCISE

Exercise 6.1 Show that the inbreeding coefficient and kinship coefficient have the following relation:

$$\Phi_{ii} = \frac{1}{2}(1 + f_i)$$

and

$$f_i = \Phi_{kl},$$

where k and l are parent of the individual i.

Exercise 6.2 Consider noninbred half sibs. Calculate the condensed coefficient of identity Δ_8.

Exercise 6.3 Calculate the kinship coefficient for the first cousins.

Exercise 6.4 Show that

$$\operatorname{cov}(g_{i1l}, g_{k1l}) = 4p_l(1-p_l)\Phi_{ik}.$$

Exercise 6.5 Show that

$$\operatorname{cov}(x_{i1l}, x_{i2l}) = f_i p_l(1-p_l) + (1-f_i)p_l(1-p_l)\theta_{M(i)P(i)},$$

where $\theta_{M(i)P(I)} = a_{M(i)}^T \Phi_N a_{P(i)}$.

Exercise 6.6 Let $\hat{f}_i = \hat{G}_{ii} = \frac{1}{L}\sum_{l=1}^{L} \frac{E\left[(g_{il} - 2\bar{p}_l)^2\right]}{2\bar{p}_l(1-\bar{p}_l)}$. Show that

$$\hat{G}_{ii} - 1 \rightarrow f_i\left[1 - \theta_{M(i)P(i)}\right] + \theta_{M(i)P(i)},$$

where $M(i)$ and $P(i)$ denote the mother and father of individual i, respectively.

Exercise 6.7 Let $\hat{f}_i = \frac{1}{L}\sum_{l=1}^{L} \frac{E\left[(g_{il} - 2\hat{p}_{il})^2\right]}{2\hat{p}_{il}(1-\hat{p}_{il})}$, where the individual-specific allele frequency $\hat{p}_{il}$ is estimated by Equation 6.35. Show that

$$\hat{f}_i \rightarrow f_i\left[1 - \frac{\theta_{M(i)P(i)} - \theta_{ii}}{1-\theta_{ii}}\right] + \frac{\theta_{M(i)P(i)} - \theta_{ii}}{1-\theta_{ii}},$$

where $\theta_{ii} = a_i^T \Phi_N a_i$.

Exercise 6.8 Show

$$E\left[(\alpha\alpha)^2\right] = \frac{1}{4}V_{AA},$$

where V_{AA} is an additive × additive interaction variance.

Exercise 6.9 If x_i is defined in Equation 6.63, then show

$$E[x_i] = 2p \quad \text{and} \quad \operatorname{var}(x_i) = 2pq.$$

Exercise 6.10 If x_i is defined in Equation 6.64 then show

$$E[x_i]=0 \quad \text{and} \quad \text{var}(x_i)=2pq.$$

Exercise 6.11 Assume that log-likelihood is given by

$$l=-\frac{n}{2}\log(2\pi)-\frac{1}{2}\log|V|-\frac{1}{2}(Y-H\gamma)^T V^{-1}(Y-H\gamma).$$

Show that

$$\frac{\partial l}{\partial V}=-\frac{1}{2}V^{-1}+\frac{1}{2}V^{-1}(Y-H\gamma)(Y-H\gamma)^T V^{-1}.$$

Exercise 6.12 Show that

$$(Y-H\hat{\gamma})^T V^{-1}(Y-H\hat{\gamma})=Y^T PY.$$

Exercise 6.13 Show that the solution to $\frac{\partial L(W,V)}{\partial V}=0$ is

$$\hat{V}=\left(\text{vec}(Y^T)-(W\otimes I_K)\text{vec}(\alpha^T)\right)\left(\text{vec}(Y^T)-(W\otimes I_K)\text{vec}(\alpha^T)\right)^T.$$

Exercise 6.14 Let $K=1$; show

$$\frac{\partial V}{\partial \sigma_A^j}=Z_j Z_j^T.$$

Exercise 6.15 Show Equation 6.252:

$$(Z\otimes I_K)(I_m\otimes\Sigma_A)(Z^T\otimes I_K)=(ZZ^T)\otimes\Sigma_A.$$

Exercise 6.16 Show

$$HZ=\frac{n_c(n-n_c)}{n}\begin{bmatrix}\bar{Z}_A^1-\bar{Z}_G^1\\ \vdots \\ \bar{Z}_A^T-\bar{Z}_G^T\end{bmatrix}.$$

Exercise 6.17 Show

$$Cov(x_i, x_j) = 2\phi_{ij}\sigma^2, \quad \text{where } \sigma^2 = p(1-p).$$

Exercise 6.18 Show that the test statistic can then be rewritten as

$$T_{CF} = \frac{T_C}{P_{corr}},$$

where

T_C is the population-based collapsing test statistic

$P_{corr} = \frac{n}{n_c n_G}\left(D_r - \frac{n_c}{n} D_p\right)^T \Phi\left(D_r - \frac{n_c}{n} D_p\right)$ is a correction factor

CHAPTER 7

Interaction Analysis

A PUBLISHED CATALOG of *Genome-Wide Association Studies* (GWAS) reported significant association of 26,791 SNPs with more than 1,704 traits in 2,337 publication on February 6, 2017 (https://www.genome.gov/page.cfm?pageid=26525384&clearquery=1#result_table). These results provide substantial information for understanding the mechanisms of the diseases. Although great progress in GWAS has been made, the significant SNP associations identified by GWAS account for only a few percent of the genetic variance (Altshuler et al. 2008; Frazer et al. 2009). Searching for the remaining genetic variance is a great challenge (Wang 2008).

One way to discover the remaining genetic variance or "missing heritability" is to study gene–gene and gene–environment interactions. Complex diseases are the consequence of the interplay of genetic and environmental factors. Modern complex theory assumes that the complexity is attributed to the interactions among the components of the system; therefore, interaction has been considered as a sensible measure of complexity of the biological systems. The interactions hold a key for dissecting the genetic structure of complex diseases and elucidating the biological and biochemical pathways underlying the diseases (Cordell 2009).

Over the past several decades, geneticists have debated intensely about how to define and measure interaction in epidemiologic studies (Ottman 1996). Many researchers indicated the importance of distinguishing biological interaction and statistical interaction (Liberman et al. 2007). Biological interaction between gene and gene or gene and environment is often defined as the interdependent operation of genetic and environmental factors that cause diseases. In contrast, statistical gene–gene or gene–environment interaction is defined as the interdependence between the effects of genetic and environmental risk factors in the context of a statistical model. The effects of disease risk factors are often measured by relative risks and odds ratios, and interaction is defined as departure from additive or multiplicative joint effects (Khoury and Wacholder 2009). Alternative to the classical definition of the statistical interaction, various new definitions of interaction that are based on interdependence among the risk factors causing disease have been proposed. Gene–gene and gene–environment interactions can generally be defined as a stochastic dependence between genetic and environmental risk factors in causing phenotypic variation among individuals. This definition does not require specifying the statistical models of the risks and is similar, although not exactly identical, to the definition of biological interaction.

In this chapter, we will cover odds ratio calculations, regression, and logistic regression analysis, which are some of the existing methods available to evaluate the gene–gene and gene–environment interactions (Xiong and Wu 2010). Functional data analysis techniques are a powerful tool for sequence-based genetic studies. We will introduce the single- and multiple-variate functional regressions and functional logistic regression for interaction analysis of quantitative and qualitative traits with next-generation sequencing (NGS) data. In Sections 4.2.2 and 5.4, we introduced canonical correlation analysis (CCA) as a statistical framework for association analysis. In this chapter, we will extend CCA from association analysis to interaction analysis and formulate CCA as a unified framework for testing gene–gene and gene–environment interactions for both quantitative and qualitative traits with both common and rare variants. As we previously pointed out, linear regression analysis can be viewed as specific CCA.

7.1 MEASURES OF GENE–GENE AND GENE–ENVIRONMENT INTERACTIONS FOR A QUALITATIVE TRAIT

7.1.1 Binary Measure of Gene–Gene and Gene–Environment Interactions

In this section, we take a common approach to gene–gene and gene–environment interactions. We consider both binary genetic and environment measures. In this case, both statistical formulations of gene–gene and gene–environment interaction measures are the same. Throughout this section, we introduce only gene–gene interaction. However, all studies of gene–gene interactions can be applied to gene–environment interaction unless otherwise stated.

Consider two binary genetic factors, G_1 and G_2. The genetic factor G_1 is coded as $G_1 = 1$ ($G_1 = 0$) if an individual carries risk-increasing genotypes (AA or Aa if A is a risk-increasing allele) (or no-risk genotypes). Similarly, we can define G_2. Let D be an indicator of disease. Two study designs, cohort study and case–control study, are often used in investigating gene–gene interaction. In the cohort study, the measures of gene–gene interaction are usually defined by relative risk, and in the case–control study, they are often defined by odds ratio. There are two types of binary measure of gene–gene interaction: additive and multiplicative measures.

7.1.1.1 The Binary Measure of Gene–Gene Interaction for the Cohort Study Design

We first introduce the measure of gene-gene interaction for the cohort study design. Define the disease risk as the conditional probability of being affected, given two particular genetic factors for each of the four possible combinations of the two genetic risk factors:

$$h_{11} = P\left(D=1 \middle| G_1=1, G_2=1\right), \quad h_{12} = P\left(D=1 \middle| G_1=1, G_2=0\right),$$
$$h_{21} = P\left(D=1 \middle| G_1=0, G_2=1\right), \quad h_{22} = P\left(D=1 \middle| G_1=0, G_2=0\right).$$

The relative risks are defined as the ratio of disease risk over baseline risk:

$$RR_{11} = \frac{h_{11}}{h_{22}}, \quad RR_{12} = \frac{h_{12}}{h_{22}}, \quad RR_{21} = \frac{h_{21}}{h_{22}}. \tag{7.1}$$

Definition 7.1

Additive and multiplicative measures of gene–gene interaction are respectively defined as

$$I_{add} = RR_{11} - RR_{12} - RR_{21} + 1 \tag{7.2}$$

and

$$I_{multiple} = \frac{RR_{11}}{RR_{12}RR_{21}}. \tag{7.3}$$

Figure 7.1 shows the gene–gene interaction where two disease risk lines are crossed and $h_{11} - h_{12} \neq h_{21} - h_{22}$. In other words, the disease risk of the genotypes at the first locus depends on the genotype at the second locus. In the absence of interaction, i.e., $I_{add} = 0$ and $I_{multiple} = 1$, we have

$$RR_{11} - RR_{12} = RR_{21} - 1 \tag{7.4}$$

and

$$RR_{11} = RR_{12} * RR_{21}. \tag{7.5}$$

Equations 7.4 and 7.5 can be reexpressed in terms of haplotype frequencies. We define the joint probability of the individual with the specified genetic factors in the general and disease populations:

$$P_{11} = P(G_1 = 1, G_2 = 1), \quad P_{12} = P(G_1 = 1, G_2 = 0), \quad P_{21} = P(G_1 = 0, G_2 = 1),$$
$$P_{22} = P(G_1 = 0, G_2 = 0),$$

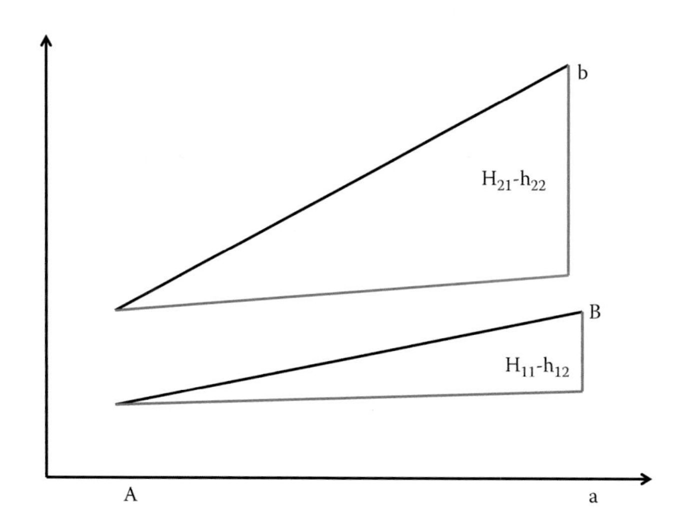

FIGURE 7.1 Illustration of gene–gene interaction.

and

$$P_{11}^{A} = P\left(G_1 = 1, G_2 = 1 \middle| D = 1\right), \quad P_{12}^{A} = P\left(G_1 = 1, G_2 = 0 \middle| D = 1\right),$$
$$P_{21}^{A} = P\left(G_1 = 0, G_2 = 1 \middle| D = 1\right), \quad P_{22}^{A} = P\left(G_1 = 0, G_2 = 0 \middle| D = 1\right).$$

Disease risk can be expressed in terms of P_{ij} and P_{ij}^{A}:

$$h_{ij} = P\left(D = 1 \middle| G_1 = i, G_2 = j\right) = \frac{P(D=1)P\left(G_1 = i, G_2 = j \middle| D = 1\right)}{P\left(G_1 = i, G_2 = j\right)}$$
$$= \frac{P(D=1)\,p_{ij}^{A}}{P_{ij}}. \tag{7.6}$$

Then, from Equations 7.4 through 7.6, we can obtain the following result.

Result 7.1

In the absence of interaction,

$$\frac{P_{11}^{A}}{P_{11}} - \frac{P_{12}^{A}}{P_{12}} = \frac{P_{21}^{A}}{P_{21}} - \frac{P_{22}^{A}}{P_{22}} \tag{7.7}$$

and

$$\frac{P_{11}^{A}P_{22}^{A}}{P_{11}P_{22}} = \frac{P_{12}^{A}P_{21}^{A}}{P_{12}P_{21}} \tag{7.8}$$

hold.

To gain understanding of the measure of gene–gene interaction, we study several special cases:

Case 1

One locus is not the disease locus. If we assume that G_1 is only a marker and will not cause disease, then we have

$$\frac{P_{11}^{A}}{P_{11}} = \frac{P\left(G_2 = 1 \middle| D\right)P\left(G_1 = 1 \middle| G_2 = 1, D\right)}{P\left(G_2 = 1\right)P\left(G_1 = 1 \middle| G_2 = 1\right)} = \frac{P\left(G_2 = 1 \middle| D\right)P\left(G_1 = 1 \middle| G_2 = 1\right)}{P\left(G_2 = 1\right)P\left(G_1 = 1 \middle| G_2 = 1\right)} = \frac{P\left(G_2 = 1 \middle| D\right)}{P\left(G_2 = 1\right)}.$$

Similarly, we have

$$\frac{P_{12}^{A}}{P_{12}} = \frac{P\left(G_2 = 0 \middle| D\right)}{P\left(G_2 = 0\right)}, \quad \frac{P_{21}^{A}}{P_{21}} = \frac{P\left(G_2 = 1 \middle| D\right)}{P\left(G_2 = 1\right)}, \quad \frac{P_{22}^{A}}{P_{22}} = \frac{P\left(G_2 = 0 \middle| D\right)}{P\left(G_2 = 0\right)},$$

which implies that

$$\frac{P_{11}^A}{P_{11}}-\frac{P_{12}^A}{P_{12}}=\frac{P(G_2=1|D)}{P(G_2=1)}-\frac{P(G_2=0|D)}{P(G_2=0)}=\frac{P_{21}^A}{P_{21}}-\frac{P_{22}^A}{P_{22}},$$

and

$$\frac{P_{11}^A P_{22}^A}{P_{11}P_{22}}=\frac{P(G_2=1|D)P(G_2=0|D)}{P(G_2=1)P(G_2=0)}=\frac{P_{12}^A P_{21}^A}{P_{12}P_{21}}.$$

Thus, we obtain $I_{add}=0$ and $I_{multiple}=1$. In other words, if one locus is a marker, there is interaction between a marker and causal locus or between two markers. Hence, both additive and multiplicative measures of gene–gene interaction correctly characterize the marker case.

The additive and multiplicative measures of gene–gene interaction have close relations with log-linear models. Consider the following log-linear model:

$$\log\left(P\left(D=1|G_1, G_2\right)\right)=\alpha+\beta_{G_1}G_1+\beta_{G_2}G_2+\beta_{G_1G_2}G_1G_2, \tag{7.9}$$

where

β_{G_1} is a genetic effect at the first locus
β_{G_2} is a genetic effect at the second locus
$\beta_{G_1G_2}$ is a gene–gene interaction effect

From this model, it follows that

$$RR_{11}=e^{\beta_{G1}+\beta_{G2}+\beta_{G1G2}},\quad RR_{12}=e^{\beta_{G1}},\quad \text{and}\quad RR_{21}=e^{\beta_{G2}}. \tag{7.10}$$

The additive and multiplicative measures of gene–gene interaction are respectively given by

$$I_{add}=e^{\beta_{G1}+\beta_{G2}+\beta_{G1G2}}-e^{\beta_{G1}}-e^{\beta_{G2}}+1$$

and

$$I_{muliple}=e^{\beta_{G1G2}}. \tag{7.11}$$

The regression coefficient $\beta_{G_1G_2}$ in the log-linear model is equal to a logarithm of the multiplicative measure of gene–gene interaction.

The additive measure and multiplicative measures are not completely overlapped. The question of which measure should be used to detect interaction has long been debated. What measure should be used to study interaction depends on the purpose of investigation (Ottman 1996).

7.1.1.2 *The Binary Measure of Gene–Gene Interaction for the Case–Control Study Design*

Next we study the measure of gene–gene interaction for the case–control study design. The gene–gene interaction for the case–control study design is usually measured by odds ratio. The genotype at the first locus, the second locus, and interaction odds ratios are defined, respectively, by

$$OR_{G_1} = \frac{\frac{h_{12}}{1-h_{12}}}{\frac{h_{22}}{1-h_{22}}}, \quad OR_{G_2} = \frac{\frac{h_{21}}{1-h_{21}}}{\frac{h_{22}}{1-h_{22}}}, \quad \text{and} \quad OR_{G_1G_2} = \frac{\frac{h_{11}}{1-h_{11}}}{\frac{h_{22}}{1-h_{22}}}. \tag{7.12}$$

Definition 7.2

Using odds ratios, we can define the additive and multiplicative measures of gene–gene interaction, respectively, as

$$I_{Oadd} = OR_{G_1G_2} - OR_{G_1} - OR_{G_2} + 1 \tag{7.13}$$

and

$$I_{Omultiple} = \frac{OR_{G_1G_2}}{OR_{G_1}OR_{G_2}}. \tag{7.14}$$

Without gene–gene interaction, Equations 7.13 and 7.14 can be simplified as

$$OR_{G_1G_2} - OR_{G_1} - OR_{G_2} + 1 = 0$$

and

$$OR_{G_1G_2} = OR_{G_1} * OR_{G_2}.$$

Define haplotype frequencies in controls as

$$P_{11}^N = P(G_1 = 1, G_2 = 1 | D = 0), \quad P_{12}^N = P(G_1 = 1, G_2 = 0 | D = 0),$$
$$P_{21}^N = P(G_1 = 0, G_2 = 1 | D = 0), \quad \text{and} \quad P_{22}^N = P(G_1 = 0, G_2 = 0 | D = 0).$$

Similar to Equations 7.7 and 7.8, we have the following result.

Result 7.2

In the absence of interaction,

$$\frac{P_{11}^A}{P_{11}^N} - \frac{P_{12}^A}{P_{12}^N} = \frac{P_{21}^A}{P_{21}^N} - \frac{P_{22}^A}{P_{22}^N} \tag{7.15}$$

and

$$\frac{P_{11}^A P_{22}^A}{P_{11}^N P_{22}^N} = \frac{P_{12}^A P_{21}^A}{P_{12}^N P_{21}^N} \tag{7.16}$$

hold.

It is interesting to know that replacing population haplotype frequencies in Equations 7.7 and 7.8 by haplotype frequencies in controls will lead to Equations 7.15 and 7.16.

Consider the following logistic model:

$$P(D=1|G_1, G_2) = \frac{e^{\alpha+\beta_{G_1}G_1+\beta_{G_2}G_2+\beta_{G_1G_2}G_1G_2}}{1+e^{\alpha+\beta_{G_1}G_1+\beta_{G_2}G_2+\beta_{G_1G_2}G_1G_2}},$$

which gives the genotypes at the first locus, second locus, and interaction odds ratios (Exercise 7.5)

$$OR_{G_1} = e^{\beta_{G_1}}, \quad OR_{G_2} = e^{\beta_{G_2}}, \quad \text{and} \quad OR_{G_1G_2} = e^{\beta_{G_1}+\beta_{G_2}+\beta_{G_1G_2}}.$$

Using these expressions for the genotypes at the first locus, second locus, and interaction odds ratios, we can obtain the following representation for the additive and multiplicative measures of gene–gene interaction:

$$I_{add} = e^{\beta_{G_1}+\beta_{G_2}+\beta_{G_1G_2}} - e^{\beta_{G_1}} - e^{\beta_{G_2}} + 1$$

and

$$I_{multiple} = e^{\beta_{G_1G_2}}, \tag{7.17}$$

which is similar to Equation 7.11.

7.1.2 Disequilibrium Measure of Gene–Gene and Gene–Environment Interactions

The concept of linkage disequilibrium (LD) and measuring its value play an essential role in genetic studies of complex diseases. We can define LD as a measure for quantifying the magnitude of gene–gene and gene–environment interactions, which allow us to borrow tools from LD studies, and it is referred to as a disequilibrium measure of gene–gene or gene–environment interaction hereafter. Let D be an indicator of the disease status and P_D be the population prevalence of disease. Treating the environmental variable as a locus, the interaction between the gene and environment can be viewed as the interaction between two loci. Similar to the measure of LD, we can also define a disequilibrium measure (covariance) between the gene and environment in the general population as follows: $\delta = P_{11}P_{22} - P_{12}P_{21}$. This mathematical form is precisely the same as the form of the measure of LD. The disequilibrium measure characterizes the dependence between the gene and gene or between gene and environment. If two genetic and environmental variables are independent, the disequilibrium measure between the gene and environment will be equal to zero.

To investigate whether the interaction between genes create the disequilibrium between them, we derived the disequilibrium measure in the disease population. Recall that the probability P_{11}^A can be expressed in terms of P_{11} and disease risk h_{11}:

$$\begin{aligned} P_{11}^A &= P(G_1 = 1, G_2 = 1 | D) = \frac{P(D=1, G_1 = 1, G_2 = 1)}{P_D} \\ &= \frac{P(G_1 = 1, G_2 = 1) P(D = 1 | G_1 = 1 | G_2 = 1)}{P_D} \\ &= \frac{P_{11} h_{11}}{P_D}. \end{aligned} \tag{7.18}$$

Similarly, we have

$$P_{12}^A = \frac{P_{12} h_{12}}{P_D}, \tag{7.19}$$

$$P_{21}^A = \frac{P_{21} h_{21}}{P_D}, \tag{7.20}$$

and

$$P_{22}^A = \frac{P_{22} h_{22}}{P_D}. \tag{7.21}$$

Now we can establish the LD relationship between the cases and controls. Combining Equations 7.18 through 7.21 gives

$$\begin{aligned} \delta^A &= P_{11}^A P_{22}^A - P_{12}^A P_{21}^A \\ &= \frac{P_{11} P_{22} h_{11} h_{22} - P_{12} P_{21} h_{12} h_{21}}{P_D^2} \\ &= \frac{h_{11} h_{22}}{P_D^2} \delta + \frac{P_{12} P_{21}}{P_D^2} (h_{11} h_{22} - h_{12} h_{21}). \end{aligned} \tag{7.22}$$

Definition 7.3

The above equation motivates us to define the following disequilibrium measure of gene–gene interaction as

$$I_\delta = h_{11} h_{22} - h_{12} h_{21}. \tag{7.23}$$

If h_{ij} in Equation 7.23 is replaced by P_{ij}, it becomes a measure of disequilibrium. Therefore, I_δ is referred to as the disequilibrium measure of gene–gene interaction. Surprisingly, we can show from Equations 7.8 and 7.23 that the disequilibrium measure I_δ is equal to zero if and only if the multiplicative measure of gene–gene interaction for the cohort study $I_{multiple}$ is equal to one. In other words, the absence of gene–gene interaction, which is detected by the disequilibrium measure of gene–gene interaction, can also be detected by multiplicative measure of gene–gene interaction.

With the aid of Equation 7.23, the measure of disequilibrium in the disease population can be expressed as

$$\delta^A = \frac{h_{11}h_{22}}{P_D^2}\delta + \frac{P_{12}P_{21}}{P_D^2} I_\delta. \tag{7.24}$$

Equation 7.24 shows that the disequilibrium between two genes in the disease population comes from two parts. One part is from the disequilibrium between two genes in the general population. Another is from the gene–gene interaction. Rewrite Equation 7.24 as

$$\delta^A - \frac{h_{11}h_{22}}{P_D^2}\delta = \frac{P_{12}P_{21}}{P_D^2} I_\delta,$$

which shows that the difference in the disequilibrium between the disease population and general population is proportional to the disequilibrium measure of gene–gene interaction. This forms the basis for formal testing for the gene–gene interaction.

7.1.3 Information Measure of Gene–Gene and Gene–Environment Interactions

In studying the information measure of gene–gene interaction, the loci G_1 and G_2 can be either coded as 0 and 1 as before or coded as 0, 1, and 2, indicating three genotypes. The environmental exposure is coded as before. Mutual information is to measure dependence between two random variables. The mutual information between two genes in the general population is defined as

$$I(G_1; G_2) = \sum_{i=0}^{2}\sum_{j=0}^{1} P(G_1 = i, G_2 = j)\log\frac{P(G_1 = i, G_2 = j)}{P(G_1 = i)P(G_2 = j)}. \tag{7.25}$$

Information theory (Cover et al. 1991) shows that mutual information $I(G_1; G_2)$ is equal to zero if and only if

$$P(G_1 = i, G_2 = j) = P(G_1 = i)P(G_2 = j), \quad (i = 0,1,2;\ j = 0,1,2),$$

i.e., two gene variables are independent.

The mutual information between two genes in the disease population is given by

$$I(G_1;G_2|D)=\sum_{i=0}^{2}\sum_{j=0}^{2}P(G_1=i,G_2=j|D=1)\log\frac{P(G_1=i,G_2=j|D=1)}{P(G_1=i|D=1)P(G_2=j|D=1)},\quad (7.26)$$

which can be reduced to

$$\begin{aligned}I(G_1;G_2|D)=&\sum_{i=0}^{2}\sum_{j=0}^{2}P(G_1=i,G_2=j|D=1)\log\frac{P(G_1=i,G_2=j)}{P(G_1=i)P(G_2=j)}\\&+\sum_{i=0}^{2}\sum_{j=0}^{2}P(G_1=i,G_2=j|D=1)\log\frac{P(D=1|G_1=i,G_2=j)/P_D}{\frac{P(D=1|G_1=i)}{P_D}\frac{P(D=1|G_2=j)}{P_D}},\end{aligned}\quad (7.27)$$

where $P_D=P(D=1)$ is the prevalence of the disease.

Equation 7.27 shows that mutual information $I(G_1;G_2|D)$ has two components. The first term in Equation 7.27 is due to dependence between two genes in the general population. The second term in Equation 7.27 is due to interaction.

Definition 7.4

Thus, we define information measure of interaction between two genes as

$$I_{G_1G_2}=\sum_{i=0}^{2}\sum_{j=0}^{2}P(G_1=i,G_2=j|D=1)\log\frac{P(D=1|G_1=i,G_2=j)/P_D}{\frac{P(D=1|G_1=i)}{P_D}\frac{P(D=1|G_2=j)}{P_D}}\quad (7.28)$$

or

$$I_{G_1G_2}=\sum_{i=0}^{2}\sum_{j=0}^{2}P(G_1=i,G_2=j|D=1)\log\frac{h_{ij}P_D}{h_ih_j},\quad (7.29)$$

which implies that $I_{G_1G_2}=0$ if and only if

$$\frac{P(D=1|G_1=i,G_2=j)}{P_D}=\frac{P(D=1|G_1=i)}{P_D}\frac{P(D=1|G_2=j)}{P_D}\quad (i=0,1,2,\ j=0,1,2)\quad (7.30)$$

or

$$h_{ij}P_D = h_i h_j, \quad (i = 0,1,2, \ j = 0,1,2), \tag{7.31}$$

where
h_{ij} is the disease risk of an individual carrying genotype $G_1 = i$ and genotype $G_2 = j$
h_i is the marginal disease risk of an individual carrying genotype $G_1 = i$
h_j is the marginal disease risk of an individual carrying genotype $G_2 = j$

Information measure of interaction has two remarkable features. First, it is defined in terms of penetrance and hence related to the cause of the disease. Second, the interaction is measured by the interdependent operation of two genes in causing disease. The absence of gene–gene interaction indicates that Equation 7.30 should hold. If G is coded as 0 or 1, then Equation 7.30 is equivalent to

$$h_{11}h_{22} = h_{12}h_{21}$$

or Equation 7.5, $RR_{11} = RR_{12} * RR_{21}$.

Equation 7.27 can be rewritten as

$$I(G_1; G_2 | D) - \sum_{i=0}^{2}\sum_{j=0}^{2} P(G_1 = i, G_2 = j | D = 1)\log\frac{P(G_1 = i, G_2 = j)}{P(G_1 = i, G_2 = j)} = I_{G_1G_2}. \tag{7.32}$$

The second term in the left side of Equation 7.32 is the mutual information between two genes in the general population if the probabilities $P(G_1 = i, G_2 = j|D)$ are replaced by $P(G_1 = i, G_2 = j)$. Equation 7.32 shows that the modified difference in mutual information between cases and the general population is proportional to the information measure of gene–gene interaction.

If we assume that two genotype variables in the general population are independent, then

$$I(G_1, G_2 | D) = I_{G_1G_2}.$$

In this case, the mutual information between two genes in the disease population is equal to the information measure of the gene–gene interaction. This provides an easy way to calculate the information measure of gene–gene interaction.

To gain an understanding of information measure of gene–gene interaction, we study several special cases.

Case 1

A locus, for example, G_1, is not the disease locus. If we assume that G_1 is only a marker and will not cause disease, then we have

$$P(D = 1 | G_1 = i, G_2 = j) = P(D = 1 | G_2 = j) \quad \text{and} \quad P(D = 1 | G_1 = i) = P_D,$$

which implies that

$$\frac{P\left(D=1\middle|G_1=i,G_2=j\right)/P_D}{\frac{P\left(D=1\middle|G_1=i\right)}{P_D}\frac{P\left(D=1\middle|G_2=j\right)}{P_D}}=1.$$

Thus, we obtain $I_{G_1G_2}=0$. In other words, if the locus G_1 is a marker, there is no interaction between the loci G_1 and G_2. The interaction measure $I_{G_1G_2}$ between two loci should be equal to zero. Hence, our information measure of gene–gene interaction correctly characterizes the marker case.

7.1.4 Measure of Interaction between a Gene and a Continuous Environment

Many environmental variables, for example, ages, incomes, and gene expressions, are continuous variables. Generally, there is more information when a risk factor is represented by a continuous variable than a categorical variable. A dichotomization of a continuous variable will lose information. Therefore, developing measures of interaction between a gene and the environment that can be applied to continuous environmental variables is indispensable in the studies of gene–environment interaction.

7.1.4.1 Multiplicative Measure of Interaction between a Gene and a Continuous Environment

To extend multiplicative measure of gene–environment interaction for binary environment to continuous environment, we first introduce point-wise risk and relative risk. We take a noncarrier of the susceptible genotype and average environment as the baseline. Denote the continuous environmental variable by e and its expectation by μ. Let the point-wise risk be defined as

$$h_{1e}=P\left(D=1\middle|G=1,E=e\right),\quad h_{1\mu}=P\left(D=1\middle|G=1,E=\mu\right),$$
$$h_{2e}=P\left(D=1\middle|G=0,E=e\right),\quad h_{2\mu}=P\left(D=1\middle|G=0,E=\mu\right).$$

Definition 7.5

Then, the point-wise relative risk can be defined as

$$RR_{1e}=\frac{h_{1e}}{h_{2\mu}},\quad RR_{1\mu}=\frac{h_{1\mu}}{h_{2\mu}},\quad \text{and}\quad RR_{2e}=\frac{h_{2e}}{h_{2\mu}}. \tag{7.33}$$

Point-wise multiplicative measure of interaction between a gene and a continuous environment is then defined by

$$I_{PMGE}=\log\frac{RR_{1e}}{RR_{1\mu}RR_{2e}}. \tag{7.34}$$

If we assume that E is an environmental variable with a normal density function of mean μ and variance σ^2 and that environmental variable E conditional on $G=i$ follows a normal density with mean μ_i and variance σ_i^2 and their corresponding densities in the disease population are a normal density function with means μ_D, μ_{Di} and variances $\sigma_D^2, \sigma_{Di}^2$, then Equation 7.34 can be reduced to

$$I_{PMGE} = \log\frac{\mu_{D0}\mu_1}{\mu_{D1}\mu_0} + \frac{1}{2}\log\frac{\sigma_{D0}^2\sigma_1^2}{\sigma_{D1}^2\sigma_0^2} + \frac{\left(e-\mu_{D0}\right)^2}{2\sigma_{D0}^2} + \frac{\left(e-\mu_0\right)^2}{2\sigma_0^2} - \frac{\left(e-\mu_{D1}\right)^2}{2\sigma_{D1}^2} - \frac{\left(e-\mu_1\right)^2}{2\sigma_1^2}. \quad (7.35)$$

We define the expectation of point-wise multiplicative measure of gene–environment interaction I_{PMGE} as the multiplicative measure of interaction between a gene and a continuous environment:

$$I_{MGE} = E\left[\log\frac{RR_{1e}}{RR_{1\mu}RR_{2e}}\right]. \quad (7.36)$$

Under the assumption of normal distribution of environmental variable, the multiplicative measure of interaction between a gene and a continuous environment is simplified to

$$I_{MGE} = \log\frac{\mu_{D0}\mu_1}{\mu_{D1}\mu_0} + \frac{1}{2}\log\frac{\sigma_{D0}^2\sigma_1^2}{\sigma_{D1}^2\sigma_0^2}. \quad (7.37)$$

In the absence of gene–environment interaction, the interaction measure I_{MGE} is equal to zero.

7.1.4.2 Disequilibrium Measure of Interaction between a Gene and a Continuous Environment

Definition 7.6

Disequilibrium measure of interaction between a binary genetic factor and a continuous environmental factor can be defined as

$$I_\delta\left(e\right) = h_{1e}h_{2\mu} - h_{1\mu}h_{2e}. \quad (7.38)$$

$I_\delta(e)$ is a function of the environmental factor. Let $P(G=1, E=\mu)$, $P(G=0, E=\mu)$, $P(G=1, E=e)$, and $P(G=0, E=e)$ be the joint probability density functions. The concept of disequilibrium measure between two random variables can be extended to measure dependence between a random variable and a random function. We first define a point-wise disequilibrium measure between the coded binary genotype and the continuous environment in the general population as

$$\delta\left(e\right) = P\left(G=1, E=e\right)P\left(G=0, E=\mu\right) - P\left(G=1, E=\mu\right)P\left(G=0, E=e\right).$$

Note that

$$
\begin{aligned}
P(G=1,E=e|D=1) &= \frac{P(G=1,E=e)h_{1e}}{P_D},\\
P(G=1,E=\mu|D=1) &= \frac{P(G=1,E=\mu)h_{1\mu}}{P_D},\\
P(G=0,E=e|D=1) &= \frac{P(G=0,E=e)h_{2e}}{P_D},\\
P(G=0,E=\mu|D=1) &= \frac{P(G=0,E=\mu)h_{2\mu}}{P_D}.
\end{aligned}
\tag{7.39}
$$

The point-wise disequilibrium measure between the coded binary genotype and the continuous environment in the disease population is

$$
\begin{aligned}
\delta^A(e) &= P(G=1,E=e|D)P(G=0,E=\mu|D) - P(G=1,E=\mu|D)P(G=0,E=e|D)\\
&= \frac{P(G=1,E=e)P(G=0,E=\mu)h_{1e}h_{2\mu} - P(G=1,E=\mu)P(G=0,E=e)h_{1\mu}h_{2e}}{P_D^2}\\
&= \frac{\delta(e)h_{1e}h_{2\mu} + P(G=1,E=\mu)P(G=0,E=e)(h_{1e}h_{2\mu} - h_{1\mu}h_{2e})}{P_D^2}.
\end{aligned}
\tag{7.40}
$$

The point-wise disequilibrium measure in the disease population is then given by

$$
\delta^A(e) = \frac{h_{1e}h_{2\mu}}{P_D^2}\delta(e) + \frac{P(G=1,E=\mu)P(G=0,E=e)}{P_D^2}I_\delta(e). \tag{7.41}
$$

7.1.4.3 Mutual Information Measure of Interaction between a Gene and a Continuous Environment

Mutual information between a gene and a continuous environmental factor E is defined as

$$
I(G;E) = \sum_{i=0}^{1}\int P(G=i,e)\log\frac{P(G=i,e)}{P(G=i)P(e)}de,
$$

where

$P(G=i,e)$ is the joint probability density function of $G=i$ and e

$P(e)$ is a density function of environmental variable

$P(G=i)$ is a probability mass function of the genotype

Definition 7.7

We define the mutual information measure of the interaction between the binary genetic factor and continuous environment as

$$I_{GCE} = \sum_{i=0}^{1} \int P\left(G=i, E=e \mid D=1\right) \log \frac{P\left(D=1 \mid G=i, E=e\right)/P_D}{\left[P\left(D=1 \mid G=i\right)/P_D\right]\left[P\left(D=1 \mid E=e\right)/P_D\right]} de. \quad (7.42)$$

The mutual information between the binary genetic factor and continuous environment in the disease population is given by (Exercise 7.8)

$$I\left(G; E \mid D\right) = \sum_{i=0}^{1} \int P\left(G=i, E=e \mid D=1\right) \log \frac{P\left(G=i, E=e\right)}{P\left(G=i\right)P\left(E=e\right)} de + I_{GCE}. \quad (7.43)$$

Equation 7.43 shows that the mutual information in the disease population consists of two terms. The first term in the right side of equation is involved in mutual information between gene and environment in the general population, and the second term is the mutual information measure of the interaction between the gene and the continuous environment.

Unlike mutual information between two discrete variables, which is easy to calculate, the mutual information between the discrete variable and continuous variable requires the calculation of an integral that may involve intensive numerical computation. However, when an environmental variable has a normal distribution with mean μ_D and variance σ_D^2 and conditional on the genotype $G=i$ and has a normal distribution with mean μ_{Di} and variance σ_{Di}^2 in the disease population, we can show that (Exercise 7.9)

$$I\left(G; E \mid D\right) = \frac{1}{2}\log \sigma_D^2 - \frac{1}{2}\left[P\left(G=0 \mid D\right)\log \sigma_{D_0}^2 + P\left(G=1 \mid D\right)\log \sigma_{D_1}^2\right]. \quad (7.44)$$

Similarly, for the general population, if we assume that the environmental variable has a normal distribution with mean μ and variance σ^2 and also has normal distributions with the conditional mean μ_i and variance σ_i^2, given the genotype $G=i$, we have

$$\begin{aligned}
&\sum_{i=0}^{1} \int P\left(G=i, E=e \mid D=1\right) \log \frac{P\left(G=i, E=e\right)}{P\left(G=i\right)P\left(E=e\right)} de \\
&= \frac{1}{2}\left[\log \sigma^2 - P\left(G=0 \mid D\right)\log \sigma_0^2 - P\left(G=1 \mid D\right)\log \sigma_1^2\right] + \frac{\sigma_{D_1}^2 + \left(\mu_D - \mu\right)^2}{2\sigma^2} \\
&- P\left(G=0 \mid D\right)\frac{\sigma_{D_0}^2 + \left(\mu_{D_0} - \mu_0\right)^2}{2\sigma_0^2} - P\left(G=1 \mid D\right)\frac{\sigma_{D_1}^2 + \left(\mu_{D_1} - \mu_1\right)^2}{2\sigma_1^2}.
\end{aligned}$$

Then, the mutual information measure of interaction between the binary genetic factor and continuous environmental variable with normal distributions is given by

$$I_{GCE} = \frac{1}{2}\log\frac{\sigma_D^2}{\sigma^2} - \frac{1}{2}\left[P(G=0|D)\log\frac{\sigma_{D_0}^2}{\sigma_0^2} + P(G=1|D)\log\frac{\sigma_{D_1}^2}{\sigma_1^2}\right] - \frac{\sigma_D^2 + (\mu_D - \mu)^2}{2\sigma^2}$$
$$+ P(G=0|D)\frac{\sigma_{D_0}^2 + (\mu_{D_0} - \mu_0)^2}{2\sigma_0^2} + P(G=1|D)\frac{\sigma_{D_1}^2 + (\mu_{D_1} - \mu_1)^2}{2\sigma_1^2}. \quad (7.45)$$

7.2 STATISTICS FOR TESTING GENE–GENE AND GENE–ENVIRONMENT INTERACTIONS FOR A QUALITATIVE TRAIT WITH COMMON VARIANTS

In the previous section, we presented four types of measures of gene–gene and gene–environment interactions that provide the basis for developing statistics to formally test for gene–gene and gene–environment interactions. In this section, we will study statistics for testing gene–gene and gene–environment interactions based on the measure of gene–gene and gene–environment interactions.

7.2.1 Relative Risk and Odds-Ratio-Based Statistics for Testing Interaction between a Gene and a Discrete Environment

Gene–gene and gene–environment interactions can be identified by a formal test. The test statistics can be developed parallel to the measure of gene–gene and gene–environment interactions. We first study statistics for testing gene–gene and gene–environment interactions in the cohort study design. The test statistics depend on the scale of the measurement of the gene–gene and gene–environment interactions. Since statistics for testing gene–environment interaction are similar to the statistics for testing gene–gene interaction, again in the following discussion, we will focus on statistics for testing the gene–gene interactions unless otherwise stated. However, statistics for testing gene–gene interactions can be easily extended to testing for gene–environment interactions.

Recall that the additive measure of gene–gene interaction is defined as

$$I_{add} = RR_{11} - RR_{01} - RR_{10} + 1$$
$$= \frac{h_{11}}{h_{22}} - \frac{h_{12}}{h_{22}} - \frac{h_{21}}{h_{22}} + 1.$$

Let n_1, n_2, n_3, and n_4 be the number of individuals carrying genotypes $G_1 = 1$ and $G_2 = 1$; $G_1 = 1$ and $G_2 = 0$, $G_1 = 0$; and $G_2 = 1$, $G_1 = 0$, and $G_2 = 0$, respectively. It can be shown that the variance of estimate of the additive interaction measure is given by (Lehmann 1983) (Exercise 7.10)

$$\hat{V}_{Radd} = \frac{h_{11}(1-h_{11})}{n_1 h_{22}^2} + \frac{h_{12}(1-h_{12})}{n_2 h_{22}^2} + \frac{h_{21}(1-h_{21})}{n_3 h_{22}^2} + \frac{(1-h_{22})(h_{12}+h_{21}-h_{11})^2}{n_4 h_{22}^3}.$$

Define the statistic to test for gene–gene interaction for the additive scale of measurement as

$$T_{Radd} = \frac{I_{add}^2}{V_{Radd}}. \tag{7.46}$$

Then, under the null hypothesis of no gene–gene interaction, the statistic T_{Radd} is asymptotically distributed as a central $\chi_{(1)}^2$ distribution.

In a case–control study design, we use odds ratio to measure gene–gene interaction. Similar to relative risk measure of gene–gene interaction in the cohort study design, there are also two odds ratio measures of gene–gene interaction in the case–control study design: additive and multiplicative measures.

The odds ratio additive measure of gene–gene interaction can be rewritten as

$$\begin{aligned} T_{Oadd} &= \mathrm{OR}_{GE} - \mathrm{OR}_G - \mathrm{OR}_E + 1 \\ &= \frac{P_{11}^A P_{22}^N}{P_{22}^A P_{11}^N} - \frac{P_{12}^A P_{22}^N}{P_{22}^A P_{12}^N} - \frac{P_{21}^A P_{22}^N}{P_{22}^A P_{21}^N} + 1. \end{aligned}$$

Similar to Appendix 7A, the variance of T_{Oadd} is given by

$$\begin{aligned} V_{Oadd} &= \frac{1}{n_A}\left(\frac{\mathrm{OR}_{G_1G_2}^2}{P_{11}^A} + \frac{\mathrm{OR}_{G_1}^2}{P_{12}^A} + \frac{\mathrm{OR}_{G_2}^2}{P_{21}^A} + \frac{\left(\mathrm{OR}_{G_1G_2} - \mathrm{OR}_{G_1} - \mathrm{OR}_{G_2}\right)^2}{P_{22}^A}\right) \\ &+ \frac{1}{n_G}\left(\frac{\mathrm{OR}_{G_1G_2}^2}{P_{11}^N} + \frac{\mathrm{OR}_{G_1}^2}{P_{12}^N} + \frac{\mathrm{OR}_{G_2}^2}{P_{21}^N} + \frac{\left(\mathrm{OR}_{G_1G_2} - \mathrm{OR}_{G_1} - \mathrm{OR}_{G_2}\right)^2}{P_{22}^N}\right). \end{aligned}$$

Therefore, for the additive measure, we define the statistic for testing gene–gene interaction as

$$T_{Oadd} = \frac{I_{Oadd}^2}{V_{Oadd}}, \tag{7.47}$$

which is asymptotically distributed as a central $\chi_{(1)}^2$ distribution under the null hypothesis of no gene–gene interaction.

The logarithm of odds-ratio multiplicative measure of gene–gene interaction can be expressed as

$$\begin{aligned} \log\left(I_{Omultiple}\right) &= \log \frac{\mathrm{OR}_{G_1G_2}}{\mathrm{OR}_{G_1}\mathrm{OR}_{G_2}} \\ &= \log \frac{P_{11}^A P_{22}^A}{P_{12}^A P_{21}^A} - \log \frac{P_{11}^N P_{22}^N}{P_{12}^N P_{21}^N}. \end{aligned}$$

The variance of the estimate of the logarithm of odds ratio is given by (Appendix 7A)

$$\hat{V}_{Rmultiple} = \frac{1}{n_A}\left(\frac{1}{P_{11}^A} + \frac{1}{P_{12}^A} + \frac{1}{P_{21}^A} + \frac{1}{P_{22}^A}\right) + \frac{1}{n_G}\left(\frac{1}{P_{11}^N} + \frac{1}{P_{12}^N} + \frac{1}{P_{21}^N} + \frac{1}{P_{22}^N}\right).$$

The statistics for testing gene–gene interaction, which is based on the odds-ratio multiplicative measure, can defined as

$$T_{Omultiple} = \frac{\left(\log \frac{P_{11}^{A} P_{22}^{A}}{P_{12}^{A} P_{21}^{A}} - \log \frac{P_{11}^{N} P_{22}^{N}}{P_{12}^{N} P_{21}^{N}}\right)^2}{\frac{1}{n_A}\left(\frac{1}{P_{11}^{A}} + \frac{1}{P_{12}^{A}} + \frac{1}{P_{21}^{A}} + \frac{1}{P_{22}^{A}}\right) + \frac{1}{n_G}\left(\frac{1}{P_{11}^{N}} + \frac{1}{P_{12}^{N}} + \frac{1}{P_{21}^{N}} + \frac{1}{P_{22}^{N}}\right)}, \quad (7.48)$$

which is again asymptotically distributed as a central $\chi^2_{(1)}$ distribution under the null hypothesis of no gene–gene interaction.

7.2.2 Disequilibrium-Based Statistics for Testing Gene–Gene Interaction

7.2.2.1 Standard Disequilibrium Measure–Based Statistics

To investigate the LD pattern generated by gene–gene interaction, we assume that two disease susceptibility loci are in Hardy–Weinberg equilibrium. Two loci can be either unlinked or linked. Let D_1 and d_1 be the two alleles at the first disease locus with frequencies P_{D_1} and P_{d_1}, respectively. Let D_2 and d_2 be the two alleles at the second disease locus with frequencies P_{D_2} and P_{d_2}, respectively. Alleles D_1 and d_1 can be indexed by 1 and 2, respectively. At the first disease locus, let D_1D_1 be genotype 11, D_1d_1 be genotype 12, and d_1d_1 be genotype 22. The genotypes at the second disease locus are similarly defined. Two-locus genotypes are simply denoted by $ijkl$ for individuals carrying the haplotypes ik and jl arranged from the left to the right. Let f_{ijkl} be the penetrance of the haplotypes ik and jl arranged from the left to the right. Let P_{11}, P_{12}, P_{21}, and P_{22} be the frequencies of the haplotypes $H_{D_1D_2}$, $H_{D_1d_2}$, $H_{d_1D_2}$, and $H_{d_1d_2}$ in the general population, respectively. Let P_{11}^A, P_{12}^A, P_{21}^A, and P_{22}^A be their corresponding haplotype frequencies in the disease population. Let $P_{D_1}^A$, $P_{d_1}^A$, $P_{D_2}^A$, and $P_{d_2}^A$ be the frequencies of alleles D_1, d_1, D_2, and d_2 in the disease population, respectively.

For ease of discussion, we introduce a concept of haplotype penetrance. Consider a haplotype with allele i at the first disease locus and allele k at the second disease locus. Then, the penetrance of haplotype h_{ik} is defined as

$$h_{ik} = P_{D_1} P_{D_2} f_{i1k1} + P_{D_1} P_{d_2} f_{i1k2} + P_{d_1} P_{D_2} f_{i2k1} + P_{d_1} P_{d_2} f_{i2k2}.$$

Let $\delta = P_{11} - P_{D_1} P_{D_2}$ be the measure of LD in the general population. The frequencies of the haplotypes in the disease population are given by

$$P_{11}^A = \frac{P_{11} h_{11}}{P_A}, \quad P_{12}^A = \frac{P_{12} h_{12}}{P_A},$$
$$P_{21}^A = \frac{P_{21} h_{21}}{P_A}, \quad P_{22}^A = \frac{P_{22} h_{22}}{P_A}, \quad (7.49)$$

where P_A is the prevalence of disease.

An interaction between two linked loci is defined in terms of penetrance of haplotype. Specifically, we define a measure of interaction between two loci, which quantifies the magnitude of interaction as

$$I = h_{11}h_{22} - h_{12}h_{21}.$$

The measure of the LD in the disease population is defined as $\delta^A = P_{11}^A P_{22}^A - P_{12}^A P_{21}^A$. In Equation 7.24, we can show that

$$\begin{aligned} \delta^A &= \frac{P_{11}P_{22}h_{11}h_{22} - P_{12}P_{21}h_{12}h_{21}}{P_A^2} \\ &= \frac{\delta h_{11}h_{22} + P_{12}P_{21}I}{P_A^2}. \end{aligned} \tag{7.50}$$

Note that

$$P_{11}^A = \frac{P_{11}h_{11}}{P_A} \quad \text{and} \quad P_{22}^A = \frac{P_{22}h_{22}}{P_A},$$

which implies that

$$\frac{h_{11}h_{22}}{P_A^2} = \frac{P_{11}^A P_{22}^A}{P_{11}P_{22}}. \tag{7.51}$$

Substituting Equation 7.51 into Equation 7.50 gives

$$\left(\delta_A - \frac{P_{11}^A P_{22}^A}{P_{11}P_{22}}\delta\right) = \frac{P_{12}P_{21}}{P_A^2}I. \tag{7.52}$$

It is known that the variances of δ_A and δ are (Equation 3.8 in Weir 1990)

$$\hat{V}_A = \mathrm{var}\left(\hat{\delta}_A\right) = \frac{\hat{P}_{D_1}^A\left(1-\hat{P}_{D_1}^A\right)\hat{P}_{D_2}^A\left(1-\hat{P}_{D_2}^A\right) + \left(1-2\hat{P}_{D_1}^A\right)\left(1-2\hat{P}_{D_2}^A\right)\hat{\delta}_A - \hat{\delta}_A^2}{2n_A}$$

and

$$\hat{V}_N = \mathrm{var}\left(\hat{\delta}\right) = \frac{\hat{P}_{D_1}\left(1-\hat{P}_{D_1}\right)\hat{P}_{D_2}\left(1-\hat{P}_{D_2}\right) + \left(1-2\hat{P}_{D_1}\right)\left(1-2\hat{P}_{D_2}\right)\hat{\delta}_N - \hat{\delta}_N^2}{2n_G},$$

where n_A and n_G are the number of sampled individuals in the cases and controls, respectively.

Motivated by Equation 7.52, we define the test statistic:

$$T_I = \frac{\left(\delta_A - \frac{P_{11}^A P_{22}^A}{P_{11}P_{22}}\delta\right)^2}{V_A + \left(\frac{P_{11}^A P_{22}^A}{P_{11}P_{22}}\right)^2 V_N}. \tag{7.53}$$

T_I will be asymptotically distributed as a central $\chi^2_{(1)}$ distribution under the null hypothesis of no interaction.

7.2.2.2 Composite Measure of Linkage Disequilibrium for Testing Interaction between Unlinked Loci

Statistical interaction models essentially treated the interaction effect as a residual term in genetic analysis and hence are likely to limit the power to detect interaction. Alternative to statistical interaction models, interactions between two loci (or genes) can be understood as the irreducible dependencies between loci causing disease (Akulin and Bratko 2003). Although LD-based statistics have demonstrated high power to detect interaction between two loci (Zhao et al. 2006), in general, linkage phase information of marker loci for unrelated individuals is unknown; only genotype data are available. Experiments for the generation of haplotype data are expensive and time consuming. Estimation of haplotypes based on genotype data inevitably incurs the errors, which in turn will lead to increasing false positives in the detection of interactions between two loci. In this section, we introduce the composite measure of LD for testing interaction between two unlinked loci when only genotype data are available (Wu et al. 2008).

Let $P_{1/1}$, $P_{1/2}$, $P_{2/1}$, and $P_{2/2}$ be the frequencies of $H_{D_1/D_2}, H_{D_1/d_2}, H_{d_1/D_2}$, and H_{d_1/d_2}, respectively, where the slash denotes that the two chromosomes in the individual, which are from different parents. Let $P_{1/1}^A$, $P_{1/2}^A$, $P_{2/1}^A$, and $P_{2/2}^A$ be their corresponding frequencies of $H_{D_1/D_2}, H_{D_1/d_2}, H_{d_1/D_2}$, and H_{d_1/d_2} in the disease population.

The nongametic frequency can be calculated by genotype frequencies (Figure 7.1). For example,

$$P_{1/1} = P_{D_1D_2}^{D_1D_2} + \frac{1}{2}\left(P_{d_1D_2}^{D_1D_2} + P_{D_1D_2}^{D_1d_2} + P_{d_1D_2}^{D_1d_2}\right). \tag{7.54}$$

In Section 7.2.2.1, the penetrance of haplotype $H_{D_1D_2}$ is defined as the probability that an individual with the haplotype $H_{D_1D_2}$ is affected. It is a weighted sum of the penetrance that contains haplotype $H_{D_1D_2}$. The penetrance h_{12}, h_{21}, and h_{22} is similarly defined.

The penetrance of two alleles at different loci on different chromosomes, H_{D_1/D_2}, can be similarly defined as

$$h_{1/1} = \left[P_{D_1D_2}^{D_1D_2} f_{1111} + \frac{1}{2}\left(P_{D_1d_2}^{D_1D_2} f_{1112} + P_{d_1D_2}^{D_1D_2} f_{1211} + P_{D_1d_2}^{d_1D_2} f_{2112}\right)\right] \Big/ P_{1/1}. \tag{7.55}$$

It is a weighted sum of genotypic penetrance. Similarly, we can define the penetrance $h_{1/2}, h_{2/1}$, and $h_{2/2}$. If we assume Hardy–Weinberg equilibrium and genotypic equilibrium in general population, then we have $h_{11} = h_{1/1}, h_{12} = h_{1/2}$, $h_{21} = h_{2/1}$, and $h_{22} = h_{2/2}$. Let $\delta_{D_1 D_2} = P_{11} - P_{D_1} P_{D_2}$ be the measure of intragametic LD that measures the association of alleles from different loci on the same haplotype (Schaid 2004) and $\delta_{D_1/D_2} = P_{1/1} - P_{D_1} P_{D_2}$ be the measure of intergametic LD that measures the association of two alleles from different loci on different haplotypes (Schaid 2004) in the general population. We can show that haplotype frequencies in disease population can be expressed as

$$P_{11}^A = \frac{P_{11} h_{11}}{P_A}, \quad P_{12}^A = \frac{P_{12} h_{12}}{P_A}, \quad P_{21}^A = \frac{P_{21} h_{21}}{P_A}, \quad P_{22}^A = \frac{P_{22} h_{22}}{P_A}, \tag{7.56}$$

and

$$P_{1/1}^A = \frac{P_{1/1} h_{1/1}}{P_A}, \quad P_{1/2}^A = \frac{P_{1/2} h_{1/2}}{P_A}, \quad P_{2/1}^A = \frac{P_{2/1} h_{2/1}}{P_A}, \quad P_{2/2}^A = \frac{P_{2/2} h_{2/2}}{P_A}, \tag{7.57}$$

where P_A denotes disease prevalence.

Now we calculate the measures of intragametic LD and intergametic LD in the disease population under a general two-locus disease model. The measure of intragametic LD and the measure of intergametic LD in the disease population are denoted by $\delta_{D_1 D_2}^A$ and δ_{D_1/D_2}^A, respectively.

Similar to Equation 7.50, we can show

$$\delta_{D_1 D_2}^A = \frac{\delta_{D_1 D_2}}{P_A} h_{11} + \frac{P_{D_1} P_{D_2}}{P_A} \left(h_{11} - \frac{h_{D_1} h_{D_2}}{P_A} \right), \tag{7.58}$$

and

$$\delta_{D_1/D_2}^A = \frac{\delta_{D_1/D_2}}{P_A} h_{1/1} + \frac{P_{D_1} P_{D_2}}{P_A} \left(h_{1/1} - \frac{h_{D_1} h_{D_2}}{P_A} \right), \tag{7.59}$$

where $h_{D_1} = P(Affected|D_1)$ and $h_{D_2} = P(Affected|D_2)$.

We define a measure of intragametic interaction that measures the interaction of two alleles from different loci on the same haplotype as $I_{intra} = h_{11} - \frac{h_{D_1} h_{D_2}}{P_A}$ and a measure of intergametic interaction that measures the interaction of two alleles from different alleles on the different haplotypes as $I_{inter} = h_{1/1} - \frac{h_{D_1} h_{D_2}}{P_A}$. Then a measure of total interaction between two loci, which consists of intragametic interaction and intergametic interaction is given by

$$I = I_{intra} + I_{inter}. \tag{7.60}$$

Equation 7.60 clearly shows that the interaction between two loci is defined by the penetrance of the two loci. Although the penetrance of the risks is not directly related to biological process, it is related to the causes of the disease. Therefore, the above definition of interaction is related to biological interaction. It follows from Equations 7.58 through 7.60 that the composite measure of LD, $\Delta_{D_1D_2}^A$ (Weir 1996), in the disease population is given by

$$\Delta_{D_1D_2}^A = \delta_{D_1D_2}^A + \delta_{D_1/D_2}^A = \frac{\delta_{D_1D_2}}{P_A} h_{11} + \frac{\delta_{D_1/D_2}}{P_A} h_{1/1} + \frac{P_{D_1} P_{D_2}}{P_A} I. \tag{7.61}$$

The absence of interaction between two loci is then defined as

$$h_{11} = \frac{h_{D_1} h_{D_2}}{P_A} \quad \text{or} \quad \frac{h_{11}}{P_A} = \frac{h_{D_1}}{P_A} \frac{h_{D_2}}{P_A} \quad \text{and} \quad h_{1/1} = \frac{h_{D_1} h_{D_2}}{P_A} \quad \text{or} \quad \frac{h_{1/1}}{P_A} = \frac{h_{D_1}}{P_A} \frac{h_{D_2}}{P_A}. \tag{7.62}$$

Equation 7.62 indicates that similar to linkage equilibrium where the frequency of a haplotype is equal to the product of the frequencies of the component alleles of the haplotype, the absence of interaction between two loci implies that the proportion of individuals carrying two alleles (either in the same chromosome or in the different chromosome) in the disease population is equal to the product of proportions of individuals carrying a single allele in the disease population if we assume that the disease is caused by only two investigated disease loci. In other words, interaction between two disease susceptibility loci occurs when the contribution of one locus to the disease depends on another locus. In contrast to the additive model for interaction, which was introduced by Fisher, the interaction model defined by Equations 7.60 and 7.62 are referred as to a multiplicative interaction model.

In the previous discussion, we showed that under the multiplicative disease model, interaction between unlinked loci will create LD. Intuitively, we can test for interaction by comparing the difference in the composite genotypic disequilibrium between two unlinked loci between cases and controls. Precisely, if we denote the estimators of the composite LD measures in cases and controls by $\hat{\Delta}_A$ and $\hat{\Delta}_N$, respectively, then the test statistic can be defined as

$$T_I = \frac{\left(\hat{\Delta}_A - \hat{\Delta}_N\right)^2}{\text{Var}\left(\hat{\Delta}_A\right) + \text{Var}\left(\hat{\Delta}_N\right)}, \tag{7.63}$$

where

$$\hat{\Delta}_A = \hat{P}_{11}^A + \hat{P}_{1/1}^A - 2\hat{P}_{D_1}^A \hat{P}_{D_2}^A, \quad \hat{\Delta}_N = \hat{P}_{11}^N + \hat{P}_{1/1}^N - 2\hat{P}_{D_1}^N \hat{P}_{D_2}^N,$$

$$\text{Var}\left(\hat{\Delta}_A\right) = \frac{1}{n_A}\left[\left(\hat{\pi}_{D_1}^A + \hat{\delta}_{D_1}^A\right)\left(\hat{\pi}_{D_2}^A + \hat{\delta}_{D_2}^A\right) + \frac{1}{2}\hat{\tau}_{D_1}^A \hat{\tau}_{D_2}^A \hat{\Delta}_A + +\hat{\tau}_{D_1}^A \hat{\delta}_{D_1D_2D_2}^A + \hat{\tau}_{D_2}^A \hat{\delta}_{D_1D_1D_2}^A + \hat{\Delta}_{D_1D_1D_2D_2}^A\right],$$

$$\operatorname{Var}\left(\hat{\Delta}_N\right)=\frac{1}{n_G}\left[\left(\hat{\pi}_{D_1}^N+\hat{\delta}_{D_1}^N\right)\left(\hat{\pi}_{D_2}^N+\hat{\delta}_{D_2}^N\right)+\frac{1}{2}\hat{\tau}_{D_1}^N\hat{\tau}_{D_2}^N\hat{\Delta}_N++\hat{\tau}_{D_1}^N\hat{\delta}_{D_1D_2D_2}^N+\hat{\tau}_{D_2}^N\hat{\delta}_{D_1D_1D_2}^N+\hat{\Delta}_{D_1D_1D_2D_2}^N\right],$$

$$\hat{\pi}_{D_1}^A=\hat{P}_{D_1}^A\left(1-\hat{P}_{D_1}^A\right),\quad \hat{\pi}_{D_2}^A=\hat{P}_{D_2}^A\left(1-\hat{P}_{D2}^A\right),\quad \hat{\delta}_{D_1}^A=\hat{P}_{D_1D_1}^A-\left(\hat{P}_{D_1}^A\right)^2,\quad \hat{\delta}_{D_2}^A=\hat{P}_{D_2D_2}^A-\left(\hat{P}_{D_2}^A\right)^2,$$

$$\hat{\tau}_{D_1}^A=\left(1-2\hat{P}_{D_1}^A\right),\quad \hat{\tau}_{D_2}^A=\left(1-2\hat{P}_{D_2}^A\right),\quad \hat{\delta}_{D_1D_1D_2}^A=\hat{P}_{D_1D_1D_2}^A-\hat{P}_{D_1}^A\hat{\Delta}_A-\hat{P}_{D_2}\hat{\delta}_{D_1}^A-\left(\hat{P}_{D_1}^A\right)^2\hat{P}_{D_2}^A,$$

$$\hat{\delta}_{D_1D_2D_2}^A=\hat{P}_{D_1D_2D_2}^A-\hat{P}_{D_2}^A\hat{\Delta}_A-\hat{P}_{D_1}\hat{\delta}_{D_2}^A-\left(\hat{P}_{D_2}^A\right)^2\hat{P}_{D_1}^A,$$

$$\begin{aligned}\hat{\Delta}_{D_1D_1D_2D_2}^A=&\hat{P}_{D_1D_1/D_2D_2}^A-2\hat{P}_{D_1}^A\hat{\delta}_{D_1D_2D_2}^A-2\hat{P}_{D_2}^A\hat{\delta}_{D_1D_1D_2}-2\hat{P}_{D_1}^A\hat{P}_{D_2}^A\hat{\Delta}_A\\&-\left(\hat{\Delta}_A\right)^2-\left(\hat{P}_{D_1}^A\right)^2\hat{\delta}_{D_2}^A-\left(\hat{P}_{D_2}^A\right)^2\hat{\delta}_{D_1}^A-\hat{\delta}_{D_1}^A\hat{\delta}_{D_2}^A-\left(\hat{P}_{D_1}^A\hat{P}_{D_2}^A\right)^2,\end{aligned}$$

$\hat{\pi}_{D_1}^N,\hat{\pi}_{D_2}^N,\hat{\tau}_{D_1}^N,\hat{\tau}_{D_2}^N,\hat{\delta}_{D_1}^N,\hat{\delta}_{D_2}^N,\hat{\delta}_{D_1D_2D_2}^N,\hat{\delta}_{D_1D_1D_2}^N$, and $\hat{\Delta}_{D_1D_1D_2D_2}^N$ are similarly defined for controls (the formula for calculations of the composite measure of LD in cases and controls is given in Weir 1996), $P_{11}^A, P_{1/1}^A, P_{D_1}^A, P_{D_2}^A, P_{11}^N P_{1/1}^N, P_{D_1}^N$, and $P_{D_2}^N$ are defined as before $\hat{P}_{11}^A,\hat{P}_{1/1}^A,\hat{P}_{D_1}^A,\hat{P}_{D_2}^A,\hat{P}_{11}^N\hat{P}_{1/1}^N,\hat{P}_{D_1}^N$, and $\hat{P}_{D_2}^N$ are their estimators, n_A and n_G denote the number of sampled individuals in cases and controls, respectively. The variance of the composite LD measure was the large-sample variance.[15] Under the null hypothesis and assumption of Hardy–Weinberg equilibrium, the variance of the composite measure of LD in cases and controls becomes $\operatorname{Var}\left(\hat{\Delta}_A\right)=\frac{\hat{\pi}_{D_1}^A\hat{\pi}_{D_2}^A}{n_A}$ and $\operatorname{Var}\left(\hat{\Delta}_N\right)=\frac{\hat{\pi}_{D_1}^N\hat{\pi}_{D_2}^N}{n_G}$. When the sample size is large enough to ensure application of large sample theory, test statistic T_I is asymptotically distributed as a central $\chi_{(1)}^2$ distribution under the null hypothesis of no interaction (both intragametic interaction and intergametic interaction) between two unlinked loci and assumption of Hardy–Weinberg equilibrium.

In theory, we can use case-only design to study interaction between two loci. However, in practice, background LD between two unlinked loci may exist in the population due to many unknown factors. Therefore, the test statistic based on case–control design is more robust than the statistic based on case-only design.

7.2.3 Information-Based Statistics for Testing Gene–Gene Interaction

In Section 7.1.3, we discussed mutual information measure of interaction. Now we study statistics for testing gene–gene interaction using mutual information measure of interaction. Consider two loci G_1 and G_2. Define

$$f_{ij}=P\left(G_1=i,\,G_2=j\right)\log\frac{P\left(G_1=i,\,G_2=j\right)}{P\left(G_1=i\right)P\left(G_2=j\right)}\quad\left(i=0,1,2,\ j=0,1,2\right)$$

and

$$f_{D_{ij}} = P\left(G_1 = i, G_2 = j \middle| D = 1\right) \log \frac{P\left(G_1 = i, G_2 = j \middle| D = 1\right)}{P\left(G_1 = i \middle| D = 1\right) P\left(G_2 = j \middle| D = 1\right)} \quad (i = 0,1,2,\ j = 0,1,2).$$

Let $f = [f_{00}, f_{01}, \ldots, f_{22}]^T$, $f_D = [f_{D00}, f_{D01}, \ldots, f_{D22}]^T$, $P_{ij} = P(G_1 = i, G_2 = j)$, and $P_{D_{ij}} = P(G_1 = i, G_2 = j | D = 1)$. Define $P = [P_{00}, P_{01}, \ldots P_{22}]^T$ and $P_D = [P_{D00}, P_{D01}, \ldots, P_{D22}]^T$. The joint probabilities of the genotype variables in both the general population and disease population follow multinomial distributions with the following covariance matrices:

$$\Sigma = \text{diag}\left(P\right) - PP^T \quad \text{and} \quad \Sigma_D = \text{diag}\left(P_D\right) - P_D P_D^T.$$

Let the Jacobian matrices of f and f_D with respect to P and P_D be $B = \left(\frac{\partial f_D}{\partial P_D^T}\right)$ and $C = \left(\frac{\partial f}{\partial P^T}\right)$, respectively. It is easy to see that

$$\frac{\partial f_{ij}}{\partial P_{ij}} = \log \frac{P_{ij}}{P_{i.} P_{.j}} - \frac{P_{ij}}{P_{i.}} - \frac{P_{ij}}{P_{.j}} + 1, \quad \underset{(l \neq j)}{\frac{\partial f_{ij}}{\partial P_{il}}} = -\frac{P_{ij}}{P_{i.}}, \quad \underset{(k \neq i)}{\frac{\partial f_{ij}}{\partial P_{kj}}} = -\frac{P_{ij}}{P_{.j}}, \quad \underset{(k \neq i, l \neq j)}{\frac{\partial f_{ij}}{\partial P_{kl}}} = 0, \tag{7.64}$$

where $P_{i.} = \sum_{j=0}^{2} P_{ij}$ and $P_{.j} = \sum_{i=0}^{2} P_{ij}$. The partial derivatives of the function $f_{D_{ij}}$ with respect to $P_{D_{kl}}$ can be similarly defined. Let n_A be the number of sampled individuals in the cases and n_G be the number of sampled individuals in the controls. Define

$$\Lambda = \frac{B \Sigma_D B^T}{n_A} + \frac{C \Sigma C^T}{n_G}.$$

The statistic for testing the gene–gene interactions is then defined as

$$T_{MIS} = \left(\hat{f}_D - \hat{f}\right)^T \hat{\Lambda}^- \left(\hat{f}_D - \hat{f}\right), \tag{7.65}$$

where

$\hat{f}$, $\hat{f}_D$, and $\hat{\Lambda}$ are the estimators of f, f_D, and Λ

$\hat{\Lambda}^-$ is a generalized inverse of matrix $\hat{\Lambda}$

When the sample size is sufficiently large to ensure application of large sample theory, the test statistic T_{MIS} is asymptotically distributed as a central $\chi^2_{(2)}$ distribution under the null hypothesis of no gene–gene interaction if we assume that two loci in the general population are in linkage equilibrium.

In many cases, two loci in the general population may not be in linkage equilibrium. Therefore, in these cases, using the statistic T_{MIS} to test gene–gene interaction is inappropriate. We extend the information-based statistic for testing gene–gene interaction to a general case where two loci may be in linkage disequilibrium. Information measure of gene–gene interaction can be rewritten as

$$I_{G_1G_2} = \sum_{i=0}^{2}\sum_{j=0}^{2} P_{Dij} \log \frac{P_{Dij}P_{i.}P_{.j}}{P_{Di.}P_{D.j}P_{ij}}. \tag{7.66}$$

To calculate the variance of the estimate of $I_{G_1G_2}$, we first calculate its partial derivatives:

$$\frac{\partial I_{G_1G_2}}{\partial P_{Dij}} = \log \frac{P_{Dij}P_{i.}P_{.j}}{P_{Di.}P_{D.j}P_{ij}} - 1$$

and

$$\frac{\partial I_{G_1G_2}}{\partial P_{ij}} = \frac{P_{Di.}}{P_{i.}} + \frac{P_{D.j}}{P_{.j}} - \frac{P_{Dij}}{P_{ij}}.$$

Then, similar to Appendix 7A, using the delta method, we can obtain the approximate variance of $I_{G_1G_2}$:

$$V_{MI} = \frac{1}{n_A}\left\{\sum_{i=0}^{2}\sum_{j=0}^{1}\left(\frac{\partial I_{G_1G_2}}{\partial P_{Dij}}\right)^2 P_{Dij} - \left(\sum_{i=0}^{2}\sum_{j=0}^{1}\frac{\partial I_{G_1G_2}}{\partial P_{Dij}} P_{Dij}\right)^2\right\}$$
$$+ \frac{1}{n_G}\left\{\sum_{i=0}^{2}\sum_{j=0}^{1}\left(\frac{\partial I_{G_1G_2}}{\partial P_{ij}}\right)^2 P_{ij} - \left(\sum_{i=0}^{2}\sum_{j=0}^{1}\frac{\partial I_{G_1G_2}}{\partial P_{ij}} P_{ij}\right)^2\right\}.$$

An information-based statistic for testing gene–gene interaction can then be defined as

$$T_{MI} = \frac{I_{G_1G_2}^2}{V_{MI}}, \tag{7.67}$$

which is asymptotically distributed as a central $\chi_{(1)}^2$ distribution under the null hypothesis of no gene–gene interaction.

If we assume that two loci in the general population are in linkage equilibrium, then under the null hypothesis of no gene–gene interaction, the variance V_{MI} will become zero. In this case, the test statistic T_{MI} will become undefined. We either use the statistic T_{MIS} defined in Equation 7.65 for testing gene–gene interaction or use the following statistic to test for gene–gene interaction:

$$T_{MII} = 2n_A \sum_{i=0}^{2}\sum_{j-0}^{1} P_{Dij} \log \frac{P_{Dij}}{P_{Di.}P_{D.j}}, \tag{7.68}$$

which is asymptotically distributed as a central $\chi^2_{(1)}$ distribution (Brillinger 2004). In practice, it is not convenient to test linkage equilibrium in the general population. Therefore, we need to develop statistics for testing gene–gene interaction, which can be applied to both linkage equilibrium and LD cases.

Let

$$x_{ij} = P_{Dij} \log \frac{P_{Dij} P_{i.} P_{.j}}{P_{Di.} P_{D.j} P_{ij}} \quad \text{and} \quad X = \left[x_{00}, x_{01}, \ldots, x_{22} \right]^T.$$

Then, its partial derivatives with respect to P_{Dij} and P_{ij} are given by

$$\frac{\partial x_{ij}}{\partial P_{Dij}} = \log \frac{P_{Dij} P_{i.} P_{.j}}{P_{Di.} P_{D.j} P_{ij}} + P_{Dij} \left(\frac{1}{P_{Dij}} - \frac{1}{P_{Di.}} - \frac{1}{P_{D.j}} \right),$$

$$\underset{l \neq j}{\frac{\partial x_{ij}}{\partial P_{Dil}}} = -\frac{P_{Dij}}{P_{Di.}}, \quad \underset{k \neq i}{\frac{\partial x_{ij}}{\partial P_{Dkj}}} = -\frac{P_{Dij}}{P_{D.j}}, \quad \underset{k \neq i, l \neq j}{\frac{\partial x_{ij}}{\partial P_{Dkl}}} = 0,$$

$$\frac{\partial x_{ij}}{\partial P_{ij}} = P_{Dij} \left(\frac{1}{P_{i.}} + \frac{1}{P_{.j}} - \frac{1}{P_{ij}} \right), \quad \underset{l \neq j}{\frac{\partial x_{ij}}{\partial P_{il}}} = \frac{P_{Dij}}{P_{i.}}, \quad \underset{k \neq i}{\frac{\partial x_{ij}}{\partial P_{kj}}} = \frac{P_{Dij}}{P_{.j}}, \quad \text{and} \quad \underset{k \neq i, l \neq j}{\frac{\partial x_{ij}}{\partial P_{kl}}} = 0.$$

Assume that the vectors P_D and P and the matrices Σ and Σ_D are defined as before. Let

$$B_I = \left(\frac{\partial X}{\partial P_D^T} \right) \quad \text{and} \quad C_I = \left(\frac{\partial X}{\partial P^T} \right).$$

Then, the covariance matrix of X is given by

$$\Lambda_I = \frac{B_I \Sigma_D B_I^T}{n_A} + \frac{C_I \Sigma C_I^T}{n_G}.$$

Define the statistic for testing gene–gene interaction as

$$T_{MIB} = \hat{X}^T \hat{\Lambda}_I^{-1} \hat{X}. \tag{7.69}$$

We can show that under null hypothesis of no gene–gene interaction, the statistic T_{MIB} is asymptotically distributed as a central $\chi^2_{(2)}$ distribution regardless of whether two loci in the general population are linkage equilibrium or not. Therefore, the statistic T_{MIB} can be used to test for gene–gene interaction in any cases.

7.2.4 Haplotype Odds Ratio and Tests for Gene–Gene Interaction

Over the last several decades, epidemiologists have debated intensely about how to define and measure interaction in epidemiologic studies. The concept of gene–gene interactions is often used but rarely specified with precision (Jakulin 2005). In general, statistical gene–gene interaction is defined as departure from additive or multiplicative joint effects of the genetic risk factors. It is increasingly recognized that statistical interactions are scale dependent (An et al. 2009). In other words, how to define the effects of a risk factor and how to measure departure from the independence of effects will greatly affect the assessment of gene–gene interaction. The most popular scale upon which risk factors are measured in case–control studies is odds-ratio. The traditional odds-ratio is defined in terms of genotypes at two loci. Similar to two-locus association analysis where only genotype information at two loci is used, odds-ratio defined by genotypes for testing interaction will not employ allelic association information. However, it is known that interaction between two loci will generate allelic associations in some cases (Zhao et al. 2006). Since they do not use allelic association information between two loci, the statistical methods based on the odds-ratio that is defined in terms of genotypes will have less power to detect interaction. To overcome this limitation, we can define odds-ratio in terms of a pseudohaplotype (which is defined as two alleles located on the same paternal or maternal chromosomes) for measuring interaction and develop a statistic based on a pseudohaplotype-defined odds-ratio for testing interaction between two loci (either linked or unlinked).

We begin with defining genotype-based, allele-based, and haplotype-based odds ratios and then define their three types of statistics for testing gene–gene interactions. Most materials are from Wu et al. (2010).

7.2.4.1 Genotype-Based Odds Ratio Multiplicative Interaction Measure

Consider two loci: G_1 and G_2. Assume that the codes $G_1=1(G_1=0)$ and $G_2=1(G_2=0)$ denote whether an individual is a carrier (or noncarrier) of the susceptible genotype at the loci G_1 and G_2, respectively. Let D denote the disease status where $D=1(D=0)$ indicates an affected (or unaffected) individual. Consider the following logistic model:

$$P\left(D=1\middle|G_1, G_2\right)=\frac{e^{\alpha+\beta_{G_1}G_1+\beta_{G_2}G_2+\beta_{G_1G_2}G_1G_2}}{1+e^{\alpha+\beta_{G_1}G_1+\beta_{G_2}G_2+\beta_{G_1G_2}G_1G_2}}. \quad (7.70)$$

The odds-ratio associated with G_1 for a nonsusceptible genotype at locus G_2. $(G_2=0)$ is defined as

$$\mathrm{OR}_{G_1}=\frac{P\left(D=1\middle|G_1=1, G_2=0\right)/P\left(D=0\middle|G_1=1, G_2=0\right)}{P\left(D=1\middle|G_1=0, G_2=0\right)/P\left(D=0\middle|G_1=0, G_2=0\right)}.$$

The odds-ratio associated with G_2 is similarly defined as

$$\mathrm{OR}_{G_2}=\frac{P\left(D=1\middle|G_1=0, G_2=1\right)/P\left(D=0\middle|G_1=0, G_2=1\right)}{P\left(D=1\middle|G_1=0, G_2=0\right)/P\left(D=0\middle|G_1=0, G_2=0\right)}.$$

The odds-ratio associated with G_1 and G_2 compared to the baseline category $G_1=0$ and $G_2=0$ is then defined as

$$\mathrm{OR}_{G_1G_2}=\frac{P(D=1|G_1=1,G_2=1)/P(D=0|G_1=1,G_2=1)}{P(D=1|G_1=0,G_2=0)/P(D=0|G_1=0,G_2=0)}.$$

The odds for baseline category $G_1=0$ and $G_2=0$ are defined as

$$\mathrm{OR}_b=\frac{P(D=1|G_1=0,G_2=0)}{P(D=0|G_1=0,G_2=0)}.$$

From Equation 7.70, we clearly have

$$\mathrm{OR}_b=e^{\alpha},\quad \mathrm{OR}_{G_1}=e^{\beta_{G1}},\quad \mathrm{OR}_{G_2}=e^{\beta_{G2}},\quad \text{and}\quad \mathrm{OR}_{G_1G_2}=\mathrm{OR}_{G_1}\mathrm{OR}_{G_2}e^{\beta_{G1G2}}.$$

Define a multiplicative interaction measure between two loci, G_1 and G_2, as

$$I_{G_1G_2}=\log\frac{\mathrm{OR}_{G_1G_2}}{\mathrm{OR}_{G_1}\mathrm{OR}_{G_2}}. \tag{7.71}$$

It is clear that

$$\beta_{G_1G_2}=I_{G_1G_2}. \tag{7.72}$$

If $\mathrm{OR}_{G_1G_2}=\mathrm{OR}_{G_1}\mathrm{OR}_{G_2}$, i.e., there is no interaction between loci G_1 and G_2, then $I_{GH}=0$. This shows that the logistic regression coefficient for interaction term $\beta_{G_1G_2}$ is equivalent to the interaction measure defined as log odds-ratio. The interaction measure $I_{G_1G_2}$ can also be written as

$$I_{G_1G_2}=\log\frac{P(G_1=1,G_2=1|D=1)P(G_1=0,G_2=0|D=1)}{P(G_1=1,G_2=0|D=1)P(G_1=0,G_2=1|D=1)}-\log\frac{P(G_1=1,G_2=1|D=0)P(G_1=0,G_2=0|D=0)}{P(G_1=1,G_2=0|D=0)P(G_1=0,G_2=1|D=0)}.$$

The values of odds-ratio defined in terms of genotypes depends on how to code indicator variables G_1 and G_2.

7.2.4.2 Allele-Based Odds Ratio Multiplicative Interaction Measure

Similar to the odds ratio for genotypes, we can define odds-ratio in terms of alleles. Let $P(D=1|G_1^i,G_2^j)$ be the probability of an individual with alleles G_1^i and G_2^j being affected.

We can similarly define $P\left(D=0 \middle| G_1^i, G_2^j\right)$. We then can determine the odds-ratio associated with allele G_1^1 at the G_1 locus and allele G_2^1 at the G_2 locus compared to the baseline G_1^2/G_2^2 as

$$\mathrm{OR}_{G_1^1/G_2^1} = \frac{\dfrac{P\left(D=1 \middle| G_1^1, G_2^1\right)}{P\left(D=0 \middle| G_1^1, G_2^1\right)}}{\dfrac{P\left(D=1 \middle| G_1^2, G_2^2\right)}{P\left(D=0 \middle| G_1^2, G_2^2\right)}}.$$

Similarly, we measure the odds-ratio associated with alleles G_1^1/G_2^2 and G_2/H_1, respectively, as

$$\mathrm{OR}_{G_1^1/G_2^2} = \frac{\dfrac{P\left(D=1 \middle| G_1^1, G_2^2\right)}{P\left(D=0 \middle| G_1^1, G_2^2\right)}}{\dfrac{P\left(D=1 \middle| G_1^2, G_2^2\right)}{P\left(D=0 \middle| G_1^2, G_2^2\right)}}$$

and

$$\mathrm{OR}_{G_1^2/G_2^1} = \frac{\dfrac{P\left(D=1 \middle| G_1^2, G_2^1\right)}{P\left(D=0 \middle| G_1^2, G_2^1\right)}}{\dfrac{P\left(D=1 \middle| G_1^2, G_2^2\right)}{P\left(D=0 \middle| G_1^2, G_2^2\right)}}.$$

Similar to genotype, we can define a multiplicative interaction measure in terms of log odds-ratio for an allele as

$$I_{G_1/G_2} = \log \frac{\mathrm{OR}_{G_1^1/G_2^1}}{\mathrm{OR}_{G_1^1/G_2^2}\mathrm{OR}_{G_1^2/G_2^1}},$$

which is equivalent to

$$I_{G_1/G_2} = \log(R) - \log(S),$$

where $R = \dfrac{P\left(G_1^1, G_2^1 \middle| D=1\right)P\left(G_1^2, G_2^2 \middle| D=1\right)}{P\left(G_1^1, G_2^2 \middle| D=1\right)P\left(G_1^2, G_2^1 \middle| D=1\right)}$ and $S = \dfrac{P\left(G_1^1, G_2^1 \middle| D=0\right)P\left(G_1^2, G_2^2 \middle| D=0\right)}{P\left(G_1^1, G_2^2 \middle| D=0\right)P\left(G_1^2, G_2^1 \middle| D=0\right)}$.

The "fast-epistasis" test statistics in PLINK (http://pngu.mgh.harvard.edu/~purcell/plink/index.shtml) is defined as

$$Z = \frac{\log(R) - \log(S)}{\sqrt{\mathrm{SE}(R) + \mathrm{SE}(S)}},$$

where SE(R) and SE(S) denote the standard deviation of R and S, respectively. Absence of interaction is if and only if

$$\frac{P\left(G_1^1, G_2^1 \middle| D=1\right) P\left(G_1^2, G_2^2 \middle| D=1\right)}{P\left(G_1^1, G_2^2 \middle| D=1\right) P\left(G_1^2, G_2^1 \middle| D=1\right)} = \frac{P\left(G_1^1, G_2^1 \middle| D=0\right) P\left(G_1^2, G_2^2 \middle| D=0\right)}{P\left(G_1^1, G_2^2 \middle| D=0\right) P\left(G_1^2, G_2^1 \middle| D=0\right)}.$$

This is the basis of the "fast-epistasis" test in PLINK.

7.2.4.3 Haplotype-Based Odds Ratio Multiplicative Interaction Measure

Suppose that locus G_1 has two alleles, G_1^1 and G_1^2, and the locus G_2 has two alleles, G_2^1 and G_2^2. Let $P_{G_1^1}^A$, $P_{G_1^2}^A$, $P_{G_2^1}^A$, $P_{G_2^2}^A$ and $P_{G_1^1}^N$, $P_{G_1^2}^N$, $P_{G_2^1}^N$, $P_{G_2^2}^N$ be the frequencies of alleles G_1^1, G_1^2, G_2^1, G_2^2 in the cases and controls, respectively. For the convenience of discussion, we introduce a terminology of "pseudohaplotype." When two loci are linked, a pseudohaplotype is defined as the regular haplotype. When two loci are unlinked, a pseudohaplotype is defined as a set of alleles that are located in the same paternal or maternal chromosomes. The frequencies of a pseudohaplotype can be estimated by the classical methods for the estimation of haplotype frequencies such as expectation–maximization (EM) algorithms. For simplicity, hereafter we will not make distinction between the haplotype and pseudohaplotype. When two loci are unlinked, a haplotype is understood as a pseudohaplotype. Let P_{11}^A, P_{12}^A, P_{21}^A, P_{22}^A, and P_{11}^N, P_{12}^N, P_{21}^N, P_{22}^N denote the frequencies of haplotypes $G_1^1G_2^1$, $G_1^1G_2^2$, $G_1^2G_2^1$, and $G_1^2G_2^2$ in the cases and controls, respectively. We define a penetrance of the haplotype as the probability of an individual with a haplotype being affected. Therefore, the penetrance of the haplotype is a weighted summation of the penetrance of all four genotypes with each genotype including the haplotype being considered. As in the previous section, h_{11}, h_{12}, h_{21}, and h_{22} are defined as the penetrance of the haplotypes $G_1^1G_2^1, G_1^1G_2^2, G_1^2G_2^1$, and $G_1^2G_2^2$, respectively. $G_1 = i$ and $G_2 = j$ represent a genotype coding scheme. Their represented genotypes depend on the specific genotype coding scheme. It should be noted that the haplotype $G_1^iG_2^j$ and $G_1 = i$ and $G_2 = j$ have different meanings. By the same idea in defining genotype-based odds ratio in terms of penetrance of combinations of genotypes, we can define the odds-ratio associated with the haplotypes $G_1^1G_2^1$ compared to the baseline haplotype $G_1^2G_2^2$ in terms of penetrance of the haplotypes as

$$\mathrm{OR}_{G_1^1G_2^1} = \frac{P\left(D=1 \middle| G_1^1G_2^1\right) / P\left(D=0 \middle| G_1^1G_2^1\right)}{P\left(D=1 \middle| G_1^2G_2^2\right) / P\left(D=0 \middle| G_1^2G_2^2\right)}.$$

Similarly, we calculate the odds-ratio associated with the haplotypes $G_1^1G_2^2$ and $G_1^2G_2^1$, respectively, as

$$\mathrm{OR}_{G_1^1G_2^2} = \frac{P\left(D=1 \middle| G_1^1G_2^2\right)/P\left(D=0 \middle| G_1^1G_2^2\right)}{P\left(D=1 \middle| G_1^2G_2^2\right)/P\left(D=0 \middle| G_1^2G_2^2\right)}$$

and

$$\mathrm{OR}_{G_1^2G_2^1} = \frac{P\left(D=1 \middle| G_1^2G_2^1\right)/P\left(D=0 \middle| G_1^2G_2^1\right)}{P\left(D=1 \middle| G_1^2G_2^2\right)/P\left(D=0 \middle| G_1^2G_2^2\right)}.$$

Again, similar to genotypes, we can compute a multiplicative interaction measure in terms of log odds-ratio for haplotypes as

$$I_{G_1G_2}^H = \log \frac{\mathrm{OR}_{G_1^1G_2^1}}{\mathrm{OR}_{G_1^1G_2^2}\mathrm{OR}_{G_1^2G_2^1}}. \tag{7.73}$$

In the absence of interaction, we have

$$\mathrm{OR}_{G_1^1G_2^1} = \mathrm{OR}_{G_1^1G_2^2}\mathrm{OR}_{G_1^2G_2^1}$$

and

$$\frac{P\left(D=1 \middle| G_1^1G_2^1\right)P\left(D=1 \middle| G_1^2G_2^2\right)}{P\left(D=1 \middle| G_1^1G_2^2\right)P\left(D=1 \middle| G_1^2G_2^1\right)} = \frac{P\left(D=0 \middle| G_1^1G_2^1\right)P\left(D=0 \middle| G_1^2G_2^2\right)}{P\left(D=0 \middle| G_1^1G_2^2\right)P\left(D=0 \middle| G_1^2G_2^1\right)}.$$

The multiplicative odds-ratio interaction measure in Equation 7.73 is defined by the penetrance of the haplotypes. From case–control data, it is difficult to calculate the penetrance of the haplotypes. However, we can show that the multiplicative odds-ratio interaction measure in Equation 7.73 can be reduced to (Appendix 7B)

$$I_{G_1G_2}^H = \log \frac{P_{11}^A P_{22}^A}{P_{12}^A P_{21}^A} - \log \frac{P_{11}^N P_{22}^N}{P_{12}^N P_{21}^N}. \tag{7.74}$$

There are many algorithms and software to infer the haplotype frequencies in cases and controls. Therefore, we can easily calculate the multiplicative odds-ratio interaction measure by Equation 7.74. It can be seen from Equation 7.74 that the absence of interaction between two loci occurs if and only if the ratio of haplotypes frequencies $\frac{P_{11}^A P_{22}^A}{P_{12}^A P_{21}^A}$ in the cases and the ratio of haplotypes frequencies $\frac{P_{11}^N P_{22}^N}{P_{12}^N P_{21}^N}$ in the controls are equal.

To gain an understanding of the multiplicative odds-ratio interaction measure, we study several special cases:

Case 1

One of two loci is a marker. If we assume that locus G_2 is a marker and is not associated with disease, then we have

$$P_{ij}^{A}=P\left(G_1^i\middle|D=1\right)P\left(G_2^j\middle|G_1^i\right) \quad \text{and} \quad P_{ij}^{N}=P\left(G_1^i\middle|D=0\right)P\left(G_2^j\middle|G_1^i\right),$$

which implies that

$$\begin{aligned}\frac{P_{11}^{A}P_{22}^{A}}{P_{12}^{A}P_{21}^{A}}&=\frac{P\left(G_1^1\middle|D=1\right)P\left(G_1^2\middle|D=1\right)P\left(G_2^1\middle|G_1^1\right)P\left(G_2^2\middle|G_1^2\right)}{P\left(G_1^1\middle|D=1\right)P\left(G_1^2\middle|D=1\right)P\left(G_2^2\middle|G_1^1\right)P\left(G_2^1\middle|G_1^2\right)}\\&=\frac{P_{11}^{N}P_{22}^{N}}{P_{12}^{N}P_{21}^{N}}.\end{aligned}$$

Thus, we obtain $I_{G_1G_2}^{H}=0$. In other words, if the locus G_2 is a marker, there is no interaction between two loci, G_1 and G_2. The interaction measure $I_{G_1G_2}^{H}$ between two loci should be equal to zero. Hence, our multiplicative odds-ratio interaction measure correctly characterizes the marker case.

Case 2

Logistic regression interpretation.

We define two indicator variables:

$$G_1=\begin{cases}1 & G_1^1\\0 & G_1^2\end{cases} \quad \text{and} \quad G_2=\begin{cases}1 & G_2^1\\0 & G_2^2\end{cases}. \tag{7.75}$$

Then four haplotypes at two loci can be coded as follows:

	G_1	G_2
$G_1^1G_2^1$	1	1
$G_1^1G_2^2$	1	0
$G_1^2G_2^1$	0	1
$G_1^2G_2^2$	0	0

$$P\left(D=1\middle|G_1,G_2\right)=\frac{e^{\alpha+\beta_{G_1}G_1+\beta_{G_2}G_2+\beta_{G_1G_2}G_1G_2}}{1+e^{\alpha+\beta_{G_1}G_1+\beta_{G_2}G_2+\beta_{G_1G_2}G_1G_2}}.$$

It follows from the logistic regression model in Equation 7.70 that

$$\mathrm{OR}_{G_1^1G_2^2} = e^{\beta_{G1}} = \mathrm{OR}_{G_1},$$
$$\mathrm{OR}_{G_1^2G_2^1} = e^{\beta_{G2}} = \mathrm{OR}_{G_2},$$
$$\mathrm{OR}_{G_1^1G_2^1} = e^{\beta_{G1}+\beta_{G2}+\beta_{G1G2}} = \mathrm{OR}_{G_1}\mathrm{OR}_{G_2}e^{\beta_{G1G2}},$$

where odds-ratios OR_{G1} and OR_{G2} are defined in terms of alleles, i.e.,

$$\mathrm{OR}_{G_1} = \frac{P\left(D=1|G_1^1\right)/P\left(D=0|G_1^1\right)}{P\left(D=1|G_1^2\right)/P\left(D=0|G_1^2\right)} \quad \text{and} \quad \mathrm{OR}_{G_2} = \frac{P\left(D=1|G_2^1\right)/P\left(D=0|G_2^1\right)}{P\left(D=1|G_2^2\right)/P\left(D=0|G_2^2\right)}.$$

Therefore, the haplotype multiplicative odds-ratio interaction measure $I_{G_1G_2}^H$ is equal to $I_{G_1G_2}^H = \beta_{G_1G_2}$, which has the same form as that in Equation 7.72. This indicates that if the coding for the genotypes in the genotype multiplicative odds-ratio interaction measure $I_{G_1G_2}$ is replaced by the coding for the haplotypes in Equation 7.75, then we can obtain the haplotype multiplicative odds-ratio interaction measure.

7.2.4.4 Haplotype-Based Odds Ratio Multiplicative Interaction Measure–Based Test Statistics

In the previous section, we defined the haplotype multiplicative odds-ratio interaction measure, which can be estimated by haplotype frequencies in cases and controls. Similar to Appendix 7A, by the delta method, we can obtain the variance of the estimator of the haplotype odds-ratio interaction measure (Exercise 7.13):

$$\mathrm{Var}\left(\hat{I}_{G_1G_2}^H\right) = \frac{1}{2n_A}\left[\frac{1}{P_{11}^A}+\frac{1}{P_{12}^A}+\frac{1}{P_{21}^A}+\frac{1}{P_{22}^A}\right]+\frac{1}{2n_G}\left[\frac{1}{P_{11}^N}+\frac{1}{P_{12}^N}+\frac{1}{P_{21}^N}+\frac{1}{P_{22}^N}\right],$$

where n_A and n_G are the number of sampled individuals in cases and controls. By the standard asymptotic theory, we can define the haplotype odds-ratio interaction measure–based statistic for testing the interaction between two loci:

$$T_{IH} = \frac{\left(I_{G_1G_2}^H\right)^2}{V\left(\hat{I}_{G_1G_2}^H\right)} = \frac{\left[\log\frac{\hat{P}_{11}^A\hat{P}_{22}^A}{\hat{P}_{12}^A\hat{P}_{21}^A} - \log\frac{\hat{P}_{11}^N\hat{P}_{22}^N}{\hat{P}_{12}^N\hat{P}_{21}^N}\right]^2}{\frac{1}{2n_A}\left[\frac{1}{\hat{P}_{11}^A}+\frac{1}{\hat{P}_{12}^A}+\frac{1}{\hat{P}_{21}^A}+\frac{1}{\hat{P}_{22}^A}\right]+\frac{1}{2n_G}\left[\frac{1}{\hat{P}_{11}^N}+\frac{1}{\hat{P}_{12}^N}+\frac{1}{\hat{P}_{21}^N}+\frac{1}{\hat{P}_{22}^N}\right]}, \quad (7.76)$$

where $\hat{P}_{11}^A, \hat{P}_{12}^A, \hat{P}_{21}^A, \hat{P}_{22}^A$ and $\hat{P}_{11}^N, \hat{P}_{12}^N, \hat{P}_{21}^N, \hat{P}_{22}^N$ are the estimators of the corresponding haplotype frequencies in cases and controls, respectively. When sample sizes are large enough to ensure application of large sample theory, T_{IH} is asymptotically distributed as a central $\chi_{(1)}^2$ distribution under the null hypothesis of no interaction between two loci. Under an alternative hypothesis of interaction between two loci being present, the statistic T_{IH} is asymptotically distributed as a noncentral $\chi^2_{(1)}$ distribution with noncentrality parameter proportional to the haplotype multiplicative odds-ratio interaction measure. This statistic can be applied to both linked and unlinked loci. For the unlinked loci, we can use the case-only design to study the interaction between two loci in which the equation is reduced to

$$T_{IH} = \frac{\left(\log \frac{P_{11}^A P_{22}^A}{P_{12}^A P_{21}^A}\right)^2}{\frac{1}{2n_A}\left[\frac{1}{\hat{P}_{11}^A} + \frac{1}{\hat{P}_{12}^A} + \frac{1}{\hat{P}_{21}^A} + \frac{1}{\hat{P}_{22}^A}\right]}. \tag{7.77}$$

7.2.5 Multiplicative Measure-Based Statistics for Testing Interaction between a Gene and a Continuous Environment

Under the assumption of normal distribution of the continuous environment, the average multiplicative measure of gene–environment interaction is defined as

$$I_{MGE} = \log \frac{\mu_{D0}\mu_1}{\mu_{D1}\mu_0} + \frac{1}{2}\log \frac{\sigma_{D0}^2\sigma_1^2}{\sigma_{D1}^2\sigma_0^2}.$$

To develop a statistic for testing interaction between a discrete genotype and a continuous environment, we first should study the distributions of the estimates of the means and variances of the environments in cases and the general population. Let n_0, n_1, n_{D_0}, and n_{D_1} be the number of individuals with the genotypes $G=0$ and $G=1$ in the general population and cases, respectively. Let $Y_n = \left[\hat{\mu}_0, \hat{\mu}_1, \hat{\mu}_{D_0}, \hat{\mu}_{D_1}, \hat{\sigma}_0^2, \hat{\sigma}_1^2, \hat{\sigma}_{D_0}^2, \hat{\sigma}_{D_1}\right]^T$. Then, from the standard asymptotic theory (Theorem 3.3.2 and Theorem 3.4.4 in Anderson 1984), Y_n is asymptotically distributed as a normal distribution with the mean $\tau = \left[\mu_0, \mu_1, \mu_{D_0}, \mu_{D_1}, \sigma_0^2, \sigma_1^2, \sigma_{D_0}^2, \sigma_{D_1}\right]^T$ and covariance matrix $\Sigma = \operatorname{diag}\left(\frac{\sigma_0^2}{n_0}, \frac{\sigma_1^2}{n_1}, \frac{\sigma_{D_0}^2}{n_{D_0}}, \frac{\sigma_{D_1}^2}{n_{D_1}}, \frac{2\sigma_0^4}{n_0 - 1}, \frac{2\sigma_1^4}{n_1 - 1}, \frac{2\sigma_{D_0}^2}{n_{D_0} - 1}, \frac{2\sigma_{D_1}^2}{n_{D_1} - 1}\right)$. Then, by the asymptotic theory of functions of asymptotically normal statistics (Lehmann 1983), we can show that under the null hypothesis of no gene–environment interaction, the estimate of the multiplicative interaction measure of gene–environment interaction is distributed as a normal distribution with a mean of zero and variance

$$V_{MGE} = \frac{\sigma_0^2}{n_0\mu_0^2} + \frac{\sigma_1^2}{n_1\mu_1^2} + \frac{\sigma_{D_0}^2}{n_{D_0}\mu_{D_0}^2} + \frac{\sigma_{D_1}^2}{n_{D_1}\mu_{D_1}^2} + \frac{1}{2(n_0 - 1)} + \frac{1}{2(n_1 - 1)} + \frac{1}{2(n_{D_0} - 1)} + \frac{1}{2(n_{D_1} - 1)}.$$

Therefore, we can define the following statistic for testing interaction between a gene and a continuous environment:

$$T_{GCE} = \frac{I_{MGE}^2}{V_{MGE}}. \tag{7.78}$$

Under the null hypothesis of no gene–environment interaction, the statistic T_{GCE} is asymptotically distributed as a central $\chi^2_{(1)}$ distribution.

7.2.6 Information Measure–Based Statistics for Testing Interaction between a Gene and a Continuous Environment

Similar to Section 7.2.5, we can also use asymptotic theory of nonlinear transformation of given statistics to develop an information measure–based statistic to test for interaction between a gene and a continuous variable. Let $n=n_0+n_1$ and $n_D=n_{D_0}+n_{D_1}$. Other quantities are defined as before. Recall that the information measure of interaction between a gene and a continuous environment is defined as

$$I_{GCE}=\frac{1}{2}\log\frac{\sigma_D^2}{\sigma^2}-\frac{1}{2}\left[P(G=0|D)\log\frac{\sigma_{D_0}^2}{\sigma_0^2}+P(G=1|D)\log\frac{\sigma_{D_1}^2}{\sigma_1^2}\right]-\frac{\sigma_D^2+(\mu_D-\mu)^2}{2\sigma^2}$$
$$+P(G=0|D)\frac{\sigma_{D_0}^2+(\mu_{D_0}-\mu_0)^2}{2\sigma_0^2}+P(G=1|D)\frac{\sigma_{D_1}^2+(\mu_{D_1}-\mu_1)^2}{2\sigma_1^2}.$$

Then, the variance of the estimate of I_{GCE} is given by

$$V_{GCE}=\frac{P_{D_0}^2(\mu_{D_0}-\mu_0)^2}{n_0\sigma_0^2}+\frac{P_{D_1}^2(\mu_{D_1}-\mu_1)^2}{n_1\sigma_1^2}+\frac{P_{D_0}^2(\mu_{D_0}-\mu_D)^2\sigma_{D_0}^2}{n_{D_0}\sigma_0^4}+\frac{P_{D_1}^2(\mu_{D_1}-\mu_1)^2\sigma_{D_1}^2}{n_{D_1}\sigma_1^4}$$
$$+\frac{(\mu_D-\mu)^2}{n\sigma^2}+\frac{(\mu_D-\mu)^2\sigma_D^2}{n_D\sigma^4}+\frac{P_{D_0}^2\sigma_0^4}{2n_0}\left[\frac{1}{\sigma_0^2}-\frac{\sigma_{D_0}^2+(\mu_{D_0}-\mu_0)^2}{\sigma_0^4}\right]^2$$
$$+\frac{P_{D_1}^2\sigma_1^4}{n_1}\left[\frac{1}{\sigma_1^2}-\frac{\sigma_{D_1}^2+(\mu_{D_1}-\mu_1)^2}{\sigma_1^4}\right]^2+\frac{P_{D_0}^2\sigma_{D_0}^4}{2n_{D_0}}\left(\frac{1}{\sigma_0^2}-\frac{1}{\sigma_{D_0}^2}\right)^2+\frac{P_{D_1}^2\sigma_{D_1}^4}{2n_{D_1}}\left(\frac{1}{\sigma_1^2}-\frac{1}{\sigma_{D_1}^2}\right)^2$$
$$+\frac{\sigma^4}{2n}\left(\frac{1}{\sigma^2}-\frac{\sigma_D^2+(\mu_D-\mu)^2}{\sigma^4}\right)^2+\frac{\sigma_D^4}{2n_D}\left(\frac{1}{\sigma_D^2}-\frac{1}{\sigma^2}\right)^2.$$

After we calculate the variance of the estimate of I_{GCE}, we can define the following information-based statistic to test for interaction between a gene and a continuous environment:

$$T_{MIGCE}=\frac{\hat{I}_{GCE}^2}{\hat{V}_{GCE}}, \tag{7.79}$$

which will be asymptotically distributed as a central $\chi_{(1)}^2$ distribution under the null hypothesis of no gene–environment interaction.

7.2.7 Real Example

To illustrate its application, the haplotype odds-ratio interaction measure–based statistic T_{IH} was applied to the coronary heart disease cohort study dataset with 469,612 SNPs typed in 1,926 cases and 2,938 controls from Wellcome Trust Case Control Consortium (WTCCC).

After qualitative control (QC), a total of 469,612 SNPs were left for analysis. Since testing for all possible pair-wise interactions between 469,612 SNPs is infeasible, 53,394 SNPs from 501 known pathways including adhesion, apoptosis, cell activation, cell cycle regulation, cell signaling, cytokines/chemokines, developmental biology, expression, hematopoiesis, immunology, metabolism, and neuroscience were used to test gene–gene interaction. The pathway information is from BioCarta and Kyoto Encyclopedia of Genes and Genomes database (BioCarta pathways; KEGG PATHWAY database). After Bonferroni correction, the *P*-value for a significant interaction between SNPs is 3.5×10^{-11}.

There are a total of 8664 interactions that were found to be significant with $P < 3.5 \times 10^{-11}$. Among the 8644 significant interactions, 111 contain at least one SNP that is in the previously identified cardiovascular disease susceptibility genes. Figure 7.2 presents the 111 interactions between SNPs, where a red dot indicates that the SNP is in a gene, which has been suggested to be associated with cardiovascular disease in previous literatures. The SNP with the largest number of interactions is rs3785579, which is in the CACNG1 (calcium channel, voltage-dependent, gamma subunit 1) gene and has 57 interactions with known cardiovascular disease susceptibility genes. This SNP is also found to be significant from the single-SNP association test. The SNP with the second largest number of interactions is rs642298, which is in the IL3RA (interleukin 3 receptor, alpha) gene and has 16 interactions with known cardiovascular disease susceptibility genes. This SNP shows marginal significance in the single-SNP association test ($P = 1.84 \times 10^{-6}$).

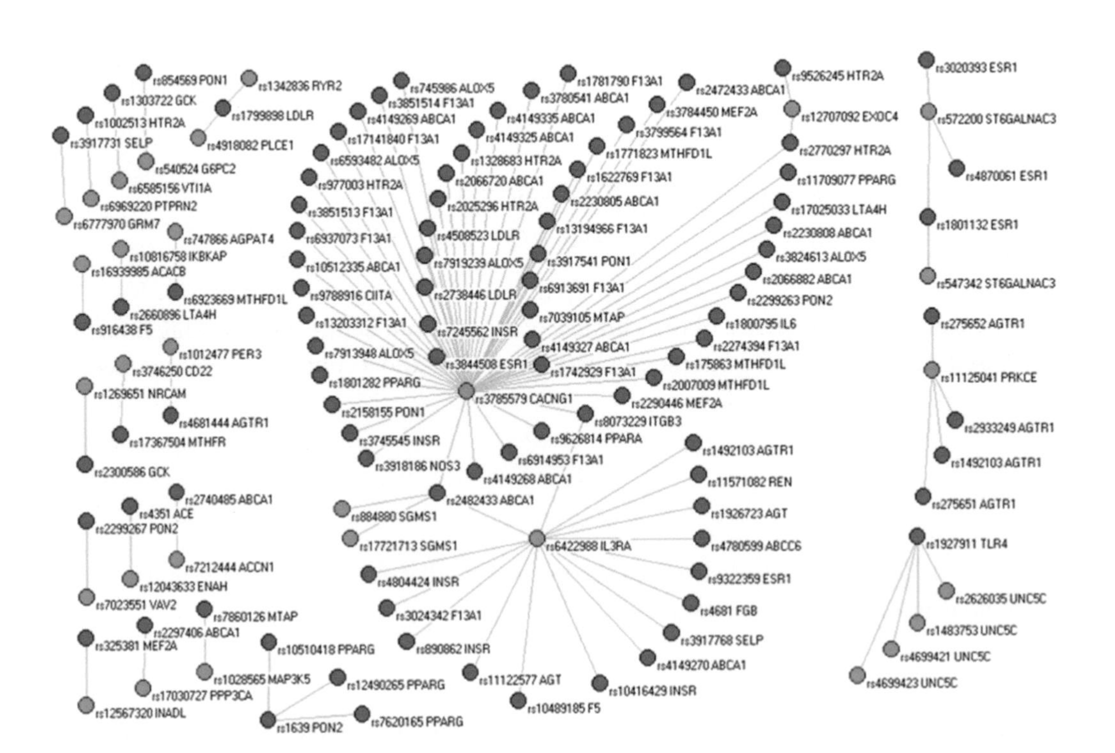

FIGURE 7.2 Interactions with known cardiovascular disease susceptibility genes shown by SNPs (a red dot indicates a previously identified cardiovascular disease susceptibility gene).

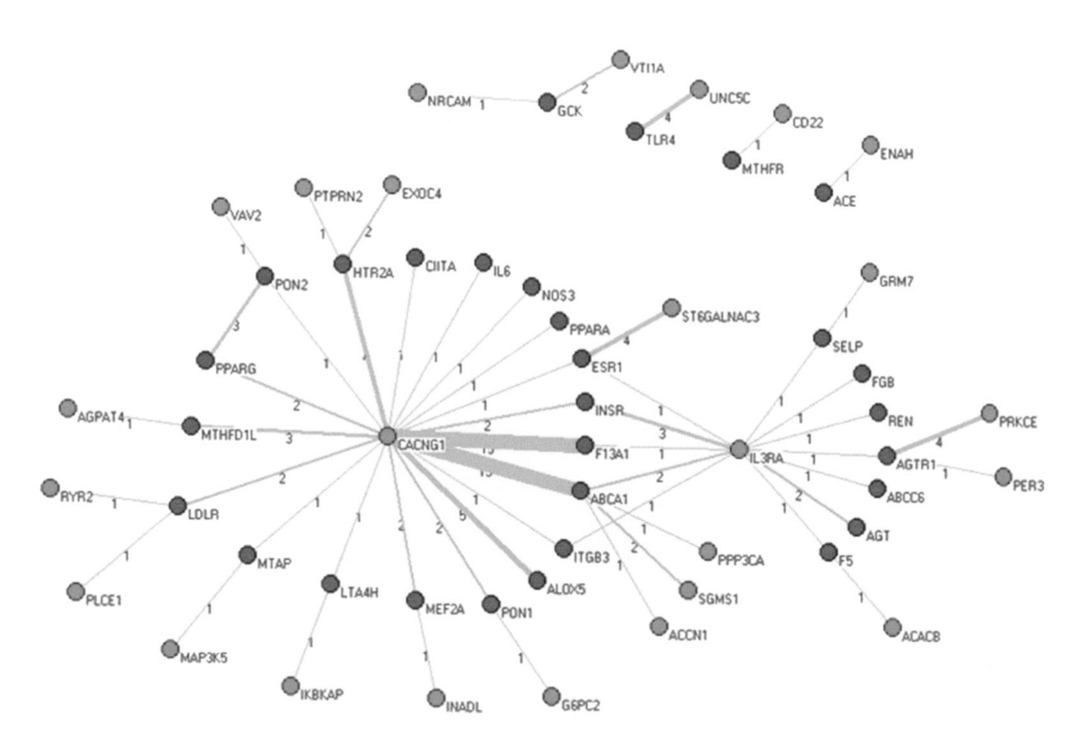

FIGURE 7.3 A total of 30 known cardiovascular disease susceptibility genes found to have significant pair-wise interactions.

There are a total of 30 known cardiovascular disease susceptibility genes found to have significant pair-wise interactions (Figure 7.3). As noted earlier, most of these cardiovascular disease susceptibility genes do not have an interaction with each other. Instead, they often interact with genes, which have not been previously reported to be related to cardiovascular disease risks. The only significant interaction between two susceptibility genes is the interaction between PON2 (paraoxonase 2) and PPARG (peroxisome proliferator-activated receptor gamma). Three SNPs in the PPARG gene (rs10510418, rs7620165, and rs12490265) are interacted with one SNP in the PON2 gene (rs1639). The PON2 gene is expressed in a variety of tissues. The PON2 gene retards cellular oxidative stress and prevents apoptosis in vascular endothelial cells, while the physiological roles of its protein product, an intracellular enzyme, are still not very clear (Horke et al. 2007). The PON2 gene has been shown to have a significant association with coronary heart disease (Sanghera et al. 1998). The protein encoded by the PPARG gene is a regulator of adipocyte differentiation. PPARG has been shown to associate with obesity, a condition closely related to coronary artery disease (Masud et al. 2003). Its interaction with apolipoportein E has also been shown to have an association with risk of coronary heart disease (Peng et al. 2003). Although both PON2 and PPARG genes have been recognized for obesity and coronary artery disease risk, their interaction has never been explored.

Of the 8664 significant interactions from the pair-wise interaction analysis, almost half of them (4327 interactions) are interactions with the CACNG1 gene. Although the CACNG1 gene has not been suggested with cardiovascular disease risk before, the pathway, which

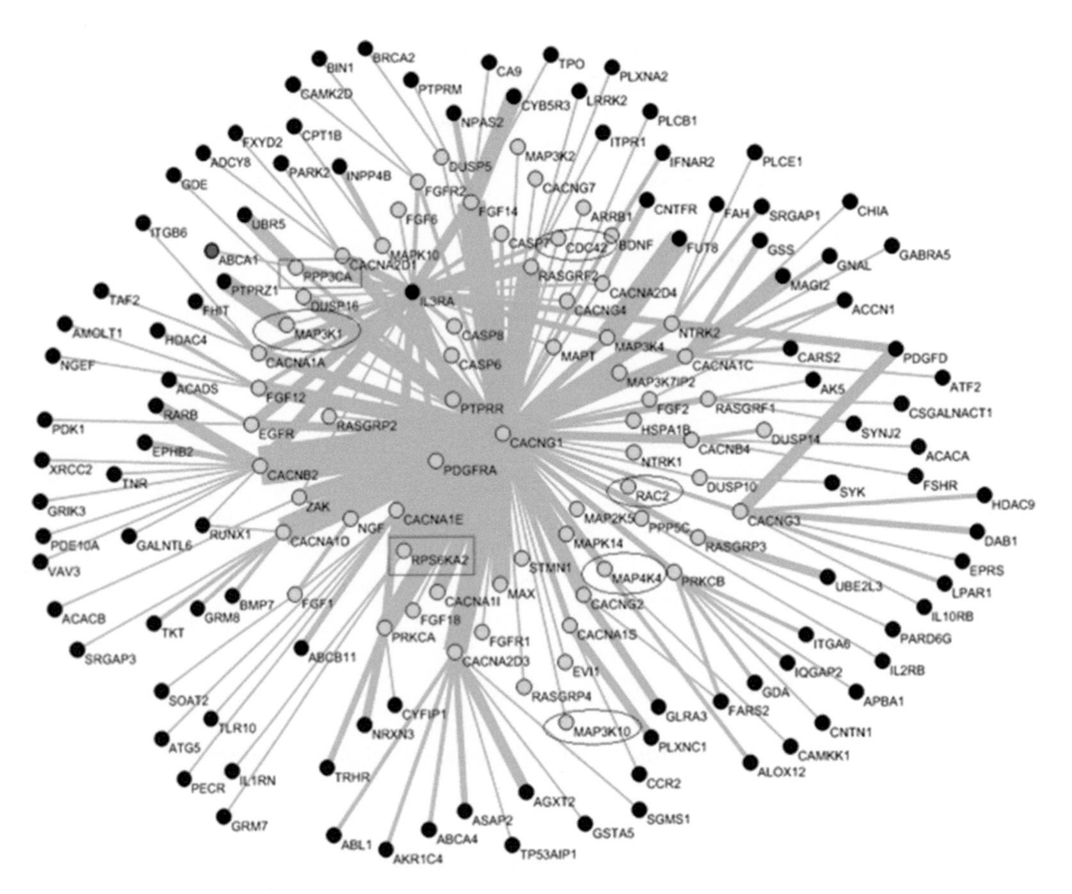

FIGURE 7.4 The CACNG1 gene interacted with genes from the MAPK signaling pathway (a yellow dot indicates the gene is involved in the MAPK signaling pathway, a red dot indicates a previously identified cardiovascular disease susceptibility gene, and the thickness of the blue line indicates the number of interactions between the two genes).

the CACNG1 gene is involved in, the MAPK (mitogen-activated protein kinase) signaling pathway, has been shown to play an important role in the pathogenesis of cardiovascular disease (Muslin 2008). MAPK represents a group of serine/threonine kinases that are believed to act downstream from protein kinase C in the smooth muscle cell regulatory cascade. Figures 7.4 and 7.5 show the interactions between the CACNG1 gene and genes that are involved in the MAPK signaling pathway.

Within the MAPK signaling pathway, the *CACNG1* gene has the most interactions with the *RPS6KA2* (ribosomal protein S6 kinase, 90 kDa, polypeptide 2) gene. There are 22 different SNPs in the RPS6KA2 gene interacting with the SNP in the CACNG1 gene. RPS6KA2 gene has been implicated in controlling cell growth and differentiation. In the MAPK signaling pathway, the CACNG1 gene is indirectly connected with the RPS6KA2 gene through the activation of Ras and ERK (extracellular signal–regulated kinase). Ras is a master regulator of intracellular signaling cascades, and it promotes the activation of MAPK and other signaling pathways (Muslin 2008). Study has shown that cardiac-specific overexpression of an activated form of some genes from the Ras family in transgenic mice

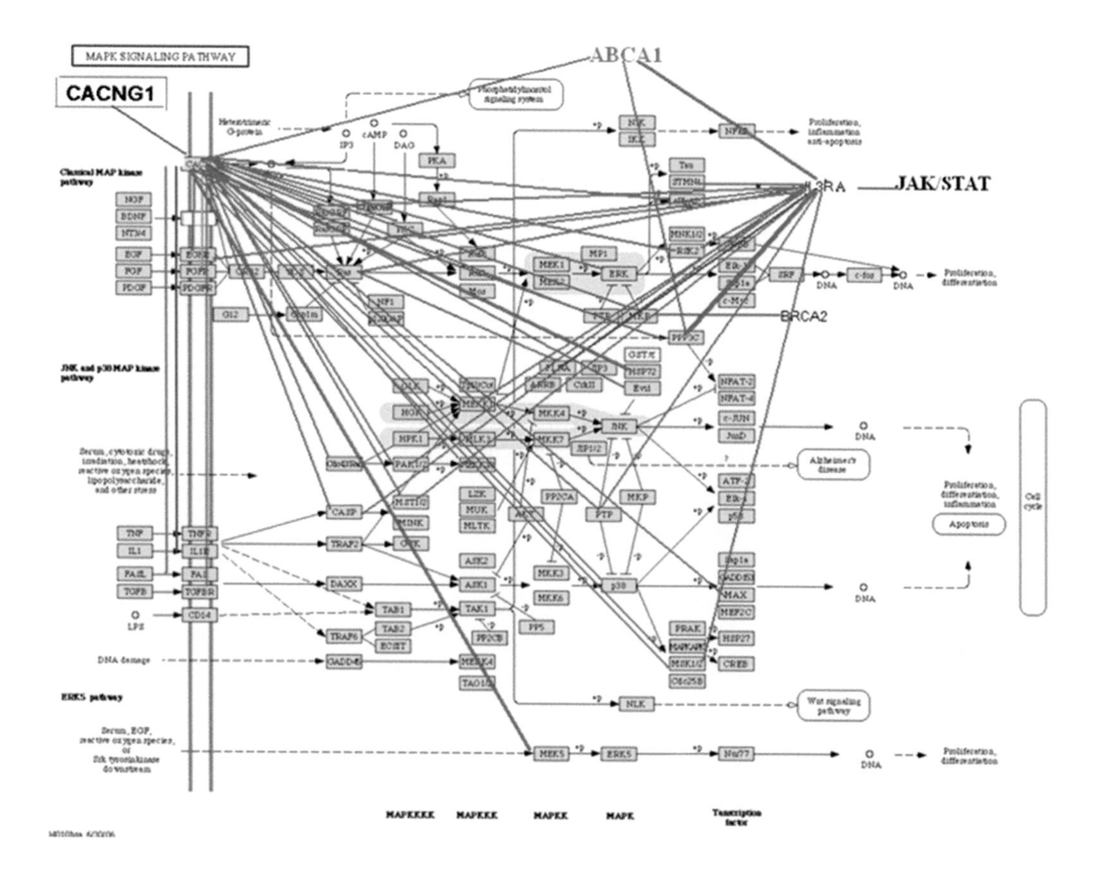

FIGURE 7.5 Gene–gene interactions in MAPK pathway.

can lead to cardiac hypertrophy and diastolic dysfunction (Hunter et al. 1995). Activation of ERK has been demonstrated in animal model systems as well as in humans with heart failure (Haq et al. 2001). Although there is no significant interaction between genes in Ras and ERK, since CACNG1 gene and RPS6K2 gene are connected through the Ras–ERK signaling cascade, the interactions between these two genes may have a substantial role in cardiovascular disease.

In the MAPK signaling pathway, in addition to Ras–ERK cascade, the JNK (c-Jun N-terminal kinase) cascade also plays an important role in the regulation of cell physiology. Overexpression studies in mice suggest that while Ras-mediated ERK activation promotes cardiac hypertrophy, JNK activation promotes cardiac dysfunction (Muslin 2008). In the pathway, the MAP3K10 (mitogen-activated protein kinase kinase kinase 10) and the MAP3K1 (mitogen-activated protein kinase kinase kinase 1) gene can indirectly activate JNK by activating mitogen-activated protein kinase kinase. One SNP in the MAP3K10 gene has a significant interaction with one SNP in the CACNG1 gene. The MAP3K1 gene has a total of 10 significant interactions: 1 with the CACNG1 gene, 1 with IL3RA gene, and 8 with the PTPRZ1 (protein tyrosine phosphatase, receptor-type, Z polypeptide 1) gene. The PTPRZ1 gene is a member of the receptor-type protein tyrosine phosphatase family. Although this gene is not in the MAPK signaling pathway, other genes from its family are found to have important functions in the MAPK signaling pathway, especially for the ERK and JNK cascades. The MAP3K1 gene can be activated by MAP4K4 (mitogen-activated protein kinase kinase kinase kinase 4), CDC42 (cell division cycle 42), and RAC2 (Ras-related C3 botulinum toxin substrate 2) genes in the MAPK signaling pathway. The MAP4K4 gene has a total of seven significant interactions: one with the CAMKK1 (calcium/calmodulin-dependent protein kinase kinase 1, alpha) gene, two with the ALOX12 (arachidonate 12-lipoxygenase) gene, and four with the CACNG1 gene. The CDC42 gene has one interaction with the CACNG1 gene and two interactions with the IL3RA gene. And RAC2 gene has just one interaction with the CACNG1 gene. Overall, the CACNG1 gene has significant interactions with genes that can indirectly activate the JNK cascade. Moreover, the CACNG1 gene also has one interaction with the MAPK10 (mitogen-activated protein kinase 10) gene, a gene that encodes the JNK family members JNK3. These results suggest that although the CACNG1 gene is not directly involved in the JNK cascade, it interacts with some key genes in the JNK cascade.

7.3 STATISTICS FOR TESTING GENE–GENE AND GENE–ENVIRONMENT INTERACTION FOR A QUALITATIVE TRAIT WITH NEXT-GENERATION SEQUENCING DATA

The critical barrier in interaction analysis for next-generation sequencing (NGS) data is that the traditional pair-wise interaction analysis that is suitable for common variants is difficult to apply to rare variants because of their prohibitive computational time, large number of tests, and low power. The great challenges for successful detection of interactions with NGS data are (1) the demands in the paradigm of changes in interaction analysis, (2) severe multiple testing, and (3) heavy computations.

The current paradigm of pair-wise interaction analysis is the lack of power to detect interaction between rare variants in a population due to the low frequencies of the rare variants. Interactions may be present in only a few samples, or even no sampled individuals at all will display the interaction effects. Although we can observe a large effect from a few pairs of interactions in a few samples, the interaction effects of the rare variants in the population are small due to their low frequencies. Large discrepancies in the number of observations between different combinations of rare variants will cause serious problems in identifying interactions in the population.

The development of novel concepts and statistics for testing interaction between rare variants and between rare and common variants, which can reduce the dimensionality of the data, the number of tests, and the time of computations and improve the power to detect interaction, are needed. To meet this challenge, we first introduce a strategy that changes a basic unit of interaction analysis from a pair of SNPs to a pair of genes (or genomic regions) (Peng et al. 2010). We take a gene as a basic unit of the interaction analysis and collectively test for interaction between all possible pairs of SNPs within two genes. This new paradigm of interaction analysis has two remarkable features. First, it uses all information in the gene to collectively test for interaction between multiple SNPs within the gene. Therefore, it not only can increase the power but also can reduce false-positive rates due to sampling variance caused by the low frequency of the rare variants. Second, it will largely reduce the number of tests and will alleviate multiple testing problems, which are more severe in interaction analysis than in association studies.

Functional data analysis techniques are then introduced as a major tool for developing statistics to implement this new paradigm for interaction analysis with NGS data. A test for interaction between two genes is formulated as an interaction test in the functional logistic regression model. In the functional logistic regression model, the genotype score functions (genetic variant profiles) are defined as a function of the genomic position of the genetic variants rather than a set of discrete genotype values, and the logit transform of the probability of an individual being affected is predicted by genotype score functions with their interaction terms. Functional logistic regression is a natural extension of the standard logistic regression for traditional interaction analysis (Zhao et al. 2016).

7.3.1 Multiple Logistic Regression Model for Gene–Gene Interaction Analysis

Recall the genotype-odds-ratio-based model for testing interaction between two SNPs. Consider two loci: G and H. Assume that the codes $G=1(G=0)$ and $H=1(H=0)$ denote whether an individual is a carrier (or noncarrier) of the susceptible genotype at the loci G and H, respectively. Let D denote disease status where $D=1(D=0)$ indicates an affected (or unaffected) individual. Recall the logistic model (7.70)

$$\pi_i = P\left(D=1 \mid G_i, H_i\right) = \frac{e^{\mu+\beta_G G_i+\beta_H H_i+\beta_{GH} G_i H_i}}{1+e^{\mu+\beta_G G_i+\beta_H H_i+\beta_{GH} G_i H_i}}$$

or

$$OR_{G_2} = \frac{P\left(D=1 \mid G_2^1\right)/P\left(D=0 \mid G_2^1\right)}{P\left(D=1 \mid G_2^2\right)/P\left(D=0 \mid G_2^2\right)}. \tag{7.80}$$

The model (7.80) for testing interaction between two loci can be extended to a multilocus interaction model. Consider two genomic regions, $[a_1, b_1]$ and $[a_2, b_2]$. Let x_{ij} be the indicator for the genotype at the jth SNP of the ith individual. The multilocus interaction model is

$$\log \frac{\pi_i}{1-\pi_i} = \mu + \sum_{m=1}^{M} v_{im}\rho_m + \sum_{j=1}^{k_1} x_{ij}\alpha_j + \sum_{j=1}^{k_1}\sum_{l=1, l\neq j}^{k_1} x_{ij}x_{il}\gamma_{jl}, \tag{7.81}$$

where

μ is an overall mean
w_{im} is the covariate
η_m is the coefficient associated with the covariate
α_j is the main genetic additive effect of the jth SNP
γ_{jl} is an interaction effect between the jth SNP and lth SNP

The right side of Equation 7.81 can be written in a vector form:

$$\log\frac{\pi_i}{1-\pi_i} = v_i\rho + x_i\alpha + z_i\gamma \tag{7.82}$$

or

$$\log\frac{\pi_i}{1-\pi_i} = H_i\beta, \tag{7.83}$$

where

$$v_i = \left[1, v_{i1},\ldots,v_{iM}\right],\quad \rho = \left[\mu, \rho_1,\ldots,\rho_M\right]^T,$$

$$x_i = \left[x_{i1},\ldots x_{ik_1}\right],\quad \alpha = \left[\alpha_1,\ldots,\alpha_{k_1}\right]^T,\quad z_i = \left[x_{i1}x_{i2},\ldots,x_{ik_1}x_{ik_1-1}\right],$$

$$\gamma = \left[\gamma_{12},\ldots,\gamma_{k_1(k_1-1)}\right]^T,\quad H_i = \left[v_i, x_i, z_i\right],\ \text{and}\ \ \beta = \begin{bmatrix}\rho\\ \alpha\\ \gamma\end{bmatrix}.$$

The logistic model (7.83) has the form (8.6). The fast parameter estimation using proximal methods will be discussed in Chapter 8.

7.3.2 Functional Logistic Regression Model for Gene–Gene Interaction Analysis

As in previous chapters, we use genetic variant profiles that will recognize information contained in the physical location of the SNP as a major data form. The densely distributed genetic variants across the genomes in large samples can be viewed as realizations of a Poisson process. The densely typed genetic variants in a genomic region for each individual are so close that these genetic variant profiles can be treated as observed data taken from curves. The genetic variant profiles are called functional.

We first define the genotypic function. Consider two genomic regions, $[a_1, b_1]$ and $[a_2, b_2]$. Let $x_i(t)$ and $x_i(s)$ be genotypic functions of the ith individual defined in the regions $[a_1, b_1]$ and $[a_2, b_2]$, respectively. Let t and s be a genomic position in the first and second genomic regions, respectively. Define a genotype profile, $x_i(t)$, of the ith individual as

$$X_i\left(t\right) = \begin{cases} 1, & MM \\ 0, & Mm, \\ -1, & mm \end{cases} \tag{7.84}$$

which is a widely used indicator variable for a genotype at an SNP. The logistic regression model (7.81) for modeling main effects and interaction effects at multiple loci can be adapted for testing interaction between two regions:

$$\log\frac{\pi_i}{1-\pi_i} = \mu + \sum_{m=1}^{M} v_{im}\rho_m + \sum_{j=1}^{k_1} x_i\left(t_j\right)\alpha_j + \sum_{l=1}^{k_2} z_i\left(s_l\right)\beta_l + \sum_{j=1}^{k_1}\sum_{l=1}^{k_2} x_i\left(t_j\right)z_i\left(s_l\right)\gamma_{jl}, \tag{7.85}$$

where μ, w_{im}, and η_m are defined as before, α_j is the main genetic additive effect of the jth SNP in the first genomic region, β_l is the main genetic additive effect of the lth SNP in the second genomic region, γ_{jl} is an interaction effect between the jth SNP in the first genomic region and the lth SNP in the second genomic region, and $x_i(t_j)$ and $z_i(s_l)$ are indicator variables for the genotypes at the jth SNP and the lth SNP as defined in Equation 7.84, respectively.

Next we extend the logistic regression model (7.85) to the functional logistic regression for modeling main and interaction effects. We begin with reviewing some functional data analysis results. By Karhunen–Loeve expansion (Ash and Gardner 1975), we have

$$x_i(t) = \sum_{j=1}^{\infty} \xi_{ij}\phi_j(t)$$

and

$$x_i(s) = \sum_{k=1}^{\infty} \eta_{ik}\psi_k(s), \tag{7.86}$$

where $\phi_j(t)$ and $\psi_k(s)$ are sequences of orthonormal basis functions. The functional principal scores are defined by

$$\xi_{ij} = \int_T x_i(t)\phi_j(t)dt$$

and

$$\eta_{ik} = \int_S x_i(s)\psi_k(s)ds.$$

We extend the traditional logistic regression model (7.81) to the following functional logistic regression model for gene–gene interaction analysis:

$$\log\frac{\pi_i}{1-\pi_i} = \mu + \sum_{m=1}^{M} v_{im}\rho_m + \int_T \alpha(t)x_i(t)dt + \int_S \beta(s)x_i(s)ds + \int_T\int_S \gamma(t,s)x_i(t)x_i(s)dtds, \tag{7.87}$$

where

- $\alpha(t)$ and $\beta(s)$ are the putative genetic additive effects of two SNPs located at the genomic positions t and s, respectively
- $\gamma(t,s)$ is the putative interaction effect between two SNPs located at the genomic positions t and s

Thus, π_i is given by

$$\pi_i = \frac{e^{\mu+\sum_{m=1}^{M} v_{im}\rho_m + \int_T \alpha(t)x_i(t)dt + \int_S \beta(s)x_i(s)ds + \int_T\int_S \gamma(t,s)x_i(t)x_i(s)dtds}}{1+e^{\mu+\sum_{m=1}^{M} v_{im}\rho_m + \int_T \alpha(t)x_i(t)dt + \int_S \beta(s)x_i(s)ds + \int_T\int_S \gamma(t,s)x_i(t)x_i(s)dtds}}.$$

Substituting Equation 7.86 into Equation 7.87, we obtain

$$\log\frac{\pi_i}{1-\pi_i}=\mu+\sum_{m=1}^{M}v_{im}\rho_m+\int_T\alpha(t)\sum_{j=1}^{\infty}\xi_{ij}\phi_j(t)dt+\int_S\beta(s)\sum_{j=1}^{\infty}\eta_{ik}\psi_k(s)ds$$

$$+\int_T\int_S\gamma(t,s)\left[\sum_{j=1}^{\infty}\xi_{ij}\phi_j(t)\right]\left[\sum_{k=1}^{\infty}\eta_{ik}\psi_k(s)\right]dtds+\varepsilon_i$$

$$=\mu+\sum_{m=1}^{M}v_{im}\rho_m+\sum_{j=1}^{\infty}\xi_{ij}\int_T\alpha(t)\phi_j(t)dt+\sum_{j=1}^{\infty}\eta_{ik}\int_S\beta(s)\Psi_k(s)ds$$

$$+\sum_{j=1}^{\infty}\sum_{k=1}^{\infty}\xi_{ij}\eta_{ik}\int_T\int_S\gamma(t,s)\phi_j(t)\Psi_k(s)dtds. \quad (7.88)$$

Let $\alpha_j=\int_T\alpha(t)\phi_j(t)dt, \beta_k=\int_S\beta(s)\psi_k(s)ds$, and $\gamma_{jk}=\int_T\int_S\gamma(t,s)\phi_j(t)\psi_k(s)dtds$. Then, Equation 7.88 can be reduced to

$$\log\frac{\pi_i}{1-\pi_i}=\mu+\sum_{m=1}^{M}v_{im}\rho_m+\sum_{j=1}^{\infty}\xi_{ij}\alpha_j+\sum_{k=1}^{\infty}\eta_{ik}\beta_k+\sum_{j=1}^{\infty}\sum_{k=1}^{\infty}\xi_{ij}\eta_{ik}\gamma_{jk}. \quad (7.89)$$

Let $v_i=[1,v_{i1},\ldots,v_{iM}]^T$, $\rho=\begin{bmatrix}1\\ \rho_1\\ \vdots\\ \rho_M\end{bmatrix}$, $\alpha=\begin{bmatrix}\alpha_1\\ \vdots\\ \alpha_J\end{bmatrix}$, $\beta=\begin{bmatrix}\beta_1\\ \vdots\\ \beta_K\end{bmatrix}$, $\gamma=\begin{bmatrix}\gamma_{11}\\ \gamma_{12}\\ \vdots\\ \gamma_{JK}\end{bmatrix}$, $b=\begin{bmatrix}\alpha_0\\ \alpha\\ \beta\\ \gamma\end{bmatrix}$,

$\xi_i=\begin{bmatrix}\xi_{i1}\\ \vdots\\ \xi_{iJ}\end{bmatrix}$, $\eta_i=\begin{bmatrix}\eta_{i1}\\ \vdots\\ \eta_{iK}\end{bmatrix}$, $\Gamma_i=\begin{bmatrix}\xi_{i1}\eta_{i1}\\ \vdots\\ \xi_{iJ}\eta_{iK}\end{bmatrix}$.

Then, we have

$$\log\frac{\pi_i}{1-\pi_i}=v_i^T\rho+\xi_i^T\alpha+\eta_i^T\beta+\Gamma_i^T\gamma$$

$$=\begin{bmatrix}v_i^T & \eta_i^T & \Gamma_i^T\end{bmatrix}\begin{bmatrix}\rho\\ \alpha\\ \beta\\ \gamma\end{bmatrix}$$

$$=W_i^Tb, \quad (7.90)$$

where $W_i=\begin{bmatrix}v_i\\ \xi_i\\ \eta_i\\ \Gamma_i\end{bmatrix}$.

The traditional odds-ratio concept defined for a locus can also be extended to the genomic region. We assume that the code $G=0$ denotes that an individual is a noncarrier of the susceptible genotypes across the first genomic region and that $G=1$ indicates that an individual carries at least one susceptible genotype in the first genomic region. We can similarly define $H=1$ ($H=0$) for the second genomic region. Let D denote disease status where $D=1$ ($D=0$) indicates an affected (unaffected) individual. For the convenience of discussion, the genotype function $x(t)$ is also coded as a binary function indicating presence ($x(t)=1$) or absence $x(t)=0$) of the risk genotype at the genomic position t. Thus, $G=0$ represents that $x(t)=0, \forall t \in [a_1, b_1]$ and $G=1$ represents that $x(t_0)=1$ at least at one genomic position, t_0. We can similarly interpret H for the second genomic region. The odds-ratio associated with the first genome region is defined as

$$\mathrm{OR}_1 = \frac{P(D=1|G=1, H=0)/P(D=0|G=1, H=0)}{P(D=1|G=0, H=0)/P(D=0|G=0, H=0)} = e^{\int_T \alpha(t)x(t)dt}. \tag{7.91}$$

Similarly, the odds-ratio associated with the second genomic region is defined as

$$\mathrm{OR}_2 = \frac{P(D=1|G=0, H=1)/P(D=0|G=0, H=1)}{P(D=1|G=0, H=0)/P(D=0|G=0, H=0)} = e^{\int_S \beta(s)x(s)ds}. \tag{7.92}$$

The odds-ratio associated with susceptibility in both the first and second genomic regions is then computed as

$$\begin{aligned}\mathrm{OR}_{12} &= \frac{P(D=1|G=1, H=1)/P(D=0|G=1, H=1)}{P(D=1|G=0, H=0)/P(D=0|G=0, H=0)} \\ &= \mathrm{OR}_1\mathrm{OR}_2 e^{\int_T\int_S \gamma(t,s)x(t)x(s)dsdt}.\end{aligned} \tag{7.93}$$

Define a multiplicative interaction measure between two genomic regions as

$$I_{12} = \log\frac{\mathrm{OR}_{12}}{\mathrm{OR}_1\mathrm{OR}_2} = \int_T\int_S \gamma(t,s)x(t)x(s)dsdt. \tag{7.94}$$

If we assume that each genomic region has only one SNP, then Equations 7.91 through 7.93 are reduced to

$$\mathrm{OR}_1 = e^{\alpha},\quad \mathrm{OR}_2 = e^{\beta},\quad \mathrm{OR}_{12} = \mathrm{OR}_1\mathrm{OR}_2 e^{\gamma},\quad \text{and}\quad I_{12} = \gamma = \int_T\int_S \gamma(t,s)x(t)x(s)dsdt,$$

which are consistent with the standard results for traditional analysis of interaction between two SNPs.

7.3.3 Statistics for Testing Interaction between Two Genomic Regions

Assume that the total number of individuals in cases and controls is n. Let $y_i, i=1,2,\ldots,n$ denote the disease status of the ith individual. A value of 1 ($y_i=1$) is used to indicate "disease" and a value of 0 ($y_i=0$) to indicate "normal." From Equation 7.90, it follows that

$$\pi_i = E\left[y_i = 1 \middle| W_i\right] = \frac{e^{W_i^T b}}{1+e^{W_i^T b}}. \tag{7.95}$$

The likelihood function is given by

$$L(y|b) = \prod_{i=1}^{n} \pi_i^{y_i}\left(1-\pi_i\right)^{1-y_i}. \tag{7.96}$$

The maximum likelihood method will be used to estimate parameters b. Let $W = \begin{bmatrix} W_1 & \cdots & W_n \end{bmatrix}^T$. The variance–covariance matrix of the estimate $\hat{b}$ is given by

$$\text{Var}\left(\hat{b}\right) = \left(W^T D W\right)^{-1}, \tag{7.97}$$

where $D = \text{diag}(\pi_1, \ldots, \pi_n)$.

Now we study to test for interaction between two genomic regions (or genes). Formally, we investigate the problem of testing the following hypothesis:

$$\gamma(t,s) = 0, \quad \forall t \in [a_1, b_1],\ s \in [a_2, b_2],$$

which is equivalent to testing the hypothesis in Equation 7.90:

$$\gamma = 0.$$

Let Λ be the matrix corresponding to the parameter γ of the variance matrix $\text{Var}(\hat{b})$ in Equation 7.97. Define the test statistic for testing the interaction between two genomic regions, $[a_1, b_1]$ and $[a_2, b_2]$, as

$$T_I = \hat{\gamma}^T \Lambda^{-1} \hat{\gamma}. \tag{7.98}$$

Then, under the null hypothesis $H_0: \gamma = 0$, T_I is asymptotically distributed as a central $\chi^2_{(JK)}$ distribution.

7.4 STATISTICS FOR TESTING GENE–GENE AND GENE–ENVIRONMENT INTERACTION FOR QUANTITATIVE TRAITS

Epistasis is a biologically important component of genetic architecture of quantitative traits (Mackay 2014). Epistasis analysis raises great challenges in statistics and computations. The statistical challenge for genome-wide interaction analysis arises from the multiple statistical tests. The computational challenge is to require a prohibitive amount of computational time. It was recently reported that the average number of SNPs per kb in the 202 drug target genes sequenced in 12,514 European subjects is about 48 SNPs (Nelson et al. 2012). The total number of all possible pairs of SNPs across the genome for large sample sizes can reach as many as 10^{16}. The dimension of whole-genome sequencing is extremely high. Suppose that 5000 pair-wise tests can be finished in 1 s. All possible pair-wise interaction tests would take about $\frac{1.04\times 10^{16}}{5000*(60*60*24*365)}=6580$ years to finish.

Although many statistical methods for epistasis analysis of quantitative traits have been developed, regression-type methods are the core methods for detecting epistasis in quantitative genetic analysis (Cordell 2009). In this section, we will first introduce the Kempthorne model for quantifying genetic additive × additive, additive × dominance, dominance × additive, and dominance × dominance epistasis effects, assuming both unlinked and linked loci. Then, we will study the single-variate and multivariate regression models for epistasis analysis of single and multiple quantitative traits.

The multivariate regression methods were originally designed to detect epistasis for common variants (Steen 2011) and are difficult to be applied to rare variants for their high type 1 error rates and low power to detect interaction between rare variants. To address a critical barrier in the detection of gene–gene interactions with NGS data, similar to interaction analysis for qualitative traits, we take a genome region (or gene) as a basic unit of interaction analysis and use all the information that can be accessed to collectively test for interaction between all possible pairs of SNPs within two genome regions (or genes). This will shift the paradigm of interaction studies from pair-wise interaction analysis to region–region (gene–gene) interaction analysis where we collectively test for interaction between two sets of loci within genomic regions or genes. The gene-based epistasis analysis methods were then mainly formulated as functional regression models (FRG) (Ramsay and Silverman 2005) with scalar response where the genotype score functions (genetic variant profiles) are defined as a function of the genomic position of the genetic variants rather than a set of discrete genotype values, and the quantitative trait is predicted by genotype score functions with their interaction terms.

7.4.1 Genetic Models for Epistasis Effects of Quantitative Traits

Consider two alleles at each SNP locus. Epistasis effects between two loci are typically modeled by a linear partition of the nine genotypic values in quantitative genetics (Mao et al. 2006). Assume that locus 1 has alleles A and a, denoted by i and j, and locus 2 has alleles B and b, denoted by k and l. Let G_{ijkl} and P_{ijkl} be the genotypic value and probability of the individual carrying genotype ij at locus 1 and kl at locus 2, respectively. The Kempthorne model of a genotypic value for two loci is

$$\begin{aligned} G_{ijkl} = {} & \mu+\alpha_i+\alpha_j+\alpha_k+\alpha_l+\delta_{ij}+\delta_{kl}+(\alpha\alpha)_{ik}+(\alpha\alpha)_{il}+(\alpha\alpha)_{jk}+(\alpha\alpha)_{jl}+(\alpha\delta)_{ikl} \\ & +(\alpha\delta)_{jkl}+(\delta\alpha)_{ijk}+(\delta\alpha)_{ijl}+(\delta\delta)_{ijkl}. \end{aligned} \tag{7.99}$$

Let μ be the mean genotypic value in the population, which can be estimated by

$$\mu = \sum_{ijkl} P_{ijkl} G_{ijkl}. \tag{7.100}$$

Similarly, we define the marginal mean of genotypic values for individuals carrying specific alleles. Let μ_i be the mean of genotypic values for individuals carrying allele $i(i = A, a)$ and μ_k be the marginal mean of genotypic values for individuals carrying allele $k(k = B, b)$.They can be estimated by

$$\mu_i = G_{i...} = \sum_{jkl} P_{ijkl} G_{ijkl}, \tag{7.101}$$

$$\mu_k = G_{..k.} = \sum_{ijl} P_{ijkl} G_{ijkl}. \tag{7.102}$$

Next we define μ_{ij} as the marginal mean of genotypic values for individuals carrying genotype ij at locus 1 $(j = AA, Aa, aa)$, μ_{ik} as the marginal mean of genotypic values for individuals carrying allele i at locus 1 and allele k at locus 2 $(i = A, a, j = B, b)$, μ_{ikl} as the marginal mean of genotypic values for individuals carrying allele i at locus 1 and genotype kl at locus 2 $(i = A, a, kl = BB, Bb, bb)$, and μ_{ijk} as the mean of genotypic value for individuals carrying genotype ij at locus 1 and allele k at locus 2 $(ij = AA, Aa, aa; k = B, b)$. These quantities can be estimated as

$$\mu_{ij} = G_{ij..} = \sum_{kl} P_{ijkl} G_{ijkl}, \tag{7.103}$$

$$\mu_{ik} = G_{i.k.} = \sum_{jl} P_{ijkl} G_{ijkl}, \tag{7.104}$$

$$\mu_{ikl} = G_{i.kl} = \sum_{j} P_{ijkl} G_{ijkl}, \tag{7.105}$$

$$\mu_{ijk} = G_{ijk.} = \sum_{l} P_{ijkl} G_{ijkl}. \tag{7.106}$$

The Kempthorne model defines genetic additive, dominance, and epistasis effects as deviations from the overall mean. Similar to ANOVA, these variations should satisfy constraints. For example, the following constraints should be imposed:

$$\begin{gathered} \sum_i P_i \alpha_i = 0, \\ \sum_j P_{ij} \delta_{ij} = 0, \\ \sum_k P_{ik} (\alpha\alpha)_{ik} = 0, \\ \sum_{kl} P_{ikl} (\alpha\delta)_{ikl} = 0, \end{gathered} \tag{7.107}$$

where P_i, P_{ij}, P_{jk}, and P_{ikl} are frequencies of allele i, genotype ij, allele i at locus 1, allele k at locus 2, allele i at locus 1, and genotype kl at locus 2, respectively. Under these constraints, we can use least square methods or simply multiply some relevant frequencies and summarize them to solve the problem (7.99).

For example, multiplying the frequencies P_{ijkl} on both sides of Equation 7.99 and summarizing the resulting equations, we obtain

$$\sum_{ijkl} P_{ijkl} G_{ijkl} = G_{....} = \mu. \tag{7.108}$$

By similar arguments, we obtain

$$P_i G_{i...} = P_i \mu + P_i \alpha_i,$$

which implies

$$\alpha_i = G_{i...} - \mu. \tag{7.109}$$

Similarly, we can obtain

$$\alpha_k = \mu_k - \mu, \tag{7.110}$$

$$\delta_{ij} = \mu_{ij} - \mu - \alpha_i - \alpha_j, \tag{7.111}$$

$$\delta_{kl} = \mu_{kl} - \mu - \alpha_k - \alpha_l, \tag{7.112}$$

$$(\alpha\alpha)_{ik} = \mu_{ik} - \mu - \alpha_i - \alpha_k, \tag{7.113}$$

$$(\alpha\delta)_{ikl} = \mu_{ikl} - \mu - \alpha_i - \alpha_k - \alpha_l - \delta_{kl} - (\alpha\alpha)_{ik} - (\alpha\alpha)_{il}, \tag{7.114}$$

$$(\delta\alpha)_{ijk} = \mu_{ijk} - \mu - \alpha_i - \alpha_j - \alpha_k - \delta_{ij} - (\alpha\alpha)_{ik} - (\alpha\alpha)_{jk}, \tag{7.115}$$

$$\begin{aligned}(\delta\delta)_{ijkl} &= G_{ijkl} - \mu - \alpha_i - \alpha_j - \alpha_k - \alpha_l - \delta_{ij} - \delta_{kl} - (\alpha\alpha)_{ik} - (\alpha\alpha)_{il} \\ &\quad - (\alpha\alpha)_{jk} - (\alpha\alpha)_{jl} - (\alpha\delta)_{ikl} - (\alpha\delta)_{jkl} - (\delta\alpha)_{ijk} - (\delta\alpha)_{ijl}.\end{aligned} \tag{7.116}$$

If we assume Hardy–Weinberg equilibrium (HWE) and linkage equilibrium, we have $P_{ij} = P_i P_j$, $P_{kl} = P_K P_l$, $P_{ik} = P_i P_k$, $P_{il} = P_i P_l$, $P_{jl} = P_j P_l$, and $P_{jk} = P_j P_k$. Under HWE, Equation 7.109 is reduced to

$$\begin{aligned}\alpha_A &= P_A \mu_{AA} + P_a \mu_{Aa} - \mu \\ &= P_A \mu_{AA} + P_a \mu_{Aa} - P_A^2 \mu_{AA} - 2 P_A P_a \mu_{Aa} - P_a^2 \mu_{aa} \\ &= P_A (1 - P_A) \mu_{AA} + P_a (1 - 2P_A) \mu_{Aa} - P_a^2 \mu_{aa} \\ &= P_a \left[P_A \mu_{AA} + (P_a - P_A) \mu_{Aa} - P_a \mu_{aa} \right] \\ &= P_a \alpha_1,\end{aligned} \tag{7.117}$$

where

$$\alpha_1 = P_A\mu_{AA} + (P_a - P_A)\mu_{Aa} - P_a\mu_{aa} \tag{7.118}$$

is often referred to as a substitution effect or additive effect.

By a similar calculation, we have

$$\alpha_a = -P_A\alpha_1. \tag{7.119}$$

Equations 7.117 through 7.119 can be extended to the second locus:

$$\begin{aligned}\alpha_2 &= P_B\mu_{BB} + (P_b - P_B)\mu_{Bb} - P_b\mu_{bb},\\ \alpha_B &= P_b\alpha_2,\\ \alpha_b &= -P_B\alpha_2.\end{aligned} \tag{7.120}$$

By the similar argument, we can show that

$$\begin{aligned}\delta_1 &= \mu_{AA} - 2\mu_{Aa} + \mu_{aa},\\ \delta_{AA} &= P_a^2\delta_1, \delta_{Aa} = -P_AP_a\delta_1, \delta_{aa} = P_A^2\delta_1,\\ \delta_2 &= \mu_{BB} - 2\mu_{Bb} + \mu_{bb},\\ \delta_{BB} &= P_b^2\delta_2, \delta_{Bb} = -P_BP_b\delta_2, \delta_{bb} = P_B^2\delta_2.\end{aligned} \tag{7.121}$$

Now we calculate the additive × additive interaction effects.

It follows from Equation 7.113 that

$$(\alpha\alpha)_{AB} = \mu_{AB} - \mu - \alpha_A - \alpha_B. \tag{7.122}$$

Note that we can have

$$\mu_{AB} = P_AP_B\mu_{AABB} + P_AP_b\mu_{AABb} + P_aP_B\mu_{AaBB} + P_aP_b\mu_{AaBb}. \tag{7.123}$$

Similarly, we have

$$\begin{aligned}\mu_{AA} &= P_B^2\mu_{AABB} + 2P_BP_b\mu_{AABb} + P_b^2\mu_{AAbb},\\ \mu_{Aa} &= P_B^2\mu_{AaBB} + 2P_BP_b\mu_{AaBb} + P_b^2\mu_{Aabb},\\ \mu_{aa} &= P_B^2\mu_{aaBB} + 2P_BP_b\mu_{aaBb} + P_b^2\mu_{aabb},\\ \mu_{BB} &= P_A^2\mu_{AABB} + 2P_AP_a\mu_{AaBB} + P_a^2\mu_{aaBB},\\ \mu_{Bb} &= P_A^2\mu_{AABb} + 2P_AP_a\mu_{AaBb} + P_a^2\mu_{aaBb},\\ \mu_{bb} &= P_A^2\mu_{AAbb} + 2P_AP_a\mu_{Aabb} + P_a^2\mu_{aabb}.\end{aligned} \tag{7.124}$$

The overall mean μ can be calculated by

$$\begin{aligned}\mu = {} & P_A^2P_B^2\mu_{AABB} + 2P_A^2P_BP_b\mu_{AABb} + P_A^2P_b^2\mu_{AAbb} + 2P_AP_aP_B^2\mu_{AaBB} + 4P_AP_aP_BP_b\mu_{AaBb}\\ & + 2P_AP_aP_b^2\mu_{Aabb} + P_a^2P_B^2\mu_{aaBB} + 2P_a^2P_BP_b\mu_{aaBb} + P_a^2P_b^2\mu_{aabb}.\end{aligned} \tag{7.125}$$

Substituting Equations 7.117, 7.120, 7.123 through 7.125 into Equation 7.122 gives

$$(\alpha\alpha)_{AB} = P_a P_b \gamma_{AA}, \tag{7.126}$$

where

$$\gamma_{AA} = P_A P_B \mu_{AABB} + P_A (P_b - P_B) \mu_{AABb} - P_A P_b \mu_{AAbb} + (P_a - P_A) P_B \mu_{AaBB} + P_a P_b \mu_{aabb}$$
$$+ (P_a - P_A)(P_b - P_B) \mu_{AaBb} - (P_a - P_A) P_b \mu_{Aabb} - P_a P_B \mu_{aaBB} - P_a (P_b - P_B) \mu_{aaBb}. \tag{7.127}$$

The measure γ_{AA} is often referred to as an additive × additive interaction effect.

Similarly, we have

$$\begin{aligned} (\alpha\alpha)_{Ab} &= -P_a P_B \gamma_{AA}, \\ (\alpha\alpha)_{aB} &= -P_A P_b \gamma_{AA}, \\ (\alpha\alpha)_{ab} &= P_A P_B \gamma_{AA}. \end{aligned} \tag{7.128}$$

We can show that additive × dominance interaction effects are (Exercise 7.15)

$$\begin{aligned} &(\alpha\delta)_{ABB} = P_b P_a^2 \gamma_{AD}, \quad (\alpha\delta)_{ABb} = -P_a P_B P_b \gamma_{AD}, \quad (\alpha\delta)_{Abb} = P_a P_B^2 \gamma_{AD}, \\ &(\alpha\delta)_{aBB} = -P_A P_b^2 \gamma_{AD}, \quad (\alpha\delta)_{aBb} = P_A P_B P_b \gamma_{AD}, \quad (\alpha\delta)_{abb} = -P_A P_B^2 \gamma_{AD}, \\ &\gamma_{AD} = P_A \mu_{AABB} - 2P_A \mu_{AABb} + P_A \mu_{AAbb} + (P_a - P_A) \mu_{AaBB} - 2(P_a - P_A) \mu_{AaBb}, \end{aligned} \tag{7.129}$$

where γ_{AD} is referred to as an additive × dominance interaction effect.

Similarly, dominance × additive interaction effects are given by

$$\begin{aligned} &(\delta\alpha)_{AAB} = P_a^2 P_b \gamma_{DA}, \quad (\delta\alpha)_{AaB} = -P_A P_a P_b \gamma_{DA}, \quad (\delta\alpha)_{aaB} = P_A^2 P_b \gamma_{DA}, \\ &(\delta\alpha)_{AAb} = -P_a^2 P_B \gamma_{DA}, \quad (\delta\alpha)_{Aab} = P_A P_a P_B \gamma_{DA}, \quad (\delta\alpha)_{aab} = -P_A^2 P_B \gamma_{DA}, \\ &\gamma_{DA} = P_B (\mu_{AABB} - 2\mu_{AaBB} + \mu_{aaBB}) + (P_b - P_B)(\mu_{AABb} - 2\mu_{AABb} + \mu_{aaBb} \\ &\quad - P_b (\mu_{AAbb} - 2\mu_{Aabb} + \mu_{aabb}), \end{aligned} \tag{7.130}$$

where γ_{DA} is called a dominance × additive interaction effect. Finally, we give the formula for the calculation of dominance × dominance interaction effects:

$$\begin{aligned} &(\delta\delta)_{AABB} = P_a^2 P_b^2 \gamma_{DD}, \quad (\delta\delta)_{AaBB} = -P_A P_a P_b^2 \gamma_{DD}, \quad (\delta\delta)_{aaBB} = P_A^2 P_B^2 \gamma_{DD}, \\ &(\delta\delta)_{AABb} = -P_a^2 P_B P_b \gamma_{DD}, \quad (\delta\delta)_{AaBb} = P_A P_a P_B P_b \gamma_{DD}, \quad (\delta\delta)_{aaBb} = -P_A^2 P_B P_b \gamma_{DD}, \\ &(\delta\delta)_{AAbb} = P_a^2 P_B^2 \gamma_{DD}, \quad (\delta\delta)_{Aabb} = -P_A P_a P_B^2 \gamma_{DD}, \quad (\delta\delta)_{aabb} = P_A^2 P_B^2 \gamma_{DD}, \\ &\gamma_{DD} = \mu_{AABB} - 2\mu_{AaBB} + \mu_{aaBB} - 2\mu_{AABb} + 4\mu_{AaBb} - 2\mu_{aaBb} + \mu_{aaBB} - 2\mu_{aaBb} + \mu_{aabb}, \end{aligned} \tag{7.131}$$

where γ_{DD} is called a dominance × dominance interaction effect.

These formulas lay foundation for regression model for interaction analysis.

7.4.2 Regression Model for Interaction Analysis with Quantitative Traits

We can extend the regression model in Section 4.1.3 to include the interaction term to implement the interaction models for a quantitative trait investigated in the previous section. Consider two loci, A with two alleles, A and a, and B with two alleles, B and b. Let P_A, P_a, P_B, and P_b be frequencies of alleles A, a, B, and b, respectively. Assume that n individuals are sampled. Let Y_i be a phenotypic value of the ith individual. A simple regression model for a quantitative trait is given by

$$\begin{aligned} Y_i &= \mu + X_{i1}\alpha_1 + X_{i2}\alpha_2 + Z_{i1}\delta_1 + Z_{i_2}\delta_2 + X_{i1}X_{i2}\gamma_{AA} \\ &+ X_{i1}Z_{i2}\gamma_{AD} + Z_{i1}X_{i2}\gamma_{DA} + Z_{i1}Z_{i2}\gamma_{DD} + \varepsilon_i, \end{aligned} \tag{7.132}$$

where μ is an overall mean, ε_i are independent and identically distributed normal variables with zero mean and variance σ_e^2

$$X_{i1} = \begin{cases} 2P_a, & AA \\ P_a - P_A, & Aa \\ -2P_A, & aa \end{cases}, \quad X_{i2} = \begin{cases} 2P_b, & BB \\ P_b - P_B, & Bb \\ -2P_B & bb \end{cases},$$

$$Z_{i1} = \begin{cases} P_a^2, & AA \\ -P_A P_a, & Aa, \\ P_A^2 & aa \end{cases} \quad \text{and} \quad Z_{i2} = \begin{cases} P_b^2, & BB \\ -P_B P_b, & Bb. \\ P_B^2 & bb \end{cases} \tag{7.133}$$

We add a constant $-P_a + P_A$ into X_{i1}. Then, the indicator variable will be transformed to

$$X_i = \begin{cases} 1, & A_1A_1 \\ 0, & A_1A_2 \\ -1, & A_2A_2 \end{cases},$$

which is a widely used indicator variable for additive effect in quantitative genetic analysis.

To show that the models (7.132) and (7.133) implement the genetic main and interaction effect models investigated in Section 7.4.1, we check $(\alpha\alpha)_{AABB}$. In the model (7.132), when the genotypes are AA and BB at locus 1 and locus 2, respectively, the coefficient before γ_{AA} is $X_{i1}X_{i2} = 4P_aP_b$ and the additive × additive interaction effect for these genotypes is $4P_aP_b$. From Equation 7.126, it follows that the additive × additive interaction effect for these genotypes is $4(\alpha\alpha)_{AB} = 4P_aP_b\gamma_{AA}$. They are exactly the same.

Next we study to use regression to model interactions between two genomic regions. Consider two genomic regions, $[a_1, b_1]$ and $[a_2, b_2]$. Assume that there are k_1 SNPs $(A_1, \ldots, A_{k_1})$ in the first genomic region and k_2 SNPs $(B_1, \ldots, B_{k_2})$ in the second genomic region. Let P_{A_j} and P_{a_j} be the frequencies of alleles A_j and a_j at the SNP A_j in the first genomic region, respectively. Similarly, let P_{B_j} and P_{b_j} be the frequencies of alleles B_j and b_j at the SNP B_j in the second genomic region, respectively. Define indicator variables

$$x_{ij}^1 = \begin{cases} 2P_{a_j} & A_jA_j \\ P_{a_j} - P_{A_j} & A_ja_j \\ -2P_{A_j} & a_ja_j \end{cases}, \quad x_{il}^2 = \begin{cases} 2P_{b_l} & B_lB_l \\ P_{b_l} - P_{B_l} & B_lb_l \\ -2P_{B_l} & b_lb_l \end{cases},$$

$$z_{ij}^1 = \begin{cases} P_{a_j}^2 & A_jA_j \\ -P_{A_j}P_{a_j} & A_ja_j \\ P_{A_j}^2 & a_ja_j \end{cases}, \text{ and } z_{il}^2 = \begin{cases} P_{b_l}^2 & B_lB_l \\ -P_{B_l}P_{b_l} & B_lb_l \\ P_{B_l}^2 & b_lb_l \end{cases}.$$

A regression model for interaction analysis is defined as

$$\begin{aligned} y_i = \mu &+ \sum_{j=1}^{k_1} x_{ij}^1\alpha_j^1 + \sum_{l=1}^{k_2} x_{il}^2\alpha_l^2 + \sum_{j=1}^{k_1} z_{ij}^1\beta_j^1 + \sum_{l=1}^{k_2} z_{il}^2\beta_l^2 + \sum_{j=1}^{k_1}\sum_{l=1}^{k_2} x_{ij}^1 x_{il}^2 \gamma_{AA}^{jl} \\ &+ \sum_{j=1}^{k_1}\sum_{l=1}^{k_2} x_{ij}^1 z_{il}^2 \gamma_{AD}^{jl} + \sum_{j=1}^{k_1}\sum_{l=1}^{k_2} z_{ij}^1 x_{il}^2 \gamma_{DA}^{jl} + \sum_{j=1}^{k_1}\sum_{l=1}^{k_2} z_{ij}^1 z_{il}^2 \gamma_{Dd}^{jl} + \varepsilon_i, \end{aligned} \tag{7.134}$$

where μ is an overall mean, α_j^1 is the main genetic additive effect of the jth SNP in the first genomic region, α_l^2 is the main genetic additive effect of the jth SNP in the second genomic region, β_j^1 is the main genetic dominance effect of the jth SNP in the first genomic region, β_l^2 is the main genetic dominance effect of the lth SNP in the second genomic region, γ_{AA}^{jl} is an additive × additive interaction effect between the jth SNP in the first genomic region and the lth SNP in the second genomic region, γ_{AD}^{jl} is the additive × dominance interaction effect between the jth SNP in the first genomic region and the lth SNP in the second genomic region, γ_{DA}^{jl} is the dominance × additive interaction effect between the jth SNP in the first genomic region and the lth SNP in the second genomic region, γ_{DD}^{jl} is the genetic dominance × dominance interaction effect between the jth SNP in the first genomic region and the lth SNP in the second genomic region, and ε_i are independent and identically distributed normal variables with a mean of zero and variance σ^2. Test statistics for testing the gene–gene interaction for the multiple regression model will be discussed in next section.

7.4.3 Functional Regression Model for Interaction Analysis with a Quantitative Trait

7.4.3.1 Model

Consider two genomic regions, $[a_1, b_1]$ and $[a_2, b_2]$. Let $x_i(t)$ and $x_i(s)$ be genotypic functions of the ith individual defined in the regions $[a_1, b_1]$ and $[a_2, b_2]$, respectively. Let y_i be the phenotypic value of a quantitative trait measured on the ith individual. Let t and s be a genomic position in the first genomic region and second genomic regions, respectively. Define a genotype profile $x_i(t_j)$ and $z_i(t_j)$ of the ith individual as

$$x_i(t_j) = \begin{cases} 2P_{a_j} & A_jA_j \\ P_{a_j} - P_{A_j} & A_ja_j \\ -2P_{A_j} & a_ja_j \end{cases} \text{ and } z_i(t_j) = \begin{cases} P_{a_j}^2 & A_jA_j \\ -P_{A_j}P_{a_j} & A_ja_j \\ P_{A_j}^2 & a_ja_j \end{cases},$$

where P_{A_j} and P_{a_j} are the frequencies of alleles A_j and a_j at the SNP A_j located in t_j of the first genomic region, respectively. We can similarly define $x_i(s_l)$ and $z_i(s_l)$ for the SNP B_l located in s_l of the second genomic region.

A functional regression model for a quantitative trait can be defined as

$$\begin{aligned} y_i = \alpha_0 &+ \int_T \alpha(t)x_i(t)dt + \int_S \beta(s)x_i(s)ds + \int_T \delta^1(t)z_i(t)dt + \int_S \delta^2(s)z_i(s)ds \\ &+ \int_T\int_S \gamma_{AA}(t,s)x_i(t)x_i(s)dtds + \int_T\int_S \gamma_{AD}(t,s)x_i(t)z_i(s)dtds \\ &+ \int_T\int_S \gamma_{DA}(t,s)z_i(t)x_i(s)dtds + \int_T\int_S \gamma_{DD}(t,s)z_i(t)z_i(s)dtds + \varepsilon_i, \end{aligned} \quad (7.135)$$

where α_0 is an overall mean; $\alpha(t)$ and $\beta(s)$ are genetic additive effects of two putative QTLs located at the genomic positions t and s, respectively; $\delta^1(t)$ and $\delta^2(s)$ are genetic dominance effects of two putative QTLs located at the genomic positions t and s, respectively; $\gamma_{AA}(t, s)$ is the additive × additive interaction effect between two putative QTLs located at the genomic positions t and s; $\gamma_{AD}(t, s)$, $\gamma_{DA}(t, s)$, and $\gamma_{DD}(t, s)$ are the additive × dominance, dominance × additive, dominance × dominance interaction effects between two putative QTLs located at the genomic positions t and s, respectively; $x_i(t)$, $x_i(s)$, $z_i(t)$, and $z_i(s)$ are genotype profiles; and ε_i are independent and identically distributed normal variables with a mean of zero and variance σ_e^2.

7.4.3.2 Parameter Estimation

We assume that both phenotypes and genotype profiles are centered. The genotype profiles $x_i(t)$ and $x_i(s)$ are expanded in terms of orthonormal basis function as

$$x_i(t) = \sum_{j=1}^{\infty} \xi_{ij}\phi_j(t), \quad x_i(s) = \sum_{k=1}^{\infty} \eta_{ik}\psi_k(s)$$

and

$$z_i(t) = \sum_{l=1}^{\infty} \zeta_{il}\varphi_l(t), \quad z_i(s) = \sum_{m=1}^{\infty} \tau_{im}\theta_m(s) \quad (7.136)$$

where $\phi_j(t)$, $\psi_k(s)$, $\varphi_l(t)$, and $\theta_m(s)$ are sequences of orthonormal eigenfunction functions. The expansion coefficients ξ_{ij} and η_{ik} are estimated by

$$\xi_{ij} = \int_T x_i(t)\phi_j(t)dt, \quad \eta_{ik} = \int_S x_i(s)\psi_k(s)ds,$$
$$\zeta_{il} = \int_T z_i(t)\varphi_l(t)dt, \quad \tau_{im} = \int_S z_i(s)\theta_m(s)ds. \tag{7.137}$$

In practice, numerical methods for integrals will be used to calculate the expansion coefficients.

Substituting Equation 7.136 into Equation 7.135, we obtain

$$\begin{aligned}
y_i =& \alpha_0 + \int_T \alpha(t)\sum_{j=1}^{\infty}\xi_{ij}\phi_j(t)dt + \int_S \beta(s)\sum_{k=1}^{\infty}\eta_{ik}\psi_k(s)ds \\
&+ \int_T \delta^1(t)\sum_{l=1}^{\infty}\zeta_{il}\varphi_l(t)dt + \int_S \delta^2(s)\sum_{m=1}^{\infty}\tau_{im}\theta_m(s)ds \\
&+ \int_T\int_S \gamma_{AA}(t,s)\left(\sum_{j=1}^{\infty}\xi_{ij}\phi_j(t)\right)\left(\sum_{k=1}^{\infty}\eta_{ik}\psi_k(s)\right)dtds + \int_T\int_S \gamma_{AD}(t,s)\sum_{j=1}^{\infty}\xi_{ij}\phi_j(t)\sum_{m=1}^{\infty}\tau_{im}\theta_m(s)dtds \\
&+ \int_T\int_S \gamma_{DA}(t,s)\sum_{l=1}^{\infty}\zeta_{il}\varphi_l(t)\sum_{k=1}^{\infty}\eta_{ik}\psi_k(s)dtds + \int_T\int_S \gamma_{DD}(t,s)\sum_{l=1}^{\infty}\zeta_{il}\varphi_l(t)\sum_{m=1}^{\infty}\tau_{im}\theta_m(s)dtds + \varepsilon_i.
\end{aligned} \tag{7.138}$$

After some algebra, Equation 7.138 can be reduced to

$$\begin{aligned}
y_i =& \alpha_0 + \sum_{j=1}^{\infty}\xi_{ij}\alpha_j + \sum_{k=1}^{\infty}\eta_{ik}\beta_k + \sum_{l=1}^{\infty}\zeta_{il}\delta_l^1 + \sum_{m=1}^{\infty}\tau_{im}\delta_m^2 + \sum_{j=1}^{\infty}\sum_{k=1}^{\infty}\xi_{ij}\eta_{ik}\gamma_{jk}^{AA} + \sum_{j=1}^{\infty}\sum_{m=1}^{\infty}\xi_{ij}\tau_{im}\gamma_{jm}^{AD} \\
&+ \sum_{l=1}^{\infty}\sum_{k=1}^{\infty}\zeta_{il}\eta_{ik}\gamma_{lk}^{DA} + \sum_{l=1}^{\infty}\sum_{m=1}^{\infty}\zeta_{il}\tau_{im}\gamma_{lm}^{DD} + \varepsilon_i,
\end{aligned} \tag{7.139}$$

where

$$\alpha_j = \int_T \alpha(t)\phi_j(t)dt, \quad \beta_k = \int_S \beta(s)\psi_k(s)ds, \quad \delta_l^1 = \int_T \delta^1(t)\varphi_l(t)dt, \quad \delta_m^2 = \int_S \delta^2(s)\theta_m(s)ds,$$

$$\gamma_{jk}^{AA} = \int_T\int_S \gamma_{AA}(t,s)\phi_j(t)\psi_k(s)dtds, \quad \gamma_{jm}^{AD} = \int_T\int_S \gamma_{AD}(t,s)\phi_j(t)\theta_m(s)dtds,$$

$$\gamma_{lk}^{DA} = \int_T\int_S \gamma_{DA}(t,s)\varphi_l(t)\psi_k(s)dtds \quad \text{and} \quad \gamma_{lm}^{DD} = \int_T\int_S \gamma_{DD}(t,s)\varphi_l(t)\theta_m(s)dtds.$$

The parameters α_j, β_k, δ_1^1, δ_1^2, γ_{jk}^{AA}, γ_{jm}^{AD}, γ_{lk}^{DA}, and γ_{lm}^{DD} are referred to as genetic additive, dominance effects at the first and second genomic regions and additive × additive, additive × dominance, and dominance × dominance interaction effect scores. These scores can also be viewed as the expansion coefficients of the genetic effect functions with respect to orthonormal basis functions:

$$\alpha(t)=\sum_j \alpha_j\phi_j(t),\quad \beta(s)=\sum_k \beta_k\psi_k(s),\quad \delta^1(t)=\sum_l \delta_l^1\varphi_l(t),\quad \delta^2(s)=\sum_m \delta_m^2\theta_m(s),$$

$$\gamma_{AA}(s,t)=\sum_j\sum_k \gamma_{jk}^{AD}\phi_j(s)\psi_k(t),\quad \gamma_{AD}(s,t)=\sum_j\sum_m \gamma_{jm}^{AD}\phi_j(t)\theta_m(s),$$

$$\gamma_{DA}(s,t)=\sum_l\sum_k \gamma_{lk}^{DA}\varphi_l(t)\psi_k(s),$$

and

$$\gamma_{DD}(s,t)=\sum_l\sum_m \gamma_{lm}^{DD}\varphi_l(t)\theta_m(s). \tag{7.140}$$

Define vectors for phenotypes and genetic effects:

$$Y=[y_1,\ldots,y_n]^T,\quad \alpha=[\alpha_0,\alpha_1,\ldots,\alpha_J]^T,\quad \beta=[\beta_1,\ldots,\beta_K]^T,\quad \delta^1=[\delta_1^1,\ldots,\delta_L^1]^T,$$

$$\delta^2=[\delta_1^2,\ldots,\delta_M^2]^T,\quad \gamma_{AA}=[\gamma_{11}^{AA},\ldots,\gamma_{JK}^{AA}]^T,\quad \gamma_{AD}=[\gamma_{11}^{AD},\ldots,\gamma_{JM}^{AD}]^T,\quad \gamma_{DA}=[\gamma_{11}^{DA},\ldots,\gamma_{LK}^{DA}]^T,$$

$$\gamma_{DD}=[\gamma_{11}^{DD},\ldots,\gamma_{LM}^{DD}]^T,\quad b=\left[\alpha^T,\beta^T,(\delta^1)^T,(\delta^2)^T,\gamma_{AA}^T,\gamma_{AD}^T,\gamma_{DA}^T,\gamma_{DD}^T\right]^T.$$

Define regression coefficient matrices:

$$\xi=\begin{bmatrix}\xi_{11} & \cdots & \xi_{1J}\\ \cdots & \cdots & \cdots\\ \xi_{n1} & \cdots & \xi_{nJ}\end{bmatrix},\quad \eta=\begin{bmatrix}\eta_{11} & \cdots & \eta_{1K}\\ \cdots & \cdots & \cdots\\ \eta_{n1} & \cdots & \eta_{nK}\end{bmatrix},$$

$$\zeta=\begin{bmatrix}\zeta_{11} & \cdots & \zeta_{1L}\\ \cdots & \cdots & \cdots\\ \zeta_{n1} & \cdots & \zeta_{nL}\end{bmatrix},\quad \tau=\begin{bmatrix}\tau_{11} & \cdots & \tau_{1M}\\ \cdots & \cdots & \cdots\\ \tau_{n1} & \cdots & \tau_{nM}\end{bmatrix},$$

$$\Gamma_{AA}=\begin{bmatrix}\xi_{11}\eta_{11} & \cdots & \xi_{1J}\eta_{1K}\\ \cdots & \cdots & \cdots\\ \xi_{n1}\eta_{n1} & \cdots & \xi_{nJ}\eta_{nK}\end{bmatrix},\quad \Gamma_{AD}=\begin{bmatrix}\xi_{11}\tau_{11} & \cdots & \xi_{1J}\tau_{1M}\\ \cdots & \cdots & \cdots\\ \xi_{n1}\tau_{n1} & \cdots & \xi_{nJ}\tau_{nM}\end{bmatrix},$$

$$\Gamma_{DA}=\begin{bmatrix}\zeta_{11}\eta_{11} & \cdots & \zeta_{1L}\eta_{1K}\\ \cdots & \cdots & \cdots\\ \zeta_{n1}\eta_{n1} & \cdots & \zeta_{nL}\eta_{nK}\end{bmatrix},\quad \text{and}\quad \Gamma_{DD}=\begin{bmatrix}\zeta_{11}\tau_{11} & \cdots & \zeta_{1L}\tau_{1M}\\ \cdots & \cdots & \cdots\\ \zeta_{n1}\tau_{n1} & \cdots & \zeta_{nL}\tau_{nM}\end{bmatrix}.$$

If we take the final terms in the model (7.139), Equation 7.139 can be reduced to

$$\begin{aligned} Y &= \xi\alpha + \eta\beta + \zeta\delta^1 + \tau\delta^2 + \Gamma_{AA}\gamma_{AA} + \Gamma_{AD}\gamma_{AD} + \Gamma_{DA}\gamma_{DA} + \Gamma_{DD}\gamma_{DD} + \varepsilon \\ &= W_1 b_1 + \Gamma\gamma + \varepsilon, \\ &= Wb + \varepsilon, \end{aligned} \tag{7.141}$$

where

$$W_1 = [\xi \quad \eta \quad \zeta \quad \tau], \quad \Gamma = \begin{bmatrix}\Gamma_{AA} & \Gamma_{AD} & \Gamma_{DA} & \Gamma_{DD}\end{bmatrix}, \quad W = \begin{bmatrix}W_1 & \Gamma\end{bmatrix},$$

$$b_1 = \begin{bmatrix} \alpha \\ \beta \\ \delta^1 \\ \delta^2 \end{bmatrix}, \quad \gamma = \begin{bmatrix} \gamma_{AA} \\ \gamma_{AD} \\ \gamma_{DA} \\ \gamma_{DD} \end{bmatrix}, \quad b = \begin{bmatrix} b_1 \\ \gamma \end{bmatrix}.$$

Therefore, the interaction models with integrals are transformed to the traditional multivariate regression models (7.140) for interaction analysis. The standard least square estimator of b is given by

$$\hat{b} = \left(W^T W\right)^{-1} W^T \left(Y - \bar{Y}\right), \tag{7.142}$$

and its variance is

$$\mathrm{Var}\left(\hat{b}\right) = \hat{\sigma}^2 \left(W^T W\right)^{-1}, \tag{7.143}$$

where

$$\hat{\sigma}^2 = \frac{1}{n - J - K - L - M - JK - JM - LK - LM} \left(Y - \bar{Y}\right)^T \left[I - W\left(W^T W\right)^{-1} W^T\right]\left(Y - \bar{Y}\right).$$

Substituting the estimated genetic effect scores $\hat{\alpha}_j, \hat{\beta}_k, \hat{\delta}_l^1, \hat{\delta}_m^2, \hat{\gamma}_{jk}^{AA}, \hat{\gamma}_{jm}^{AD}, \hat{\gamma}_{lk}^{DA}$, and $\hat{\gamma}_{lm}^{DD}$ into Equation 7.140 yields the estimated genetic additive effect and additive × additive interaction effect functions $\hat{\alpha}(t), \hat{\beta}(s), \hat{\delta}^1(t), \hat{\delta}^2(s), \hat{\gamma}_{AA}(t,s), \hat{\gamma}_{AD}(t,s), \hat{\gamma}_{DA}(t,s)$, and $\hat{\gamma}_{DD}(t,s)$.

7.4.3.3 *Test Statistics*

An essential problem in genetic interaction studies of the quantitative trait is to test the interaction between two genomic regions (or genes). Formally, we investigate the problem of testing the following hypothesis:

$$\begin{aligned} &\gamma_{AA}(t,s) = 0, \quad \gamma_{AD}(t,s) = 0, \quad \gamma_{DA}(t,s) = 0, \quad \gamma_{DD}(t,s) = 0, \\ &\forall t \in [a_1, b_1], \quad s \in [a_2, b_2], \end{aligned}$$

which is equivalent to testing the hypothesis

$$\gamma_{AA}=0, \gamma_{AD}=0, \gamma_{DA}=0, \quad \text{and} \quad \gamma_{DD}=0, \quad \text{or} \quad \gamma=0,$$

where γ is defined as in Equation 7.141.

Let Λ be the matrix corresponding to the parameter γ of the variance matrix $\text{Var}(\hat{b})$ in Equation 7.143. Define the test statistic for testing the interaction (including additive × additive, additive × dominance, dominance × additive, and dominance × dominance interaction effects) between two genomic regions, $[a_1, b_1]$ and $[a_2, b_2]$, as

$$T_I=\hat{\gamma}^T\Lambda^{-1}\hat{\gamma}. \tag{7.144}$$

Then, under the null hypothesis $H_0:\gamma=0$, T_I is asymptotically distributed as a central $\chi^2_{(JK+JM+LK+LM)}$ distribution if $JK+JM+LK+LM$ components are taken in the expansion Equation 7.139.

7.4.3.4 Simulations and Applications to Real Example

For simplicity, throughout this section, we consider only the additive effect and the additive × additive interaction effect (Zhang et al. 2014).

7.4.3.4.1 Null Distribution of Test Statistics In the previous section, we have shown that the test statistics T_I is asymptotically distributed as a central $\chi^2_{(JK)}$ distribution when we consider only the additive × additive interaction effect. To examine the null distribution of test statistics, we performed a series of simulation studies to compare their empirical levels with the nominal ones (Zhang et al. 2014).

We calculated the type 1 error rates for rare variants and both common and rare variants. We assumed the model to generate a phenotype.

Model (with marginal effects of two genes):

$$y_i=\mu+\sum_{j=1}^{k_1}x_{ij}\alpha_j+\sum_{l=1}^{k_2}z_{il}\beta_l+\varepsilon_i, \tag{7.145}$$

where

x_{ij} and z_{il} are indicator variables for the genotype at the jth SNP in the first gene and at the l the SNP in the second gene, respectively

$\alpha_j=(1-p_j)(r_1-1)$

$\beta_l=(1-p_l)(r_2-1)$

p_j and p_l are the frequencies of minor allele at the jth SNP in the first gene and at the lth SNP in the second gene, respectively

r_1 and r_2 are risk parameters and are equal to 1.2 and 1.4, respectively

We assume 20% of variants to be risk variants.

TABLE 7.1 Description of the Four Genes

	Total Number of SNPs	Number of Rare Variants	Median MAF	Coding Length
PLCH2	129	114	2.62E−04	29,212
PANK4	101	94	2.62E−04	18,062
TNFRSF14	25	23	2.62E−04	7,462
KANK4	69	55	2.62E−04	83,247

TABLE 7.2 Average Type 1 Error Rates of the Statistics for Testing for Interaction between Two Genes with Rare Variants and Marginal Effects at Both Genes over Six Pairs of Genes

Sample Size	0.001	0.01	0.05
500	0.0012	0.0093	0.0450
1000	0.0011	0.0102	0.0474
2000	0.0010	0.0105	0.0505
3000	0.0010	0.0094	0.0465
4000	0.0010	0.0114	0.0492
5000	0.0010	0.0092	0.0499

A total of 1,000,000 chromosomes by resampling from 1911 individuals with variants in four genes selected from the NHLBI's Exome Sequencing Project (ESP), of which the descriptions of the four genes are summarized in Table 7.1, were generated. The number of sampled individuals range from 500 to 5000, and 5000 simulations were repeated. Table 7.2 summarized the average type 1 error rates of the statistics for testing interaction between two genes with rare variants and marginal effects at both genes over six pairs of genes at the nominal levels $\alpha = 0.05$, $\alpha = 0.01$, and $\alpha = 0.001$. Table 7.2 showed that the type 1 error rates of the test statistics for testing for interaction between two genes with marginal effects were not appreciably different from the nominal α levels. These results are also true for the genes without marginal effects (Zhang et al. 2014).

7.4.3.4.2 Power Simulations Simulations can also be used to evaluate the power of the functional regression models for testing the interaction between two genes or genomic regions for a quantitative trait. A true quantitative genetic model is given as follows. Consider H pairs of quantitative trait loci (QTL) from two genes (genomic regions). Let Q_{h_1} and q_{h_1} be two alleles at the first QTL and Q_{h_2} and q_{h_2} be two alleles at the second QTL for the H pair of QTLs. Let u_{ijkl} be the genotypes of the uth individual with $ij = Q_{h_1}Q_{h_1}, Q_{h_1}q_{h_1}, q_{h_1}q_{h_1}$ and $kl = Q_{h_2}Q_{h_2}, Q_{h_2}q_{h_2}, q_{h_2}q_{h_2}$ and g_{uijkl} be its genotypic value. The following multiple linear regression is used as a genetic model for a quantitative trait:

$$y_u = \sum_{h=1}^{H} g_{uijkl}^{h} + \varepsilon_u, \quad u = 1,2,\ldots,n,$$

where

g_{uijkl}^{h} is a genotypic value of the hth pair of QTLs
ε_u is distributed as a standard normal distribution $N(0,1)$

Here the result is presented for the Dominant or Dominant (Dom ∪ Dom) interaction model. Readers who are interested in other interaction models such as Dominant and Dominant (Dom ∩ Dom), Recessive or Recessive (Rec ∪ Rec), and Threshold model are referred to the paper (Zhang et al. 2014). We assume that the parameter r varies from 0 to 1.

Figure 7.6 showed power curves of three statistics: the function regression, the regression on PCA, and the pair-wise interaction tests where permutations were used to adjust for multiple testing for testing interaction between two genomic regions that consist of rare variants for a quantitative trait as a function of the relative risk parameter r at the significance level $\alpha = 0.05$ under the Dom ∪ Dom model, assuming sample sizes of 2000. Figure 7.6 showed that the functional regression model had more power to detect interaction than the regression on PCA and traditional pair-wise interaction tests.

7.4.3.4.3 Real Example To illustrate the application of functional regression for testing the interaction, we present a real example. The dataset that included 2225 individuals of European origin was from the NHLBI's Exome Sequencing Project. The trait we considered was high-density lipoprotein (HDL). The logarithm of HDL was taken as a trait value. The total number of genes being tested for interactions, which included both common and rare variants, was 14,503. A P-value for declaring significant interaction after applying the

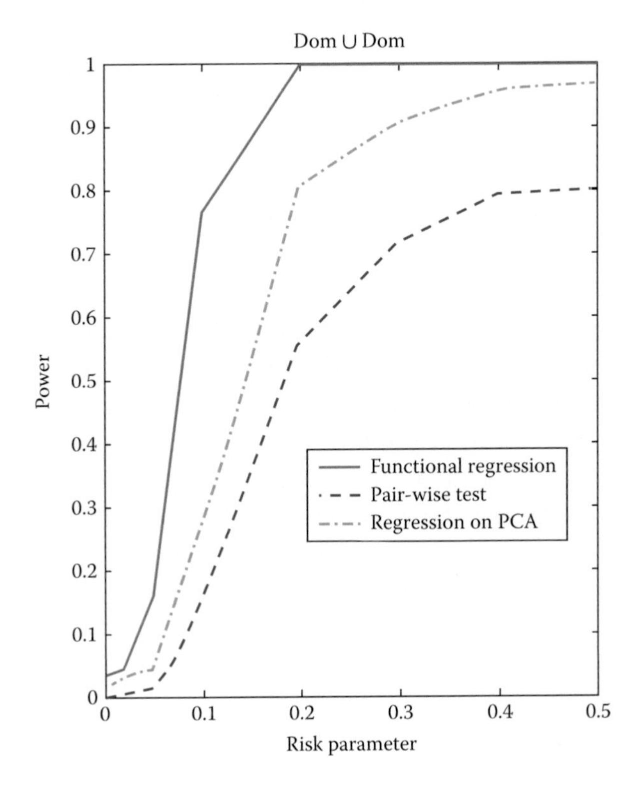

FIGURE 7.6 Power curves of three statistics: the function regression, the regression on PCA, and the pair-wise interaction tests.

Bonferroni correction for multiple tests was 4.75×10^{-10}. We identified 27 pairs of genes showed significant evidence of interaction with P-values $<4.58 \times 10^{-10}$ by the functional regression additive × additive interaction model (Zhang et al. 2014). We also identified 130 pairs of genes with P-values $<9.87 \times 10^{-9}$. The pair-wise interactions between rare and rare variants and between rare and common variants were often observed, but the significant pair-wise interactions between common and common variants were less observed. We also observed interacting genes formed an interaction network. A total of 31 genes—*GALNT2, RPA2, GALNT2, PCSK9, GALNT3, APOB, GCKR, HMGCR, MLXIPL, BAZ1B, TBL2, LPL, ABCA1, APOA5, APOA4, APOA1, FADS2, FADS3, MADD, FOLH1, MVK, MMAB, LIPC, CETP, CTCF, PRMT7, GALNT1, LIPG, LDLR, DNAH11*, and *APOE*—that were associated with serum lipid levels in recent GWAS (Aulchenko et al. 2009) were tested for additive × additive interactions. All 31 genes showed mild interaction with more than one gene (P-value range from 7.81×10^{-8} to 9.97×10^{-5}). The interacting genes formed networks (Figure 7.7) (Zhang et al. 2014).

7.4.4 Functional Regression Model for Interaction Analysis with Multiple Quantitative Traits

7.4.4.1 Model

Assume that n individuals are sampled. Let y_{ik}, $k = 1, 2, \ldots, K$, be the kth trait values of the ith individual. Consider two genomic regions $[a_1, b_1]$ and $[a_2, b_2]$. Let $x_i(t)$, $z_i(t)$, $x_i(s)$, and $z_i(s)$ be genotypic functions for additive and dominance effects of the ith individual defined in the regions $[a_1, b_1]$ and $[a_2, b_2]$, respectively, and defined as that in Section 7.4.3.1. Let $y_i = [y_{i1}, \ldots, y_{iK}]^T$ be the vector of the trait values measured on the ith individual. Let t and s be a genomic position in the first and second genomic regions, respectively.

The functional regression model for a quantitative trait can be defined as (Zhang et al. 2016)

$$\begin{aligned} y_{ik} = {} & \alpha_{0k} + \sum_{h=1}^{H} \upsilon_{ih}\lambda_{kh} + \int_T \alpha_k(t)x_i(t)dt + \int_S \beta_k(s)x_i(s)ds + \int_T \delta_k^1(t)z_i(t)dt + \int_S \delta_k^2(s)z_i(s)ds \\ & + \int_T\int_S \gamma_{AAk}(t,s)x_i(t)x_i(s)dtds + \int_T\int_S \gamma_{ADk}(t,s)x_i(t)z_i(s)dtds + \int_T\int_S \gamma_{DAk}(t,s)z_i(t)x_i(s)dtds \\ & + \int_T\int_S \gamma_{DDk}(t,s)z_i(t)z_i(s)dtds + \varepsilon_{ik}, \end{aligned} \tag{7.146}$$

where α_{0k} is an overall mean; υ_{ih} is a covariate; λ_{kh} is the regression coefficient associated with the covariate; $\alpha_k(t)$ and $\beta_k(s)$ are genetic additive effects of two putative QTLs located at the genomic positions t and s, respectively; $\delta_k^1(t)$ and $\delta_k^2(s)$ are genetic dominance effects of two putative QTLs located at the genomic positions t and s, respectively; $\gamma_{AAk}(t,s)$, $\gamma_{ADk}(t,s)$, $\gamma_{DAk}(t,s)$, and $\lambda_{DDk}(t,s)$ are the additive × additive, additive × dominance, dominance × additive, and dominance × dominance interaction effects between two putative QTLs located at the genomic positions t and s for the kth trait, $k = 1, \ldots, K$, respectively; and ε_{ik}

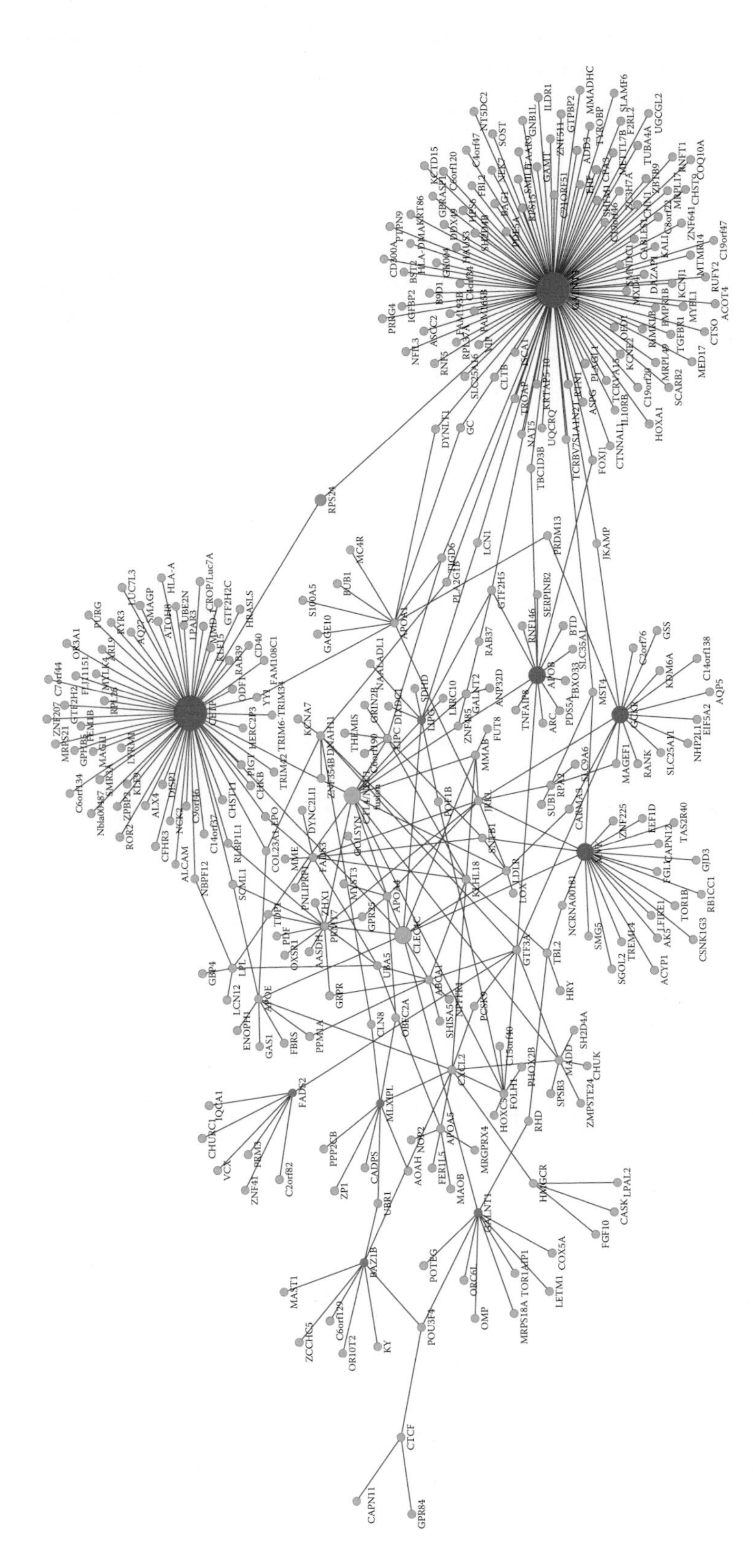

FIGURE 7.7 Networks of 384 pairs of mild interactions between 31 genes that influenced lipid levels, discovered in previous GWAS, and other genes in our analysis.

are independent and identically distributed normal variables with a mean of zero and covariance matrix Σ. Consider that covariates in the model (7.146) allow incorporating PCA scores for population stratification, sex, age, BMI, and other biomarkers into the model.

7.4.4.2 Parameter Estimation

We assume that both phenotypes and genotype profiles are centered. The genotype profiles $x_i(t)$, $x_i(s)$, $z_i(t)$, and $z_i(s)$ are expanded in terms of the orthonormal basis function as that in Equation 7.136. The larger the number of variants in the genes, the more accurate the eigenfunction expansion. If the number of variants is less than three, the eigenfunction expansion of the genotypic profiles is impossible. The multivariate functional regression (MFRG) can only be used for a gene with more than three variants.

Substituting an expansion into Equation 7.136 gives

$$y_{ik} = \alpha_{0k} + \sum_{h=1}^{H} \upsilon_{ih}\lambda_{kh} + \int_T \alpha_k(t)\sum_{j=1}^{\infty}\xi_{ij}\phi_j(t)dt + \int_S \beta_k(s)\sum_{l=1}^{\infty}\eta_{il}\psi_l(s)ds + \int_T \delta_k^1(t)\sum_{u=1}^{\infty}\zeta_{iu}\varphi_u(t)dt$$
$$+ \int_S \delta_k^2(s)\sum_{m=1}^{\infty}\tau_{im}\theta_m(s)ds + \int_T\int_S \gamma_{AAk}(t,s)\left(\sum_{j=1}^{\infty}\xi_{ij}\phi_j(t)\right)\left(\sum_{l=1}^{\infty}\eta_{il}\psi_l(s)\right)dtds +$$
$$+ \int_T\int_S \gamma_{ADk}(t,s)\sum_{j=1}^{\infty}\xi_{ij}\phi_j(t)\sum_{m=1}^{\infty}\tau_{im}\theta_m(s)dtds + \int_T\int_S \gamma_{DAk}(t,s)\sum_{u=1}^{\infty}\zeta_{iu}\varphi_u(t)\sum_{l=1}^{\infty}\eta_{il}\psi_l(s)dtds$$
$$+ \int_T\int_S \gamma_{DDk}(t,s)\sum_{u=1}^{\infty}\zeta_{iu}\varphi_u(t)\sum_{m=1}^{\infty}\tau_{im}\theta_m(s)dtds + \varepsilon_i. \tag{7.147}$$

After some algebra, Equation 7.147 can be further reduced to

$$y_{ik} = \alpha_{0k} + \sum_{h=1}^{H}\upsilon_{ih}\lambda_{kh} + \sum_{j=1}^{\infty}\xi_{ij}\alpha_{kj} + \sum_{l=1}^{\infty}\eta_{il}\beta_{kl} + \sum_{u=1}^{\infty}\zeta_{iu}\delta_{ku}^1 + \sum_{m=1}^{\infty}\tau_{im}\delta_{km}^2$$
$$+ \sum_{j=1}^{\infty}\sum_{l=1}^{\infty}\xi_{ij}\eta_{il}\gamma_{kjl}^{AA} + \sum_{j=1}^{\infty}\sum_{m=1}^{\infty}\xi_{ij}\tau_{im}\gamma_{kjm}^{AD} + \sum_{u=1}^{\infty}\sum_{l=1}^{\infty}\zeta_{iu}\eta_{il}\gamma_{kul}^{DA} + \sum_{u=1}^{\infty}\sum_{m=1}^{\infty}\zeta_{iu}\tau_{im}\gamma_{kum}^{DD} + \varepsilon_{ik}, \tag{7.148}$$

where

$$\alpha_{kj} = \int_T \alpha_k(t)\phi_j(t)dt,\quad \beta_{kl} = \int_S \beta_k(s)\psi_l(s)ds,\quad \delta_{ku}^1 = \int_T \delta_k^1(t)\varphi_u(t)dt,\quad \delta_{km}^2 = \int_S \delta_k^2(s)\theta_m(s)ds,$$
$$\gamma_{kjl}^{AA} = \int_T\int_S \gamma_{AAk}(t,s)\phi_j(t)\psi_l(s)dtds,\quad \gamma_{kjm}^{AD} = \int_T\int_S \gamma_{ADk}(t,s)\phi_j(t)\theta_m(s)dtds,$$
$$\gamma_{kul}^{DA} = \int_T\int_S \gamma_{DAk}(t,s)\varphi_u(t)\psi_l(s)dtds,\quad \gamma_{kum}^{DD} = \int_T\int_u \gamma_{DDk}(t,s)\varphi_u(t)\theta_m(s)dtds.$$

The parameters $\alpha_{kj}, \beta_{kl}, \delta^1_{ku}, \delta^2_{km}, \gamma^{AA}_{kjl}, \gamma^{AD}_{kjm}, \gamma^{DA}_{kul}$, and γ^{DD}_{kum} are referred to as genetic additive, dominance, and additive × additive, additive × dominance, dominance × additive, and dominance × dominance interaction effect scores for the kth trait. These scores can also be viewed as the expansion coefficients of the genetic effect functions with respect to orthonormal basis functions:

$$\alpha_k(t)=\sum_j \alpha_{kj}\phi_j(t),\quad \beta_k(s)=\sum_l \beta_{kl}\psi_l(s),\quad \delta^1_k(t)=\sum_u \delta^1_{ku}\varphi_u(t),\quad \delta^2_k(s)=\sum_m \delta^2_{km}\theta_m(s),$$

$$\gamma_{AAk}(t,s)=\sum_j\sum_l \gamma^{AA}_{kjl}\phi_j(t)\psi_l(s),\quad \gamma_{ADk}(t,s)=\sum_j\sum_m \gamma^{AD}_{kjm}\phi_j(t)\theta_m(s),$$

$$\gamma_{DAk}(t)=\sum_u\sum_l \gamma^{DA}_{kul}\varphi_u(t)\psi_l(s),\quad \text{and}\quad \gamma_{DDk}(t)=\sum_u\sum_m \gamma^{DD}_{kum}\varphi_u(t)\theta_m(s).$$

Equation 7.148 can be approximated by (Appendix 7C)

$$\begin{aligned} Y &= e\alpha_0+\upsilon\lambda+\xi\alpha+\eta\beta+\zeta\delta^1+\tau\delta^{2+}\Gamma\gamma+\varepsilon \\ &= WB+\varepsilon, \end{aligned} \tag{7.149}$$

where $W=\begin{bmatrix} e & \upsilon & \xi & \eta & \zeta & \tau & \Gamma \end{bmatrix}$ and $B=\begin{bmatrix} \alpha_0 \\ \lambda \\ \alpha \\ \beta \\ \delta^1 \\ \delta^2 \\ \gamma \end{bmatrix}$.

Therefore, we transform the original functional regression interaction model into the classical multivariate regression interaction model by eigenfunction expansions. Multivariate regression analysis can directly be used for solving problem (7.149). The standard least square estimators of B and the variance covariance matrix Σ are respectively given by

$$\hat{B}=\left(W^TW\right)^{-1}W^TY, \tag{7.150}$$

$$\hat{\Sigma}=\frac{1}{n}(Y-WB)^T(Y-WB). \tag{7.151}$$

Denote the last $(J+U)(L+M)$ row of the matrix $(W^TW)^{-1}W^T$ by A. Then, the estimator of the parameter γ is given by

$$\hat{\gamma}=AY. \tag{7.152}$$

The vector of matrix γ can be written as

$$\text{vec}(\hat{\gamma}) = (I \otimes A)\text{vec}(Y). \quad (7.153)$$

By the assumption of the variance matrix of Y, we obtain the variance matrix of vec(Y):

$$\text{var}(\text{vec}(Y)) = \Sigma \otimes I_n. \quad (7.154)$$

Thus, it follows from Equations 7.153 and 7.154 that

$$\begin{aligned} \Lambda = \text{var}(\text{vec}(\gamma)) &= (I \otimes A)(\Sigma \otimes I_n)(I \otimes A^T) \\ &= \Sigma \otimes (AA^T). \end{aligned} \quad (7.155)$$

7.4.4.3 Test Statistics

An essential problem in genetic interaction studies of the quantitative traits is to test the interaction between two genomic regions (or genes). Formally, we investigate the problem of testing the following hypothesis:

$$\gamma_{AAk}(t,s) = 0, \quad \gamma_{ADk}(t,s) = 0, \quad \gamma_{DAk}(t,s) = 0, \quad \gamma_{DDk}(t,s) = 0,$$
$$\forall t \in [a_1, b_1], \quad s \in [a_2, b_2], \quad k = 1,2,\ldots,K,$$

which is equivalent to testing the null hypothesis:

$$H_0 : \gamma = 0.$$

Define the test statistic for testing the interaction between two genomic regions, $[a_1, b_1]$ and $[a_2, b_2]$, with K quantitative traits as

$$T_I = (\text{vec}(\hat{\gamma})^T \Lambda^{-1} \text{vec}(\hat{\gamma}).$$

Then, under the null hypothesis $H_0: \gamma = 0$, T_I is asymptotically distributed as a central $\chi^2_{(K(J+U)(L+M))}$ distribution if $(J+U)(L+M)$ components are taken in the expansion Equation 7.136.

Other test statistics for multivariate regression analysis introduced in Chapter 5 can be applied here. For example, we can also develop likelihood-ratio-based statistics for testing interaction.

Setting $W = [W_1 \quad W_2]$, we can write the model as

$$E[Y] = W_1 \begin{bmatrix} \alpha \\ \beta \\ \delta^1 \\ \delta^2 \end{bmatrix} + W_2 \gamma.$$

Under $H_0 : \gamma = 0$, we have the model:

$$Y = W_1 \begin{bmatrix} \alpha \\ \beta \\ \delta^1 \\ \delta^2 \end{bmatrix} + \varepsilon.$$

The estimators will be

$$\begin{bmatrix} \hat{\alpha} \\ \hat{\beta} \\ \hat{\delta}^1 \\ \hat{\delta}^2 \end{bmatrix} = \left(W_1^T W_1\right)^{-1} W_1^T Y \quad \text{and} \quad \hat{\Sigma}_1 = \frac{1}{n}\left(Y - W_1 \begin{bmatrix} \hat{\alpha} \\ \hat{\beta} \\ \hat{\delta}^1 \\ \hat{\delta}^2 \end{bmatrix}\right)^T \left(Y - W_1 \begin{bmatrix} \hat{\alpha} \\ \hat{\beta} \\ \hat{\delta}^1 \\ \hat{\delta}^2 \end{bmatrix}\right).$$

The likelihood for the full model and reduced model are, respectively, given by

$$L\left(\hat{\alpha}, \hat{\beta}, \hat{\delta}^1, \hat{\delta}^2, \hat{\gamma}, \hat{\Sigma}\right) = \frac{e^{-nK/2}}{\left(2\pi\right)^{nK/2} \left|\hat{\Sigma}\right|^{n/2}}$$

and

$$L\left(\hat{\alpha}, \hat{\beta}, \hat{\delta}^1, \hat{\delta}^2, \hat{\Sigma}_1\right) = \frac{e^{-nK/2}}{\left(2\pi\right)^{nK/2} \left|\hat{\Sigma}_1\right|^{n/2}}.$$

The likelihood-ratio-based statistic for testing interaction between two genomic regions with multivariate traits is defined as

$$T_{I\Lambda} = -n \log\left(\frac{\left|\hat{\Sigma}\right|}{\left|\hat{\Sigma}_1\right|}\right).$$

Again, under the null hypothesis $H_0 : \gamma = 0$, $T_{I\Lambda}$ is asymptotically distributed as a central $\chi^2_{(K(J+U)(L+M))}$ distribution if $(J+U)(L+M)$ components are taken in the expansion Equation 7.136.

By similar arguments, we can also test individual additive × additive, additive × dominance, dominance × additive, and dominance × dominance interactions.

7.4.4.4 Simulations and Real Example Applications

For simplicity, throughout this section, we consider only additive effect and additive × additive interaction effect with multiple traits (Zhang et al. 2016).

7.4.4.4.1 Type 1 Error Rates To validate the null distribution of test statistics, we calculated the type 1 error rates for rare alleles. We assume the null model with marginal genetic effects (additive model) at both genes:

$$y_{ik} = \mu_k + \sum_{j=1}^{J} x_{ij}\alpha_{kj} + \sum_{l=1}^{L} z_{il}\beta_{kl} + \varepsilon_{ik},$$

where

$$x_{ij} = \begin{cases} 2(1-P_j) & A_jA_j \\ 1-2P_j & A_ja_j \\ -2P_j & a_ja_j \end{cases},\ z_{il} = \begin{cases} 2(1-q_l) & B_lB_l \\ 1-2q_l & B_lb_l \\ -2q_l & b_lb_l \end{cases},\ \alpha_{kj} = \alpha_k = (r_{pk}-1)f_0,\ \beta_{kl} = \beta_k = (r_{qk}-1)f_0,$$

P_j and q_l are frequencies of alleles A_j and B_l, respectively; r_{pk} and r_{qk} are risk parameters of the kth trait for the SNPs in the first and second genes, respectively, and randomly selected from 1.1 to 1.6; f_0 is a baseline penetrance and set to 1; and ε are defined as before, ε_i is distributed as

$$\begin{bmatrix} \varepsilon_1 & \cdots & \varepsilon_k \end{bmatrix} \sim N\left(\begin{bmatrix} 0 & \cdots & 0 \end{bmatrix}, \begin{pmatrix} 1 & \cdots & 0.5 \\ \vdots & \ddots & \vdots \\ 0.5 & \cdots & 1 \end{pmatrix} \right).$$

A total of 1,000,000 chromosomes by resampling from 2018 individuals with variants in five genes (*TNFRSF14, GBP3, KANK4, IQGAP3, GALNT2*) selected from the NHLBI's Exome Sequencing Project (ESP) were generated. We randomly selected 20% of SNPs as causal variants. The number of sampled individuals from populations of 1,000,000 chromosomes ranged from 1000 to 5000. A total of 5000 simulations were repeated. Table 7.3 presented average type 1 error rates of the test statistics for testing the interaction between two genes

TABLE 7.3 Average Type 1 Error Rates of the Statistic for Testing Interaction between Two Genes with Marginal Effects at Two Genes Consisting of Only Rare Variants with 10 Traits over 10 Pairs of Genes

Sample Size	0.05	0.01	0.001
1000	0.0604	0.0125	0.0158
2000	0.0512	0.0105	0.0114
3000	0.0486	0.0101	0.0011
4000	0.0477	0.0098	0.0011
5000	0.0458	0.0093	0.0009

with marginal effect at both genes consisting of only rare variants with 10 traits over 10 pairs of genes selected from the above five genes at the nominal levels $\alpha = 0.05$, $\alpha = 0.01$, and $\alpha = 0.001$. The statistics for testing interaction between two genomic regions have the similar type 1 error rates in the other two scenarios: with marginal genetic effect at one gene or without marginal genetic effects at two genes (data not shown). These results clearly showed that the type 1 error rates of the MFRG-based test statistics for testing interaction between two genes with multiple traits with or without marginal effects were not appreciably different from the nominal α levels.

7.4.4.4.2 Power Simulations To provide some guidance for interaction analysis with multiple traits, we used simulated data to estimate the power of the test statistics for detecting the interaction between two genes for two quantitative traits (Zhang et al. 2016). Consider H pairs of quantitative trait loci (QTL) from two genes (genomic regions). Let Q_{h_1} and q_{h_1} be two alleles at the first QTL and Q_{h_2} and q_{h_2} be two alleles at the second QTL for the H pair of QTLs. Let u_{ijkl} be the genotypes of the uth individual with $ij = Q_{h_1}Q_{h_1}, Q_{h_1}q_{h_1}, q_{h_1}q_{h_1}$ and $kl = Q_{h_2}Q_{h_2}, Q_{h_2}q_{h_2}, q_{h_2}q_{h_2}$ and $g_{mu_{ijkl}}$ be its genotypic value for the mth trait. The following multiple regression is used as a genetic model for the mth quantitative trait:

$$y_{mu} = \sum_{h=1}^{H} g_{mu_{ijkl}}^{h} + \varepsilon_{mu}, \quad u = 1,2,\ldots,n, \; m = 1,\ldots,M,$$

where $g_{mu_{ijkl}}^{h}$ is a genotypic value of the hth pair of QTLs for the mth quantitative trait and ε_{mu} are distributed as $\begin{bmatrix}\varepsilon_1 & \cdots & \varepsilon_m\end{bmatrix} \sim N\left(\begin{bmatrix}0 & \cdots & 0\end{bmatrix}, \begin{pmatrix} 1 & \cdots & 0.5 \\ \vdots & \ddots & \vdots \\ 0.5 & \cdots & 1 \end{pmatrix}\right)$.

Consider interactions model: (1) Dominant or Dominant (Table 7.4). Power simulations results for the other three models, (2) Dominant and Dominant, (3) Recessive or Recessive, and (4) Threshold model, were given in Zhang et al. (2016). A total of 1,000,000 individuals by resampling from 2,018 individuals of European origin with variants in two genes, *IQGAP3* and *ACTN2*, selected from the ESP dataset were generated. Twenty percent of the variants were selected as causal variants. A total of 2000 individuals were sampled from the generated dataset. A total of 1000 simulations were repeated for the power calculation.

The power of the proposed MFRG model is compared with the single trait functional regression (SFRG) model, the multivariate pair-wise interaction test, and the regression on principal components (PCs). Figure 7.8 presented the power of three statistics for testing the interaction between two genomic regions (or genes) with only rare variants as a function of sample sizes under the Dominant or Dominant model, assuming 20% of the risk is due to rare variants and the risk parameter $r = 0.05$ (Zhang et al. 2016). The results showed that (1) the functional regression model had more power than the multivariate point-wise regression model and (2) the power of tests with multiple traits was higher than that of single trait.

TABLE 7.4 The Interaction Models

Models	First Locus	Second Locus		
		$Q_{h_2}Q_{h_2}$	$Q_{h_2}q_{h_2}$	$q_{h_2}q_{h_2}$
Dominant or Dominant	$Q_{h_1}Q_{h_1}$	r	r	r
	$Q_{h_1}q_{h_1}$	r	r	r
	$q_{h_1}q_{h_1}$	r	r	0
Dominant and Dominant	$Q_{h_1}Q_{h_1}$	r	r	0
	$Q_{h_1}q_{h_1}$	r	r	0
	$q_{h_1}q_{h_1}$	0	0	0
Recessive or Recessive	$Q_{h_1}Q_{h_1}$	r	r	r
	$Q_{h_1}q_{h_1}$	r	0	0
	$q_{h_1}q_{h_1}$	r	0	0
Threshold	$Q_{h_1}Q_{h_1}$	r	r	0
	$Q_{h_1}q_{h_1}$	r	0	0
	$q_{h_1}q_{h_1}$	0	0	0

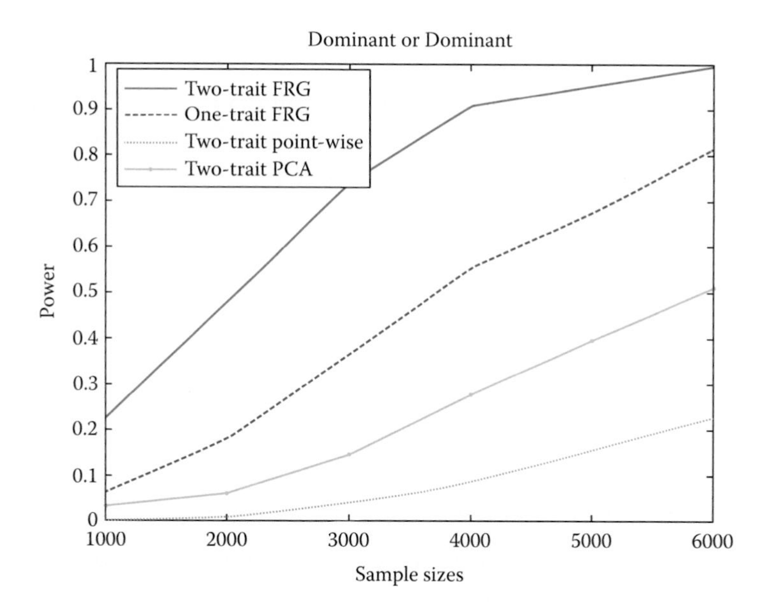

FIGURE 7.8 Power curves of three statistics for testing interaction of two genes with only rare variants under Dominant or Dominant as a function of sample sizes.

7.4.4.4.3 Real Example Application To further illustrate its applications, the multivariate functional regression models for testing additive × additive interaction were applied to data from the NHLBI's ESP Project. We consider five phenotypes: HDL, LDL, total cholesterol, SBP, and DBP. A total of 2018 individuals of European origin from 15 different cohorts in the ESP Project was used in analysis. The rank-based inverse normal transformation of the phenotypes was taken as trait values. The total number of genes tested for interactions,

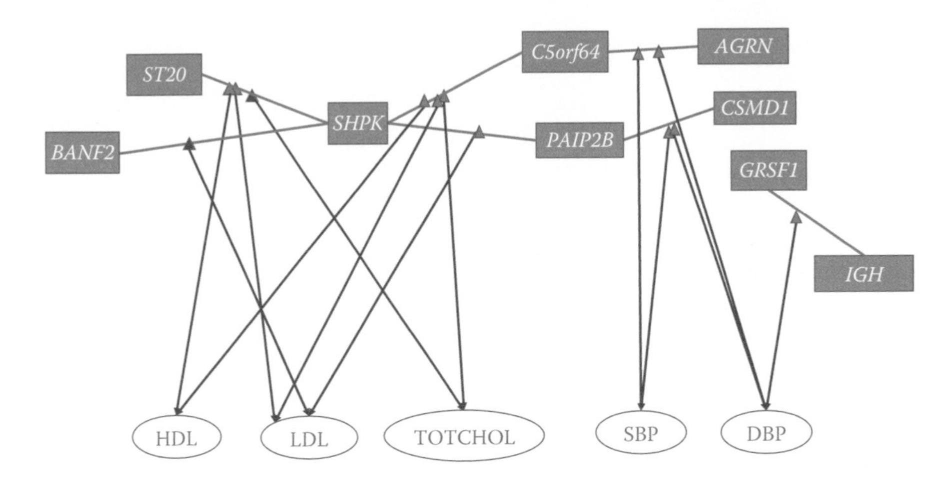

FIGURE 7.9 Observed 7 pairs of genes showing pleiotropic epistasis effects by univariate of epistasis analysis individually.

which included both common and rare variants, was 18,498. Therefore, a P-value for declaring significant interaction after applying the Bonferroni correction for multiple tests was 2.92×10^{-10}. A total of 267 pairs of genes, which were derived from 160 genes, showed significant evidence of epistasis influencing five traits with P-values $<1.96 \times 10^{-10}$ and formed a large interaction network (Zhang et al. 2016). Among 267 significantly interacted genes identified by joint analysis with five traits, we observed only one pair of genes: ST20 and SHPK showed epistasis influencing LDL at the genome-wide significance level by univariate analysis of epistasis with the LDL individually (P-value $<6.48 \times 10^{-11}$). However, if we release the significance level to $P < 5.0 \times 10^{-8}$, we observed seven pairs of genes showing pleiotropic epistasis effects by univariate of epistasis analysis individually (Figure 7.9). This demonstrated that although by each individual trait analysis they only showed mild evidence of epistasis, by simultaneous epistasis analysis of multiple correlated traits, the genes showed strong evidence of epistasis influencing multiple traits. The results imply that the genetic analysis of multiple traits can reveal the complicated genetic structures of the complex traits, which may be missed by univariate genetic analysis.

7.5 MULTIVARIATE AND FUNCTIONAL CANONICAL CORRELATION AS A UNIFIED FRAMEWORK FOR TESTING FOR GENE–GENE AND GENE–ENVIRONMENT INTERACTION FOR BOTH QUALITATIVE AND QUANTITATIVE TRAITS

In Sections 4.2.2 and 5.4, we introduced canonical correlation analysis (CCA) as a statistical framework for testing the association of a gene or genomic region with a quantitative trait or multiple quantitative traits. In this section, we will extend CCA from association analysis to interaction analysis. We will introduce CCA as a unified framework for testing gene–gene and gene–environment interaction for both quantitative and qualitative traits with both common and rare variants. As we previously pointed out, linear regression analysis can be viewed as specific CCA.

7.5.1 Data Structure of CCA for Interaction Analysis

Consider two genomic regions. There are p SNPs and q SNPs in the first and second genomic regions, respectively. Assume that n individuals are sampled. For the ith individual, two vectors of the genetic variation data (genotypes or functional principal component scores) for the first and second genomic regions are denoted by $x_i = [x_{i1}, \ldots, x_{ip}]$ and $z_i = [z_{i1}, \ldots, z_{iq}]$, respectively. Let $\xi_i = [x_{i1}z_{i1}, \ldots, x_{i1}z_{iq}, \ldots, x_{ip}z_{i1}, \ldots, x_{ip}z_{iq}]$. The single-variate regression (for a single quantitative trait), multivariate regression (for multiple quantitative traits), and logistic regression (for a qualitative trait) will be used to preprocess the genotype data for removing the genetic main effects of two genomic regions before taking their residuals for interaction analysis. We start with interaction analysis of a single quantitative trait.

7.5.1.1 Single Quantitative Trait

Before performing single quantitative trait interaction analysis, we first regress the single trait on the genotypes of two genomic regions. Let y_i be a trait value of the ith individual. Consider a regression model:

$$y_i = \mu + \sum_{m=1}^{M} \tau_{im}\theta_m + \sum_{j=1}^{p} x_{ij}\alpha_j + \sum_{l=1}^{q} z_{il}\beta_l + \varepsilon_i, \tag{7.156}$$

where

- μ is an overall mean
- τ_{im} are covariates such as age, sex, and principal component (PC) scores for removing the impact of population structure
- θ_m are their corresponding regression coefficients
- α_j and β_l are genetic main effects for the first and second genomic regions, respectively

After the model fits the data, we calculate the residual for each individual:

$$\eta_i = y_i - \hat{\mu} - \sum_{m=1}^{M} \tau_{im}\hat{\theta}_m - \sum_{j=1}^{p} x_{ij}\hat{\alpha}_j - \sum_{l=1}^{q} z_{il}\hat{\beta}_l. \tag{7.157}$$

Define a vector of a residual and a genetic data matrix:

$$\eta = \begin{bmatrix} \eta_1 \\ \vdots \\ \eta_m \end{bmatrix} \text{ and } \xi = \begin{bmatrix} \xi_{11} & \cdots & \xi_{1pq} \\ \vdots & \vdots & \vdots \\ \xi_{n1} & \vdots & \xi_{npq} \end{bmatrix}. \tag{7.158}$$

7.5.1.2 Multiple Quantitative Trait

Consider K traits. Let y_{ik} be the kth trait values of the ith individual. The multivariate regression model is

$$y_{ik} = \mu_k + \sum_{m=1}^{M} \tau_{im}\theta_{mk} + \sum_{j=1}^{J_1} x_{ij}\alpha_{kj} + \sum_{l=1}^{J_2} z_{il}\beta_{kl} + \varepsilon_{ik}, \tag{7.159}$$

where

- μ_k is an overall mean of the kth trait
- θ_{mk} is the regression coefficient associated with the covariate τ_{im}
- α_{kj} is the main genetic additive effect of the jth genetic variant in the first genomic region for the kth trait
- β_{kl} is the main genetic additive effect of the lth genetic variant in the second genomic region for the kth trait

Let $\hat{y}_{ik}$ be the predicted value by the fitted model:

$$\hat{y}_{ik} = \hat{\mu}_k + \sum_{m=1}^{M} \tau_{im}\hat{\theta}_{mk} + \sum_{j=1}^{J_1} x_{ij}\hat{\alpha}_{kj} + \sum_{l=1}^{J_2} z_{il}\hat{\beta}_{kl}.$$

The residual is defined as

$$\eta_{ik} = y_{ik} - \hat{y}_{ik}.$$

Define the residual matrix:

$$\eta = \begin{bmatrix} \eta_{11} & \cdots & \eta_{1K} \\ \vdots & \vdots & \\ \eta_{n1} & \cdots & \eta_{nK} \end{bmatrix}.$$

Data matrix ξ is defined as before.

7.5.1.3 A Qualitative Trait

To investigate the interaction analysis for a qualitative trait using CCA, we will use logistic regression to preprocess data. Consider a logistic regression model:

$$\log \frac{\pi_i}{1-\pi_i} = \mu + \sum_{m=1}^{M} \tau_{im}\theta_m + \sum_{j=1}^{p} x_{ij}\alpha_j + \sum_{l=1}^{q} z_{il}\beta_l, \tag{7.160}$$

where π_i is the probability of the ith individual being affected and other parameters are defined as in Equation 7.156.

After we fit the data to the logistic regression model (7.160), we define the residual:

$$\eta_i = \log\frac{\pi_i}{1-\pi_i} - \hat{\mu} - \sum_{m=1}^{M}\tau_{im}\hat{\theta}_m - \sum_{j=1}^{p}x_{ij}\hat{\alpha}_j - \sum_{l=1}^{q}z_{il}\hat{\beta}_l. \tag{7.161}$$

Finally, we define the vector of residuals:

$$\eta = \begin{bmatrix} \eta_1 \\ \vdots \\ \eta_m \end{bmatrix}.$$

The genetic data matrix ξ is defined as before.

7.5.2 CCA and Functional CCA

For the convenience, we assume that $K \le pq$. Define the covariance matrix:

$$\Sigma = \begin{bmatrix} \Sigma_{\xi\xi} & \Sigma_{\xi\eta} \\ \Sigma_{\eta\xi} & \Sigma_{\eta\eta} \end{bmatrix}.$$

Matrix Σ will be estimated by

$$\hat{\Sigma} = \begin{bmatrix} \hat{\Sigma}_{\xi\xi} & \hat{\Sigma}_{\xi\eta} \\ \hat{\Sigma}_{\eta\xi} & \hat{\Sigma}_{\eta\eta} \end{bmatrix} = \frac{1}{n}\begin{bmatrix} \xi^T\xi & \xi^T\eta \\ \eta^T\xi & \eta^T\eta \end{bmatrix}. \tag{7.162}$$

The solution to CCA starts with defining the R^2 matrix (Equation 1.221):

$$R^2 = \hat{\Sigma}_{\eta\eta}^{-1/2}\hat{\Sigma}_{\eta\xi}\hat{\Sigma}_{\xi\xi}^{-1}\hat{\Sigma}_{\xi\eta}\hat{\Sigma}_{\eta\eta}^{-1/2}. \tag{7.163}$$

Let

$$W = \hat{\Sigma}_{\xi\xi}^{-1/2}\hat{\Sigma}_{\xi\eta}\hat{\Sigma}_{\eta\eta}^{-1/2}. \tag{7.164}$$

Suppose that the singular value decomposition (SVD) of the matrix is given by

$$W = U\Lambda V^T, \tag{7.165}$$

where $\Lambda = \text{diag}(\lambda_1, \ldots, \lambda_d)$ and $d = \min(K, pq)$. It is clear that

$$W^T W = R^2 = V\Lambda^2 V^T. \tag{7.166}$$

The matrices of canonical covariates are defined as

$$A = \Sigma_{\xi\xi}^{-1/2} U, \\ B = \Sigma_{\eta\eta}^{-1/2} V. \tag{7.167}$$

The vector of canonical correlations is

$$CC = \left[\lambda_1, \ldots, \lambda_d\right]^T. \tag{7.168}$$

Canonical correlations between the interaction terms and phenotypes measure the strength of the interaction. The CCA produces multiple canonical correlations. But we wish to use a single number to measure the interaction. We propose to use the summation of the square of the singular values as a measure to quantify the interaction:

$$r = \sum_{i=1}^{d} \lambda_i^2 = \mathrm{Tr}\left(\Lambda^2\right) = \mathrm{Tr}\left(R^2\right). \tag{7.169}$$

To test the interaction between two genomic regions is equivalent to testing independence between ξ and η or testing the hypothesis that each variable in the set ξ is uncorrelated with each variable in the set η. The null hypothesis of no interaction can be formulated as

$$H_0 : \Sigma_{\xi\eta} = 0.$$

The likelihood ratio for testing $H_0 : \Sigma_{\xi\eta} = 0$ is

$$\Lambda_r = \frac{\Sigma|}{\Sigma_{\xi\xi}\left|\Sigma_{\eta\eta}\right|} = \prod_{i=1}^{d}\left(1 - \lambda_i^2\right), \tag{7.170}$$

which is equal to the Wilks' lambda Λ defined in the multivariate linear regression model in Chapter 5.

This demonstrates that testing for interaction using multivariate linear regression can be treated as special case of CCA.

We usually define the likelihood ratio test statistic for testing the interaction as

$$T_{CCA} = -N \sum_{i=1}^{d} \log\left(1 - \lambda_i^2\right). \tag{7.171}$$

For small λ_i^2, T_{CCA} can be approximated by $N \sum_{i=1}^{d} \lambda_i^2 = Nr$, where r is the measure of interaction between two genomic regions. The stronger the interaction, the higher the power with which the test statistic can test the interaction.

Under the null hypothesis $H_0:\Sigma_{\xi\eta}=0$, T_{CCA} is asymptotically distributed as a central χ^2_{Kpq}. When the sample size is large, Bartlett (1939) suggests using the following statistic for hypothesis testing:

$$T_{CCA} = -\left[N - \frac{(d+3)}{2}\right]\sum_{i=1}^{d}\log\left(1-\lambda_i^2\right). \tag{7.172}$$

If the functional principal component scores are taken as genetic variants in matrix ξ, then the multivariate CCA becomes functional CCA. All previous discussion for the multivariate CCA can be applied to functional CCA.

7.5.3 Kernel CCA

Kernel CCA is a nonlinear extension of canonical correlation analysis with positive definite kernels (Fukumizu et al. 2007). In Section 5.4.2, we study the kernel CCA. Here, we adapt it to the interaction analysis. Define two kernels, $K(.,\eta)$ and $K(.,\xi)$, and two kernel matrices:

$$K_\eta = \begin{bmatrix} K(\eta_1,\eta_1) & \cdots & K(\eta_1,\eta_n) \\ \vdots & \vdots & \vdots \\ K(\eta_n,\eta_1) & \cdots & K(\eta_n,\eta_n) \end{bmatrix} \quad \text{and} \quad K_\xi = \begin{bmatrix} K(\xi_1,\xi_1) & \cdots & K(\xi_1,\xi_n) \\ \vdots & \vdots & \vdots \\ K(\xi_n,\xi_1) & \cdots & K(\xi_n,\xi_n) \end{bmatrix}.$$

Let $H = I_n - \frac{1}{n}\mathbf{1}\mathbf{1}^T$, $\mathbf{1}=[1,1,\ldots,1]^T$, I_n be an $n \times n$ dimensional identity matrix. Define the centralized kernel matrices:

$$\tilde{K}_\xi = HK_\xi H \quad \text{and} \quad \tilde{K}_\eta = HK_\eta H.$$

Define the corresponding kernel matrix of Σ:

$$K_\Sigma = \begin{bmatrix} \tilde{K}_\xi\tilde{K}_\xi & \tilde{K}_\xi\tilde{K}_\eta \\ \tilde{K}_\eta\tilde{K}_\xi & \tilde{K}_\eta\tilde{K}_\eta \end{bmatrix}. \tag{7.173}$$

Define the R^2 matrix:

$$R^2 = \left(\tilde{K}_\eta\tilde{K}_\eta\right)^{-1/2}\tilde{K}_\eta\tilde{K}_\xi\left(\tilde{K}_\xi\tilde{K}_\xi\right)^{-1}\left(\tilde{K}_\xi\tilde{K}_\eta\right)\left(\tilde{K}_\eta\tilde{K}_\eta\right)^{-1/2}. \tag{7.174}$$

Let

$$W = \left(\tilde{K}_\xi\tilde{K}_\xi\right)^{-1/2}\left(\tilde{K}_\xi\tilde{K}_\eta\right)\left(\tilde{K}_\eta\tilde{K}_\eta\right)^{-1/2}. \tag{7.175}$$

The SVD of the matrix

$$W = U\Lambda V^T, \tag{7.176}$$

where $\Lambda = \text{diag}(\lambda_1, \ldots, \lambda_d)$ and $d = \min(K, pq)$.

Then, the above discussion in Section 7.5.2 can be applied here. The test statistic is defined as

$$T_{CCA} = -N\sum_{i=1}^{d}\log\left(1-\lambda_i^2\right), \quad (7.177)$$

or

$$T_{CCA} = -\left[N-\frac{(d+3)}{2}\right]\sum_{i=1}^{d}\log\left(1-\lambda_i^2\right). \quad (7.178)$$

Under the null hypothesis of no interaction, the statistic T_{CCA} is asymptotically distributed as a central χ^2_{Kpq}.

SOFTWARE PACKAGE

The paper (Koo et al. 2015) listed a number of software package for gene–gene and gene–environment interaction analysis. Software based on functional regression models for a single trait and multiple traits can be found in the website https://sph.uth.edu/hgc/faculty/xiong/software-A.html.

APPENDIX 7A: VARIANCE OF LOGARITHM OF ODDS RATIO

We first consider individuals in controls. Let $\theta = \left[P_{11}^N, P_{12}^N, P_{21}^N, P_{22}^N\right]^T$. Using multinomial distribution, we can obtain the variance–covariance matrix of θ:

$$\Sigma = \frac{1}{n_G}\left\{\begin{bmatrix} \frac{1}{P_{11}^N} & 0 & 0 & 0 \\ 0 & \frac{1}{P_{12}^N} & 0 & 0 \\ 0 & 0 & \frac{1}{P_{21}^N} & 0 \\ 0 & 0 & 0 & \frac{1}{P_{22}^N} \end{bmatrix} - \begin{bmatrix} P_{11}^N \\ P_{12}^N \\ P_{21}^N \\ P_{22}^N \end{bmatrix}\begin{bmatrix} P_{11}^N & P_{12}^N & P_{21}^N & P_{22}^N \end{bmatrix}\right\}.$$

Let $h(\theta) = \log\frac{P_{11}^N P_{22}^N}{P_{12}^N P_{21}^N} = \log P_{11}^N + \log P_{22}^N - \log P_{12}^N - \log P_{21}^N$. Then, we have

$$\frac{\partial h(\theta)}{\partial\theta} = \left[\frac{1}{P_{11}^N}, -\frac{1}{P_{12}^N}, -\frac{1}{P_{21}^N}, \frac{1}{P_{22}^N}\right]^T.$$

Note that

$$\begin{bmatrix} p_{11}^N & p_{12}^N & p_{21}^N & p_{22}^N \end{bmatrix} \frac{\partial h(\theta)}{\partial \theta} = 1 - 1 - 1 + 1 = 0. \tag{7A.1}$$

According to the delta method, the variance of $h(\theta)$ can be approximated by

$$\begin{aligned}
\operatorname{Var}\left(h\left(\hat{\theta}\right)\right) &= \frac{\partial h^T(\theta)}{\partial \theta} \Sigma \frac{\partial h(\theta)}{\partial \theta} \\
&= \frac{1}{n_G} \begin{bmatrix} \frac{1}{P_{11}^N} & -\frac{1}{P_{12}^N} & -\frac{1}{P_{21}^N} & \frac{1}{P_{22}^N} \end{bmatrix} \begin{bmatrix} \frac{1}{P_{11}^N} & 0 & 0 & 0 \\ 0 & \frac{1}{P_{12}^N} & 0 & 0 \\ 0 & 0 & \frac{1}{P_{21}^N} & 0 \\ 0 & 0 & 0 & \frac{1}{P_{22}^N} \end{bmatrix} \begin{bmatrix} \frac{1}{P_{11}^N} \\ -\frac{1}{P_{12}^N} \\ -\frac{1}{P_{21}^N} \\ \frac{1}{P_{22}^N} \end{bmatrix} - 0 \\
&= \frac{1}{n_G} \begin{bmatrix} 1 & -1 & -1 & 1 \end{bmatrix} \begin{bmatrix} \frac{1}{P_{11}^N} \\ -\frac{1}{P_{12}^N} \\ -\frac{1}{P_{21}^N} \\ \frac{1}{P_{22}^N} \end{bmatrix} \\
&= \frac{1}{n_G} \left(\frac{1}{P_{11}^N} + \frac{1}{P_{12}^N} + \frac{1}{P_{21}^N} + \frac{1}{P_{22}^N} \right).
\end{aligned}$$

Similarly, we have

$$\operatorname{Var}\left(\log \frac{P_{11}^A P_{22}^A}{P_{12}^A P_{21}^A} \right) = \frac{1}{n_A} \left(\frac{1}{P_{11}^A} + \frac{1}{P_{12}^A} + \frac{1}{P_{21}^A} + \frac{1}{P_{22}^A} \right).$$

Therefore, combining the above two equations, we obtain

$$\hat{V}_{Rmultiple} = \frac{1}{n_A} \left(\frac{1}{P_{11}^A} + \frac{1}{P_{12}^A} + \frac{1}{P_{21}^A} + \frac{1}{P_{22}^A} \right) + \frac{1}{n_G} \left(\frac{1}{P_{11}^N} + \frac{1}{P_{12}^N} + \frac{1}{P_{21}^N} + \frac{1}{P_{22}^N} \right).$$

APPENDIX 7B: HAPLOTYPE ODDS-RATIO INTERACTION MEASURE

Note that

$$P\left(D=1\middle|G_1^1G_2^1\right)=\frac{P\left(D=1,G_1^1G_2^1\right)}{P\left(G_1^1G_2^1\right)}=\frac{P(D=1)P_{11}^A}{P_{11}}. \tag{7B.1}$$

Similarly, we have

$$P\left(D=0\middle|G_1^1G_2^1\right)=\frac{P(D=0)P_{11}^N}{P_{11}}, \tag{7B.2}$$

$$P\left(D=1\middle|G_1^2G_2^2\right)=\frac{P(D=1)P_{22}^A}{P_{22}}, \tag{7B.3}$$

$$P\left(D=0\middle|G_1^2G_2^2\right)=\frac{P(D=0)P_{22}^N}{P_{22}}. \tag{7B.4}$$

Combining Equations 7B.1 through 7B.4, the odds-ratio $\mathrm{OR}_{G_1^1G_2^1}$ is reduced to

$$\mathrm{OR}_{G_1^1G_2^1}=\frac{P_{11}^AP_{22}^N}{P_{11}^NP_{22}^A}. \tag{7B.5}$$

By the similar arguments, we obtain

$$\mathrm{OR}_{G_1^1G_2^2}=\frac{P_{12}^AP_{22}^N}{P_{12}^NP_{22}^A}, \tag{7B.6}$$

$$\mathrm{OR}_{G_1^2G_2^1}=\frac{P_{21}^AP_{22}^N}{P_{21}^NP_{22}^A}. \tag{7B.7}$$

Substituting Equations 7B.5 through 7B.7 into Equation 7.73, finally we have

$$\begin{aligned} I_{G_1G_2}^H &= \log\frac{P_{11}^AP_{22}^N}{P_{11}^NP_{22}^A}-\log\frac{P_{12}^AP_{22}^N}{P_{12}^NP_{22}^A}-\log\frac{P_{21}^AP_{22}^N}{P_{21}^NP_{22}^A} \\ &= \log\frac{P_{11}^AP_{22}^A}{P_{12}^AP_{21}^A}-\log\frac{P_{11}^NP_{22}^N}{P_{12}^NP_{21}^N}. \end{aligned}$$

APPENDIX 7C: PARAMETER ESTIMATION FOR MULTIVARIATE FUNCTIONAL REGRESSION MODEL

Substituting expansion into Equation 7.136 gives

$$y_{ik}=\alpha_{0k}+\sum_{h=1}^{H}\upsilon_{ih}\lambda_{kh}+\int_T\alpha_k(t)\sum_{j=1}^{\infty}\xi_{ij}\phi_j(t)dt+\int_S\beta_k(s)\sum_{l=1}^{\infty}\eta_{il}\psi_l(s)ds+\int_T\delta_k^1(t)\sum_{u=1}^{\infty}\zeta_{iu}\varphi_u(t)dt$$
$$+\int_S\delta_k^2(s)\sum_{m=1}^{\infty}\tau_{im}\theta_m(s)ds+\int_T\int_S\gamma_{AAk}(t,s)\left(\sum_{j=1}^{\infty}\xi_{ij}\phi_j(t)\right)\left(\sum_{l=1}^{\infty}\eta_{il}\psi_l(s)\right)dtds+$$
$$+\int_T\int_S\gamma_{ADk}(t,s)\sum_{j=1}^{\infty}\xi_{ij}\phi_j(t)\sum_{m=1}^{\infty}\tau_{im}\theta_m(s)dtds+\int_T\int_S\gamma_{DAk}(t,s)\sum_{u=1}^{\infty}\zeta_{iu}\varphi_u(t)\sum_{l=1}^{\infty}\eta_{il}\psi_l(s)dtds$$
$$+\int_T\int_S\gamma_{DDk}(t,s)\sum_{u=1}^{\infty}\zeta_{iu}\varphi_u(t)\sum_{m=1}^{\infty}\tau_{im}\theta_m(s)dtds+\varepsilon_i. \tag{7C.1}$$

After some algebra, Equation 7C.1 can be further reduced to

$$y_{ik}=\alpha_{0k}+\sum_{h=1}^{H}\upsilon_{ih}\lambda_{kh}+\sum_{j=1}^{\infty}\xi_{ij}\alpha_{kj}+\sum_{l=1}^{\infty}\eta_{il}\beta_{kl}+\sum_{u=1}^{\infty}\zeta_{iu}\delta_{ku}^1+\sum_{m=1}^{\infty}\tau_{im}\delta_{km}^2$$
$$+\sum_{j=1}^{\infty}\sum_{l=1}^{\infty}\xi_{ij}\eta_{il}\gamma_{kjl}^{AA}+\sum_{j=1}^{\infty}\sum_{m=1}^{\infty}\xi_{ij}\tau_{im}\gamma_{kjm}^{AD}+\sum_{u=1}^{\infty}\sum_{l=1}^{\infty}\zeta_{iu}\eta_{il}\gamma_{kul}^{DA}+\sum_{u=1}^{\infty}\sum_{m=1}^{\infty}\zeta_{iu}\tau_{im}\gamma_{kum}^{DD}+\varepsilon_{ik}, \tag{7C.2}$$

where

$$\alpha_{kj}=\int_T\alpha_k(t)\phi_j(t)dt,\ \beta_{kl}=\int_S\beta_k(s)\psi_l(s)ds,\ \delta_{ku}^1=\int_T\delta_k^1(t)\varphi_u(t)dt,\ \delta_{km}^2=\int_S\delta_k^2(s)\theta_m(s)ds,$$
$$\gamma_{kjl}^{AA}=\int_T\int_S\gamma_{AAk}(t,s)\phi_j(t)\psi_l(s)dtds,\quad \gamma_{kjm}^{AD}=\int_T\int_S\gamma_{ADk}(t,s)\phi_j(t)\theta_m(s)dtds,$$
$$\gamma_{kul}^{DA}=\int_T\int_S\gamma_{DAk}(t,s)\varphi_u(t)\psi_l(s)dtds,\quad \gamma_{kum}^{DD}=\int_T\int_u\gamma_{DDk}(t,s)\varphi_u(t)\theta_m(s)dtds.$$

The parameters $\alpha_{kj},\beta_{kl},\delta_{ku}^1,\delta_{km}^2,\gamma_{kjl}^{AA},\gamma_{kjm}^{AD},\gamma_{kul}^{DA}$, and γ_{kum}^{DD} are referred to as genetic additive, dominance, additive × additive, additive × dominance, dominance × additive, and dominance × dominance interaction effect scores for the kth trait. These scores can also be viewed as the expansion coefficients of the genetic effect functions with respect to orthonormal basis functions:

$$\alpha_k(t)=\sum_j\alpha_{kj}\phi_j(t),\quad \beta_k(s)=\sum_l\beta_{kl}\psi_l(s),\quad \delta_k^1(t)=\sum_u\delta_{ku}^1\varphi_u(t),\quad \delta_k^2(s)=\sum_m\delta_{km}^2\theta_m(s),$$
$$\gamma_{AAk}(t,s)=\sum_j\sum_l\gamma_{kjl}^{AA}\phi_j(t)\psi_l(s),\quad \gamma_{ADk}(t,s)=\sum_j\sum_m\gamma_{kjm}^{AD}\phi_j(t)\theta_m(s),$$
$$\gamma_{DAk}(t)=\sum_u\sum_l\gamma_{kul}^{DA}\varphi_u(t)\psi_l(s),\quad\text{and}\quad \gamma_{DDk}(t)=\sum_u\sum_m\gamma_{kum}^{DD}\varphi_u(t)\theta_m(s).$$

Let

$$Y=[Y_1,\ldots,Y_K]=\begin{bmatrix} Y_{11} & \cdots & Y_{1K} \\ \vdots & \ddots & \vdots \\ Y_{n1} & \cdots & Y_{nK} \end{bmatrix},\quad e=\begin{bmatrix} 1 \\ \vdots \\ 1 \end{bmatrix},\quad \alpha_0=[\alpha_{01},\ldots,\alpha_{0K}],\quad \upsilon=\begin{bmatrix} \upsilon_{11} & \cdots & \upsilon_{1M} \\ \vdots & \ddots & \vdots \\ \upsilon_{n1} & \cdots & \upsilon_{nM} \end{bmatrix},$$

$$\lambda_k=\begin{bmatrix} \lambda_{k1} \\ \vdots \\ \lambda_{kM} \end{bmatrix},\quad \lambda=[\lambda_1,\ldots,\lambda_K],\quad \xi=\begin{bmatrix} \xi_{11} & \cdots & \xi_{1J} \\ \vdots & \ddots & \vdots \\ \xi_{n1} & \cdots & \xi_{nJ} \end{bmatrix},\quad \eta=\begin{bmatrix} \eta_{11} & \cdots & \eta_{1L} \\ \vdots & \ddots & \vdots \\ \eta_{n1} & \cdots & \eta_{nL} \end{bmatrix},\quad \xi_i=\begin{bmatrix} \xi_{i1} \\ \vdots \\ \xi_{iJ} \end{bmatrix},$$

$$\zeta=\begin{bmatrix} \zeta_{11} & \cdots & \zeta_{1U} \\ \vdots & \ddots & \\ \zeta_{n1} & \cdots & \zeta_{nU} \end{bmatrix},\quad \tau=\begin{bmatrix} \tau_{11} & \cdots & \tau_{1M} \\ \vdots & \ddots & \vdots \\ \tau_{n1} & \cdots & \tau_{nM} \end{bmatrix},\quad \zeta_i=\begin{bmatrix} \zeta_{i1} \\ \vdots \\ \zeta_{iU} \end{bmatrix},\quad \tau_i=\begin{bmatrix} \tau_{i1} \\ \vdots \\ \tau_{iM} \end{bmatrix},\quad \eta_i=\begin{bmatrix} \eta_{i1} \\ \vdots \\ \eta_{iL} \end{bmatrix},$$

$$\Gamma_{AA}=\begin{bmatrix} \xi_1^T\otimes\eta_1^T \\ \vdots \\ \xi_n^T\otimes\eta_n^T \end{bmatrix}=\begin{bmatrix} \xi_{11}\eta_{11} & \cdots & \xi_{11}\eta_{1L} & \cdots & \xi_{1J}\eta_{11} & \cdots & \xi_{1J}\eta_{1L} \\ \cdots\ \cdots & \cdots & & \cdots & & \cdots\ \cdots & \cdots \\ \xi_{n1}\eta_{n1} & \cdots & \xi_{n1}\eta_{nL} & \cdots & \xi_{nJ}\eta_{n1} & \cdots & \xi_{nJ}\eta_{nL} \end{bmatrix},\quad \Gamma_{AD}=\begin{bmatrix} \xi_1^T\otimes\tau_1^T \\ \vdots \\ \xi_n^T\otimes\tau_n^T \end{bmatrix},$$

$$\Gamma_{DA}=\begin{bmatrix} \zeta_1^T\otimes\eta_1^T \\ \vdots \\ \zeta_n^T\otimes\eta_n^T \end{bmatrix},\quad \Gamma_{DD}=\begin{bmatrix} \zeta_1^T\otimes\tau_1^T \\ \vdots \\ \zeta_n^T\otimes\tau_n^T \end{bmatrix},\quad \Gamma=[\Gamma_{AA},\Gamma_{AD},\Gamma_{DA},\Gamma_{DD}],$$

$$\gamma_{AAk}=[\gamma_{k11}^{AA},\ldots,\gamma_{k1L}^{AA},\ldots,\gamma_{kJ1}^{AA},\ldots,\gamma_{kJL}^{AA}]^T,\quad \gamma_{ADk}=[\gamma_{k11}^{AD},\ldots,\gamma_{k1M}^{AD},\ldots,\gamma_{kJ1}^{AD},\ldots,\gamma_{kJM}^{AD}]^T,$$

$$\gamma_{DAk}=[\gamma_{k11}^{DA},\ldots,\gamma_{k1L}^{DA},\ldots,\gamma_{kU1}^{DA},\ldots,\gamma_{kUL}^{DA}]^T,\quad \gamma_{DDk}=[\gamma_{k11}^{DD},\ldots,\gamma_{k1M}^{DD},\ldots,\gamma_{kU1}^{DD},\ldots,\gamma_{kUM}^{DD}]^T,$$

$$\gamma_k=[\gamma_{AAk}^T,\gamma_{ADk}^T,\gamma_{DAk}^T,\gamma_{DDk}^T]^T,\quad \gamma=\begin{bmatrix}\gamma_1 & \cdots & \gamma_K\end{bmatrix},\quad \delta_k^1=\begin{bmatrix} \delta_{k1}^1 \\ \vdots \\ \delta_{kU}^1 \end{bmatrix},\quad \delta^1=[\delta_1^1,\ldots,\delta_K^1],$$

$$\alpha_k=\begin{bmatrix} \alpha_{k1} \\ \vdots \\ \alpha_{kJ} \end{bmatrix},\quad \alpha=[\alpha_1,\ldots,\alpha_K],\quad \beta_k=\begin{bmatrix} \beta_{k1} \\ \vdots \\ \beta_{kL} \end{bmatrix},\quad \beta=[\beta_1,\ldots,\beta_K],\quad \delta_k^2=\begin{bmatrix} \delta_{k1}^2 \\ \vdots \\ \delta_{kM}^2 \end{bmatrix},\quad \delta^2=[\delta_1^2,\ldots\delta_K^2],$$

$$\varepsilon=\begin{bmatrix} \varepsilon_{11} & \cdots & \varepsilon_{1K} \\ \cdots & \cdots & \cdots \\ \varepsilon_{n1} & \cdots & \varepsilon_{nK} \end{bmatrix}.$$

Equation 7C.2 can be written in a matrix form:

$$Y = e\alpha_0 + \upsilon\lambda + \xi\alpha + \eta\beta + \zeta\delta^1 + \tau\delta^{2+}\Gamma\gamma + \varepsilon$$
$$= WB + \varepsilon, \quad (7C.3)$$

where $W = \begin{bmatrix} e & \upsilon & \xi & \eta & \zeta & \tau & \Gamma \end{bmatrix}$ and $B = \begin{bmatrix} \alpha_0 \\ \lambda \\ \alpha \\ \beta \\ \delta^1 \\ \delta^2 \\ \gamma \end{bmatrix}$.

EXERCISE

Exercise 7.1 Consider two loci, G_1 and G_2. Define the disease risk due to genetics as $h_{ij} = P(D|G_1 = i, G_2 = j)$, the quasi-haplotype frequencies in the general and disease populations as

$$P_{ij} = P(G_1 = i, G_2 = j) \quad \text{and} \quad P_{ij}^A = P(G_1 = i, G_2 = j | D = 1).$$

Show that in the absence of interaction

$$\frac{P_{11}^A P_{22}^A}{P_{11} P_{22}} = \frac{P_{12}^A P_{21}^A}{P_{12} P_{21}}$$

holds.

Exercise 7.2 Show that if an SNP is a marker, then both additive and multiplicative interaction measures between a marker and causal variant will be zero.

Exercise 7.3 Show that

$$\text{RR}_{11} = e^{\beta_G + \beta_E + \beta_{GE}}, \; \text{RR}_{12} = e^{\beta_G}, \quad \text{and} \quad \text{RR}_{21} = e^{\beta_E}.$$

Exercise 7.4 Extend Equation 7.9 to two genetic loci. Show that

$$I_{muliple} = e^{\beta_{GE}}.$$

Exercise 7.5 Show that

$$\text{OR}_G = e^{\beta_g}, \quad \text{OR}_E = e^{\beta_e}, \quad \text{and} \quad \text{OR}_{GE} = e^{\beta_g + \beta_e + \beta_{ge}}.$$

Exercise 7.6 Show that the disequilibrium measure I_δ is equal to zero if and only if the multiplicative measure of gene–environment interaction for the cohort study $I_{multiple}$ is equal to one.

Exercise 7.7 Show that $I_{GE}=0$ is equivalent to $h_{11}h_{22}=h_{12}h_{21}$.

Exercise 7.8 Show that the mutual information between the binary genetic factor and continuous environment in disease population is given by

$$I(G;E|D)=\sum_{i=0}^{1}\int P(G=i,E=e|D=1)\log\frac{P(G=i,E=e)}{P(G=i)P(E=e)}de+I_{GCE}.$$

Exercise 7.9 When an environmental variable has a normal distribution with mean μ_D and variance σ_D^2 and conditional on the genotype $G=i$, it has normal distribution with mean μ_{Di} and variance σ_{Di}^2 in the disease population. Show that

$$I(G;E|D)=\frac{1}{2}\log\sigma_D^2-\frac{1}{2}\left[P(G=0|D)\log\sigma_{D_0}^2+P(G=1|D)\log\sigma_{D_1}^2\right].$$

Exercise 7.10 Show that

$$\hat{V}_{Radd}=\frac{h_{11}(1-h_{11})}{n_1h_{22}^2}+\frac{h_{12}(1-h_{12})}{n_2h_{22}^2}+\frac{h_{21}(1-h_{21})}{n_3h_{22}^2}+\frac{(1-h_{22})(h_{12}+h_{21}-h_{11})^2}{n_4h_{22}^3}.$$

Exercise 7.11 Show that

$$\delta_{D_1D_2}^{A}=\frac{\delta_{D_1D_2}}{P_A}h_{11}+\frac{P_{D_1}P_{D_2}}{P_A}\left(h_{11}-\frac{h_{D_1}h_{D_2}}{P_A}\right)$$

and

$$\delta_{D_1/D_2}^{A}=\frac{\delta_{D_1/D_2}}{P_A}h_{1/1}+\frac{P_{D_1}P_{D_2}}{P_A}\left(h_{1/1}-\frac{h_{D_1}h_{D_2}}{P_A}\right).$$

Exercise 7.12 Show that

$$V_{MI}=\frac{1}{n_A}\left\{\sum_{i=0}^{2}\sum_{j=0}^{1}\left(\frac{\partial I_{G_1G_2}}{\partial P_{Dij}}\right)^2P_{Dij}-\left(\sum_{i=0}^{2}\sum_{j=0}^{1}\frac{\partial I_{G_1G_2}}{\partial P_{Dij}}P_{Dij}\right)^2\right\}$$
$$+\frac{1}{n_G}\left\{\sum_{i=0}^{2}\sum_{j=0}^{1}\left(\frac{\partial I_{G_1G_2}}{\partial P_{ij}}\right)^2P_{ij}-\left(\sum_{i=0}^{2}\sum_{j=0}^{1}\frac{\partial I_{G_1G_2}}{\partial P_{ij}}P_{ij}\right)^2\right\}.$$

Exercise 7.13 Show that

$$\operatorname{Var}\left(\hat{I}^{H}_{G_1G_2}\right)=\frac{1}{2n_A}\left[\frac{1}{P^A_{11}}+\frac{1}{P^A_{12}}+\frac{1}{P^A_{21}}+\frac{1}{P^A_{22}}\right]+\frac{1}{2n_G}\left[\frac{1}{P^N_{11}}+\frac{1}{P^N_{12}}+\frac{1}{P^N_{21}}+\frac{1}{P^N_{22}}\right].$$

Exercise 7.14 Show that dominance effects can be expressed as

$$\delta_1=\mu_{AA}-2\mu_{Aa}+\mu_{aa},$$
$$\delta_{AA}=P_a^2\delta_1,\quad \delta_{Aa}=-P_AP_a\delta_1,\quad \delta_{aa}=P_A^2\delta_1,$$
$$\delta_2=\mu_{BB}-2\mu_{Bb}+\mu_{bb},$$
$$\delta_{BB}=P_b^2\delta_2,\quad \delta_{Bb}=-P_BP_b\delta_2,\quad \delta_{bb}=P_B^2\delta_2.$$

Exercise 7.15 Show that additive × dominance interaction effects are

$$\left(\alpha\delta\right)_{ABB}=P_bP_a^2\gamma_{AD},$$

where $\gamma_{AD}=P_A\mu_{AABB}-2P_A\mu_{AABb}+P_A\mu_{AAbb}+(P_a-P_A)\mu_{AaBB}-2(P_a-P_A)\mu_{AaBb}$.

CHAPTER 8

Machine Learning, Low-Rank Models, and Their Application to Disease Risk Prediction and Precision Medicine

Supervised learning attempts to learn from already labeled data how to predict the class of unlabeled data. There are two general data types: labeled data and unlabeled data. Unlabeled data are often called predictors, or features. In a typical supervised learning task, data are represented as a table of examples. Each example is described by a fixed number of measurements or features along with a label that denotes its class. Features are typically SNPs, gene expression levels, sex, age, and environmental variables such as drug dosages. Status of the disease and classes of the subject are labeled data. We usually use X to denote unlabeled data and Y to denote labeled data. In supervised learning, we divide datasets into a training dataset and test dataset. Suppose that in the training dataset, N individuals are sampled. We assume that in the training dataset, both unlabeled data and labeled data are known. From given data, $\{x_n \in R^d, y_n \in (1, \ldots, k)\}$, $n = 1, \ldots, N$, in the training dataset, we learn a predict model, $y = f(x)$ (Figure 8.1). We use the trained model to classify a new feature vector, x, into one of the existing classes, y. This new data is often referred to as the testing dataset.

Machine learning is a core area in big data analysis. Machine learning has been widely used in robotics, driverless cars, space exploration, web search, E-commerce, and finance. Machine learning is also a powerful tool for disease risk prediction, diagnosis, management, treatment selection, and precision medicine.

Many algorithms have been developed for supervised learning. Widely used algorithms include logistic regression, discriminant analysis, support vector machine, deep learning, and decision tree, among others. This chapter focuses on logistic regression, discriminant analysis, and support vector machines.

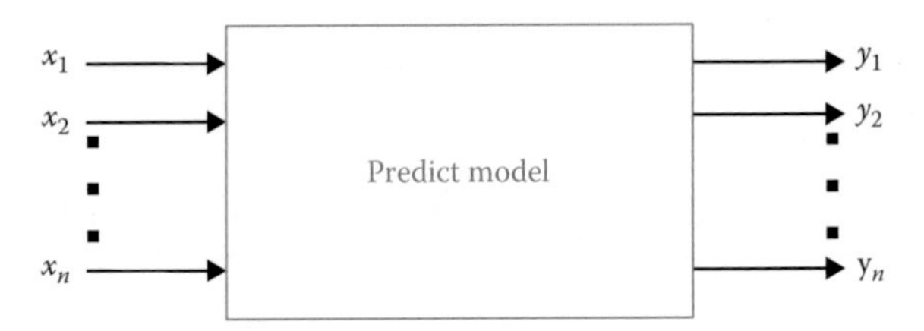

FIGURE 8.1 A generic system for supervised learning.

8.1 LOGISTIC REGRESSION

8.1.1 Two-Class Logistic Regression

Consider the binary output variable Y that can take values $\{1, 0\}$. For example, $Y=1$ indicates a disease status of an individual and $Y=0$ indicates a healthy status of an individual. Let $X \in R^P$ be a d-dimensional vector of predictors or features. Given a vector of predictors X, we want to calculate the conditional expectation $E[Y|X]=P(Y=1|X]$ of Y on the predictors. We assume that we observe pairs of predictors and binary output values $\{(x_i, y_i), i=1, \ldots, N\}$ and that binary output values follow a Bernoulli distribution. The likelihood for observing the binary output values y_i is (McCullagh and Nelder 1989)

$$L(p, y) = \prod_{i=1}^{n} p^{y_i} (1-p)^{1-y_i}. \tag{8.1}$$

The maximum likelihood estimator of p is $\hat{p} = \frac{1}{n} \sum_{i=1}^{n} y_i$ (Exercise 8.1).

In many scenarios, each trial has its own success probability, p_i. The likelihood in Equation 8.1 becomes

$$L(p, y) = \prod_{i=1}^{n} p_i^{y_i} (1-p_i)^{1-y_i}. \tag{8.2}$$

The maximum likelihood estimator of the parameter is $\hat{p}_i = y_i$. The parameter estimator does not provide any useful information. Only when we link the success probability p_i with observed predictors will the model and estimator become useful. In other words, the interesting quantity is a conditional probability, $P(Y=1|X=x)$. We make the logistic transformation $\log \frac{p(Y=1|X)}{1-p(Y=1|X)}$ as a linear function of predictors X. In other words, model logistic transformation of the conditional probability as

$$\log \frac{p(Y_i=1|x_i)}{1-p(Y_i=1|x_i)} = x_i^T \beta, \tag{8.3}$$

where $x_i = [1, x_{i2}, \ldots, x_{id}]^T$, $\beta = [\beta_1, \beta_2, \ldots, \beta_d]^T$.

For the convenience of presentation, denote $p(Y_i = 1|x_i)$ as $p(y_i|x_i, \beta)$. Then, Equation 8.3 is reduced to

$$\log \frac{p(y_i|x_i,\beta)}{1-p(y_i|x_i,\beta)} = x_i^T\beta. \tag{8.4}$$

In genetic studies, we will include covariates, indicator variables for genetic additive, and dominance effects into the model. Consider k SNPs and $l-1$ covariates. The model (8.4) can be expanded as

$$\log \frac{p(y_i|w_i,x_i,z_i,\eta,\alpha,\delta)}{1-p(y_i|w_i,x_i,z_i,\eta,\alpha,\delta)} = w_i\eta + x_i\alpha + z_i\delta, \tag{8.5}$$

where

$w_i = [w_{i1} = 1, w_{i2}, \ldots, w_{il}]$ is a vector of covariates
$\eta = [\eta_1, \eta_2, \ldots, \eta_l]^T$ is a vector of coefficients associated with the covariates
$x_i = [x_{i1}, x_{i2}, \ldots, x_{ik}]$ is a vector of genotype indicator variables
$\alpha = [\alpha_1, \alpha_2, \ldots, \alpha_k]^T$ is a vector of genetic additive effects
$z_i = [z_{i1}, z_{i2}, \ldots, z_{ik}]$ is a vector of genotype indicator variables for the genetic dominance effects
$\delta = [\delta_1, \delta_2, \ldots, \delta_k]^T$ is a vector of genetic dominance effects

Let $H_i = [w_i \; x_i \; z_i]$ and $\beta = \begin{bmatrix} \eta \\ \alpha \\ \delta \end{bmatrix}$. Equation 8.5 can be further reduced to

$$\log \frac{p(y_i|H_i,\beta)}{1-p(y_i|H_i,\beta)} = H_i\beta. \tag{8.6}$$

Solving Equation 8.6 for the conditional probability gives

$$p(y_i|H_i,\beta) = \frac{e^{H_i\beta}}{1+e^{H_i\beta}} = \frac{1}{1+e^{-H_i\beta}}. \tag{8.7}$$

The binary output values or class labels are predicted by conditional probability. We should predict $y=1$ when $p \geq 0.5$ and predict $y=0$ when $p<0.5$. It is clear from Equation 8.7 that when $H\beta \geq 0$, we have $p \geq 0.5$ or $y=1$ and when $H\beta < 0$, we have $p<0.5$ or $y=0$. The decision boundary that separates two classes is the solution of $H\beta = 0$.

Example 8.1 (Figure 8.2)

Consider $\eta = [-3 \quad 1 \quad 1]$. The decision boundary is a line, $w_1 + w_2 = 3$. When $w_1 + w_2 \geq 3$, $P \geq 0.5$ and $y=1$; when $w_1 + w_2 < 3$, $p<0.5$ and $y=0$. The line $\beta_1 + \beta_2 = 3$ separates the points with a triangle and points with circle in the plane.

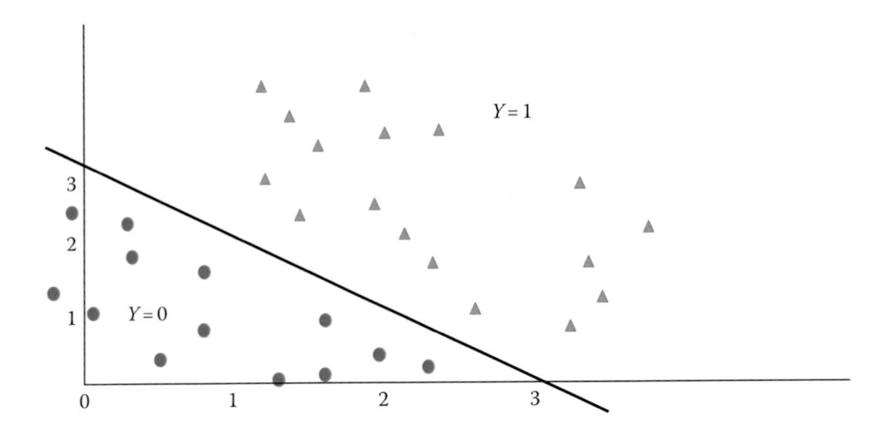

FIGURE 8.2 Boundary line.

Example 8.2 (Figure 8.3)

Consider the model $H\beta = -2 + w_1^2 + w_2^2$. The decision boundary is a circle, $w_1^2 + w_2^2 = 2$. When $w_1^2 + w_2^2 \geq 2$, then $y = 1$; when $w_1^2 + w_2^2 < 2$, we predict $y = 0$. The circle $w_1^2 + w_2^2 = 2$ separates two classes of data.

For each training point, we have a vector of observed covariates and genotypes, H_i, and an observed class, y_i. The class variable y_i follows a Bernoulli distribution with the conditional probability $p(y_i|H_i, \beta)$ as its parameter. Similar to Equation 8.2, the likelihood and log-likelihood are, respectively, given by

$$L(\beta) = \prod_{i=1}^{n} p\left(y_i \middle| H_i, \beta\right)^{y_i} \left(1 - p\left(Y_i \middle| H_i, \beta\right)\right)^{1-y_i}, \tag{8.8}$$

and

$$\begin{aligned} l(\beta) &= \sum_{i=1}^{n} \left[y_i \log p\left(y_i \middle| H_i, \beta\right) + (1 - y_i) \log\left(1 - p\left(y_i \middle| H_i, \beta\right)\right) \right] \\ &= \sum_{i=1}^{n} \left[y_i \log \frac{p\left(y_i \middle| H_i, \beta\right)}{1 - p\left(y_i \middle| H_i, \beta\right)} + \log\left(1 - p\left(y_i \middle| H_i, \beta\right)\right) \right] \\ &= \sum_{i=1}^{n} \left[y_i \left(H_i \beta\right) - \log\left(1 + e^{H_i \beta}\right) \right]. \end{aligned} \tag{8.9}$$

8.1.2 Multiclass Logistic Regression

It is often observed that the number of classes is more than two. For example, in medical diagnosis, the results of diagnosis can be type 2 diabetes, hypertension, heart disease, lung cancer, healthy individual, and others. As shown in Figure 8.4, multiclass logistic regression attempts to find a boundary to separate three classes. Suppose that we have C classes.

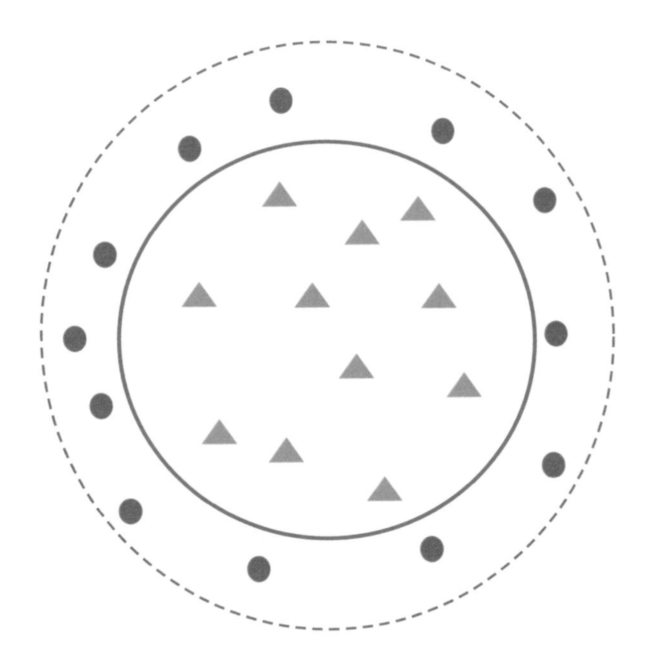

FIGURE 8.3 Boundary circle.

The output variable Y can take one of the set of values, $\{0, 1, \ldots, C-1\}$. We can assume that the variable Y follows a multinomial distribution. The predicted conditional probability $P(y_i = c|H_i, \beta)$ is defined as

$$p\left(y_i = c\middle|H_i,\beta\right) = \frac{e^{H_i\beta^{(c)}}}{1+\sum_{c=1}^{C-1} e^{H_i\beta^{(c)}}}, \tag{8.10}$$

where $\beta^{(c)} = \begin{bmatrix} \delta^{(c)} \\ \eta^{(c)} \\ \alpha^{(c)} \end{bmatrix}$, $c = 1, 2, \ldots, C-1$.

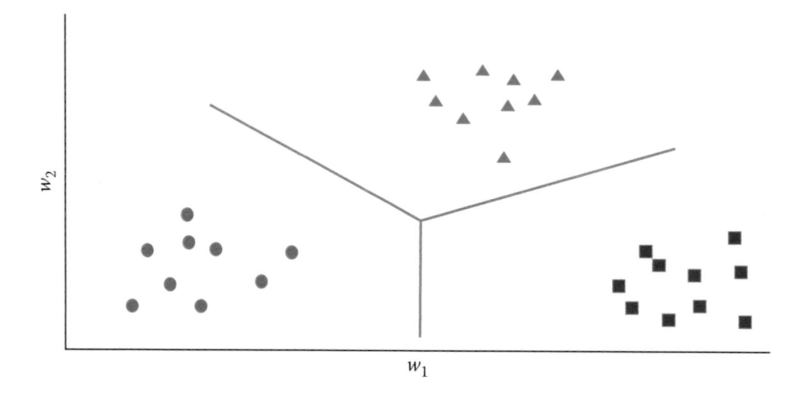

FIGURE 8.4 Multiclass classification.

The log-likelihood is defined as

$$
\begin{aligned}
l(\beta) &= \sum_{i=1}^{n}\sum_{c=1}^{C-1}\left[I(y_i = c)\log\frac{e^{H_i\beta^{(c)}}}{1+\sum_{c=1}^{C-1} e^{H_i\beta^{(c)}}}\right] \\
&= \sum_{i=1}^{n}\sum_{c=1}^{C-1}\left[I(y_i = c)\left(H_i\beta^{(c)} - \log\left(1+\sum_{c=1}^{C-1} e^{H_i\beta^{(c)}}\right)\right)\right],
\end{aligned} \tag{8.11}
$$

where $I(y_i = c) = \begin{cases} 1 & y_i = c \\ 0 & y_i \neq c \end{cases}$.

After a vector of new predictors, H_*, is available, the "maximum of a posteriori" decision rule is used to assign the subject to the class c_* with the highest conditional probability, e.g.,

$$
c_* = \arg\max_c \frac{e^{H_{*i}\beta^{(c)}}}{1+\sum_{c=}^{C-1} e^{H_*\beta^{(c)}}}.
$$

8.1.3 Parameter Estimation

The log-likelihood functions (8.9) and (8.59) are a nonlinear function of the parameters. Closed forms for the maximum likelihood estimators of the parameters are not available. Numerical optimization methods are used to search the maximum likelihood estimators. A typical optimization method is the Newton–Raphson method. Newton–Raphson approximates the likelihood function by a second-order Taylor series and optimizes that approximation. We first discuss the parameter estimation for a two-class logistic regression. The maximum likelihood estimators of the parameters must satisfy the equation

$$
\frac{\partial l(\beta)}{\partial \beta} = 0. \tag{8.12}
$$

The gradient of the log-likelihood function $\frac{\partial l(\beta)}{\partial \beta}$ can be approximated by

$$
\frac{\partial l(\beta)}{\partial \beta} \approx \frac{\partial l(\beta_0)}{\partial \beta} + \frac{\partial^2 l(\beta_0)}{\partial \beta \partial \beta^T}(\beta - \beta_0). \tag{8.13}
$$

Substituting Equation 8.13 into Equation 8.12, we obtain

$$
\frac{\partial l(\beta_0)}{\partial \beta} + \frac{\partial^2 l(\beta_0)}{\partial \beta \partial \beta^T}(\beta - \beta_0) = 0. \tag{8.14}
$$

Matrix $\frac{\partial^2 l(\beta_0)}{\partial\beta\partial\beta^T}$ is often called the Hessian matrix of the function $l(\beta)$. Solving Equation 8.14 for the vector of parameters β gives an approximate solution to Equation 8.12:

$$\beta = \beta_0 - \left[\frac{\partial^2 l(\beta_0)}{\partial\beta\partial\beta^T}\right]^{-1} \frac{\partial l(\beta_0)}{\partial\beta}. \tag{8.15}$$

Let $\xi_i = H_i\beta$. Then, Equation 8.9 can be written as

$$l(\beta) = \sum_{i=1}^{n} \left[y_i \xi_i - \log\left(1 + e^{\xi_i}\right) \right]. \tag{8.16}$$

Using Equation 8.16, we can obtain the gradient of the log-likelihood function

$$\begin{aligned} \frac{\partial l(\beta)}{\partial\beta} &= \sum_{i=1}^{n} \left[y_i \frac{\partial \xi_i}{\partial\beta} - \frac{e^{\xi_i}}{1+e^{\xi_i}} \frac{\partial \xi_i}{\partial\beta} \right] \\ &= \sum_{i=1}^{n} (y_i - \pi_i) \frac{\partial \xi_i}{\partial\beta} \\ &= \sum_{i=1}^{n} (y_i - \pi_i) H_i^T, \end{aligned} \tag{8.17}$$

where $\pi_i = P\left(y_i \middle| H_i, \beta\right) = \frac{e^{\xi_i}}{1+e^{\xi_i}}$.

Let

$$Y = \begin{bmatrix} y_1 \\ \vdots \\ y_n \end{bmatrix}, \quad \pi = \begin{bmatrix} \pi_1 \\ \vdots \\ \pi_n \end{bmatrix}, \quad H = \begin{bmatrix} H_1 \\ \vdots \\ H_n \end{bmatrix}.$$

Then, Equation 8.17 can be written in a matrix form:

$$\frac{\partial l(\beta)}{\partial\beta} = H^T (Y - \pi). \tag{8.18}$$

Now calculate the Hessian matrix. Note that

$$\frac{\partial \pi_i}{\partial\beta^T} = \frac{e^{\xi_i}}{\left(1+e^{\xi_i}\right)^2} H_i = \pi_i (1 - \pi_i) H_i$$

and

$$\frac{\partial \pi}{\partial \beta^T} = \Pi H, \tag{8.19}$$

where $\Pi = \text{diag}(\pi_1(1-\pi_1), \ldots, \pi_n(1-\pi_n))$.

Therefore, taking the partial derivatives with respect to β^T on both sides of Equation 8.18 gives the Hessian matrix of the log-likelihood:

$$\frac{\partial^2 l(\beta)}{\partial \beta \partial \beta^T} = -H^T \Pi H. \tag{8.20}$$

Result 8.1

The Newton–Raphson algorithm is summarized as follows.

Step 1. Initialization. Compute the initial value:

$$\beta^{(0)} = \left(H^t H\right)^{-1} H^T Y.$$

Step 2. Compute the conditional probabilities:

$$\xi_i^{(t)} = H_i \beta^{(t)}, \quad \pi_i^{(t)} = \frac{e^{\xi_i}}{1+e^{\xi_i}} \quad \text{and} \quad \pi^{(t)} = \left[\pi_1^{(t)}, \ldots, \pi_n^{(t)}\right]^T.$$

Step 3. Compute the gradient of the log-likelihood:

$$\left.\frac{\partial l}{\partial \beta}\right|_{\beta^{(t)}} = H^T\left(Y - \pi^{(t)}\right).$$

Step 4. Compute the Hessian matrix of the log-likelihood:

$$\left.\frac{\partial^2 l}{\partial \beta \partial \beta^T}\right|_{\beta^{(t)}} = -H^T \Pi^{(t)} H, \quad \Pi^{(t)} = \text{diag}\left(\pi_1^{(t)}\left(1-\pi_1^{(t)}\right), \ldots, \pi_n^{(t)}\left(1-\pi_n^{(t)}\right)\right).$$

Step 5. Update the parameters β:

$$\beta^{(t+1)} = \beta^{(t)} + \left[H^T \Pi^{(t)} H\right]^{-1} H^T\left(Y - \pi^{(t)}\right).$$

Step 6. Check convergence:
If $\|\beta^{(t+1)} - \beta^{(t)}\|_2 \le \varepsilon$, stop. Otherwise, if $\beta^{(t)} \leftarrow \beta^{(t+1)}$, go to step 2.

Now we study the parameter estimation in multiclass logistic regression. Recall that the log-likelihood function for multiclass logistic regression is

$$l(\beta)=\sum_{i=1}^{n}\sum_{c=1}^{C-1}\left[I(y_i=c)\left(H_i\beta^{(c)}-\log\left(1+\sum_{c=1}^{C-1}e^{H_i\beta^{(c)}}\right)\right)\right]. \tag{8.21}$$

The partial derivative $\frac{\partial l(\beta)}{\partial \beta^{(c)}}$ is

$$\begin{aligned}\frac{\partial l(\beta)}{\partial \beta^{(c)}}&=\sum_{i=1}^{n}I(y_i=c)\left(H_i^T-\frac{e^{\xi_i^{(c)}}}{1+\sum_{c=1}^{C-1}e^{\xi_i^{(c)}}}H_i^T\right)\\&=\sum_{i=1}^{n}I(y_i=c)\left(1-\pi_i^{(c)}\right)H_i^T,\end{aligned} \tag{8.22}$$

where $\xi_i^{(c)}=H_i\beta^{(c)}$ and $\pi_i^{(c)}=\frac{e^{\xi_i^{(c)}}}{1+\sum_{c=1}^{C-1}e^{\xi_i^{(c)}}}$.

When $j\neq c$, similarly, we obtain

$$\begin{aligned}\frac{\partial l(\beta)}{\partial \beta^{(j)}}&=\sum_{i=1}^{n}I(y_i=c)\left(0-\frac{e^{\xi_i^{(j)}}}{1+\sum_{c=1}^{C-1}e^{\xi_i^{(c)}}}H_i^T\right)\\&=-\sum_{i=1}^{n}I(y_i=c)\pi_i^{(j)}H_i^T.\end{aligned} \tag{8.23}$$

Combining Equations 8.22 and 8.23 gives

$$\frac{\partial l(\beta)}{\partial \beta^{(c)}}=\sum_{i=1}^{n}\left[I(y_i=c)-\pi_i^{(c)}\right]H_i^T. \tag{8.24}$$

Now compute $\frac{\partial^2 l(\beta)}{\partial \beta^{(c)}\left(\partial \beta^{(j)}\right)^T}$. When $j=c$, we have

$$\frac{\partial^2 l(\beta)}{\partial \beta^{(c)}\left(\partial \beta^{(c)}\right)^T}=-\sum_{i=1}^{n}H_i^T\frac{\partial \pi_i^{(c)}}{\partial\left(\beta^{(c)}\right)^T}=-\sum_{i=1}^{n}H_i^TH_i\pi_i^{(c)}\left(1-\pi_i^{(c)}\right). \tag{8.25}$$

When $j \neq c$, we obtain

$$\frac{\partial^2 l(\beta)}{\partial \beta^{(c)} \left(\partial \beta^{(j)}\right)^T} = -\sum_{i=1}^{n} H_i^T \frac{\partial \pi_i^{(c)}}{\partial \left(\beta^{(j)}\right)^T} = -\sum_{i=1}^{n} H_i^T H_i \pi_i^{(c)} \pi_i^{(j)}. \tag{8.26}$$

Combining Equations 8.25 and 8.26, we have

$$\frac{\partial^2 l(\beta)}{\partial \beta^{(c)} \left(\partial \beta^{(j)}\right)^T} = -\sum_{i=1}^{n} \pi_i^{(c)} \left(I(c=j) - \pi_i^{(j)}\right) H_i^T H_i. \tag{8.27}$$

Define

$$\nabla l(\beta) = \begin{bmatrix} \frac{\partial l(\beta)}{\partial \beta^{(1)}} & \cdots & \frac{\partial l(\beta)}{\partial \beta^{(C-1)}} \end{bmatrix} \text{ and } \Pi(\beta) = \begin{bmatrix} \frac{\partial^2 l}{\partial \beta^{(1)} \partial \left(\beta^{(1)}\right)^T} & \cdots & \frac{\partial^2 l}{\partial \beta^{(1)} \partial \left(\beta^{(C-1)}\right)^T} \\ \vdots & \vdots & \vdots \\ \frac{\partial^2 l}{\partial \beta^{(C-1)} \partial \left(\beta^{(1)}\right)^T} & \cdots & \frac{\partial^2 l}{\partial \beta^{(C-1)} \partial \left(\beta^{(C-1)}\right)^T} \end{bmatrix}.$$

Define the matrix for the output class label variables:

$$\mathbf{Y} = \begin{bmatrix} I(y_1 = 1) & \cdots & I(y_1 = C-1) \\ \vdots & \vdots & \vdots \\ (Iy_n = 1) & \cdots & I(y_n = C-1) \end{bmatrix}$$

and the matrix of parameters:

$$B = \begin{bmatrix} \beta^{(1)} & \cdots & \beta^{(C-1)} \end{bmatrix}.$$

The algorithm for the parameter estimation is summarized as follows.

Result 8.2

The Newton–Raphson algorithm for estimation of parameters in the multiclass logistic regression is given below.

Step 1. Initialization. Compute the initial value:

$$B^{(0)} = \left(H^t H\right)^{-1} H^T Y.$$

Step 2. Compute the conditional probabilities:

$$\xi_i^{(t)} = \left[\xi_i^{(1)} \quad \cdots \quad \xi_i^{(C-1)}\right] = H_i B^{(t)}, \quad \pi_i^{(c)} = \frac{e^{\xi_i^{(c)}}}{1+\sum_{c=1}^{C-1} e^{\xi_i^{(c)}}} \quad \text{and} \quad \pi_i^{(t)} = \left[\pi_i^{(1)} \quad \cdots \quad \pi_i^{(C-1)}\right].$$

Step 3. Compute the gradient of the log-likelihood:

$$\left.\frac{\partial l(\beta)}{\partial \beta^{(c)}}\right|_{\beta^{(t)}} = \sum_{i=1}^{n}\left[I(y_i = c) - \left(\pi_i^{(c)}\right)^{(t)}\right] H_i^T,$$

$$\left(\nabla l(\beta)\right)^{(t)} = \left[\left.\frac{\partial l(\beta)}{\partial \beta^{(1)}}\right|_{\beta^{(t)}} \quad \cdots \quad \left.\frac{\partial l(\beta)}{\partial \beta^{(C-1)}}\right|_{\beta^{(t)}}\right].$$

Step 4. Compute the Hessian matrix of the log-likelihood:

$$\left.\frac{\partial^2 l(\beta)}{\partial \beta^{(c)}\left(\partial \beta^{(j)}\right)^T}\right|_{\beta^{(t)}} = -\sum_{i=1}^{n}\left(\pi_i^{(c)}\right)^{(t)}\left(I(c=j) - \left(\pi_i^{(j)}\right)^{(t)}\right) H_i^T H_i,$$

$$\Pi\left(\beta^{(t)}\right) = \begin{bmatrix} \left.\frac{\partial^2 l}{\partial \beta^{(1)} \partial\left(\beta^{(1)}\right)^T}\right|_{\beta^{(t)}} & \cdots & \left.\frac{\partial^2 l}{\partial \beta^{(1)} \partial\left(\beta^{(C-1)}\right)^T}\right|_{\beta^{(t)}} \\ \vdots & \vdots & \vdots \\ \left.\frac{\partial^2 l}{\partial \beta^{(C-1)} \partial\left(\beta^{(1)}\right)^T}\right|_{\beta^{(t)}} & \cdots & \left.\frac{\partial^2 l}{\partial \beta^{(C-1)} \partial\left(\beta^{(C-1)}\right)^T}\right|_{\beta^{(t)}} \end{bmatrix}.$$

Step 5. Update the parameters β:

$$B^{(t+1)} = B^{(t)} - \left[\Pi\left(\beta^{(t)}\right)\right]^{-1}\left(\nabla l(\beta)\right)^{(t)}.$$

Step 6. Check convergence:

If $\|B^{(t+1)} - B^{(t)}\|_F \leq \varepsilon$, stop. Otherwise, if $B^{(t)} \leftarrow B^{(t+1)}$, go to step 2.

8.1.4 Test Statistics

First consider association tests for two-class logistic regression. Suppose that there are K loci. The null hypothesis for no association is

$$H_0 : \alpha = [\alpha_1,\ldots,\alpha_K]^T = 0 \quad \text{and} \quad \delta = [\delta_1,\ldots,\delta_K]^T = 0.$$

The alternative hypothesis is

$$H_a : \alpha \neq 0 \quad \text{and} \quad \delta \neq 0.$$

Let $l(\beta_0)$ denote the log-likelihood of the logistic model under the null hypothesis of no association: let $\beta_0 = \eta$ and $l(\beta)$ be the log-likelihood of the logistic model under the full model. Define the statistic

$$T_{\log} = -2\left(l(\beta_0) - l(\beta)\right). \tag{8.28}$$

Then, under the null hypothesis of no association, $T_{\log}$ is asymptotically distributed as a central $\chi^2_{(2K)}$ distribution.

Next consider the multiclass logistic regression model. The null hypothesis of no association for multiclass logistic regression model is given by

$$H_0 : \alpha^{(c)} = \left[\alpha_1^{(c)},\ldots,\alpha_K^{(c)}\right]^T = 0 \quad \text{and} \quad \delta^{(c)} = \left[\delta_1^{(c)},\ldots,\delta_K^{(c)}\right]^T = 0, \quad c = 1,2,\ldots,C-1.$$

The parameter β_0 is now given by

$$\beta_0 = \begin{bmatrix} \eta^{(1)} \\ \vdots \\ \eta^{(C-1)} \end{bmatrix}.$$

Let $l_m(\beta_0)$ denote the log-likelihood of the multiclass logistic model under the null hypothesis of no association: let $\beta_0 = \eta$ and $l_m(\beta)$ be the log-likelihood of the multiclass logistic model under the full model.

Define the statistic

$$T_{m\log} = -2\left(l_m(\beta_0) - l_m(\beta)\right). \tag{8.29}$$

Then, under the null hypothesis of no association, $T_{m\log}$ is asymptotically distributed as a central $\chi^2_{(2(C-1)K)}$ distribution.

8.1.5 Network-Penalized Two-Class Logistic Regression

8.1.5.1 Model

Logistic regression is used for disease and drug response prediction. The next generation of genomic, sensing, and imaging technologies will produce a deluge of DNA sequencing, transcriptomic, imaging, behavioral, and clinical multiple phenotypic data with millions of features. An essential issue for the success of prediction is feature selection. When the number of predictors exceeds the number of samples, the traditional logistic analysis breaks down. Parameter penalization is a popular technique for feature selection and high dimension reduction in fitting the logit models (Huang et al. 2015; Min et al. 2016; Wu et al. 2016). We consider three types of penalization. The first one is L_1-norm penalization on the covariates. The second one is group LASSO for genes that consists of multiple SNPs or number of principal components or functional principal components, in which case all the genetic effects of the SNPs associated with a gene are required to shrink to zero simultaneously. The third one is network-based penalization. Regulatory relationships between genes exist. Genes form gene regulatory networks. Multiple phenotypes are highly correlated and may form causal networks. Many SNPs are in linkage disequilibrium. The dependence relationships between SNPs can be modeled by undirected graphs. Incorporation of such structure information underlying data generation can improve the smoothness of the estimated parameters over the network and improve the accuracy of prediction.

For simplicity, we only consider genetic additive effects. Predictors consist of three parts: covariates including environments, genotypes, and phenotypes (or gene expressions). Assume L covariates including 1, G genes, and K phenotypes. Recall that the logistic model is given by

$$\pi_i = p\left(y_i = 1 \middle| H_i, \beta\right) = \frac{e^{\xi_i}}{1 + e^{\xi_i}}, \tag{8.30}$$

where $\xi_i = H_i\beta$, $H_i = \begin{bmatrix} z_i & w_i & x_i \end{bmatrix}$, and $\beta = \begin{bmatrix} \delta \\ \eta \\ \alpha \end{bmatrix}$. Specifically, the vector of phenotypes z and their coefficients are, respectively, denoted by $z_i = [z_{i1}, \ldots, z_{iK}]$ and $\delta = [\delta_1, \delta_2, \ldots, \delta_K]^T$. The genotype indicator variables or functional principal component scores are denoted by $x_i = \left[x_i^{(1)}, \ldots, x_i^{(G)}\right]$, $x_i^{(g)} = \left[x_{i1}^{(g)}, \ldots, x_{ik_g}^{(g)}\right]$, $g = 1, \ldots, G$. Correspondingly, we denote the genetic effects by

$$\alpha = \begin{bmatrix} \alpha_1 \\ \vdots \\ \alpha_G \end{bmatrix}, \quad \alpha_g = \begin{bmatrix} \alpha_{g1} \\ \vdots \\ \alpha_{gk_g} \end{bmatrix}, \quad g = 1, \ldots, G, \quad m = \sum_{g=1}^{G} k_g.$$

We can simply write $\alpha = [\alpha_1, \alpha_2, \ldots, \alpha_m]^T$. The covariates w_i and the vector of their logistic regression coefficients η are defined as before.

Now we consider the first type of penalty on the phenotype coefficients δ and covariate coefficients η. The L_1-norm regularization on δ and η is defined as

$$J_1 = \sum_{u=1}^{K} \|\delta_u\|_1 + \sum_{l=2}^{L} \|\eta_l\|_1 . \tag{8.31}$$

The second type of penalty is group LASSO and is defined as

$$J_2(\alpha) = \sum_{g=1}^{G} \phi_g \|\alpha_g\|_2 . \tag{8.32}$$

The third type of penalty is a network penalty in which constraints are posed on graphs (Figure 8.5). Consider a graph, $\acute{G} = (V, E)$, where V is the node set and E is the set of edges. We assume that the graph $\acute{G}$ consists of three subnetworks: phenotype subnetworks $G_1 = (V_1, E_1)$, genotype–phenotype connect subnetworks $G_2 = (V_2, E_2)$, and environment–phenotype connect subnetworks $G_3 = (V_3, E_3)$. In the phenotype subnetwork, the node denotes the phenotype variable and $|V_1| = K$. The environment–phenotype subnetwork has edges connecting a covariate and a phenotype. The total number of nodes is $|V_2| = L + K - 1$. Similarly, the genotype–phenotype subnetwork G_3 is characterized by edges. An edge connects two nodes: an SNP (or functional principal component) and a phenotype node. The total number of nodes is $|V_3| = m + K$. For the subnetwork, we use edge penalty as a way to penalize the network. We penalize the difference between the variables at adjacent nodes. First, we consider the edges in the phenotype subnetwork. Consider the edge between nodes u and v with weight s_{uv}. The constraint posed on the edge is $\bar{s}_{uv} \|\delta_u - \delta_v\|_2^2$. Thus, the penalty for the phenotype subnetwork is

$$\sum_{(u,v) \in E_1} \bar{s}_{uv} \|\delta_u - \delta_v\|_2^2 . \tag{8.33}$$

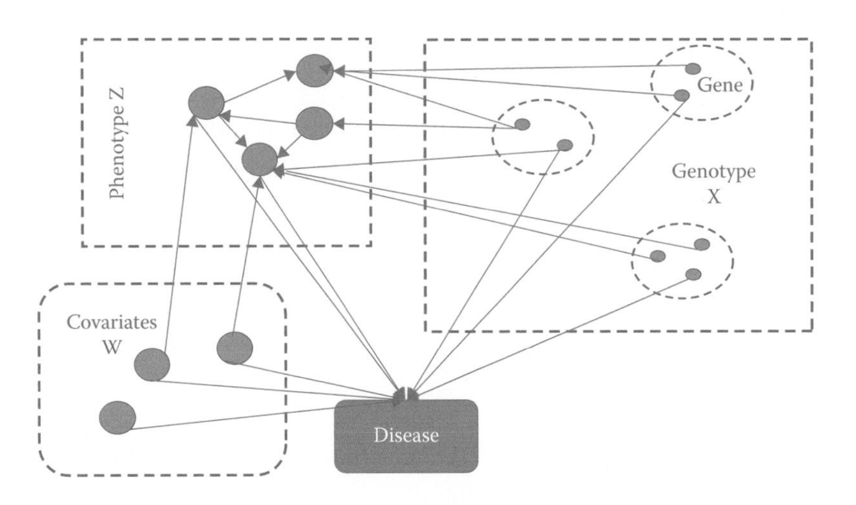

FIGURE 8.5 Genotype–environment–phenotype–disease networks

Then, consider the penalty for the environment–phenotype subnetwork. The penalty for the edge connecting a covariate, η_l, and a phenotype, δ_u, is $\bar{s}_{ul}\left\|\eta_l-\delta_u\right\|_2^2$. The penalty for

$$\sum_{(l,u)\in E_2}\bar{s}_{ul}\left\|\eta_l-\delta_u\right\|_2^2. \tag{8.34}$$

Finally, consider the penalty for the genotype–phenotype connect subnetwork. The penalty for the edge connecting an SNP, α_j, and a phenotype, δ_u, is $\bar{s}_{uj}\left\|\alpha_j-\delta_u\right\|_2^2$. The penalty for the genotype–phenotype connect subnetwork is

$$\sum_{(j,u)\in E_3}\bar{s}_{uj}\|\alpha_j-\delta_u\|_2^2. \tag{8.35}$$

Combining Equations 8.33 through 8.35 gives the penalty for the whole network:

$$J_3(\beta)=\sum_{(u,v)\in E_1}\bar{s}_{uv}\|\delta_u-\delta_v\|_2^2+\sum_{(l,u)\in E_2}\bar{s}_{ul}\|\eta_l-\delta_u\|_2^2+\sum_{(j,u)\in E_3}\bar{s}_{uj}\|\alpha_j-\delta_u\|_2^2. \tag{8.36}$$

We construct an adjacency matrix for the whole network as follows. First define the elements of the adjacency matrix:

$$s_{uv}=\begin{cases}\bar{s}_{uv} & (u,v)\in E_1\\ 0 & (u,v)\notin E_1\end{cases},\quad s_{ul}=\begin{cases}\bar{s}_{ul} & (u,l)\in E_2\\ 0 & (u,l)\notin E_2\end{cases},\quad s_{uj}=\begin{cases}\bar{s}_{uj} & (u,j)\in E_3\\ 0 & (u,j)\in E_3\end{cases}.$$

Then, we define the adjacency matrix:

$$S_{\delta\delta}=\begin{bmatrix}0 & s_{u_1u_2} & \cdots & s_{u_1u_K}\\ \vdots & \vdots & & \vdots\;\vdots\\ s_{u_Ku_1} & s_{u_Ku_2} & \cdots & 0\end{bmatrix}_{K\times K},\quad S_{\delta\eta}=\begin{bmatrix}s_{u_1l_1} & \cdots & s_{u_1l_L}\\ \vdots & \vdots & \vdots\\ s_{u_Kl_1} & \cdots & s_{u_Kl_L}\end{bmatrix}_{K\times L},\quad S_{\delta\alpha}=\begin{bmatrix}s_{u_1j_1} & \cdots & s_{u_1j_m}\\ \vdots & \vdots & \vdots\\ s_{u_kj_1} & \cdots & s_{u_kj_m}\end{bmatrix}_{K\times m},$$

and

$$S=\begin{bmatrix}S_{\delta\delta} & S_{\delta\eta} & S_{\delta\alpha}\\ S_{\delta\eta}^T & 0 & 0\\ S_{\delta\alpha}^T & 0 & 0\end{bmatrix}_{(K+L+m)\times(K+L+m)}.$$

Let **1** be a $(K+L+m)$-dimensional vector with all the elements of one. Define a $(K+L+m)$-dimensional degree vector, $d=S\mathbf{1}$, and a degree matrix, $D=\text{diag}(d_1,d_2,\ldots,d_{(K+L+m)})$. The Laplacian matrix associated with the whole network is $D-S$. We can use the Laplacian matrix to rewrite Equation 8.36 in a matrix form:

$$J_3(\beta)=\beta^T(D-S)\beta. \tag{8.37}$$

Therefore, the penalized log-likelihood function is defined as

$$l_p(\beta) = -l(\beta) + \lambda_1 \sum_{u=1}^{K} ||\delta_u||_1 + \lambda_2 \sum_{l=2}^{L} ||\eta_l||_1] + \lambda_3 \sum_{g=1}^{G} \phi_g ||\alpha_g||_2 + \lambda_4 \beta^T (D - S)\beta, \quad (8.38)$$

where $l(\beta) = \sum_{i=1}^{n} \left[y_i \xi_i - \log\left(1 + e^{\xi_i}\right) \right]$, $\xi_i = H_i \beta$, $H_i = [z_i \quad w_i \quad x_i]$, and $\beta = \begin{bmatrix} \delta \\ \eta \\ \alpha \end{bmatrix}$.

Now we decompose the penalized log-likelihood function into a differential part and a nondifferential part:

$$l_p(\beta) = f(\beta) + \Omega(\beta), \quad (8.39)$$

where $f(\beta) = -l(\beta) + \lambda_4 \beta^T (D - S)\beta$ and $\Omega(\beta) = \lambda_1 \sum_{u=1}^{K} \|\delta_u\|_1 + \lambda_2 \sum_{l=2}^{L} \|\eta_l\|_1] + \lambda_3 \sum_{g=1}^{G} \phi_g \|\alpha_g\|_2$.

The nonsmooth penalty $\Omega(\beta)$ can be further written as

$$\Omega(\beta) = \Omega_1(\delta) + \Omega_2(\eta) + \Omega_3(\alpha),$$

where

$$\Omega_1(\delta) = \lambda_1 \sum_{u=1}^{K} \|\delta_u\|_1, \quad \Omega_2(\eta) = \lambda_2 \sum_{l=2}^{L} \|\eta_l\|_1], \quad \text{and} \quad \Omega_3(\alpha) = \lambda_3 \sum_{g=1}^{G} \phi_g \|\alpha_g\|_2.$$

For easy implementation, $\Omega(\beta)$ can be replaced by

$$\Omega_0(\beta) = \lambda_1 \sum_{u=1}^{K} \|\delta_u\|_1 + \lambda_2 \sum_{l=2}^{L} \|\eta_l\|_1] + \lambda_3 \sum_{g=1}^{G} \|\alpha_g\|_2.$$

However, since all predictors are penalized to the same degree, this will lead to estimation bias. To overcome this limitation, we can use an adaptive group LASSO. First, we solve the optimization problem

$$\tilde{\beta} = \arg\min_{\beta} l_p(\beta) = f(\beta) + \Omega_0(\beta),$$

where $\tilde{\beta} = \begin{bmatrix} \tilde{\delta} \\ \tilde{\eta} \\ \tilde{\alpha} \end{bmatrix}$.

Then, taking $\phi_g = \frac{1}{\|\tilde{\alpha}_g\|_2}$, we define

$$\Omega(\beta) = \lambda_1 \sum_{u=1}^{K} ||\delta_u||_1 + \lambda_2 \sum_{l=2}^{L} ||\eta_l||_1] + \lambda_3 \sum_{g=1}^{G} \frac{1}{||\tilde{\alpha}_g||_2} ||\alpha_g||_2 \tag{8.40}$$

and solve the optimization problem (8.39).

8.1.5.2 Proximal Method for Parameter Estimation

An optimization problem (8.39) is a nonlinear and nonsmooth optimization problem and is difficult to solve. Recently, proximal methods have been developed as powerful tools for solving nonsmooth convex optimization problems (Parikh and Boyd 2013 and Chapter 1). We adapt the proximal method to solve the problem (8.39). The proximal point algorithms for solving network-penalized logistic regression are summarized as follows (see Appendix 8A).

Result 8.3

Step 1. Given S, D, $v, \lambda_1, \lambda_2, \lambda_3$, and ϕ_g, $g = 1, 2, \ldots, G$. Initial values $\beta^{(0)}$ are obtained by the ordinary logistic regression without parameter penalization.

Step 2. Calculate $\pi_i = \frac{e^{\xi_i^t}}{1 + e^{\xi_i^t}}$, $\xi_i^t = H_i \beta^t$, $H_i = [z_i \quad w_i \quad x_i]$, $\beta^t = \begin{bmatrix} \delta^t \\ \eta^t \\ \alpha^t \end{bmatrix}$. Calculate the gradient of the log-likelihood:

$$u^t = \beta^t - v\nabla f(\beta^t) = \begin{bmatrix} u_\delta^t \\ u_\eta^t \\ u_\alpha^t \end{bmatrix}$$

$$= \begin{bmatrix} \delta^t + v \sum_{i=1}^{n} (y_i - \pi_i) z_i^T - 2v\lambda_4 (D_{\delta\delta} - S_{\delta\delta}) \delta^t + 2v\lambda_4 S_{\delta\eta} \eta^t + 2v\lambda_4 S_{\delta\alpha} \alpha^t \\ \eta^t + v \sum_{i=1}^{n} (y_i - \pi_i) w_i^T - 2v\lambda_4 D_{\eta\eta} \eta^t + 2v\lambda_4 S_{\delta\eta}^T \delta^t \\ \alpha^t + v \sum_{i=1}^{n} (y_i - \pi_i) x_i^T - 2v\lambda_4 D_{\alpha\alpha} \alpha^t + 2v\lambda_4 S_{\delta\alpha}^T \delta^t \end{bmatrix}.$$

Step 3. Update

$$\delta^{t+1} = \begin{bmatrix} \text{sign}\left(u_{\delta_1}^t\right)\left(\left|u_{\delta_1}^t\right| - \lambda_1 v\right)_+ \\ \vdots \\ \text{sign}\left(u_{\delta_K}^t\right)\left(\left|u_{\delta_K}^t\right| - \lambda_1 v\right)_+ \end{bmatrix},$$

$$\eta^{t+1} = \begin{bmatrix} u_{\eta_1}^t \\ \text{sign}\left(u_{\eta_2}^t\right)\left(\left|u_{\eta_2}^t\right| - \lambda_2 v\right)_+ \\ \vdots \\ \text{sign}\left(u_{\eta_L}^t\right)\left(\left|u_{\eta_L}^t\right| - \lambda_2 v\right)_+ \end{bmatrix},$$

$$\alpha^{t+1} = \begin{bmatrix} \alpha_1^{t+1} \\ \vdots \\ \alpha_G^{t+1} \end{bmatrix} = \begin{bmatrix} \left(1 - \frac{v\lambda_3\phi_1}{\left\|u_{\alpha_1}^t\right\|_2}\right)_+ u_{\alpha_1}^t \\ \vdots \\ \left(1 - \frac{v\lambda_3\phi_g}{\left\|u_{\alpha_G}^t\right\|_2}\right)_+ u_{\alpha_G}^t \end{bmatrix}.$$

Step 4. Check convergence:

If $\|\delta^{t+1} - \delta^t\|_2 + \|\eta^{t+1} - \eta^t\|_2 + \|\alpha^{t+1} - \alpha^t\|_2 < \varepsilon$, stop. Otherwise, if $\delta^t \leftarrow \delta^{t+1}, \eta^t \leftarrow \eta^{t+1}, \alpha^t \leftarrow \alpha^{t+1}$, go to step 2.

8.1.6 Network-Penalized Multiclass Logistic Regression

8.1.6.1 Model

Now we study the network-penalized multiclass logistic regression. Again, for simplicity, we only consider genetic additive effect. Predictors consist of three parts: covariates including environments, genotypes, and phenotypes (or gene expressions). Assume L covariates including 1, G genes, and K phenotypes. The multiclass logistic regression is given by (Tian et al. 2015)

$$p\left(y_i = c \middle| H_i, \beta\right) = \frac{e^{H_i\beta^{(c)}}}{1 + \sum_{c=1}^{C-1} e^{H_i\beta^{(c)}}}, \tag{8.41}$$

where $\xi_i = H_i\beta$, $H_i = \begin{bmatrix} z_i & w_i & x_i \end{bmatrix}$, and

$$\beta^{(c)} = \begin{bmatrix} \delta^{(c)} \\ \eta^{(c)} \\ \alpha^{(c)} \end{bmatrix}, \quad \delta^{(c)} = \begin{bmatrix} \delta_1^{(c)} \\ \vdots \\ \delta_K^{(c)} \end{bmatrix}, \quad \eta^{(c)} = \begin{bmatrix} \eta_1^{(c)} \\ \vdots \\ \eta_L^{(c)} \end{bmatrix}, \quad \alpha^{(c)} = \begin{bmatrix} \alpha_1^{(c)} \\ \vdots \\ \alpha_G^{(c)} \end{bmatrix}, \quad \alpha_g^{(c)} = \begin{bmatrix} \alpha_{g1}^{(c)} \\ \vdots \\ \alpha_{gk_g}^{(c)} \end{bmatrix} \quad c = 1, 2, \ldots, C-1.$$

The log-likelihood is defined as

$$l(\beta)=\sum_{i=1}^{n}\sum_{c=1}^{C-1}\left[I(y_i=c)\log\frac{e^{H_i\beta^{(c)}}}{1+\sum_{c=1}^{C-1}e^{H_i\beta^{(c)}}}\right]$$

$$=\sum_{i=1}^{n}\sum_{c=1}^{C-1}\left[I(y_i=c)\left(H_i\beta^{(c)}-\log\left(1+\sum_{c=1}^{C-1}e^{H_i\beta^{(c)}}\right)\right)\right], \tag{8.42}$$

where $I(y_i=c)=\begin{cases}1 & y_i=c\\ 0 & y_i\neq c.\end{cases}$

We can rewrite the parameters corresponding to their associated variables:

$$\delta_k=\begin{bmatrix}\delta_k^{(1)}\\ \vdots\\ \delta_k^{(C-1)}\end{bmatrix},\quad \eta_l=\begin{bmatrix}\eta_l^{(1)}\\ \vdots\\ \eta_l^{(C-1)}\end{bmatrix},\quad \alpha_g=\begin{bmatrix}\alpha_g^{(1)}\\ \vdots\\ \alpha_g^{(C-1)}\end{bmatrix},\quad \alpha_g^{(c)}=\begin{bmatrix}\alpha_{g1}^{(c)}\\ \vdots\\ \alpha_{gk_g}^{(c)}\end{bmatrix}.$$

The parameters for all classes should be shrunk to zero if a variable is not selected in the model. To achieve this, a group LASSO penalty should be used to penalize the parameters. Specifically, we define three group LASSO penalties $\Omega_1(\delta)$, $\Omega_2(\eta)$, and $\Omega_3(\alpha)$ as follows.

We first impose constraints $\Omega_1(\delta)$ for the parameters associated with phenotypes:

$$\Omega_1(\delta)=\lambda_1\sum_{u=1}^{K}\phi_{1k}\|\delta_k\|_2, \tag{8.43}$$

where $\|\delta_k\|_2=\sqrt{\sum_{c=1}^{C-1}\left(\delta_k^{(c)}\right)^2}$.

Then, penalize the parameters associated with covariates $\Omega_2(\eta)$:

$$\Omega_2(\eta)=\lambda_2\sum_{l=1}^{L}\phi_{2l}\|\eta_l\|_2, \tag{8.44}$$

where $\|\eta_l\|_2=\sqrt{\sum_{c=1}^{C-1}\left(\eta_l^{(c)}\right)^2}$.

Now consider the penalization on the genetic additive effects. We should penalize the genetic effects contributed with C classes of all SNPs with a gene. The penalty $\Omega_3(\alpha)$ for the genetic effects is defined as

$$\Omega_3(\alpha)=\lambda_3\sum_{g=1}^{G}\phi_{3g}\|\alpha_g\|_2, \tag{8.45}$$

where $\|\alpha_g\|_2=\sqrt{\sum_{c=1}^{C-1}\sum_{u=1}^{k_g}\left(\alpha_{gu}^{(c)}\right)^2}$.

Now we consider the network penalty. For each fixed c class, similar to Equation 8.37, the network-constrained penalty is $(\beta^{(c)})(D-S)\beta^{(c)}$. The network-constrained penalty for all classes is defined as

$$\sum_{c=1}^{C-1}\left(\beta^{(c)}\right)(D-S)\beta^{(c)}, \tag{8.46}$$

where D and S are defined as before.

Therefore, the penalized log-likelihood function is defined as

$$l_p(\beta)=f(\beta)+\Omega(\beta), \tag{8.47}$$

where $f(\beta)=-l(\beta)+\sum_{c=1}^{C-1}\left(\beta^{(c)}\right)(D-S)\beta^{(c)}$ and $\Omega(\beta)=\Omega_1(\delta)+\Omega_2(\eta)+\Omega_3(\alpha)$. The goal is to minimize

$$\min_{\beta} l_p(\beta). \tag{8.48}$$

8.1.6.2 Proximal Method for Parameter Estimation in Multiclass Logistic Regression

Similar to Section 8.1.5.2, we can use the proximal method (Parikh and Boyd 2014) to solve the optimization problem (8.48). The proximal point algorithms for network-penalized multiclass logistic regression are summarized as Result 8.4.

Result 8.4 Proximal Point Algorithm

Step 1: Given $S, D, v, \lambda_1, \lambda_2, \lambda_3, \phi_{1k}, k=1,2,\ldots,K, \phi_{2l}, l=1,2,\ldots,L$, and $\phi_{3g}, g=1,2,\ldots,G$. Initial values $(\beta^{(c)})^{(0)}, c=1,2,\ldots,C-1$ are obtained by the ordinary multiclass logistic regression without parameter penalization.

Step 2: Calculate $H_i=\begin{bmatrix} z_i & w_i & x_i \end{bmatrix}$.

Step 3: Calculate $\left(\xi_i^{(c)}\right)^{(t)}=H_i\left(\beta^{(c)}\right)^{(t)}$ and $\left(\pi_i^{(c)}\right)^{(t)}=\dfrac{e^{\left(\xi_i^{(c)}\right)^{(t)}}}{1+\sum_{c=1}^{C-1}e^{\left(\xi_i^{(c)}\right)^{(t)}}}$, $c=1,2,\ldots,C-1$,

$$\text{where } \left(\beta^{(c)}\right)^{(t)}=\begin{bmatrix}\left(\delta^{(c)}\right)^{(t)}\\ \left(\eta^{(c)}\right)^{(t)}\\ \left(\alpha^{(c)}\right)^{(t)}\end{bmatrix}.$$

Calculate the gradient of the log-likelihood:

$$\left(u^{(c)}\right)^{(t)} = \left(\beta^{(c)}\right)^{(t)} - v\nabla_{\beta^{(c)}} f\left(\left(\beta^{(c)}\right)^{(t)}\right) = \begin{bmatrix} \left(u_{\delta}^{(c)}\right)^{(t)} \\ \left(u_{\eta}^{(c)}\right)^{(t)} \\ \left(u_{\alpha}^{(c)}\right)^{(t)} \end{bmatrix}$$

$$= \begin{bmatrix} \left(\delta^{(c)}\right)^{(t)} + v\sum_{i=1}^{n}\left[I(y_i = c) - \left(\pi_i^{(c)}\right)^{(t)}\right] z_i^T - 2v\lambda_4\left(D_{\delta\delta} - S_{\delta\delta}\right)\left(\delta^{(c)}\right)^{(t)} \\ + 2v\lambda_4 S_{\delta\eta}\left(\eta^{(c)}\right)^{(t)} + 2v\lambda_4 S_{\delta\alpha}\left(\alpha^{(c)}\right)^{(t)} \\ \left(\eta^{(c)}\right)^{(t)} + v\sum_{i=1}^{n}(I(y_i = c) - \left(\pi_i^{(c)}\right)^{(t)} w_i^T - 2v\lambda_4 D_{\eta\eta}\left(\eta^{(c)}\right)^{(t)} + 2v\lambda_4 S_{\delta\eta}^T\left(\delta^{(c)}\right)^{(t)} \\ \left(\alpha^{(c)}\right)^{(t)} + v\sum_{i=1}^{n}\left(I(y_i = c) - \left(\pi_i^{(c)}\right)^{(t)}\right) x_i^T - 2v\lambda_4 D_{\alpha\alpha}\left(\alpha^{(c)}\right)^{(t)} + 2v\lambda_4 S_{\delta\alpha}^T\left(\delta^{(c)}\right)^{(t)} \end{bmatrix}.$$

Let

$$u_{\delta_k}^{(t)} = \begin{bmatrix} \left(u_{\delta_k}^{(1)}\right)^{(t)} \\ \vdots \\ \left(u_{\delta_k}^{(C-1)}\right)^{(t)} \end{bmatrix}, \quad k = 1,\ldots,K, \quad u_{\eta_l}^{(t)} = \begin{bmatrix} \left(u_{\eta_l}^{(1)}\right)^{(t)} \\ \vdots \\ \left(u_{\eta_l}^{(C-1)}\right)^{(t)} \end{bmatrix}, \quad l = 1,\ldots,L$$

$$u_{\alpha_g}^{(t)} = \begin{bmatrix} \left(u_{\alpha_g}^{(1)}\right)^{(t)} \\ \vdots \\ \left(u_{\alpha_g}^{(C-1)}\right)^{(t)} \end{bmatrix}, \quad \left(u_{\alpha_g}^{(c)}\right)^{(t)} = \begin{bmatrix} \left(u_{\alpha_{g1}}^{(c)}\right)^{(t)} \\ \vdots \\ \left(u_{\alpha_{gk_g}}^{(c)}\right)^{(t)} \end{bmatrix}, \quad g = 1,2,\ldots,G.$$

Step 4: Update

$$\delta^{(t+1)} = \begin{bmatrix} \delta_1^{(t+1)} \\ \vdots \\ \delta_K^{(t+1)} \end{bmatrix} = \begin{bmatrix} \left(1 - \frac{v\lambda_1\phi_{11}}{\|u_{\delta_1}^{(t)}\|_2}\right)_+ u_{\delta_1}^{(t)} \\ \vdots \\ \left(1 - \frac{v\lambda_1\phi_{1K}}{\|u_{\delta_K}^{(t)}\|_2}\right)_+ u_{\delta_K}^{(t)} \end{bmatrix},$$

$$\eta^{(t+1)} = \begin{bmatrix} \eta_1^{(t+1)} \\ \vdots \\ \eta_L^{(t+1)} \end{bmatrix} = \begin{bmatrix} \left(1 - \frac{\nu\lambda_2\phi_{21}}{\|u_{\eta_1}^{(t)}\|_2}\right)_+ u_{\eta_1}^{(t)} \\ \vdots \\ \left(1 - \frac{\nu\lambda_2\phi_{2L}}{\|u_{\eta_L}^{(t)}\|_2}\right)_+ u_{\eta_L}^{(t)} \end{bmatrix},$$

$$\alpha^{t+1} = \begin{bmatrix} \alpha_1^{t+1} \\ \vdots \\ \alpha_G^{t+1} \end{bmatrix} = \begin{bmatrix} \left(1 - \frac{\nu\lambda_3\phi_{31}}{\|u_{\alpha_1}^{(t)}\|_2}\right)_+ u_{\alpha_1}^{(t)} \\ \vdots \\ \left(1 - \frac{\nu\lambda_3\phi_{3g}}{\|u_{\alpha_G}^{(t)}\|_2}\right)_+ u_{\alpha_G}^{(t)} \end{bmatrix}.$$

Step 5: Check convergence:

If $\|\delta^{t+1}-\delta^t\|_2+\|\eta^{t+1}-\eta^t\|_2+\|\alpha^{t+1}-\alpha^t\|_2<\varepsilon$, stop. Otherwise, if $\delta^t \leftarrow \delta^{t+1}, \eta^t \leftarrow \eta^{t+1}, \alpha^t \leftarrow \alpha^{t+1}$, go to step 3.

The penalties ϕ_{1k}, ϕ_{2l}, and ϕ_{3g} can be determined by two steps. At the first step, assume that they are all equal to 1 and then use proximal methods to solve the problem. At the second step, we define $\phi_{1k} = \frac{1}{\|\tilde{\delta}_k\|_2}$, $\phi_{2l} = \frac{1}{\|\tilde{\eta}_l\|_2}$, and $\phi_{3g} = \frac{1}{\|\tilde{\alpha}_g\|_2}$, where $\tilde{\delta}$, $\tilde{\eta}$, and $\tilde{\alpha}$ are the solutions at the first step.

8.2 FISHER'S LINEAR DISCRIMINANT ANALYSIS

8.2.1 Fisher's Linear Discriminant Analysis for Two Classes

Fisher's linear discriminant analysis (LDA) has been a widely used tool for classification in machine learning. Due to its simplicity and high computational speed, LDA was our first choice for classification. Fisher's linear discriminant is a classification method that projects high-dimensional data onto a line and performs classification in this one-dimensional space (Figure 8.6). The projection maximizes the distance between the means of the two classes while minimizing the variance within each class.

Consider two populations, which are to be separated. Let x be a vector of observations and $y=a^Tx$ be a linear combination of the observations, which is usually referred to as a discriminant direction. Let $\bar{x}_1$ and $\bar{x}_2$ be the means of the observations in populations 1 and 2, respectively. Thus, the projection of the means $\bar{x}_1$ and $\bar{x}_2$ to the discriminant direction is (Figure 8.6)

$$\bar{y}_1 = a^T\bar{x}_1, \quad \bar{y}_2 = a^T\bar{x}_2.$$

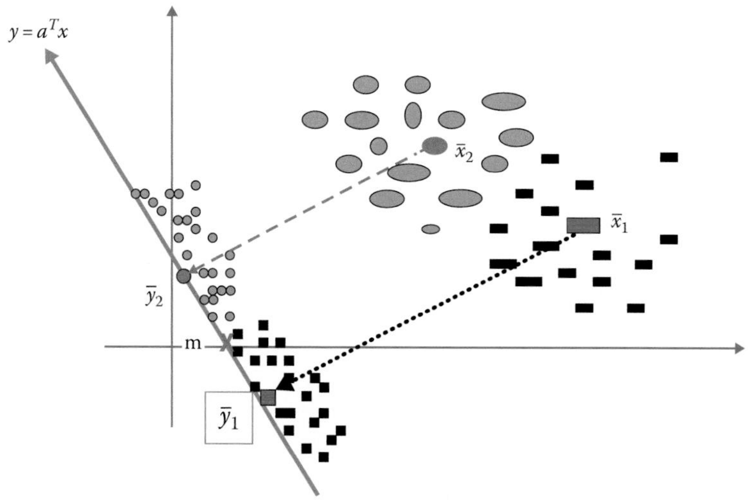

FIGURE 8.6 Scheme of linear discriminant analysis.

The square of the difference between $\bar{y}_1$ and $\bar{y}_2$ is given by

$$\begin{aligned}(\bar{y}_1 - \bar{y}_2)^2 &= a^T(\bar{x}_1 - \bar{x}_2)(\bar{x}_1 - \bar{x}_2)^T a \\ &= a^T dd^T a,\end{aligned}$$

where $d = \bar{x}_1 - \bar{x}_2$ is the difference between the two centers of the populations.

Assume that the covariance matrices of populations 1 and 2 are equal and denoted by Σ. Then, the variance of $\bar{y}_1 - \bar{y}_2$ is given by

$$\begin{aligned}\mathrm{Var}(\bar{y}_1 - \bar{y}_2) &= a^T \mathrm{cov}(\bar{x}_1 - \bar{x}_2) a \\ &= a^T\left(\frac{1}{n_1}\Sigma + \frac{1}{n_2}\Sigma\right)a \\ &= \frac{n_1 + n_2}{n_1 n_2} a^T \Sigma a,\end{aligned}$$

where n_1 and n_2 are the number of individuals sampled from populations 1 and 2.

The estimate of variance of $\bar{y}_1 - \bar{y}_2$ is given by $\frac{n_1 + n_2}{n_1 n_2} a^T S a$, where S is the pooled estimate of the covariance matrix Σ. Our goal is to select the vector a to achieve separation of the sample means $\bar{y}_1$ and $\bar{y}_2$. To reach this goal, maximize the ratio of the square of the difference between $\bar{y}_1$ and $\bar{y}_2$ to its variance:

$$\mathrm{Max} \frac{|\bar{y}_1 - \bar{y}_2|^2}{\mathrm{var}(\bar{y}_1 - \bar{y}_2)} = \frac{a^T dd^T a}{a^T S a}, \tag{8.49}$$

where $d = \bar{x}_1 - \bar{x}_2$.

Since a constant will not affect the results of maximization in Equation 8.49, $\frac{n_1+n_2}{n_1n_2}a^TSa$ is replaced by a^TSa.

To solve the optimization problem (8.49), we pose the following constraint to normalize the data:

$$a^TSa=1. \quad (8.50)$$

Therefore, the optimization problem (8.49) can be reduced to the following optimization problem:

$$\begin{aligned} &\text{Max} \quad a^Tdd^Ta \\ &\text{s.t.} \quad a^TSa=1. \end{aligned} \quad (8.51)$$

The constrained optimization problem (8.51) can be reduced to the unconstrained optimization problem by the Lagrange multiplier method:

$$\text{Max} \quad F=a^Tdd^Ta+\lambda\left(1-a^TSa\right).$$

Taking the derivative of F with respect to yields

$$dd^Ta-\lambda Sa=0 \quad (8.52)$$

and multiplying a^T on both sides of the equation and applying Equation 8.50, we have

$$\lambda=\left(a^Td\right)^2. \quad (8.53)$$

Substituting Equation 8.53 into Equation 8.52 gives

$$a=\frac{1}{a^Td}S^{-1}d. \quad (8.54)$$

Multiplying a^Tdd^T on both sides of Equation 8.54, we obtain

$$a^Tdd^Ta=d^TS^{-1}d. \quad (8.55)$$

Therefore, the projection of point x on the discriminant direction is

$$y=\left(\bar{x}_1-\bar{x}_2\right)^TS^{-1}x. \quad (8.56)$$

The square of the distance between two populations, D^2, is defined as $\left(\bar{y}_1-\bar{y}_2\right)^2$. Recall that

$$\left(\bar{y}_1-\bar{y}_2\right)^2=a^Tdd^Ta. \quad (8.57)$$

Substituting Equation 8.55 into Equation 8.57 yields

$$D^2 = d^T S^{-1} d. \tag{8.58}$$

Let X_1 and X_2 denote observations from populations 1 and 2, respectively. Fisher's idea is to project the high-dimensional multivariate observations x_1 and x_2 on the discriminant direction such that their projections, y_1 and y_2, are separated as much as possible. Fisher suggested taking linear combinations of the x's to generate y's, which can be easily manipulated mathematically. The midpoint $\hat{m}$ between the two projections of the sample means on the discriminant direction, $\bar{y}_1 = (\bar{x}_1 - \bar{x}_2)^T S^{-1} \bar{x}_1$ and $\bar{y}_2 = (\bar{x}_1 - \bar{x}_2)^T S^{-1} \bar{x}_2$, is given by

$$\hat{m} = \frac{1}{2}(\bar{x}_1 - \bar{x}_2)^T S^{-1} (\bar{x}_1 + \bar{x}_2).$$

Allocation Rule

The classification rule based on Fisher's linear discrimination function for an unknown sample, x_0, is as follows:

Assign x_0 to population 1, if $(\bar{x}_1 - \bar{x}_2)^T S^{-1} x_0 \geq \hat{m}$.
Assign x_0 to population 2, if $(\bar{x}_1 - \bar{x}_2)^T S^{-1} x_0 \leq \hat{m}$.

Example 8.3

Suppose that the mean values in normal and disease populations are, respectively, given by

$$\bar{X}_1 = \begin{bmatrix} -0.0150 \\ -0.080 \end{bmatrix}, \quad \bar{X}_2 = \begin{bmatrix} -0.300 \\ 0.0400 \end{bmatrix}$$

and the inverse of the sampling matrix is given by

$$S_{pooled}^{-1} = \begin{bmatrix} 131.158 & -90.423 \\ -90.423 & 108.147 \end{bmatrix}.$$

The discriminant direction a is

$$a = S^{-1}(\bar{x}_1 - \bar{x}_2) = \begin{bmatrix} 48.2308 \\ -38.7482 \end{bmatrix}.$$

The midpoint $\hat{m}$ is

$$\hat{m} = \frac{1}{2}(\bar{x}_1 - \bar{x}_2)^T S^{-1} (\bar{x}_1 + \bar{x}_2) = -6.8214.$$

Assume that a new observation is

$$x_0 = \begin{bmatrix} -0.210 \\ -0.44 \end{bmatrix}.$$

Since

$$\left(\bar{x}_1 - \bar{x}_2\right)^T S^{-1} x_0 = 6.9207 > \hat{m} = -6.8214,$$

then the new observation x_0 is from population 1.

8.2.2 Multiclass Fisher's Linear Discriminant Analysis

Fisher's discriminant analysis can be extended to multiple classes. Its primary purpose is to separate populations. It can, however, also be used to classify the subjects. It does not need to specify a normal distribution, but it indeed needs to assume that population covariance matrices are equal and of full rank. Linear discriminant analysis seeks projections such that high-dimensional data can be mapped into the most discriminative low-dimensional subspace (Johnson and Wichern 2002).

Consider K classes and the p-dimensional observation vector x. Let μ_i and $\bar{\mu}$ be the means of the ith population ($i = 1, 2, \ldots, K$) and combined populations, respectively.

Define between-class scatter matrix:

$$B_\mu = \sum_{i=1}^{K} \left(\mu_i - \bar{\mu}\right)\left(\mu_i - \bar{\mu}\right)^T. \tag{8.59}$$

Instead of one discriminant direction in a two-class linear discriminant analysis, the multiclass discriminant analysis will seek multiple discriminant directions.

Consider the projection of the observation x from the ith population on a discriminant direction

$$y_i = a^T x_i,$$

and calculate the expected values:

$$\mu_{yi} = E\left[y_i\right] = a^T \mu_i. \tag{8.60}$$

If population covariance matrices are equal, then we have

$$\operatorname{var}\left(y_i\right) = \operatorname{var}\left(y\right) = a^T \operatorname{cov}\left(x\right) a = a^T \Sigma a, \tag{8.61}$$

where $\operatorname{cov}(x) = \Sigma$.

The overall mean is defined as

$$\bar{\mu}_Y = a^T \bar{\mu}.$$

Then, the sum of squared distances from populations to the overall mean of Y is

$$\sum_{i=1}^{K} \left(\mu_{yi} - \bar{\mu}_y\right)^2 = a^T B_\mu a. \tag{8.62}$$

The first discriminant direction, a_1, is obtained by maximizing the ratio of the between-class scatter matrix to the variance of Y:

$$\max_a \frac{a^T B_\mu a}{a^T \Sigma a}. \tag{8.63}$$

Set

$$a^T \Sigma a = 1.$$

Then, the optimization problem (8.63) is reduced to solving the following optimization problem:

$$\begin{aligned} &\max_a \quad a^T B_\mu a \\ &\text{s.t.} \quad a^T \Sigma a = 1. \end{aligned} \tag{8.64}$$

Using the Lagrange multiplier, the constrained optimization problem (8.64) is reduced to the following unconstrained optimization problem:

$$F = a^T B_\mu a + \lambda\left(1 - a^T \Sigma a\right).$$

Setting the partial derivative $\frac{\partial F}{\partial a} = 0$, we obtain

$$B_\mu a - \lambda \Sigma a = 0,$$

which implies

$$\Sigma^{-1} B_\mu a = \lambda a. \tag{8.65}$$

The vector a_1 is the eigenvector of the matrix $\Sigma^{-1} B_\mu$ and the ratio is equal to

$$\lambda_1 = D^2 = \frac{a_1^T B_\mu a_1}{a_1^T \Sigma a_1}. \tag{8.66}$$

The ith discriminant direction maximizes the ratio (8.63) subject to being orthogonal to all previous $i-1$ directions. In general, the number of discriminant directions will be $\min(p, K-1)$.

Now we study how to classify objects. Let A be the matrix consisting of eigenvectors with nonzero eigenvalues:

$$A = \begin{bmatrix} a_1 & \cdots & a_s \end{bmatrix}^T,$$

where s is the number of the nonzero eigenvalues.

Assume that x_0 is a new observation. The projection of the new observations and means of the populations to the discriminant directions are equal to

$$y_0 = Ax_0, \quad y_1 = A\bar{x}_1, \ldots, y_k = A\bar{x}_K.$$

Define the distance between the new observation and populations as

$$d_i = \|y_0 - y_i\|^2, \quad i = 1, \ldots, K.$$

Find a j such that

$$j = \arg\ \min(d_i).$$

Then, the new observation is classified to the class j.

8.2.3 Connections between Linear Discriminant Analysis, Optimal Scoring, and Canonical Correlation Analysis (CCA)

8.2.3.1 Matrix Formulation of Linear Discriminant Analysis

Now we consider a general formulation of LDA. Consider an $n \times p$ design matrix:

$$X = \begin{bmatrix} x_{11} & \cdots & x_{1p} \\ \vdots & \vdots & \vdots \\ x_{n1} & \cdots & x_{np} \end{bmatrix}. \tag{8.67}$$

For convenience, assume that each column of the matrix X is centered to have zero mean unit variance.

Define an $n \times K$ matrix of dummy variables for the K classes:

$$Y = \begin{bmatrix} y_{11} & \cdots & y_{1K} \\ \vdots & \vdots & \vdots \\ y_{n1} & \cdots & y_{nK} \end{bmatrix}, \tag{8.68}$$

where y_{ik} is an indicator variable for whether the ith observation $x_i = [x_{i1}, \ldots, x_{ip}]$ belongs to the kth class.

Let $\Omega_k = \{i | y_{ik} = 1\}$, $k = 1, \ldots, K$. The total sample covariance matrices Σ can be partitioned into the sum of the between-class covariance Σ_b and within-class covariance Σ_w:

$$\begin{aligned}\Sigma &= \frac{1}{n} X^T X \\ &= \frac{1}{n} \sum_{i=1}^{n} x_i^T x_i \\ &= \frac{1}{n} \sum_{k=1}^{K} \sum_{i\in\Omega_k} x_i^T x_i \\ &= \frac{1}{n} \sum_{k=1}^{K} \sum_{i\in\Omega_k} \left(x_i - \bar{x}_k + \bar{x}_k\right)^T \left(x_i - \bar{x}_k + \bar{x}_k\right) \\ &= \frac{1}{n} \sum_{k=1}^{K} \sum_{i\in\Omega_k} \left(x_i - \bar{x}_k\right)^T \left(x_i - \bar{x}_k\right) + \frac{1}{n} \sum_{k=1}^{K} n_k \bar{x}_k^T \bar{x}_k \\ &= \Sigma_w + \Sigma_b, \end{aligned} \tag{8.69}$$

where $\Sigma_w = \frac{1}{n} \sum_{k=1}^{K} \sum_{i\in\Omega_k} \left(x_i - \bar{x}_k\right)^T \left(x_i - \bar{x}_k\right)$, $\Sigma_b = \frac{1}{n} \sum_{k=1}^{K} n_k \bar{x}_k^T \bar{x}_k$, $\bar{x}_k = \frac{1}{n_k} \sum_{i\in\Omega_k} x_i$, and n_k is the sample size of the kth class.

The diagonal matrix of the sample sizes can be expressed in terms of Y:

$$Y^T Y = \begin{bmatrix} \sum_{i=1}^{n} y_{i1}^2 & \cdots & \sum_{i=1}^{n} y_{i1} y_{iK} \\ \vdots & \vdots & \vdots \\ \sum_{i=1}^{n} y_{iK} y_{i1} & \cdots & \sum_{i=1}^{n} y_{iK}^2 \end{bmatrix} = \begin{bmatrix} n_1 & \cdots & 0 \\ \vdots & \vdots & \vdots \\ 0 & \cdots & n_k \end{bmatrix}. \tag{8.70}$$

Similarly, the between-class covariance matrix and within-class covariance matrix can be expressed in terms of the matrices X and Y:

$$Y^T X = \begin{bmatrix} \sum_{i\in\Omega_1} x_{i1} & \cdots & \sum_{i\in\Omega_1} x_{ip} \\ \vdots & \vdots & \vdots \\ \sum_{i\in\Omega_k} x_{i1} & \cdots & \sum_{i\in\Omega_k} x_{ip} \end{bmatrix} = \begin{bmatrix} n_1 \bar{x}_1 \\ \vdots \\ n_K \bar{x}_K \end{bmatrix}, \tag{8.71}$$

$$\frac{1}{n}\left(Y^TX\right)^T\left(Y^TY\right)^{-1}Y^TX=\left[n_1\bar{x}_1^T \quad \cdots \quad n_K\bar{x}_K^T\right]\begin{bmatrix}1/n_1 & \cdots & 0\\ \vdots & \vdots & \vdots\\ 0 & \cdots & 1/n_K\end{bmatrix}\begin{bmatrix}n_1\bar{x}_1\\ \vdots\\ n_K\bar{x}_K\end{bmatrix}$$

$$=\frac{1}{n}\sum_{k=1}^{K}n_k\bar{x}_k^T\bar{x}_k=\Sigma_b, \tag{8.72}$$

and

$$\begin{aligned}\Sigma_w&=\Sigma-\Sigma_b\\ &=\frac{1}{n}X^T\left(1-P_Y\right)X,\end{aligned} \tag{8.73}$$

where $P_Y = Y(Y^TY)^{-1}Y^T$.

Let $A=[a_1, \ldots, a_s]$ be the matrix of discriminant vectors (directions) and be referred to as the discriminant matrix. The LDA problem can be formulated as solving the following constrained optimization problem (Wu et al. 2015):

$$\begin{aligned}&\max_A \ \operatorname{Tr}\left(A^T\Sigma_bA\right)\\ &\text{s.t.} \ \ A^T\Sigma_wA=I.\end{aligned} \tag{8.74}$$

The Lagrangian function for solving the optimization problem (8.74) is

$$L(A,\Lambda)=\operatorname{Tr}\left(A^T\Sigma_bA\right)+\operatorname{Tr}\left(\Lambda\left(I-A^T\Sigma_wA\right)\right). \tag{8.75}$$

Using Equation 1.164, we obtain

$$\frac{\partial L(A,\Lambda)}{\partial A}=2\left(\Sigma_bA-\Sigma_wA\Lambda\right)=0,$$

which implies the eigenequation

$$\Sigma_bA=\Sigma_wA\Lambda. \tag{8.76}$$

Eigenequation (8.76) can be transformed to

$$\Sigma_w^{-1/2}\Sigma_b\Sigma_w^{-1/2}B=B\Lambda, \tag{8.77}$$

where $B=\Sigma_w^{1/2}A$.

8.2.3.2 Optimal Scoring and Its Connection with Linear Discriminant Analysis

Optimal scoring is another formulation of the LDA (Clemmensen et al. 2011; Hastie et al. 1994). We can formulate the discriminant analysis problem as a regression with categorical response Y. Let $\theta_1, \ldots, \theta_K$ be the scores for the categories $1, 2, \ldots, K$. Taking $Y\theta$ as response, we can regress $Y\theta$ on the observations X. The regression coefficients are the discriminant vectors. In other words, the LDA can be formulated as the following optimal scoring problem:

$$\begin{aligned} &\min_{B,\theta} \quad \|Y\theta - XB\|_F^2 \\ &\text{s.t.} \quad \frac{1}{n}\theta^T Y^T Y\theta = I, \end{aligned} \tag{8.78}$$

where $\|.\|_F$ denotes the Frobenius norm. We can show that the optimal scoring problem is equivalent to the LDA (Appendix 8A). The scoring $\theta = [\theta_1, \ldots, \theta_L]$, $L \leq K - 1$ is referred to as the scoring matrix and $B = [B_1, \ldots, B_L]$ is referred to as the discriminant direction matrix. As a consequence, LDA can be reformulated as a multioutput regression problem. All regression techniques can be applied to LDA.

8.2.3.3 Connection between LDA and CCA

In this section, we show the equivalence of CCA and LDA. Consider two datasets: the observation matrix X and categorical response matrix Y. In a matrix form, CCA can be formulated as

$$\begin{aligned} &\max_{\theta,A} \quad \mathrm{Tr}\left(\theta^T Y^T XA\right) \\ &\text{s.t.} \quad \theta^T Y^T Y\theta = I, \quad A^T X^T XA = I. \end{aligned} \tag{8.79}$$

The Lagrangian multiplier algorithm for solving a constrained optimization problem (8.79) gives

$$\max_{\theta,A,\Lambda,\Pi} \quad F = \mathrm{Tr}\left(\theta^T Y^T XA\right) + \frac{1}{2}\mathrm{Tr}\left[\Lambda\left(I - \theta^T Y^T Y\theta\right)\right] + \frac{1}{2}\mathrm{Tr}\left[\Pi\left(I - A^T X^T XA\right)\right].$$

Taking a derivative of function, F, with respect to the matrices θ and A, and using Equation 1.164, we obtain

$$\frac{\partial F}{\partial \theta} = Y^T XA - Y^T Y\theta\Lambda = 0, \tag{8.80}$$

$$\frac{\partial F}{\partial A} = X^T Y\theta - X^T XA\Pi = 0. \tag{8.81}$$

Multiplying Equation 8.80 by θ^T and Equation 8.81 by A^T and using Equation 8.79 gives

$$\Lambda = \theta^T Y^T XA = A^T X^T Y\theta = \Pi.$$

Consequently, Equations 8.80 and 8.81 are reduced to

$$Y^T XA - Y^T Y\theta\Lambda = 0, \tag{8.82}$$

$$X^T Y\theta - X^T XA\Lambda = 0. \tag{8.83}$$

Solving Equation 8.82 for θ gives

$$\theta = \left(Y^T Y\right)^{-1} Y^T XA\Lambda^{-1}. \tag{8.84}$$

Substituting Equation 8.84 into Equation 8.83, we obtain

$$X^T Y\left(Y^T Y\right)^{-1} Y^T XA\Lambda^{-1} - X^T XA\Lambda = 0. \tag{8.85}$$

Substituting Equations 8.69 and 8.72 into Equation 8.85 yields

$$\Sigma_b A = \left(\Sigma_b + \Sigma_w\right) A\Lambda^2. \tag{8.86}$$

Recall that LDA in (8.74) can also be formulated as

$$\begin{aligned} &\max_A \quad \operatorname{Tr}\left(A^T \Sigma_b A\right) \\ &\text{s.t.} \quad A^T\left(\Sigma_b + \Sigma_w\right) A = I. \end{aligned} \tag{8.87}$$

Solving the optimization problem (8.87) leads to the following eigenequation:

$$\Sigma_b A = \left(\Sigma_b + \Sigma_w\right) A\Lambda_0. \tag{8.88}$$

Taking $\Lambda_0 = \Lambda^2$, Equations 8.86 and 8.88 are equivalent. The canonical directions and the discriminant directions are the same. The R^2 matrix is given by

$$R^2 = \Sigma_{xx}^{-1/2} \Sigma_b \Sigma_{xx}^{-1/2}.$$

8.3 SUPPORT VECTOR MACHINE

Support vector machines (SVMs) are a set of related supervised learning methods used for classification and regression and are perhaps one of the most popular algorithms in machine learning (Burges 1998; Christianini and Shawe-Taylor 2000; Vapnik 1998). The basic idea that drives the initial development of SVMs is that for a given learning task, with a given finite amount of training data, the best generalization performance will be achieved by the balance between the accuracy attained on any dataset and the ability of the machine to learn the particular training set without error. In this section, the basic SVM learning algorithms will be introduced and their recent development, penalized SVMs to make the material self-contained, will be presented.

8.3.1 Introduction

The SVMs are theoretically motivated by statistical learning theory and are successfully applied in many fields. We start the introduction with the binary-class SVMs. Recall that in geometry, each data point can be represented by a p-dimensional vector (a list of p numbers) and belongs to only one of two classes. Classify binary-class points by assigning them to one of two disjoint half spaces either in the original input space of the problem for linear classifiers or in a higher-dimensional feature space for nonlinear classifiers (Fung and Mangasarian 2004). Similar to linear discriminant analysis, the question is whether we can separate them with a $p-1$ dimensional hyperplane. There are many hyperplanes that might satisfy this property (Figure 8.7). However, our task is to find out if we can achieve maximum separation (margin) between the two classes. In other words, we attempt to find the hyperplane so that the distance from the hyperplane to the nearest data point is maximized. Define such hyperplanes as the maximum-margin hyperplane and such a linear classifier is known as a maximum-margin classifier (Figure 8.8).

8.3.2 Linear Support Vector Machines

8.3.2.1 Separable Case

Begin with the simplest case: linear SVMs trained on separable data. Given a training set $\{x_i, y_i\}_{i=1}^{n}$ with input feature patterns x_i and output values $y_i \in \{-1, +1\}$ for class labels, a hyperplane can be mathematically represented by an algebra equation (Figure 8.9):

$$w^T x + b = 0, \tag{8.89}$$

where w is the normal vector perpendicular to the separating hyperplane. Adding the offset parameter b allows the margin to be increased. In its absence, the hyperplane is forced to

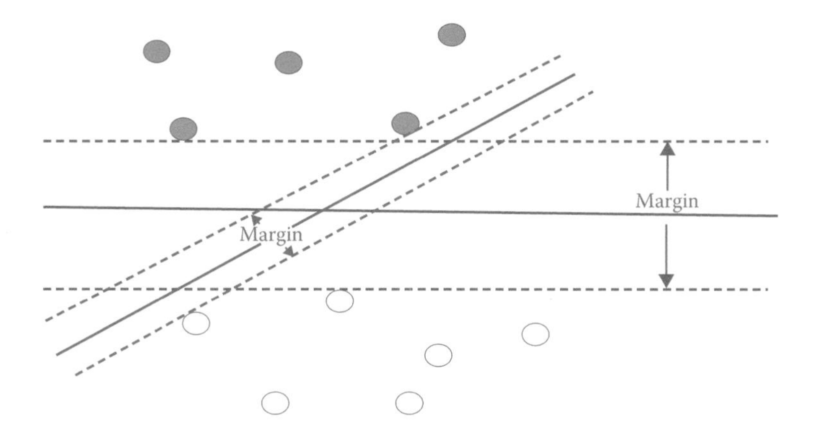

FIGURE 8.7 Multiple planes that can separate two classes of observations.

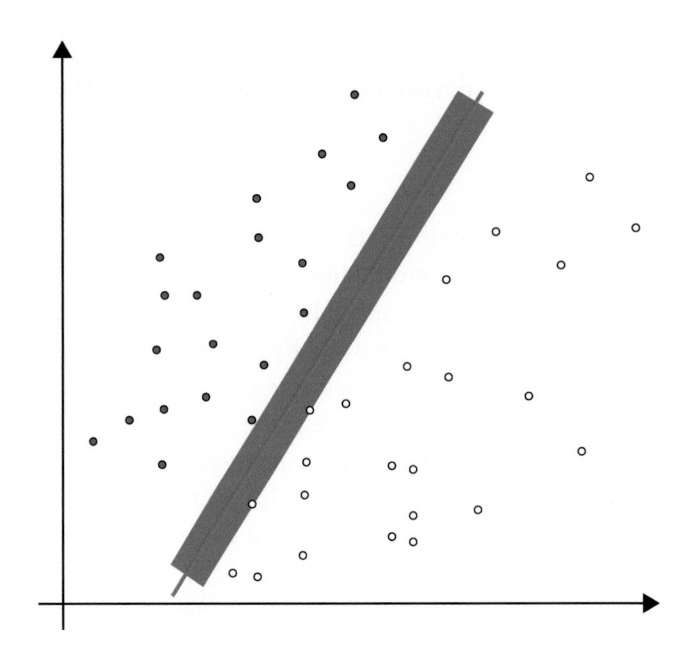

FIGURE 8.8 Maximum-margin hyperplane.

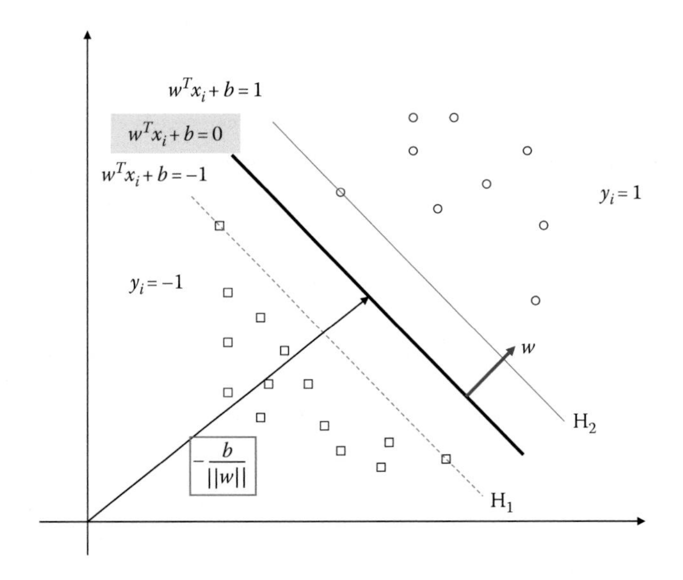

FIGURE 8.9 Linear separating hyperplane.

pass through the origin. Let $||w||$ be the Euclidean norm of w. Then, $\frac{|b|}{||w||}$ is the perpendicular distance from the hyperplane to the origin. Let x_0 be a point in the space. We can show that the distance from x_0 to the plane is equal to (Appendix 8C)

$$\frac{\left|w^T x_0 + b\right|}{||w||}. \tag{8.90}$$

In the linear separable case, the purpose of the SVMs is to seek the separating hyperplane with largest margin. Suppose that all training data satisfy the following constraints:

$$w^T x_i + b \geq 1 \quad \text{for } y_i = 1$$

$$w^T x_i + b \leq -1 \quad \text{for } y_i = -1,$$

which can be combined as

$$y_i \left(w^T x_i + b \right) - 1 \geq 0. \tag{8.91}$$

When the points lie on the hyperplane H_1, we have $w^T x_i + b = -1$, and when the points lie on the hyperplane H_2, the constraint $w^T x_i + b = 1$ holds. The perpendicular distances from the origin to the hyperplane H_1 and the hyperplane H_2 are $\frac{|-1-b|}{||w||}$ and $\frac{|1-b|}{||w||}$, respectively. Thus, the distance between two hyperplanes, H_1 and H_2, is $\frac{2}{||w||}$.

The hyperplane H_1 and the hyperplane H_2 are parallel and parallel to the separating plane, and there are no training points between them. The margin of the separating hyperplane is defined as the distance between the hyperplane H_1 and the hyperplane H_2, which is equal to $\frac{2}{||w||}$. The goal is to find a hyperplane (defined by w and b) with the largest margin. Mathematically, attempt to find a normal vector w of a separating hyperplane maximizing $\frac{2}{||w||}$ or minimizing $||w||$. Formally, to find a separating hyperplane with the largest margin can be expressed as

$$\begin{aligned} &\min \quad \frac{1}{2} w^T w \\ &\text{s.t.} \quad -y_i \left(w^T x_i + b \right) + 1 \leq 0, \ \forall i. \end{aligned} \tag{8.92}$$

Using the Lagrangian multiplier, the optimization problem (8.92) can be reformulated as

$$L_P = \frac{1}{2} w^T w - \sum_{i=1}^{l} \alpha_i \left[y_i \left(w^T x_i + b \right) - 1 \right], \tag{8.93}$$

where $\alpha_i \geq 0, \forall i$.

Problem (8.93) is a convex quadratic programming problem. The problem (8.93) is referred to as the primary problem. We can equivalently solve the following dual problem:

$$\max_{\alpha} \min_{w,b} L_P == \frac{1}{2} w^T w - \sum_{i=1}^{l} \alpha_i \left[y_i \left(w^T x_i + b \right) - 1 \right]. \tag{8.94}$$

To minimize L_P, set partial derivatives of L_p with respect to w and b to be equal to zero, which leads to

$$\begin{aligned}\frac{\partial L_P}{\partial w} &= w - \sum_{i=1}^{l} \alpha_i y_i x_i = 0 \\ \frac{\partial L_P}{\partial b} &= \sum_{i=1}^{l} \alpha_i y_i = 0.\end{aligned} \tag{8.95}$$

Substituting Equation 8.95 into Equation 8.93 yields

$$L_D = \sum_{i=1}^{l} \alpha_i - \frac{1}{2} \sum_{i=1}^{l} \sum_{j=1}^{l} \alpha_i \alpha_j y_i y_j x_i^T x_j. \tag{8.96}$$

Note that we give the Lagrangian different labels: P for primal and D for dual. The problem (8.94) is then transformed to

$$\begin{aligned}&\max \quad L_D \\ &\text{s. t.} \quad \alpha_i \geq 0, \quad \sum_{i=1}^{l} \alpha_i y_i = 0, \ \forall i.\end{aligned} \tag{8.97}$$

Next we need to develop an algorithm to solve the dual problem. The sequential minimal optimization algorithm can be used to solve the dual problem and will be introduced in Section 8.3.2.4.

8.3.2.2 Nonseparable Case

In many cases, as shown in Figure 8.10, the data from different classes are nonseparable. The above algorithm for separable cases cannot be applied to a nonseparable case. However, we can modify the above algorithms to include a nonseparable case by relaxing the constraints (8.91). By introducing the positive slack variable ξ_i in constraints (8.91) (Figure 8.11), we obtain

$$\begin{aligned}&w^T x_i + b \geq 1 - \xi_i \quad \text{for } y_i = 1 \\ &w^T x_i + b \leq -1 + \xi_i \quad \text{for } y_i = -1 \\ &\xi_i \geq 0, \ \forall i,\end{aligned} \tag{8.98}$$

where ξ_i is used to measure errors. The constraints (8.98) can be rewritten as

$$y_i \left(w^T x_i + b \right) \geq 1 - \xi_i. \tag{8.99}$$

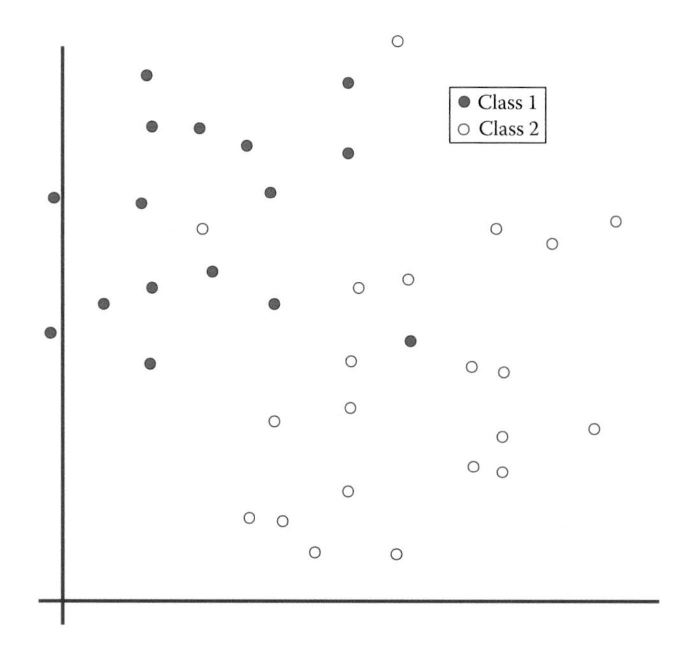

FIGURE 8.10 Nonseparable case.

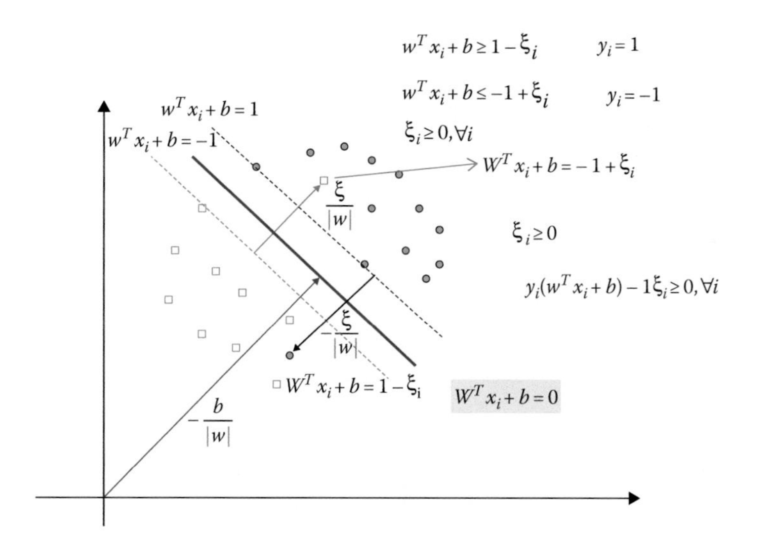

FIGURE 8.11 Constraints for the nonseparable case.

Taking the errors into consideration, an extra cost for errors needs to be added into the objective function for the separable case. Thus, for a nonseparable case, the objective function becomes

$$\frac{1}{2} w^T w + c \sum_i \xi_i, \tag{8.100}$$

where c is a positive number and chosen by users for penalty of the errors. A large c corresponds to a high penalty to errors. Again, this is also a convex quadratic programming problem. By the same argument as before, using the Lagrange multipliers, the following primal Lagrange is

$$L_p = \frac{1}{2} w^T w + c \sum_i \xi_i - \sum_i \alpha_i \left[y_i \left(w^T x_i + b \right) - 1 + \xi_i \right] - \sum_i \mu_i \xi_i, \tag{8.101}$$
$$\alpha_i \geq 0, \quad \mu_i \geq 0, \quad \xi_i \geq 0.$$

Here, the α_is and μ_is are Lagrange multipliers and are constrained to be nonnegative. The dual Lagrange function is obtained by setting the partial derivatives of L_p with respect to w, b, ξ_i to be equal to zero:

$$\frac{\partial L_P}{\partial w} = w - \sum \alpha_i y_i x_i = 0,$$
$$\frac{\partial L_P}{\partial b} = -\sum_i \alpha_i y_i = 0,$$
$$\frac{\partial L}{\partial \xi_i} = c - \alpha_i - \mu_i = 0.$$

In other words, we obtain

$$w = \sum_i \alpha_i y_i x_i, \quad \sum_i \alpha_i y_i = 0, \quad \text{and} \quad c = \alpha_i + \mu_i.$$

Substituting the above equalities into Equation 8.101 results in the following dual problem of SVMs for a nonseparable case:

$$\text{Maximize} \quad L_D = \sum_i \alpha_i - \frac{1}{2} \sum_i \sum_j \alpha_i \alpha_j y_i y_j x_i^T x_j \tag{8.102}$$
$$\text{s.t.} \quad 0 \leq \alpha_i \leq c, \quad \sum_i \alpha_i y_i = 0.$$

8.3.2.3 The Karush–Kuhn–Tucker (KKT) Conditions

The Karush–Kuhn–Tucker (KKT) conditions are necessary conditions for an optimal point of the optimization problem (8.102) and play an important role in the constrained optimization theory and algorithm development. Consider a general constrained optimization problem:

$$\min_x \ f(x)$$
$$\text{s.t.} \quad g(x) \leq 0, \quad g(x) = \left[g_1(x), \ldots, g_m(x) \right]^T. \tag{8.103}$$

Let u be the Lagrange multipliers. The KKT conditions for the nonlinear optimization problem (8.103) are

1. $f(\bar{x})+\bar{u}^T g(\bar{x})=\min_x\{f(x)+\bar{u}g(x)\}$
2. Primal feasibility: $g(\bar{x})\le 0, i=1,\ldots,m$
3. Dual feasibility: $\bar{u}\ge 0$
4. Complementary slackness: $\bar{u}_i g_i(\bar{x})=0, i=1,\ldots,m$

The KKT dual-complementarity conditions can be used to test for the convergence of the SMO algorithms. Now we study the KKT dual-complementary conditions for the SVMs. The constraints for the primal feasibility for the SVMs are

$$y_i\left(w^T x_i+b\right)-1+\xi_i\ge 0, \tag{8.104}$$

$$\xi_i\ge 0, \quad i=1,\ldots,m. \tag{8.105}$$

Let $u=[\alpha^T,\mu^T]^T$. Then, their dual-complementarity conditions are

$$\alpha_i\left[y_i\left(w^T x_i+b\right)-1+\xi_i\right]=0, \tag{8.106}$$

$$\mu_i\xi_i=0, \quad i=1,\ldots,m. \tag{8.107}$$

Now consider three cases: (1) $\alpha_i=0$, (2) $0<\alpha_i<c$, and (3) $\alpha_i=c$ (ftp://www.ai.mit.edu).

1. If $\alpha_i=0$, then $\mu_i=c$. It follows from Equation 8.107 that $\xi_i=0$. Therefore, from Equation 8.104, we obtain

$$y_i\left(w^T x_i+b\right)-1\ge 0. \tag{8.108}$$

2. If $0<\alpha_i<c$, then from Equation 8.106, we have

$$y_i\left(w^T x_i+b\right)-1+\xi_i=0. \tag{8.109}$$

However, $\mu_i=c-\alpha_i>0$, which implies $\xi_i=0$. Therefore, from Equation 8.109, we have

$$y_i\left(w^T x_i+b\right)-1=0. \tag{8.110}$$

3. If $\alpha_i=c$, again from Equation 8.106, we have

$$y_i\left(w^T x_i+b\right)-1+\xi_i=0. \tag{8.111}$$

But $\mu_i = c - \alpha_i = 0$, which implies $\xi_i \geq 0$. Thus, we have

$$y_i\left(w^T x_i + b\right) - 1 \leq 0. \tag{8.112}$$

From the above discussion, observe that the KKT conditions depend on a key quantity, $y_i(w^T x_i + b) - 1$. Define

$$\begin{aligned} R_i &= y_i\left(w^T x_i + b\right) - 1 \\ &= y_i\left(w^T x_i + b\right) - y_i^2 \\ &= y_i\left(\left(w^T x_i + b\right) - y_i\right) = y_i E_i, \end{aligned} \tag{8.113}$$

where $E_i = (w^T x_i + b) - y_i$ is defined as the prediction error. In summary, the KKT dual-complementary conditions are

$$\alpha_i = 0 \Rightarrow R_i \geq 0, \tag{8.114}$$

$$\alpha_i = c \Rightarrow R_i \leq 0, \tag{8.115}$$

$$0 < \alpha_i < c \Rightarrow R_i = 0. \tag{8.116}$$

8.3.2.4 Sequential Minimal Optimization (SMO) Algorithm

The SMO is a popular algorithm for solving the dual problem (8.102) (Platt 1998). The SMO is motivated by the coordinate ascent algorithm. The original coordinate ascent algorithm is to optimize function $L_D(\alpha_1, \ldots, \alpha_m)$ with respect to just one selected variable, α_i, while holding all the other variables, $\alpha_j, j = 1, \ldots, m, \ j \neq i$, fixed. However, since all the dual variables must satisfy the constraint

$$\sum_j \alpha_j y_j = 0, \tag{8.117}$$

it implies that

$$\alpha_i y_i = -\sum_{j \neq i} \alpha_j y_j. \tag{8.118}$$

When all the other variables, $\alpha_j, j = 1, \ldots, m, j \neq i$, are fixed, it follows from Equation 8.118 that the selected variable α_i is also fixed. Therefore, no changes can be made to α_i without violating the constraint (8.117). This indicates that if we want to update the dual variables α_i, we must update at least two variables simultaneously in order to satisfy the constraint (8.117).

The SMO algorithm optimizes two dual variables (coordinates) at a time. A simple version of the SMO algorithm iterates over all $\alpha_i, i=1, \ldots, m$. If α_i does not fulfill the KKT conditions, an α_j is randomly selected from the remaining $m-1$ αs. The SMO algorithm then jointly optimizes α_i and α_j. If none of the αs are changed after a few iterations over all the αs, then the algorithm stops.

Without loss of generality, suppose we are optimizing α_1, α_2 while holding $\alpha_3, \ldots, \alpha_m$ fixed. To maximize the dual function $L_D(\alpha_1, \alpha_2, \alpha_3, \ldots, \alpha_m)$ in (8.102), first write the dual function L_D in terms of the dual variable α_2 while keeping the constraint (8.118) satisfied. Let

$$\gamma = -\sum_{j=3}^{m} \alpha_j y_j. \tag{8.119}$$

Then,

$$y_1\alpha_1 + y_2\alpha_2 = y_1\alpha_1^{(0)} + y_2\alpha_2^{(0)} = \gamma. \tag{8.120}$$

Let $K_{11} = x_1^T x_1$, $K_{12} = x_1^T x_2$, $K_{22} = x_2^T x_2$, and $\eta = 2K_{12} - K_{11} - K_{12}$. After a lengthy calculation, we can show that

$$L_D = \frac{1}{2}\eta\alpha_2^2 + \left(y_2\left(E_1^{(0)} - E_2^{(0)}\right) - \eta\alpha_2^{(0)}\right)\alpha_2 + \text{Const}, \tag{8.121}$$

where $E_1^{(0)} = \left(w^{(0)}\right)^T x_1 + b - y_1$ and $E_2^{(0)} = \left(w^{(0)}\right)^T x_2 + b - y_2$.

To find α_2 that maximizes the objective function L_D, set the first derivative of L_D with respect to α_2 equal to zero:

$$\frac{dL_D}{d\alpha_2} = \eta\alpha_2 + \left(y_2\left(E_1^{(0)} - E_2^{(0)}\right) - \eta\alpha_2^{(0)}\right) = 0. \tag{8.122}$$

Solving Equation 8.122 for α_2 gives

$$\alpha_2 = \alpha_2^{(0)} + \frac{y_2\left(E_2^{(0)} - E_1^{(0)}\right)}{\eta}. \tag{8.123}$$

To assess whether α_2 reaches maximum of the function L_D, take the second derivative of L_D with respect to α_2:

$$\frac{d^2 L_D}{d\alpha_2^2} = \eta. \tag{8.124}$$

Since $\eta = 2K_{12} - K_{11} - K_{22} = -(x_2 - x_1)^T(x_2 - x_1) \leq 0$, the L_D is a convex function and hence α_2 reaches maximum of the function L_D.

Next check whether the estimated optimal value α_2 is in the feasible region. Let $s = y_1 y_2$ and $\gamma = \alpha_1^{(0)} + s\alpha_2^{(0)}$. Consider two cases:

1. $s = 1$. Then, we have $\alpha_1 + \alpha_2 = \gamma$.
   If $\gamma > c$, from Figure 8.12, we clearly observe $\gamma - c \leq \alpha_2 \leq c$.
   If $\gamma < c$, then Figure 8.13 shows $0 \leq \alpha_2 \leq \gamma$.

Similarly, we have

2. $s = -1$. Then, we have $\alpha_1 - \alpha_2 = \gamma$.
   If $\gamma > 0$, then $0 \leq \alpha_2 \leq c - \gamma$.
   If $\gamma < 0$, then $-\gamma \leq \alpha_2 \leq c$.

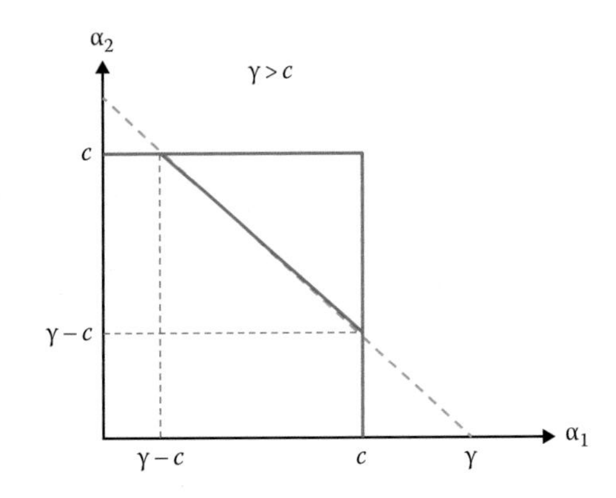

FIGURE 8.12 Range of the variables α_1 and α_2 for $\gamma > c$.

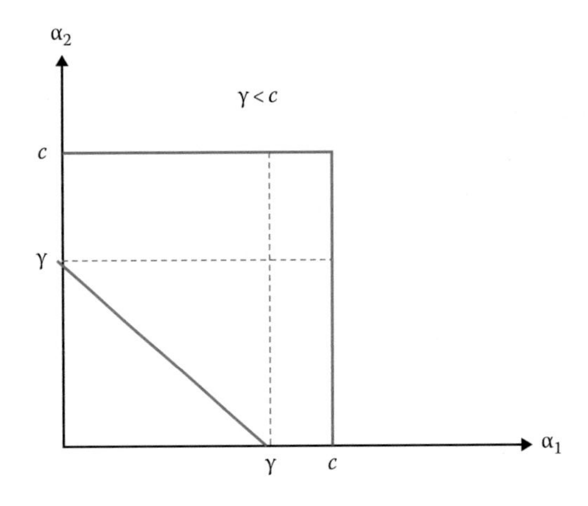

FIGURE 8.13 Range of the variables α_1 and α_2 for $\gamma < c$.

Let L be the lower bound and H be upper bound, i.e., $L \le \alpha_2 \le H$. In summary, if $y_i = y_j$, then $L = \max(0, \gamma - c)$ and $H = \min(c, \gamma)$; if $y_i \ne y_j$, then $L = \max(0, -\gamma)$ and $H = \min(c, c - \gamma)$. Therefore, if $\eta < 0$, the solutions to the dual problem are

$$\alpha_2 = \begin{cases} H & \text{if } H < \alpha_2^{(0)} + \dfrac{y_2\left(E_2^{(0)} - E_1^{(0)}\right)}{\eta} \\ \alpha_2^{(0)} + \dfrac{y_2\left(E_2^{(0)} - E_1^{(0)}\right)}{\eta} & \text{if } L \le \alpha_2^{(0)} + \dfrac{y_2\left(E_2^{(0)} - E_1^{(0)}\right)}{\eta} \le H \\ L & \text{if } \alpha_2^{(0)} + \dfrac{y_2\left(E_2^{(0)} - E_1^{(0)}\right)}{\eta} < L \end{cases} \tag{8.125}$$

and

$$\alpha_1 = \alpha_1^{(0)} - s\frac{y_2\left(E_2^{(0)} - E_1^{(0)}\right)}{\eta}. \tag{8.126}$$

If $\eta = 0$, then L_D is a linear function of α_2. We evaluate the objective function L_D at the two endpoints, 0 and c, and set α_2 to be the one with the larger objective function value L_D.

Finally, update the threshold b using the KKT dual-complementarity conditions. Let

$$b_1 = b^{(0)} - E_1^{(0)} - \left(\alpha_1 - \alpha_1^{(0)}\right) y_1 x_1^T x_1 - \left(\alpha_2 - \alpha_2^{(0)}\right) y_2 x_2^T x_1 \tag{8.127}$$

and

$$b_2 = b^{(0)} - E_2^{(0)} - \left(\alpha_1 - \alpha_1^{(0)}\right) y_1 x_1^T x_2 - \left(\alpha_2 - \alpha_2^{(0)}\right) y_2 x_2^T x_2, \tag{8.128}$$

where $b^{(0)}$ is the threshold in the previous iteration.

Let $E(x, y) = \sum_{j=1}^{m} \alpha_j y_j x_j^T x + b - y_i$. Then, the change in $E(x,y)$ due to changes in α_i is

$$\Delta E(x,y) = \Delta\alpha_1 y_1 x_1^T x + \Delta\alpha_2 y_2 x_2^T x + \Delta b. \tag{8.129}$$

Now consider three scenarios:

1. If $0 < \alpha_1 < c$, it follows from Equation 8.116 that

$$0 = R_1 = y_1 E_1,$$

which implies that

$$E_1 = E_1^{(0)} + \Delta\alpha_1 y_1 x_1^T x_1 + \Delta\alpha_2 y_2 x_2^T x_1 + \Delta b = 0. \tag{8.130}$$

Combining Equations 8.127 and 8.130 gives

$$b = b_1.$$

Similarly, if $0 < \alpha_2 < c$, then $b = b_2$ and $b_1 = b_2$ (Exercise 8.6).

2. If $\alpha_1 = 0$, then it follows from Equation 8.114 that

$$y_1 E_1 \geq 0. \tag{8.131}$$

If $y_1 > 0$, then $E_1 \geq 0$. It follows from Equation 8.129 that

$$E_1 = E_1^{(0)} + \Delta E(x_1, y) = E_1^{(0)} + \Delta\alpha_1 y_1 x_1^T x_1 + \Delta\alpha_2 y_2 x_2^T x_1 + \Delta b \geq 0,$$

or

$$E_1^{(0)} + \Delta\alpha_1 y_1 x_1^T x + \Delta\alpha_2 y_2 x_2^T x + b - b^{(0)} \geq 0, \tag{8.132}$$

which gives

$$b \geq b^{(0)} - E_1^{(0)} - \left(\alpha_1 - \alpha_1^{(0)}\right) y_1 x_1^T x_1 - \left(\alpha_2 - \alpha_2^{(0)}\right) y_2 x_2^T x_1 = b_1. \tag{8.133}$$

By similar arguments, we can show that if $\alpha_2 = 0$ and $y_2 > 0$, then $b \geq b_2$.

If $y_1 < 0$ and $y_2 < 0$ similarly, show that

$$b \leq b_1 \quad \text{and} \quad b \leq b_2. \tag{8.134}$$

3. If $\alpha_i = c$, then it follows from Equation 8.115 that

$$y_1 E_1 \leq 0.$$

If $y_1 > 0$, similarly, show that

$$b \leq b_1. \tag{8.135}$$

If $y_1 < 0$, then $E_1 > 0$. Therefore, we obtain

$$b > b_1.$$

Similarly, show that if $y_2 > 0$, then $b \le b_2$ and if $y_2 < 0$, then $b > b_2$. Assume that $\gamma \ne 0$ and if both α_1 and α_2 are at the bounds, then one will be 0 and the other one will be c. Therefore, when both α_1 and α_2 are at the bounds, all the thresholds between b_1 and b_2 satisfy the KKT conditions. Therefore, the threshold is

$$b = \begin{Bmatrix} b_1 & 0 < \alpha_1 < c \\ b_2 & 0 < \alpha_2 < c \\ (b_1 + b_2)/2 & \text{otherwise} \end{Bmatrix}. \tag{8.136}$$

8.3.3 Nonlinear SVM

Nonlinear SVM formulations start from the assumption that all the training data satisfy the following constraints:

$$\begin{cases} w^T \phi(x_i) + b \ge +1, & \text{if } y_i = +1, \\ w^T \phi(x_i) + b \le -1, & \text{if } y_i = -1. \end{cases} \tag{8.137}$$

Here, the nonlinear mapping $\phi(\cdot)$ maps the input data into a higher-dimensional space, and w is a normal to the hyperplane. Note that the dimension of w is not specified (it can be infinite dimensional). Suppose we have some hyperplane that separates the positive from the negative examples (a "separating hyperplane"). Define the "margin" of a separating hyperplane to be the summation of shortest distance from the separating hyperplane to the closest positive and negative examples. It can be shown that the margin is simply $2/\sqrt{w^T w}$. Our goal is to find the pair of hyperplanes, which gives the maximum margin. This can be accomplished by minimizing $w^T w$, subject to the above constraints.

8.3.4 Penalized SVMs

The original linear SVMs that were discussed in the previous sections attempt to find a hyperplane separating the two classes of data points:

$$\begin{aligned} &\frac{1}{2} w^T w + c \sum_i \xi_i \\ &\text{s.t.} \quad y_i \left(w^T x_i + b \right) \ge 1 - \xi_i, \quad \xi_i \ge 0. \end{aligned} \tag{8.138}$$

However, a major limitation for the traditional formulation of the SVMs is that the SVM cannot perform the automatic feature selection. To overcome this limitation, introduce hinge loss SVMs including the squared hinge loss function and huberized hinge loss function (Wang et al. 2008; Xu et al. 2015; Yang and Zou 2015). Begin with the squared hinge SVMs. It is clear that the optimization problem (8.138) can also be formulated as

$$\min_{w,b} \frac{1}{n} \sum_{i=1}^{n} \left[\left(1 - y_i \left(w^T x_i + b \right) \right)_+ \right]^2 + \frac{\lambda_1}{2} \|w\|_2^2 + \frac{\lambda_2}{2} b^2, \tag{8.139}$$

where $L(t)=(1-t)_+$ is the hinge loss. When $1-t\le 0$, i.e., $y_i(w^Tx_i+b)\ge 1$, the constraints $y_i(w^Tx_i+b)\ge 1-\xi_i$, $\xi_i\ge 0$ are satisfied, we do not need to change the parameters for enforcing constraints. When the constraints are violated, i.e., $y_i(w^Tx_i+b)<1$ or $1-t>0$, minimization of the hinge loss aims to reduce the overlap between the two classes. Similar to the squared hinge loss, we can also introduce the huberized hinge loss SVMs. The huberized hinge loss function measures misclassification. The huberized hinge loss is defined as

$$\Phi_H(t)=\begin{Bmatrix} 0 & t>1 \\ \dfrac{(1-t)^2}{2\delta} & 1-\delta<t\le 1 \\ 1-t-\dfrac{\delta}{2} & t\le 1-\delta \end{Bmatrix}. \tag{8.140}$$

The huberized hinge SVM is then formulated as

$$\min_{w,b}\frac{1}{n}\sum_{i=1}^{n}\Phi_H\left(y_i\left(w^Tx_i+b\right)\right)+\frac{\lambda_1}{2}||w||_2^2+\frac{\lambda_2}{2}b^2. \tag{8.141}$$

The unified formulation of the squared hinge SVMs and the huberized hinge SVMs is

$$\min_{w,b}\frac{1}{n}\sum_{i=1}^{n}\Phi\left(y_i\left(w^Tx_i+b\right)\right)+\frac{\lambda_1}{2}||w||_2^2+\frac{\lambda_2}{2}b^2, \tag{8.142}$$

where $\Phi(t)=\begin{cases} L(t) & \text{squared hinge SVM} \\ \Phi_H(t) & \text{huberized hinge SVM} \end{cases}.$

In Section 8.1.5, we introduced a general form of penalty functions including network, L_1, and group penalties:

$$\Omega_0(\beta)=\lambda_1\sum_{u=1}^{K}||\delta_u||_1+\lambda_2\sum_{l=2}^{L}||\eta_l||_1]+\lambda_3\sum_{g=1}^{G}\phi_g||\alpha_g||_2+\frac{\lambda_4}{2}\beta^T(D-S)\beta+\frac{\lambda_5}{2}b^2. \tag{8.143}$$

A general form for the penalized SVMs is defined as

$$\min_{b,\beta}\frac{1}{n}\sum_{i=1}^{n}\Phi\left(y_i\left(H_i\beta+b\right)\right)+\frac{\lambda_4}{2}\beta^T(D-S)\beta+\frac{\lambda_5}{2}b^2+\Omega(\beta), \tag{8.144}$$

where $\Omega(\beta)=\lambda_1\sum_{u=1}^{K}||\delta_u||_1+\lambda_2\sum_{l=2}^{L}||\eta_l||_1]+\lambda_3\sum_{g=1}^{G}\phi_g||\alpha_g||_2$.

For convenience, write Equation 8.144 as

$$\min_{b,\beta} F(b,\beta) = f(b,\beta) + \frac{\lambda_5}{2} b^2 + \Omega(\beta), \tag{8.145}$$

where $f(b,\beta) = \frac{1}{n}\sum_{i=1}^{n} \Phi(y_i(H_i\beta + b)) + \frac{\lambda_4}{2}\beta^T(D-S)\beta.$

Using the chain rule, we have

$$\begin{aligned} \nabla_b f &= \frac{1}{n}\sum_{i=1}^{n} \Phi'(y_i(H_i\beta + b))y_i \\ \nabla_\beta f &= \sum_{i=1}^{n} \Phi'(y_i(H_i\beta + b))y_i H_i^T + \lambda_4(D-S)\beta, \end{aligned} \tag{8.146}$$

where

$$\Phi'(t) = \begin{cases} L'(t) & \text{squared hinge SVM} \\ \Phi'_H(t) & \text{huberized hinge SVM} \end{cases}, \quad L'(t) = \begin{cases} 0 & t > 1 \\ -2(1-t) & t \le 1 \end{cases}$$

and

$$\Phi'_H(t) = \begin{cases} 0 & t > 1 \\ \dfrac{t-1}{\delta} & 1-\delta < t \le 1. \\ -1 & t \le 1-\delta \end{cases}$$

Let $u = \begin{bmatrix} b \\ \beta \end{bmatrix}$ and $v_i = \begin{bmatrix} y_i \\ y_i H_i^T \end{bmatrix}$. Then, $y_i(H_i\beta + b) = v_i^T u$, $f(b,\beta) = f(u)$, and

$$\nabla_u f(u) = \begin{bmatrix} \nabla_b f(u) \\ \nabla_\beta f(u) \end{bmatrix} = \frac{1}{n}\sum_{i=1}^{n} \Phi'(v_i^T u)v_i + \begin{bmatrix} 0 \\ \lambda_4(D-S)\beta \end{bmatrix}. \tag{8.147}$$

Now we derive the Lipschitz constant. It follows from Equation 8.147 that

$$\begin{aligned} \|\nabla_u f(u) - \nabla_u f(u)\|_2 &\le \frac{1}{n}\sum_{i=1}^{n} \|(\Phi'(v_i^T u) - \Phi'(v_i^T u))v_i\|_2 + \lambda_4 \|D-S\| \|u-u\|_2 \\ &\le \frac{L_0}{n}\sum_{i=1}^{n} \|v_i\|_2^2 \|u-\hat{u}\|_2 + \lambda_4\|D-S\|\|u-\hat{u}\|_2 \\ &\le \left(\frac{L_0}{n}\sum_{i=1}^{n} \|v_i\|_2^2 + \lambda_4\|D-S\|\right)\|u-\hat{u}\|_2 \\ &\le L\|u-\hat{u}\|_2, \end{aligned} \tag{8.148}$$

where

$$L_0 = \begin{cases} 2 & \text{squared hinge SVM} \\ \frac{1}{\delta} & \text{huberized hinge SVM} \end{cases}$$

and

$$\begin{aligned} L &= \frac{L_0}{n}\sum_{i=1}^{n}||v_i||_2^2 + \lambda_4||D-S|| \\ &= \frac{L_0}{n}\sum_{i=1}^{n} y_i^2\left(1+||H_i^2||_2\right) + \lambda_4||D-S||. \end{aligned} \quad (8.149)$$

Let $\hat{u}^{k-1} = u^{k-1} + \omega_{k-1}\left(u^{k-1} - u^{k-2}\right)$. Then, it follows from Equations 8.145 through 8.148 that

$$F\left(u,\hat{u}^{k-1}\right) \le f\left(\hat{u}^{k-1}\right) + \left(\nabla_u f\left(\hat{u}^{k-1}\right)\right)^T\left(u-\hat{u}^{k-1}\right) + \frac{L}{2}||u-u^{k-1}||_2^2 + \frac{\lambda_5}{2}b^2 + \Omega(\beta). \quad (8.150a)$$

A general form for the accelerated proximal gradient method for solving the optimization problem (8.145) is

$$\begin{aligned} \hat{u}^{k-1} &= u^{k-1} + \omega_{k-1}\left(u^{k-1} - u^{k-2}\right), \\ u^k &= \text{prox}_{l_k g(u)}\left(\hat{u}^{k-1} - l_k \nabla_u f\left(\hat{u}^{k-1}\right)\right), \end{aligned} \quad (8.150b)$$

where $g(u) = \frac{\lambda_5}{2}b^2 + \Omega(\beta)$.

Specifically, the algorithm is given below (Fung and Mangasarian 2004; Parikh and Boyd 2014).

Accelerated Proximal Gradient Algorithm

1. Input data (x_i, y_i), $i = 1, \ldots, n$, penalty parameters $\lambda_1, \lambda_2, \lambda_3, \lambda_4, \lambda_5$, and δ for the huberized SVM.
2. Initialization: choose $u^0 = [b^0, (\beta^0)^T]^T$, $u^{-1} = u^0$, compute L using Equation 8.149, and select $l_0 \in \left(0, \frac{1}{L}\right)$. Set $k = 1$.
3. **While** not converged do
4. Let $\hat{u}^{k-1} = u^{k-1} + \omega_{k-1}\left(u^{k-1} - u^{k-2}\right)$ for some $\omega_{k-1} \le 1$.

5. Compute

$$b^k = \frac{l_{k-1}\hat{b}^{k-1} - \nabla_b f\left(\hat{u}^{k-1}\right)}{l_{k-1} + \lambda_5}, \tag{8.151}$$

where $\nabla_b f\left(\hat{u}^{k-1}\right) = \frac{1}{n}\sum_{i=1}^{n} \Phi'\left(v_i^T \hat{u}^{k-1}\right) y_i$.

6. Compute

$$\beta^t = \hat{\beta}^{k-1} - l_k \nabla_\beta f\left(\hat{u}^{k-1}\right) = \begin{bmatrix} u_\delta^t \\ u_\eta^t \\ u_\alpha^t \end{bmatrix}$$

$$= \begin{bmatrix} \hat{\delta}^{k-1} - \frac{l_k}{n}\sum_{i=1}^{n}\Phi'\left(v_i^T \hat{u}^{k-1}\right) y_i Z_i^T + 2l_k\lambda_4\left(D_{\delta\delta} - S_{\delta\delta}\right)\hat{\delta}^{k-1} + 2l_k\lambda_4 S_{\delta\eta}\hat{\eta}^{k-1} + 2l_k\lambda_4 S_{\delta\alpha}\hat{\alpha}^{k-1} \\ \hat{\eta}^{k-1} - \frac{l_{k-1}}{n}\sum_{i=1}^{n}\Phi'\left(v_i^T \hat{u}^{k-1}\right) y_i w_i^T - 2l_{k-1}\lambda_4 D_{\eta\eta}\hat{\eta}^{k-1} + 2l_{k-1}\lambda_4 S_{\delta\eta}^T\hat{\delta}^{k-1} \\ \hat{\alpha}^{k-1} - \frac{l_{k-1}}{n}\sum_{i=1}^{n}\Phi'\left(v_i^T \hat{u}^{k-1}\right) y_i x_i^T + 2l_{k-1}\lambda_4 D_{\alpha\alpha}\hat{\alpha}^{k-1} + 2l_{k-1}\lambda_4 S_{\delta\alpha}^T\hat{\delta}^{k-1} \end{bmatrix},$$

$$\delta^k = \begin{bmatrix} \operatorname{sign}\left(u_{\delta_1}^t\right)\left(\left|u_{\delta_1}^t\right| - \lambda_1 l_{k-1}\right)_+ \\ \vdots \\ \operatorname{sign}\left(u_{\delta_K}^t\right)\left(\left|u_{\delta_K}^t\right| - \lambda_1 l_{k-1}\right)_+ \end{bmatrix}, \tag{8.152}$$

$$\eta^k = \begin{bmatrix} \operatorname{sign}\left(u_{\eta_1}^t\right)\left(\left|u_{\eta_1}^t\right| - \lambda_2 l_{k-1}\right)_+ \\ \operatorname{sign}\left(u_{\eta_2}^t\right)\left(\left|u_{\eta_2}^t\right| - \lambda_2 l_{k-1}\right)_+ \\ \vdots \\ \operatorname{sign}\left(u_{\eta_L}^t\right)\left(\left|u_{\eta_L}^t\right| - \lambda_2 l_{k-1}\right)_+ \end{bmatrix}, \tag{8.153}$$

$$\alpha^k = \begin{bmatrix} \left(1 - \frac{l_{k-1}\lambda_3\phi_1}{||u_{\alpha_1}^t||_2}\right)_+ u_{\alpha_1}^t \\ \vdots \\ \left(1 - \frac{l_{k-1}\lambda_3\phi_g}{||u_{\alpha_G}^t||_2}\right)_+ u_{\alpha_G}^t \end{bmatrix}. \tag{8.154}$$

7. Update for l_k.

Given $\hat{u}^{k-1}$, l_{k-1} and $\tau \in (0,1)$,

$$\lambda := l_{k-1}.$$

Repeat

a. Let $\hat{u} := \text{prox}_{\lambda g(u)}\left(\hat{u}^{k-1} - \lambda \nabla_u f\left(\hat{u}^{k-1}\right)\right)$.
b. Break if $f(u) \le f_\lambda\left(u, \hat{u}^{k-1}\right)$,

where $f(u) = f(b,\beta) = \frac{1}{n}\sum_{i=1}^{n} \Phi\left(y_i\left(H_i\beta + b\right)\right) + \frac{\lambda_4}{2}\beta^T \ (D-S)\beta$ and

$$f_\lambda\left(u,\hat{u}^{k-1}\right) = f\left(\hat{u}^{k-1}\right) + \left(\nabla_u f\left(\hat{u}^{k-1}\right)\right)^T\left(u - \hat{u}^{k-1}\right) + \frac{1}{2\lambda}\left\|u - u^{k-1}\right\|_2^2.$$

c. Update $\lambda := \tau\lambda$.

Return $l_k := \lambda$, $\hat{u}^k := u$.

8. Let $k = k+1$.

9. End **while.**

8.4 LOW-RANK APPROXIMATION

In genomic and epigenomic analysis, data are often highly dimensional and complex and consist of various heterogeneous data types, with noise and many missing entries. We need to dissect complex data structures, remove noisy data points, and fill in missing entries. For the numerical dataset, PCA that finds a low-rank matrix to minimize the approximation error to the original dataset under the Euclidean distance measure is a widely used tool for dimension reduction and exploratory data analysis.

Our data include multiple data types; we can extend PCA to the generalized low-rank models in which the Euclidean distance measure is replaced by other loss functions and regularization is added on the low-dimensional factors. In other words, we extend the PCA to the generalized low-rank model (GLRM) that projects high-dimensional data into low-dimensional space to minimize a loss function on the approximation error subject to regularization of the low-dimensional factors (Udell et al. 2016). The GLRM emerges as a useful tool for dimension reduction and integration of heterogeneous data.

8.4.1 Quadratically Regularized PCA

8.4.1.1 Formulation

In Section 1.5.1.2, we show that the variance of the first principal component is the largest eigenvalue of the covariance matrix Σ of X. Therefore, the proportion of the variance of the first principal component over the total variance is

$$\frac{\lambda_1}{\lambda_1 + \lambda_2 + \cdots + \lambda_k}.$$

To increase the proportion of the first principal component in the total variation of the low-dimensional space, we reduce every eigenvalue by a constant, γ. We can show that

$$\frac{\lambda_1-\gamma}{\lambda_1-\gamma+\lambda_2-\gamma+\cdots+\lambda_k-\gamma}>\frac{\lambda_1}{\lambda_1+\lambda_2+\cdots+\lambda_k}.$$

To achieve this goal, we extend the PCA to quadratically regularized PCA. In Section 1.5.1.1, we introduced the traditional PCA. Now we add quadratic regularization on two factor matrices, U and V, to the objective.

Consider the data matrix $A\in R^{m\times n}$ and two matrix factors: $X\in R^{m\times k}$ and $Y\in R^{k\times n}$. The quadratically regularized PCA problem is mathematically formulated as

$$\min_{X,Y} \quad ||A-XY||_F^2+\gamma||X||_F^2+\gamma||Y||_F^2, \tag{8.155}$$

where $\gamma\geq 0$ is the regularization parameter. Since we restrict $||X||_F^2\leq\alpha$ and $||Y||_F^2\leq\alpha$, we can expect that the single values of the product matrix XY will be reduced. When $\gamma=0$, the quadratically regularized PCA problem is reduced to the traditional PCA.

Assume that the single value decomposition (SVD) of A is given by

$$A=U\Lambda V^T=\left[u_1,\ldots,u_r\right]\begin{bmatrix}\lambda_1 & \cdots & 0\\ \vdots & \ddots & \vdots\\ 0 & \cdots & \lambda_r\end{bmatrix}\begin{bmatrix}v_1^T\\ \vdots\\ v_r^T\end{bmatrix}, \tag{8.156}$$

where $U\in R^{m\times r}$ and $V\in R^{n\times r}$ have orthonormal columns, $\lambda_1\geq\lambda_2\geq\cdots\geq\lambda_r>0$, $r=\text{Rank}(A)$, U and V are referred to as the left and right singular vectors of the matrix A, respectively, and $\lambda_1,\ldots,\lambda_r$ are the singular values of A. Using matrix calculus, find the solution to the optimization problem (8.155) (Udell et al. 2016) (Appendix 8D),

$$X=U_\Omega\left(\Lambda_\Omega-\gamma I\right)^{1/2}, \tag{8.157}$$

where U_Ω and V_Ω denote the submatrix of U, V with columns indexed by Ω, respectively, $|\Omega|\leq k$ with $\lambda_i\geq\gamma$ for $i\in\Omega$, and similarly, denote Λ_Ω. For example, if $\Omega=\{i:1,2,3\}$, then we have $U_\Omega=[u_1,u_2,u_3]$, $\Lambda_\Xi=\text{diag}(\lambda_1-\gamma,\lambda_2-\gamma,\lambda_3-\gamma)$, and $V_\Omega=[v_1,v_2,v_3]$. In general, Equation 8.51 can also be written as

$$X=U_k\tilde{\Lambda}_k^{1/2},\quad Y=\tilde{\Lambda}_k^{1/2}V_k^T, \tag{8.158}$$

where $U_k=\left[u_1,\ldots,u_k\right]$, $V_k=\left[v_1,\ldots,v_k\right]$, $\tilde{\Lambda}_k=\text{diag}\left(\left(\lambda_1-\gamma\right)_+,\ldots,\left(\lambda_k-\gamma\right)_+\right)$, $(a)_+=\max(a,0)$.

8.4.1.2 Interpretation

The PCA for high-dimensional data identifies a new set of orthogonal basis in the low-dimensional space in which the original dataset is re-expressed. From Equation 8.52, it can be easily seen that matrix Y of the principal components can be expressed as

$$Y = \begin{bmatrix} Y_1 \\ \vdots \\ Y_k \end{bmatrix} = \begin{bmatrix} (\lambda_1 - \gamma)_+^{1/2} V_1^T \\ \vdots \\ (\lambda_k - \gamma)_+^{1/2} V_k^T \end{bmatrix}, \quad (8.159)$$

which implies that

$$Y_i Y_j^T = 0, \; i \neq j \quad \text{and} \quad \|Y_i\|^2 = (\lambda_i - \gamma)_+ .$$

In other words, the principal components are orthogonal. The principal components $Y_i, i = 1, \ldots, k$ form a new set of basis. In PCA, the data A can be approximated by

$$A \approx XY$$

or

$$\begin{bmatrix} A_1 \\ \vdots \\ A_m \end{bmatrix} \approx \begin{bmatrix} X_1 \\ \vdots \\ X_m \end{bmatrix} \begin{bmatrix} Y_1 \\ \vdots \\ Y_k \end{bmatrix}, \quad (8.160)$$

where

$$A_i = \begin{bmatrix} a_{i1} & \cdots & a_{in} \end{bmatrix}, \; X_i = \begin{bmatrix} x_{i1} & \cdots & x_{in} \end{bmatrix}, \; Y_j = \begin{bmatrix} y_{j1} & \cdots & y_{jn} \end{bmatrix}.$$

Therefore, from Equation 8.160, we can express the data of the ith sample A_i as

$$A_i = \sum_{j=1}^{k} x_{ij} Y_j . \quad (8.161)$$

Equation 8.161 indicates that the original data can be expressed as a linear combination of new basis vectors.

Multiplying Y_l^T on both sides of Equation 8.161, we obtain

$$x_{ij} = A_i Y_j^T .$$

The coefficient x_{ij} is called the principal score or loading of sample i on the principal component j.

The original data with n features can be compressed to $k < n$ new features. The row vector X_i is associated with sample i and can be viewed as a feature vector for the sample i using the compressed set of $k < n$ features. The vector X_i is the representation of sample i in terms of the principal components Y_j, e.g., new bases (axes). The principal component scores x_{ij} can also be viewed as the coordinates in the new system of axes. Therefore, Equation 8.161 implies that the original data A_i for sample i can be projected into the low-dimensional space that is defined by the principal components as a system of bases, and its coordinates are the principal scores x_{ij}. We often use a score plot to represent the original data.

8.4.2 Generalized Regularization

8.4.2.1 Formulation

A serious limitation of the PCA is that each principal component uses a linear combination of all features. This will increase the cost of measuring additional features and difficulty to interpret the results. The interpretation of the principal components would be facilitated if the principal components involve very few nonzero features. It is desirable to select the most informative subset of the features for defining principal components to eliminate the cost of measuring the additional less informative features and facilitate the interpretation of the results. To achieve this, we can extend the PCA to allow arbitrary regularization on the rows x_i of principal component score matrix X and columns y_j of the principal component matrix Y.

Mathematically form the generally regularized PCA problem as (Udell et al. 2016)

$$\text{minimize} \quad \sum_{(i,j)\in\Omega}\left(A_{ij}-x_i y_j\right)^2+\sum_{i=1}^{m} r_i\left(x_i\right)+\sum_{j=1}^{n}\tilde{r}_j\left(y_j\right), \tag{8.162}$$

where $r_i(x_i)$ and $\tilde{r}_j\left(y_j\right)$ are regularizers. When $r_i\left(x_i\right)=\gamma\|x_i\|_2^2, \tilde{r}_j\left(y_j\right)=\gamma\|y_j\|_2^2$, regularized PCA (8.56) reduces to the quadratically regularized PCA.

Regularized PCA problem (8.56) can also be written in a matrix form:

$$\min_{X,Y} \quad \|A-XY\|_F^2+r\left(X\right)+\tilde{r}\left(Y\right), \tag{8.163}$$

where $r\left(X\right)=\sum_{i=1}^{m} r\left(x_i\right)$ and $\tilde{r}\left(Y\right)=\sum_{j=1}^{n}\tilde{r}\left(r_j\right)$.

8.4.2.2 Sparse PCA

Sparse PCA is an important specific case of the regularized PCA. We can easily interpret and understand each basis and each sample if a small number of features and small number of bases are used. Recall that

$$A=XY, \tag{8.164}$$

and

$$X = U_k \Lambda_k^{1/2}, \quad Y = \Lambda_k^{1/2} V_k^T. \tag{8.165}$$

The matrix of principal components Y can be approximated by

$$Y = \begin{bmatrix} Y_1 \\ \vdots \\ Y_k \end{bmatrix} = \Lambda_k^{-1/2} U_k^T A = \begin{bmatrix} \frac{1}{\sqrt{\lambda_1}} U_1^T A \\ \vdots \\ \frac{1}{\sqrt{\lambda_k}} U_k^T A \end{bmatrix}, \tag{8.166}$$

which implies that the principal component basis Y_k can be expressed as a linear combination of all n features, $Y_i = \frac{1}{\sqrt{\lambda_i}} U_i^T A$, regarding features are informative or not. To remove noninformative features, we impose L_1-norm on the matrix of principal components Y: $\tilde{r}(Y) = |Y|_1 = \sum_{j=1}^{k} |Y_j|_1$, where $|Y_j|_1 = \sum_{i=1}^{n} |Y_{ji}|$. We then formulate the sparse PCA as follows:

$$\min_{X,Y} \; ||A - XY||_F^2 + \gamma ||Y||_1. \tag{8.167}$$

The sparse PCA problem (8.167) can have alternative formulation. If we impose the orthonormality constraint $YY^T = I$, then from Equation 8.164, we obtain $X = AY^T$. A PCA problem can be formulated as a regression:

$$\begin{aligned} \min_{Y} \quad & ||A - XY||_F^2 = ||A - AY^T Y||_F^2 \\ \text{subject to} \quad & YY^T = I_k, \end{aligned} \tag{8.168}$$

where $Y^T = \begin{bmatrix} Y_1 & \cdots & Y_k \end{bmatrix}$ and its columns form an orthonormal basis. Each principal component is derived from a linear combination of all feature variables. Let $B = Y^T$. The PCA problem (8.168) can be rewritten as

$$\begin{aligned} \min_{Y,B} \quad & ||A - ABY||_F^2 \\ \text{subject to} \quad & YY^T = I_k, \end{aligned} \tag{8.169}$$

where $B = [\beta_1, \beta_2, \ldots, \beta_k]$, and the columns of B, which minimize (8.168), define the PCA basis V. In order to make the PCA basis sparse, a regularization constraint should be posed in the PCA regression formulation (8.169). Incorporation of a sparse penalty in the PCA regression formulation reduces the number of features.

8.5 GENERALIZED CANONICAL CORRELATION ANALYSIS (CCA)

CCA is a commonly used supervised dimension reduction tool. CCA finds the correlations between two sets of multidimensional variables including multiple phenotypes and multi-locus genotypes. Generalization techniques developed for unsupervised dimension reduction in Section 8.2 can also be extended to CCA. The generalized CCA will increase the power for genomic, epigenomic, and imaging data analysis. Similar to Section 8.2, extension from the traditional CCA to generalized CCA consists of two categories: (1) imposing constraints and (2) enlarging types of cost function.

8.5.1 Quadratically Regularized Canonical Correlation Analysis

We first extend the unregularized traditional CCA to quadratically regularized CCA. The easiest way for CCA extension is to first approximate the original data using quadratically regularized PCA and then apply the CCA algorithm to the approximated dataset. Consider two matrices, $X \in R^{n\times p}$ and $Y \in R^{n\times q}$. We assume $p \le q$. Suppose that the SVD of matrices X and Y are, respectively, given by

$$X = u_1 s_1 v_1^T, \quad Y = u_2 s_2 v_2^T, \tag{8.170}$$

where $u_1 = \left[u_1^{(1)},\ldots,u_p^{(1)}\right]$, $v_1 = \left[v_1^{(1)},\ldots,v_p^{(1)}\right]$, $u_2 = \left[u_1^{(2)},\ldots,u_q^{(2)}\right]$, $v_2 = \left[v_1^{(2)},\ldots,v_q^{(2)}\right]$, $s_1 = \text{diag}(\lambda_1, \ldots, \lambda_p)$, $s_2 = \text{diag}(\rho_1, \ldots, \rho_q)$, $\lambda_1 \ge \cdots \ge \lambda_p > 0$, and $\rho_1 \ge \cdots \ge \rho_q > 0$ are the single values of matrices X and Y, rank $(X) = p$, and rank $(Y) = q$.

To derive quadratically regularized CCA, we apply the quadratically regularized PCA to the data matrices X and Y separately. Matrices X and Y can then be approximated by their quadratically regularized PCA as follows:

$$X \approx \tilde{X} = \tilde{u}_1 \tilde{s}_1 \tilde{v}_1^T, \quad Y \approx \tilde{Y} = \tilde{u}_2 \tilde{s}_2 \tilde{v}_2^T, \tag{8.171}$$

where $\tilde{u}_1 = \left[u_1^{(1)},\ldots,u_k^{(1)}\right]$, $\tilde{v}_1 = \left[v_1^{(1)},\ldots,v_k^{(1)}\right]$, $\tilde{u}_2 = \left[u_1^{(2)},\ldots,u_l^{(2)}\right]$, $\tilde{v}_2 = \left[v_1^{(2)},\ldots,v_l^{(2)}\right]$, $\tilde{s}_1 = \text{diag}\left((\lambda_1-\gamma_1)_+,\ldots,(\lambda_k-\gamma_1)_+\right)$, $\tilde{s}_2 = \text{diag}\left((\rho_1-\gamma_2)_+,\ldots,(\rho_l-\gamma_2)_+\right)$, γ_1, γ_2 are penalty parameters.

Next the standard CCA techniques discussed in Section 1.6.3 are applied to the dataset:

$$\begin{bmatrix} \tilde{X} \\ \tilde{Y} \end{bmatrix}.$$

We denote linear combinations of the matrices $\tilde{X}$ and $\tilde{Y}$ by

$$\tilde{u} = \tilde{X}\tilde{A}, \quad \tilde{v} = \tilde{Y}\tilde{B}. \tag{8.172}$$

We calculate the SVD of the matrix $\tilde{u}_1^T \tilde{u}_2$ as

$$\tilde{u}_1^T \tilde{u}_2 = \tilde{U}\tilde{\Lambda}\tilde{V}^T. \tag{8.173}$$

Using Equation 1.236, we obtain the transformation matrices:

$$\tilde{A} = \tilde{v}_1 \tilde{s}_1^{-1} \tilde{U}, \quad \tilde{B} = \tilde{v}_2 \tilde{s}_2^{-1} \tilde{V}. \tag{8.174}$$

Substituting Equation 8.174 into Equation 8.172, we obtain the matrices of canonical covariates $\tilde{u}$ and $\tilde{v}$.

To derive the canonical correlations, we calculate

$$\begin{aligned}
\tilde{u}^T \tilde{v} &= \tilde{A}^T \tilde{X}^T \tilde{Y} \tilde{B} \\
&= \tilde{A}^T \tilde{v}_1 \tilde{s}_1 \tilde{u}_1^T \tilde{u}_2 \tilde{s}_2 \tilde{v}_2^T \tilde{B} && \text{by Equation 8.171} \\
&= \tilde{A}^T \tilde{v}_1 \tilde{s}_1 \tilde{U} \tilde{\Lambda} \tilde{V}^T \tilde{s}_2 \tilde{v}_2^T \tilde{B} && \text{by Equation 8.173} \\
&= \tilde{U}^T \tilde{s}_1^{-1} \tilde{v}_1^T \tilde{v}_1 \tilde{s}_1 \tilde{U} \tilde{\Lambda} \tilde{V}^T \tilde{s}_2 \tilde{v}_2^T \tilde{v}_2 \tilde{s}_2^{-1} \tilde{V} && \text{by Equation 8.174} \\
&= \tilde{\Lambda}.
\end{aligned} \tag{8.175}$$

8.5.2 Sparse Canonical Correlation Analysis

The standard CCA assumes that the sample size is larger than the number of variables. However, the genomic and epigenomic data are often high dimensional where the number of variables exceeds the sample size. The standard CCA is difficult to apply to large genomic and epigenomic problems. Sparse CCA produces linear combinations of only a small number of variables from each dataset, which is much smaller than the sample size, thereby making the standard CCA appropriate (Wilms and Croux 2015). In addition, the sparse CCA facilitates the interpretability of the results.

8.5.2.1 Least Square Formulation of CCA

Regression is a well-studied technique for wide-range application. The least square formulation of the CCA allows easy extension of the standard CCA to sparse CCA using well-developed penalized regression techniques. It also serves the purpose of applying the CCA to large-scale multiclass classification and disease risk prediction problems (Sun et al. 2011).

8.5.2.1.1 Reduced Rank Regression Regression formulation of CCA often uses reduced rank regression techniques (Izenman 2008). Before we study reduced rank regression, we introduce two theorems, the Eckart–Young Theorem for matrix approximation and the Poincare Separation Theorem for estimating the low bound of eigenvalues of the symmetric matrix.

The Eckart–Young Theorem (Appendix 8E). Consider two matrices: $A \in R^{n \times p}$ and $B \in R^{n \times p}$. Let $r = \min\ (n, p)$. Assume that the SVD of matrix A is $A = U \Lambda V^T$, where

$$U = \begin{bmatrix} U_k & U_{r-k} \end{bmatrix}, \quad \Lambda = \begin{bmatrix} \Lambda_k & 0 \\ 0 & \Lambda_{r-k} \end{bmatrix}, \quad V = \begin{bmatrix} V_k & V_{r-k} \end{bmatrix}, \quad U_k \in R^{n \times k}, \quad U_{r-k} \in R^{n \times (r-k)},$$

$$\Lambda_k = \text{diag}(\sigma_1, \ldots, \sigma_k), \quad \Lambda_{r-k} = \text{diag}(\sigma_{k+1}, \ldots, \sigma_r), \quad V_k \in R^{p \times k}, \quad V_{r-k} \in R^{p \times (r-k)}, \quad k < r.$$

For any matrix of rank at most k,

$$||A - A_k|| \leq ||A - B||, \tag{8.176}$$

where $A_k = U_k \Lambda_k V_k^T$, $||.||$ is either the Frobenius norm $||.||_F$ or the Euclidean norm $||.||_2$ of the matrix.

Poincare Separation Theorem (Appendix 8F). Let $A \in R^{n \times n}$ be a symmetric matrix and let $G \in R^{n \times k}$ be a semiorthonormal matrix such that $G^T G = I_k$. The eigenvalues of matrices A and $G^T AG$ are denoted by $\sigma_1(A) \geq \sigma_2(A) \geq \cdots \geq \sigma_n(A)$ and $\sigma_1(G^T AG) \geq \sigma_2(G^T AG) \geq \cdots \geq \sigma_k(G^T AG)$, respectively. Then,

$$\sigma_j(A) \geq \sigma_j\left(G^T AG\right), \quad j = 1, \ldots, k, \tag{8.177}$$

with equality if matrix G is formed by the first k eigenvectors of A.

Difference between the standard regression and the reduced rank regression is that in the standard regression model, we assume the full rank of the regression coefficient matrix, but in the reduced rank regression, we allow that the rank of the regression coefficient matrix is not full. Consider the following reduced rank regression model:

$$Y = \mu + CX + \varepsilon, \tag{8.178}$$

where $Y \in R^k$, $\mu \in R^k$, $C \in R^{k \times r}$, $X \in R^r$, $\varepsilon \in R^k$. We assume that the error ε has mean $E[\varepsilon] = 0$ and covariance matrix $\text{cov}(\varepsilon) = \Sigma_{\varepsilon\varepsilon}$ and is uncorrelated with X. We assume that

$$\text{rank}\left(C\right) = l \leq \min\left(k, r\right). \tag{8.179}$$

When the rank of the regression coefficient matrix is deficient, some rows or columns are dependent. In other words, some constraints among regression coefficients exist. The statistical methods for the full-rank regression coefficient matrix cannot be applied since the ordinary derivatives of the objective function minimizing the square of difference between the observed response phenotypes and predicted responses with respect to the regression coefficients will not consider constraints.

The reduced rank regression coefficients can be estimated by the least square estimation methods. For the generality, we consider the weighted least square estimates. For the convenience of discussion, we assume that both Y and X are centered. To estimate matrix C, we consider a weighted sum of square of errors:

$$\begin{aligned}
W(l) &= E\left[(Y - CX)^T \Gamma (Y - CX)\right] \\
&= E\left[Y^T \Gamma Y - Y^T \Gamma CX - X^T C^T \Gamma Y + X^T C^T \Gamma CX\right] \\
&= E\left[\text{Trace}\left(Y^T \Gamma Y - Y^T \Gamma CX - X^T C^T \Gamma Y + X^T C^T \Gamma CX\right)\right] \\
&= E\left[\text{Trace}\left(\Gamma^{\frac{1}{2}} YY^T \Gamma^{\frac{1}{2}} - \Gamma^{\frac{1}{2}} CXY \Gamma^{\frac{1}{2}} - \left(\Gamma^{\frac{1}{2}} C\right)^T \Gamma^{\frac{1}{2}} YX^T + \left(\Gamma^{\frac{1}{2}} C\right)^T \Gamma^{\frac{1}{2}} CXX^T\right)\right].
\end{aligned} \tag{8.180}$$

Let

$$\Sigma_{YY}^{*}=\Gamma^{\frac{1}{2}}E\left[YY^{T}\right]\Gamma^{\frac{1}{2}}=\Gamma^{\frac{1}{2}}\Sigma_{YY}\Gamma^{\frac{1}{2}},\quad C^{*}=\Gamma^{\frac{1}{2}}C,\quad \Sigma_{XY}^{*}=E\left[XY\right]\Gamma^{\frac{1}{2}}=\Sigma_{XY}\Gamma^{\frac{1}{2}},\quad \Sigma_{YX}^{*}=\left(\Sigma_{XY}^{*}\right)^{T}.$$

Using these notations, Equation 8.180 can be reduced to

$$W(l)=\text{Trace}\left(\Sigma_{YY}^{*}-C^{*}\Sigma_{XY}^{*}-\left(C^{*}\right)^{T}\Sigma_{YX}^{*}+C^{*}\Sigma_{XX}\left(C^{*}\right)^{T}\right). \tag{8.181}$$

Our goal is to find C^* that minimizes $W(l)$. However, the rank-deficient matrix C^* implies that elements in C^* are constrained. This prevents application of the traditional calculus to searching minimum of $W(l)$. Alternative algebra methods can be used to solve minimization problem (8.180). After some algebra, Equation 8.181 can be reduced to

$$W(l)=\text{Trace}\left(\left(C^{*}\Sigma_{XX}^{1/2}-\Sigma_{YX}^{*}\Sigma_{XX}^{-1/2}\right)\left(C^{*}\Sigma_{XX}^{1/2}-\Sigma_{YX}^{*}\Sigma_{XX}^{-1/2}\right)^{T}+\Sigma_{YY}^{*}-\Sigma_{YX}^{*}\Sigma_{XX}^{-1}\Sigma_{XY}^{*}\right). \tag{8.182}$$

Since only the first term in Equation 8.182 involves the unknown regression coefficient matrix C^*, minimizing $W(l)$ is equivalent to minimizing the following objective function:

$$F=\left(C^{*}\Sigma_{XX}^{1/2}-\Sigma_{YX}^{*}\Sigma_{XX}^{-1/2}\right)\left(C^{*}\Sigma_{XX}^{1/2}-\Sigma_{YX}^{*}\Sigma_{XX}^{-1/2}\right)^{T}. \tag{8.183}$$

Let $A=\Sigma_{YX}^{*}\Sigma_{XX}^{-1/2}$ and $B=C^{*}\Sigma_{XX}^{1/2}$. Minimizing F in Equation 8.183 again is reduced to the problem (8.176).

Let $m=\min(k,r)$. The SVD of matrix A is given by

$$A=U\Lambda V^{T}, \tag{8.184}$$

where $U=\left[U_{l}\quad U_{m-l}\right]$, $\Lambda=\begin{bmatrix}\Lambda_{l} & 0\\ 0 & \Lambda_{m-l}\end{bmatrix}$, $V=\left[V_{l}\quad V_{m-l}\right]$, $U_{l}\in R^{k\times l}$, $U_{r-l}\in R^{k\times(r-l)}$, $V_{l}\in R^{r\times l}$, $V_{m-l}\in R^{r\times(m-l)}$.

We can use the Eckart–Young Theorem to solve the optimization problem (8.183). In other words, the optimization problem (8.183) can be solved by

$$C^{*}\Sigma_{XX}^{1/2}=U_{l}\Lambda_{l}V_{l}^{T}. \tag{8.185}$$

Solving Equation 8.185, we obtain

$$C=\Gamma^{-1/2}U_{l}\Lambda_{l}V_{l}^{T}\Sigma_{XX}^{-1/2}. \tag{8.186}$$

Next show that we can only use the left singular vector U_l to express C. It follows from Equation 8.184 that

$$A=U_{l}\Lambda_{l}V_{l}^{T}+U_{m-l}\Lambda_{m-l}V_{m-l}^{T}. \tag{8.187}$$

Using Equation 8.187, we obtain

$$U^T A = \begin{bmatrix} U_l^T \\ U_{m-l}^T \end{bmatrix} U_l \Lambda_l V_l^T + U_{m-l} \Lambda_{m-l} V_{m-l}^T$$
$$= \begin{bmatrix} \Lambda_l V_l^T \\ \Lambda_{m-l} V_{m-l}^T \end{bmatrix}. \tag{8.188}$$

However, $U^T A$ can also be expressed as

$$U^T A = \begin{bmatrix} U_l^T A \\ U_{m-l}^T A \end{bmatrix}. \tag{8.189}$$

Comparing Equations 8.188 and 8.189 yields

$$U_l^T A = \Lambda_l V_l^T. \tag{8.190}$$

It follows from Equations 8.185 and 8.190 that

$$C^* \Sigma_{XX}^{1/2} = U_l U_l^T A$$

or

$$\Gamma^{1/2} C \Sigma_{XX}^{1/2} = U_l U_l^T \Sigma_{YX}^* \Sigma_{XX}^{-1/2}$$
$$= U_l U_l^T \Gamma^{1/2} \Sigma_{YX} \Sigma_{XX}^{-1/2}. \tag{8.191}$$

Multiplying $\Gamma^{-1/2}$ from the left and multiplying $\Sigma_{XX}^{-1/2}$ from the right on both sides of the Equation 8.191, we obtain

$$C = \Gamma^{-1/2} U_l U_l^T \Gamma^{1/2} \Sigma_{YX} \Sigma_{XX}^{-1}. \tag{8.192}$$

Matrix C is referred to as the *reduced rank regression coefficient matrix* with rank l, and matrix Γ is referred to as the *weight matrix.*

Matrix U_l is also the matrix of eigenvectors associated with the matrix

$$AA^T = \Gamma^{1/2} \Sigma_{YX} \Sigma_{XX}^{-1} \Sigma_{XY} \Gamma^{1/2}. \tag{8.193}$$

Define

$$H = \Gamma^{-1/2} U_l, \tag{8.194}$$

$$B = U_l^T \Gamma^{1/2} \Sigma_{YX} \Sigma_{XX}^{-1}. \tag{8.195}$$

It is clear that $C=HB$. The reduced rank regression model (8.178) can be rewritten as

$$Y=\mu+HBX+\varepsilon, \tag{8.196}$$

where μ is estimated by

$$\hat{\mu}=\bar{Y}-HB\bar{X} \tag{8.197}$$

where H and B are estimated by Equations 8.194 and 8.195.

Now we calculate the mean square of errors $W_{\min}(l)$. From Equation 8.180, we know that

$$\begin{aligned} W(l) &= E\left[(Y-CX)^T\Gamma(Y-CX)\right] \\ &= E\left[\text{Trace}\left((Y-CX)^T\Gamma(Y-CX)\right)\right] \\ &= E\left[\text{Trace}\left((Y-CX)\Gamma(Y-CX)^T\Gamma\right)\right] \\ &= \text{Trace}\left(\left(\Sigma_{YY}-C\Sigma_{XY}-\Sigma_{YX}C^T+C\Sigma_{XX}C^T\right)\Gamma\right). \end{aligned} \tag{8.198}$$

Using Equations 8.184 and 8.193, we can easily see that

$$\Gamma^{1/2}\Sigma_{YX}\Sigma_{XX}^{-1}\Sigma_{XX}\Gamma^{1/2}=U\Lambda^2U^T. \tag{8.199}$$

Combining Equations 8.192 and 8.199 yields

$$\begin{aligned} C\Sigma_{XY} &= \Gamma^{-1/2}U_lU_l^T\Gamma^{1/2}\Sigma_{YX}\Sigma_{XX}^{-1}\Sigma_{XY} \\ &= \Gamma^{-1/2}U_lU_l^T\Gamma^{1/2}\Sigma_{YX}\Sigma_{XX}^{-1}\Sigma_{XY}\Gamma^{-1/2}\Gamma^{1/2} \\ &= \Gamma^{-1/2}U_lU_l^TU\Lambda^2U^T\Gamma^{1/2}. \end{aligned} \tag{8.200}$$

It is clear that

$$\begin{aligned} U_lU_l^TU\Lambda^2U^T &= U_lU_l^T\begin{bmatrix} U_l & U_{m-l}\end{bmatrix}\begin{bmatrix}\Lambda_l^2 & 0 \\ 0 & \Lambda_{m-l}^2\end{bmatrix}\begin{bmatrix}U_l^T \\ U_{m-l}^T\end{bmatrix} \\ &= U_l\Lambda_l^2U_l^T. \end{aligned} \tag{8.201}$$

Substituting Equation 8.201 into Equation 8.200, we obtain

$$C\Sigma_{XY}=\Gamma^{-1/2}U_l\Lambda_l^2U_l^T\Gamma^{-1/2}. \tag{8.202}$$

Similarly, we can obtain

$$\Sigma_{YX} C^T = \Gamma^{-1/2} U_l \Lambda_l^2 U_l^T \Gamma^{-1/2}, \tag{8.203}$$

$$C \Sigma_{XX} C^T = \Gamma^{-1/2} U_l \Lambda_l^2 U_l^T \Gamma^{-1/2}. \tag{8.204}$$

Combing Equations 8.202 through 8.204, we have

$$-C\Sigma_{XY} - \Sigma_{YX} C^T + C\Sigma_{XX} C^T = -\Gamma^{-1/2} U_l \Lambda_l^2 U_l^T \Gamma^{-1/2}. \tag{8.205}$$

Substituting Equation 8.205 into Equation 8.198 results in

$$\begin{aligned} W_{\min}(l) &= \text{Trace}\left(\left(\Sigma_{YY} - \Gamma^{-1/2} U_l \Lambda_l^2 U_l^T \Gamma^{-1/2}\right)\Gamma\right) \\ &= \text{Trace}\left(\Sigma_{YY}\Gamma\right) - \text{Trace}\left(U_l \Lambda_l^2 U_l^T\right) \\ &= \text{Trace}\left(\Sigma_{YY}\Gamma\right) - \Sigma_{j=1}^{l} \sigma_j^2. \end{aligned} \tag{8.206}$$

From Equation 8.206, when the rank of the reduced rank regression coefficient matrix increases, the mean square of errors decreases.

8.5.2.1.2 Least Square Formulation Recall that in Section 1.6.1, we consider vectors $X \in R^p$ and $Y \in R^q$. Let

$$G = \begin{bmatrix} a_1^T \\ \vdots \\ a_p^T \end{bmatrix}, \quad H = \begin{bmatrix} b_1^T \\ \vdots \\ b_p^T \end{bmatrix}, \tag{8.207}$$

where
G is a $p \times p$ dimensional matrix
H is a $p \times q$ dimensional matrix

Two vectors of the canonical variables are given by

$$\xi = GX, \quad \omega = HY. \tag{8.208}$$

From Equations 1.220, 1.221, and 1.228, we know that G, H are given by

$$G = V^T \Sigma_{YY}^{-1/2} \Sigma_{YX} \Sigma_{XX}^{-1}, \quad H = V^T \Sigma_{YY}^{-1/2}, \tag{8.209}$$

where

$$\begin{aligned} R &= \Sigma_{YY}^{-1/2} \Sigma_{YX} \Sigma_{XX}^{-1} \Sigma_{XY} \Sigma_{YY}^{-1/2} \\ &= V \Lambda^2 V^T \end{aligned} \tag{8.210}$$

and

$$\Sigma_{XX}^{-1/2}\Sigma_{XY}\Sigma_{YY}^{-1/2}=U\Lambda V^{T}. \tag{8.211}$$

Now we show that CCA defined by Equations 8.208 through 8.211 can be formulated by regression (Izenman 2008):

$$HY=\nu+GX+\varepsilon. \tag{8.212}$$

Proof

To estimate the parameters ν, H, and G in the regression model (8.212), least square methods are used to find ν, H, and G to minimize the $p\times p$ matrix:

$$F=E\left[(HY-\nu-GX)(HY-\nu-GX)^{T}\right], \tag{8.213}$$

where we assume that the covariance matrix of ω is

$$\Sigma_{\omega\omega}=H\Sigma_{YY}H^{T}=I_{p}. \tag{8.214}$$

For the convenience of discussion, the data are first centered. Let $\boldsymbol{\omega}_c=\boldsymbol{\omega}-\boldsymbol{\mu}_x$ and $X_c=X-\boldsymbol{\mu}_x$. Then, Equation 8.213 is transformed to

$$F=E\left[(\omega_c-GX_c)(\omega_c-GX_c)^{T}\right]. \tag{8.215}$$

Equation 8.215 can be further reduced to

$$\begin{aligned}F&=E\left[(\omega_c-GX_c)(\omega_c-GX_c)^{T}\right]\\&=\Sigma_{\omega\omega}-\Sigma_{\omega X}G^{T}-G\Sigma_{X\omega}+G\Sigma_{XX}G^{T}.\end{aligned} \tag{8.216}$$

The least square method is used to minimize the average sum of square of errors:

$$\begin{aligned}&\min_{G}\ \text{Trace}\left(\Sigma_{\omega\omega}-\Sigma_{\omega X}G^{T}-G\Sigma_{X\omega}+G\Sigma_{XX}G^{T}\right)\\&=\min_{G}\ \text{Trace}\left(\Sigma_{\omega\omega}-\Sigma_{\omega X}\Sigma_{XX}^{-1}\Sigma_{X\omega}+\left(G\Sigma_{XX}^{1/2}-\Sigma_{\omega X}\Sigma_{XX}^{-1/2}\right)\left(G\Sigma_{XX}^{1/2}=\Sigma_{\omega X}\Sigma_{XX}^{-1/2}\right)^{T}\right)\\&=\text{Trace}\left(\Sigma_{\omega\omega}-\Sigma_{\omega X}\Sigma_{XX}^{-1}\Sigma_{X\omega}\right)+\min_{G}\ \text{Trace}\left(\left(G\Sigma_{XX}^{1/2}-\Sigma_{\omega X}\Sigma_{XX}^{-1/2}\right)\left(G\Sigma_{XX}^{1/2}=\Sigma_{\omega X}\Sigma_{XX}^{-1/2}\right)^{T}\right).\end{aligned} \tag{8.217}$$

Taking $G = \Sigma_{\omega X} \Sigma_{XX}^{-1}$, Equation 8.217 is reduced to

$$\begin{aligned} &\min_{G} \quad \text{Trace}\left(\Sigma_{\omega\omega} - \Sigma_{\omega X} G^T - G\Sigma_{X\omega} + G\Sigma_{XX} G^T\right) \\ &\geq \text{Trace}\left(\Sigma_{\omega\omega} - \Sigma_{\omega X} \Sigma_{XX}^{-1} \Sigma_{X\omega}\right). \end{aligned} \tag{8.218}$$

Using Equations 8.209 and 8.214, we obtain

$$\begin{aligned} \text{Trace}\left(\Sigma_{\omega\omega} - \Sigma_{\omega X} \Sigma_{XX}^{-1} \Sigma_{X\omega}\right) &= \text{Trace}\left(I_p - H\Sigma_{YX} \Sigma_{XX}^{-1} \Sigma_{XY} H^T\right) \\ &= p - \sum_{j=1}^{p} \sigma_j\left(H\Sigma_{YX} \Sigma_{XX}^{-1} \Sigma_{XY} H^T\right). \end{aligned} \tag{8.219}$$

Substituting Equation 8.219 into Equation 8.218, we obtain

$$\begin{aligned} &\min_{G} \quad \text{Trace}\left(\Sigma_{\omega\omega} - \Sigma_{\omega X} G^T - G\Sigma_{X\omega} + G\Sigma_{XX} G^T\right) \\ &\geq p - \max_{G} \sum_{j=1}^{p} \sigma_j\left(H\Sigma_{YX} \Sigma_{XX}^{-1} \Sigma_{XY} H^T\right). \end{aligned} \tag{8.220}$$

To use the Poincare Separation Theorem for maximizing the second term in the inequality (8.219), we define $D = \Sigma_{YY}^{1/2} H^T$. Then, the matrix product $H\Sigma_{YX} \Sigma_{XX}^{-1} \Sigma_{XY} H^T$ is reduced to

$$H\Sigma_{YX} \Sigma_{XX}^{-1} \Sigma_{XY} H^T = D^T \Sigma_{YY}^{-1/2} \Sigma_{YX} \Sigma_{XX}^{-1} \Sigma_{XY} \Sigma_{YY}^{-1/2} D. \tag{8.221}$$

Using Equation 8.211, we can reduce Equation 8.221 to

$$H\Sigma_{YX} \Sigma_{XX}^{-1} \Sigma_{XY} H^T = D^T RD. \tag{8.222}$$

The Poincare Separation Theorem ensures that

$$\sigma_j\left(D^T RD\right) \leq \sigma_j\left(R\right)$$

and

$$\sum_{j=1}^{p} \sigma_j\left(D^T RD\right) \leq \sum_{j=1}^{p} \sigma_j\left(R\right). \tag{8.223}$$

When $D = V$, we have

$$\sum_{j=1}^{p} \sigma_j \left(V^T R V\right) = \sum_{j=1}^{p} \sigma_j (R) = \sum_{j=1}^{p} \sigma_j^2.$$

Therefore,

$$\begin{aligned} \min_G \quad & \text{Trace}\left(\Sigma_{\omega\omega} - \Sigma_{\omega X} G^T - G \Sigma_{X\omega} + G \Sigma_{XX} G^T\right) \\ & \geq p - \sum_{j=1}^{p} \sigma_j^2. \end{aligned} \tag{8.224}$$

Recall that

$$D = \Sigma_{YY}^{1/2} H^T. \tag{8.225}$$

When $D = V$, solving Equation 8.225 for the matrix H, we obtain

$$D = V^T \Sigma_{YY}^{-1/2}, \tag{8.226}$$

$$G = V^T \Sigma_{YY}^{-1/2} \Sigma_{YY} \Sigma_{XX}^{-1}. \tag{8.227}$$

Matrices H, G in Equations 8.226 and 8.227 are exactly the same as that in Equation 8.128. This shows that the CCA can be transformed into regression analysis.

Substituting Equations 8.226 and 8.227 into Equation 8.208 leads to

$$\xi = V^T \Sigma_{YY}^{-1/2} \Sigma_{YX} \Sigma_{XX}^{-1} X, \quad \omega = V^T \Sigma_{YY}^{-1/2} Y. \tag{8.228}$$

The covariance matrix $\text{cov}(\xi, \omega)$ is given by

$$\begin{aligned} \text{cov}(\xi,\omega) &= V^T \Sigma_{YY}^{-1/2} \Sigma_{YX} \Sigma_{XX}^{-1} \Sigma_{XY} \Sigma_{YY}^{-1/2} V \\ &= V^T R V \\ &= \Lambda^2. \end{aligned}$$

Similarly, we have $Var(\xi) = \Lambda^2$, $\text{var}(\omega) = I$. Then, the canonical correlation coefficient matrix is

$$\begin{aligned} corr(\xi, \omega) &= \left(\text{var}(\xi)\right)^{-1} \text{cov}(\xi, \omega) \left(\text{var}(\omega)\right)^{-1/2} \\ &= \Lambda^{-1} \Lambda^2 = \Lambda. \end{aligned} \tag{8.229}$$

If we assume G is an $l \times p$ matrix and H is an $l \times q$ matrix, then we have

$$H = V_l^T \Sigma_{YY}^{-1/2}, \tag{8.230}$$

$$G = V_l^T \Sigma_{YY}^{-1/2} \Sigma_{YX} \Sigma_{XX}^{-1}, \tag{8.231}$$

$$\min_G \ \text{Trace}\left(\Sigma_{\omega\omega} - \Sigma_{\omega X} G^T - G\Sigma_{X\omega} + G\Sigma_{XX} G^T\right)$$
$$\geq p - \sum_{j=1}^{l} \sigma_j^2. \tag{8.232}$$

8.5.2.2 CCA for Multiclass Classification

One widely used application of CCA is supervised learning. CCA projects the data onto a low-dimensional space directed by class label information (Sun et al. 2011). However, CCA and sparse CCA involve a computationally expensive generalized eigenvalue problem. To directly design sparse CCA algorithms poses a great challenge. In Section 8.2.3, we showed that CCA is equivalent to optimal scoring. Optimal scoring is a biregression problem and needs to iteratively solve two regression problems. To overcome the computational difficulty of CCA and sparse CCA for multiclass classification, in this section, we show that finding canonical directions can be formulated as a single regression problem without iteration. This will save large computations.

8.5.2.2.1 Least Squares for Regression and Classification Suppose that the observation data matrix $X \in R^{n\times p}$ and categorical response matrix $Y \in R^{n\times K}$ are as defined in (8.67) and (8.68), respectively. Recall that the optimization problem (8.79) for CCA can be reduced to

$$\begin{aligned} \max_A \quad & \text{Tr}\left[A^T X^T Y \left(Y^T Y\right)^{-1} Y^T XA\right] \\ \text{s.t.} \quad & A^T X^T XA = I, \end{aligned} \tag{8.233}$$

where A is called a canonical direction matrix.

Using the Lagrangian multiplier method, we obtain the eigenequation

$$\left[X^T Y\left(Y^T Y\right)^{-1} Y^T X\right] A = \left[X^T X\right] A\Lambda$$

or

$$\left(X^T X\right)^{-1/2}\left[X^T Y\left(Y^T Y\right)^{-1} Y^T X\right]\left(X^T X\right)^{-1/2} B = B\Lambda, \tag{8.234}$$

where

$$B = \left(X^T X\right)^{1/2} A. \tag{8.235}$$

Let the SVD of Y be

$$Y = U_y \Sigma_y V_y^T, \quad (8.236)$$

where $U_y \in R^{n \times K}$ and $V_y \in R^{K \times K}$ are orthogonal matrices and $\sum_y$ is a diagonal matrix. Define the regression

$$U_y = XW + \varepsilon, \quad (8.237)$$

where $U_y = [u_1, \ldots, u_K], u_k \in R^n$ and $W = [W_1, \ldots, W_K], W_k \in R^P$.

Then, we can show Result 8.5.

Result 8.5

The regression coefficient matrix W is equal to the canonical direction matrix A (Appendix 8G).

Therefore, a canonical direction matrix can be found by solving the regression (8.238):

$$\min_W \; ||U_y - XW||_F^2 = \sum_{k=1}^{K} ||u_k - XW_k||^2. \quad (8.238)$$

8.5.3 Sparse Canonical Correlation Analysis via a Penalized Matrix Decomposition

The CCA is a powerful tool for genomic and epigenomic analysis. However, in genomic data analysis, the number of variables under consideration is much larger than the number of samples. When the number of SNPs is larger than the number of samples, the CCA is not well defined. A solution is sparse CCA (Witten and Tibshirani 2009) that identifies sparse linear combinations of two sets of highly correlated variables.

8.5.3.1 Sparse Singular Value Decomposition via Penalized Matrix Decomposition

The sparse CCA can be formulated as a penalized matrix decomposition problem or can be recasted a regression framework (Wilms and Croux 2015). Here, we mainly introduce the sparse CCA via a penalized matrix decomposition. Recall that the algorithm for CCA is implemented by solving the eigenvalue problem of the following matrix:

$$R = \Sigma_{yy}^{-1/2} \Sigma_{yx} \Sigma_{xx}^{-1} \Sigma_{xy} \Sigma_{yy}^{-1/2}. \quad (8.239)$$

Define the matrix

$$K = \Sigma_{xx}^{-1/2} \Sigma_{xy} \Sigma_{yy}^{-1/2}. \quad (8.240)$$

Then, the right eigenvector of the singular value decomposition (SVD) of the matrix K is the solution of eigenequation (8.239). In fact, suppose that the SVD of matrix K is given by

$$K = UDV^T. \tag{8.241}$$

Thus,

$$K^T K = R = VDU^T UDV^T = VD^2V^T. \tag{8.242}$$

Equation 8.242 shows that the right eigenvector of the SVD of the matrix K is the eigenvector of the matrix R. Let

$$U = \left[u_1,\ldots,u_q\right] \text{ and } V = \left[v_1,\ldots,v_q\right]. \tag{8.243}$$

Then,

$$\Sigma_{xx}^{-1/2} \Sigma_{xy} \Sigma_{yy}^{-1/2} V = UD, \tag{8.244}$$

which implies that

$$d_1 u_1 = \Sigma_{xx}^{-1/2} \Sigma_{xy} \Sigma_{yy}^{-1/2} v_1. \tag{8.245}$$

Therefore, the canonical vectors are given by

$$b = \Sigma_{yy}^{-1/2} v_1 \text{ and } a = \Sigma_{xx}^{-1} \Sigma_{xy} \Sigma_{yy}^{-1/2} v_1 = \Sigma_{xx}^{-1/2} u_1. \tag{8.246}$$

The SVD of matrix K problem can be formulated as the following optimization problem:

$$\begin{aligned} &\min_{U,V} \ \frac{1}{2}||K - UDV^T||_F^2 \\ &\text{s.t. } \ U^TU = I, \ \ V^TV = I, \end{aligned} \tag{8.247}$$

where $||\cdot||_F$ is the Frobenius norm of the matrix.

However, by expanding out the squared Frobenius norm (Witten et al. 2009), we have

$$\begin{aligned} ||K - UDV^T||_F^2 &= \text{Tr}\left(\left(K - UDV^T\right)^T \left(K - UDV^T\right)\right) \\ &= ||K||_F^2 - 2\text{Tr}\left(VDU^TK\right) + \text{Tr}\left(VD^2V^T\right) \\ &= |K_F^2|| - 2\text{Tr}\left(VDU^TK\right) + \text{Tr}\left(D^2\right) \\ &= -2\sum_{i=1}^{q} \lambda_i u_i^T K v_i + |K_F^2|| + \text{Tr}\left(D^2\right). \end{aligned} \tag{8.248}$$

It follows from Equation 8.248 that minimizing $\|K - UDV^T\|_F^2$ is equivalent to maximizing $\sum_{i=1}^{q} \lambda_i u_i^T K v_i$. Therefore, to make the left and right eigenvectors u and v of the SVD of matrix K sparse, we formulate the following optimization problem:

$$\begin{aligned} &\max_{u,v} \quad u^T K v \\ &\text{s.t.} \quad u^T u = 1, \quad v^T v = 1, \quad \|u\|_1 \le c_1, \quad \|v\|_1 \le c_2. \end{aligned} \tag{8.249}$$

The objective function in (8.249) is bilinear. When v is fixed, it is linear in u and when u is fixed, it is linear in v. Therefore, we can iteratively solve the optimization problem (8.249). With v fixed, the optimization problem becomes

$$\begin{aligned} &\max_{u} \quad u^T K v \\ &\text{s.t.} \quad u^T u = 1, \quad \|u\|_1 \le c_1. \end{aligned} \tag{8.250}$$

Using the Lagrange multiplier, we have

$$L(u,\lambda,\mu) = u^T K v + \frac{\lambda}{2}\left(1 - u^T u\right) + \frac{\mu}{2}\left(c_1 - \|u\|_1\right).$$

Setting a derivative of the Lagrangian function $L(u, \lambda, \mu)$ with respect to μ to 0, we obtain

$$\frac{\partial L(u,\lambda,\mu)}{\partial u} = Kv - \lambda u - \mu \partial \|u\|_1 = 0, \tag{8.251}$$

where

$$\partial \|u\|_1 = \begin{cases} \text{sgn}(u) & u \ne 0 \\ [-1,1] & u = 0 \end{cases}.$$

Consider two scenarios:

1. $\lambda > 0$
   The Karush–Kuhn–Tucker (KKT) conditions: $\lambda\left(\|u\|_2^2 - 1\right) = 0$ implies that $\|u\|_2^2 = 1$ holds. Thus, from Equation 8.251, we obtain

$$u = \frac{1}{\lambda}\left(Kv - \mu \partial \|u\|_1\right). \tag{8.252}$$

   Solving Equation 8.252 for u gives

$$u = \frac{S(Kv,\mu)}{\|S(Kv,\mu\|_2}, \tag{8.253}$$

   where $S(Kv, \mu) = \text{sgn}(Kv)(|Kv| - \mu)_+$ is the soft threshold operator and $(x)_+ = \max(0, x)$.

Now we determine μ. By the KKT condition, $\mu(c_1 - \|u\|_1) = 0$, we obtain that

i. Select $\mu = 0$ if $\|u\|_1 \le c_1$.

ii. Otherwise, we choose μ such that $\|u\|_1 = c_1$.

2. $\lambda = 0$
   Set $u = S(Kv, \mu)$. Again, we have

iii. Select $\mu = 0$ if $\|u\|_1 \le c_1$.

iv. Otherwise, we choose μ such that $\|u\|_1 = c_1$.

In summary, the algorithm for the sparse SVD of matrix K is given by

1. Select sparseness parameters λ_u and λ_v.
2. Let $K^1 \leftarrow K$.
3. For $j = 1, \ldots, q$
   a. Initialize v_j and normalize $v_j \leftarrow \frac{v_j}{\|v_j\|_2}$.
   b. Iterate until convergence:
      i. $u_j \leftarrow \frac{S(K^j v_j, \lambda_u)}{\|S(K^j v_j, \lambda_u)\|_2}$, where $\lambda_u = 0$ if this results in $\|u_j\|_1 \le c_1$; otherwise, λ_u is chosen to be a positive constant such that $\|u_j\|_1 = c_1$.
      ii. $v_j \leftarrow \frac{S(K^j u_j, \lambda_v = 0)}{\|S(K^j u_j, \lambda_v)\|_2}$, where $\lambda_v = 0$ if this results in $\|v_j\|_1 \le c_2$; otherwise, λ_v is chosen to be a positive constant such that $\|v_j\|_1 = c_2$.
   c. $d_j \leftarrow u_j^T K^j v_j$.
   d. $K^{j+1} \leftarrow K^j - d_j u_j v_j^T$.

Finally, we set

$$b_j = \Sigma_{yy}^{-1/2} v_j \quad \text{and} \quad a_j = \Sigma_{xx}^{-1} \Sigma_{xy} \Sigma_{yy}^{-1/2} v_j = \Sigma_{xx}^{-1/2} u_j.$$

8.5.3.2 Sparse CCA via Direct Regularization Formulation

The problem in the penalized matrix decomposition for the sparse CCA is that although eigenvectors u and v are sparse, the canonical vectors a and b may not be sparse. To overcome this limitation, we introduce another algorithm for the sparse CCA (Wilms and Croux 2016; Witten et al. 2009).

By directly employing L_1-norm regularization, we can formulate sparse CCA as

$$\max_{w_1,w_2} \; w_1^T \sum_{xy} w_2 \qquad (8.254)$$
$$\text{subject to } \; \|w_1\|_2^2 \le 1, \; \|w_2\|_2^2 \le 1, \; \|w_1\|_1 \le c_1 \|w_2\|_1 \le c_2.$$

Following similar arguments as in the previous sections, single-factor sparse CCA algorithms for solving problem (8.254) is given by the following:

1. Initialize w_2 to have L_2 norm 1.
2. Iterate the following two steps until convergence:

a. $w_1 \leftarrow \frac{S(\Sigma_{xy} w_2, \lambda_1)}{\|S(\Sigma_{xy} w_2, \lambda_1)\|_2}$, $\lambda_1 = 0$, if this results in $\|w_1\|_1 \le c_1$; otherwise, $\lambda_1 > 0$ is chosen so that

$$\|w_1\|_1 = c_1. \qquad (8.255a)$$

b. $w_2 \leftarrow \frac{S(\Sigma_{xy} w_1, \lambda_2)}{\|S(\Sigma_{xy} w_1, \lambda_2\|_2}$, $\lambda_2 = 0$, if this results in $\|w_2\|_1 \le c_2$; otherwise, $\lambda_2 > 0$is chosen so that

$$\|w_2\|_1 = c_2. \qquad (8.255b)$$

The single-factor sparse CCA algorithm can be easily extended to obtain multiple canonical vectors.

Algorithm for Obtaining Sparse CCA Factors

1. Let $Z^1 \leftarrow X^T Y$.
2. For $j = 1, 2, \ldots, J$
   i. Initialize w_2^j to have L_2 norm 1.
   ii. Iterate the following two steps until convergence:
      a. $w_1^j \leftarrow \frac{S(Z^j w_2^j, \lambda_1)}{\|S(Z^j w_2^j, \lambda_1\|_2}$, $\lambda_1 = 0$, if this results in $\|w_1^j\|_1 \le c_1$; otherwise, $\lambda_1 > 0$ is chosen so that $\|w_1^j\|_1 \le c_1$.
      b. $w_2^j \leftarrow \frac{S(Z^j w_1^j, \lambda_2)}{\|S(Z^j w_1^j, \lambda_2\|_2}$, $\lambda_2 > 0$, if this results in $\|w_2^j\|_1 \le c_2$; otherwise, $\lambda_2 > 0$ is chosen so that $\|w_2^j\|_1 \le c_2$.
   iii. $Z^{j+1} \leftarrow Z^j - \left[\left(w_1^j\right)^T Z^j w_2^j \right] w_1^j \left(w_2^j\right)^T$, where w_1^j and w_2^j are the jth canonical vectors.

8.6 INVERSE REGRESSION (IR) AND SUFFICIENT DIMENSION REDUCTION

8.6.1 Sufficient Dimension Reduction (SDR) and Sliced Inverse Regression (SIR)

A large number of variables in genomic, epigenomic, physiological, and imaging datasets, the well-known "curse of dimensionality" poses a great challenge to regression, classification, and cluster analysis. Dimension reduction is an essential tool for reducing the impact of noise and irrelevant predictors on regression and risk prediction. Dimension reduction is used to identify a linear or nonlinear combination of the original set of variables while preserving relevant information (Nileson et al. 2007). We have two classes of dimension reduction: unsupervised dimension reduction and supervised dimension reduction. Principal component analysis (PCA) is a typical method for unsupervised dimension reduction, which projects predictor data onto a linear space without response variable information. Supervised dimension reduction is to discover the best subspace that maximally reduces the dimension of the input while preserving the information necessary to predict the response variable. Let Y be a univariate response variable (phenotype), which can be a continuous or discrete variable, and X a p-dimensional vector of predictors (genotypes, functional principal component scores, and other feature variables). The current popular supervised dimension reduction method is SDR, which aims to find a linear subspace S such that the response Y is conditionally independent of the covariate vector X, given the projection of X on S (Cook 2004):

$$Y \perp X \big| P_S X, \tag{8.256}$$

where

$\perp$ indicates independence

P_S represents a projection operator in the standard inner product on S

In other words, all the information of X about Y is contained in the space S. The subspace S is referred to as a dimension reduction subspace. The projection $P_S X$ can be expressed as a linear combination of the original predictors, $\beta_1^T X, \ldots, \beta_k^T X$, where $[\beta_1, \ldots, \beta_k]$ form a basis for the dimension reduction subspace S. Using the projections on to the dimension reduction subspace S, we can regress the response variable Y on the predictors through their predictions on the dimension reduction subspace without loss of information:

$$y = f\left(\beta_1^T X, \ldots, \beta_k^T X, \varepsilon\right), \tag{8.257}$$

where

f is an unknown function

ε is a random error independent of X

After the dimension reduction subspace is estimated, we can plot Y versus the new predictors $\beta_1^T X, \ldots, \beta_k^T X$. The model (8.156) implies that most of relevant information in X about Y is included in $\left\{\beta_1^T X, \ldots, \beta_k^T X\right\}$.

The subspace S may not be unique. To uniquely describe dimension reduction subspace, we introduce central subspace (CS) that is defined as the intersections of all dimension reduction subspaces S, if it is also a dimension reduction subspace (Cook 1994). The CS is denoted by $S_{Y|X}$. The dimension k of the central subspace $S_{y|x}$ is far less than the number of available variants p. Let $B=[\beta_1, \ldots, \beta_k]$ form a basis of central subspace $S_{y|x}$.

For the convenience of discussion, we often standardize the predictors. Let Σ_x be a covariance matrix of X. We define the standardized predictors as

$$Z=\Sigma_x^{-1/2}\left(X-E(X)\right). \tag{8.258}$$

It is clear from Equation 8.258 that $\text{var}(Z)=1$. Working on the standardized predictors Z, we can easily show that the central subspace $S_{Y|Z}$ is equal to $\Sigma_x^{1/2}S_{Y|X}$ (Exercise 8.14). Let $H=[\eta_1, \ldots, \eta_k]$ form a basis of $S_{Y|Z}$, then $\eta_k=\Sigma_x^{1/2}\beta_k$.

Many methods have been developed for identifying CS. These methods are based on inversion regression (Li 1991). Intuitively, if in the forward regression (8.257), data in the central space predict the response and contain all information of predictors X involving the response variable Y, then the conditional mean $E[Z|Y]$ must lie in the central space. To further reduce the dimension of the data, we perform PCA on the central space. The principal component can be found by maximizing the variance of the following variable: $g=\eta^T E[Z|Y]$. In other words, we maximize

$$\text{var}(g)=\eta^T \text{cov}\left(E\left[Z|Y\right]\right)\eta. \tag{8.259}$$

If we normalize the vector in the central space by imposing the constraints

$$\eta^T\eta=1, \tag{8.260}$$

then our goal is to solve the following constrained optimization problem:

$$\begin{aligned}&\max_{\eta_k\in R^p} \quad \eta_k^T \text{cov}\left(E\left[Z|Y\right]\right)\eta_k\\&\text{subject to} \quad \eta_k^T\eta_k=1, \quad \eta_k^T\eta_l=0, \quad l=1,\ldots,k-1.\end{aligned} \tag{8.261}$$

Substituting $\eta_k=\Sigma_x^{1/2}\beta_k$ into Equation 8.261, we obtain

$$\begin{aligned}&\max_{\beta_k\in R^p} \beta_k^T \quad \text{cov}\left(E[X|Y]\right)\beta_k\\&\text{Subject to} \quad \beta_k^T\Sigma_x\beta_x=1, \quad \beta_k^T\Sigma_k\beta_l=0, \quad l=1,\ldots,k-1.\end{aligned} \tag{8.262}$$

The solutions to the optimization problem (8.262) give the orthogonal directions such that the central inverse regression function has the largest variations.

To make a mathematically rigorous presentation and to use the generalized eigenequations to estimate the CS, we make the following assumptions (Cook 2004):

1. *Linearity condition*: $E(Z|P_{SY|Z}Z) = P_{SY|Z}Z$. Let a $p \times k$ dimensional matrix $\gamma = [\gamma_1, \ldots, \gamma_k]$ form the basis matrix of the CS $S_{Y|Z}$. Then, assumption (1) requires that $E(Z|\gamma^T Z)$ be a linear function of $\gamma^T Z$. The linearity condition holds for the elliptically symmetric distribution including the normal distribution (Li 1991). We can show that under the linearity condition and model (8.156), the inverse regression curve $E[Z|Y]$ is contained in the linear subspace spanned by the vectors η_l, $l = 1, \ldots, k$. In fact, assume that a vector, b, is in the orthogonal complement of the space spanned by the basis $\{\eta_1, \ldots, \eta_k\}$, which implies that $\eta_l^T b = 0$, $l = 1, \ldots, k$. To show that $E[Z|Y]$ is contained in the linear subspace spanned by the vectors η_l, $l = 1, \ldots, k$, we only need to show that $b^T E[Z|Y] = 0$. However, using conditioning on expression, we have $b^T E\left[Z|Y\right] = E\left[E\left[b^T Z|\eta_l^T Z\right]\right]Y$. It suffices to show that $E\left[b^T Z|\eta_l^T Z\right] = 0$ or $E\left[\left(E\left[b^T Z + \eta_l^T Z\right]\right)^2\right] = 0$. By conditioning, $E\left[\left(E\left[b^T Z|\eta_l^T Z\right]\right)^2\right] = E\left[E\left[b^T Z|\eta_l^T Z\right]Z^T b\right]$, and linearity condition, we have $E\left[b^T Z|\eta_l^T Z\right] = c_0 + \sum_{l=1}^{k} c_l \eta_l^T Z$. Therefore, we obtain $E\left[E\left[b^T Z|\eta_l^T Z\right]Z^T b\right] = c_0 E\left[Z\right] + \sum_{l=1}^{k} c_l \eta_l^T E\left[ZZ^T\right]b = \sum_{l=1}^{k} c_l \eta_l^T b = 0$, i.e., $b^T E[Z|Y] = 0$. This implies that $E[Z|Y]$ lies in the CS.
2. *Coverage condition*: Span $\{E(Z|Y = y)|y = 1, \ldots h\} = S_{Y|Z}$, where we assume that Y is discretized into h slices. In other words, the space spanned by the inverse conditional mean defines the CS. Therefore, for each y of Y, the inverse conditional mean $E(Z|Y = y)$ is a linear combination of bases of $S_{Y|X}$:

$$E\left(Z|Y = y\right) = \gamma \rho_y, \tag{8.263}$$

   where γ is the basis matrix for $S_{Y|Z}$ and $\rho_y \in R^k$ is a k-dimensional vector.
3. *Constant covariance condition*: $\text{Var}(Z|P_{S_{Y|Z}}Z) = Q_{S_{Y|Z}}$, where $Q_{S_{Y|Z}} = I_p - P_{S_{Y|Z}}$. In other words, $\text{Var}(Z|P_{SY|Z}Z)$ should be a nonrandom matrix.

To characterize the basis in the CS for Z, we define a linear combination of variables in $E(Z|Y)$ as $\gamma_0^T E\left(Z|Y\right)$. To maximally employ information in the $S_{Y|Z}$, CS, we maximize the variance of $\gamma_0^T E\left(Z|Y\right)$:

$$\text{cov}\left(\gamma_0^T E\left(Z|Y\right)\right) = \gamma_0^T \text{cov}\left(E\left(Z|Y\right)\right)\gamma_0, \tag{8.262a}$$

under the constraints $\gamma_0^T\gamma_0 = 1$ or

$$\begin{aligned}\operatorname{cov}\left(\gamma_0^T E(Z|Y)\right) &= \gamma_0^T \operatorname{cov}\left(E(Z|Y)\right)\gamma_0 \\ &= \left(\Sigma^{-1/2}\gamma_0\right)^T \operatorname{cov}\left(E\left(X - E(X)|Y\right)\right)\Sigma^{-1/2}\gamma_0 \\ &= \beta^T \operatorname{cov}\left(E\left(X - E(X)|Y\right)\right)\beta, \end{aligned} \tag{8.262b}$$

under the constraints

$$\gamma_0^T\gamma_0 = 1 \quad \text{or} \quad \beta^T \Sigma \beta = 1,$$

where $\beta_1, \ldots, \beta_k$ is the basis for $S_{Y|X}$. Solving the optimization problem (8.262a) or (8.262b) leads to the following eigenequation:

$$\operatorname{cov}\left(E(Z|Y)\right)\gamma = \lambda_z \gamma, \tag{8.263a}$$

or

$$\operatorname{cov}\left(E\left(X - E(X)|Y\right)\right)\beta = \lambda_x \Sigma_x \beta, \tag{8.263b}$$

where λ_z and λ_x are eigenvalues and γ and β are eigenvectors, respectively. Solutions to eigenequation (8.263a) and (8.263b) yield the basis matrices $\eta = [\gamma_1, \ldots, \gamma_d]$ for $S_{Y|Z}$ and $B = [\beta_1, \ldots, \beta_k]$ for $S_{Y|X}$, respectively.

Eigenequation problem (8.263) can be efficiently solved by the sliced inverse regression (SIR) method (Li 1991). Suppose that the response variable y is discretized and partitioned into J slices. Let $\Sigma_{E(X|Y_J)}$ be the between-slice sample covariance matrix defined by

$$\hat{\Sigma}_{E(X|Y_J)} = \frac{1}{n}\sum_{j=1}^{J} n_j \left(\hat{m}_j - \hat{m}\right)\left(\hat{m}_j - \hat{m}\right)^T, \tag{8.264}$$

where

$\hat{m}$ is the sample overall mean

$\hat{m}_j = \frac{1}{n_j}\sum_{i \in S_j} x_j$ is the sample mean for the jth slice

S_j is the index set for the jth slice

Assume that Σ_x is estimated by sampling matrix $\hat{\Sigma}_x = \frac{1}{n}\sum_{i=1}^{n}(X_i - \hat{m})(X_i - \hat{m})^T$.

Equation 8.263b is then reduced to

$$\hat{\Sigma}_{E(X|Y_j)}\beta = \lambda_x \hat{\Sigma}_x \beta. \tag{8.265}$$

8.6.2 Sparse SDR

8.6.2.1 Coordinate Hypothesis

Although it is a powerful tool for dimension reduction, SDR also has serious limitations. SDR uses all of the original predictors to estimate the CS. The results are difficult to interpret and cannot be used for the discovery of predictive variants. To overcome these limitations, we can remove irrelevant predictors while preserving all information about the response variable. This idea can be formulated by introducing the concept of coordinate hypothesis, which assumes that some coordinate variables (components) in the basis vectors for the CS are zero, i.e., their corresponding original variables will make no contribution to the projection of the original predictors on the CS. Let H be a selected r-dimensional subspace of the predictor space that specifies the hypothesis of a set of components on the basis vectors being zero. We test the coordinate hypothesis of the form

$$P_H S_{Y|X} = O_p, \tag{8.266}$$

where O_p represents the origin in R^p. For example, by arranging the order of variables in the dataset, we can always partition the predictor dataset X into two parts: $X = \left[X_1^T, X_2^T\right]^T$. The corresponding basis matrix for CS can also be partitioned as $B = \left[\beta_1^T, \beta_2^T\right]^T$. Let $H = \text{Span}((0, I_r)^T)$. Equation 8.266 implies

$$P_H S_{Y|X} = (0, I_r)\begin{bmatrix}\beta_1 \\ \beta_2\end{bmatrix} = \beta_2 = 0.$$

Equation 8.266 provides a general framework for SDR-based variable selection. Since the number of variables involved in genome-based disease risk prediction may reach as high as millions or ten millions of variables, in practice, it is difficult to solve such large SDR-based variable selection problems. Fortunately, we can use a split-and-conquer approach to solve this problem. We can show that SDR for a whole genome can be partitioned into a number of small sub-SDR problems defined for divided small genomic regions (see Supplemental Note B). The combined sub-SDR solutions for genomic regions will globally solve the SDR for a whole genome.

8.6.2.2 Reformulation of SIR for SDR as an Optimization Problem

The SIR for estimation of the CS can be formulated as an eigenvalue problem as shown in Equation 8.265. The eigenvalue problem can also be formulated as a constrained optimization problem (Chen and Li 1998; Wang and Zhu 2013):

$$\begin{aligned} &\min_{T_i, b_i} \; E\left[\left(T_i(Y) - E(T_i(Y)) - (X - E[X])^T b_i\right)^2\right] \\ &\text{s.t. } \operatorname{var}(T_i(Y)) = 1, \; \operatorname{cov}(T_i(Y), T_j(Y)) = 0, \; j = 1,\ldots,i-1, \; i = 1,\ldots,d, \end{aligned} \tag{8.267}$$

where $T_i(y)$ is a set of transformation function of the response variable y. The transformation function can be expanded in terms of basis functions:

$$T_i(y) - E(T_i(y)) = \sum_{k=1}^{K} \xi_{ik}\phi_k(y) = \xi_i^T \phi(y), \quad i = 1,\ldots,d, \tag{8.268}$$

where $\phi_k(y)$ are known basis functions, ξ_{ik} are coefficients of expansion, $\phi(y) = [\phi_1(y), \ldots, \phi_K(y)]^T$, and $\xi_i = [\xi_{i1}, \ldots \xi_{ik}]^T$. For convenience, we assume that $\phi_1(y) \equiv 1$. Assume that the predictors are centered. We then denote $x = X - E(X)$. The optimization problem (8.267) can be written in terms of expansion coefficients:

$$\begin{aligned} &\min_{\xi_i, b} \; E\left[\left(\xi_i^T \phi(y) - x^T b_i\right)^2\right] \\ &\text{s.t. } \xi_i^T \operatorname{cov}(\phi(y))\xi_i = 1, \; \xi_i^T \operatorname{cov}(\phi(y))\xi_j = 0, \; j = 1\ldots,i-1, \; i = 1,\ldots,d. \end{aligned} \tag{8.269}$$

Assume that $(X_1, Z_1), \ldots, (X_n, Z_n)$ are sampled and that predictors X_j are centered. Let $Z = [\phi(z_1), \ldots, \phi(z_n)]^T$ and $X = [x_1, \ldots, x_n]^T$.

Then, the sampling formulas for the expectation and covariance in problem (8.269) are given by

$$E\left[\left(\xi_i^T \phi(z) - x^T b_i\right)^2\right] \approx \frac{1}{n}||Z\xi_i - Xb_i||_2^2 \quad \text{and} \quad \operatorname{cov}(\phi(z)) \approx \frac{1}{n} Z^T Z.$$

Therefore, the problem (8.269) can be approximated by

$$\begin{aligned} &\min_{\theta_i \in R^k, b_i \in R^P} \; ||Z\theta_i - X\beta_i||^2 \\ &\text{s. t. } \theta_i^T D\theta_i = 1, \; \theta_i^T D\theta_j = 0, \; j = 1,\ldots,i-1, \; i = 1,\ldots,d, \end{aligned} \tag{8.270}$$

where $D = \frac{Z^T Z}{n}$, $\theta_i = \xi_i$, and $\beta_i = \frac{1}{\sqrt{n}} b_i$.

In other words, the SIR is reformulated as a biconvex optimization problem, or an optimal scoring problem (Clemmensen et al. 2011; Wang and Zhu 2013).

8.6.2.3 Solve Sparse SDR by Alternative Direction Method of Multipliers

To systematically search the genetic variants of prediction value across the genome based on SDR, penalized techniques should be used to solve the optimal scoring problem (8.270). Furthermore, in the genome-based disease risk prediction, the number of genetic variants is much larger than the number of sampled individuals. As a consequence, the sample covariance of matrix of genetic variants is singular and its inverse does not exist. Finding solutions to the optimal scoring is problematic. The sparse optimal scoring or SDR algorithms are needed.

Let $B=[\beta_1,\ldots,\beta_i]=\left[\beta_1^*,\ldots,\beta_p^*\right]^T$, $i=1,\ldots,d$ be a $p\times i$ matrix, which forms the basis matrix of the CS. For a simplified discussion, β_{ij}, $i=1,2,\ldots,d$ is referred to as the jth coordinate in the CS. In the sparse optimal scoring formulation of SDR by Wang and Zhu, the penalty is imposed separately for each vector in the CS. Consequently, a coordinate in some vectors in the CS will be penalized toward zero, but the same coordinate in other vectors in the CS may not be penalized to zero. Therefore, it is difficult to use their sparse optimal scoring formulation of SDR for variable screening. To overcome this limitation and develop a sparse SDR that can simultaneously reduce the dimension and the number of predictors, we introduce a coordinate-independent penalty function. We introduce a coordinate-independent penalty function to penalize the coordinate in all reduction directions (vectors forming the CS) toward zero (Chen et al. 2010):

$$\rho(B)=\sum_{l=1}^{p}\lambda_l\sqrt{\beta_l^{*T}\beta_l^*}=\sum_{l=1}^{p}\lambda_l\|\beta_l^*\|_2.$$

For simplicity of computation, we define $\lambda_l=\lambda\|\beta_l^*\|_2^{-r}$ and $r\geq 0$ is a prespecified parameter. Thus, the penalty function can be simplified to

$$\rho(B)=\lambda\sum_{l=1}^{p}\|\beta_l^*\|_2^{1-r},\tag{8.271}$$

where λ is a penalty parameter.

After introducing the penalty function, the sparse version of optimal scoring problem (8.270) for penalizing the variable can be defined as

$$\min_{\theta_i\in R^k,\beta_i\in R^p}\ \|Z\theta_i-X\beta_i\|_2^2+\lambda\sum_{l=1}^{p}\|\beta_l^*\|_2^{1-r}\tag{8.272}$$
$$\text{s. t.}\quad \theta_i^TD\theta_i=1,\ \theta_i^TD\theta_j=0,\quad j<i,\ i=1,\ldots,d.$$

The problem (8.272) is a biconvex problem. It is convex in θ for each β and convex in β for each θ. It can be solved by a simple iterative algorithm. The iterative process consists of two steps: (1) for fixed θ_i we optimize with respect to β_i and for fixed β_i we optimize with respect to θ_i. The algorithms are given below.

Step 1: Initialization.

Let $D = \frac{Z^T Z}{n}$ and $Q_1 = [1, 1, \ldots, 1]^T$. We first initialize for $\theta_i^{(0)}, i = 1, \ldots, d$

$$\tilde{\theta}_i^{(0)} = \left(I - Q_i Q_i^T D\right)\theta_*, \quad \theta_i^{(0)} = \frac{\tilde{\theta}_i^{(0)}}{\sqrt{\tilde{\theta}_i^{T(0)} D \tilde{\theta}_i^{(0)}}}, \quad Q_{i+1} = \left[Q_i : Q_i\right],$$

where θ_* is a random k vector.

Step 2: Iterate between $\theta^{(s)}$ and $\beta^{(s)}$ until convergence or until a specified maximum number of iterations ($s = 1, 2, \ldots$) is reached:

Step A: For fixed $\theta_i^{(s-1)}, i = 1, \ldots, d$, we solve the following minimization problem:

$$\min_{\beta_i^{(s)} \in R^P} \sum_{i=1}^{d} \|Z\theta_i^{(s-1)} - X\beta_i^{(s)}\|_2^2 + \lambda\left((1-\delta)\sum_{j=1}^{p} \|\beta_j^{*(s)}\|_2^2 + \delta \sum_{j=1}^{p} \|\beta_j^{*(s)}\|_2^{1-r}\right), \tag{8.273a}$$

where $B^{(s)} = \left[\beta_1^{(s)}, \ldots, \beta_d^{(s)}\right] = \left[\beta_1^{*(s)}, \ldots, \beta_p^{*(s)}\right]^T$.

Step B: For fixed $\beta_i^{(s)}$, $i = 1, \ldots, d$, we seek $\theta_i^{(s)}$, $i = 1, \ldots, d$, which solve the following unconstrained optimization problem:

$$\begin{aligned} &\min_{\theta_i^{(s)} \in R^k} \| \left\| Z\theta_i^{(s)} - X\beta_i^{(s)} \right\|_2^2 \\ &\text{s. t. } \ \theta_i^{(s)T} D \theta_i^{(s)} = 1, \quad \theta_i^{(s)T} D \theta_j^{(s)} = 0, \quad j = 1, \ldots, i-1. \end{aligned} \tag{8.273b}$$

A solution to the above optimization leads to a nonlinear equation (Appendix 8H):

$$\theta_i^{(s)} = \frac{\left(I - Q_i^{(s)} Q_i^{(s)T} D\right) D^{-1} Z^T X \beta_i^{(s)}}{\theta_i^{T(s)} Z^T X \beta_i^{(s)}}, \tag{8.274}$$

where $Q_1^{(0)} = Q_1^{(1)} = \cdots = Q_1^{(s)} = [1, 1, \ldots, 1]^T$.

By Newton's method or simple iteration, we obtain a solution, $\tilde{\theta}_i^{(s)}$, to Equation 8.273b. Set

$$\theta_i^{(s)} = \frac{\tilde{\theta}_i^{(s)}}{\sqrt{\tilde{\theta}_i^{T(s)} D \tilde{\theta}_i^{(s)}}}, \quad Q_{i+1} = \left[Q_i : \theta_i\right], \quad i = 1, \ldots, d.$$

If $||\theta_i^{(s+1)} - \theta_i^{(s)}||_2 < \varepsilon$, $||\beta_i^{(s+1)} - \beta_i^{(s)}||_2 < \varepsilon$, $i = 1, \ldots, d$, then stop; otherwise, $s := s + 1$, in which case, go to step A.

Now we study how to use ADMM to solve the optimization problem (8.273a). The algorithm for ADMM to solve optimization problem (8.273a) is given below (Appendix 8H).

Initial value ($m = 0$):

$$\begin{aligned} \beta_j^{(s)(0)} &= \left(X^T X + \frac{\rho}{2} I\right)^{-1} X^T Z \theta_j^{(s)} \\ \alpha_j^{(s)(0)} &= \beta_k^{(s)(0)} \\ u_j^{(s)(0)} &= 0, \quad j = 1, \ldots, i. \end{aligned} \tag{8.275}$$

For fixed $\theta_j^{(s)}$, $j = 1, \ldots, i$, iterate with m until convergence:

Step (i): $m := m + 1$.

Step (ii):

$$\beta_j^{(s)(m+1)} = \left(X^T X + \frac{\rho}{2} I\right)^{-1} \left[X^T Z \theta_j^{(s)} + \frac{\rho}{2}\left(\alpha_j^{(m)} - u_j^{(m)}\right)\right], \quad j = 1, \ldots, d. \tag{8.276a}$$

Step (iii): Let

$$B = \left[\beta_1^{(s)(m+1)}, \ldots, \beta_i^{(s)(m+1)}\right] = \left[\beta_1^{*(s)}, \ldots, \beta_p^{*(s)}\right]^T, \quad \alpha^{(s)(m)} = \left[\alpha_1^{(s)(m)}, \ldots, \alpha_i^{(s)(m)}\right] = \left[\alpha_p^{*(s)(m)}\right]^T$$

and

$$u^{(s)(m)} = \left[u_1^{(s)(m)}, \ldots, u_i^{(s)(m)}\right] = \left[u_1^{*(s)(m)}\right]^T.$$

Then

$$\alpha_i^{*(s)(m+1)} = \left(\beta_i^{*(s)(m+1)} + u_i^{*(s)(m)}\right) \left(\frac{\left(\dfrac{||\beta_i^{*(s)(m+1)} + u_i^{*(s)(m)}||_2^{(1+r)} - \dfrac{\lambda\delta\left(1 - r^2\right)}{\rho}}{1 + \dfrac{2\lambda(1-\delta)(1+r)}{\rho}} \right)^{\frac{1}{1+r}}}{||\beta_i^{*(s)(m+1)} + u_i^{*(s)(m)}||_2} \right), \quad l = 1, \ldots, p.$$

and

$$\alpha^{(s)(m+1)} = \left[\alpha_1^{*(s)(m+1)},\ldots,\alpha_p^{*(s)(m+1)}\right]^T = \left[\alpha_1^{(s)(m+1)},\ldots,\alpha_d^{(s)(m+1)}\right]. \quad (8.276b)$$

Step (iv):

$$u_j^{(s)(m+1)} = u_j^{(s)(m)} + \beta_j^{(s)(m+1)} - \alpha_j^{(s)(m+1)}. \quad (8.276c)$$

When $||\beta_j^{(s)(m+1)} - \beta_j^{(s)(m)}||_2 \le \varepsilon$, $||\alpha_j^{(s)(m+1)} - \alpha_j^{(s)(m)}||_2 \le \varepsilon$, $||u_j^{(s)(m+1)} - u_j^{(s)(m)}||_2 \le \varepsilon$, $j = 1,\ldots,i$, stop and go to step B; otherwise, go to step (i).

Since the number of SNPs selected for risk prediction by the sparse SDR method is usually less than 50,000 SNPs, a split-and-conquer algorithm is used to search clinically valuable SNPs for disease risk prediction. Briefly, we first divide the whole genome into K subgenomic regions. The sparse SDR method is then applied to each subgenomic region to search SNPs with some optimal criterion in the training dataset. Then, we collect all selected SNPs in each subgenomic region to generate a new dataset for final classification. The sparse SDR method is again applied to the new generated dataset to search SNPs and predict disease. In the previous section, we proved that the proposed split-and-conquer algorithm can reach global optimal classification or disease risk prediction.

8.6.2.4 Application to Real Data Examples

The first clinical use of genetic variants is disease risk prediction that can discriminate between individuals who will develop the disease of interest and those who will not. To examine whether it can systematically search clinically valuable genetic variants for disease prediction, the proposed sparse SDR method was first applied to the GWAS data of the Wellcome Trust Case Control Consortium (WTCCC) for coronary artery disease (CAD) study where 1,929 cases and 2,938 controls were sampled and the total number of SNP markers is 393,473 (Frayling et al. 2007).

To reduce bias in the genotypes, we removed all SNPs that were excluded in the original WTCCC CAD study (Frayling et al. 2007). To unbiasedly evaluate the performance of the sparse SDR method for disease risk prediction, a 5-fold cross-validation (CV) was first used to select SNPs for disease risk prediction. Specifically, the original sample was randomly partitioned into five equal-size subsamples. Of the five subsamples, a single subsample was retained as the test dataset, another single sample was used as the validation dataset, and the remaining three subsamples were used as a training dataset. The cross-validation process is then repeated five times (the folds), with each of the five subsamples used exactly once as the test dataset. The five results from the folds can then be averaged to produce a single estimation of sensitivity, specificity, and class. The whole genome was first partitioned into 20 subgenomic regions; each subgenomic region included 20,000 SNPs except for the 20th subgenomic region that included 13,473 SNPs. To evaluate its performance for disease risk prediction without bias, the sparse SDR was applied to each subgenomic region in the training dataset. After convergence of the iteration process in the optimal scoring algorithm, the top 2000 SNPs with the largest contribution to the classification (the largest absolution values of β in Equation 8.273a) were selected for first variable screening. The results from each of the 20 subgenomic regions were then combined. We totally selected 40,000 SNPs that were used for the second variable screening. We then split the 40,000 SNPs into four equal numbers of SNP subdatasets. Again, the sparse SDR was applied to each of the four subdatasets. The top 1500 SNPs with the largest contribution to the classification were then selected for the third (final) variable screening and classification accuracy calculation. Variable screening was performed by the sparse SDR method for each of the five sets of training datasets. Finally, the selected SNPs were then used to learn a predictive discriminative model on training individuals, which was in turn used to classify CAD on the validation dataset. We finally selected the SNPs with the best classification accuracy that was evaluated in the validation dataset. The selected SNPs and the trained model in the validation dataset were finally used to evaluate sensitivity, specificity, and classification accuracy in the test dataset.

The final classification performance results of the sparse SDR method were compared with the software implementing sparse logistic regression (Friedman et al. 2010) and the SNP ranking method based on the GWAS P-values of the χ^2 association test and logistic regression. The P-value of the SNPs was calculated in the training dataset. The SNP ranking method did not need to split the genome region into subgenomic regions. The P-value of each SNP was calculated once in the training dataset. For the SNP ranking method, we selected the SNPs with the best classification accuracy that was evaluated in the validation dataset. The selected SNPs and the trained model in the validation dataset were then used to evaluate sensitivity, specificity, and accuracy. The results were summarized in Tables 8.1 through 8.3 where CV1–CV5 represent five partitioned datasets in 5-fold cross-validations. We observed that accuracy of classifying CAD using the proposed sparse SDR method in the test dataset was much higher than that from using the sparse logistic regression and SNP ranking method. Using the sparse SDR method compared to the sparse logistic regression and SNP ranking significantly increased classification accuracy at least by ~15%.

SOFTWARE PACKAGE

R package for discriminant analysis can be found in "Quick-R: Discriminant Function Analysis." A good package for network-penalized logistic regression is package "penalized" (https://cran.r-project.org/web/packages/penalized/penalized.pdf). A widely used software for implementing SVM is LIBSVM in package e1071 (https://cran.r-project.org/web/packages/e1071/vignettes/svmdoc.pdf). Package "PMA" (https://cran.r-project.org/web/packages/PMA/PMA.pdf) implements sparse CCA. Software for sparse SDR that is designed to select features can be downloaded from our website (http://www.sph.uth.tmc.edu/hgc/faculty/xiong/index.htm).

TABLE 8.1 Accuracy of the CAD Classifications Using the Sparse SDR Method

Sample	Training Dataset			Validation Dataset			Test Dataset					
Folder	Sensitivity	Specificity	Accuracy	Sensitivity	Specificity	Accuracy	Sensitivity	Specificity	Accuracy	λ	α	# SNPs
CV-1	0.623	0.934	0.812	0.617	0.927	0.801	0.545	0.925	0.775	3	0.04	13
CV-2	0.610	0.954	0.819	0.577	0.922	0.786	0.605	0.929	0.797	3	0.02	16
CV-3	0.599	0.975	0.824	0.514	0.956	0.788	0.537	0.932	0.776	10	0.06	23
CV-4	0.575	0.972	0.814	0.537	0.965	0.792	0.494	0.967	0.787	10	0.02	21
CV-5	0.576	0.941	0.797	0.567	0.942	0.794	0.542	0.928	0.772	8	0.12	13
Average	0.597	0.955	0.813	0.562	0.943	0.792	0.544	0.936	0.782			

TABLE 8.2 Accuracy of the CAD Classifications Using the Sparse Logistic Regression

Sample	Training Dataset			Validation Dataset			Test Dataset				
Folder	**Sensitivity**	**Specificity**	**Accuracy**	**Sensitivity**	**Specificity**	**Accuracy**	**Sensitivity**	**Specificity**	**Accuracy**	**α**	**# SNPs**
CV-1	0.283	0.968	0.698	0.207	0.962	0.655	0.211	0.957	0.663	0.024	63
CV-2	0.270	0.969	0.695	0.235	0.943	0.664	0.214	0.950	0.651	0.024	61
CV-3	0.247	0.958	0.673	0.206	0.940	0.661	0.238	0.953	0.671	0.025	30
CV-4	0.273	0.962	0.687	0.249	0.951	0.668	0.203	0.956	0.669	0.024	50
CV-5	0.241	0.975	0.685	0.204	0.974	0.670	0.186	0.963	0.650	0.024	66
Average	0.263	0.966	0.688	0.220	0.954	0.663	0.210	0.956	0.661		

TABLE 8.3 Accuracy of the CAD Classifications Using SNP Ranking

Sample Folder	Training Dataset			Validation Dataset			Test Dataset			# SNPs
	Sensitivity	Specificity	Accuracy	Sensitivity	Specificity	Accuracy	Sensitivity	Specificity	Accuracy	
CV-1	0.419	0.855	0.684	0.393	0.867	0.674	0.359	0.868	0.667	28
CV-2	0.333	0.917	0.688	0.332	0.903	0.677	0.340	0.912	0.679	6
CV-3	0.368	0.903	0.689	0.337	0.882	0.675	0.384	0.890	0.690	7
CV-4	0.392	0.880	0.686	0.424	0.884	0.699	0.346	0.851	0.658	14
CV-5	0.319	0.950	0.701	0.295	0.962	0.699	0.306	0.961	0.698	4
Average	0.366	0.901	0.689	0.356	0.900	0.685	0.347	0.896	0.678	

APPENDIX 8A: PROXIMAL METHOD FOR PARAMETER ESTIMATION IN NETWORK-PENALIZED TWO-CLASS LOGISTIC REGRESSION

We aim at solving the optimization problem:

$$l_p(\beta)=f(\beta)+\Omega(\beta). \tag{8A.1}$$

Function $f(\beta)$ can be expanded in terms of a Taylor expansion (Equation 1.38):

$$f(\beta)=f(\beta^t)+\nabla f(\beta^t)^T(\beta-\beta^t)+\frac{1}{2v}||\beta-\beta^t||_2^2, \tag{8A.2}$$

where

$$\nabla f(\beta^t)=-\sum_{i=1}^{n}(y_i-\pi_i)H_i^T+2\lambda_3(D-S)\beta^t, \tag{8A.3}$$

$\pi_i=\frac{e^{\xi_i^t}}{1+e^{\xi_i^t}}$, $\xi_i^t=H_i\beta^t$, and $\frac{1}{v}$ is an upper bound on the Lipschitz constant of gradient ∇f.
Substituting Equation 8A.2 into the optimization problem (8A.1), we obtain the reduced optimization problem:

$$\min_{\beta}\ f(\beta^t)+\nabla f(\beta^t)^T(\beta-\beta^t)+\frac{1}{2v}||\beta-\beta^t||_2^2+\Omega(\beta). \tag{8A.4}$$

Equation 8A.4 can be reformulated as

$$\min_{\beta}\frac{1}{2v}\beta-\left\|\left(\beta^t-v\nabla f(\beta^t)\right)\right\|_2^2+\Omega(\beta). \tag{8A.5a}$$

Recall that the proximal operator for function $v\Omega(\beta)$ is defined as (see Equation 1.43)

$$\text{Prox}_{v\Omega}(u)=\arg\min_{\beta}\left(\Omega(\beta)+\frac{1}{2v}||\beta-u||_2^2\right).$$

Therefore, the iteration of proximal method for solving β is

$$\beta^{t+1}=\mathbf{Prox}_{v\Omega}\left(\beta^t-v\nabla f(\beta^t)\right). \tag{8A.5b}$$

Let

$$\beta^{t+1}=\begin{bmatrix}\delta^{t+1}\\ \eta^{t+1}\\ \alpha^{t+1}\end{bmatrix}$$

and

$$u^t = \beta^t - v\nabla f\left(\beta^t\right) = \begin{bmatrix} u_\delta^t \\ u_\eta^t \\ u_\alpha^t \end{bmatrix} = \begin{bmatrix} \delta^t + v\sum_{i=1}^{n}\left(y_i - \pi_i\right)z_i^T - 2v\lambda_4\left(D_{\delta\delta} - S_{\delta\delta}\right)\delta^t + 2v\lambda_4 S_{\delta\eta}\eta^t \\ +2v\lambda_4 S_{\delta\eta}\eta^t + 2v\lambda_4 S_{\delta\alpha}\alpha^t \\ \eta^t + v\sum_{i=1}^{n}\left(y_i - \pi_i\right)w_i^T - 2v\lambda_4 D_{\eta\eta}\eta^t + 2v\lambda_4 S_{\delta\eta}^T\delta^t \\ \alpha^t + v\sum_{i=1}^{n}\left(y_i - \pi_i\right)x_i^T - 2v\lambda_4 D_{\alpha\alpha}\alpha^t + 2v\lambda_4 S_{\delta\alpha}^T\delta^t \end{bmatrix}. \quad (8A.6)$$

Then, function $\Omega(\beta) + \frac{1}{2v}\|\beta - u^t\|_2^2$ can be written as

$$\Omega_1(\delta) + \Omega_2(\eta) + \Omega_3(\alpha) + \frac{1}{2v}\left[\|\delta - u_\delta^t\|_2^2 + \|\eta - u_\eta^t\|_2^2 + \|\alpha - u_\alpha^t\|_2^2\right].$$

Then,

$$\min_\beta \quad \Omega(\beta) + \frac{1}{2v}\|\beta - u^t\|_2^2$$
$$= \min_\delta\left[\left(\Omega_1(\delta) + \frac{1}{2v}\|\delta - u_\delta^t\|_2^2\right)\right] + \min_\eta\left[\Omega_2(\eta) + \frac{1}{2v}\|\eta - u_\eta^t\|_2^2\right] + \min_\alpha\left[\Omega_3(\alpha) + \frac{1}{2v}\|\alpha - u_\alpha^t\|_2^2\right],$$

which implies

$$\beta^{t+1} = \arg\min_\beta \ \Omega(\beta) + \frac{1}{2v}\|\beta - u^t\|_2^2$$
$$= \begin{bmatrix} \delta^{t+1} \\ \eta^{t+1} \\ \alpha^{t+1} \end{bmatrix} = \begin{bmatrix} \arg\min_\delta\left[\Omega_1(\delta) + \frac{1}{2v}\|\delta - u_\delta^t\|_2^2\right] \\ \arg\min_\eta\left[\Omega_2(\eta) + \frac{1}{2v}\|\eta - u_\eta^t\|_2^2\right] \\ \arg\min_\alpha\left[\Omega_3(\alpha) + \frac{1}{2v}\|\alpha - u_\alpha^t\|_2^2\right] \end{bmatrix}$$
$$= \begin{bmatrix} \mathbf{Prox}_{v\Omega_1}\left(u_\delta^t\right) \\ \mathbf{Prox}_{v\Omega_2}\left(u_\eta^t\right) \\ \mathbf{Prox}_{v\Omega_3}\left(u_\alpha^t\right) \end{bmatrix}. \quad (8A.7)$$

Now we first calculate

$$\mathbf{Prox}_{\nu\Omega_1}(u_\delta^t) = \arg\min_{\delta}\left[\lambda_1 \sum_{u=1}^{K} \left\|\delta_u\right\|_1 + \frac{1}{2\nu}\left\|\delta - u_\delta^t\right\|_2^2\right]. \tag{8A.8}$$

Let

$$u_\delta^t = \begin{bmatrix} u_{\delta_1}^t \\ \vdots \\ u_{\delta_K}^t \end{bmatrix}.$$

Then, it follows from Equations 1.44b and 8A.8 that

$$\mathbf{Prox}_{\nu\Omega_1}\left(u_\delta^t\right) = \begin{bmatrix} \mathbf{Peox}_{\lambda_1\nu\|\delta_1\|_1}\left(u_{\delta_1}^t\right) \\ \vdots \\ \mathbf{Prox}_{\lambda_1\nu\|\delta_K\|_1}\left(u_{\delta_K}^t\right) \end{bmatrix}. \tag{8A.9}$$

Equation 1.137 gives

$$\mathbf{Prox}_{\lambda_1\nu\|\delta_k\|_1}\left(u_{\delta_k}^t\right) = \mathbf{sign}\left(u_{\delta_k}^t\right)\left(\left|u_{\delta_k}^t\right| - \lambda_1\nu\right)_+, \quad k = 1,\ldots,K, \tag{8A.10}$$

where $(x)_+ = \max(0,x)$.

Combining Equations 8A.7, 8A.9, and 8A.10, we obtain

$$\delta^{t+1} = \begin{bmatrix} \mathbf{sign}\left(u_{\delta_1}^t\right)\left(\left|u_{\delta_1}^t\right| - \lambda_1\nu\right)_+ \\ \vdots \\ \mathbf{sign}\left(u_{\delta_K}^t\right)\left(\left|u_{\delta_K}^1\right| - \lambda_1\nu\right)_+ \end{bmatrix}. \tag{8A.11}$$

We also can directly derive formula (8A.11). Our goal is to find solution

$$\arg\min_{\delta_k}\left[\lambda_1\left|\delta_k\right| + \frac{1}{2\nu}\|\delta_k - u_{\delta_k}^t\|_2^2\right]. \tag{8A.12}$$

Let $F = \lambda_1\left|\delta_k\right| + \frac{1}{2\nu}\|\delta_k - u_{\delta_k}^t\|_2^2$. Then, its subdifferential ∂F is

$$\partial F = \lambda_1\Gamma_k + \frac{1}{\nu}\left(\delta_k - u_{\delta_k}^t\right), \tag{8A.13}$$

where

$$\Gamma_k = \begin{cases} 1 & \delta_k > 0 \\ [-1,1] & \delta_k = 0. \\ -1 & \delta_k < 0 \end{cases}$$

A necessary condition for δ_k^* to minimize the function F is

$$0 \in \sigma F$$

or

$$0 \in \lambda_1 \Gamma_k + \frac{1}{v}\left(\delta_k - u_{\delta_k}^t\right),$$

which implies

$$\delta_k \in u_{\delta_k}^t - \lambda_1 v \Gamma_k. \tag{8A.14}$$

We first show that δ_k and $u_{\delta_k}^t$ have the same sign. Otherwise, if $u_{\delta_k}^t > 0$, then $\delta_k < 0$. But this implies $\Gamma_k < 0$ and $u_{\delta_k}^t - \lambda_1 v \Gamma > 0$. This leads to violation of condition (8A.14). Therefore, it must be that $\delta_k > 0$, i.e.,

$$\begin{aligned} \delta_k &= \max\left(0, u_{\delta_k}^t - \lambda_1 v\right) \\ &= \mathbf{sign}\left(u_{\delta_k}^t\right)\left(u_{\delta_k}^t - \lambda_1 v\right)_+ . \end{aligned} \tag{8A.15}$$

Similarly, we can show that if $u_{\delta_k}^t < 0$, then $\delta_k < 0$, i.e.,

$$\begin{aligned} \delta_k &= u_{\delta_k}^t + \lambda_1 v \\ &= -\left|u_{\delta_k}^t\right| + \lambda_1 v \\ &= -\left(\left|u_{\delta_k}^t\right| - \lambda_1 v\right) \\ &= \mathbf{sign}\left(u_{\delta_k}^t\right)\max\left(0, \left|u_{\delta_k}^t\right| - \lambda_1 v\right) \\ &= \mathbf{sign}\left(u_{\delta_k}^t\right)\left(\left|u_{\delta_k}^t\right| - \lambda_1 v\right)_+ . \end{aligned} \tag{8A.16}$$

If $u_{\delta_k}^t = 0$, then $\delta_k = 0$. Otherwise, if $\delta_k > 0$, then $u_{\delta_k}^t - \lambda_1 v \Gamma_k = -\lambda_1 v < 0$, which violates the condition (8A.14). Similarly, we can show that $\delta_k < 0$ is also impossible. In this case,

$$\left(\left|u_{\delta_k}^t\right| - \lambda_1 v\right)_+ = \left(-\lambda_1 v\right)_+ = 0 = \delta_k. \tag{8A.17}$$

Combining Equations 8A.15 through 8A.17 gives

$$\delta_k = \mathbf{sign}\left(u_{\delta_k}^t\right)\left(\left|u_{\delta_k}^t\right| - \lambda_1 \nu\right)_+, \quad k = 1,\ldots,K.$$

Similarly, we can obtain

$$\eta^{t+1} = \begin{bmatrix} u_{\eta_1}^t \\ \mathbf{sign}\left(u_{\eta_2}^t\right)\left(\left|u_{\eta_2}^t\right| - \lambda_2 \nu\right)_+ \\ \vdots \\ \mathbf{sign}\left(u_{\eta_L}^t\right)\left(\left|u_{\eta_L}^t\right| - \lambda_2 \nu\right)_+ \end{bmatrix}. \tag{8A.18}$$

Finally, we calculate $\mathbf{Prox}_{\nu\Omega_3}\left(u_\alpha^t\right)$ using group LASSO. Recall that

$$\mathbf{Prox}_{\nu\Omega_3}\left(u_\alpha^t\right) = \begin{bmatrix} \mathbf{Prox}_{\nu\lambda_3\phi_1}\left(u_{\alpha_1}^t\right) \\ \vdots \\ \mathbf{Prox}_{\nu\lambda_3\phi_G}\left(u_{\alpha_G}^t\right) \end{bmatrix}. \tag{8A.19}$$

Using Equation 1.151 gives

$$\mathbf{Prox}_{\nu\lambda_3\phi_g}\left(u_{\alpha_g}^t\right) = \left(1 - \frac{\nu\lambda_3\phi_g}{\left\|u_{\alpha_g}^t\right\|_2}\right)_+ u_{\alpha_g}^t, \quad g = 1,2,\ldots,G. \tag{8A.20}$$

Therefore, the update formula for α^{t+1} is

$$\alpha^{t+1} = \begin{bmatrix} \alpha_1^{t+1} \\ \vdots \\ \alpha_G^{t+1} \end{bmatrix} = \begin{bmatrix} \left(1 - \frac{\nu\lambda_3\phi_1}{\left\|u_{\alpha_{1g}}^t\right\|_2}\right)_+ u_{\alpha_1}^t \\ \vdots \\ \left(1 - \frac{\nu\lambda_3\phi_g}{\left\|u_{\alpha_G}^t\right\|_2}\right)_+ u_{\alpha_G}^t \end{bmatrix}. \tag{8A.21}$$

Group LASSO can also be directly solved. Specifically, we want to obtain the solution to

$$\min_{\alpha_g} \left\|\alpha_g\right\|_2 + \frac{1}{2\nu\lambda_3\phi_g}\left\|\alpha_g - u_{\alpha_g}^t\right\|_2^2. \tag{8A.22}$$

Let $F=||\alpha_g||_2+\frac{1}{2\nu\lambda_3\phi_g}||\alpha_g-u^t_{\alpha_g}||_2^2$. Then, when $\alpha_g\neq 0$, we set $\frac{\partial F}{\partial \alpha_g}=0$:

$$\frac{\partial F}{\partial \alpha_g}=\frac{\alpha_g}{\left\|\partial_g\right\|_2}+\frac{1}{\nu\lambda_3\phi_g}\left(\alpha_g-u^t_{\alpha_g}\right)=0, \tag{8A.23}$$

which implies

$$\left[1+\frac{\nu\lambda_3\phi_g}{\left\|\alpha_g\right\|_2}\right]\alpha_g=u^t_{\alpha_g}. \tag{8A.24}$$

Taking an L_2 norm on both sides of Equation 8A.24 gives

$$\left(1+\frac{\nu\lambda_3\phi_g}{\left\|\alpha_g\right\|_2}\right)\left\|\alpha_g\right\|_2=\left\|u^t_{\alpha_g}\right\|_2. \tag{8A.25}$$

Solving Equation 8A.25 for $||\alpha_g||_2$, we obtain

$$\left\|\alpha_g\right\|_2=\left\|u^t_{\alpha_g}\right\|_2-\nu\lambda_3\phi_g. \tag{8A.26}$$

Substituting Equation 8A.26 into Equation 8A.24 yields

$$\alpha_g=\left(1-\frac{\nu\lambda_3\phi_g}{\left\|u^t_{\alpha_g}\right\|_2}\right)u^t_{\alpha_g}. \tag{8A.27}$$

Now we investigate $\alpha_g=0$. Denote the subgradient of α_g at $\alpha_g=0$ by s_g. Then, we have

$$\left\|s_g\right\|_2\leq 1. \tag{8A.28}$$

From Equation 8A.23, we obtain that

$$s_g=\frac{u^t_{\alpha_g}}{\nu\lambda_3\phi_g}, \tag{8A.29}$$

which implies

$$\frac{||u^t_{\alpha_g}||_2}{\nu\lambda_3\phi_g}\leq 1$$

or it must be

$$1-\frac{\nu\lambda_3\phi_g}{\left\|u^t_{\alpha_g}\right\|_2}\leq 0. \tag{8A.30}$$

In other words, when $u^t_{\alpha_g} = 0$, if $1 - \frac{\nu\lambda_3\phi_g}{\|u^t_{\alpha_g}\|_2} \le 0$, then we must set $\alpha_g = 0$, and Equation 8A.27 becomes

$$\alpha_g = \left(1 - \frac{\nu\lambda_{3\phi_g}}{\left\|u^t_{\alpha_g}\right\|_2}\right)_+ u^t_{\alpha_g}. \tag{8A.31}$$

APPENDIX 8B: EQUIVALENCE OF OPTIMAL SCORING AND LDA

We use a Lagrangian multiplier to solve the optimal scoring problem (8.78):

$$\min_{B,\theta,\lambda} \quad F = \mathrm{Tr}\left[Y\theta - XB\right)^T \left(Y\theta - XB\right) + \mathrm{Tr}\left[\Lambda\left(I - \frac{1}{n}\theta^T Y^T \theta\right)\right]. \tag{8B.1}$$

Using Equation 1.164, we obtain

$$\frac{\partial F}{\partial B} = -2X^T\left(Y\theta - XB\right) = 0$$
$$\frac{\partial F}{\partial \theta} = -2Y^T\left(Y\theta - XB\right) - 2\frac{1}{n}Y^T Y\theta\Lambda = 0,$$

which can be reduced to

$$X^T Y\theta - X^T XB = 0, \tag{8B.2}$$

$$Y^T Y\theta\left(I - \frac{1}{n}\Lambda\right) - Y^T XB = 0. \tag{8B.3}$$

Solving Equation 8B.3 for θ gives

$$\theta = \left(Y^T Y\right)^{-1} Y^T XB\left(I - \frac{1}{n}\Lambda\right)^{-1}. \tag{8B.4}$$

Recall that

$$\left(I - \frac{1}{n}\Lambda\right)^{-1} = I + \frac{1}{n}\left(\Lambda^{-1} - \frac{1}{n}I\right)^{-1}. \tag{8B.5}$$

Substituting Equation 8B.5 into Equation 8B.4, we obtain

$$\theta = \left(Y^T Y\right)^{-1} Y^T XB + \frac{1}{n}\left(Y^T Y\right)^{-1} Y^T XB\left(\Lambda^{-1} - \frac{1}{n}I\right)^{-1}. \tag{8B.6}$$

It follows from Equations 8.69 and 8.72 that

$$X^T Y (Y^T Y)^{-1} Y^T X = n \Sigma_b, \tag{8B.7}$$

and

$$X^T X = n \Sigma_b + n \Sigma_w. \tag{8B.8}$$

Substituting Equations 8B.6 through 8B.8 into Equation 8B.2 gives

$$X^T Y \left(Y^T Y\right)^{-1} Y^T XB + \frac{1}{n} X^T Y \left(Y^T Y\right)^{-1} Y^T XB \left(\Lambda^{-1} - \frac{1}{n} I\right)^{-1} - X^T XB = 0,$$

which will be reduced to

$$n \Sigma_b B + \Sigma_b B \left(\Lambda^{-1} - \frac{1}{n} I\right)^{-1} - n \Sigma_b B - n \Sigma_w B = 0$$

or

$$\Sigma_b B = \Sigma_w B (n\Lambda^{-1} - I). \tag{8B.9}$$

Let $\Lambda_0 = n\Lambda^{-1} - I$. Then, Equation 8B.9 can be written as

$$\Sigma_b B = \Sigma_w B \Lambda_0. \tag{8B.10}$$

Equation 8B.10 is equivalent to Equation 8.76. This shows that the optimal score is equivalent to LDA.

APPENDIX 8C: A DISTANCE FROM A POINT TO THE HYPERPLANE

Assume that x_0 is a point in the space and x is its projection to the plane, $w^T x + b = 0$. Figure 8C.1 shows that the vector $x_0 - x$ is parallel to w. So we have

$$x_0 - x = kw, \tag{8C.1}$$

where the point x is in the plane. We normalize the normal vector in Equation 8.89. Thus, the point x satisfies the following equation:

$$\frac{w^T x + b}{||w||} = 0.$$

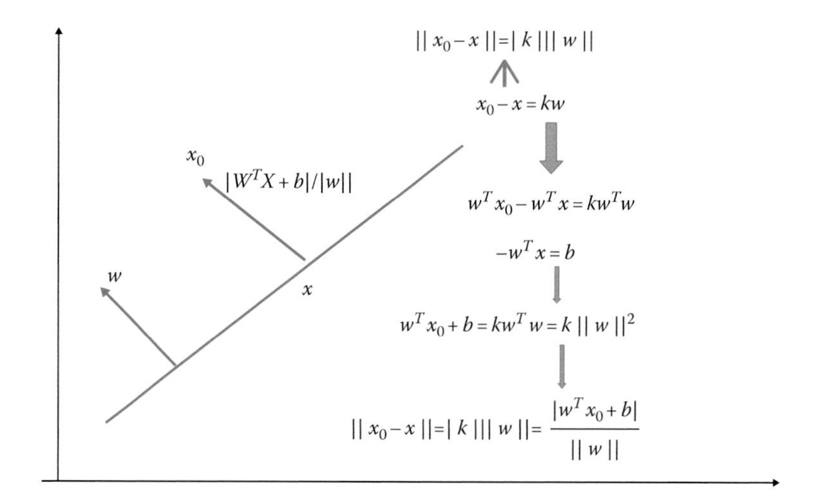

FIGURE 8C.1 Distance from a point to the hyperplane.

Multiplying both sides of Equation 8C.1 by w, we obtain

$$w^T x_0 - w^T x = kw^T w. \tag{8C.2}$$

However, it follows from Equation 8.89 that

$$-w^T x = b,$$

which implies that

$$w^T x_0 + b = kw^T w = k\|w\|^2. \tag{8C.3}$$

From Equation 8C.1, we know that the distance between point x_0 and the plane is equal to

$$\|x_0 - x\| = |k|\|w\|. \tag{8C.4}$$

Combining Equations 8C.3 and 8C.4 yields

$$||x_0 - x|| = |k|||w|| = \frac{|w^T x_0 + b|}{||w||}$$

Thus, the distance from point x_0 to the hyperplane is

$$\frac{|wx_0 + b|}{||w||}.$$

APPENDIX 8D: SOLVING A QUADRATICALLY REGULARIZED PCA PROBLEM

Let

$$\begin{aligned} F &= ||A - XY||_F^2 + \gamma||Y||_F^2 \\ &= \mathrm{Tr}\left(\left(A - XY\right)^T \left(A - XY\right)\right) + \gamma\mathrm{Tr}\left(X^T X\right) + \gamma\mathrm{Tr}\left(Y^T Y\right). \end{aligned}$$

The optimal conditions for the problem (8.155) are

$$\frac{\partial F}{\partial X} = 0 \quad \text{and} \quad \frac{\partial F}{\partial Y} = 0. \tag{8D.1}$$

Using formula (1.164) for calculation of the derivative of the trace with respect to matrices, we obtain

$$\frac{\partial F}{\partial X} = -2\left(A - XY\right)Y^T + 2\gamma X = 0$$

or

$$-\left(A - XY\right)Y^T + \gamma X = 0 \tag{8D.2}$$

and

$$\frac{\partial F}{\partial Y} = -2X^T\left(A - XY\right) + 2\gamma Y = 0$$

or

$$-X^T\left(A - XY\right) + \gamma Y = 0. \tag{8D.3}$$

Taking transpose on both sides of Equation 8D.3, we obtain

$$-\left(A - XY\right)^T X + \gamma Y^T = 0. \tag{8D.4}$$

Now we show that the solutions

$$X = U_k \tilde{\Lambda}_k^{1/2} \quad \text{and} \quad Y = \tilde{\Lambda}_k^{1/2} V_k^T \tag{8D.5}$$

satisfy Equations 8D.2 and 8D.4.

Recall

$$A = U\Lambda V^T \tag{8D.6}$$

and

$$\begin{aligned}\tilde{\Lambda}_k &= \text{diag}\left((\lambda_1-\gamma)_+,\ldots,(\lambda_1-\gamma)_+,\ldots,(\lambda_k-\gamma)_+\right)\\ &= \text{diag}\left(\lambda_1-\lambda,\ldots,\lambda_l-\gamma,0,\ldots,0\right)\\ &= \begin{bmatrix}\tilde{\Lambda}_l & 0\\ 0 & 0\end{bmatrix}.\end{aligned} \tag{8D.7}$$

Therefore, using Equations 8D.5 and 8D.6, we obtain

$$-(A-XY)Y^T+\gamma X = -\left(U\Lambda V^T - U_k\tilde{\Lambda}_k V_k^T\right)V_k\tilde{\Lambda}_k^{1/2}+\gamma U_k\tilde{\Lambda}_k^{1/2}. \tag{8D.8}$$

Note that

$$-U\Lambda V^T V_k = -U\Lambda\begin{bmatrix}I_k\\ 0_{r-k}\end{bmatrix} = -U\begin{bmatrix}\Lambda_k\\ 0\end{bmatrix} = -U_k\Lambda_k \tag{8D.9}$$

and

$$\begin{aligned}U_k\tilde{\Lambda}_k V_k^T V_k &= U_k\tilde{\Lambda}_k\\ &= \begin{bmatrix}U_l & U_{k-l}\end{bmatrix}\begin{bmatrix}\tilde{\Lambda}_l & 0\\ 0 & 0\end{bmatrix} = \begin{bmatrix}U_l\tilde{\Lambda}_l & 0\end{bmatrix}.\end{aligned} \tag{8D.10}$$

It follows from Equations 8D.7 and 8D.8 that

$$-U\Lambda V^T V_k\tilde{\Lambda}_k^{1/2} = -\begin{bmatrix}U_l & U_{k-l}\end{bmatrix}\begin{bmatrix}\Lambda_l & 0\\ 0 & \Lambda_{k-l}\end{bmatrix}\begin{bmatrix}\tilde{\Lambda}_l^{1/2} & 0\\ 0 & 0\end{bmatrix} = \begin{bmatrix}-U_l\Lambda_l\tilde{\Lambda}_k^{1/2} & 0\end{bmatrix}. \tag{8D.11}$$

Similarly, from Equations 8D.7 and 8D.10, we obtain

$$U_k\tilde{\Lambda}_k V_k^T V_k\tilde{\Lambda}_k^{1/2} = \begin{bmatrix}U_l\tilde{\Lambda}_l & 0\end{bmatrix}\begin{bmatrix}\tilde{\Lambda}_l^{1/2} & 0\\ 0 & 0\end{bmatrix} = \begin{bmatrix}U_l\tilde{\Lambda}_l\tilde{\Lambda}_k^{1/2} & 0\end{bmatrix}. \tag{8D.12}$$

It is easy to see that

$$U_k\tilde{\Lambda}_k^{1/2} = \begin{bmatrix}U_l\tilde{\Lambda}_l^{1/2} & 0\end{bmatrix}. \tag{8D.13}$$

Combining Equations 8D.8, 8D.11 through 8D.13, we obtain

$$\begin{aligned}-(A-XY)Y^T+\gamma X&=\left[-U_l\Lambda_l\tilde{\Lambda}_l^{1/2}+\gamma U_l\tilde{\Lambda}_l^{1/2}+U_l\tilde{\Lambda}_l\tilde{\Lambda}_l^{1/2}\ \ 0\right]\\&=\left[-U_l\left(\Lambda_l-\gamma I_l\right)\tilde{\Lambda}_l^{1/2}+U_l\tilde{\Lambda}_l\tilde{\Lambda}_l^{1/2}\ \ 0\right]\\&=\left[-U_l\tilde{\Lambda}_l\tilde{\Lambda}_l^{1/2}+U_l\tilde{\Lambda}_l\tilde{\Lambda}_l^{1/2}\ \ 0\right]\\&=0.\end{aligned}\tag{8D.14}$$

APPENDIX 8E: THE ECKART–YOUNG THEOREM

For the self-containment of this book, we give the proof here, following the directions in https://www.cs.princeton.edu/courses/archive/spring12/cos598C/svdchapter.pdf. The celebrated Eckart–Young Theorem discovered the best. The goal is to find matrix A_k with the rank at most k, which is a solution to the optimization problem

$$\begin{aligned}&\min_B \ \|A-B\|\\&\text{subject to } \operatorname{rank}(B)\le k.\end{aligned}\tag{8E.1}$$

A tool for deriving the best k rank approximation of a matrix is the SVD. The SVD provides a set of basis vector for matrix and vector representation. We first prove the Frobenius norm case. Assume that the SVD of the matrix A is $A = U\Lambda V^T$, where

$$U=\begin{bmatrix}U_k & U_{r-k}\end{bmatrix},\ \Lambda\begin{bmatrix}\Lambda_k & 0\\ 0 & \Lambda_{r-k}\end{bmatrix},\ V=\begin{bmatrix}V_k & V_{r-k}\end{bmatrix},\ U_k\in R^{n\times k},\ U_{r-k}\in R^{n\times(r-k)},$$

$$\Lambda_k=\operatorname{diag}(\sigma_1,\ldots,\sigma_k),\ \Lambda_{r-k}=\operatorname{diag}(\sigma_{k+1},\ldots,\sigma_r),\ V_k\in R^{p\times k},\ V_{r-k}\in R^{p\times(r-k)},\ k<r.$$

Let $A_k = U_k\Lambda_k V_k^T$. We can show that the rows of A_k are the projections of the rows of A onto the subspace S_k spanned by the first k right singular vectors of A. Indeed, suppose that a is a row vector of the matrix A and v_i, $i = 1, \ldots, k$ are the basis vectors of the subspace S_k. Then, the projection of the vector a can be expressed as

$$a\approx\sum_{i=1}^{k}\xi_i v_i^T.\tag{8E.2}$$

Multiplying v_j from the left on both sides of Equation 8E.2 and using the orthonormality $v_i^T v_j = 0$ of the vectors v_i, we obtain

$$\xi_i = av_i.\tag{8E.3}$$

Substituting Equation 8E.3 into Equation 8E.2 yields

$$a\approx\sum_{i=1}^{k}(av_i)v_i^T\tag{8E.4}$$

Therefore, the set of projections of matrix A onto subspace S_k is

$$A = \begin{bmatrix} a_1 \\ \vdots \\ a_n \end{bmatrix} \approx \sum_{i=1}^{k} \begin{bmatrix} a_1 v_i \\ \vdots \\ a_n v_i \end{bmatrix} v_i^T = \sum_{i=1}^{k} (Av_i) v_i^T. \tag{8E.5}$$

Recall that the SVD of A is given by

$$AV = U \Lambda V^T,$$

which implies

$$AV = U\Lambda,$$

or

$$Av_i = \sigma_i u_i. \tag{8E.6}$$

Substituting Equation 8E.6 into Equation 8E.5, we obtain

$$A \approx \sum_{i=1}^{k} \sigma_i u_i v_i^T = U_k \Lambda_k V_k^T = A_k. \tag{8E.7}$$

Next we prove that for any matrix B of rank at most k, we have

$$\left\| A - A_k \right\|_F \le \left\| A - B \right\|_F. \tag{8E.8}$$

Let B be any matrix of rank at most k. The dimension of B is at most k. Let $V_k = \begin{bmatrix} v_1 & \cdots & v_k \end{bmatrix}$ be the space with dimension k. The matrix B can be represented by

$$B = \sum_{i=1}^{k} (Bv_i) v_i^T. \tag{8E.9}$$

The Frobenius norm of the difference matrix $A - B$ can be calculated as follows. The SVD of A is

$$\begin{aligned} A &= \sum_{i=1}^{k} \sigma_j u_j v_j^T + \sum_{i=k+1}^{n} \sigma_i u_i v_i^T \\ &= \sum_{i=1}^{k} (Av) v_j^T + \sum_{i=k+1}^{n} \sigma_i u_i v_i^T. \end{aligned} \tag{8E.10}$$

Combining Equations 8E.9 and 8E.10, we have

$$\|A-B\|_F^2 = \left\| \sum_{i=1}^{k} (Av_i - Bv_i)v_i^T + \sum_{j=k+1}^{n} \sigma_j u_j v_j^T \right\|_F^2$$

$$= \left\| \sum_{i=1}^{k} (Av_i - Bv_i)v_i^T \right\|_F^2 + \left\| \sum_{j=k+1}^{n} \sigma_j u_j v_j^T \right\|_F^2, \tag{8E.11}$$

where the second equality is due to $v_i^T v_j = 0$.

Recall that

$$\left\| \sum_{j=k+1}^{n} \sigma_j u_j v_j^T \right\|_F^2 = \|U_{n-k}\Lambda_{n-k}V_{n-k}^T\|_F^2$$

$$= \text{Trace}\left(\left(U_{n-k}\Lambda_{n-k}V_{n-k}^T\right)^T \left(U_{n-k}\Lambda_{n-k}V_{n-k}^T\right)\right)$$

$$= \text{Trace}\left(\text{V}_{n-k}\Lambda_{n-k}V_{n-k}^T\right)$$

$$= \text{Trace}\left(\Lambda_{n-k}V_{n-k}^T V_{n-k}\right)$$

$$= \text{Trace}\left(\Lambda_{n-k}\right)$$

$$= \sum_{j=k+1}^{n} \sigma_j^2, \tag{8E.12}$$

where $U_{n-k} = \begin{bmatrix} u_{k+1} & \cdots & u_n \end{bmatrix}$, $\Lambda_{n-k} = \text{diag}\left(\sigma_{k+1},\ldots,\sigma_n\right)$, $V_{n-k} = \begin{bmatrix} v_{k+1} & \cdots & v_n \end{bmatrix}$.

Since $\left\| \sum_{i=1}^{k} (Av_i - Bv_i)v_i^T \right\|_F^T \geq 0$, it is clear from Equations 8E.11 and 8E.12 that

$$||A-A_k||_F^2 = ||U_{n-k}\Lambda_{n-k}V_{n-k}^T||_F^2 = \sum_{j=k+1}^{n} \sigma_j u_j v_j^T \leq ||A-B||_F^2,$$

where equality occurs when $Bv_i = Av_i$, $i = 1, \ldots, k$, i.e.,

$$B = \sum_{i=1}^{k} \left(Bv_i\right)v_i^T = \sum_{i=1}^{k} \left(Av_i\right)v_i^T = \sum_{i=1}^{k} \sigma_i u_i v_i^T = A_k.$$

Next we study the Euclidean norm case. By contradiction, we will show that

$$||A-A_k||_2 \leq ||A-B||_2.$$

Suppose that there is some matrix B of rank at most k such that

$$\|A-B\|_2 \le \|A-A_k\|_2 . \tag{8E.13}$$

Since the rank of the matrix B is at most k, the null space of B, denoted by Null (B), has a dimension of at least $n-k$. Consider space V_{k+1}, spanned by the first $k+1$ right singular vectors $v_1, \ldots, v_{k+1}$. The intersection between Null (B) and V_{k+1} is not empty. Let $z \neq 0$ with $\|z\|_2 = 1$ be in the intersection space:

$$\text{Null}\left(B\right) \cap V_{k+1}.$$

By definition of Euclidean norm of the matrix, we have

$$\|A-B\|_2^2 \ge \|(A-B)z\|_2^2 . \tag{8E.14}$$

Since z lies in the Null (B) and Null $(B) \perp B$, we have $Bz = 0$, which implies

$$\left(A-B\right)z = Az. \tag{8E.15}$$

Since by assumption z is in the space V_{k+1}, z is orthogonal to the vectors $v_{k+2}, \ldots, v_n$, i.e.,

$$v_i^T z = 0, \quad i = k+2, \ldots, n. \tag{8E.16}$$

Combining Equations 8E.14 through 8E.16 we obtain

$$\begin{aligned} \|(A-B)z\|_2^2 &= \|Az\|_2^2 = z^T A^T Az = z^T V \Lambda V^T z = z^T \left[\sum_{i=1}^{k+1} \sigma_i^T v_i v_i^T + \sum_{i=k+2}^{n} \sigma_i^T v_i v_i^T \right] z \\ &= \sum_{i=1}^{k+1} \sigma_i^2 \left(z^T v_i v_i^T z \right) \\ &\ge \sigma_{k+1}^2 \sum_{i=1}^{k+1} \left(v_i^T z \right)^2 . \end{aligned} \tag{8E.17}$$

Since z is in the space V_{k+1} and $\|z\|_2 = 1$, we have

$$\|z\|_2^2 = \left\| \sum_{i=1}^{k+1} (v_i^T z) v_i \right\|_2^2 = \sum_{i=1}^{k+1} (v_i^T z)^2 = 1. \tag{8E.18}$$

Substituting Equation 8E.17 into Equation 8E.16, we obtain

$$\|(A-B)z\|_2^2 \ge \sigma_{k+1}^2 . \tag{8E.19}$$

Therefore, it follows from Equations 8.183 and 8E.18 that

$$\|A-B\|_2^2 \geq \sigma_{k+1}^2. \tag{8E.20}$$

Using (3) in Example 1.5, we have $\|U_{n-k}\Lambda_{n-k}V_{n-k}^T\|_2 = \sigma_{k+1}$.

Thus,

$$\|A-A_k\|_2 = \|U_{n-k}\Lambda_{n-k}V_{n-k}^T\|_2 = \sigma_{k+1} \leq \|A-B\|_2,$$

which contradicts with assumption (8E.13):

$$\|A-B\|_2 \leq \|A-A_k\|_2.$$

Therefore, we prove that

$$\|A-A_k\|_2 \leq \|A-B\|_2$$

and equality occurs when $B=A_k$.

APPENDIX 8F: POINCARE SEPARATION THEOREM

Let $A \in R^{n\times n}$ be a symmetric matrix with eigenvalues $\sigma_1 \geq \cdots \geq \sigma_n$ and associated eigenvectors $v_1, \ldots, v_n$. The subspace spanned by $v_1, \ldots, v_k$ is denoted by S_k, and its orthogonal complement is denoted by $S_{k-1}^{\perp}$. We first show that

$$\sigma_k = \max_{\substack{\|x\|_2=1 \\ x\in S_{k-1}^{\perp}}} x^T Ax. \tag{8F.1}$$

Let $A = V\Lambda V^T$ be the eigendecomposition of A, where $\Lambda = \text{diag}(\sigma_1, \ldots, \sigma_n)$. The quadratic form x^TAx is $x^TV\Lambda V^Tx = y^T\Lambda y$, where $y = V^Tx$ and $\|y\|_2 = \|x\|_2$. If $x \in S_{k-1}^{\perp}$, then $y \in S_{k-1}^{\perp}$. In fact, the vector x can be expressed in terms of basis vectors $\{v_1, \ldots, v_n\}$:

$$\sum_{i=1}^{k-1} \alpha_i v_i + \sum_{i=k}^{n} \alpha_i v_i,$$

or

$$x = \begin{bmatrix} x_1 \\ 0 \\ x_n \end{bmatrix} = \begin{bmatrix} v_1^T \\ 0 \\ v_n^T \end{bmatrix} \alpha. \tag{8F.2}$$

The assumption of $x \in S_{k-1}^{\perp}$ implies that

$$x = \begin{bmatrix} 0 \\ \vdots \\ 0 \\ v_k^T \alpha \\ \vdots \\ v_n^T \alpha \end{bmatrix}. \tag{8F.3}$$

Since A is a symmetric matrix, we have $V^T = V$. Thus,

$$\begin{aligned} v_j^T y &= v_j^T V x \\ &= v_j^T \begin{bmatrix} v_1 & \cdots & v_j & \cdots & v_k & \cdots & v_n \end{bmatrix} x \\ &= \begin{bmatrix} 0 & \cdots & 1 & \cdots & 0 & \cdots & 0 \end{bmatrix} x \\ &= x_j, \quad j = 1, \ldots, k-1. \end{aligned}$$

But, from Equation 8F.3, we know that $x_j = 0$, which implies that $v_j^T y = 0,\ j = 1, \ldots, k-1$, or $y \in S_{k-1}^{\perp}$.

If we express y in terms of $\{v_1, \ldots, v_n\}$, then we have

$$y = \begin{bmatrix} y_1 \\ \vdots \\ y_{k-1} \\ y_k \\ \vdots \\ y_n \end{bmatrix} = V\beta.$$

Similar to Equation 8F.3, we can show that $y_1 = \cdots = y_{k-1} = 0$. Therefore,

$$x^T A x = \begin{bmatrix} 0 & \cdots & 0 & y_k & \cdots & y_n \end{bmatrix} \begin{bmatrix} \sigma_1 & & & & & \\ & \ddots & & & & \\ & & \sigma_{k-1} & & & \\ & & & \sigma_k & & \\ & & & & \ddots & \\ & & & & & \sigma_n \end{bmatrix} \begin{bmatrix} 0 \\ \vdots \\ 0 \\ y_k \\ \vdots \\ y_n \end{bmatrix} = \sum_{i=k}^{n} \sigma_i y_i^2,$$

which implies that

$$x^T A x \le \sigma_k \sum_{i=k}^{n} y_i^2 = \sigma_k, \tag{8F.4}$$

where we use the factor $\sum_{i=k}^{n} y_i^2 = \|y\|_2^2 = \|x\|_2^2 = 1$.

Taking $y_k = 1$, $y_j = 0$, $\forall_j \neq k$, or $x_* = v_k$, we obtain

$$x_*^T A x_* = \sigma_k.$$

This shows that

$$\sigma_j = \max_{\substack{\|x\|_2=1 \\ x \in S_{k-1}^{\perp}}} x^T A x. \tag{8F.5}$$

Let $\mu_1 \geq \mu_2 \geq \cdots \mu_k$ be the eigenvalues of the matrix $G^T A G$. Since $A = V\mathbf{\Lambda} V^T$, we have $G^T A G = G^T V \mathbf{\Lambda} V^T G$. The eigenvectors of the matrix $G^T A G$ are $G^T V$. Let W_j be a subspace spanned by the eigenvectors $\{G^T v_i, \ldots, G^T v_j\}$. Let $x = Gy$. Then, we have $\|x\|_2 = \|y\|_2$. Let $S_j = \{v_1, \ldots, V_j\}$. If $x \in S_{j-1}^{\perp}$, then $v_i^T x = 0$, $i = 1, \ldots, j-1$. Thus, $v_i^T x = v_i^T G y = \left(G^T v_i\right)^T y = 0$, $\forall i = 1, \ldots, j-1$, which implies that $y \in W_{j-1}^{\perp}$.

Therefore,

$$\sigma_j = \max_{\substack{\|x\|_2=1 \\ x \in S_{j-1}^{\perp}}} x^T A x \geq \max_{\substack{\|x\|_2=1 \\ x \in S_{j-1}^{\perp}, x = Gy}} x^T A x = \max_{\substack{\|y\|_2=1 \\ y \in W_{j-1}^{\perp}}} y^T G^T A G y = \mu_j, \quad j = 1, \ldots, k. \tag{8F.6}$$

When $G = V_k$, the first k eigenvectors of A, then we have

$$\begin{aligned} G^T A G &= V_k^T \Lambda V^T V_k \\ &= V_k^T \begin{bmatrix} V_k & V_{n-k} \end{bmatrix} \begin{bmatrix} \Lambda_k & 0 \\ 0 & \Lambda_{n-k} \end{bmatrix} \begin{bmatrix} V_k^T \\ V_{n-k}^T \end{bmatrix} V_k \\ &= \begin{bmatrix} I_k & 0 \end{bmatrix} \begin{bmatrix} \Lambda_k & 0 \\ 0 & \Lambda_{n-k} \end{bmatrix} \begin{bmatrix} 0 \\ 0 \end{bmatrix} \\ &= \Lambda_k. \end{aligned}$$

This shows that the eigenvalues of matrix $G^T A G$ are the first k eigenvalues of A.

APPENDIX 8G: REGRESSION FOR CCA

Now we show that the regression coefficient matrix W in Equation 8.237 is equal to the canonical direction matrix A. We first use the SVD of the matrices X and Y to compute the canonical direction matrix A.

Let the SVD of the matrix X be

$$X = U_x \Sigma_x V_x^T. \tag{8G.1}$$

Then, we have

$$\left(X^T X\right)^{-1} = V_x \Sigma_x^{-2} V_x^T \tag{8G.2}$$

and

$$\left(X^T X\right)^{-1/2} = V_x \Sigma_x^{-1} V_x^T. \tag{8G.3}$$

Let the SVD of the matrix Y be

$$Y = U_y \Sigma_y V_y^T. \tag{8G.4}$$

Then, again we have

$$\left(Y^T Y\right)^{-1} = V_y \Sigma_y^{-2} V_y^T \tag{8G.5}$$

and

$$\left(Y^T Y\right)^{-1/2} = V_y \Sigma_y^{-1} V_y^T. \tag{8G.6}$$

Therefore, using the above SVD, we obtain the decomposing of the between-class covariance matrix:

$$\begin{aligned}\Sigma_b &= X^T Y \left(Y^T Y\right)^{-1} X \\ &= V_x \Sigma_x U_x^T U_y \Sigma_y V_y^T V_y \Sigma_y^{-2} V_y^T V_y \Sigma_y U_y^T U_x \Sigma_x V_x^T \\ &= V_x \Sigma_x U_x^T U_y U_y^T U_x \Sigma_x V_x^T.\end{aligned} \tag{8G.7}$$

Using Equations 8G.3 and 8G.7 gives

$$\begin{aligned}\left(X^T x\right)^{-1/2} \Sigma_b \left(X^T X\right)^{-1/2} &= V_x \Sigma_x^{-1} V_x^T V_x \Sigma_x U_x^T U_y U_y^T U_x V_x^T V_x \Sigma_x^{-1} V_x^T \\ &= V_x U_x^T U_y U_x^T U_x V_x^T \\ &- BIB^T,\end{aligned} \tag{8G.8}$$

where

$$B = V_x U_x^T U_y. \tag{8G.9}$$

Using Equations 8G.9, 8G.3, and 8.237, we obtain

$$\begin{aligned}A &= \left(X^T X\right)^{-1/2} B \\ &= V_x \Sigma_x^{-1} V_x^T V_x U_x^T U_y \\ &= V_x \Sigma_x^{-1} U_x^T U_y.\end{aligned} \tag{8G.10}$$

The least square estimator of the regression (8.237) is

$$\begin{aligned} W &= \left(X^T X\right)^{-1} X^T U_y \\ &= V_x \Sigma_x^{-2} V_x^T V_x \Sigma_x U_x^T U_y \\ &= V_x \Sigma_x^{-1} U_x^T U_y. \end{aligned}$$

This shows that $A = W$.

APPENDIX 8H: PARTITION OF GLOBAL SDR FOR A WHOLE GENOME INTO A NUMBER OF SMALL REGIONS

Suppose that genome is divided into d genomic regions. For the jth genomic region, we assume that some components of the basis vector are zero. The basis vector in the jth genomic region can be denoted by

$$\gamma_j = \begin{bmatrix} \xi_j \\ 0 \end{bmatrix}.$$

From Equation 8.263, we have

$$E\left\{\begin{bmatrix} Z_1^j \\ Z_2^j \end{bmatrix} \middle| Y = y\right\} = \begin{bmatrix} \xi_j \\ 0 \end{bmatrix} \rho_y, \quad j = 1, 2, \ldots, d. \tag{8H.1}$$

By arranging the order of variables, the vector Z and γ can be written as

$$Z = \begin{bmatrix} Z_1^1 \\ \vdots \\ Z_d^1 \\ 0 \end{bmatrix} = \begin{bmatrix} Z_1 \\ Z_2 \end{bmatrix} \quad \text{and} \quad \gamma = \begin{bmatrix} \xi_1 \\ \vdots \\ \xi_d \\ 0 \end{bmatrix} = \begin{bmatrix} \xi \\ 0 \end{bmatrix}, \tag{8H.2}$$

where $Z_1 = \begin{bmatrix} Z_1^1 \\ \vdots \\ Z_d^1 \end{bmatrix}$ and $\xi = \begin{bmatrix} \xi_1 \\ \vdots \\ \xi_d \end{bmatrix}$.

Combining Equations 8H.1 and 8H.2, we obtain

$$\begin{aligned} E\left[Z_1 \middle| Y = y\right] &= \xi \rho_y \\ E\left[Z_2 \middle| Y = y\right] &= 0. \end{aligned} \tag{8H.3}$$

But

$$\operatorname{cov}\left(E\left(Z \middle| Y\right)\right) = \begin{bmatrix} \operatorname{cov}\left(E\left(Z_1 \middle| Y\right)\right) & 0 \\ 0 & 0 \end{bmatrix}. \tag{8H.4}$$

It follows from Equation 8H.4 that

$$\operatorname{cov}\left(E\left(Z_1 \mid Y\right)\right)\xi = \lambda_z \xi, \tag{8H.5}$$

which implies that zero components can be removed by solving an eigenequation for each genomic region.

Recall that

$$Z = \Sigma^{-1/2}\left(X - E(X)\right)$$

and

$$E\left(Z \mid Y = y\right) = \gamma \rho_y,$$

which implies that

$$E\left(X - E(X) \mid Y\right) = \Sigma_x \beta \rho_y. \tag{8H.6}$$

For the jth genomic region, we assume that some components of the basis vector for $S_{Y|X}$ are zero. The basis vector in the jth genomic region can be denoted by

$$\beta_j = \begin{bmatrix} B_j \\ 0 \end{bmatrix}.$$

From Equation 8H.6, we have

$$E\left\{\begin{matrix} X_1^j - E\left(X_1^j\right) \\ X_2^j - E\left(X_2^j\right) \end{matrix} \middle| Y = y\right\} = \begin{bmatrix} \Sigma_{11}^j & \Sigma_{12}^j \\ \Sigma_{21}^j & \Sigma_{22}^j \end{bmatrix}\begin{bmatrix} B_j \\ 0 \end{bmatrix}\rho_y, \quad j = 1, 2, \ldots, m.$$

By arranging the order of variables, the vector X and β can be written as

$$X = \begin{bmatrix} X_1^1 \\ \vdots \\ X_m^1 \\ X_1^2 \\ \vdots \\ X_m^2 \end{bmatrix} = \begin{bmatrix} X_1 \\ X_2 \end{bmatrix} \quad \text{and} \quad \beta = \begin{bmatrix} B_1 \\ \vdots \\ B_m \\ 0 \end{bmatrix} = \begin{bmatrix} B \\ 0 \end{bmatrix}, \tag{8H.7}$$

where $X_1 = \begin{bmatrix} X_1^1 \\ \vdots \\ X_m^1 \end{bmatrix}$, $X_2 = \begin{bmatrix} X_1^2 \\ \vdots \\ X_m^2 \end{bmatrix}$ and $B = \begin{bmatrix} b_1 \\ \vdots \\ b_m \end{bmatrix}$.

Combining Equations 8H.6 and 8H.7, we obtain

$$\begin{bmatrix} E\left(X_1 - E\left(X_1\right) \middle| Y = y\right) \\ E\left(X_2 - E\left(X_2\right) \middle| Y = y\right) \end{bmatrix} = \begin{bmatrix} \Sigma_{11} & \Sigma_{12} \\ \Sigma_{21} & \Sigma_{22} \end{bmatrix} \begin{bmatrix} B \\ 0 \end{bmatrix} \rho_y. \tag{8H.8}$$

From Equation 8H.8, we can obtain the following covariance matrix:

$$\operatorname{cov}\left(E\left(X - E(X) \middle| Y\right)\right) = \begin{bmatrix} \Sigma_{11} B E\left(\rho_y \rho_y^T\right) B^T \Sigma_{11} & \Sigma_{11} B E\left(\rho_y \rho_y^T\right) B^T \Sigma_{12} \\ \Sigma_{21} B E\left(\rho_y \rho_y^T\right) B^T \Sigma_{11} & \Sigma_{21} B E\left(\rho_y \rho_y^T\right) B^T \Sigma_{12} \end{bmatrix}. \tag{8H.9}$$

Then, we have

$$\operatorname{cov}\left(E\left(X_1 - E\left(X_1\right) \middle| Y\right)\right) = \Sigma_{11} B E\left(\rho_y \rho_y^T\right) B^T \Sigma_{11}.$$

Let $\beta = \begin{bmatrix} e \\ 0 \end{bmatrix}$. The eigenequation for the vector of variable X_1 is given by

$$\operatorname{cov}\left(E\left(X_1 - E\left(X_1\right) \middle| Y\right)\right) e = \Sigma_{11} B E\left(\rho_y \rho_y^T\right) B^T \Sigma_{11} e = \lambda_x \Sigma_{11} e, \tag{8H.10}$$

which implies that

$$\Sigma_{21} B E\left(\rho_y \rho_y^T\right) B^T \Sigma_{11} e = \lambda_x \Sigma_{21} e. \tag{8H.11}$$

It follows from Equation 8H.9 that

$$\begin{aligned} \operatorname{cov}\left(E\left(X - E(X) \middle| Y\right)\right)\beta &= \begin{bmatrix} \Sigma_{11} B E\left(\rho_y \rho_y^T\right) B^T \Sigma_{11} & \Sigma_{11} B E\left(\rho_y \rho_y^T\right) B^T \Sigma_{12} \\ \Sigma_{21} B E\left(\rho_y \rho_y^T\right) B^T \Sigma_{11} & \Sigma_{21} B E\left(\rho_y \rho_y^T\right) B^T \Sigma_{12} \end{bmatrix} \begin{bmatrix} e \\ 0 \end{bmatrix} \\ &= \begin{bmatrix} \Sigma_{11} B E\left(\rho_y \rho_y^T\right) B^T \Sigma_{11} e \\ \Sigma_{21} B E\left(\rho_y \rho_y^T\right) B^T \Sigma_{11} e \end{bmatrix}. \end{aligned} \tag{8H.12}$$

Combining Equations 8H.10 through 8H.12 leads to

$$\begin{aligned} \operatorname{cov}\left(E\left(X - E(X) \middle| Y\right)\right)\beta &= \lambda_x \begin{bmatrix} \Sigma_{11} e \\ \Sigma_{21} e \end{bmatrix} \\ &= \lambda_x \begin{bmatrix} \Sigma_{11} & \Sigma_{12} \\ \Sigma_{21} & \Sigma_{22} \end{bmatrix} \begin{bmatrix} e \\ 0 \end{bmatrix} \\ &= \lambda_x \Sigma \beta, \end{aligned} \tag{8H.13}$$

which implies that a solution to the eigenequation for subvector of predictors X_1 also satisfies the eigenequation (8H.13) for the whole vector of predictors X.

APPENDIX 8I: OPTIMAL SCORING AND ALTERNATIVE DIRECTION METHODS OF MULTIPLIERS (ADMM) ALGORITHMS

We first introduce alternative direction methods of multipliers (ADMM) for solving the constrained convex optimization problem (Boyd et al. 2011). We consider the following general optimization problem:

$$\begin{aligned} &\text{minimize} \quad f(x)+g(z) \\ &\text{s.t.} \qquad\quad Ax+Bz=c \end{aligned} \tag{8I.1}$$

We form the augmented Lagrangian

$$L\rho(x,z,y)=f(x)+g(z)+y^T(Ax+Bz-c)+\frac{\rho}{2}\|Ax+Bz-c\|_2^2. \tag{8I.2}$$

The ADMM algorithm is given by

$$\begin{aligned} x^{k+1} &= \arg\min_x L_\rho\left(x,z^k,y^k\right) \\ z^{k+1} &= \arg\min_x L\rho\left(x^{k+1},z,y^k\right) \\ y^{k+1} &= y^k+\rho\left(Ax^{k+1},Bz-c\right). \end{aligned} \tag{8I.3}$$

Combining the linear and quadratic terms in the augmented Lagrangian and scaling the dual variable yields

$$y^T(Ax+Bz-c)+\frac{\rho}{2}||Ax+Bz-c||_2^2=\frac{\rho}{2}||r+u||_2^2\frac{\rho}{2}||u||_2^2, \tag{8I.4}$$

where $r=Ax+Bz-c, u=\frac{1}{\rho}y$ is the scaled dual variable.

The scaled form of ADMM in algorithm (8I.3) can be expressed as

$$x^{k+1}=\arg\min_x\left(f(x)\frac{\rho}{2}||Ax^{k+1}+Bz^k-c+u^k||_2^2\right),$$

$$z^{k+1}=\arg\min_z\left(g(z)\frac{\rho}{2}||Ax^{k+1}+Bz-c+u^k||_2^2\right),$$

$$u^{k+1}-u^k+Ax^{k-1}+Bz^{k-1}-c. \tag{8I.5}$$

Let $f_j\left(\beta_1^{(s)}\right)=\|Z\theta_j^{(s-1)}-X\beta_1^{(s)}\|_2^2$, $B^{(s)}=\left[\beta_1^{(s)},\ldots,\beta_d^{(s)}\right]=\left[\beta_1^{*(s)},\ldots,\beta_p^{*(s)}\right]^T$, and $g\left(\alpha_1^{(s)},\ldots,\alpha_d^{(s)}\right)=g\left(\alpha_1^{*(s)},\ldots,\alpha_p^{*(s)}\right)=\lambda\sum_{i=1}^{p}\|\alpha_1^{*(s)}\|_2^{1-r}$, where $\alpha^{(s)}=\left[\alpha_1^{(s)},\ldots,\alpha_d^{(s)}\right]=\left[\alpha_1^{*(s)},\ldots,\alpha_p^{*(s)}\right]^T$. The problem (8.273a) in the text can be rewritten as

$$\min \sum_{j=1}^{d} f_j\left(\beta_j^{(s)}\right)+g\left(\alpha_1^{(s)},\ldots,\alpha_d^{(s)}\right) \tag{8I.6}$$

$$\text{s.t. } \beta_j^{(s)}-\alpha_j^{(s)}=0, \quad j=1,\ldots,d.$$

The scaled ADMM for solving problem (8I.6) is given by

$$\begin{aligned}\beta_j^{(s)(m+1)}&=\underset{\beta}{\arg\min}\left(f_j\left(\beta_j^{(s)}\right)\right)+\frac{\rho}{2}\left\|\beta_j^{(s)}-\alpha_j^{(s)(m+1)}-u_j^{(s)(m)}\right\|_2^2\Big)\\ &=\left(X^TX+\frac{\rho}{2}I\right)^{-1}\left[X^TZ\theta_j^{(s)}+\frac{\rho}{2}\left(\alpha_j^{(s)(m)}-u_j^{(s)(m)}\right)\right],\end{aligned} \tag{8I.7}$$

$$\alpha^{*(s)(m+1)}=\underset{\alpha^{(s)}}{\arg\min}\ (g(\alpha^{(s)})+\frac{\rho}{2}\sum_{j=1}^{d}\left\|\alpha_j^{(s)}-\beta_j^{(s)(m+1)}-u_j^{(s)(m)}\right\|_2^2\Big), \tag{8I.8}$$

$$u_j^{(s)(m+1)}=u_j^{(s)(m)}+\beta_j^{(s)(m+1)}-\alpha_j^{(s)(m+1)}. \tag{8I.9}$$

Now we study how to solve the nonsmooth optimization problem (8I.8).

First, we assume that $\|\alpha_1^{*(s)}\|_2\neq 0,\ i=1,\ldots,p$. By definition, we have

$$\frac{\partial g(a)}{\partial \alpha_j^{(s)}}=\lambda(1-r)\left[\frac{\alpha_{1j}^{(s)}}{\left\|\alpha_1^{*(s)}\right\|_2^{1+r}},\ldots,\frac{\alpha_{pj}^{(s)}}{\left\|\alpha_p^{*(s)}\right\|_2^{1+r}}\right]^T. \tag{8I.10}$$

Minimization problem (8I.8) can be solved by the following equation:

$$\frac{\partial g(a)}{\partial \alpha_j^{(s)}}+\rho\left(\alpha_j^{(s)}-\beta_j^{(s)(m+1)}-u_j^{(s)(m)}\right)=0, \quad j=1,\ldots,d. \tag{8I.11}$$

Let $B^{(s)}=\left[\beta_1^{(s)},\ldots,\beta_d^{(s)}\right]=\left[\beta_1^{*(s)},\ldots,\beta_p^{(s)}\right]^T$, $\alpha^{(s)}=\left[\alpha_1^{(s)},\ldots,\alpha_d^{(s)}\right]=\left[\alpha_1^{*(s)},\ldots,\alpha_p^{*(s)}\right]^T$, and $u^{(s)(m)}=\left[u_1^{(s)(m)},\ldots,\beta_d^{(s)(m)}\right]=\left[u_1^{*(s)(m)},\ldots,u_p^{*(s)(m)}\right]^T$.

Changing column vectors in Equation 8I.11 to row vector, we obtain

$$\lambda(1-r)\frac{\alpha_i^{*(s)}}{||\alpha_i^{*(s)}||_2^{1+r}}+\rho\left(\alpha_i^{*(s)}-\beta_i^{*(s)(m+1)}-u_i^{*(s)(m)}\right)=0, \quad i=1,\ldots,p,$$

which implies that

$$\left(\frac{\lambda(1-r)}{||\alpha_i^{*(s)}||_2^{1+r}}+\rho\right)\alpha_i^{*(s)}=\rho\left(\beta_i^{*(s)(m+1)}+u_i^{*(s)(m)}\right), \quad i=1,\ldots,p. \tag{8I.12}$$

Dividing Equation 8I.12 by ρ, we obtain

$$\left(1+\frac{\lambda(1-r)}{\rho||\alpha_i^{*(s)}||_2^{1+r}}\right)\alpha_i^{*(s)}=\beta_i^{*(s)(m+1)}+u_i^{*(s)(m)}. \tag{8I.13}$$

Taking norm $||\cdot||_2$ on both sides of Equation 8I.13, we have

$$\left[\left(1+\frac{\lambda(1-r)}{\rho||\alpha_i^{*(s)}||_2^{1+r}}\right)\right]^{(1+r)}||\alpha_i^{*(s)}||_2^{1+r}=||\beta_i^{*(s)(m+1)}+u_i^{*(s)(m)}||_2^{(1+r)}. \tag{8I.14}$$

By approximation, we have

$$\left[\left(1+\frac{\lambda(1-r)}{\rho||\alpha_i^{*(s)}||_2^{1+r}}\right)\right]^{(1+r)}=1+(1+r)\frac{\lambda(1-r)}{\rho||\alpha_i^{*(s)}||_2^{1+r}}.$$

Substituting the above equation into Equation 8I.14, we obtain

$$||\alpha_i^{*(s)}||_2^{1+r}=||\beta_i^{*(s)(m+1)}+u_i^{*(s)(m)}||_2^{(1+r)}-\frac{\lambda(1-r^2)}{\rho}. \tag{8I.15}$$

Therefore, we have

$$||\alpha_i^{*(s)}||_2=\left[||\beta_i^{*(s)(m+1)}+u_i^{*(s)(m)}||_2^{(1+r)}-\frac{\lambda(1-r^2)}{\rho}\right]^{\frac{1}{1+r}}. \tag{8I.16}$$

Substituting Equation 8I.16 into Equation 8I.14 results in

$$1+\frac{\lambda(1-r)}{\rho||\alpha_i^{*(s)}||_2^{1+r}}=\frac{||\beta_i^{*(s)(m+1)}+u_i^{*(s)(m)}||_2}{||\alpha_i^{*(s)}||_2}$$

$$=\frac{||\beta_i^{*(s)(m+1)}+u_i^{*(s)(m)}||_2}{\left[||\beta_i^{*(s)(m+1)}+u_i^{*(s)(m)}||_2^{(1+r)}-\frac{\lambda\left(1-r^2\right)}{\rho}\right]^{\frac{1}{1+r}}}. \tag{8I.17}$$

Substituting Equation 8I.17 into Equation 8I.13, we obtain

$$\alpha_i^{*(s)(m+1)}=\frac{\beta_i^{*(s)(m+1)}+u_i^{*(s)(m)}}{1+\frac{\lambda(1-r)}{\rho||\alpha_i^{*(s)}||_2^{1+r}}}$$

$$=\left(\beta_i^{*(s)(m+1)}+u_i^{*(s)(m)}\right)\left[1-\frac{\lambda\left(1-r^2\right)}{\rho||\beta_i^{*(s)(m+1)}+u_i^{*(s)(m)}||_2^{(1+r)}}\right]_+^{\frac{1}{1+r}}. \tag{8I.18}$$

Let

$$\alpha^{(s)(m+1)}=\left[\alpha_1^{*(s)(m+1)},\ldots,\alpha_p^{*(s)(m+1)}\right]^T=\left[\alpha_1^{(s)(m+1)},\ldots,\alpha_d^{(s)(m+1)}\right]. \tag{8I.19}$$

Substituting Equation 8I.19 into Equation 8I.9, we can obtain $u_i^{(s)(m+1)}$, $j=1,\ldots,d$.

Next we briefly discuss how to obtain solution (12b) (Clemmensen et al. 2011). By the Lagrangian multiplier method, the constrained optimization problem (8.273b) can be formulated as the following unconstrained optimization problem:

$$L(\theta_i,\lambda_\theta,\mu)=\left(Z\theta_i^{(s)}-X\beta_i^{(s)}\right)^T\left(Z\theta_i^{(s)}-X\beta_i^{(s)}\right)+\lambda_\theta\left(1-\theta_i^{(s)T}D\theta_i^{(s)}\right)+\sum_{j=1}^{i-1}\mu_j\theta_j^{(s)T}D\theta_i^{(s)}. \tag{8I.20}$$

Differential $L(\theta_i, \lambda_\theta, \mu)$ with respect to $\theta_i^{(s)}$ and setting it to be equal to zero, we obtain

$$Z^T\left(Z\theta_i^{(s)}-X\beta_i^{(s)}\right)\lambda_\theta D\theta_i^{(s)}+\sum_{j=1}^{i-1}\mu_jD\theta_j^{(s)}=0, \tag{8I.21}$$

which implies that

$$\lambda_\theta=\left[\theta_j^{(s)T}Z^TZ\theta_i^{(s)}-\theta_i^{(s)T}Z^TX\beta_i^{(s)}\right]=n-\theta_j^{(s)T}Z^TX\beta_i^{(s)} \tag{8I.22}$$

and

$$\mu_j = \theta_j^{(s)} - Z^T X\beta_i^{(s)}, \quad j = 1,\ldots,i-1. \tag{8I.23}$$

Substituting Equations 8I.22 and 8I.23 into Equation 8I.21, we obtain

$$(n-\lambda_\theta)D\theta_i^{(s)} - Z^T X\beta_i^{(s)} + D\sum_{j=1}^{i-1}\theta_j^{(s)}\theta_j^{(s)T} Z^T X\beta_i^{(s)} = 0. \tag{8I.24}$$

Let $Q_1^{(s)} = \left[\theta_1^{(s)},\ldots,\theta_{i-1}^{(s)}\right]$, where $Q_1^{(s)} = [1,1,\ldots,1]^T$. Then, we have $Q_i^{(s)}Q_i^{(s)T} = \sum_{j=1}^{i=1}\theta_j^{(s)}\theta_j^{(s)T}$. The solution to equation (8I.24) is given by

$$\theta_i^{(s)} = a\left(I - Q_i^{(s)}Q_i^{(s)T}D\right)D^{-1}Z^T X\beta_i^{(s)}), \quad \text{where } a = \frac{1}{n-\lambda_\theta}. \tag{8I.25}$$

Equation 8I.25 is a nonlinear equation. We can use Newton's method to solve it.

The recursive formula for solving it is given by

$$\theta_i^{(s)(m+1)} = \theta_i^{(s)(m)} - \left[\left(Z^T X\beta\right)\otimes\theta_i^{(s)(m)}\right]^{-1}\left[\theta_i^{T(s)(m)}Z^T X\beta_i^{(s)}\theta_i^{(s)(m)} - \left(I - Q_i^{(s)}Q_i^{(s)T}D\right)D^{-1}Z^T X\beta_i^{(s)}\right].$$

We can also use iterative algorithm to Equation 8I.25. Just iteratively,

$$\theta_i^{(s)T(m+1)} = \frac{\left(I - Q_i^{(s)}Q_i^{(s)T}D\right)D^{-1}Z^T X\beta_i^{(s)}}{\theta_i^{(s)T(m)}Z^T X\beta_i^{(s)}}. \tag{8I.26}$$

Finally, we set

$$\theta_i^{(s)} = \frac{\theta_i^{(s)T(m+1)}}{\sqrt{\theta_i^{(s)T(m+1)}D\theta_i^{(s)T(m+1)}}}, \quad Q_{i+1} = \left[Q_i : \theta_i\right], \quad i = 1,\ldots,d. \tag{8I.27}$$

EXERCISES

Exercise 8.1 Assume that the likelihood for observing the binary output values y_i is

$$L(p,y) = \prod_{i=1}^{n} p^{y_i}(1-p)^{y_i}.$$

Find the maximum likelihood estimator of the parameter p.

Exercise 8.2 Derive the formula

$$\frac{\partial^2 l(\beta)}{\partial \beta^{(c)} \left(\partial \beta^{(j)}\right)^T}.$$

Exercise 8.3 Show that Equation 8.36 can be rewritten as Equation 8.37:

$$J_3(\beta) = \beta^T (D - S)\beta.$$

Exercise 8.4 Show that

$$\alpha^{t+1} = \begin{bmatrix} \alpha_1^{t+1} \\ \vdots \\ \alpha_G^{t+1} \end{bmatrix} = \begin{bmatrix} \left(1 - \frac{v\lambda_3 \phi_{31}}{\|u_{\alpha_1}^{(t)}\|_2}\right)_+ u_{\alpha_1}^{(t)} \\ \vdots \\ \left(1 - \frac{v\lambda_3 \phi_{3g}}{\|u_{\alpha_G}^{(t)}\|_2}\right)_+ u_{\alpha_G}^{(t)} \end{bmatrix}.$$

Exercise 8.5 Show that

$$L_D = \frac{1}{2}\eta\alpha_2^2 + \left(y_2\left(E_1^{(0)} - E_2^{(0)}\right) - \eta\alpha_2^{(0)}\right)\alpha_2 + \text{Const},$$

$$E_1^{(0)} = \left(w^{(0)}\right)^T x_1 + b - y_1 \text{ and } E_2^{(0)} = \left(w^{(0)}\right)^T x_2 + b - y_2.$$

Exercise 8.6 Show that

If $0 < \alpha_1 < c$ and $0 < \alpha_2 < c$, then $b_1 = b_2$.

Exercise 8.7 Let Σ_x be a covariance matrix of X. We define the standardized predictors as $Z = \Sigma_x^{-1/2}\left(X - E(X)\right)$. Prove that var($Z$) = 1.

Exercise 8.8 Show that $S_{Y|Z} = \Sigma_x^{-1/2} S_{Y|Z}$.

Exercise 8.9 Show that

$$w_1^j \leftarrow \frac{S\left(Z^j w_2^j, \lambda_1\right)}{\|S\left(Z^j w_2^j, \lambda_1\right)\|_2},$$

$\lambda_1 = 0$, if this results in $\|w_1^j\|_1 \le c_1$; otherwise, $\lambda_1 > 0$ is chosen so that $\|w_1^j\|_1 = c_1$.

Exercise 8.10 Let $f(x) = \|x\|_1$. Derive the formula $\text{Prox}_{\lambda f}(v)$.

Exercise 8.11 Show that $C = \Gamma^{-1/2} U_l \Lambda_l V_l^T \Sigma_{XX}^{-1/2}$ in Equation 8.186.

Exercise 8.12 Show that $A = U_l \Lambda_l V_l^T + U_{m-l} \Lambda_{m-l} V_{m-l}^T$.

Exercise 8.13 Show that the matrix U_l is also the matrix of eigenvectors associated with the matrix $AA^T = \Gamma^{1/2} \Sigma_{YX} \Sigma_{XX}^{-1} \Sigma_{XY} \Gamma^{1/2}$.

Exercise 8.14 Let Σ_x be a covariance matrix of X. We define the standardized predictors as $Z = \Sigma_x^{-1/2} \left(X - E(X) \right)$. Prove that var($Z$) = 1.

Exercise 8.15 Show that $S_{Y|Z} = \Sigma_x^{-1/2} S_{Y|X}$.

References

Akey, J., Jin, L., Xiong, M. (2001). Haplotypes vs single marker linkage disequilibrium tests: What do we gain? *Eur J Hum Genet* 9:291–300.

Akulin, A., Bratko, I. (2003). Analyzing attribute dependencies. *Lect Notes Artif Intell* 2838:229–240.

Altshuler, D., Clark, A.G. (2005). Genetics. Harvesting medical information from the human family tree. *Science* 307:1052–1053.

Altshuler, D., Daly, M.J., Lander, E.S. (2008). Genetic mapping in human disease. *Science* 322(5903):881–888.

An, P., Mukherjee, O., Chanda, P., Yao, L., Engelman, C.D. et al. (2009). The challenge of detecting epistasis (G G interactions): Genetic analysis workshop 16. *Genet Epidemiol* 33(Suppl. 1): S58–S67.

Anderson, T.W. (1984). *An Introduction to Multivariate Statistical Analysis*, 2nd edition. John Wiley & Sons, New York.

Aschard, H., Vilhjálmsson, B.J., Greliche, N., Morange, P.E., Trégouët, D.A., Kraft, P. (2014). Maximizing the power of principal-component analysis of correlated phenotypes in genome-wide association studies. *Am J Hum Genet* 94(5):662–676.

Ash, R.B., Gardner, M.F. (1975). *Topics in Stochastic Processes*. Academic Press, New York.

Aulchenko, Y.S., Ripatti, S., Lindqvist, I., Boomsma, D., Heid, I.M. (2009). Loci influencing lipid levels and coronary heart disease risk in 16 European population cohorts. *Nat Genet* 41:47–55.

Bacanu, S.A., Nelson, M.R., Whittaker, J.C. (2011). Comparison of methods and sampling designs to test for association between rare variants and quantitative traits. *Genet Epidemiol* 35:226–235.

Bach, F. (2010). Structured sparsity-inducing norms through submodular functions. *Adv Neural Inf Process Syst* 23:118–126.

Bach, F.R., Jenatton, R., Mairal, J., Obozinski, G. (2011). Convex optimization with sparsity-inducing norms. *Optim Mach Learn* 5:19–53.

Bach, F., Jenatton, R., Mairal, J., Obozinski, G. (2012). Optimization with sparsity-inducing penalties. *Found Trends Mach Learn* 4(1):1–106.

Bansal, V., Libiger, O., Torkamani, A., Schork, N.J. (2010). Statistical analysis strategies for association studies involving rare variants. *Nat Rev Genet* 11:773–785.

Barrett, J.C., Fry, B., Maller, J., Daly, M.J. (2005). Haploview: Analysis and visualization of LD and haplotype maps. *Bioinformatics* 21(2):263–265.

Bartlett, M.S. (1939). A note on tests of significance in multivariate analysis. *Proc Cambr Philos Soc* 35:180–185.

Beck, A. (2014). *Introduction to Nonlinear Optimization: Theory, Algorithms, and Applications with MATLAB*. SIAM, Philadelphia, PA.

Bonnet, A., Gassiat, E., Lévy-Leduc, C. (2015). Heritability estimation in high dimensional sparse linear mixed models. *Electron J Stat* 9:2099–2129.

Borwein, J.M., Lewis, A.S. (2006). *Convex Analysis and Nonlinear Optimization: Theory and Examples*. Springer-Verlag, New York.

Boyd, S., Parikh, N., Chu, E., Peleato, B., Eckstein, J. (2011). Distributed optimization and statistical learning via the alternating direction method of multipliers. *Found Trends Mach Learn* 3(1):1–122.

Brillinger, D.R. (2004). Some data analyses using mutual information. *Braz J Probab Stat* 18:163–183.

Broadaway, K.A., Cutler, D.J., Duncan, R., Moore, J.L., Ware, E.B. et al. (2016). Statistical approach for testing cross-phenotype effects of rare variants. *Am J Hum Genet* 98(3):525–540.

Brookes, A.J., Robinson, P.N. (2015). Human genotype-phenotype databases: Aims, challenges and opportunities. *Nat Rev Genet* 16(12):702–715.

Burges, C.J.C. (1998). A tutorial on support vector machines for pattern recognition. *Data Min Knowl Discov* 2:121–167.

Cardot, H., Ferraty, F., Sarda, P. (1999). Functional linear model. *Stat Probab Lett* 45:11–22.

Carlson, C.S., Eberle, M.A., Kruglyak, L., Nickerson, D.A. (2004). Mapping complex disease loci in whole-genome association studies. *Nature* 429:446–452.

Chakravarti, A. (2011). Genomics is not enough. *Science* 334:15.

Chapman, N.H., Wijsman, E.M. (1998). Genome screens using linkage disequilibrium tests: Optimal marker characteristics and feasibility. *Am J Hum Genet* 63:1872–1885.

Chen, C.-H., Li, K.-C. (1998). Can SIR be as popular as multiple linear regression? *Stat Sin* 8:289–316.

Chen, W., Chen, D., Zhao, M., Zou, Y., Zeng, Y., Gu, X. (2015). Genepleio software for effective estimation of gene pleiotropy from protein sequences. *Biomed Res Int* 2015:269150.

Chen, X., Zou, C., Cook, R.D. (2010). Coordinate-independent sparse sufficient dimension reduction and variable selection. *Ann Stat* 38:3696–3723.

Christianini, N., Shawe-Taylor, J. (2000). *An Introduction to Support Vector Machines and Other Kernel-Based Learning Methods*. Cambridge University Press, Cambridge, U.K.

Clark, C.W. (1990). *Mathematical Bioeconomics: The Optimal Management of Renewable Resources*, 2nd edition. John Wiley & Sons, New York.

Clemmensen, L., Hastie, T., Witten, D., Ersbøll, B. (2011). Sparse discriminant analysis. *Technometrics* 53(4):406–413.

Collins, F.S. (2004). The case for a US prospective cohort study of genes and environment. *Nature* 429:475–477.

Conomos, M.P., Reiner, A.P., Weir, B.S., Thornton, T.A. (2016). Model-free estimation of recent genetic relatedness. *Am J Hum Genet* 98(1):127–148.

Cook, R.D. (1994). On the interpretation of regression plots. *J Am Stat Assoc* 89:177–189.

Cook, R.D. (2004). Testing predictor contributions in sufficient dimension reduction. *Ann Stat* 32:1062–1092.

Cordell, H.J. (2009). Detecting gene-gene interactions that underlie human diseases. *Nat Rev Genet* 10:392–404.

Cover, T.M., Thomas, J.A. (1991). *Elements of Information Theory*. John Wiley & Sons Inc, New York.

Dahl, A., Iotchkova, V., Baud, A., Johansson, Å., Gyllensten, U., Soranzo, N., Mott, R., Kranis, A., Marchini, J. (2016). A multiple-phenotype imputation method for genetic studies. *Nat Genet* 48(4):466–472.

Davies, R. (1980). The distribution of a linear combination of chi-square random variables. *Appl Stat* 29:323–333.

Dawy, Z., Goebel, B., Hagenauer, J., Andreoli, C., Meitinger, T., Mueller, J.C. (2006). Gene mapping and marker clustering using Shannon's mutual information. *IEEE/ACM Trans Comput Biol Bioinform* 3(1):47–56.

Delaneau, O., Howie, B., Cox, A.J., Zagury, J.F., Marchini, J. (2013). Haplotype estimation using sequencing reads. *Am J Hum Genet* 93(4):687–696.

Falconer, D.S., Mackay, T.F.C. (1996). *Introduction to Quantitative Genetics*, 4th edition. Longman, Harlow, England.

Fan, R., Wang, Y., Mills, J.L., Wilson, A.F., Bailey-Wilson, J.E., Xiong, M. (2013). Functional linear models for association analysis of quantitative traits. *Genet Epidemiol* 37(7):726–742.

Febrero-Bande, M., Galeano, P., Gonzalez-Manteiga, W. (2010). Measures of influence for the functional linear model with scalar response. *J Multivar Anal* 101:327–339.

Frank, I.E., Friedman, J.H. (1993). A statistical view of some chemometrics regression tools (with discussion). *Technometrics* 35:140–143.

Frayling, T.M., Timpson, N.J., Weedon, M.N., Zeggini, E., Freathy, R.M., Lindgren, C.M., Perry, J.R., Elliott, K.S., Lango, H., Rayner, N.W. (2007). A common variant in the FTO gene is associated with body mass index and predisposes to childhood and adult obesity. *Science* 316:889–894.

Frazer, K.A., Murray, S.S., Schork, N.J., Topol, E.J. (2009). Human genetic variation and its contribution to complex traits. *Nat Rev Genet* 10(4):241–251.

Friedman, J., Hastie, T., Tibshirani, R. (2010). Regularization paths for generalized linear models via coordinate descent. *J Stat Softw* 33:1–22.

Fukumizu, K., Bach, F.R., Gretton, A. (2007). Statistical consistency of kernel canonical correlation analysis. *J Mach Learn Res* 8:361–383.

Fung, G.M., Mangasarian, O.L. (2004). Multicategory proximal support vector machine. *Mach Learn* 59:77–97.

Furlotte, N.A., Eskin, E. (2015). Efficient multiple-trait association and estimation of genetic correlation using the matrix-variate linear mixed model. *Genetics* 200(1):59–68.

Gao, H., Wu, Y., Zhang, T., Wu, Y., Jiang, L., Zhan, J., Li, J., Yang, R. (2015). Multiple-trait genome-wide association study based on principal component analysis for residual covariance matrix. *Heredity (Edinb)* 113(6):526–532.

Gilmour, A.R., Thompson, R., Cullis, B.R. (1995). Average information REML: An efficient algorithm for variance parameter estimation in linear mixed models. *Biometrics* 51:1440–1450.

Golan, D., Lander, E.S., Rosset, S. (2014). Measuring missing heritability: Inferring the contribution of common variants. *Proc Natl Acad Sci USA* 111(49):E5272–E5281.

Graybill, F.A. (1976). *Theory and Application of the Linear Model.* Wadsworth & Brooks/Cole Advanced Books & Software, Pacific Grove, CA.

Graybill, F.A. (1983). *Matrices with Applications in Statistics*, 2nd edition. Wadsworth International Group, Belmont, CA.

Gretton, A. (2015). Introduction to RKHS, and some simple kernel algorithms. http://www.gatsby.ucl.ac.uk/~gretton/coursefiles/lecture4_introToRKHS.pdf, accessed March, 2016.

Gumedze, F.N., Dunne, T.T. (2011). Parameter estimation and inference in the linear mixed model. *Linear Algebra Appl* 435:1920–1944.

Haq, S., Choukroun, G., Lim, H., Tymitz, K.M., del Monte, F. et al. (2001). Differential activation of signal transduction pathways in human hearts with hypertrophy versus advanced heart failure. *Circulation* 103(5):670–677.

Hartville, D.A. (1977). Maximum likelihood approaches to variance component estimation and to related problems. *J Am Stat Assoc* 72:320–338.

Hartwell, L. (2004). Genetics. Robust interactions. *Science* 303:774–775.

Hastie, T., Tibshirani, R., Buja, A. (1994). Flexible discriminant analysis by optimal scoring. *J Am Stat Assoc* 89:1250–1270.

Heckerman, D., Gurdasani, D., Kadie, C., Pomilla, C., Carstensen, T. et al. (2016). Linear mixed model for heritability estimation that explicitly addresses environmental variation. *Proc Natl Acad Sci USA* 113(27):7377–7382.

Henderson, D., Plaschko, P. (2006). *Stochastic Differential Equations in Science and Engineering.* World Scientific, Hackensack, NJ.

Hill, W.G., Zhang, X.S. (2012). On the pleiotropic structure of the genotype-phenotype map and the evolvability of complex organisms. *Genetics* 190(3):1131–1137.

Horikawa, Y., Oda, N., Cox, N.J., Li, X., Orho-Melander, M. et al. (2000). Genetic variation in the gene encoding calpain-10 is associated with type 2 diabetes mellitus. *Nat Genet* 26:163–175.

Horke, S., Witte, I., Wilgenbus, P., Krüger, M., Strand, D., Förstermann, U. (2007). Paraoxonase-2 reduces oxidative stress in vascular cells and decreases endoplasmic reticulum stress-induced caspase activation. *Circulation* 115(15):2055–2064.

Hotelling, H. (1931). The generalization of student's ratio. *Ann Math Stat* 2:360–378.

Huang, H.H., Liang, Y., Liu, X.Y. (2015). Network-based logistic classification with an enhanced L 1/2 solver reveals biomarker and subnetwork signatures for diagnosing lung cancer. *Biomed Res Int* 2015:713953.

Hudson, R.R. (2002). Generating samples under a wright-fisher neutral model of genetic variation. *Bioinformatics* 18:337–338.

Hunter, J.J., Tanaka, N., Rockman, H.A., Ross, J., Jr., Chien, K.R. (1995). Ventricular expression of a MLC-2v-ras fusion gene induces cardiac hypertrophy and selective diastolic dysfunction in transgenic mice. *J Biol Chem* 270(39):23173–23178.

Izenman, A.J. (2008). *Modern Multivariate Statistical Techniques: Regression, Classification, and Manifold Learning*. Springer, New York.

Jakulin, A. (2005). Machine learning based on attribute interaction. PhD dissertation. University of Ljubljana, Ljubljana, Slovenia.

Jenatton, R., Audibert, J.-Y., Bach, F.R. (2011). Structured variable selection with Sparsity-inducing norms. *J Mach Learn Res* 12:2777–2824.

Jimenez-Sanchez, G., Childs, B., Valle, D. (2001). Human disease genes. *Nature* 409:853–855.

Johnson, R.A., Wichern, D.W. (2002). *Applied Multivariate Statistical Analysis*, 5th edition. Prentice Hall, Upper Saddle River, NJ.

Jorde, L.B. (2000). Linkage disequilibrium and the search for complex disease genes. *Genome Res* 10:1435–1444.

Joyce, P., Tavare, S. (1995). The distribution of rare alleles. *J Math Biol* 33:602–618.

Kang, H.M., Sul, J.H., Service, S.K., Zaitlen, N.A., Kong, S.-y. et al. (2010). Variance component model to account for sample structure in genome-wide association studies. *Nat Genet* 42:348–354.

Khoury, M.J., Wacholder, S. (2009). Invited commentary: From genome-wide association studies to gene-environment-wide interaction studies-challenges and opportunities. *Am J Epidemiol* 169:227–230.

Kim, J.M., Lee, K.H., Jeon, Y.J., Oh, J.H., Jeong, S.Y. et al. (2006). Identification of genes related to Parkinson's disease using expressed sequence tags. *DNA Res* 13:275–286.

Koo, C.L., Liew, M.J., Mohamad, M.S., Salleh, A.H.M., Deris, S., Ibrahim, Z., Susilo, B., Hendrawan, Y., Wardani, A.K. (2015). Software for detecting gene-gene interactions in genome wide association studies. *Biotechnol Bioprocess Eng* 20:662–676.

Korte, A., Farlow, A. (2013). The advantages and limitations of trait analysis with GWAS: A review. *Plant Methods* 9:29.

Krishna Kumar, S., Feldman, M.W., Rehkopf, D.H., Tuljapurkar, S. (2016). Limitations of GCTA as a solution to the missing heritability problem. *Proc Natl Acad Sci USA* 113(1):E61–E70.

Kullback, S. (1959). *Information Theory and Statistics*. Wiley, New York.

Larson, N.B., Jenkins, G.D., Larson, M.C., Vierkant, R.A., Sellers, T.A. et al. (2014). Kernel canonical correlation analysis for assessing gene-gene interactions and application to ovarian cancer. *Eur J Hum Genet* 22(1):126–131.

Lee, S., Abecasis, G.R., Boehnke, M., Lin, X. (2014). Rare-variant association analysis: Study designs and statistical tests. *Am J Hum Genet* 95(1):5–23.

Lee, S., Wu, M.C., Lin, X. (2012). Optimal tests for rare variant effects in sequencing association studies. *Biostatistics* 13(4):762–775.

Lee, S.H., Wray, N.R., Goddard, M.E., Visscher, P.M. (2011). Estimating missing heritability for disease from genome-wide association studies. *Am J Hum Genet* 88(3):294–305.

Lehmann, E.L. (1983). *Theory of Point Estimation*. John Wiley & Sons, New York.

Lewontin, R.C. (1964). The interaction of selection and linkage. II. Optimum models. *Genetics* 50:757–782.

Li, B., Leal, S.M. (2008). Methods for detecting associations with rare variants for common diseases: Application to analysis of sequence data. *Am J Hum Genet* 83:311–321.

Li, K.-C. (1991). Sliced inverse regression for dimension reduction. *J Am Stat Assoc* 86:316–327.

Liberman, U., Puniyani, A., Feldman, M.W. (2007). On the evolution of epistasis II: A generalized Wright-Kimura framework. *Theor Popul Biol* 71(2):230–238.

Lin, D., Zhang, J., Li, J., Calhoun, V.D., Deng, H.W., Wang, Y.P. (2013). Group sparse canonical correlation analysis for genomic data integration. *BMC Bioinform* 14:245.

Lin, D.Y., Tang, Z.Z. (2011). A general framework for detecting disease associations with rare variants in sequencing studies. *Am J Hum Genet* 89(3):354–367.

Lin, S., Cutler, D.J., Zwick, M.E., Chakravarti, A. (2002). Haplotype inference in random population samples. *Am J Hum Genet* 71:1129–1137.

Lippert, C., Listgarten, J., Liu, Y., Kadie, C.M., Davidson, R.I., Heckerman, D. (2011). FaST linear mixed models for genome-wide association studies. *Nat Methods* 8:833–835.

Listgarten, J., Kadie, C., Schadt, E.E., Heckerman, D. (2010). Correction for hidden confounders in the genetic analysis of gene expression. *Proc Natl Acad Sci USA* 107:16465–16470.

Liu, D., Ghosh, D., Lin, X. (2008). Estimation and testing for the effect of a genetic pathway on a disease outcome using logistic kernel machine regression via logistic mixed models. *BMC Bioinform* 9:292.

Liu, G., Lin, Z., Yan, S., Sun, J., Yu, Y., Ma, Y. (2013). Robust recovery of subspace structures by low rank representation. *IEEE Trans Pattern Anal Mach Intell* 35(1):171–184.

Liu, T., Thalamuthu, A. (2011). Identify by descent and association analysis of dichotomous traits based on large pedigrees. *BMC Proc* 5:S31.

Liu, Z., Lin, S. (2005). Multilocus LD measure and tagging SNP selection with generalized mutual information. *Genet Epidemiol* 29:353–364.

Luo, L., Boerwinkle, E., Xiong, M.M. (2011). Association studies for next-generation sequencing. *Genome Res* 21(7):1099–1108.

Luo, L., Zhu, Y., Xiong, M. (2012). Quantitative trait locus analysis for next-generation sequencing with the functional linear models. *J Med Genet* 49(8):513–524.

Luo, L., Zhu, Y., Xiong, M.M. (2013). Smoothed functional principal component analysis for testing association of the entire allelic spectrum of genetic variation. *Eur J Hum Genet* 21(2):217–224.

Lupski, J.R., Belmont, J.W., Boerwinkle, E., Gibbs, R.A. (2011). Clan genomics and the complex architecture of human disease. *Cell* 147:32–43.

Lynch, M., Walsh, B. (1998). *Genetics and Analysis of Quantitative Traits*. Sinauser Associates, Inc. Publishers, Sunderland, MA.

Mackay, T.F. (2014). Epistasis and quantitative traits: Using model organisms to study gene-gene interactions. *Nat Rev Genet* 15(1):22–33.

Madsen, B.E., Browning, S.R. (2009). A groupwise association test for rare mutations using a weighted sum statistic. *PLoS Genet* 5:e1000384.

Mao, Y., London, N.R., Ma, L., Dvorkin, D., Da, Y. (2006). Detection of SNP epistasis effects of quantitative traits using an extended Kempthorne model. *Physiol Genomics* 28(1):46–52.

Masud, S., Ye, S., SAS Group (2003). Effect of the peroxisome proliferator activated receptor-gamma gene Pro12Ala variant on body mass index: A meta-analysis. *J Med Genet* 40(10):773–780.

Matsuda, H. (2000). Physical nature of higher-order mutual information: Intrinsic correlations and frustration. *Phys Rev E* 62(3 Pt A):3096–3102.

McCullagh, P., Nelder, J.A. (1989). *Generalized Linear Models*, 2nd edition. Chapman & Hall/CRC, New York.

McCulloch, C.E., Searle, S.R. (2001). *Generalized, Linear and Mixed Models*. John Wiley & Sons Inc., New York.

Min, W., Liu, J., Zhang, S. (2016). Network-regularized sparse logistic regression models for clinical risk prediction and biomarker discovery. *IEEE/ACM Trans Comput Biol Bioinform*. doi: 10.1109/TCBB.2016.2640303 [Epub ahead of print].

Mitteroecker, P., Cheverud, J.M., Pavlicev, M. (2016). Multivariate analysis of genotype-phenotype association. *Genetics* 202(4):1345–1363.

Molenberghs, G., Verbeke, G. (2007). Likelihood ratio, score, and Wald tests in a constrained parameter space. *Am Stat* 61:22–27.

Montgomery, S.B., Sammeth, M., Gutierrez-Arcelus, M., Lach, R.P., Ingle, C., Nisbett, J., Guigo, R., Dermitzakis, E.T. (2010). Transcriptome genetics using second generation sequencing in a Caucasian population. *Nature* 464(7289):773–777.

Mordukhovich, B., Nam, N.M. (2017). Geometric approach to subdifferential calculus. *Optimization* 66(6):839–873. arXiv:1504.06004.

Morris, A.P., Zeggini, E. (2010). An evaluation of statistical approaches to rare variant analysis in genetic association studies. *Genet Epidemiol* 34:188–193.

Muller, K., Peterson, B. (1984). Practical methods for computing power in testing the multivariate general linear hypothesis. *Comput Stat Data Anal* 2(2):143–158.

Muslin, A.J. (2008). MAPK signalling in cardiovascular health and disease: Molecular mechanisms and therapeutic targets. *Clin Sci* 115(7):203–218.

Najmabadi, H., Hu, H., Garshasbi, M., Zemojtel, T., Abedini, S.S. et al. (2011). Deep sequencing reveals 50 novel genes for recessive cognitive disorders. *Nature* 478:57–63.

National Human Genome Research Institute. (2015). A catalog of published genome-wide association studies. https://www.genome.gov/page.cfm?pageid=26525384&clearquery=1#result_table, accessed July, 2015.

Neale, B.M., Sham, P.C. (2004). The future of association studies: Gene-based analysis and replication. *Am J Hum Genet* 75:353–362.

Nelson, M.R., Wegmann, D., Ehm, M.G., Kessner, D., St Jean, P. et al. (2012). An abundance of rare functional variants in 202 drug target genes sequenced in 14,002 people. *Science* 337(6090):100–104.

Nilsson, J., Sha, F., Jordan, M.I. (2007). Regression on manifolds using kernel dimension reduction. *Proceedings of the 24th International Conference on Machine Learning*. ACM, New York, pp. 697–704.

Nothnagel, M., Furst, R., Rohde, K. (2002). Entropy as a measure for linkage disequilibrium over multilocus haplotype blocks. *Hum Hered* 54(4):186–198.

Obozinski, G., Jacob, L., Vert, J.P. (2011). Group lasso with overlaps: The latent group lasso approach. arXiv preprint arXiv:1110.0413.

Ott, J., Kamatani, Y., Lathrop, M. (2011). Family-based designs for genome-wide association studies. *Nat Rev Genet* 12:465–474.

Ott, J., Wang, J., Leal, S.M. (2015). Genetic linkage analysis in the age of whole-genome sequencing. *Nat Rev Genet* 16(5):275–284.

Ottman, R. (1996). Theoretic epidemiology. Gene-environment interaction: Definitions and study designs. *Prev Med* 25:764–770.

Pagano, M., Gauvreau, K. (1993). *Principles of Biostatistics*. Duxbury Press, Belmont, CA.

Pan, W., Kim, J., Zhang, Y., Shen, X., Wei, P. (2014). A powerful and adaptive association test for rare variants. *Genetics* 197(4):1081–1095.

Parikh, N., Boyd, S. (2014). Proximal algorithms. *Found Trends Optimization* 1(3):123–231.

Pavone, M. (1994). On the Riesz representation theorem for bounded linear functionals. *Proc R Irish Acad Sect A: Math Phys Sci* 94A(1):133–135.

Peng, D.Q., Zhao, S.P., Nie, S., Li, J. (2003). Gene-gene interaction of PPARgamma and ApoE affects coronary heart disease risk. *Int J Cardiol* 92(2–3):257–263.

Peng, Q., Zhao, J., Xue, F. (2010). A gene-based method for detecting gene-gene co-association in a case-control association study. *Eur J Hum Genet* 18:582–587.

Platt, J. (1998). Fast training of support vector machines using sequential minimal optimization, in *Advances in Kernel Methods—Support Vector Learning*, B. Scholkopf, C. Burges, A. Smola, eds. MIT Press, Cambridge, MA.

Press, W.H. (2011). Canonical correlation clarified by singular value decomposition. http://numerical.recipes/whp/notes/CanonCorrBySVD.pdf, accessed May, 2016.

Price, A.L., Kryukov, G.V., de Bakker, P.I., Purcell, S.M., Staples, J. et al. (2010a). Pooled association tests for rare variants in exon-resequencing studies. *Am J Hum Genet* 86:832–838.

Price, A.L., Zaitlen, N.A., Reich, D., Patterson, N. (2010b). New approaches to population stratification in genome-wide association studies. *Nat Rev Genet* 11:459–463.

Pritchard, J.K., Cox, N.J. (2002). The allelic architecture of human disease genes: Common disease-common variant… or not? *Hum Mol Genet* 11:2417–2423.

Qin, Z.S., Niu, T., Liu, J.S. (2002). Partition-ligation-expectation-maximization algorithm for haplotype inference with single-nucleotide polymorphisms. *Am J Hum Genet* 71:1242–1247.

Ramsay, J.O., Silverman, B.W. (2005). *Functional Data Analysis*, 2nd edition. Springer, New York.

Ray, D., Pankow, J.S., Basu, S. (2016). USAT: A unified score-based association test for multiple phenotype-genotype analysis. *Genet Epidemiol* 40(1):20–34.

Reich, D.E., Gabriel, S.B., Altshuler, D. (2003). Quality and completeness of SNP databases. *Nat Genet* 33:457–458.

Risch, N., Merikangas, K. (1996). The future of genetic studies of complex human diseases. *Science* 273:1516–1517.

Romeo, S., Pennacchio, L.A., Fu, Y., Boerwinkle, E., Tybjaerg-Hansen, A. et al. (2007). Population-based resequencing of ANGPTL4 uncovers variations that reduce triglycerides and increase HDL. *Nat Genet* 39:513–516.

Sagan, H. (1969). *Introduction to the Calculus of Variation*. Dover Publications, Inc., New York.

Sanghera, D.K., Aston, C.E., Saha, N., Kamboh, M.I. (1998). DNA polymorphisms in two paraoxonase genes (PON1 and PON2) are associated with the risk of coronary heart disease. *Am J Hum Genet* 62(1):36–44.

Schaid, D.J. (2004). Linkage disequilibrium testing when linkage phase is unknown. *Genetics* 166:505–512.

Schölkopf, B., Smola, A., Müller, K.R. (1997). Kernel principal component analysis. *Artificial Neural Networks—ICANN'97*, Lausanne, Switzerland, pp. 583–588.

Searle, S.R., Casella, G., McCulloch, C.E. (1992). *Variance Components*. John Wiley & Sons, New York.

Serfling, R.J. (1980). *Approximation Theorems of Mathematical Statistics*. John Wiley & Sons, New York.

Shannon, C.E. (1948). A mathematical theory of communication. *Bell Syestems Tech J* 27:379–423.

Shawe-Taylor, J., Cristianini, N. (2004). *Kernel Methods for Pattern Analysis*. Cambridge University Press, New York.

Sherwin, W.B. (2015). Genes are information, so information theory is coming to the aid of evolutionary biology. *Mol Ecol Resour* 15:1259–1261.

Shimada, M., Miyagawa, T., Kawashima, M., Tanaka, S., Honda, Y. et al. (2010). An approach based on a genome-wide association study reveals candidate loci for narcolepsy. *Hum Genet* 128:433–441.

Shugart, Y.Y., Zhu, Y., Guo, W., Xiong, M.M. (2012). Weighted pedigree-based statistics for testing the association of rare variants. *BMC Genomics* 13:667.

Sing, C.F., Stengard, J.H., Kardia, S.L. (2003). Genes, environment, and cardiovascular disease. *Arterioscler Thromb Vasc Biol* 23:1190–1196.

Solovieff, N., Cotsapas, C., Lee, P.H., Purcell, S.M., Smoller, J.W. (2013). Pleiotropy in complex traits: Challenges and strategies. *Nat Rev Genet* 14:483–495.

Speed, D., Balding, D.J. (2014). MultiBLUP: Improved SNP-based prediction for complex traits. *Genome Res* 24(9):1550–1557.

Steen, K.V. (2011). Travelling the world of gene-gene interactions. *Brief Bioinform* 13:1–19.

Stephens, J.C., Schneider, J.A., Tanguay, D.A., Choi, J., Acharya, T. et al. (2001). Haplotype variation and linkage disequilibrium in 313 human genes. *Science* 293:489–493.

Stephens, M. (2013). A unified framework for association analysis with multiple related phenotypes. *PLoS One* 8:e65245.

Stram, D.O., Lee, J.W. (1994). Variance components testing in the longitudinal mixed effects model. *Biometrics* 50:1171–1177.

Sun, L., Ji, S., Ye, J.P. (2011). Canonical correlation analysis for mulilabel classification: A least-squres formulation, extensions, and analysis. *IEEE Trans Pattern Anal Mach Intell* 33(1):194–200.

Tian, X., Wang, X., Chen, J. (2015). Network-constrained group lasso for high-dimensional multinomial classification with application to cancer subtype prediction. *Cancer Inform* 13(Suppl. 6):25–33.

Tran, M.N. (2008). An introduction to theoretical properties of functional principal component analysis. Department of Mathematics and Statistics, The University of Melbourne, Victoria, Australia.

Udell, M., Horn, C., Zadeh, R., Boyd, S. (2016). Generalized low rank models. *Found Trends Mach Learn* 9(1):1–118.

Vapnik, V. (1998). *Statistical Learning Theory*. John Wiley & Sons, Inc., New York.

Verbeke, G., Molenberghs, G. (2003). The use of score tests for inference on variance components. *Biometrics* 59:254–262.

Vilhjálmsson, B.J., Yang, J., Finucane, H.K., Gusev, A., Lindström, S., Ripke, S. et al. (2015). Modeling linkage disequilibrium increases accuracy of polygenic risk scores. *Am J Hum Genet* 97(4):576–592.

Wagner, G.P., Zhang, J. (2011). The pleiotropic structure of the genotype–phenotype map: The evolvability of complex organisms. *Nat Rev Genet* 12:204–213.

Wakefield, J. (2008). Inference for variance components by REML. *Stat/Biostat* 571:53–130.

Wang, K. (2008). Genetic association tests in the presence of epistasis or gene-environment interaction. *Genet Epidemiol* 32:606–614.

Wang, L., Zhu, J., Zou, H. (2008). Hybrid huberized support vector machines for microarray classification and gene selection. *Bioinformatics* 24(3):412–419.

Wang, T., Zhu, L. (2013). Sparse sufficient dimension reduction using optimal scoring. *Comput Stat Data Anal* 57(1):223–232.

Weir, B. (1990). *Genetic Data Analysis*. Sinauer Associates, Sunderland, MA.

Weir, B.S. (1979). Inferences about linkage disequilibrium. *Biometrics* 35:235–254.

Weir, B.S. (1996). *Genetic Data Analysis II: Methods for Discrete Population Genetic Data*. Sinauer Associates, Sunderland, MA.

Weir, B.S., Cockerham, C.C. (1989). Complete characterization of disequilibrum at two loci, in *Mathematical Evolutionary Theory*, M.W. Feldman ed., pp. 86–110. Princeton University Press, Princeton, NJ.

Weir, B.S., Hill, W.G. (2002). Estimating F-statistics. *Annu Rev Genet* 36:721–750.

Weissbrod, O., Geiger, D., Rosset, S. (2016). Multikernel linear mixed models for complex phenotype prediction. *Genome Res* 26(7):969–979.

Wijsman, E.M. (2016). Family-based approaches: Design, imputation, analysis, and beyond. *BMC Genet* 17(Suppl. 2):9.

Wilms, I., Croux, C. (2015). Sparse canonical correlation analysis from a predictive point of view. *Biom J* 57(5):834–851.

Wilms, I., Croux, C. (2016). Robust sparse canonical correlation analysis. *BMC Syst Biol* 10(1):72.

Witten, D.M., Tibshirani, R., Hastie, T. (2009). A penalized matrix decomposition, with applications to sparse principal components and canonical correlation analysis. *Biostatistics* 10(3):515–534.

Witten, D.M., Tibshirani, R.J. (2009). Extensions of sparse canonical correlation analysis with applications to genomic data. *Stat Appl Genet Mol Biol* 8: Article 28.

Wu, M.C., Lee, S., Cai, T., Li, Y., Boehnke, M., Lin, X. (2011). Rare-variant association testing for sequencing data with the sequence kernel association test. *Am J Hum Genet* 89(1):82–93.

Wu, M.Y., Zhang, X.F., Dai, D.Q., Ou-Yang, L., Zhu, Y., Yan, H. (2016). Regularized logistic regression with network-based pairwise interaction for biomarker identification in breast cancer. *BMC Bioinform* 17:108. doi: 10.1186/s12859-016-0951-7.

Wu, X., Dong, H., Luo, L., Zhu, Y., Peng, G., Reveille, J.D., Xiong, M. (2010). A novel statistic for genome-wide interaction analysis. *PLoS Genet* 6(9):e1001131.

Wu, X., Jin, L., Xiong, M.M. (2008). Composite measure of linkage disequilibrium for testing interaction between unlinked loci. *Eur J Hum Genet* 16(5):644–651.

Wu, Y., Wipf, D., Yun, J.M. (2015). Understanding and evaluating sparse linear discriminant analysis. Proceedings of the Eighteenth International Conference on Artificial Intelligence and Statistics. *PMLR* 38:1070–1078.

Xiong, M., Zhao, J., Boerwinkle, E. (2002). Generalized T2 test for genome association studies. *Am J Hum Genet* 70:1257–1268.

Xiong, M.M., Wu, X. (2010). Chapter 3: Statistics for testing gene-environment interaction, in *Environmental Factors, Genes, and the Development of Human Cancers*, D. Roy, M.T. Dorak, eds., pp. 53–95. Springer Science+Business Media, New York.

Xiong, M.M., Zhao, J., Boerwinkle, E. (2003). Haplotype block linkage disequilibrium mapping. *Front Biosci* 8:85–93.

Xu, Y., Akrotirianakis, I., Chakraborty, A. (2015). Proximal gradient method for huberized support vector machine. arXiv:1511.09159, pages 1–17.

Yang, J., Benyamin, B., McEvoy, B.P., Gordon, S., Henders, A.K. et al. (2010). Common SNPs explain a large proportion of the heritability for human height. *Nat Genet* 42(7):565–569.

Yang, J., Lee, S.H., Goddard, M.E., Visscher, P.M. (2011). GCTA: A tool for genome-wide complex trait analysis. *Am J Hum Genet* 88:76–82.

Yang, J., Zaitlen, N.A., Goddard, M.E., Visscher, P.M., Price, A.L. (2014). Advantages and pitfalls in the application of mixed-model association methods. *Nat Genet* 46(2):100–106.

Yang, Y., Zou, H. (2015). A fast unified algorithm for solving group-lasso penalize learning problems. *Stat Comput* 25:1129.

Yu, J., Pressoir, G., Briggs, W.H., Bi, I.V., Yamasaki, M. et al. (2006). A unified mixed-model method for association mapping that accounts for multiple levels of relatedness. *Nat Genet* 38:203–208.

Zaitlen, N., Kraft, P. (2012). Heritability in the genome-wide association era. *Hum Genet* 131(10):1655–1664.

Zhang, F., Boerwinkle, E., Xiong, M.M. (2014). Epistasis analysis for quantitative trait with next-generation sequencing data. *Genome Res* 24(6):989–998.

Zhang, F., Xie, D., Liang, M., Xiong, M. (2016). Functional regression models for epistasis analysis of multiple quantitative traits. *PLoS Genet* 12(4):e1005965.

Zhang, J.T. (2005). Approximate and asymptotic distributions of chi-squared-type mixtures with applications. *J Am Stat Assoc* 100:273–285.

Zhang, K., Peters, J., Janzing, D., Schölkopf, B. (2012). Kernel-based conditional independence test and application in causal discovery. arXiv:1202.3775.

Zhang, Z., Ersoz, E., Lai, C.Q., Todhunter, R.J., Tiwari, H.K. et al. (2010). Mixed linear model approach adapted for genome-wide association studies. *Nat Genet* 42:355–360.

Zhao, J., Boerwinkle, E., Xiong, M.M. (2005). An entropy-based statistic for genome-wide association studies. *Am J Hum Genet* 77:27–40.

Zhao, J., Jin, L., Xiong, M. (2006). Test for interaction between two unlinked loci. *Am J Hum Genet* 79(5):831–845.

Zhao, J., Zhu, Y., Xiong, M. (2016). Genome-wide gene-gene interaction analysis for next-generation sequencing. *Eur J Hum Genet* 24(3):421–428.

Zhao, T., Liu, Y., Wang, P., Li, S., Zhou, D. et al. (2009). Positive association between the PDLIM5 gene and bipolar disorder in the Chinese Han population. *J Psychiatry Neurosci* 34:199–204.

Zheng, X., Weir, B.S. (2016). Eigenanalysis of SNP data with an identity by descent interpretation. *Theor Popul Biol* 107:65–76.

Zhou, J.J., Cho, M.H., Lange, C., Lutz, S., Silverman, E.K., Laird, N.M. (2015). Integrating multiple correlated phenotypes for genetic association analysis by maximizing heritability. *Hum Hered* 79(2):93–104.

Zhou, X., Stephens, M. (2012). Genome-wide efficient mixed-model analysis for association studies. *Nat Genet* 44:821–826.

Zhu, Y., Xiong, M.M. (2012). Family-based association studies for next-generation sequencing. *Am J Hum Genet* 90(6):1028–1045.

Zou, H., Hastie, T. (2005). Regularization and variable selection via the elastic net. *J R Stat Soc Series B Stat Methodology* 67:301–320.

Index

M

N

O

R

T

U

V

W